THE ROUTLEDGE HANDBOOK OF PHILOSOPHY AND SCIENCE OF ADDICTION

The problem of addiction is one of the major challenges and controversies confronting medicine and society. It also poses important and complex philosophical and scientific problems. What is addiction? Why does it occur? And how should we respond to it, as individuals and as a society?

The Routledge Handbook of Philosophy and Science of Addiction is an outstanding reference source to the key topics, problems and debates in this exciting subject. It spans several disciplines and is the first collection of its kind. Organized into three clear parts, the forty-five chapters by a team of international contributors examine key areas, including:

- the meaning of addiction to individuals
- conceptions of addiction
- varieties and taxonomies of addiction
- methods and models of addiction
- evolution and addiction
- history, sociology and anthropology
- population distribution and epidemiology
- developmental processes
- vulnerabilities and resilience
- psychological and neural mechanisms
- prevention, treatment and spontaneous recovery
- public health and the ethics of care
- social justice, law and policy.

Essential reading for students and researchers in addiction research and in philosophy, particularly the philosophy of mind and psychology and ethics, *The Routledge Handbook of Philosophy and Science of Addiction* will also be of great interest to those in related fields, such as medicine, mental health, social work, and social policy.

Hanna Pickard is Professor in Philosophy of Psychology at the University of Birmingham, UK, and a Visiting Research Scholar in the Program of Cognitive Science, Princeton University, 2017–19.

Serge H. Ahmed is a Research Director at the Centre National de la Recherche Scientifique (CNRS). He currently works at the Centre Broca Nouvelle-Aquitaine, Bordeaux Neurocampus, Université de Bordeaux, France.

ROUTLEDGE HANDBOOKS IN PHILOSOPHY

Routledge Handbooks in Philosophy are state-of-the-art surveys of emerging, newly refreshed, and important fields in philosophy, providing accessible yet thorough assessments of key problems, themes, thinkers, and recent developments in research.

All chapters for each volume are specially commissioned, and written by leading scholars in the field. Carefully edited and organized, *Routledge Handbooks in Philosophy* provide indispensable reference tools for students and researchers seeking a comprehensive overview of new and exciting topics in philosophy. They are also valuable teaching resources as accompaniments to textbooks, anthologies, and research-orientated publications.

Also available:

THE ROUTLEDGE HANDBOOK OF COLLECTIVE INTENTIONALITY
Edited by Marija Jankovic and Kirk Ludwig

THE ROUTLEDGE HANDBOOK OF SCIENTIFIC REALISM
Edited by Juha Saatsi

THE ROUTLEDGE HANDBOOK OF PACIFISM AND NON-VIOLENCE
Edited by Andrew Fiala

THE ROUTLEDGE HANDBOOK OF CONSCIOUSNESS
Edited by Rocco J. Gennaro

THE ROUTLEDGE HANDBOOK OF PHILOSOPHY AND SCIENCE OF ADDICTION
Edited by Hanna Pickard and Serge H. Ahmed

THE ROUTLEDGE HANDBOOK OF MORAL EPISTEMOLOGY
Edited by Karen Jones, Mark Timmons, and Aaron Zimmerman

For more information about this series, please visit: https://www.routledge.com/Routledge-Handbooks-in-Philosophy/book-series/RHP

THE ROUTLEDGE HANDBOOK OF PHILOSOPHY AND SCIENCE OF ADDICTION

*Edited by Hanna Pickard and
Serge H. Ahmed*

LONDON AND NEW YORK

First published 2019
by Routledge
2 Park Square, Milton Park, Abingdon, Oxon OX14 4RN

and by Routledge
52 Vanderbilt Avenue, New York, NY 10017

First issued in paperback 2020

Routledge is an imprint of the Taylor & Francis Group, an informa business

© 2019 selection and editorial matter, Hanna Pickard and Serge H. Ahmed; individual chapters, the contributors

The right of Hanna Pickard and Serge H. Ahmed to be identified as the authors of the editorial material, and of the authors for their individual chapters, has been asserted in accordance with sections 77 and 78 of the Copyright, Designs and Patents Act 1988.

All rights reserved. No part of this book may be reprinted or reproduced or utilised in any form or by any electronic, mechanical, or other means, now known or hereafter invented, including photocopying and recording, or in any information storage or retrieval system, without permission in writing from the publishers.

Trademark notice: Product or corporate names may be trademarks or registered trademarks, and are used only for identification and explanation without intent to infringe.

British Library Cataloguing-in-Publication Data
A catalogue record for this book is available from the British Library

Library of Congress Cataloging-in-Publication Data
Names: Pickard, Hanna, editor.
Title: The Routledge handbook of philosophy and science of addiction /
edited by Hanna Pickard and Serge H. Ahmed.
Description: 1 [edition]. | New York : Routledge, 2018. | Series:
Routledge handbooks in philosophy | Includes bibliographical
references and index.
Identifiers: LCCN 2017058207| ISBN 9781138909281 (hardback : alk.
paper) | ISBN 9781315689197 (e-book)
Subjects: LCSH: Compulsive behavior—Philosophy.
Classification: LCC RC533 .R68 2018 | DDC 616.85/84—dc23
LC record available at https://lccn.loc.gov/2017058207

ISBN 13: 978-0-367-57150-4 (pbk)
ISBN 13: 978-1-138-90928-1 (hbk)

Typeset in Bembo
by Swales & Willis Ltd, Exeter, Devon, UK

CONTENTS

Notes on contributors   *x*
Acknowledgements   *xx*

   Introduction   1
   Hanna Pickard and Serge H. Ahmed

PART I
What is addiction?   5

SECTION A
Conceptions of addiction   7

1 The puzzle of addiction   9
   Hanna Pickard

2 Deriving addiction: an analysis based on three elementary features of making choices   23
   Gene M. Heyman

3 The picoeconomics of addiction   34
   George Ainslie

4 Addiction as a disorder of self-control   45
   Edmund Henden

5 Addiction: the belief oscillation hypothesis 54
   Neil Levy

6 Addiction and moral psychology 63
   Chandra Sripada and Peter Railton

7 Identity and addiction 77
   Owen Flanagan

8 The harmful dysfunction analysis of addiction: normal brains
   and abnormal states of mind 90
   Jerome C. Wakefield

9 The evolutionary significance of drug toxicity over reward 102
   Edward H. Hagen and Roger J. Sullivan

SECTION B
Varieties, taxonomies, and models of addiction 121

10 Defining addiction: a pragmatic perspective 123
   Walter Sinnott-Armstrong and Jesse S. Summers

11 Diagnosis of addictions 132
   Marc Auriacombe, Fuschia Serre, Cécile Denis, and Mélina Fatséas

12 Reconsidering addiction as a syndrome: one disorder with
   multiple expressions 145
   Paige M. Shaffer and Howard J. Shaffer

13 Developing general models and theories of addiction 160
   Robert West, Simon Christmas, Janna Hastings, and Susan Michie

14 Gambling disorder 173
   Seth W. Whiting, Rani A. Hoff, and Marc N. Potenza

15 Food addiction 182
   Ashley Gearhardt, Michelle Joyner, and Erica Schulte

16 "A walk on the wild side" of addiction: the history and
   significance of animal models 192
   Serge H. Ahmed

PART II
Explaining addiction: culture, pathways, mechanisms 205

SECTION A
Anthropological, historical, and socio-psychological perspectives 207

17 Power and addiction 209
    Jim Orford

18 Sociology of addiction 220
    Richard Hammersley

19 The fuzzy boundaries of illegal drug markets and why they matter 229
    Lee D. Hoffer

20 Multiple commitments: heterogeneous histories of neuroscientific
    addiction research 240
    Nancy D. Campbell

SECTION B
Developmental processes, vulnerabilities, and resilience 251

21 The epidemiological approach: an overview of methods and models 253
    James C. Anthony

22 A genetic framework for addiction 275
    Philip Gorwood, Yann Le Strat, and Nicolas Ramoz

23 Choice impulsivity: a drug-modifiable personality trait 286
    *Annabelle M. Belcher, Carl W. Lejuez, F. Gerard Moeller,
    Nora D. Volkow, and Sergi Ferré*

24 Stress and addiction 299
    Rajita Sinha

SECTION C
Psychological and neural mechanisms 313

25 Mechanistic models for understanding addiction as a
    behavioural disorder 315
    Dominic Murphy and Gemma Lucy Smart

26	Controlled and automatic learning processes in addiction Lee Hogarth	325
27	Decision-making dysfunctions in addiction Antonio Verdejo-Garcia	339
28	The current status of the incentive sensitization theory of addiction Mike J.F. Robinson, Terry E. Robinson, and Kent C. Berridge	351
29	Resting-state and structural brain connectivity in individuals with stimulant addiction: a systematic review Anna Zilverstand, Rafael O'Halloran, and Rita Z. Goldstein	362
30	Imaging dopamine signaling in addiction Diana Martinez and Felipe Castillo	380
31	The neurobiology of placebo effects Elisa Frisaldi, Diletta Barbiani, and Fabrizio Benedetti	392
32	Brain mechanisms and the disease model of addiction: is it the whole story of the addicted self? A philosophical-skeptical perspective Şerife Tekin	401

PART III
Consequences, responses, and the meaning of addiction 411

SECTION A
Listening and relating to addicts 413

33	The Outcasts Project: humanizing heroin users through documentary photography and photo-elicitation Aaron Goodman	415
34	Our stories, our knowledge: the importance of addicts' epistemic authority in treatment Peg O'Connor	431
35	Reactive attitudes, relationships, and addiction Jeanette Kennett, Doug McConnell, and Anke Snoek	440

SECTION B
Prevention, treatment, and spontaneous recovery — 453

36 Contingency management approaches — 455
    Kristyn Zajac, Sheila M. Alessi, and Nancy M. Petry

37 Twelve-step fellowship and recovery from addiction — 464
    John F. Kelly and Julie V. Cristello

38 Opioid substitution treatment and harm minimization approaches — 478
    Mark K. Greenwald

39 Self-change: genesis and functions of a concept — 490
    Harald Klingemann and Justyna I. Klingemann

SECTION C
Ethics, law, and policy — 499

40 Addiction: a structural problem of modern global society — 501
    Bruce K. Alexander

41 Don't be fooled by the euphemistic language attesting to a gentler war on drugs — 511
    Carl L. Hart

42 Drug legalization and public health: general issues, and the case of cannabis — 518
    Robin Room

43 Addiction and drug (de)criminalization — 531
    Douglas Husak

44 Criminal law and addiction — 540
    Stephen J. Morse

45 Addiction and mandatory treatment — 554
    Steve Matthews

Index — *564*

CONTRIBUTORS

Serge H. Ahmed is a Research Director at the Centre National de la Recherche Scientifique (CNRS). He currently works at the Centre Broca Nouvelle-Aquitaine, Bordeaux Neurocampus, Université de Bordeaux, France. He has a long-standing interest in understanding addiction from a broad range of disciplines, including philosophy, economics, sociology, psychology, evolutionary biology, and neurobiology. His experimental work mainly focuses on developing valid animal models of addiction for neurobiological research. His theoretical work mainly consists of contributing to the formulation of computational models of addiction.

George Ainslie is a Research Psychiatrist at the Veterans Affairs Medical Center, Coatesville, PA, USA, and Professor at the School of Economics, University of Cape Town, South Africa. His field is defined by the title of his first book: *Picoeconomics: The Strategic Interaction of Successive Motivational States within the Person* (Cambridge University Press 1992).

Sheila M. Alessi, PhD, is an Associate Professor at the University of Connecticut School of Medicine, Farmington, CT, USA. Her research focuses on improving substance use and other behavioral health outcomes, including use of mobile technologies to target, support, and incentivize behavior change.

Bruce K. Alexander is Professor Emeritus at Simon Fraser University in Burnaby, BC, Canada. He has been a researcher in the field of addiction since 1970, beginning with the "Rat Park" experiments and culminating with *The Globalization of Addiction* (Oxford University Press 2008/2010).

James C. (Jim) Anthony, MSc, PhD, is Professor of Epidemiology at Michigan State University College of Human Medicine in East Lansing, MI, USA. He enjoys teaching and mentoring enthusiastic research fellows and graduate trainees.

Marc Auriacombe is a Professor of Psychiatry and Addiction Medicine at the Medical School of the University of Bordeaux, Bordeaux, France and an Adjunct Associate Professor of Psychiatry at the University of Pennsylvania, Philadelphia, PA, USA. He is the Director of the addiction research team at CNRS USR 3413 Sanpsy and Medical Director of the Addiction Treatment

Services of the Ch. Perrens Hospital and University Hospital of Bordeaux (CHU), France. He was appointed by the American Psychiatric Association as a member of the Substance-Related Disorders DSM-5 Workgroup (2007–2013).

Diletta Barbiani received a MSc in Cognitive Psychology at the University of Bologna, Italy, and is now a PhD student at the Department of Neuroscience of the University of Turin Medical School, Italy. Her research focuses on higher cognitive functions, such as expectations and beliefs, particularly in relation to the placebo response in Parkinson's disease and performance in extreme environments.

Annabelle M. (Mimi) Belcher, PhD, is an Assistant Professor in the Department of Psychiatry at the University of Maryland School of Medicine, Baltimore, MD, USA. Her work focuses on understanding the neurobiological and behavioral repercussions of substance use disorders using preclinical models of exposure to drugs of abuse.

Fabrizio Benedetti, MD, is Professor of Neurophysiology and Human Physiology at the University of Turin Medical School, Italy, and Director of the Center for Hypoxia at the Plateau Rosà Laboratories in Breuil-Cervinia, Italy, and Zermatt, Switzerland. He has been nominated to the Academy of Europe and the European Dana Alliance for the Brain. Recent awards include the British Medical Association Award in 2009 for his book *Placebo Effects* (2nd edition, Oxford University Press 2014) and the Seymour Solomon Award of the American Headache Society in 2012.

Kent C. Berridge is James Olds Distinguished University Professor of Psychology and Neuroscience at the University of Michigan, Ann Arbor, MI, USA. He received his Bachelor of Science from UC Davis in 1979, followed by a Masters and PhD from the University of Pennsylvania in 1980 and 1983. His research examines the neurobiology of pleasure and desire, with implications for motivational disorders such as drug addiction and eating disorders.

Nancy D. Campbell is Professor of Science and Technology Studies and Associate Dean of Graduate Studies and Research in the School of Humanities, Arts, and Social Sciences at Rensselaer Polytechnic Institute in Troy, New York. She is a historian of science who focuses on addiction research. Her books include *Discovering Addiction: The Science and Politics of Substance Abuse Research* (University of Michigan Press 2007); *Using Women: Gender, Drug Policy, and Social Justice* (Routledge 2000); *Gendering Addiction: The Politics of Drug Treatment in a Neurochemical World* (co-authored with Elizabeth Ettorre: Palgrave 2011); and *The Narcotic Farm: The Rise and Fall of America's First Prison for Drug Addicts* (co-authored with J.P. Olsen and Luke Walden; Harry N. Abrams Inc. 2008).

Felipe Castillo is a resident physician in psychiatry, currently training in the Bronx, New York. His interest is to bridge the roles of scientific research and community service. His long-term career goals include being able to discover and deliver better treatment for psychiatric illnesses such as drug addiction.

Simon Christmas is an independent insight expert with a reputation for solving problems and moving debates forward through the application of clear thinking and design and delivery of qualitative research. He is a Visiting Senior Research Fellow at the Centre for Public Policy Research at King's College London and an Associate Consultant at the Centre for Behaviour Change at University College London.

Contributors

Julie V. Cristello graduated from Saint Anselm College in 2012. She is currently a Senior Clinical Research Coordinator at the Center for Addiction Medicine and Recovery Research Institute at Massachusetts General Hospital/Harvard Medical School, Boston, MA, USA.

Cécile Denis received a PhD in addiction psychology research from the University of Bordeaux, Bordeaux, France. In 2010 she received a NIDA INVEST/CTN fellowship and she is currently a research associate at the Center for Studies of Addiction, University of Pennsylvania, Philadelphia, PA, USA, and an adjunct researcher at Sanpsy CNRS USR 3413, University of Bordeaux, Bordeaux, France.

Mélina Fatséas is an addiction psychiatrist with a PhD in addiction psychology research and an HDR in addiction psychiatry. She is a senior researcher at the Addiction Team of Sanpsy CNRS USR 3413 and a Senior Lecturer at the Medical School of the University of Bordeaux, Bordeaux, France, as well as staff addiction psychiatrist within the Addiction Treatment Services of the Ch. Perrens Hospital and University Hospital of Bordeaux (CHU), Bordeaux, France.

Sergi Ferré, MD, PhD, is a Senior Investigator and Chief of the Integrative Neurobiology Section at the National Institute on Drug Abuse Intramural Research Program (NIDA-IRP), Baltimore, MD, USA. His work investigates the use of receptor heteromers as targets for drug development in neuropsychiatric disorders, including substance use disorders.

Owen Flanagan is James B. Duke University Professor of Philosophy at Duke University, Durham, NC, USA. He works in philosophy of mind and psychology, and ethics. His most recent book is *The Geography of Morals: Varieties of Moral Possibility* (Oxford University Press 2017).

Elisa Frisaldi, PhD, is a human geneticist and a postdoctoral fellow in neuroscience at the Department of Neuroscience of the University of Turin Medical School, Turin, Italy. Her research focuses on higher cognitive functions, particularly in relation to the placebo response in pain, Parkinson's disease and myasthenia gravis. Her work places particular emphasis on the integration between neurobiological mechanisms underlying the placebo effect and their potential ethical and social implications and clinical applications. In 2015 she received the William S. Kroger Award from the American Society of Clinical Hypnosis and the Best Presentation Award at the 9th Congress of the European Pain Federation (EFIC).

Ashley Gearhardt, PhD, is an Assistant Professor of Psychology in the Clinical Science Area at the University of Michigan, Ann Arbor, MI, USA.

Rita Z. Goldstein is a tenured Professor of Psychiatry and Neuroscience at the Icahn School of Medicine at Mount Sinai, New York, USA. She directs the Neuroimaging of Addictions and Related Conditions (NARC) Research Program, internationally recognized for its use of innovative multimodality neuroimaging methods (including MRI, PET, EEG/ERP) for translational studies of the cognitive and emotional processes underlying human drug addiction and related disorders. Her research interests include the elucidation of the role of dopamine and prefrontal cortical deficits in iRISA (Impairments in Response Inhibition and Salience Attribution) in drug addiction and other disorders of self-control.

Contributors

Aaron Goodman is a faculty member in the Journalism and Communication Studies Department at Kwantlen Polytechnic University in Surrey, BC, Canada. His research involves interactive documentary production, multimedia storytelling and documentary photography. He specializes in collecting, archiving and disseminating testimonies of survivors of genocide, conflict and human rights violations. Website: www.aarongoodman.com.

Philip Gorwood, MD, PhD, is Professor of Psychiatry at Sainte-Anne Hospital and Head of the CMME Department [Clinique des Maladies Mentales et de L'Encéphale]. He is also Head of the Team 1 at INSERM (Institut National de la Santé et de la Recherche Médicale) research unit 894 (Center of Psychiatry and Neuroscience), Paris, France. He has published over 240 scientific articles (h-index=46) and 24 book chapters. He is Editor-in-Chief of the journal *European Psychiatry* (IF=3.9); the Founder and Director of the Recovery Research Institute at the Massachusetts General Hospital (MGH), Boston, MA, USA; the Program Director of the Addiction Recovery Management Service (ARMS); and the Associate Director of the Center for Addiction Medicine at MGH.

Mark K. Greenwald, PhD, is Professor and Associate Chair for Research in the Department of Psychiatry and Behavioral Neurosciences, and Adjunct Professor of Pharmacy Practice, at Wayne State University, Detroit, MI, USA. He directs the Substance Abuse Research Division, including its Human Pharmacology Laboratory and outpatient opioid treatment clinic. His research focuses on (1) pharmacological, environmental and individual difference determinants of drug seeking/demand and adverse consequences, primarily with opioids but also cocaine, marijuana and nicotine; (2) using brain-imaging techniques to understand the clinical neurobiology of substance use disorders; and (3) developing medication and behavioral treatments to reduce the harms from substance use disorders.

Edward H. Hagen is a Professor in the Department of Anthropology at Washington State University, Vancouver, WA, USA. He specializes in evolutionary approaches to health, social behavior, and cognition.

Richard Hammersley is Professor of Health Psychology at the University of Hull, UK. He has researched addiction for thirty years, often in multidisciplinary collaborations, and is best known for his work on drugs and crime.

Carl L. Hart is the Chairperson at the Department of Psychology and Dirk Ziff Professor of Psychology at the Departments of Psychology and Psychiatry at Columbia University, New York, USA. He has published dozens of scientific articles in the area of neuropsychopharmacology and is co-author of the textbook *Drugs, Society and Human Behavior* (with Charles Ksir: 16th edition, McGraw-Hill Education 2018). His book *High Price: A Neuroscientist's Journey of Self-Discovery that Challenges Everything You Know about Drugs and Society* (Harper 2013) was the 2014 winner of the PEN/E. O. Wilson Literary Science Writing Award.

Janna Hastings is an expert in the development and application of biomedical ontologies for scientific research. With a background in Computer Science and Philosophy, she has many years of experience developing resources to organize scientific data and resolve data integration, interpretation and translation challenges across disciplinary boundaries. She is currently completing a PhD in Biological Sciences at the University of Cambridge, UK.

Contributors

Edmund Henden is Professor of Professional Studies at Oslo Metropolitan University, Norway. His primary areas of research are in applied ethics and the philosophy of psychology. He is currently researching issues concerning informed consent and manipulation.

Gene M. Heyman is a Senior Lecturer in the Department of Psychology at Boston College, Boston, MA, USA. His research has focused on choice, drug self-administration, and the history and correlates of drug use. He is the author of *Addiction: A Disorder of Choice* (Harvard University Press 2009). Currently he is working on the social and economic correlates of drug overdoses in the US and a method for quantifying the allocation of covert attention.

Rani A. Hoff is a Professor of Psychiatry at the Yale School of Medicine, New Haven, CT, USA. She is also the Associate Director of the Robert Wood Johnson Clinical Scholars Program at Yale and the Director of the Northeast Program Evaluation Center for VA Connecticut Health Care System.

Lee D. Hoffer is an Associate Professor of Medical Anthropology at Case Western Reserve University (CWRU) in Cleveland, Ohio, USA, with a secondary appointment of Professor in the Department of Psychiatry at CWRU School of Medicine. He has conducted survey and ethnographic research on drug addiction since 1995.

Lee Hogarth is an Associate Professor of Psychology at the University of Exeter, UK. His work focuses on individual differences in vulnerability to addiction.

Douglas Husak is Distinguished Professor of Philosophy and Law at Rutgers University, New Brunswick, NJ, USA. He holds both a PhD and a JD from Ohio State University. He specializes in the philosophy of criminal law with an emphasis on drug policy. His major books include *Ignorance of Law* (Oxford University Press 2017); *The Philosophy of Criminal Law: Collected Essays* (Oxford University Press 2011); *Overcriminalization* (Oxford University Press 2008); and *Drugs and Rights* (Cambridge University Press 1992). He is the Editor-in-Chief of *Criminal Law and Philosophy* and the former Editor-in-Chief of *Law and Philosophy*.

Michelle Joyner, MS, is a doctoral student of psychology in the clinical science area at the University of Michigan, Ann Arbor, MI, USA.

John F. Kelly is the Elizabeth R. Spallin Associate Professor of Psychiatry in Addiction Medicine at Harvard Medical School, Boston, MA, USA – the first endowed professor in addiction medicine at Harvard. He is also the Founder and Director of the Recovery Research Institute at the Massachusetts General Hospital (MGH), the Program Director of the Addiction Recovery Management Service (ARMS), and the Associate Director of the Center for Addiction Medicine at MGH.

Jeanette Kennett joined Macquarie University, Sydney, NSW, Australia, in 2009 as a CoRE joint appointment between the Philosophy Department and the Macquarie Centre for Cognitive Science. Previously she was Principal Research Fellow in The Centre for Applied Philosophy and Public Ethics at the Australian National University and also at Charles Sturt University (2008–9). Her current research focus is on the contribution the sciences can make to our understanding of the cognitive and affective underpinnings of moral reasoning, moral judgment, and moral agency, and the implications for meta-ethics and moral psychology.

Contributors

Harald Klingemann is a Senior Research Professor at the Bern University of Applied Sciences, Switzerland. His research focuses on the natural history of addiction and treatment systems, for which he received an Honorary Doctorate from the University of Stockholm in 2003.

Justyna I. Klingemann is a Medical Sociologist working at the Department of Studies on Alcohol and Drug Dependence at the Institute of Psychiatry and Neurology in Warsaw, Poland. She conducted the first study on self-change from alcohol addiction in Poland.

Carl W. Lejuez, PhD, is Dean of the College of Liberal Arts and Sciences at the University of Kansas, Lawrence, KS, USA. His research interests cover substance use disorders, personality disorders and mood disorders, with the goal of using findings from laboratory-based studies in the development of novel assessment and treatment strategies.

Yann Le Strat is Deputy Head of the Department of Psychiatry and Addiction Medicine at Louis Mourier Hospital, Colombes, France, and Full Professor of Psychiatry at Paris Diderot University. He holds a medical degree in psychiatry and a PhD in genetics and neuroscience from Paris 6 University. He is a researcher at INSERM *(Institut National de la Santé et de la Recherche Médicale)* research unit 894 (Center of Psychiatry and Neuroscience), with a focus on addiction and mental disorders, and epidemiology and genetics. He has published over 60 scientific articles and 15 book chapters.

Neil Levy is Professor of Philosophy at Macquarie University, Sydney, NSW, Australia, and a Research Fellow at the Oxford Centre for Neuroethics, Oxford, UK. He has published widely on self-control, moral responsibility, philosophy of mind and related topics. His most recent book is *Consciousness and Moral Responsibility* (Oxford University Press 2014).

Diana Martinez is Associate Professor of Psychiatry at Columbia University Medical Center, New York, USA. She specializes in Positron Emission Tomography (PET) imaging research, which can be used to image brain neurochemistry. Her research focuses on using PET to direct better treatments for drug addiction by investigating the underlying chemistry and using neuroscience to develop treatment approaches.

Steve Matthews is Senior Research Fellow at the Plunkett Centre for Ethics, St Vincent's Hospital, and also at the Centre for Moral Philosophy and Applied Ethics, Australian Catholic University in Sydney, NSW, Australia. His previous positions include appointments to Macquarie University, Charles Sturt University, and Monash University. He works primarily in applied ethics, moral psychology and the philosophy of psychiatry.

Doug McConnell is a Postdoctoral Research Fellow at Charles Sturt University, Sydney, NSW, Australia. His research interests include moral psychology, bioethics, and applied philosophy, particularly in relation to addiction and psychiatry. His recent work, "Narrative self-constitution and recovery from addiction" in *American Philosophical Quarterly*, investigates the effect of self-conceptual content and structure on self-governance.

Susan Michie is Professor of Health Psychology and Director of the Centre for Behaviour Change at University College London, UK (www.ucl.ac.uk/behaviour-change). Her research focuses on behavior change in relation to health: how to understand it theoretically and apply theory to intervention development, evaluation and implementation. She has also developed innovative methods for intervention reporting and evidence synthesis, working across disciplines such as information science and computer science. She leads the Human Behaviour-Change Project (www.humanbehaviourchange.org).

F. Gerard Moeller, MD, is Professor in the Departments of Psychiatry and Pharmacology and Toxicology at Virginia Commonwealth University, Richmond, VA, USA, where he also serves as the Division Chair for Addiction Psychiatry, the Director of the Institute for Drug and Alcohol Studies, and the Director of Addiction Medicine. His main research interest is in the identification of the neurobiological substrates of individual differences (with particular emphasis on impulsive behaviors) to guide therapeutic treatment strategies.

Stephen J. Morse is Ferdinand Wakeman Hubbell Professor of Law, Professor of Psychology and Law in Psychiatry and Associate Director of the Center for Neuroscience and Society at the University of Pennsylvania, Philadelphia, PA, USA. He was Co-Director of the MacArthur Foundation Law and Neuroscience Project and is a Diplomate in Forensic Psychology of the American Board of Professional Psychology. He is also a past president of Division 41 of the American Psychological Association; a recipient of the American Academy of Forensic Psychology's Distinguished Contribution Award; a recipient of the American Psychiatric Association's Isaac Ray Award for distinguished contributions to forensic psychiatry and the psychiatric aspects of jurisprudence; a member of the MacArthur Foundation Research Network on Mental Health and Law; and a trustee of the Bazelon Center for Mental Health Law.

Dominic Murphy is Associate Professor of History and Philosophy of Science at the University of Sydney, NSW, Australia. He is the author of *Psychiatry in the Scientific Image* (MIT Press 2006).

Peg O'Connor is Professor of Philosophy at Gustavus Adolphus College, St. Peter, MN, USA. Her most recent book is *Life on the Rocks: Finding Meaning in Addiction and Recovery* (Central Recovery Press 2016). She writes a blog, "Philosophy Stirred, Not Shaken," for psychologytoday.com.

Rafael O'Halloran is a Senior Scientist at Hyperfine Research (Guilford, CT, USA) where he is developing novel MRI applications and specializes in methods development in diffusion-weighted MRI for the advancement of our understanding of neuropsychiatric disorders.

Jim Orford trained in clinical psychology and obtained his PhD at the Institute of Psychiatry in London before moving to a joint University/Health Service post in Exeter, and subsequently to the University of Birmingham, UK, where he is now Emeritus Professor of Clinical and Community Psychology in the School of Psychology. He has researched and written extensively about alcohol, drug and especially gambling problems. Probably his best-known work is *Excessive Appetites: A Psychological View of Addictions* (2nd edition, Blackwell-Wiley 2001), and his most recent work on the subject is *Power, Powerlessness and Addiction* (Cambridge University Press 2013). In 2012 he set up the Gambling Watch UK website, campaigning for a public health approach to gambling.

Nancy M. Petry, PhD, is a Professor at the University of Connecticut School of Medicine, Farmington, CT, USA, and the Editor of *Psychology of Addictive Behaviors*. Her research focuses on behavioral treatments for addictive disorders, ranging from substance use disorders to pathological gambling.

Contributors

Hanna Pickard is Professor in Philosophy of Psychology at the University of Birmingham, UK, and a Visiting Research Scholar in the Program of Cognitive Science, Princeton University, Princeton, NJ, USA, 2017–19. In addition to her academic work, from 2007–17 she worked in a therapeutic community for people with personality and related disorders. Website: www.hannapickard.com.

Marc N. Potenza is Professor of Psychiatry, Child Study and Neuroscience at the Yale School of Medicine, New Haven, CT, USA, and a Senior Scientist at the National Center on Addiction and Substance Abuse. He is Director of the Yale Center of Excellence in Gambling Research.

Peter Railton is the G.S. Kavka Distinguished University Professor and J.S. Perrin Professor of Philosophy at the University of Michigan, Ann Arbor, MI, USA. His primary areas of research are ethics, metaethics, moral psychology, and the philosophy of science. He is a fellow of the American Academy of Arts and Sciences, and his work has appeared in the *Philosophical Review*, *Ethics*, *Philosophy of Science*, *Philosophy and Public Affairs*, *Cognition*, and *Perspectives in Psychological Science*. He is the author of *Facts, Values, and Norms* (Cambridge University Press 2003) and co-author of *Homo Prospectus* (Oxford University Press 2016).

Nicolas Ramoz is a researcher, INSERM tenure track position, and expert in genetics, molecular biology and pharmacogenetics in psychiatric and addictive disorders. He also works on human genetics and epigenetics. He has published 60 papers (h-index=23) and 10 book chapters.

Mike J.F. Robinson is Assistant Professor of Psychology and Assistant Professor of Neuroscience & Behavior at Wesleyan University, Middletown, CT, USA. He graduated from the University of Sussex with a First Class Honors degree in Neuroscience, and was granted a Master's degree and PhD from McGill in Experimental Psychology. His research examines forms of excessive motivation and desire, and risky decision-making in drug addiction, gambling and diet-induced obesity.

Terry E. Robinson is Elliot S. Valenstein Distinguished University Professor of Psychology and Neuroscience at the University of Michigan, Ann Arbor, MI, USA. He received his undergraduate degree from the University of Lethbridge, a Master's degree from the University of Saskatchewan and his PhD in Psychology (Biopsychology) from the University of Western Ontario. His research concerns the persistent behavioral and neurobiological consequences of drug use, and the implications of these for addiction and relapse.

Robin Room is a Professor at the Centre for Social Research on Alcohol and Drugs, Stockholm University, Stockholm, Sweden, and the Centre for Alcohol Policy Research, La Trobe University, Melbourne, Australia.

Erica Schulte, MS, is a doctoral student of psychology in the clinical science area at the University of Michigan, Ann Arbor, MI, USA.

Fuschia Serre graduated with a Master's degree in neuroscience and a PhD in addiction psychology research from the University of Bordeaux, France. She is a research engineer at the Addiction research team of Sanpsy CNRS USR 3413 at University of Bordeaux, Bordeaux, France.

Contributors

Paige M. Shaffer, MPH, is an Epidemiologist for the Department of Public Health in Massachusetts, USA. She conducts research and writes on topics related to epidemiology, public health, homelessness, psychiatry, and addiction. She has consulted on numerous addiction-related projects with Laval University, the World Health Organization, and Harvard Medical School.

Howard J. Shaffer, PhD, is the Morris E. Chafetz Associate Professor of Psychiatry in the Field of Behavioral Sciences at Harvard Medical School, Boston, MA, USA; he is also the Director of the Division on Addiction at the Harvard-affiliated Cambridge Health Alliance. His research, writing, and teaching have shaped how the health care field conceptualizes and treats the full range of addictive behaviors.

Rajita Sinha, PhD, is the Foundations Fund Endowed Professor in Psychiatry, Neuroscience and Child Study at the Yale University School of Medicine, New Haven, CT, USA. She is a licensed Clinical Psychologist and Clinical Neuroscientist, Chief of the Psychology Section in Psychiatry and Co-Director of Education for the Yale Center for Clinical Investigation (home of Yale's NIH-supported Clinical Translational Science Award). She is also the Founding Director of the Yale Interdisciplinary Stress Center that focuses on the stress, reward and self-control mechanisms that promote addiction and chronic disease risk, and the development of interventions to address these mechanisms. She has published over 250 scientific peer-reviewed publications, and serves as the North American Senior Editor for *Addiction Biology* and is on the Editorial Board of *Neuropsychopharmacology*.

Walter Sinnott-Armstrong is Stillman Professor at Duke University, Durham, NC, USA, in the Philosophy Department, the Kenan Institute for Ethics, the Center for Cognitive Neuroscience, and the Law School. He publishes widely in ethics, moral psychology and neuroscience, philosophy of law, epistemology, informal logic, and philosophy of religion.

Gemma Lucy Smart is a MSc Candidate at the University of Sydney, NSW, Australia. Her current research interests lie in the philosophy of psychiatry and neuroscience, with a particular focus on addiction. Her other research interests include science and ethics, and human geography. She has published on the social application and implications of agricultural science in remote developing regions, and homelessness among urban Indigenous Australians.

Anke Snoek is a Postdoctoral Researcher at the University of Maastricht, Maastricht, Netherlands, where she works on a project on parental alcohol abuse. Her work combines empirical and theoretical approaches, and is situated at the intersection of philosophy, neuroethics, and qualitative interviews. She obtained her PhD at Macquarie University, and her Master's degree at the University of Humanistic Studies.

Chandra Sripada is Associate Professor of Philosophy and Psychiatry at the University of Michigan, Ann Arbor, MI, USA. His research employs cross-disciplinary approaches to study agency, self-control, and free will. His recent work appears in leading journals in philosophy, psychology, and neuroscience.

Roger J. Sullivan is a Professor of Anthropology at California State University, Sacramento, CA, USA. He is a biological anthropologist with interests in mental illnesses, social cognition and substance use. He conducts field work on mental health in Oceania.

Contributors

Jesse S. Summers is an Academic Dean in Trinity College of Arts & Sciences, Kenan Fellow in the Kenan Institute for Ethics, and Adjunct Assistant Professor of Philosophy at Duke University, Durham, NC, USA. His research focuses on philosophical issues surrounding irrationality, including rationalization, anxiety and anxiety disorders, addiction, and compulsion.

Şerife Tekin is an Assistant Professor of Philosophy at the University of Texas at San Antonio, TX, USA and an Associate Fellow of the Center for Philosophy of Science, at the University of Pittsburgh, Pittsburgh, PA, USA. Her work has appeared in journals such as *Synthese*; *Philosophy, Psychiatry and Psychology*; *Public Affairs Quarterly*; *Journal of Medical Ethics*; *Philosophical Psychology*; *The American Journal of Bioethics*.

Antonio Verdejo-Garcia is Associate Professor in the School of Psychological Sciences and the Monash Institute of Cognitive and Clinical Neurosciences (MICCN) at Monash University, VIC, Australia. He has an Adjunct Honorary Appointment in Turning Point (Eastern Health). His research focuses on the cognitive and neural mechanisms underpinning executive control and decision-making, and their implications for addiction and obesity. His work has attracted over 7,000 citations. He sits on the Editorial Board of the journals *Addiction*, *PLoS One*, *The American Journal of Drug and Alcohol Abuse*, *Current Addiction Reports*, and *Adicciones*.

Nora D. Volkow, MD, PhD, has served as the Director of the National Institute on Drug Abuse (NIDA), Bethesda, MD, USA, since 2003. Her pioneering use of brain imaging to investigate the toxic effects and addictive properties of drugs was instrumental in demonstrating that substance use disorder is a disease of the brain.

Jerome C. Wakefield, PhD, DSW, is University Professor, Professor of Social Work, Professor of the Conceptual Foundations of Psychiatry in the School of Medicine, and Associate Faculty in the Center for Bioethics at New York University, New York, USA. He is co-author of *The Loss of Sadness: How Psychiatry Transformed Normal Sorrow into Depressive Disorder* (Oxford University Press 2007), named best psychology book of 2007 by the Association of Professional and Scholarly Publishers.

Robert West is Professor of Health Psychology at University College London. He is Editor-in-Chief of the journal *Addiction* and has published more than 600 academic works. His research includes population studies of smoking cessation and the development and evaluation of smoking cessation interventions. He has co-authored the book *Theory of Addiction* (2nd edition, Wiley-Blackwell 2013).

Seth W. Whiting is a Faculty Member in the Department of Psychology and the Assistant Director at the Central Autism Treatment Center at Central Michigan University, Mount Pleasant, MI, USA. His main research interests are in problem gambling and other behavioral addictions, impulsivity, and choice.

Kristyn Zajac, PhD, is an Assistant Professor at the University of Connecticut School of Medicine, Farmington, CT, USA. Her research focuses on the development and refinement of treatments for substance use and mental health problems among high-risk adolescents and young adults.

Anna Zilverstand, PhD, is an Assistant Professor at the Department of Psychiatry at the Icahn School of Medicine at Mount Sinai, New York, NY, USA, who investigates drug addiction through clinical, neurocognitive and multimodal neuroimaging methods.

ACKNOWLEDGEMENTS

We would like to thank all of the contributors to this *Handbook* for their patience and perseverance in working with us to see the project through to completion, and for writing such excellent chapters. We are also grateful to Tony Bruce and Adam Johnson at Routledge for their encouragement and advice, and to Joy Shim for her dedication and meticulousness as a research assistant. Hanna Pickard would like to thank Ian Phillips for his constant support and intellectual companionship, Serge Ahmed for being a wonderful collaborator and source of intellectual inspiration, and, finally, the many people who have shared their experience of drug use and addiction with her over the years. Her work on this project was supported by The Wellcome Trust, The University of Birmingham, and Princeton University. Serge Ahmed would like to thank Sallouha Aidoudi for her unwavering support, the members of his lab for their dedication and inspiration, and, last but not least, Hanna Pickard for having been the perfect collaborator – generous, brilliant, and funny – throughout this intellectual adventure. His work on this project was supported by the French Research Council (CNRS), the Université de Bordeaux, and the Fondation pour la Recherche Médicale (FRM DPA20140629788).

INTRODUCTION

Hanna Pickard and Serge H. Ahmed

This *Handbook* collects together original, substantial, and in places provocative perspectives on addiction, written by leading researchers in the field from a variety of disciplines. In today's world, addiction is frequently discussed in the media, especially in relation to politics and policy debates bearing on criminal justice, public health, and society at large; as well as widely represented in literature, film, music, and other forms of art and culture. It also touches many of our lives personally in one way or another. Perhaps this is one reason why addiction research is a field shaped by questions as much as answers, and disagreement as much as consensus. It is inevitable that all of us with an interest in addiction – whatever the reason, academic or other – are potentially influenced by popular media and cultural portrayals, as well by our own personal experience and any corresponding attempt we have made to make sense of and come to terms with addiction for ourselves. In addition, what addiction researchers say *matters*. It stands to affect policy debates, public perception, and ultimately addicts themselves – many of whom are the most vulnerable and marginalized people in our communities, and who experience tremendous distress and suffering. As a result, when it comes to addiction, the objectivity required for genuine progress in understanding is not always easy to come by.

We have therefore encouraged the contributors to this *Handbook* to write from their own disciplinary and, where relevant, personal perspectives, but also not to shy away from challenging received wisdom any more than they would shy away from challenging self-evident hyperbole or myth. We have also included an extremely wide range of disciplinary perspectives encompassing animal models, anthropology, behavioral economics, clinical research and treatment, cognitive psychology, documentary photography, health and personal ethics, epidemiology, evolutionary biology, genetics, history, law, neuroscience, philosophy, psychiatry, public policy, social justice studies, and sociology; in addition, many of the chapters themselves encompass more than one of these disciplines. Our hope is that, in collecting together this diversity of bold, state-of-the-art, yet accessible chapters, disagreements can be made visible, debates about how to resolve them can begin, and new and better questions will emerge. Like the parable of the blind men and the elephant, addiction cannot be understood from one perspective alone. It is only when these multiple disciplinary perspectives are accorded their due and intelligently pieced together that a credible and complete understanding of addiction is possible.

The *Handbook* is divided into three Parts. Part I addresses the question of what addiction is. Here, as elsewhere in the field, there is no consensus. Researchers disagree about the very nature of addiction: how it should be defined; how it should be studied; what about it requires explanation; how it can be delineated from non-addictive drug use; what kind of thing it is; what kinds of things represent instances of it; which species can in theory be addicted; and which individuals in fact are. The chapters in Part I Section A **Conceptions of addiction** represent some of the more abstract and conceptual contributions to the *Handbook* in their attempt to grapple with these questions. They collectively explore the importance of ideas of value, preference, choice, control, compulsion, belief, identity, explanation, disorder, dysfunction and evolution to the nature of addiction. The chapters in Part I Section B **Varieties, taxonomies, and models of addiction** approach the question of what addiction is from a more concrete orientation. They consider our practices of diagnosis, classification, and research methods, as well as two particular kinds of behavioral addictions (gambling and food) and one particularly unusual kind of addict (non-human animals).

Part II addresses the question of why addiction occurs. As this Part amply demonstrates, this question is not best approached by searching for a single cause or kind of explanation. Rather, the chapters in this Part each explore one part of the answer – some adopt a broader, large-scale perspective while others take a narrower, small-scale one. The chapters in Part II Section A **Anthropological, historical, and socio-psychological perspectives** consider some of the wider cultural variables in the explanation of addiction. They detail how political and socio-economic forces create conditions for the development and entrenchment of addiction – as well as charting the history of addiction research itself and investigating how this history has shaped our understanding of the phenomenon to be explained. The chapters in Part II Section B **Developmental processes, vulnerabilities, and resilience** trace some of the pathways to addiction involving the interaction between biological, genetic, personality, population and environmental factors. The chapters in Part II Section C **Psychological and neural mechanisms** review and assess our increasing knowledge of the cognitive psychology and neuroscience underpinning addiction, addressing the nature of learning processes, decision-making, motivation, reward expectancy, and brain connectivity in addiction. In addition, two chapters in this Section offer a philosophical perspective on the brain sciences of addiction, one complementary and the other more skeptical.

Part III addresses the question of the meaning of addiction in today's world and what we should do about it. Part III Section A **Listening and relating to addicts** contains chapters that aim to bring the voices of addicts themselves and those who care about them into the dialogue created by this *Handbook*. In one way or another, these chapters all emphasize the ethics of care and the importance of maintaining humanity in our relations with addicts, as well as bringing into focus the terrible distress and suffering addiction can involve. The chapters in Part III Section B **Prevention, treatment, and spontaneous recovery** describe and assess evidence-based interventions to help addicts abstain from drug use or to manage it in less damaging ways, alongside the phenomenon of natural remission, with an eye to how this knowledge can shed light on what addiction is. The chapters in Part III Section C **Ethics, law, and policy** offer political and socio-economic critiques relevant to drug policy, as well as addressing a host of pressing legal issues pertaining to drug legalization and regulation, criminal prosecution, and mandatory treatment both in criminal justice contexts and in cases of civil commitment.

In dividing the *Handbook* this way, we aim to offer a structure that helps organize and systematize the different elements necessary for a credible and complete understanding of addiction – what it is, why it occurs, what it means to people and how we should respond to

Introduction

it both as individuals and as a society – and display how different disciplinary perspectives not only can but must contribute to this understanding. But the divisions are nonetheless somewhat arbitrary. Many if not most of the chapters in this *Handbook* contain material relevant to more than one Section or Part. We hope that readers may come to challenge this structure and find connections between the disparate elements, so as to overcome what remains of our collective blindness in understanding "the elephant" of addiction.

PART I

What is addiction?

SECTION A

Conceptions of addiction

1
THE PUZZLE OF ADDICTION

Hanna Pickard

The orthodox conception of drug addiction[1] within science and medicine is a neurobiological disease characterized by compulsive drug use despite negative consequences (cf. NIDA 2009; WHO 2004). This conception depends on three core ideas: disease, compulsion, and negative consequences. Yet the meaning of the ideas of disease and compulsion, and the significance of negative consequences, is rarely made explicit. I argue that it is only when the significance of negative consequences is appreciated that the puzzle of addiction comes clearly into view; and I suggest that there are both conceptual and empirical grounds for skepticism about the claim that addiction is a form of compulsion, and agnosticism about the claim that addiction is a neurobiological disease. Addiction is better characterized as involving choices which, while on the surface puzzling, can be explained by recognizing the multiple functions that drugs serve, and by contextualizing them in relation to a host of interacting factors, including psychiatric co-morbidity, limited socio-economic opportunities, temporally myopic decision-making, denial, and self-identity.[2]

The significance of negative consequences

As characterized by the orthodox conception, codified in diagnostic manuals, and of course widely known, drug addiction has severe negative consequences. These typically include the neglect of other pleasures and interests; the inability to fulfil important social and occupational roles and responsibilities; ruined relationships; the loss of social standing and community; cognitive impairment and mental health problems; physical disability and disease; and, lastly, death (cf. APA 2013; WHO 1992). In addition, addiction can be a source of terrible shame, self-hatred, and low self-worth (Flanagan 2013 and in this volume). From an ethical and public policy perspective, such pain and suffering matters straightforwardly, simply because it demands our help. However, from a theoretical perspective, negative consequences matter because they pinpoint what it is about addiction that demands explanation.

Common sense suggests that if a person knows that an action of theirs will bring about negative consequences and they are able to avoid doing it, then they do. We act, so far as we can, in our own best interests and the interests of others we care for. This is a basic folk psychological rule of thumb for explaining and predicting human action, ubiquitous in our ordinary interaction with and understanding of each other. But this is what addicts seem not to do. Although addiction has severe negative consequences, addicts continue to use drugs. This is the puzzle of addiction: *why do addicts keep using drugs despite negative consequences?*

The orthodox conception of addiction offers a parsimonious and powerful solution to this puzzle. To use a common metaphor, the explanation is that addiction "hijacks" the brain, so that addicts lose all control and cannot help taking drugs, despite the consequences and against their best interests. Hence the puzzle of why addicts keep using drugs despite negative consequences can be straightforwardly explained. If addicts could avoid using drugs, they would – but they can't, so they don't. The reason is simple: they suffer from a neurobiological disease that renders use compulsive.

No doubt there are many reasons why the orthodox conception of addiction has become so dominant. These include socio-historical, political, and economic forces (Heyman 2009; Satel and Lillienfeld 2013), arguably alongside a widespread belief that framing addiction as a disease is crucial for fighting blame and stigma and getting addicts the help they need (Volkow *et al.* 2016; but for critical discussion, see Hall *et al.* 2015 and Lewis 2015; Pickard 2017b articulates how choice models can combat blame and stigma). But, from a theoretical perspective, the orthodox conception's explanatory power is strong evidence in its support: it appears to solve the puzzle of addiction.

Compulsion

The orthodox conception's solution to the puzzle has two parts. The first appeals to compulsion to explain use in face of negative consequences. The second appeals to neurobiological disease to explain compulsion.

Consider first the idea of compulsion. There is no agreed definition. But it is standardly understood to mean an irresistible desire: a desire so strong that it is impossible for it not to lead to action. From a folk psychological perspective, we do not ordinarily conceive of our desires as irresistible. Desires may be strong and persistent. It may require sustained effort and concentration not to act on them. Meanwhile, the alternative actions genuinely available to us may be limited and the costs of not acting may be high. As a result, our desires may be hard to resist. In addition, in many circumstances, it may be justifiable not to resist, given the balance of costs for and against acting. But this is not the same as irresistibility. Desires that are hard to resist yet leave us some power to do other than what we desire should we choose: it is possible not to act on them. This possibility is what compulsion removes. Compulsion strips a person of all choice and power to do otherwise. If the desire for drugs is irresistible, then it is impossible for addicts not to use drugs. As Carl Elliott expresses this claim, an addict "must go where the addiction leads her, because the addiction holds the leash" (Elliott 2002: 48).

The appeal to compulsion understood as irresistible desire is key to the orthodox conception's explanation of persistent use in the face of negative consequences. Suppose that, even if the desire to use is hard to resist, it is not irresistible. Then the question of why use persists in the face of negative consequences remains. For, given the severity of these consequences, the *difficulty* of resisting – as opposed to the *impossibility* of resisting – is not by itself explanatory. We need to know more. The point is not that this cannot be explained; indeed, my aim in what follows is to explain it. The point is rather that the parsimony and power of the orthodox conception to explain the puzzle of addiction depends on an appeal to compulsion understood as irresistible desire. Softening the meaning of compulsion costs the orthodox conception its explanatory force.

Are addictive desires irresistible? Cravings are of course a central component of addiction (Auriacombe *et al.* in this volume; Robinson *et al.* in this volume). When access to drugs is limited, the desire for them can be psychologically encompassing and distressing. When there is in addition a state of dependence, withdrawal can cause physical suffering. No one should deny that the desire to use drugs is extremely strong or minimize the very real struggle addicts face not using (for a discussion of self-control, see Henden in this volume). But there is increasing evidence that addicts are not compelled to use. They are responsive to incentives, suggesting that the desire to use is not irresistible.

Here is a brief review of the evidence. Anecdotal and first-person reports abound of addicts who are diagnosed as dependent (and so suffer withdrawal) going "cold turkey" (cf. Heyman 2009, 2013a). Large-scale epidemiological studies suggest that the majority of addicts "mature out" without clinical intervention in their late twenties and early thirties, as the responsibilities and opportunities of adulthood, such as parenthood and employment, increase (for a review of these findings see Heyman 2009 and in this volume; but for criticism of this interpretation of the data, see Anthony in this volume). Rates of use are cost-sensitive: indeed, some addicts choose to undergo withdrawal in order to decrease tolerance, thereby reducing the cost of future use (Ainslie 2000). There is increasing evidence that contingency management treatment improves abstinence and treatment-compliance, compared with standard forms of treatment such as counselling and cognitive-behavioral therapy, by offering a reward structure of alternative goods, such as modest monetary incentives and small prizes, on condition that addicts produce drug-free urine samples (Zajac *et al.* in this volume). Experimental studies show that, when given a choice between small sums of money and taking drugs then and there in a laboratory setting, addicts will often choose money over drugs (Hart *et al.* 2000; Hart 2013). Finally, since Bruce Alexander's classic "Rat Park" experiment (Alexander *et al.* 1978, 1985), animal research on addiction has convincingly demonstrated that, although the majority of cocaine-addicted rats will escalate self-administration if offered no alternative goods, they will forego cocaine and choose alternative goods, such as sugar, saccharin, or same-sex snuggling, if these are available (Ahmed 2010; Zernig *et al.* 2013).

This evidence is strong, but we need nonetheless to be careful in drawing conclusions. There is a basic, common-sense distinction between what a person can do but won't (because they are not motivated) and what a person wants to do but can't (because they lack the ability) (Pickard 2012, 2017a). The evidence shows that the majority of addicts have the ability to refrain from use in many ordinary circumstances. But it does not demonstrate they have the ability in all possible circumstances. The attribution of an ability to refrain from use is consistent with there being occasions where, due to any variety of constraints, it cannot be exercised. Nor does the evidence demonstrate beyond doubt that the minority of addicts who do not respond to incentives have the ability to refrain but don't exercise it, rather than not having the ability at all. In the absence of any clear marker between different sub-groups of addicts that would explain the difference between the majority who refrain and the minority who don't, the evidence suggests the latter are like the former in having the ability and unlike them in not exercising it, but it does not conclusively establish this. Finally, there is the important question of how to understand conflicting self-reports from addicts, who often oscillate both intra- and inter-personally between using the language of compulsion and the language of choice (for discussion see Booth Davies 1992; Pickard 2012, 2017a). For all these reasons, caution is needed in interpreting the evidence. Nonetheless, our understanding of addiction should reflect what the evidence clearly does show, namely that, for many addicts, on many occasions, they are not compelled to use. For this reason, an appeal to compulsion understood as irresistible desire cannot be the fundamental explanation of the puzzle of persistent use despite negative consequences. It is simply not true, of too many addicts, too much of the time.

Neurobiological disease

Consider now the idea of disease. What does this mean? Our ordinary concept of disease is complicated, as well as having important social and personal consequences in our culture, including a claim to care and a removal of responsibilities in virtue of occupying "the sick role" (Parsons 1951). But, in simple terms, it typically invokes the idea of underlying pathology as the cause of observable surface-level symptoms and suffering. For example, consider the way core symptoms of Parkinson's Disease, like tremor and slow movement, are caused by brain degeneration. With respect to addiction, the surface-level symptom is drug use, and the suffering is the negative consequences thereby caused. In characterizing addiction as a neurobiological disease, the orthodox conception explains this symptom (and hence the consequent suffering) by appeal to underlying brain pathology. Addicts use drugs because something is wrong with their brains.

The ordinary concept of disease therefore invites an appeal to compulsion because prototypical symptoms of diseases are passive occurrences – things that happen to us rather than things we do. But it is possible to reject the claim that drug use is compulsive while yet maintaining that addiction is nonetheless a neurobiological disease. Addiction could be a "disease of choice" if the neural changes and processes underlying drug choices that are found in addicts are pathological (Berridge 2017).

It is important to be clear that long-term heavy drug use has chronic effects on the brain (Zilverstand *et al.* in this volume). Drugs directly affect levels of synaptic dopamine as opposed to affecting them only indirectly via the neural states and processes sub-serving learning and reward. This can explain why cues associated with drugs trigger a desire that over-estimates their anticipated reward and hence is unusually strong in its motivational strength (Redish *et al.* 2008; see too Levy in this volume). Over time, *wanting* drugs may even come apart from *liking* them: cues may trigger cravings and strongly motivate drug-seeking and drug-taking, even though drug experience offers less pleasure than it initially did or than appears commensurate with the desire to use (Robinson *et al.* in this volume; cf. Holton and Berridge 2013). In line with what was argued above, these neural changes and processes do not establish that the desire for drugs is irresistible and use is compulsive. Rather, they explain (among other things) the intensity of the desire. But are they pathological?

The answer is that we do not yet know. Just as we cannot infer irresistibility and impossibility from descriptions of underlying neural states and processes, so too we cannot infer pathology. On the one hand, from a theoretical perspective, there is no agreed understanding in philosophy or in medicine of what makes a state or process pathological. However, this much is clear: deviation, however extreme, from the statistically average states and processes characteristic of any relevant level of explanation, whether that is personal-level, cognitive-psychological, or neurobiological, is not enough. Atypicality is neither necessary nor sufficient for pathology, as there is tremendous variation between individuals and some pathologies are near universal (cf. Boorse 1977). Rather, we need an account of *the natural or proper function* of a state or process relative to a level of explanation in order to judge whether or not the difference in question counts as pathological – not just as atypical but as *dysfunctional* relative to that level. How are the processes sub-serving learning and reward supposed to function at the neurobiological level, and does their functioning in response to drugs constitute a pathology? Although it is tempting to answer yes, the truth is that it is not possible at present to settle these questions (cf. Stephens and Graham 2009; Levy 2013; for an argument that the neural changes and processes underlying addiction represent normal learning, see Lewis 2015). On the other hand, from an empirical perspective, although our knowledge of the chronic effects of drugs on the brain is ever-increasing, we do not yet have animal or human studies directly comparing dopamine responses in addicts caused by drug versus non-drug reward cues (e.g., sex or food), or directly comparing dopamine responses to drug cues in addicted subjects to dopamine response to non-drug reward cues in non-addicted subjects, in order to establish the difference. In other words, we do not yet know even how atypical an addicted subject's neural response to drug cues is.

The orthodox conception explains the puzzle of why addicts use drugs despite negative consequences by appeal to the claim that use is both compulsive and caused by a neurobiological disease. But the evidence is strong that use is not compulsive. And it is at present unclear whether the neural changes and processes underlying drug choices are correctly considered pathological. To avoid any possible confusion, I want to emphasise that adopting an agnostic attitude towards the claim that addiction is a *neurobiological disease* in no way entails that the cognitive and brain sciences cannot explain many aspects of addiction. They clearly can – as they can explain many aspects of the human mind and behavior more generally. Many of the factors I adduce below to solve the puzzle of addiction can be illuminated by scientific investigation. And some of the neural changes and processes underlying drug choices may ultimately prove to be pathological, as our theoretical understanding of pathology, and our empirical knowledge of the effects of drugs on the brain, increases. The point is rather that the question of disease is more delicate than the orthodox conception has acknowledged, and that, however this question is ultimately resolved, in absence of an appeal to compulsion, the orthodox conception cannot explain the puzzle of addiction.

The puzzle of addiction

Why then do addicts use drugs despite negative consequences? Given that use is not compelled, the puzzle of addiction is a puzzle of choice. We need to understand why addicts use drugs despite negative consequences *when they have choice*. To answer this question, consider first a more basic one. Why do people use drugs at all?

Strikingly, there is no puzzle at all with respect to this question. Alongside factors such as cultural expectations (Flanagan in this volume) and drug availability, drugs offer means to fulfilling many self-evidently valuable ends. Christian Muller and Gunter Schumann (2011) delineate the following seven clearly documented functions of drugs, identifying the common types of psychoactive substances and neuropharmacological mechanisms relevant to each: (1) improved social interaction; (2) facilitated mating and sexual behavior; (3) improved cognitive performance and counteracting fatigue; (4) facilitated recovery and coping with psychological stress; (5) self-medication for mental health problems; (6) sensory curiosity – expanded experiential horizons; and, finally (and in ways most self-evidently), (7) euphoria and hedonia. In addition, arguably drugs can offer socially isolated and ostracized individuals a sense of self-identity and a community to belong to (cf. Dingle *et al.* 2015; Flanagan in this volume). The relationships and reciprocal bonds between members of highly vulnerable and marginalized drug users are striking and strong (cf. Bourgois and Schonberg 2009). I discuss self-identity and community further below. The point here is that drugs not only bring pleasure but in addition serve many other valuable functions: drugs have multiple benefits.

The importance of this point is often overlooked. It is routinely emphasized in addiction research that addicts sometimes report no longer liking their drug of choice even when they persist in taking it (cf. Robinson *et al.* in this volume; Volkow *et al.* 2016); and it is no doubt the case that pleasure typically decreases as tolerance increases. Yet many, if not indeed most, addicts continue to find pleasure in drug use despite their addiction. Moreover, few if any of the other functions served by drugs are mediated by pleasure. For example, drugs can numb anxiety and other negative emotions, remove sexual and social inhibitions, counteract fatigue and stress, relieve boredom, and provide a sense of identity and community, without inducing pleasure. In other words, whether or not pleasure persists in addiction, other valuable drug functions do (see Flanagan in this volume for first-person reports that speak to this point).

This means that there is only a puzzle surrounding use when drugs come to have significant costs, such as the severe consequences characteristic of addiction, alongside the benefits. There is nothing notable about the choice to use drugs unless the balance between costs and benefits has tipped. And, in absence of any clear underlying neurobiological pathology, there is no sharp line determining when problem use becomes addiction. Costs and benefits can only be weighed in relation to values, which differ between people, including addicts (cf. Flanagan 2016; Sinnott-Armstrong and Summers in this volume). For example, people can care less or more about the loss of relationships and social standing, as weighed against whatever functions drug use is serving. Equally, context-specific external factors, from national practices of criminalization and policing, to socio-economic status (Hammersley in this volume; Orford in this volume), can both create and protect against costs. For example, in countries where drugs are not criminalized, addicts cannot be criminally charged, convicted, and sentenced for possession; or consider how, in contrast to a poor parent, a wealthy parent may be able to protect their child from some of the consequences of their addiction by employing a live-in nanny, thereby ensuring that more of their parental responsibilities are met and their relationship with the child better preserved. The lack of a sharp line dividing addiction from problem use is reflected in diagnostic criteria (APA 2013), but it has led some theorists to claim that the negative consequences of use are ancillary as opposed to core features of addiction, and should be removed from the construct (Martin *et al.* 2014). The difficulty with this suggestion is that, given the benefits of drugs, and, again, in absence of any clear underlying neurobiological pathology, it is only when costs exceed benefits that there is any puzzle of addiction – any reason to think that something is *wrong*.

So why then do addicts choose to use drugs when doing so has costs that look from the outside to outweigh the benefits? In addition to the facts that the desire to use is strong and persistent, and drug use is habitual and so requires concentration and effort to resist (Pickard 2012), there are at least five factors that are relevant to solving the puzzle.

Self-hatred and self-harm

Some addicts may use drugs not despite negative consequences but in part because of them. The basic folk psychological rule of thumb for explaining and predicting human action that creates the puzzle of addiction in the first place – namely, that people act, so far as they can, in their own best interests and the interests of others they care for – is only a rule of thumb. Human psychology also has a self-destructive streak, often found in people from backgrounds characterized by childhood adversity and mistreatment, and who may struggle with a negative self-concept alongside a range of mental health problems associated with addiction, especially personality disorders (Maté 2009; Pickard and Pearce 2013). People with such complex needs may deliberately and directly self-harm – through self-directed violence, such as cutting and burning, but also by other means, such as sexual and other forms of risk-taking behavior, overdosing, and, arguably, drug abuse quite generally. Addicts who share this mindset may not care about themselves enough to care about the negative consequences of use – indeed, they may, consciously or unconsciously, embrace these consequences, in keeping with a self-concept as worthless and deserving of suffering. Negative consequences only offer an incentive not to use drugs *if* a person values and cares about themselves. For people who don't, the costs of drug use may to some degree count as benefits, thereby solving the puzzle.

Human misery, limited socio-economic opportunities, and poor mental health

Some addicts may choose to continue to use drugs, notwithstanding the negative consequences, because the benefits outweigh the costs given a realistic appreciation of their circumstances and the options available (Pickard 2012). As noted above, the majority of addicts "mature out" in their late twenties and early thirties. Those for whom addiction remains a chronic problem are typically people from underprivileged backgrounds who also suffer from co-morbid psychiatric disorders, particularly anxiety, mood, and personality disorders, and who of course must equally face the stigma, stress, and other problems associated with long-term poor mental health (Compton *et al.* 2007; Regier *et al.* 1990) and lack of psychosocial integration (Alexander in this volume). The "self-medication" hypothesis has long been a staple of clinical understanding of drug use (Khantzian 1985, 1997). It is common knowledge that drugs offer relief from psychological distress. This is one of the well-documented functions of drugs listed above: we "drown our sorrows". For many chronic addicts, drugs may provide a habitual and, in the short-term, effective way of relieving suffering, caused by negative emotions alongside many other symptoms and problems typically experienced by people with mental health problems living in impoverished circumstances. Put crudely, drugs and alcohol offer a way of coping with stress, pain, and misery, when there is little possibility for genuine hope or improvement. For addicts in such circumstances there is no puzzle of addiction: the cost of abstinence is likely to be very high, while the benefits of drug use are many, and the alternative goods available or ways of relieving suffering are few.

Temporally myopic decision-making

Some addicts may choose to use drugs because, at the moment of choice, they value drugs more than they value a possible but uncertain future reward, such as improved wellbeing with respect to health, relationships, or opportunities, which is consequent on long-term abstinence. The disposition to discount the future relative to the present is a common feature of human psychology, standardly considered rational to the extent that, adjusting for the relative value of the rewards, the present reward is certain while the future reward is uncertain. But, in addition, human discount curves are typically hyperbolic, so that as a reward nears in time, its expected value increases sharply, creating shifts in preferences over time simply in response to current availability (Ainslie in this volume; cf. Heyman 2009). Addicts have steeply hyperbolic discount rates compared with the norm (Bickel and Marsch 2001; Bickel *et al.* 2014). When the drug is within immediate reach, addicts may prefer use to abstinence, even if, when the drug is not within reach, they prefer abstinence to use.

Ambivalence is characteristic of many cases of addiction. Addicts often report fluctuating desires and resolutions, alongside vacillating hope and despair, which lends a sense of psychological reality to hyperbolic discounting models. Moreover, the success of contingency management treatment (Zajac *et al.* in this volume) testifies to the role of discounting in explaining drug choices. It is remarkable that a small amount of money or a prize can provide sufficient incentive for addicts to forgo drugs, when the consequences of their addiction do not. However, the money or prize is directly and reliably available according to a fixed schedule upon the delivery of a drug-free urine sample. There is no significant delay in reward, and there is no significant uncertainty as to delivery. In comparison, the rewards of abstinence are not only temporally delayed, but also, for many addicts, extremely uncertain.

Unlike contingency management treatment rewards, the good life does not spring forth ready-made simply because an addict quits. There may be long-term physical and mental health problems that cannot be fixed simply through forgoing drugs. Equally, ruined relationships do not just snap back into shape, communities do not quickly forget, and jobs that were lost are not automatically regained. For those addicts who come from underprivileged backgrounds of poor opportunity, housing, education, and employment, opportunities do not simply materialize overnight. The creation of a life worth living requires work, and, for many addicts, the cards are stacked against them even if they kick their addiction.

Moreover, for addicts with complex needs, a "suicide option" may function to rationalize the discounting of any possible future reward consequent on a drug-free life, given the costs of abstinence in the present. The option of committing suicide can be very important to people who live with long-term psychological pain and distress, because it offers an escape that lies within their control if life becomes unbearable (Pickard 2015). In so far as drug use functions for an addict to offer relief from suffering, the cost of abstinence is very high unless and until alternative means of coping are available. The person must bear not only withdrawal and other drug-related effects of abstinence, but also the psychological pain and distress that the drugs were functioning to relieve. Hence, for addicts committed to a "suicide option", there is a serious question whether undergoing the costs of abstinence could ever be worth it. For if life becomes too unbearable, they will take the option, ensuring that there is no possible future reward for suffering in the present, and thereby eradicating its potential relevance to present decision-making. In this respect, death is the ultimate trump.

For this reason, even if the myopic temporal horizon characteristic of addiction may be in part pathological (Verdejo-Garcia in this volume), it may also be in part rational, taking into account the life circumstances and options realistically available to many addicts (Heyman 2013b; Pickard 2017a). But, either way, discounting models can explain why addicts choose to use: the future benefits of abstinence (alongside the future costs of drug use) only provide incentive *not* to use *if* they are represented as outweighing the present benefits of drug use in decision-making. There is no puzzle of addiction if addicts are temporally myopic.

Denial

Despite the fact that addicts are notoriously prone to denial, it has received surprisingly little attention in both philosophical and scientific research on addiction. Denial is a psychological defense mechanism. It can be understood as a species of motivated belief or self-deception, whereby a person fails to believe the truth of a proposition because doing so would cause psychological pain and distress, and despite evidence in its favor that would ordinarily suffice for its acceptance (Pickard 2016). Denial can explain why addicts choose to use drugs despite negative consequences. If, despite the evidence, addicts are in denial that their drug use is causing negative consequences, then the disincentive to use that negative consequences constitute is effectively removed from their psychology, and cannot guide decision-making. There is no puzzle why drug use persists if the costs associated with it are not known, and denial blocks this knowledge.

How do people learn that their drug use has negative consequences? Although it can initially seem as if this is self-evident, it is not. One way or another, addicts have to *discover* that it does. The fact that one's drug use is causing negative consequences is not immediately manifest in experience, but requires acquiring *causal knowledge*.

There are at least two kinds of causal knowledge relevant to addiction, typically acquired by two corresponding routes. On the one hand, there are large-scale generalizations, such as the knowledge that smoking causes disease. Acquiring knowledge of large-scale generalizations typically depends on equally large-scale collective research efforts involving data collection and hypothesis testing and confirmation. For example, the causal link between smoking and disease was established by extensive longitudinal comparisons of smoking versus non-smoking populations, and confirmatory evidence from animal models. Once such large-scale generalizations are known in the research community, they can be disseminated to the public through channels such as the media and public education initiatives, and become available for use in individual decision-making. So, armed with the knowledge that smoking causes disease, one can choose not to smoke to reduce the risk of disease.

On the other hand, there are small-scale individual generalizations, pertaining to our own actions and their outcomes. We can often acquire this knowledge on the basis of our experience alone. If we observe an association between two events, such as an action of ours and an outcome, we can test the possibility of a causal relation, by intervening and manipulating the hypothesized cause (our action) while monitoring the effect (the outcome). For example, although we cannot discover that smoking causes disease on our own, we can potentially discover that, in our own case, smoking causes headache. We can do this by first noticing the association and then testing the hypothesis by controlling our actions: smoke a cigarette, then observe the effects; don't smoke, then observe the effects. Once this causal knowledge is acquired, it is available for use in individual decision-making, allowing us to achieve outcomes by means of interventions such as our own actions. So, armed with the knowledge that, in one's own case, smoking causes headache, one can choose not to smoke to avoid headache.

Acquiring causal knowledge of the negative consequences of drug use, and in addition putting this knowledge to work in individual decision-making, must therefore be seen as *an achievement*. With respect to large-scale generalizations such as health risks, individuals are dependent on scientific discovery and dissemination. In addition, for decision-making to be successfully guided by these generalizations, people must believe the information they are given, possess and exercise the capacity to reason probabilistically in order to assess individual risk, and overcome any tendency towards personal exceptionalism. With respect to small-scale individual generalizations, one's experience may not offer clear confirmation. Given that the causal network of relations is likely complicated and thickly interwoven, and drugs may well be contributory as opposed to single causes, interventions and manipulations may not yield knowledge. Suppose, for example, that you are an addict who opts not to use drugs on some occasion: you refrain from use. That is unlikely to mean that your problems, including those that may initially have been caused or exacerbated by drugs, disappear. For instance, the damage to your body is unlikely to be immediately reversed; the damage to your relationships is unlikely to immediately heal. Indeed, things may get worse before they get better, as life without drugs may be more of a struggle and contain more suffering than life with them. So an intervention (foregoing drugs) may not produce the effect (the disappearance of negative consequences of use) that would support the acquisition of knowledge of a causal relationship between them.

Because it is an achievement to acquire and deploy knowledge that drug use is causing negative consequences, the ground is ripe for denial to take root. There are multiple opportunities for information-processing biases and motivational and affective influences on cognition to interfere with knowledge acquisition and its use in decision-making (Pickard 2016).

Needless to say, there are also many reasons why addicts may be motivated to deny the negative consequences of drug use. These consequences are themselves frightening and upsetting. It can be shaming to acknowledge the harm one has done by one's addiction to oneself and also potentially to others one cares for (cf. Flanagan 2013). Finally, and in ways most obviously, acknowledging the negative consequences of use creates a demand, namely, *to desist from the behavior causing them* – that is, to quit drugs. Given the strength of the desire to use and the many valuable functions drugs serve, addicts are clearly motivated to use drugs, and, as a result, to deny that there are reasons not to, namely, the negative consequences of use. But, once in denial, there is no puzzle as to why addicts choose to use despite negative consequences.[3]

Self-identity

Some addicts are not in denial. Rather, they self-identify as addicts. This self-identification can be part of why addicts use drugs despite negative consequences. They use because they are addicts. Who else would they be?

There are two parts to this explanation. The first invokes the consequences of self-labelling or self-categorization, whereby a person identifies themselves as a member of a social group. Labelling and categorizing people is informative. Social groups are typically defined in part by sets of beliefs and standards of behavior. These are the norms determining what it means to be a member of that kind or category and to which individuals are expected to conform in virtue of their membership. Labelling or categorizing a person as a member of a social group therefore leads others to form expectations of them, based on the defining group beliefs and behavior – the group norms (Leslie 2017). Self-labelling or self-categorizing as a member of a social group provides norms by which to self-regulate (Turner 1987; Hacking 1996). We act as members of the social groups with which we self-identify are expected to act, conforming our beliefs and behavior to group norms. This can have both an explicit and implicit dimension. Self-regulation can be deliberate and controlled, but over time may become more ingrained and automatic.

People who self-identify as addicts are therefore likely to persist in drug use almost by default – after all, that is what it is to be an addict. However, this may be further compounded if they view addiction according to the orthodox conception, as a neurobiological disease of compulsion. If addicts think of themselves as powerless over their desire to use, then the possibility of not using is unlikely to be considered let alone pursued – we cannot rationally aim to do the impossible (Pickard 2012).

Self-categorization can have consequences for people's beliefs and behavior whether or not the self-identity it provides offers a positive sense of self. Addiction can be experienced as an *identity loss* – as destructive of all that was meaningful in life before drugs dominated (Mackintosh and King 2012; Flanagan in this volume). In such cases, recovery often involves rediscovering a past self that addiction has "spoiled" (Goffman 1963). But it is important to recognize that addiction can also be experienced as an *identity gain* (Dingle et al. 2015). This is the second part of the explanation.

As noted above, drug user communities can offer individuals a sense of self-identity and belonging, when they are otherwise socially isolated and ostracized. Self-identifying as a member of a social group does not only provide group norms by which to self-regulate. It can also provide a positive sense of self if one values the social group with which one identifies (Tajfel 1982; cf. Becker 1963). If one's self-esteem is derived largely from membership in a social group, one is all the more motivated to conform to its norms, on pain of rejection from the group and the loss of self-identity and self-worth this would engender. In this respect, drug use may represent an identity gain in so far as it brings meaning and community that is otherwise lacking into a person's life and so has genuine value, while it is quitting drugs that, at least initially, represents the identity loss.[4] In such cases, recovery requires fashioning what we might think of as an "aspirational" self (Dingle et al. 2015) – facing the question of who, if not an addict, one will be. This may be one of the many reasons why abstinence is aided by membership of recovery support groups (Buckingham et al. 2013). These can help to create and sustain a positive new self-identity, based on identification with group norms that do not support drug use, together with the value of peer acceptance, positive regard, and belonging that comes with group membership.

Hence addicts may continue to use despite negative consequences not only because they self-identify as addicts, but, in addition, because this self-identification is of value. It can provide a positive sense of self and a community of rich and complicated relationships – never mind a set of daily routines and structure. Without this self-identification, addicts may not know who they would be.

Conclusion

The puzzle of addiction is a puzzle of choice. Why do addicts choose to use drugs when doing so has costs that look from the outside to outweigh the benefits? We can solve the puzzle by recognising the multiple functions that drugs serve, and contextualizing them in relation to factors including, but not necessarily limited to, psychiatric co-morbidity, limited socio-economic opportunities, temporally myopic decision-making, denial, and self-identity. In other words, there is no single and unified explanation of addiction. All addicts may have a strong and persistent desire to use drugs, but people make choices relative to the psychological and socio-economic conditions they find themselves in, which are vastly diverse. Many addicts use drugs to gain relief from suffering, misery, and chronic mental health problems, especially when they face limited socio-economic opportunities and have no real alternative means for addressing these needs. They may feel hopeless and despairing for many reasons, including but not limited to their addiction, and so do not look towards the future when acting but remain focused only on the present. They may be in denial that they have a problem. They may feel lonely and lost without the identity and structure that drug use and the social bonds of a drug community can provide. Recognizing the multiple functions served by drugs and the need to contextualize drug choices reveals how some of the benefits of consumption may be hidden to us from the outside, as well as how some of the costs of consumption may be hidden to addicts from the inside. To understand addiction, we need to move beyond the orthodox conception of it as a neurobiological disease of compulsion, and acknowledge the importance of these many, diverse factors. To address it, we need to change them.

Acknowledgements

I would like to thank Serge Ahmed, David Lind, Shaun Nichols and Ian Phillips for extremely helpful comments and discussion.

Notes

1. I include alcohol as well as criminalized and pharmaceutical psychoactive substances open to abuse in the referent of the term "drugs."
2. Although my focus in this chapter is confined to drug addiction, the framework presented is potentially explanatory of behavioral addictions, despite the differences between them.
3. For a more detailed discussion of the nature of denial and its role in addiction see Pickard 2016 and Pickard and Ahmed 2016.
4. As an illustration, consider this self-report from a recovered addict: "Just as a person can feel loss of identity when they lose a long-standing job, or their children have grown and left home, it is also very common, I believe, to feel loss of identity when recovering from a drug-addicted lifestyle ... I had established myself as a druggie. My friends and family knew me as such, and in a way I was proud of my varied life experiences and my street-smarts. I'd had an older boyfriend who had introduced me to the drug scene, and who I learnt a lot of drug-taking practices from. I took pride in the fact that I knew

more about drug taking than most my own age ... At age 18 I already knew how to cook and filter different drugs for IV use, and how to prepare poppies to extract the opium, I knew dosages and strengths for illicit use of prescription meds, I knew all sorts about scoring and smoking dope and lots of quirky little tricks for increasing your buzz.... Seeing as I'd not done much else with myself over those formative years of early adulthood, I didn't have a heck of a lot else going on with my sense of identity ... I began to leave my drug identity behind, but felt like I didn't have much else to equate myself with, there was a real void.... I felt not so much like I missed the druggie lifestyle, but that I was starting to lose my grip on who I was, and was finding it hard to function." From www.stuff.co.nz/stuff-nation/assignments/how-have-drugs-affected-your-life/9513619/Drugs-were-the-only-life-I-knew, quoted in McConnell (2016), who offers a discussion of self-narrative in addiction complementary to the analysis of self-identity presented here.

References

Ahmed, S. H. (2010) "Validation crisis in animal models of drug addiction: beyond non-disordered drug use toward drug addiction", *Neuroscience Biobehavioral Review* 35: 172–184.
Ainslie, G. (2000) "A research-based theory of addictive motivation", *Law and Philosophy* 19: 77–115.
Ainslie, G. (this volume, pp. 34–44) "The picoeconomics of addiction".
Alexander, B. K. (this volume, pp. 501–510) "Addiction: a structural problem of modern global society".
Alexander, B. K., Coambs, R. B., and Hadaway, P. F. (1978) "The effect of housing and gender on morphine self-administration in rats", *Psychopharmacology* 58(2): 175–179.
Alexander, B. K., Peele, S., Hadaway, P. F., Morse, S. J., Brodsky, A., and Beyerstein, B. L. (1985) "Adult, infant, and animal addiction", in S. Peele (ed.), *The Meaning of Addiction*, Lexington, MA: Lexington Books, pp. 77–96.
American Psychiatric Association (APA) (2013) *Diagnostic and Statistical Manual of Mental Disorders*, 5th edition, Washington, DC: American Psychiatric Association.
Anthony, J. C. (this volume, pp. 253–274) "The epidemiological approach: an overview of methods and models".
Auriacombe, M., Serre, F., Denis, C., and Fatséas, M. (this volume) "Diagnosis of addictions".
Becker, H. S. (1963) *Outsiders: Studies in the Sociology of Deviance*, New York, NY: Free Press.
Berridge, K. C. (2017) "Is addiction a brain disease?" *Neuroethics* 10(1): 29–33.
Bickel, W. K. and Marsch, L. A. (2001) "Toward a behavioral economic understanding of drug dependence: delay discounting processes", *Addiction* 96(1): 73–86.
Bickel, W. K., Koffarnus, M. N., Moody, L., and Wilson, A. G. (2014) "The behavioral- and neuro-economic process of temporal discounting: a candidate behavioral marker of addiction", *Neuropharmacology* 76(B): 518–527.
Boorse, C. (1977) "Health as a theoretical concept", *Philosophy of Science* 44(4): 542–573.
Booth Davies, J. (1992) *The Myth of Addiction*, Amsterdam: Harwood Academic Publishers.
Bourgois, P. and Schonberg, J. (2009) *Righteous Dopefiend*, Berkeley, CA: University of California Press.
Buckingham, S., Albery, I. P., and Frings, D. (2013) "Group membership and social identity in addiction and recovery", *Psychology of Addictive Behaviors* 27(4): 1132–1140.
Compton, W. M., Thomas, Y. F., Stinson, F. S., and Grant, B. F. (2007) "Prevalence, correlates, disability, comorbidity of DSM-IV drug abuse and dependence in the United States: results from the national epidemiologic survey on alcohol and related conditions", *Archives of General Psychiatry* 64(5): 566–576.
Dingle, G., Cruwys, T., and Frings, D. (2015) "Social identities as pathways into and out of addiction", *Frontiers in Psychiatry* 6: 1795.
Elliott, C. (2002) "Who holds the leash?" *American Journal of Bioethics* 2(2): 48.
Flanagan, O. (2013) "The shame of addiction", *Frontiers in Psychiatry* 4: 120.
Flanagan, O. (2016) "Willing addicts? Drinkers, dandies, druggies, and other Dionysians", in N. Heather and G. Segal (eds), *Addiction and Choice: Rethinking the Relationship*, Oxford, UK: Oxford University Press, pp. 66–81.
Flanagan, O. (this volume, pp. 77–89) "Identity and addiction".
Frings, D. and Albery, I. (2015) "The social identity model of cessation maintenance: formulation and initial evidence", *Addictive Behaviors* 44: 35–42.
Goffman, I. (1963) *Stigma*, Englewood Cliffs, NJ: Prentice Hall Inc.

Hacking, I. (1996) "The looping effects of human kinds", in D. Sperber, D. Premack, and A. James Premack (eds), *Causal Cognition: A Multidisciplinary Debate*, Oxford, UK: Oxford University Press, pp. 351–395.
Hall, W., Carter, A., and Forlini, C. (2015) "The brain disease model of addiction: is it supported by the evidence and has it delivered on its promise?", *Lancet Psychiatry* 2(1): 105–10.
Hammersley, R. (this volume, pp. 220–228) "Sociology of addiction".
Hart, C. (2013) *High Price*, New York, NY: Harper Collins Publishing.
Hart, C. L., Haney, M., Foltin, R. W., and Fischman, M. W. (2000) "Alternative reinforcers differentially modify cocaine self-administration by humans", *Behavioural Pharmacology* 11(1): 87–91.
Henden, E. (this volume, pp. 45–53) "Addiction as a disorder of self-control".
Heyman, G. M. (2009) *Addiction: A Disorder of Choice*, Cambridge MA: Harvard University Press.
Heyman, G. M. (2013a) "Quitting drugs: quantitative and qualitative features", *Annual Review of Clinical Psychology* 9: 29–59.
Heyman, G. M. (2013b) "Addiction and choice: theory and new data", *Frontiers in Psychiatry* 4: 31.
Heyman, G. M. (this volume, pp. 23–33) "Deriving addiction: an analysis based on three elementary features of making choices".
Holton, R. and Berridge, K. (2013) "Addiction between choice and compulsion", in N. Levy (ed.), *Addiction and Self-Control: Perspectives from Philosophy, Psychology, and Neuroscience*, New York, NY: Oxford University Press, pp. 239–268.
Khantzian, E. J. (1985) "The self-medication hypothesis of addictive disorders: focus on heroin and cocaine dependence", *American Journal of Psychiatry* 142: 1259–1264.
Khantzian, E. J. (1997) "The self-medication hypothesis of substance use disorders: a reconsideration and recent application", *Harvard Review of Psychiatry* 4(5): 231–244.
Leslie, S.-J. (2017) "The original sin of cognition: fear, prejudice, and generalization", *Journal of Philosophy* 114(8): 393–421.
Levy, N. (2013) "Addiction is not a brain disease (and it matters)", *Frontiers in Psychiatry* 4: 24.
Levy, N. (this volume, pp. 54–62) "Addiction: the belief oscillation hypothesis".
Lewis, M. (2015) *The Biology of Desire: Why Addiction is not a Disease*, New York, NY: Perseus Books Group.
Mackintosh, V. and Knight, T. (2012) "The notion of self in the journey back from addiction", *Qualitative Health Research* 22(8): 1094–1101.
Martin, C. S., Langenbucher, J. W., Chung, T., and Sher, K. J. (2014) "Truth or consequences in the diagnosis or substance use disorders", *Addiction* 109(11): 1773–1778.
Maté, G. (2009) *In the Realm of Hungry Ghosts: Close Encounters with Addiction*, Toronto: Vintage Canada.
McConnell, D. (2016) "Narrative self-constitution and recovery from addiction", *American Philosophical Quarterly* 53(3): 307–322.
Muller, C. P. and Schumann, G. (2011) "Drugs as instruments: a new framework for non-addictive psychoactive drug use", *Behavioural and Brain Sciences* 34(6): 293–310.
National Institute on Drug Abuse (2009) *Principles of Drug Addiction Treatment: A Research-based Guide*, Bethesda, MD: National Institute on Drug Abuse.
Orford, J. (this volume, pp. 209–219) "Power and addiction".
Parsons, T. (1951) *The Social System*, London, UK: Routledge & Kegan Paul Ltd.
Pickard, H. (2012) "The purpose in chronic addiction", *American Journal of Bioethics Neuroscience* 3(2): 30–39.
Pickard, H. (2015) "Choice, deliberation, violence: mental capacity and criminal responsibility in personality disorder", *International Journal of Law and Psychiatry* 40: 15–24.
Pickard, H. (2016) "Denial in addiction", *Mind & Language* 31(3): 277–299.
Pickard, H. (2017a) "Addiction" in K. Timpe, M. Griffith, and N. Levy (eds), *The Routledge Companion to Free Will*, New York, NY: Routledge, pp. 454–467.
Pickard, H. (2017b) "Responsibility without blame for addiction", *Neuroethics* 10(1): 169–180.
Pickard, H. and Ahmed, S. (2016) "How do you know you have a drug problem? The role of knowledge of negative consequences in explaining drug choice in humans and rats", in N. Heather and G. Segal (eds), *Addiction and Choice: Rethinking the Relationship*, Oxford: Oxford University Press, pp. 29–48.
Pickard, H. and Pearce, S. (2013) "Addiction in context: philosophical lessons from personality disorder clinic", in N. Levy (ed.), *Addiction and Self-Control: Perspectives from Philosophy, Psychology, and Neuroscience*, Oxford: Oxford University Press, pp. 165–184.

Redish, A. D., Jensen, S., and Johnson, A. (2008) "A unified framework for addiction: vulnerabilities in the decision process", *Behavioural Brain Science* 31(4): 415–437.

Regier, D. A., Farmer, M. E., Rae, D. S., Locke, B. Z., Keith, S. J., Judd, L., and Frederick, K. G. (1990) "Comorbidity of mental disorders with alcohol and other drug abuse. Results from the epidemiological catchment area (ECA) study", *JAMA: The Journal of the American Medical Association* 264(19): 2511–2518.

Robinson, M. J. F., Robinson, T. E., and Berridge, K. C. (this volume, pp. 351–361) "The current status of the incentive sensitization theory of addiction".

Satel, S. and Lilienfeld, S. O. (2013) "Addiction and the brain-disease fallacy", *Frontiers in Psychiatry* 4: 141.

Sinnott-Armstrong, W. and Summers, J. (this volume, pp. 123–131) "Defining addiction: a pragmatic perspective".

Stephens, G. L. and Graham, G. (2009) "An addictive lesson: a case study in psychiatry as cognitive neuroscience", in M. R. Broome and L. Bortolotti (eds), *Psychiatry as Cognitive Neuroscience*, Oxford: Oxford University Press, pp. 203–220.

Tajfel, H. (1982) "The social psychology of intergroup relations", *Annual Review of Psychology* 33(1): 1–39.

Turner, J. C. (1987) *Rediscovering the Social Group: A Self-Categorization Theory*, Oxford: Basil Blackwell.

Verdejo-Garcia, A. (this volume, pp. 339–350) "Decision-making dysfunctions in addiction".

Volkow, N. D., Koob, G. F., and McLellan, A. T. (2016) "Neurobiologic advances from the brain disease model of addiction", *The New England Journal of Medicine* 374(4): 363–371.

World Health Organization (WHO) (1992) *ICD-10 Classifications of Mental and Behavioural Disorder: Clinical Descriptions and Diagnostic Guidelines*. Geneva, Switzerland: World Health Organization.

World Health Organization (WHO) (2004) *Neuroscience of Psychoactive Substance Use and Dependence*, Geneva, Switzerland: World Health Organization.

Zajac, K., Alessi, S. M., and Petry, N. M. (this volume, pp. 455–463) "Contingency management approaches".

Zernig, G., Kummer, K. K., and Prast, J. M. (2013) "Dyadic social interaction as an alternative reward to cocaine", *Frontiers in Psychiatry* 4: 100.

Zilverstand, A., O'Halloran, R., and Goldstein, R. Z. (this volume, pp. 362–379) "Resting-state and structural brain connectivity in individuals with stimulant addictions: a systematic review".

2
DERIVING ADDICTION
An analysis based on three elementary features of making choices

Gene M. Heyman

Introduction

Individuals make choices according to quantifiable behavioral principles. Depending on specifiable conditions, these principles produce optimal outcomes, near optimal outcomes, or seriously sub-optimal outcomes, which involve compulsive-like, excessive levels of consumption of a highly preferred substance or activity (Heyman 2009). In this chapter I focus on three elementary features of how people make choices. Although they are perfectly ordinary and are active in all decision making, they can result in drug binges, excessive drug use, and the pattern of remission and relapse that characterizes addiction. By analogy meteorology textbooks teach us that the physics that governs everyday weather is the same physics that foments typhoons. My analysis begins with a brief overview of the topic to be explained: addiction.

What is addiction like?

Clinicians and researchers rely on the *Diagnostic and Statistical Manual of Mental Disorders* (*DSM*) to distinguish drug addicts from drug users so that this manual provides a useful starting place for characterizing addiction. The authors of the 4th edition (APA 1994), which is the reference for much of the research cited in this chapter, state:

> The essential feature of Substance Dependence is a cluster of cognitive, behavioral, and physiological symptoms indicating that the individual continues use of the substance despite significant substance-related problems. There is a pattern of repeated self-administration that usually results in tolerance, withdrawal, and compulsive drug-taking behavior.
>
> *p. 176*

Following this passage, the *DSM* lists several clinically significant behavioral features of drug use: tolerance, withdrawal, using more drug than initially intended, failing to stop using after vowing to do so, and spending excessive amounts of time chasing, procuring, and consuming drugs. If three or more of these symptoms are present in the previous twelve months then the drug user is considered "drug dependent." However, the APA's account of addiction leaves out essential information and is misleading if this information is not included. (There is a 5th edition of the *DSM* (APA 2013), but I believe that it will prove less useful as a research tool than earlier editions because it relaxes the criteria for distinguishing between drug users and drug addicts.)

Common properties of addictive drugs

The temporal profile of costs and benefits. The positive hedonic effects of drug use are virtually immediate, whereas the costs are delayed and probabilistic. For instance, many smokers do not get cancer, and the delay from the onset of smoking to smoking-related illnesses is measured in decades. In contrast, the rewarding pharmacological and sensory effects of smoking are immediate and certain, taking place in seconds.

Sites of action, dose levels, psychological implications. With the exception of alcohol, addictive drugs bind to subcellular neuronal components that mediate neuronal communication. Consequently, they are highly potent; miniscule (milligram) doses produce dramatic changes in thought, feeling, and action. Nevertheless, the dose levels for self-administered drugs are several orders of magnitude greater than their naturally occurring counterparts (e.g., Comer *et al.* 2010; Dole 1980). This difference has important psychological implications. First, drugs are not satiating in the sense that food and drink are. This means that there are no naturally occurring self-inhibiting processes as in food consumption. Rather, the user has to judge whether he or she has taken enough drug. Second, addictive drugs can produce psychological effects that have no peer. For instance, a common theme in memoirs and interviews with addicts is that their drug highs are unlike anything else. Typical accounts of heroin include the following: "filling me up with a sensation I never felt before. . . . [i]t was the most intense nothingness there ever was. . . . There's no right, no wrong. Everything's beautiful." Addictive drugs corner the market on intoxication.

The natural history of drug addiction

The consequences of drug use change over time. Initially, there is a "honeymoon" period in which few, if any, of the costs of getting high have had time to emerge. But, as drug use continues and becomes more frequent, the consequences begin to include direct negative effects, such as tolerance and withdrawal, and indirect, socially mediated penalties, such as criticism from friends and family, legal problems, financial problems, workplace problems, and the like. Over time, the negative effects accumulate and the positive effects weaken (e.g., tolerance) so that, eventually, there is a period in which drug users often claim they want to stop using but keep using anyway. This is often described as loss of control.

Day-to-day patterns of use. Regular drug use requires planning and subterfuge. Drugs are often hard to come by and, if illegal, purchases must be made with care, planning and duplicity. To a lesser extent, legal addictive drug use comes with the same challenges. Implicit in these observations is that even the heaviest users are not high all the time. Indeed, many addicts have regular or part-time jobs and families that they tend to (e.g., Courtwright *et al.* 1989; Hanson 1985). Thus, addicts are not without opportunities for non-intoxicated reflection on whether to continue using drugs (e.g., Toneatto *et al.* 1999).

Filling in the gaps in the DSM account of addiction

This overview is based on observations and reflections on drug use in addicts (e.g., Biernacki 1986; Waldorf *et al.* 1991). Although the reports were not aimed at providing diagnoses, they are consistent with the APA account—with one exception: the word "compulsive." However, the *DSM* authors (APA 1994: 178) go on to say that what they mean by "compulsive" is that drug users take more drug than they initially intended or for longer than they intended and tried to cut back or quit but didn't. However, taking more drugs than intended does not imply compulsion; it could mean that there were consumption-dependent changes in preference. For example, it would be surprising if preferences did not change while intoxicated. More generally, all of the *DSM*'s examples of compulsion serve equally well as examples of ambivalence or changes in preference. Addictive drugs are uniquely attractive as well as uniquely perverse.

Three elementary features of choice

This next section identifies three elementary features of making choices. They are in play whenever anyone chooses between two or more substances or activities. Although they are perfectly ordinary, and usually lead to adaptive if not optimal outcomes, they can produce seriously suboptimal, pathological outcomes under certain conditions. Addictive drugs provide these conditions.

(1) *Preferences are dynamic.* The value of a substance or activity changes as a function of previous choices and/or the passage of time. In most instances the relationship is negative, as when eating reduces the value of food because of satiation. The positive instances are less common but familiar. Salty, sweet, and fatty foods whet appetite, at least at first, and activities that foster greater skill and knowledge grow in value as a function of practice.

(2) *Individuals always choose the better option.* This is true by definition. However, what is the best choice is ambiguous because there is usually more than one way to frame the options. As described below, and as is shown in Figures 2.1 and 2.2, it is possible for the best choice from a "local" perspective to be the worst choice from a "global" perspective.

(3) *In a series of choices between two or more items, it is possible to aggregate the options in different ways, which, in turn, yield different choices.* For instance, given a series of choices between two items, say, "A" and "B," consumers may choose more of "B" if they frame their choices as a series of independent trials, but then choose more of "A" if they frame their choices as bundles composed of different proportions of "A" and "B." It is convenient to label individuals who frame their choices as independent trials "local bookkeepers" and to label individuals who frame their choices as bundles "global bookkeepers."

Figures 2.1 and 2.2 are based on these three principles. They mimic graphs that researchers use to illustrate quantitative models of the relationship between reward and choice in laboratory studies of choice (e.g., Herrnstein *et al.* 1993). In the lab, the equations accurately predict the relationship between choice and reward rates. However, as models of addiction they are necessarily highly schematic simplifications that leave much out. Nevertheless, Figures 2.1 and 2.2 predict the essentials of addiction as well as some of the details that distinguish addiction from other psychiatric disorders.

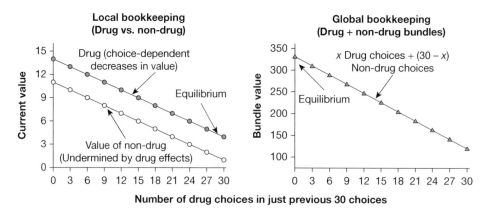

Figure 2.1 Choice for drug and non-drug days. In a highly schematic way the graph mimics the *DSM's* account of addiction.

On the *x*-axis of the left and right panels of Figure 2.1 is the number of choices for the drug out of the most recent 30 choices. The *y*-axis in the left-side panel represents the value of each item at the moment of choice. For instance, if 15 of the just previous 30 choices are for the drug then the current values of the drug and non-drug are "9" and "6" (arbitrary units). The *y*-axis for the global bookkeeper (right-side panel) lists the value of every possible combination of 30 drug and non-drug days. For instance, the global bookkeeper asks "what is the best combination of the two items: 15 drug days and 15 non-drug days or maybe 10 drug days and 20 non-drug days?" Put another way, global bookkeepers frame their options as rates of consumption, just as someone might choose between two or more weekly meal plans composed of different proportions of meats, vegetables, salads, and soups, or someone might choose between two or more schedules of going to the gym, working in the garden, spending time in service activities, shopping, cooking, etc. According to this analysis, a key component of how individuals make choices is how they frame them.

Binging on drugs, remission, and relapse as shifts in how choices are framed

According to the *DSM*, addicts are compulsive drug users, e.g., binge on drugs, take more drugs than initially intended, quit using but then relapse. Figures 2.1 and 2.2 illustrate these patterns, although the underlying principles are the general properties of all choices, not compulsion. Consider Figure 2.1 first.

The negative slope for the drug represents its decline in value due to tolerance. The negative slope for the non-drugs represents the toxic effects of drugs on other aspects of life, e.g., drug-related problems at home, in the neighborhood, and at work (what the *DSM* refers to as "significant substance-related problems"). Since by definition individuals always choose what is best, the local bookkeeper keeps choosing the drug over the non-drug even though this is making everything worse. The result is an all-out binge, and, according to the right-side panel, the lowest possible combination of drug and non-drugs days.

Now, imagine a dramatic change in circumstances that turns the local bookkeeper into a global bookkeeper. For instance, a common theme in the ethnographic literature is an event that precipitates taking stock of one's life. The examples are witnessing an overdose or the

Deriving addiction

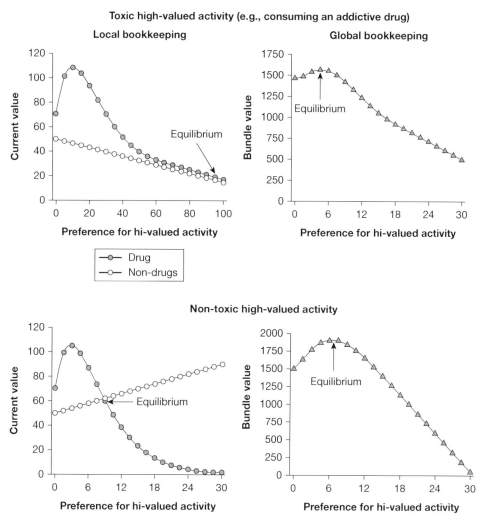

Figure 2.2 The relationship between choice and changes in value for a behaviorally toxic activity (e.g., drug taking, the top panel) and non-toxic but highly valued activity (bottom panel). The differences provide clues as to why certain substances are addictive and why they are usually drugs.

Source: the author.

realization that drugs have become more important than family (e.g., Biernacki 1986; Jorquez 1983; Premack 1970; Waldorf *et al.* 1991). According to the value functions in Figure 2.1, the best bundle contains exactly zero drug days, thereby bringing a halt to drug use. In other words, Figure 2.1 provides a schematic account of "hitting rock bottom" motivated recoveries. But, of course, events can also invite local bookkeeping. According to the same sources that provide first-hand accounts of remission, there is a pattern to the events that trigger relapse. They are "special occasions" and so-called "one last times." For instance, running into an old drug buddy, a recent setback, or, perhaps, just the opposite, a recent success, can trigger the myopic "today is an exception" and a return to heavy drug use. (See below for further discussion of the psychology of "special occasions.") Thus, depending on the frame of reference, simple choice principles predict either an all-out binge or abstinence, as well as some of the language that accompanies them. Also notice that the graph illustrates the AA saying that "for an alcoholic, one drink is too many and a thousand are not enough."

In the top panel of Figure 2.2, local bookkeeping also leads to binging on the higher valued activity. However, the comparison with the value functions of the lower panel provides clues as to why most highly valued activities are not addictive. Figure 2.2 also helps make the point that the conflicting predictions of local and global bookkeeping are not restricted to the highly simplified value functions of Figure 2.1.

What makes a substance or activity addictive?

The three principles that generated drug binging in Figure 2.1 and in the top panel of Figure 2.2 say nothing about drugs. But of the many activities and substances that people find highly rewarding, relatively few are the focus of addictions, and of those few, most—or all—are drugs. These two observations invite two questions: what makes something addictive and why are drugs the most likely addictive substances? Figures 2.1 and 2.2 suggest two answers. First, the comparison of the top and bottom panels of Figure 2.2 says that substances and activities that undermine the value of the competing choices are more likely to become addictive. For instance, preference for the high valued activity in the bottom panel of Figure 2.2 does not lead to binging because the non-drug activity maintains its value. Indeed, as is typically the case for conventional rewarding activities, the value functions in the bottom panel of Figure 2.2 follow the rule of diminishing marginal returns. As shown in the top panel, addictive drug consumption leads to violations of this rule for the competing goods. Second, regardless of how toxic a substance and activity might be, Figures 2.1 and 2.2 say that global bookkeeping will prevent addiction. Thus, substances that undermine or challenge cognitive capacities necessary for global bookkeeping are likely candidates for addiction. Drugs do this in three ways. Intoxication undermines judgement; the combination of immediate rewards and greatly delayed costs that are typical of addictive drugs make estimating the drug's net value difficult; and that drugs do not satiate robs the user of simple cues as to how much is enough. Hence drugs end up as the most likely candidates for supporting addictive behavior.

Addictive drugs may also induce residual cognitive deficits that interfere with global bookkeeping. Given their potent intoxicating effects, this is highly plausible. However, the evidence on this question is inconsistent (Heyman 2009; Toomey *et al.* 2003). For instance, Figure 2.3, presented below, suggests that if there are serious drug-induced cognitive deficits, they do not appear to prevent most addicts from remitting.

Local and global bookkeeping provide guides to the unique vocabulary of addiction

The terms "hitting rock bottom," kicking the habit," "going cold turkey" are specific to addiction. No one goes "cold turkey" from depression or diabetes. There are also addiction-specific excuses, as intimated above. A theory of addiction should be able to predict the unique features of addiction.

Relapse and other forms of backsliding are often precipitated by the verbal formulas "this is a special occasion," or "this is the last time," or "tomorrow I turn over a new leaf." According to Figures 2.1 and 2.2, these are perfect excuses. On the last choice in a series of choices, the conflicting dictates of local and global choice disappear. The global perspective requires future choices whose values are affected by the present choice. When it is the last of a series of choices, the only possible framework is the present, which is to say the local perspective. If drug use makes the "last" few moments the best possible last few moments, there is no reason to abstain. Of course, it may turn out that there are more "last" times and more "new leafs" to turn.

Idioms emerge "spontaneously" as a function of experience. The idioms "kicking the habit" and "going cold turkey" refer to the fact that heroin addicts sometimes quit drugs all at once and on their own. Figures 2.1 and 2.2 provide mechanisms for such sudden shifts in symptoms. These idioms are now common parlance for all drug addictions but have not become part of the conversation in regard to other psychiatric disorders. This suggests that local and global bookkeeping, although part and parcel of all decision making, play more of a role in addiction than in other psychiatric disorders. In any case, the graphs help explain drug binging, remission, relapse, and the psychology and conversation that accompanies these fundamental features of addiction.

What is the evidence for the role of local and global bookkeeping in making choices?

Local and global bookkeeping are well represented in experiments on choice, in economic analyses of consumer choice, and every-day observations. Consider local bookkeeping first.

The equilibrium points in the local bookkeeping panels of Figures 2.1 and 2.2 are identical to experimental psychology's "matching law" equilibrium points (Herrnstein 1970). This result, as the label "matching law" suggests, has been observed in hundreds of experiments, conducted both in and outside of laboratory settings (Davison and McCarthy 1988). In experiments with infra-human subjects, matching is the expected result, despite reliably producing substantially suboptimal returns under some conditions (e.g., those shown in Figure 2.1). However, in a choice experiment with pigeons in which the experimenters introduced stimuli that explicitly signaled global bookkeeping reward bundles, the pigeons deviated from matching as predicted by global bookkeeping thereby increasing reward rate (Heyman and Tanz 1995)—a result that underscores the cognitive dimensions of local and global bookkeeping.

In experiments with human subjects the results vary. In many studies the subjects routinely match choice and reward proportions as in the animal research, whereas, in others, some subjects make choices that approximate the maximizing global bookkeeping predictions (Davison and McCarthy 1988; Herrnstein *et al.* 1993). These differences may reflect individual difference in the capacity to detect higher-order relations in the structure of the reward contingencies. For example, in a study in which the choice trials occurred every ten seconds or occurred in a pattern of three closely bunched trials separated by 30 second gaps, choices shifted from the pattern predicted by local bookkeeping towards the pattern predicted by global bookkeeping (Kudadjie-Gyamfi and Rachlin 1996). In other words, when the researchers introduced a cue that highlighted the ways in which the apparently competing choices could be treated as bundles, the subjects switched to the more profitable bundle approach.

Global bookkeeping is how economists depict consumer choice. For example, in a widely used text that has gone through at least 13 editions, Baumol and Blinder (1994) whimsically provide the example of consumers choosing between the best combinations of packs of rubber bands and pounds of cheese, e.g., "do I prefer two packs of rubber bands and three pounds of cheese or three packs of rubber bands and two pounds of cheese?" Although professors Baumol and Blinder may be alone when it comes to aggregates of rubber bands and cheese, people routinely aggregate their options. Competing bundles are what is at stake when we decide between different menus that each list different diets, when we decide between different schedules that each list different arrangements of a set of events, and even when we decide among different lifestyles and identities. For instance, a workaholic is someone who does not schedule in leisure time, and a "wild and crazy guy" is someone who does not schedule. One implication of these observations is that life styles and prudential rules (e.g., "dessert comes after the main dish," "don't drink before noon") are a culture's way of signaling the advantages of bundling choices, just as researchers arrange stimuli that encourage choice bundling when they want to teach their subjects to maximize reward.

In summary, the research suggests that local bookkeeping prevails to the extent that the items and activities in question are perceived in terms of their "natural" perceptual differences, whereas conditions that promote the capacity to form more abstract option categories, composed of combinations of the substances and activities, promote global bookkeeping.

Voluntary drug use predicts that addiction is a limited disorder

Voluntary actions and involuntary activities differ in the degree to which they are maintained by their consequences. Examples include winks versus blinks, kicking a ball versus the patellar reflex, putting on rouge versus blushing, and so on. Elsewhere the importance of these contrasts has been discussed at some length (e.g., Heyman 2009). For the purposes of the next prediction, it is sufficient to point out that if voluntary behavior is guided by its consequences then it should be self-correcting, granted that (1) the consequences are deleterious and (2) there is a better alternative at hand. Applying this rule to addiction, the implications are that (1) addiction should not occur or (2), if it does occur, it should not persist for long (assuming better alternatives are available), and (3) addicts should be able to turn away from drugs on their own without the benefits of interventions.

The logic is correct but it leaves out the natural history of addiction. As described above, there is an initial honeymoon period in which drug use is, on balance, beneficial, followed by a period in which the costs and benefits may more or less balance each other out. Nevertheless, if addiction is a disorder and addicts really do remain voluntary drug users, as assumed in the discussion of Figures 2.1 and 2.2, dependence must voluntarily come to an end. In contrast, among many experts and the informed public, addiction is widely understood to be a chronic relapsing disease and that only treatment can help addicts stay clean.

Do most addicts quit using drugs as predicted by the claim that they remain voluntary drug users?

Figure 2.3 shows the cumulative frequency of remission as a function of the onset of dependence in a nation-wide representative sample of addicts (Lopez-Quintero *et al.* 2011). The researchers recruited approximately 42,000 individuals, with the goal that their sample would mimic the demographic characteristics of individuals between the ages of 18 and 64 living in the US. Once enrolled, the participants were interviewed according to a questionnaire designed to produce an APA diagnosis, when so warranted. On the *x*-axis is the amount of time since the onset of dependence. On the *y*-axis is the proportion of individuals who, according to the interviews, met the criteria for dependence for a year or more in the past, but did not meet these same criteria in the year prior to the interview or longer. The fitted curves are negative exponentials. They were drawn according to the assumption that each year a constant proportion of those still addicted remitted independently of how long they had been using drugs.

The cumulative frequency of remission increased each year for each drug, and the good fit of the equations to the frequencies of quitting says that addicts did so at a constant rate, regardless of time since the onset of dependence. By year 4, half of those who were ever addicted to cocaine had stopped using cocaine at clinically significant levels; the half-life of dependence on marijuana was six years; and for alcohol, the half-life of dependence was considerably longer, sixteen years. Taking into consideration typical onset ages for addiction (Kessler *et al.* 2005a), the results say that most of those

Deriving addiction

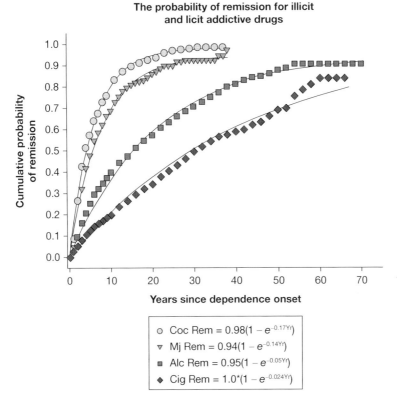

Figure 2.3 The cumulative frequency of remission as a function of time since the onset of dependence, based on Lopez-Quintero *et al.*'s report (2011). The proportion of addicts who quit each year was approximately constant. The smooth curves are based on the negative exponential equations listed in the figure.

Source: the author.

who became addicted to an illicit drug were "ex-addicts" by age thirty. Of course, many may have switched drugs rather than quit drugs, but other considerations indicate that this does not explain the trends displayed in Figure 2.3. For example, dependence on all drugs decreased as a function of age, which would not have been possible if addicts switched drugs rather than quit drugs. However, it is possible that one of the reasons that dependence on alcohol persisted so much longer than did cocaine and marijuana addiction is that a significant number of drug addicts continued to drink after stopping illicit drug use. Nevertheless, most alcoholics eventually became ex-alcoholics.

How did addicts quit?

The idea that voluntary behavior is self-correcting predicts that addicts can quit without the assistance of explicit clinical interventions. As predicted, researchers invariably report that most addicts do not seek treatment (e.g., Anthony and Helzer 1991; Robins 1993; Stinson *et al.* 2005). But there must be reasons for dependence ending. In the absence of clinical interventions, it is reasonable to suppose that the motivation was in response to the events of everyday life, such as economic issues, family pressures, concerns about staying out of jail, concerns about health, and the difficulties that attend any illegal and/or stigmatized pattern of behavior. In support of this inference, when addicts talk about why they quit drugs they point to the sort of necessities of life just listed. In addition, accounts of quitting drugs often include moral concerns. With some frequency, ex-addicts explain that they wanted to regain the respect of family members, to better meet their image of how a parent should behave, and to better approximate their image of a person who is competent and in control of their life (e.g., Biernacki 1986; Jorquez 1983; Premack 1970; Waldorf et al. 1991).

The logic of voluntary behavior predicts that dependence is a limited disorder and that it can end without the benefits of interventions. Figure 2.3 and the finding that most addicts were not in treatment support these two predictions. However, it is not unreasonable to question the reliability of the results summarized by Figure 2.3 given that they disagree so with received understandings of the nature of addiction. This issue has been discussed in detail in previous publications (e.g., Heyman 2013). Some of the key findings are that (1) the results in Figure 2.3 are not new, but replicate the findings of the three previous US surveys of psychiatric disorders (Anthony and Helzer 1991; Kessler et al. 2005b; Warner et al. 1995). (2) Unreliable self-reports cannot explain the high remission rates. And (3) missing addicts due to mortality or unwillingness to cooperate with the researchers also cannot explain Figure 2.3. The good fit of the exponential model also deserves attention. It says that the likelihood of quitting was constant regardless of how long dependence had persisted. This is a surprising result, but one that has been reported elsewhere (Vaillant 1973). See Heyman 2013 for further discussion of this finding and Cantin et al. 2010 for analogous results.

Before concluding, it should also be pointed out that unassisted quitting is not an argument against helping addicts quit drugs. It is routine and sensible to use coaches and teachers to help facilitate mastery of difficult voluntary activities. Similarly, there are proven programs for accelerating the rate at which addicts quit using drugs (Davis et al. 2016). Given the great damage that addiction causes, effective treatment programs should be strongly supported.

Conclusion

Figures 2.1, 2.2, and 2.3 strongly support the idea that how individuals frame their options plays an important role in their welfare, the welfare of their family and the welfare of their community. Yet, the experiential and biological determinants of local and global bookkeeping remain largely unexplored. What is needed is a program of research that focuses on the factors that encourage global bookkeeping. The results may not only help us understand addiction better but also help us understand better the too familiar, often disastrous, shortcomings of human decision making in general.

References

American Psychiatric Association (APA) (1994) *Diagnostic and Statistical Manual of Mental Disorders*, 4th Edition, Washington, DC: American Psychiatric Association.
American Psychiatric Association (APA) (2013) *Diagnostic and Statistical Manual of Mental Disorders*, 5th Edition, Washington, DC: American Psychiatric Association.
Anthony, J. and Helzer, J. (1991) "Syndromes of drug abuse and dependence", in L. Robins and D. Regier (eds), *Psychiatric Disorders in America: The Epidemiologic Catchment Area Study*, New York, NY: Free Press, pp. 116–154.
Baumol, W. and Blinder, A. (1994) *Economics: Principles and Policy*, Fort Worth, TX: The Dryden Press, Harcourt Brace & Company.
Biernacki, P. (1986) *Pathways from Heroin Addiction: Recovery without Treatment*. Philadelphia, PA: Temple University Press.
Cantin, L., Lenoir, M., Augier, E., Vanhille, N., Dubreucq, S., Serre, F. and Ahmed, S. H. (2010) "Cocaine is low on the value ladder of rats: possible evidence for resilience to addiction", *PLOS One* 5(7): e11592.

Comer, S., Cooper, Z., Kowalczyk, W., Sullivan, M., Evans, S., Bisaga, A. and Vosburg, S. (2010) "Evaluation of potential sex differences in the subjective and analgesic effects of morphine in normal, healthy volunteers", *Psychopharmacology* 208(1): 45–55.

Courtwright, D., Joseph, H. and Des Jarlais, D. (1989) *Addicts Who Survived: An Oral History of Narcotic Use in America, 1923–1965*, Knoxville, TN: University of Tennessee Press.

Davis, D., Kurti, A., Skelly, J., Redner, R., White, T. and Higgins, S. (2016) "A review of the literature on contingency management in the treatment of substance use disorders, 2009–2014", *Preventive Medicine* 92: 36–46.

Davison, M. and McCarthy, D. (1988) *The Matching Law: A Research Review*, Hillsdale, NJ: Lawrence Erlbaum Associates.

Dole, V. (1980) "Addictive behavior", *Scientific American* 243(6): 138–154.

Hanson, B. (1985) *Life with Heroin: Voices from the Inner City*, Lexington, MA: Lexington Books.

Herrnstein, R. (1970) "On the law of effect", *Journal of the Experimental Analysis of Behavior* 13(2): 243–266.

Herrnstein, R., Loewenstein, G., Prelec, D. and Vaughan, W. (1993) "Utility maximization and melioration: internalities in individual choice", *Journal of Behavioral Decision Making* 6(3): 149–185.

Heyman, G. (2009) *Addiction: A Disorder of Choice*, Cambridge, MA: Harvard University Press.

Heyman, G. (2013) "Quitting drugs: quantitative and qualitative features", *Annual Review of Clinical Psychology* 9: 29–59.

Heyman, G. and Tanz, L. E. (1995) "How to teach a pigeon to maximize overall reinforcement rate", *Journal of the Experimental Analysis of Behavior* 64, 277–297.

Jorquez, J. (1983) "The retirement phase of heroin using careers", *Journal of Drug Issues* 13(3): 343–365.

Kessler, R., Berglund, P., Demler, O., Jin, R., Merikangas, K. and Walters, E. (2005a) "Lifetime prevalence and age-of-onset distributions of DSM-IV disorders in the national comorbidity survey replication", *Archives of General Psychiatry* 62(6): 593–602.

Kessler, R., Chiu, W., Demler, O., Merikangas, K. and Walters, E. (2005b) "Prevalence, severity, and comorbidity of 12-month DSM-IV disorders in the national comorbidity survey replication", *Archives of General Psychiatry* 62(6): 617–627.

Kudadjie-Gyamfi, E. and Rachlin, H. (1996) "Temporal patterning in choice among delayed outcomes", *Organizational Behavior and Human Decision Processes* 65(1): 61–67.

Lopez-Quintero, C., Hasin, D., de Los Cobos, J., Pines, A., Wang, S., Grant, B. and Blanco, C. (2011) "Probability and predictors of remission from lifetime nicotine, alcohol, cannabis or cocaine dependence: results from the National Epidemiologic Survey on Alcohol and Related Conditions", *Addiction* 106(3): 657–669.

Premack, D. (1970) "Mechanisms of self-control", in W. Hunt (ed.), *Learning Mechanisms in Smoking*, Chicago, IL: Aldine, pp. 107–123.

Robins, L. N. (1993) "Vietnam veterans' rapid recovery from heroin addiction: a fluke or normal expectation?" *Addiction* 88(8): 1041–1954.

Stinson, F., Grant, B., Dawson, D., Ruan, W., Huang, B. and Saha, T. (2005) "Comorbidity between DSM-IV alcohol and specific drug use disorders in the United States: results from the National Epidemiological Survey on Alcohol and Related Conditions", *Drug and Alcohol Dependence* 80(1): 105–116.

Toneatto, T., Sobell, L., Sobell, M. and Rubel, E. (1999) "Natural recovery from cocaine dependence", *Psychology of Addictive Behaviors: Journal of the Society of Psychologists in Addictive Behaviors* 13(4): 259–268.

Toomey, R., Lyons, M. J., Eisen, S. A., Xian, H., Chantarujikapong, S., Seidman, L. J., Faraone, S. V. and Tsuang, M. T. (2003) "A twin study of the neuropsychological consequences of stimulant abuse", *Archives of General Psychiatry* 60(3): 303–310.

Vaillant, G. E. (1973) "A 20-year follow-up of New York narcotic addicts", *Archives of General Psychiatry* 29(2): 237–241.

Waldorf, D., Reinarman, C. and Murphy, S. (1991) *Cocaine Changes: The Experience of Using and Quitting*, Philadelphia, PA: Temple University Press.

Warner, L., Kessler, R., Hughes, M., Anthony, J. and Nelson, C. (1995) "Prevalence and correlates of drug use and dependence in the United States. Results from the National Comorbidity Survey", *Archives of General Psychiatry* 52(3): 219–229.

3
THE PICOECONOMICS OF ADDICTION

George Ainslie

The core paradox of addiction is a persistent tendency of the addict to choose what she believes she doesn't want. Addiction is sometimes characterized by other attributes, such as dependence on a substance, withdrawal (sickness on quitting), a preoccupation that squeezes out "normal" activities, or tolerance (a progressive insensitivity to the relevant reward), but none of the first three is necessary for the core paradox (witness gambling, cocaine dependence, and smoking, respectively), and the last is not specific: There is probably no rewarding activity that does not habituate with repetition. Activities with one or more of the other attributes may properly be called addictions, of course—dependency without awareness, or even "willing addictions" despite a knowledge of negative consequences (Flanagan 2016; Pickard & Ahmed 2016); but these do not entail the puzzling feature of choosing what one consciously doesn't want. The nub of the problem is this apparent paradox of choice. For centuries it has remained a scientific puzzle and a moral/legal quandary. The problem has become more urgent for two reasons: not, perhaps, because the human susceptibility to addiction has increased, but because (1) it endures while other causes of premature death and disability have dropped away, and (2) human craft has developed addictive activities faster than it has developed protections against them. There was a great acceleration when the growth of trade permitted worldwide sharing of opium, coca, tobacco and other natural substances, another when new techniques extracted or synthesized concentrated ingredients from them (Crocq 2007), and, arguably, a current expansion into fast-paying interactive patterns that do not depend on substances—video games, internet gambling, internet porn, and absorption in the internet itself.

What explains the paradox of unwanted behavior? Theorists have long been drawn to the answer that thinking is divided into two kinds, a far-seeing process that plans consistently over time and a myopic process left over from our evolutionary roots. Modern authors discern "visceral," "hot," or "type 1" thinking (Loewenstein 1996; Metcalf & Mischel 1999; Kahneman 2011, respectively) impinging on a deliberate, rational process (type 2, etc.), and have reported separate areas in the brain that might govern them (Luerssen et al. 2015; McClure et al. 2004). The implication is that we have a steady, rational self that is occasionally attacked by an evolutionarily primitive process, which pressures us to choose costly short-term options before returning us to our rational state. But, although there are clearly divisions of labor in the brain, this is not an adequate explanation.

Such dualism is intuitively appealing, but it has limitations. For instance, there is sometimes an urge that is impulsive with respect to another impulsive urge, e.g. laziness in executing a plan that is itself short-sighted (see Ainslie 2009). There are also familiar instances where we give in to short-sighted goals that do not arouse visceral or hot thinking, such as simple procrastination. Sometimes the short-sighted goal is too distant to be based on arousal, as when we knowingly save too little for retirement over months or years. Furthermore, the oft-reported individuals who prefer a hypothetical $50 now to $100 in three years, but do not prefer $50 in six years to $100 in nine years, meet the definition of myopia but are unlikely to have been emotionally aroused by the immediate option—as demonstrated by the persistence of the pattern when a month is added to both earlier and later delivery times (Green et al. 2005). Appetites and emotions are certainly arousable, and increase the reward value of their objects when aroused; but many short-term preferences cannot be attributed to such arousal. There is evidence that a more general mechanism promotes addictive behavior, and that this behavior may be more widespread than is usually recognized.

A marketplace model of addiction

We are just beginning to learn how brain activity corresponds to the judgments involved in self-control, but two striking findings may help us frame the problem: (1) When human subjects face the prospect of getting real money after delays of days to weeks, the activity in their reward-sensitive regions is the inverse of the delays (Kable & Glimcher 2007), thus confirming the inverse relationship of value and delay found in many behavioral experiments. (2) There are many anatomical and functional connections in the brain that are associated with choice-making (Haber & Knutson 2010), but they do not converge on any region that is likely to house the faculty of "self," as philosophers from Descartes to Fodor have imagined it. Choice seems to be governed by a bidding process among any options that can replace each other, based on a common currency that is best called reward—a marketplace, not a homunculus. As Daniel Dennett has said, "all the work done by the imaginary homunculus in the Cartesian Theater must be distributed *in time and space* in the brain" (2003, p. 123—his emphasis).

(1) Psychophysical experiments have long found that changes in sensations such as brightness, loudness, and heat are proportional to the change in the strength and/or proximity of the physical stimulus. Over the last forty years experiments have surprisingly found that the same phenomenon occurs with delay to expected rewards (Green & Myerson 2004). Surprisingly because, although value has been known since Socrates to be discounted for delay, the form of the discounting has been assumed not to result in changes of preference over time. That is, if you preferred a larger, later (LL) reward to a smaller, sooner (SS) one when both were distant, you were assumed to prefer it when both were close, if the lag between them stayed the same and you learned no new information. The only formula that permits this constant relative preference is the one that banks use for interest, widely accepted after the economist Samuelson rather belatedly drew it (1937):

$$\text{present value} = \text{value if immediate} \times \text{discount rate}^{\text{delay}}$$

The surprise is that, in contrast to the bankers' *exponential* formula, both humans and other animals tend to value delayed rewards in inverse proportion to that delay, in accordance with their perception of other kinds of stimuli, but not rationally for future planning.

$$\text{present value} = \frac{\text{value if immediate}}{1 + (\text{impatience factor} \times \text{delay})}$$

Such *hyperbolic* delay discounting often leads to preference for an LL reward when a pair are distant, but an SS reward when they are close, as in the $50 vs $100 example above. People usually learn to discount the future like banks when dealing with banks or with another smart dealer, because otherwise the dealer will *money pump* them—buy their winter coat cheap in the spring and sell it back to them higher in the fall, again and again until they learn consistent valuation. But much human folly suggests that this learning is tenuous; and the presence of hyperbolic delay discounting even in rats and pigeons (albeit over periods of seconds—Ainslie & Monterosso 2003; Ainslie & Herrnstein 1981) shows that this pattern dates back to the evolutionary origin of psychophysical laws, probably too fundamental a part of animal design to have been selected out to facilitate human planning. Exponential discounting never becomes a basic process, but depends on learned self-control.

(2) Whether there are one or two or ten brain centers that take part in the decision process, there must be a mechanism—a reward process—by which any option that you can learn to substitute for any other option is weighed, one against the other (Shizgal & Conover 1996; Levy & Glimcher 2012). Still more fundamentally, "you" comprise the substitution process, rather than an entity that stands outside of it and supervises it from above.

Implications of hyperbolic delay discounting

If choice is entirely governed by the marketplace of reward, we might expect it to be the simple matter of calculation, the way utilitarians have long imagined. But the finding of hyperbolic delay discounting literally throws a curve into this picture. If you expect to change your preference away from an LL reward before you get to it, any plan for it must include a way to forestall this change—by modifying the expected SS or LL values or making the change impossible. The ordinary process of learning paths to reward produces *interests*—the sets of paths that lead to and are rewarded by particular alternatives. This is a trivial concept except to the extent that one interest has incentive to undermine another, for instance a long-term interest in staying sober that must deal strategically with a short-term interest in getting drunk. On a different time scale, someone with post-traumatic stress disorder (PTSD) has a mid- to long-term interest in avoiding intrusive memories, and a very short-term interest in rehearsing them.[1] Short-term interests are based on the power of rewards that are close; interests based on more distant rewards have the opportunity to act earlier, but must forestall the short-term interests before they become stronger. Neither interest will simply prevail as long as each has a prospect of succeeding, a situation that is apt to endure because hyperbolic discount curves give each interest a period of dominance (see the right-hand pair of Figure 3.1). Thus a marketplace model with hyperbolic delay discounting does result in dualisms of a sort, but not between different kinds of motivation. Any alternative rewards that are available at different delays are apt to become the bases of long- and short-term interests, whether one, both, or neither evoke "type 1 thinking," and over timelines ranging from split seconds to years. Furthermore, the long-range interest in one dualism can be the short-range interest in another: A shoplifter in action needs to avoid distractions—short-term interests with respect to her plan—but a still longer-term interest is to forestall the shorter-term interest in shoplifting itself (Ainslie 2009).

The first lesson to be drawn from hyperbolic delay discounting is that a pattern of temporary preferences and regrets is not an unusual process imposed on a person by some pathological factor. Rather this is the normal baseline of motivation written over time. We naturally overvalue the imminent future, and learn the ability to maintain long-term plans only gradually and imperfectly. Everyone struggles with bad habits. If addiction is defined with a low threshold, half the people in America are addicted to something (Sussman et al. 2011). Those of us who have avoided the named addictive diagnoses are nevertheless apt to suffer from habitual overvaluation of the present moment, as in chronic procrastination, overuse of credit, or unrealistic future time commitment. So the problem for the science of addiction is not an addict's susceptibility to temptation, but why she fails to use her culture's shared knowledge to counteract it in specific areas over part of her life—and then, usually, succeeds again. (Most substance addicts eventually recover without treatment—Heyman 2009.)

The next lesson is that the mechanisms often proposed for addictions all depend on motivation (Ainslie 2016). Here I describe the three prominent proposals for which hyperbolic discounting has the greatest implications.

Three explanations for addiction re-interpreted

1. Weak will

A universal tendency to form temporary preferences makes commitment necessary for consistent behavior. Acting in one's long-term interest, there are simple ways to forestall temptations: Keep your attention away from them so they do not enter the marketplace, abandon the relevant appetite or emotion before it gets too strong, or find external influences (for instance Alcoholics Anonymous) or commitments (for instance disulfiram=Antabuse). But all of these have serious drawbacks: Attention is hard to divert for long; appetites and emotions are rewarding in their own right; other people have their own agendas, and neither they nor physical commitments may be available when you need them.

The tendency of hyperbolic curves to level out at long delays permits a more sophisticated strategy: The influence of the LL rewards (height of their curves) grows faster than that of the SS rewards when a series of choices is summed, so an arrangement to make a whole series of SS/LL choices at once favors the LL options (Figure 3.1; details in Ainslie 2005). Such an arrangement is apt to form spontaneously in anyone who notices that her current SS/LL choice predicts how she will make similar choices in the future: Then her expectation of getting LL outcomes in the set of similar choices will be somewhat determined by her current choice. This perception is apt in turn to become a factor in making the current choice, and each subsequent choice that looks similar. Each LL choice increases, or preserves, her expectation of getting the whole series—or *bundle*—of LL rewards, and each SS choice decreases this expectation, a process of *recursive self-prediction*.

The stake of LL reward may be aggregated from the evident consequences of each choice, as in binges with hangovers, or the stake may be a more distant anticipated condition such as good health or adequate savings. The necessary element for the self-prediction process is that the person's expectation of getting a category of LL reward as a whole is put at stake when each opportunity within a definable set of SS rewards occurs. This self-prediction process is recursive, in that each estimate of future self-control is fed back into the estimating process.

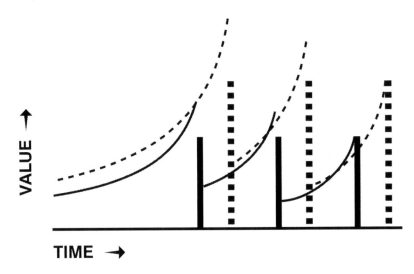

Figure 3.1 Summed, hyperbolically discounted values of three expected SS rewards versus three expected LL rewards (vertical bars), if chosen as bundles: The LL bundle will be preferred at all choice points in advance of the first pair. If the choice were only between the last pair, the SS option would be preferred when it was nearly due.

Once someone has noticed this phenomenon she is apt to propose bundles deliberately: resolutions, or diets, or tests of character. In effect she is defining a game resembling a repeated prisoner's dilemma, *intertemporal bargaining* with her expected future selves, stabilized by self-enforcing contracts in which single defections (SS choices) have more impact than single cooperations (LL choices—modeled in Monterosso et al. 2002). The reason the dieter doesn't eat the single chocolate éclair is not that it will cause a noticeable gain in weight, but that it will damage the credibility of her diet. The struggle between impulse and control now turns not so much on how close she gets to a temptation—although this remains a factor—but on whether she expects a later self to see her current choice as a defection and thus have less reason in turn not to defect. Her self-prediction has become the basis of *personal rules*. The bundling tactic has the greatest impact when a person sees the stake as her belief about her character—whether she is "the kind of person who" abuses her children, is the slave of an addiction, or has a disgusting paraphilia. The game theory of such high-stake "self-signaling" was formally worked out by Ronit Bodner and Drazen Prelec (2003).

Intertemporal bargaining forms willpower—a weapon against addictions—but also raises the stakes of choice in ways that can strengthen an addiction:

> *"Loss of control."* A person may have recruited great differential motivation in an important set of choices, so a single defection can cause a spectacular collapse—the "abstinence violation effect" (Curry et al. 1987). Then, to save her expectation of controlling herself generally, she will be strongly motivated to find a line excluding the kind of choice where her will failed from her larger rule. This means attributing the lapse to a particular aspect of that situation, even though it will make self-control more difficult when that aspect is present in the future. She may decide that she *can't* resist the urge to eat jelly donuts, or to smoke after meals, or accept cocaine when offered by a friend—or resist that particular modality of temptation at all. The result is a circumscribed area of dyscontrol.

Cognitive blocks. Since personal rules organize great amounts of motivation, they create an incentive to suborn the self-perception process. The potential damage from lapses creates an incentive not to see them, giving rise to a motivated unconscious: suppression, repression, denial.

Compulsiveness. Choices may become more important as test cases than for their own sakes, making it hard to live in the here-and-now. Awareness that damaging a personal rule may threaten to de-stabilize a larger network of intertemporal bargains may lead to an unwillingness risk modifying the rule.[2]

Thus intertemporal bargaining stabilizes not only long-term plans but also ways around them. It lays down a history of choices that have turned out either well or badly, leading to modifications in your personal rules. The logic is the same as for how court decisions over time lay down the English and American common laws. Most psychological therapies aim at unpicking the tangle of overgrown resolutions and lapses that individuals have developed in their various attempts at rule-making: "cognitive maps" (Gestalt), "conditions of worth" (client-centered), "musturbation" [sic] (rational-emotive), "overgeneralization" (cognitive behavioral), and of course the punitive superego (summarized in Corsini & Wedding 2011). Unfortunately, these therapies have not had a good record against addictions.

Short-term interests based on addictions are aggressive bargainers by definition. They may make it necessary to cut the Gordian knot of intertemporal bargains and focus concretely on a single target. The Anonymous organizations' concept of helplessness should be understood not as a negation of reward bundling, but as a warning against the compromises that ordinary self-control invites (see Monterosso & Ainslie 2007, p. S107). How to maximize the effect of a bundle of LL choices while leaving open the idea of recovery from a possible abstinence violation effect is a conundrum for all addiction therapists—logically, anticipating the latter weakens the former. For rebuilding trust AA cautiously adopts the insight of some religions in invoking the influence of a "higher power," which cannot be presumed upon to grant grace and thus remains credible after failures (see the Saint X effect in Ainslie 2001, pp. 107–108).

2. Habit

The persistence of addictions despite deteriorating reward has suggested the relevance of behavioral experiments on habit—the persistence of a choice after it is no longer rewarded. Some kinds of brain lesions make rats unable to change overlearned responses when reward contingencies change (Everitt & Robbins 2013). Addiction to stimulants in particular has been accompanied by suggestive changes in human brain activity. However, although addicts have been observed in laboratory choice-making tasks to respond less well than nonaddicts to changed information, this difference has usually been moderate, as it has been even in patients with gross brain lesions (Fellows & Farah 2005). Furthermore, overlearning is of questionable relevance to choices about whether or not to consume substances; addictive "habits" have very little to do with mindless repetition, but on the contrary require a high degree of flexible, goal-directed behavior to evade a hostile society.

In addition to the tangle of personal rules just described, hyperbolic discounting predicts a simple mechanism for apparent habit. Devalued addictive reward does not lose its power evenly over a consumption episode; initial phases stay rewarding, at least relative to the alternatives still available, but no longer last as long. Given its reduced average value the consumption may look like robotic repetition, while hyperbolic overvaluation of the near future keeps it the dominant choice. A recurring urge followed by disappointment has the time pattern of an itch, but the path of least resistance is still to scratch the itch—light another cigarette, eat another snack—rather than tax a depleted willpower fighting recurrent urges. The activity is still based on reward (see Ainslie, 2017a).

3. Ignorance of the true contingencies

Motivational science has necessarily studied the control of behavior by external rewards, and so has not permitted explanations that do without such rewards. Addiction was long assumed to arise from the action of a drug. The obvious exception of gambling addiction was attributed to delusions about how the laws of probability govern the ostensibly rational goal of getting money. However, the power of internet-assisted and video gaming makes it hard not to notice how bets become rewarding in their own right—without promising to deliver anything but outcomes *per se*. The importance that people attach to purely symbolic prizes is familiar, of course, but a motivational model in which learned ("secondary") rewards must predict some innately rewarding ("primary") event has seemed necessary, as has the location of that event beyond the subject's control (Baum 2005, pp. 277–286). However, such a mechanism does not credibly deal with the problem of activities that we do not expect to lead to an ultimate product. This is a topic where the problem of addiction reveals the need to greatly broaden the conventional theory of reward, with implications for many non-addictive activities as well. Mathematical modelers have begun to explore the notion that reward can be generated internally, but they still assume that such rewards come from innately programmed outcomes, "hard-wired from the start of the agent's life" (Singh et al. 2010, p. 73).

The theoretical problem has been that coining your own reward might be expected to produce short circuits in which people reward themselves autistically—indeed this is literally seen in the stereotyped behaviors of severe autism and, at a higher level, in the preoccupations of fantasy-prone personalities (Rhue & Lynn 1987). However, the ability to reward oneself at will is familiar enough. With a physical reward like food, availability at will makes its effect depend on appetite. Appetite builds potential payoffs disproportionately to their delays, so although hyperbolic discount curves create urges to harvest reward as soon as possible, your long-term interest is to cultivate the appetite—to "work it up," or make a personal rule not to eat between meals. The same principle is true of self-generated reward. Fiction, games, even daydreams require an appetite factor such as suspense or curiosity to be satisfying. You can harvest this appetite, as it were, at will—look ahead to the end, cheat at solitaire. But the earlier you harvest the appetite, the less well it pays. Appetite builds as a function of time, and is accelerated by challenges as in thrill-seeking, gambling, or video games. The combination of challenge and rapid success might make any behavior addicting, as Foddy has pointed out (2016). Just assuming, for illustration, that appetite builds linearly and is consumed linearly over time, the hyperbolically discounted values simply of unobstructed versus obstructed consumption patterns can take the familiar SS/LL form (Figure 3.2). In this example, the obstructed pattern is worth waiting for from the perspective of distance, but requires commitment if it is to stay chosen when the unobstructed pattern becomes imminently available.

Where the harvesting depends on physical objects such as books or cards, personal rules about it are easy to enforce. Where the harvesting is entirely mental, ways to pace it are more conjectural. Rules for such harvesting necessarily begin in discerning criteria—*occasions*—for payoffs that are either external or hard to create. Adoption of such criteria might best be called *betting*: on occasions in a book or card game, on whether a sports team will win on TV, or on

The picoeconomics of addiction

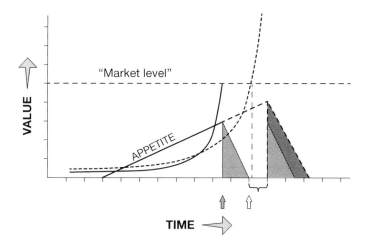

Figure 3.2 Comparison of two consumption patterns of the same expected reward, as determined by the appetite for it (stippled areas) and chosen when the hyperbolically discounted values of these areas reach market level: earlier consumption beginning at the moment of choice (solid arrow) versus consumption available later and obstructed by an obligatory delay (bracket from the clear arrow).

whether you can master a task—which can be a mental task, as long as benchmarks for mastering it are clear. The personal rule is to maintain the appetite until the occasions occur, instead of redefining the occasions, or withdrawing your attention to alternative bets or to the expectation of physical rewards. A mental action that spoils a bet is less conspicuous than turning a page or card. If you withdraw your investment in a movie when it gets too scary, you may or may not be conscious of saying to yourself, "it's only a story;" but in either case you have reduced your potential to be rewarded by its subsequent events. Likewise you may or may not be conscious of rooting less for a sports team, or valuing a friendship less, but you are apt to notice a change in your relevant payoffs. As with other personal rules, seeing yourself pull investment from a movie will make you less likely to resist the urge to do so later, and not just in this movie but in other movies and perhaps other projects.

The potential endogenous reward that depends on a bet could be called its *hedonic importance*, as opposed to an instrumental importance based on external reward. The great weakness of hedonic bets is their arbitrariness, the fact that they can as well be based on any number of scenarios leading to any number of outcomes, as in daydreams. This weakness can be overcome by betting on *singular* occasions—those that stand out from other possible occasions by being infrequent and easily distinguishable from those that are more frequent. Such bets win out in the marketplace of choice. Singularity may consist in being a remote coincidence, a round anniversary, a great exercise of skill, an event close to home or to the present moment, and so on (Ainslie 2013a). Two sources of singularity make occasions especially effective in pacing endogenous reward: a history of your having bet on them, and, perhaps confusingly, their realistic prediction of external rewards.

A history of betting recruits importance recursively. You *find* a problem or a book or a sports team important from having spent attention on them, and you *make* them more important by spending more attention on them. This recursive process has obvious similarities to intertemporal bargaining, except that the alternatives do not necessarily serve long- vs short-term interests; a withdrawal of importance may represent just a change in taste rather than a defection. However, accumulated importance is an investment, "consumption capital" (Becker & Murphy 1988), which may bring satisfactions in art or philosophy but may also create obsessions or dares—absorption in gambling, increasingly risky shoplifting, building a video game score, self-destructive mountain climbing (Leamer 1999). For want of a motivational explanation the role of hedonic importance in sustaining addictions has often been dismissed as habit.

Goals that are objectively important are often singular as well, so their instrumental benchmarks may serve as occasions for hedonic reward in addition to being secondary external rewards—"Getting there is half the fun." It is difficult to distinguish the two processes by observation (Lea & Webley 2006), which means that ostensibly productive activities that are good pacers of endogenous reward are apt to become potent process addictions in disguise, such as "skilled" gambling, day trading of stocks, or dealing in collectibles. Arguably the crossed purposes of increasing instrumental productivity versus cultivating appetite underlie much misinterpretation of economic activity in general, but that is another topic (Ainslie 2013b).

The implication of an endogenous reward process is that strong motivations can arise from nothing but a source of occasions, and, unlike secondary rewards, grow substantially with practice (see Ainslie 2017b). Sometimes the result is long-term artistic pleasure or sublimation, but occasions can also be structured to invite hedonic importance onto faster-paying patterns that can become addictive. This would seem to be the object of modern gambling technology and the video game industry (again, see Foddy 2016). The availability of big data to evaluate patterns of occasions can only make this kind of addiction grow more inviting.

So is addiction a disease?

A third, but probably not final, lesson from hyperbolic delay discounting is that "motivated" does not mean "voluntary." Will is a learned executive function—intertemporal bargaining—that reinterprets patterns of expected reward to reveal incentive for consistent choice. This is the skill that society holds us responsible for maintaining. But sometimes a personal history of bargaining has left very little expected reward to be bargained with, raising the question of whether some addicts "can" stop their activity. That is the crux of whether hopeless addiction can be called a disease. I have proposed elsewhere that it is better seen as a bankruptcy (Ainslie 2011), although, unlike financial bankruptcy, it sometimes responds to sudden shifts in bargaining that re-focus motives (e.g., Miller & C'de Baca 2001). If we want to call it a disease, we need to recognize that it is a disease of motivation, that is, one that does not bypass the mechanism of choice. Identifying it as such is only the start of identifying its incentives.

Acknowledgements

This material is the result of work supported with resources and the use of facilities at the Department of Veterans' Affairs Medical Center, Coatesville, PA, USA. The opinions expressed are not those of the Department of Veterans' Affairs or of the US Government.

Notes

1 Hyperbolic discounting suggests how negative experiences can compete with positive ones: the negative experience must have a positive phase, experienced as an urge, long enough to attract attention; so intrusive memories, rages, fears, and even physical pains must lure you into participation.

2 Compulsions should be distinguished from impulses and addictions. I recommend that "compulsive" be reserved for the case of being strictly confined by personal rules, rather than being overwhelmingly motivated in general, or even being beyond motivation, as when people talk about "compulsive drinking." Similarly, rule-bound, compulsive behavior is sometimes called addictive, as in workaholism, perfectionism, anorexia nervosa, and obsessive-compulsive personality disorder; but this usage is confusing, since the motivational dynamic is quite different—overcontrol rather than failure of control. As a group, compulsive traits have much in common with each other and differ from substance- and thrill-based activities not only in their lack of arousal but in being more consistently preferred and integrated with your values (Ainslie, 2009). Compulsions should not be called addictions or impulses, and vice versa.

References

Ainslie, G. (2001) *Breakdown of Will*, New York, NY: Cambridge University Press.
Ainslie, G. (2005) "Précis of *Breakdown of Will*", *Behavioral and Brain Sciences* 28(5): 635–673.
Ainslie, G. (2009) "Pleasure and aversion: challenging the conventional dichotomy", *Inquiry* 52(4): 357–377.
Ainslie, G. (2010) "The core process in addictions and other impulses: hyperbolic discounting versus conditioning and cognitive framing", in D. Ross, H. Kincaid, D. Spurrett, and P. Collins (eds), *What Is Addiction?* Cambridge, MA: The MIT Press, pp. 211–245.
Ainslie, G. (2011) "Free will as recursive self-prediction: does a deterministic mechanism reduce responsibility?" in J. Poland and G. Graham (eds), *Addiction and Responsibility*, Cambridge, MA: The MIT Press, pp. 55–87.
Ainslie, G. (2013a) "Grasping the impalpable: the role of endogenous reward in choices, including process addictions", *Inquiry* 56(5): 446–469.
Ainslie, G. (2013b) "Money as MacGuffin: a factor in gambling and other process addictions", in N. Levy (ed.), *Addiction and Self-Control: Perspectives from Philosophy, Psychology, and Neuroscience*, Oxford, UK: Oxford University Press, pp. 16–37.
Ainslie, G. (2016) "Palpating the elephant: current theories of addiction in the light of hyperbolic delay discounting", in N. Heather and G. Segal (eds), *Addiction and Choice: Rethinking the Relationship*, Oxford, UK: Oxford University Press.
Ainslie, G. (2017a) "Intertemporal bargaining in habit", *Neuroethics* 10, 143–153, doi: 10.1007/s12152-0169294-3.
Ainslie, G. (2017b) "*De gustibus disputare*: hyperbolic delay discounting integrates five approaches to choice", *Journal of Economic Methodology* 24(2), 166–189. http://dx.doi.org/10.1080/1350178X.2017.1309748
Ainslie, G. and Herrnstein, R. (1981) "Preference reversal and delayed reinforcement", *Animal Learning and Behavior* 9(4): 476–482.
Ainslie, G. and Monterosso, J. (2003) "Building blocks of self-control: increased tolerance for delay with bundled rewards", *Journal of the Experimental Analysis of Behavior* 79(1): 83–94.
Baum, W. M. (2005) *Understanding Behaviorism*, 2nd Edition, Malden, MA: Blackwell.
Becker, G. and Murphy, K. (1988) "A theory of rational addiction", *Journal of Political Economy* 96(4): 675–700.
Bodner, R. and Prelec, D. (2003) "The diagnostic value of actions in a self-signaling model", in I. Brocas and J. D. Carillo (eds), *The Psychology of Economic Decisions Vol. 1: Rationality and Well-Being*, Oxford, UK: Oxford University Press, pp. 105–126.
Conklin, C. A. and Tiffany, S. T. (2002) "Applying extinction research and theory to cue-exposure addiction treatments", *Addiction*, 97(2): 155–167.
Corsini, R. J. and Wedding, D. (2011) *Current Psychotherapies*, 9th Edition, Belmont, CA: Brooks/Cole.
Crocq, M.-A. (2007). "Historical and cultural aspects of man's relationship with addictive drugs", *Dialogues in Clinical Neuroscience*, 9(4): 355–361.
Curry, S., Marlatt, A. and Gordon, J. R. (1987) "Abstinence violation effect: validation of an attributional construct with smoking cessation", *Journal of Consulting and Clinical Psychology* 55(2): 145–149.
Dennett, D. C. (2003) *Freedom Evolves*, New York, NY: Viking Penguin.
Everitt, B. J. and Robbins, T. W. (2013) "From the ventral to the dorsal striatum: devolving views of their roles in drug addiction", *Neuroscience & Biobehavioral Reviews*, 37(9): 1946–1954.
Fellows, L. K. and Farah, M. J. (2005) "Different underlying impairments in decision-making following ventromedial and dorsolateral frontal lobe damage in humans", *Cerebral Cortex* 15(1), 58–63.

Flanagan, O. (2016) "Willing addicts? Drinkers, dandies, druggies and other Dionysians", in Heather, N. and Segal, G. (eds), *Addiction and Choice: Rethinking the Relationship*, Oxford, UK: Oxford University Press, pp. 66–81.

Foddy, B. (2016) "The pleasures and perils of operant behavior", in Heather, N. and Segal, G. (eds), *Addiction and Choice: Rethinking the Relationship*, Oxford, UK: Oxford University Press, pp. 49–65.

Green, L. and Myerson, J. (2004) "A discounting framework for choice with delayed and probabilistic rewards", *Psychological Bulletin* 130(5), 769–792.

Green, L., Myerson, J. and Macaux, E. W. (2005) "Temporal discounting when the choice is between two delayed rewards", *Journal of Experimental Psychology: Learning, Memory, & Cognition* 31(1): 1121–1133.

Haber, S. N. and Knutson, B. (2010) "The reward circuit: linking primate anatomy and human imaging", *Neuropsychopharmacology*, 35(5): 4–26.

Heyman, G. M. (2009) *Addiction: A Disorder of Choice*, Cambridge, MA: Harvard University Press.

Kable, J. W. and Glimcher, P. W. (2007) "The neural correlates of subjective value during intertemporal choice", *Nature Neuroscience* 10: 1625–1633.

Kahneman, D. (2011) *Thinking, Fast and Slow*, New York, NY: Farrar, Straus, and Giroux.

Lea, S. E. G. and Webley, P. (2006) "Money as tool, money as drug: the biological psychology of a strong incentive", *Behavioral and Brain Sciences* 29(2): 161–209.

Leamer, L. (1999) *Ascent: The Spiritual and Physical Quest of Legendary Mountaineer Willi Unsoeld*, Minot, ND: Quill.

Levy, D. J. and Glimcher, P. W. (2012) "The root of all value: a neural common currency for choice", *Current Opinion in Neurobiology* 22(6): 1027–1038.

Loewenstein, G. (1996) "Out of control: visceral influences on behavior", *Organizational Behavior and Human Decision Processes* 65(3): 272–292.

Luerssen, A., Gyurak, A., Ayduk, O., Wendelken, C. and Bunge, S. A. (2015) "Delay of gratification in childhood linked to cortical interactions with the nucleus accumbens", *Social Cognitive and Affective Neuroscience*, 10(12), 1769–1776.

McClure, S. M., Laibson, D. I., Loewenstein, G. and Cohen, J. D. (2004) "The grasshopper and the ant: separate neural systems value immediate and delayed monetary rewards", *Science* 306(5695): 503–507.

Metcalfe, J. and Mischel, W. (1999) "A hot/cool-system analysis of delay of gratification: dynamics of willpower", *Psychological Review* 106(1): 3–19.

Miller, W. R. and C'de Baca, J. (2001) *Quantum Change: When Epiphanies and Sudden Insights Transform Ordinary Lives*, New York, NY: Guilford.

Monterosso, J. and Ainslie, G. (2007) "The behavioral economics of will in recovery from addiction", *Drug and Alcohol Dependence* 90(1): S100–S111.

Monterosso, J. R., Ainslie, G., Toppi-Mullen, P. and Gault, B. (2002) "The fragility of cooperation: a false feedback study of a sequential iterated prisoner's dilemma", *Journal of Economic Psychology* 23(4): 437–448.

Pickard, H. and Ahmed, S. (2016) "How do you know you have a drug problem? The role of knowledge of negative consequences in explaining drug choice in humans and rats", In Heather, N. and Segal, G. (eds), *Addiction and Choice: Rethinking the Relationship*, Oxford: Oxford University Press, pp 29–48.

Rhue, J. W. and Lynn, S. J. (1987) "Fantasy proneness: the ability to hallucinate 'as real as real'", *British Journal of Experimental and Clinical Hypnosis* 4(3): 173–180.

Samuelson, P. A. (1937) "A note on measurement of utility", *Review of Economic Studies* 4(2): 155–161.

Shizgal, P. and Conover, K. (1996) "On the neural computation of utility", *Current Directions in Psychological Science* 5(2): 37–43.

Singh, S., Lewis, R. L., Barto, A. G. and Sorg, J. (2010) "Intrinsically motivated reinforcement learning: an evolutionary perspective", *IEEE Transactions on Autonomous Mental Development* 2(2): 70–82.

Sussman, S., Lisha, N. and Griffiths, M. (2011) "Prevalence of the addictions: a problem of the majority or the minority?" *Evaluation & the Health Professions* 34(1): 3–56.

Voon, V., Derbyshire, K., Rück, C., Irvine, M. A., Worbe, Y., Enander, J., Schreiber, L. Gillan, C., Fineberg, N. A., Sahakian, B. J., Robbins, T. W., Harrison, N. A., Woodk, J., Daw, N. D., Dayan, P., Grant, J. E. and Bullmore, E. T. (2015) "Disorders of compulsivity: a common bias towards learning habits", *Molecular Psychiatry* 20(3): 345–352.

4
ADDICTION AS A DISORDER OF SELF-CONTROL

Edmund Henden

Although impairment of self-control is often assumed to be a defining feature of addiction, many addicts nonetheless display what appears to be a considerable amount of control over their addictive behavior. Not only do they act intentionally and frequently in a deliberate manner, there is evidence suggesting that they are responsive to many ordinary incentives and counter-incentives (Heyman 2009; Pickard 2015). Given the great variety of addicts and their ability to control their drug use, how can impairment of self-control still be taken as *a defining feature* of addiction?

Clearly, if we want to hold on to this idea (as I think we should), a lot depends on how we understand the nature of self-control. In this chapter I shall argue, first, that a difficulty with some standard views of self-control in the philosophical literature is that they see the impairment of addicts' self-control only in terms of the role addiction plays in causing a certain kind of self-control conflict, either between the addict's drug-oriented behavior and her all things considered judgment, or between this behavior and her long-term values. But addicts, problematically, also display impaired self-control in the absence of any such conflicts. Second, there is plenty of empirical evidence suggesting that diminished attentional and impulse control is an important part of the explanation of addictive behavior. I argue that while such diminished control may produce self-control conflicts of the kind assumed in the philosophical literature, it need not do so. Still, it provides evidence that addicts have impaired self-control. By creating an inflexible and stimulus-bound practical perspective, diminished attention and impulse control mean that addicts' decision-making gets shaped by their addiction rather than by their reasons. While this makes it very difficult for addicts to revise or abandon their drug-oriented behavioral pattern given good and sufficient reasons to do so, it does not rule out the possibility of their often exercising substantial control over their drug-oriented actions.

Judgment-based vs value-based self-control

Self-controlled persons are governed by motives that are, in some sense, constitutive of their "self." In philosophical discussions of self-control, two kinds of motives tend to be highlighted. First, there are the motives that are expressed by the agent's "all things considered" or "better" judgments. Second, there are the motives that are expressed by what the agent "genuinely values." Since an agent's all things considered judgement at a particular time need not reflect what she genuinely values, these motives are not necessarily the same. It is possible, therefore, to distinguish between a judgment-based and a value-based form of self-control.

Consider first judgment-based self-control. Alfred Mele has argued that self-control is the contrary of what Aristotle called *akrasia* (incontinence, weakness of will), "a trait of character exhibited in uncompelled, intentional behavior that goes against the agent's best or better judgment" (Mele 2002: 531). A best or better judgment is "a judgment to the effect that, on the whole, it would be best to A, or (instead) better to A than to B" (Mele 2010: 392). As Mele also points out, "Someone's *making a judgment* [. . .] is an event," that is "at least suggestive of a belief arrived at on the basis of conscious deliberation" (Mele 2009: 6, 7). According to a judgment-based view, then, self-control is a trait of character exhibited in behavior that conforms with one's best or better judgment in the face of temptation to act to the contrary.

If self-control is the contrary of akrasia, it may seem that failures of self-control cannot occur in the absence of akrasia. That is, a person cannot fail to exercise self-control if he acts in conformity with his all things considered, or better judgment. However, this view is problematic. Consider the following example. An alcoholic wishing to stay abstinent may now judge that, all things considered, it would be better to stay home and watch television than go to the pub and have a drink with a friend. But as evening draws near, the temptation to drink increases, making him change his mind about what would be the better course of action – by giving too much weight to certain considerations that appear to provide reasons to meet with his friend, for example – and then revising his all things considered judgment accordingly. Is this a failure of self-control? Not if self-control is the contrary of akrasia. The alcoholic (unlike the *akrates*) does not act *against* his all things considered judgment when he chooses to meet up with his friend. Nevertheless, he plausibly exhibits a lack of self-control. This is because he revises his all things considered judgment *unreasonably*: he revises it, not because of the considerations that give him reasons to join his friend, but because he, at the time of making the choice, has a strong desire to drink – precisely the desire his judgment ruled out as outweighed by his better reasons.

In attempting to diagnose the difficulty here, it might be pointed out that the "self" of self-control must involve more than what is contained in the agent's all things considered judgment at the time of choice. While a self needs, for example, a certain stability over time – a self is not typically understood as the sort of thing that frequently changes – an agent's all things considered judgment may very well change from one moment to the next. Oscillations in judgment, or "judgment shifts" as they are called, typically occur when influences external to the self (such as impulses, desires, or strong feelings) interfere with or disrupt an agent's deliberative process at the time of choice. An alternative approach, therefore, one that aims to preserve the stability of the self, might be to see this notion as essentially involving *values* the agent herself has endorsed or endowed with normative force (Kennett 2013). Although values, of course, also may change, they seem inherently to have more stability than ordinary judgments. As Gary Watson suggests, they express what the agent "in a cool and non-deceptive moment – articulates as definitive of the good, fulfilling and defensible life" (Watson 2004: 25). Clearly, an agent's all things considered judgment at one particular moment need not express such values. According to value-based self-control, then, self-controlled persons govern themselves in accordance with what they genuinely value even in the face of strong competing motivation. Since it is common to see an agent's values as expressed by the stable set of attitudes she has relative to future behavior, value-based self-control is *diachronic*: it enables the agent to govern her actions across time, to maintain coherence between her long-term attitudes and her actions (Levy 2006). The alcoholic in the example might be said, then, to exhibit a lack of value-based self-control.

Judgment-based and value-based conceptions of self-control imply different views of the fault involved in *failures* of self-control. Suppose addicts typically lack judgment-based self-control.

That means they typically take drugs while judging at the same time that, all things considered, it would be better to abstain. Since they take drugs knowingly and in full awareness of choosing an inferior option, their fault cannot be cognitive. It must be *volitional*. It must consist in a failure to be motivated to do what they judge best. But this might not be the correct view of the failure of self-control in addiction. Suppose instead that addicts typically lack value-based self-control. As the opportunity for consumption presents itself, they typically succumb to a temptation to rationalize their reluctance to abstain by changing their mind about what would be best. At that moment, they always judge it best, all things considered, to take drugs. This would suggest that their fault is *cognitive*. It consists in a failure to judge, at the time of action, that the option that represents their genuine values is the best option.

So, is the failure of self-control in addiction typically cognitive or volitional? Let me start by considering a recent philosophical argument for the view that it is cognitive.

Levy on the impairment of self-control in addiction

The most important evidence of a self-control problem in addiction stems from addicts' own reports of failed attempts to exercise restraint. This restraint typically involves effort and a conscious act of will. Based on the further observation that these failures tend to have tremendous negative consequences for the addicts' themselves, it has been common to infer that addictive desires must be so powerful as to be literally irresistible. Clearly, this means, if correct, that addicts' impairment of self-control must involve a volitional fault, i.e., addicts are perfectly aware of the best course of action – to decline the offer of a drink, to throw away the packet of cigarettes, to abstain from heroin – but they are just unable to motivate or incentivize themselves to act accordingly. There are, however, serious problems with this inference. As several studies have shown, financial concerns, fear of arrest, values regarding parenthood, and many other factors increasing the cost of drug-taking often persuade addicts to desist (Heyman 2009; Pickard 2015). In fact, addiction often appears to involve a choice process, a conscious weighing up of the costs and benefits of different options. This is evidence that addictive desires need not be irresistible.

For some authors, self-control failure in addiction is therefore a cognitive rather than a volitional failure, involving the addict's loss of control over the valuational contents of her all things considered judgments rather than over her motivation to act in line with these judgments (Levy 2014). Based on empirical evidence that addicts tend to be "hyperbolic discounters" (Ainslie 2001) – they tend to discount the utility of future rewards by a proportion that declines as the length of the delay increases – Neil Levy has suggested that addiction creates strong temptations that induce regular and uncontrollable judgment shifts (Levy 2006). Such judgment shifts undermine self-control, he argues, because the addict fails to integrate her life sufficiently to pursue what she genuinely values. In other words, addiction impairs *value-based* self-control. Recently, Levy has supported this account with an argument purporting to show that the neural adaptations characteristic of addiction in fact provide *a mechanism* for such judgment shifts (Levy 2014 and in this volume). While addictive drugs are widely known to increase dopamine by stimulating its release or decreasing its reuptake in the nucleus accumbens, the information carried by increased levels of dopamine is open to different interpretations. Of particular relevance here is a series of well-known experiments by Wolfram Schultz and colleagues (1992), measuring firing activity of putative midbrain dopamine neurons in monkeys before they learned to associate a light cue with a reward. They found that the dopamine signal went up when the reward was delivered, but once the association between cue and reward was learned, the dopamine signal went up when the light appeared but remained flat when the reward was delivered.

Schultz and colleagues took this to show that a momentary increase in dopamine neuron firing activity indicates that the world contains more rewards than expected, or that an unexpected reward is available.

If one of the functions of midbrain dopamine neurons is to help the organism update its idea of the world's abundance of rewards, as Schultz and colleagues suggest, addiction, Levy claims, could plausibly be seen as a pathology of reward-based learning because drugs, unlike natural rewards, drive up the dopamine signal by direct chemical action on the brain. Drugs are therefore likely to generate a large prediction error: every time they are consumed (i.e., the reward is delivered) the addict's reward system tells her that the world contains more rewards than expected, rather than being exactly as expected (which would have been the normal learning response). Based on evidence linking dysfunctions of the dopamine system with pathologies of belief-formation, Levy argues that the reward system's treatment of the drug as something of ever-increasing value puts pressure on addicts to adopt and endorse an available causal model of the world that minimizes prediction error. Since the most accessible model is one in which drug use is judged better than abstention, addicts are likely to adopt and endorse this model, he claims. So when dopamine neuron firing activity increases in the presence of drugs or cues predicting drug availability, addicts will shift from judging abstention, all things considered, better than use to judging use as better, all things considered, than abstention. However, once the drug is consumed and the dopamine neuron firing activity has decreased, a rival model of the world will be triggered, where abstention is judged better than use. The result is oscillation in all things considered judgments.

Now, it is well documented that addicts generally experience difficulties in attempting to control future behavior and pursue long-term goals of abstinence. Levy is probably correct that judgment shifts occur more frequently in addicts than in non-addicts. In the next section, however, I shall argue that there still are good reasons to resist the claim that loss of control over all things considered judgment is the only, or even the typical, form the impairment of self-control takes in addiction.

Cravings and failure of impulse control

If one of the functions of midbrain dopamine neurons is indeed error prediction, these error predictions most likely lie outside of consciousness (Schroeder 2011). This means that a lot hinges on Levy's assumption that the error is passed on to higher levels of the processing hierarchy. According to Levy, minimizing the prediction error creates pressure on these higher levels to adopt and endorse the model that constitutes the judgment that drug use is, all things considered, best. It is "the model," Levy writes, "of the drug's place in the world, which the person endorsed in the earlier stages of drug use when drug use was controlled and chosen for its rewards" (Levy 2014: 348). It will be adopted and endorsed because it is the "most easily accessible" and most "easily available for recall" (ibid.: 348). This claim raises difficult questions about what makes a model "easily accessible." Clearly, addicts differ immensely in terms of psychological and social circumstances, personal resources, severity of addiction and so on. Is *the same* model of the world equally "accessible" to *all* addicts in spite of these differences? Levy could be right that the model that drug use is, all things considered, best might be the "most easily accessible" model to a young cocaine addict trying to quit for the first time. But why assume this model is equally accessible to the middle-aged nicotine addict who has desperately and unsuccessfully struggled with addiction for many years? She might not get any real pleasure out of smoking anymore. She might even wonder why she ever started given how little she now values smoking. Will the model of smoking's place in the world that she endorsed in the early stages of her addiction (possibly decades ago!) for its rewards be the one "most easily accessible" every time some smoke-related cue boosts her dopamine levels and she feels a craving to smoke? I confess I find this hard to believe.

The main problem with the judgment shift view, however, is that it ignores the role of *cravings* in the development and maintenance of addiction. It is widely agreed that dopaminergic activity in addicts is linked to experiences of drug cravings. A natural place to start, then, in interpreting the information carried by this activity, would be to focus on the role cravings play in deliberation and action guidance. Cravings appear to belong to the same category of simple motivation as "impulses" and "urges." Such motivations differ in important ways from more complex forms of cognitive motivation involving beliefs, desires, or judgments. For example, they can be very hard to separate from the actions they produce – such as the impulse to scratch produced by an itch, or the impulse to laugh in response to a joke, can be hard to separate from the experience of scratching or laughing. Usually, it is when an impulse for some reason is not immediately translated into action that it comes into conscious awareness and is experienced as an "urge" or a "craving" (West 2006).

Now, if craving is a form of impulse or urge, "yielding to temptation" might mean something quite different when it is caused by a craving than by a more complex form of cognitive motivation. Consider cases where loss of control is mediated by judgment shifts. Here, an agent yields to temptation by failing to restrain herself from re-evaluating her options when the tempting object becomes available. Plausibly, one explanation of what drives this process is an anticipation that she is likely to succumb, combined with a desire to avoid the experience of dissonance by choosing an option she believes is less valuable than the alternative (Holton 2009). Since she decides to succumb after weighing the costs and benefits of the alternative options, she will – as Levy points out (quoting Austin) – "succumb to temptation with calm and even finesse" (Levy 2014: 352). But even if loss of control can often be quite deliberate, involving the weighing of costs and benefits, it is surely, sometimes, *not* very deliberate! The best evidence of this comes from cases where people appear to be *taken by surprise* by their own failures to resist their urges or impulses. I am thinking here of actions performed under the influence of sudden anger, sexual arousal, or, perhaps, hunger. What drives such failures does not seem to be any process involving anticipation and re-evaluation. Not only do they happen too quickly for that (often in the midst of great emotional turmoil), it is not unusual for people acting under impulse to find themselves regretting the action *as* they are carrying it out, sometimes even trying to exercise restraint *at that very moment* ("I can't believe I am doing this! I should stop right now"). While they may well realize they are about to do something they judge it would be better not to do, the reason yielding to temptation in such cases is not mediated by judgment shifts is that the actions occur *before* the weighing of pros and cons. Clearly, an agent doesn't fail to control an impulse by *first* judging that performing the impulsive action is, all things considered, the best thing to do! Before discussing some of the evidence suggesting that diminished impulse control is prevalent among addicts, let me first briefly consider a different conception of self-control, one that is common in the psychological literature.

Broad psychological self-control

One reason for thinking that addiction impairs self-control might be that it seems close to being a conceptual truth that a person who is actively addicted to a drug must have impaired control over her consumption of that drug. Consider, for example, the suggestion that a certain nicotine addict who smokes 60 cigarettes a day is in "full control" of her smoking behavior. I think most people would be inclined to find this claim odd, indeed, even contradictory ("An addict who smokes three packs a day? Obviously, her consumption is out of control!"). If this is correct, it suggests that some notion of impaired control might be inherent in the common-sense notion of addiction. Is it equally odd to assume that the nicotine addict genuinely believes smoking in general to be fine, indeed even *better*, all things considered, than not smoking? Clearly this assumption is less odd. Many addicts don't make a serious effort to abstain from drugs. Some don't even want to. It seems perfectly possible to imagine many of them acting, most of the time, in accordance with their consciously held all things considered judgments when they choose to use. Still, insofar as they are "addicted," there seems to be a sense in which they cannot be in "full control" of their consumption. Is that because they are behaving contrary to what they *genuinely value* as seen from their own long-term perspective?

There doesn't seem to be anything odder about the claim that smoking is something the nicotine addict genuinely values from the vantage point of her own long-term perspective. Not all addicts are ambivalent, vacillating, or conflicted about their drug use (Flanagan 2016). Imagine, for example, a non-addicted recreational smoker who genuinely values smoking, perhaps because she associates it with a cool lifestyle. Suppose her smoking escalates and becomes an addiction. Although this means her smoking becomes excessive, she may have no desire to quit. Her long-term perspective and value system remain the same throughout and beyond this transitional phase and she regulates her life in accordance with her values. Still, when she does become addicted, most people would be inclined, I think, to deny that she is in full control of her cigarette smoking.

In fact, there seem to be many cases where it is at best unclear whether the addicts have lost value-based self-control. Consider, for example, those described in the clinical literature as "severe, chronic addicts," i.e., people who have exhibited signs of dependency for many years (sometimes decades). The search for and use of drugs may now be the single most important activity in their lives, completely dominating their thinking, emotions, and behavior. For some of them, we may suppose, it might even define who they think and feel they *are* and become part of their practical identity. If this is the case, drug-oriented values may be precisely what imposes a cross-temporal structure on their thoughts and actions. It is difficult to see in what sense these addicts should be failing to regulate their life in accordance with their own values. But, even so, most people, I think, would be inclined to say that they are not in full control of their drug-oriented behavior.

What all these examples have in common is the suggestion of a sense in which addicts seem to have lost an important form of self-control even when engaging in drug-seeking or drug-taking behavior does not give rise to the kinds of normative conflict associated with failures of value-based or judgment-based self-control. In fact, one might reasonably wonder how often people – even *self-controlled* people – really act on the basis of conscious judgments informed by comparisons of all the relevant available reasons for and against the alternative options, or in accordance with coherent sets of values they in a cool and non-deceptive moment would articulate as definitive of the good, fulfilling and defensible life. It is quite plausible, I think, that both exercises in self-control, as well as failures of self-control, can occur independently of whether these rather demanding conditions obtain. Imagine again the nicotine addict – let us call her Beth – who "in a cool and non-deceptive moment" has decided, based, we can assume, on a careful weighing of the benefits and risks of smoking (pleasant feelings, enjoyable rituals, probability of lung cancer, and so on), that she wants to continue smoking one more year before she quits. Suppose Beth is at a party and is overcome by a strong urge to smoke another cigarette. However, at that moment she suddenly remembers a TV program featuring a terminally ill lung cancer patient. Even though she discounted the risk of lung cancer in her previous "cool" assessment, the disturbing image of this person now triggers an emotional response causing her, without any consideration of pros and cons, to try to override her urge to smoke another cigarette. Unfortunately, her attempt at restraint fails. Is this a failure of *self-control*? It seems very plausible. A sudden feeling of anxiety activated a goal (not to smoke another cigarette) that caused her to make an effort to refrain from lighting another cigarette. Yet although Beth fails in that goal, she does not act contrary to an all things considered judgment. Indeed, she did not make any, and if she *had* made one at that moment, it would likely have been that smoking another cigarette was, all things considered, better than not smoking one (once she has lit the cigarette she may be content that she failed!). Neither does Beth act contrary to what she genuinely values either.

In the psychological literature Beth's failure would be treated as a paradigmatic example of a failure of self-control. Although it is difficult to find a very precise definition of self-control in this literature, it tends to be used in the sense of involving a conscious effort to override or inhibit competing urges, behaviors, or desires in order to attain particular goals. Since Beth makes a conscious effort to act in accordance with a mental goal representation of herself refraining from smoking another cigarette (a representation that *does not* include a comprehensive assessment of relevant available reasons to choose or reject any of the alternative options), but fails to override the urge to smoke, she fails to behave in conformity with her mental goal representation, and hence to exercise self-control. She is not, however, guilty of any value-based or judgment-based failure of self-control. Often, the term self-control is used, in the psychological literature, interchangeably with "self-regulation," which refers, even more broadly, to all higher-order (i.e., executive) control of lower-order processes that adapt thought, emotions, or behavior to the demands of the situation and the agent's goal(s) within that situation, including cognitive and motivational operations that are performed automatically and unconsciously (Fitzsimons and Bargh 2004). In the next section I shall argue that malfunctions in capacities associated with self-control in this broad psychological sense undermine addicts' self-control also in the absence of failures of value-based or judgment-based self-control.

Addiction as a disorder of self-control

It is widely believed that one of the most important higher-order-control functions is associated with directed attention, i.e., the capacity to voluntarily focus or shift attention (Baumeister and Heatherton 1996). Directed attention represents a common regulatory mechanism for emotion, cognition, and behavior. As MacCoon and colleagues (2004) note, it is a top-down mechanism "capable of enhancing appropriate cognitions, emotions, or behaviors, and suppressing inappropriate cognitions, emotions, or behaviors" (ibid., 422). It is crucial for self-control not only because it is necessary for maintaining focus on longer-range goals, concerns or values, but also because it is required for making conscious efforts to override or inhibit competing motivations. Thus, measures of effortful control often include indices of attention-regulation (Eisenberg 2004).

In a much-cited article from 2008, Matt Field and W. Miles Cox review evidence suggesting that addicts' attention is biased toward drug-associated stimuli. Much of this evidence comes from experiments measuring impairment in attention and impulse control such as the addiction Stroop task, where the addict must name the print color of a drug-related word and inhibit the stronger tendency to read the name of a color itself. While addicts exhibit significantly slower reaction times and are more prone to error when naming the color of drug-related words, control participants do not exhibit this pattern. The standard interpretation of this Stroop interference is that drug-related words capture addicts' attention causing excessive processing of the semantic content of these words, thereby disrupting their color naming. What it seems to suggest is that addicts find it particularly difficult to *ignore* salient, drug-related stimuli, or *exercise* directed attention (which, as noted above, is required for both planning and inhibition) in the presence of drugs or cues predicting drug availability. This hypothesis appears to be corroborated by neuroscientific evidence showing that repeated drug use "sensitizes" certain regions in the brain involved in the motivation of behavior, making them more easily activated by drug-related cues or circumstances. As the neuroscientists Robinson and Berridge note, such sensitization "produces a bias of attentional processing towards drug-associated stimuli" (Robinson and Berridge 2008).

Now, it should be noted that not all researchers are convinced of the cogency of the evidence for attentional bias in addiction (see, e.g., Hogarth this volume). However, in what follows I shall assume it is correct. Given this hypothesis, it would be reasonable to infer that cravings are triggered by cues or circumstances via entrenched patterns of attention. A likely effect of such entrenched patterns is a dramatic increase in cognitive load on higher-order functions requiring directed attention, such as inhibition, reasoning, or planning. That is because, unlike non-addicts, every time addicts engage in these activities, they have to make an effort *not* to attend disproportionally to drug-related cues or considerations. If all higher-order functions draw on the same limited resource (as many psychologists believe), and the more of this resource is consumed the more depleted it becomes, then the more depleted addicts will become compared with non-addicts (which is, of course, exactly what the addiction Stroop task shows). Entrenched patterns of attention combined with subsequent depletion reduce the capacity to switch thinking and attention among different tasks or operations in response to changing goals or circumstances. This must have important normative consequences because it plausibly results in a more inflexible and stimulus-bound practical perspective. Depending on individual differences between addicts (e.g., differences in personal and social resources), such a perspective might be hypothesized to affect their behavior in a variety of ways. For example, by restricting or altering the *reasons* that are salient to them, it might act as an option-limiting cause in some cases. In other words, reasons to abstain from drugs may no longer be recognized as reasons in the presence of drugs or cues predicting drug availability, resulting in a systematic biasing of their practical deliberation, e.g., an over-appreciation of drug-associated features of situations, or blindness to longer-range goals. This, of course, would explain why judgment shifts occur more frequently in addicts than in non-addicts (see page 48). However, as I have argued, self-control failures in addiction might take other forms than loss of control over all things considered judgments. Many addicts plausibly recognize the harmful effects of drug use on their lives even as they are seeking or taking them. For them the greatest difficulty might not be *to see* the salience of alternative reasons at the time of choice, or *to judge* what the best course of action is, all things considered, at that point in time, but to get themselves *to act* on these reasons or judgments. An inflexible and stimulus-bound practical perspective makes this very difficult because the frequency, cue-dependency, and computational speed of cravings produced via entrenched patterns of attention are likely to deplete their inhibitory mechanisms, hence making it harder for them to override these cravings – even if overriding them is precisely what they believe they should be doing.

Together, the biasing and depletion effects associated with an inflexible and stimulus-bound practical perspective are likely to make it, overall, very difficult for addicts to revise or abandon their drug-oriented behavioral pattern even if they are given good and sufficient reasons to do so. This, I contend, is the sense in which impaired self-control is a defining feature of addiction. Difficulties associated with revising or abandoning such a pattern even in the presence of good and sufficient reasons to do so, do not imply that addicts' cravings for drugs must be irresistible, nor that addicts cannot often take themselves to have reasons – even good reasons – to use drugs, reasons which, on many occasions, might play a part in the explanation of their addictive behavior. First, an inflexible and stimulus-bound practical perspective does not rule out a capacity to resist cravings. It just makes it harder over time to do so, requiring a more sustained effort, and hence increasing the likelihood of failure. Second, there is plenty of evidence that many addicts use drugs to cope with stressful or traumatic experiences. Such experiences may give them reasons to take drugs, reasons that form an important part of the explanation of their addictive behavior (Pickard 2015). Using drugs for such reasons, however, does not rule out the importance of diminished attentional and impulse control in this explanation. My central claim in this chapter is that if a person is genuinely addicted to a particular drug, that implies that she is disposed to suffer from impaired self-control with respect to its use.

References

Ainslie, G. (2001) *Breakdown of Will*, Cambridge, UK: Cambridge University Press.
Baumeister, R. F and Heatherton, T. F. (1996) "Self-regulation failure: An overview", *Psychological Inquiry* 7(1): 1–15.
Eisenberg, N., Smith, C. L., Sadovsky, A. and Spinrad, T. L. (2004) "Effortful control relations with emotions regulation, adjustment, and socialization in childhood", in R. F. Baumeister and K. D. Vohs (eds), *Handbook of Self-Regulation*, New York, NY: The Guilford Press, pp. 259–282.
Field, M. and Cox, W. M. (2008) "Attentional bias in addictive behaviors: a review of its development, causes, and consequences", *Drug and Alcohol Dependence* 97(1–2): 1–20.
Fitzsimons, G. M. and Bargh, J. A. (2004) "Automatic self-regulation", in R. F. Baumeister and K. D. Vohs (eds), *Handbook of Self-Regulation*, New York, NY: The Guilford Press, pp. 151–170.
Flanagan, O. (2016) "Willing addicts? Drinkers, dandies, druggies, and other Dionysians", in N. Heather and G. Segal (eds), *Addiction & Choice. Rethinking the Relationship*, New York, NY: Oxford University Press, pp. 66–81.
Heyman, G. M. (2009) *Addiction, a Disorder of Choice*, Cambridge, MA: Harvard University Press.
Hogarth, L. (this volume, pp. 325–338) "Controlled and automatic learning processes in addiction".
Holton, R. (2009) *Willing, Wanting, Waiting*, New York, NY: Oxford University Press.
Kennett, J. (2013) "Just say no? Addiction and the elements of self-control", in N. Levy (ed.), *Addiction and Self-Control: Perspectives from Philosophy, Psychology, and Neuroscience*, New York, NY: Oxford University Press, pp. 144–164.
Levy, N. (2006) "Autonomy and addiction", *Canadian Journal of Philosophy* 36(3): 427–448.
Levy, N. (2014) "Addiction as a disorder of belief", *Biology and Philosophy* 29(3): 337–355.
Levy, N. (this volume, pp. 54–62) "Addiction: the belief oscillation hypothesis".
MacCoon, D. G., Wallace, J. F. and Newman, J. P. (2004) "Self-regulation: context-appropriate balanced attention", in R. F. Baumeister and K. D. Vohs (eds), *Handbook of Self-Regulation*, New York, NY: The Guilford Press, pp. 422–444.
Mele, A. (2002) "Autonomy, self-control, and weakness of will", in R. Kane (ed.), *The Oxford Handbook of Free Will*, New York, NY: Oxford University Press, pp. 529–548.
Mele, A. (2009) *Backsliding. Understanding Weakness of Will*, New York, NY: Oxford University Press.
Mele, A. (2010) "Weakness of will and akrasia", *Philosophical Studies* 150(3): 391–404.
Pickard, H. (2015) "Psychopathology and the ability to do otherwise", *Philosophy and Phenomenological Research* 90(1): 135–163.
Robinson, T. E. and Berridge, K. C. (2008) "The incentive sensitization theory of addiction: some current issues", *Philosophical Transactions of the Royal Society* 363(1507): 3137–3146.
Schroeder, T. (2011) "Irrational action and addiction", in D. Ross, H. Kincaid, D. Spurrett and P. Collins (eds), *What is Addiction?* Cambridge, MA: The MIT Press, pp. 391–407.
Schultz, W., Apicella, P., Scanati, E. and Ljungberg, T. (1992) "Neuronal activity in monkey ventral striatum related to the expectation of reward", *Journal of Neuroscience* 12(1): 4595–4610.
Watson, G. (2004) "Free agency", in *Agency and Answerability*, New York, NY: Oxford University Press, pp. 13–32.
West, R. (2006) *Theory of Addiction*, Oxford, UK: Blackwell.

5
ADDICTION
The belief oscillation hypothesis

Neil Levy

In popular, philosophical and many scientific accounts of addiction, strong desires and other affective states carry a great deal of the explanatory burden. Much less of a role is given to cognitive states than to affective. But as Pickard and Ahmed (2016; see also Pickard 2016) note, addiction may be as much or more a disorder of cognition as of compulsion or desire. Pickard's focus is on denial. In this chapter my focus will be different. I will argue that in many cases at least, we can explain the lapses of abstinent addicts by way of processes that do not involve motivated reasoning (as denial or self-deception plausibly do). Mechanisms that have the role of updating beliefs in response to evidence may alter addicts' judgments concerning what they have most reason to do (in the precise circumstances in which they find themselves), and thereby cause them to act accordingly.[1]

My focus is on abstinent addicts, understood here as addicts who sincerely resolve to refrain from consuming the drug to which they are addicted and who are committed to this resolution. For many of these abstinent addicts, the resolution is not an idle wish. They may act on it in myriad ways. They may make changes, large and small, in their lives to try to make it more likely that they will abide by it. They may move homes, jobs, withdraw from their social circle, and so on, because they believe (on good grounds) that so doing will help them to remain abstinent. They may enter treatment programs, and they may commit resources to being abstinent (they may, for instance, pay for expensive treatments). We should not underestimate the efficacy of these kinds of steps. Most addicts do succeed in giving up, and many do so without any external help at all (Heyman 2009). Nevertheless, and notoriously, many abstinent addicts relapse, some after long periods of abstinence.

I am concerned with explaining these relapses and, more proximally, with explaining lapses. A *lapse* is an episode of consumption by a hitherto abstinent addict. A *relapse* is a return to regular substance abuse. Contrary to a popular myth, lapses do not always result in relapses. Addicts may lapse and then return, with more or less difficulty, to durable abstinence. However, relapses begin with lapses. Explaining lapses is not sufficient for explaining relapses, but it will contribute very significantly to an explanation of relapse.

The kinds of explanations of lapses that circulate in popular thinking about addiction do not seem to come close to explaining all of them. Commonly, these explanations cite irresistible urges, or the duress that withdrawal involves. On the first account, addicts lapse because they are gripped by a desire to consume that overcomes their powers of resistance. On the second, withdrawal is so awful that they are powerfully motivated to do almost anything to alleviate it. No doubt the strength of desires plays a role in explaining some lapses, and no doubt addicts are motivated to alleviate the pains of withdrawal. But there is plenty of evidence, direct and indirect, that neither explanation, alone or combined, is adequate.

First, there is extensive evidence that addicted people possess a great deal of control over their behaviour, even when they are in the grip of strong desires. Consumption of drugs, even among addicts, is price sensitive (Elster 1999; Neale 2002), which suggests a responsiveness to perfectly ordinary incentives. Consumption remains price sensitive among alcoholics even after a priming drink (Fingarette 1988). Second, the epidemiological evidence – the fact that the majority of addicts 'mature out' of their addiction – is hard to square with an irresistible desire account. But the withdrawal account does even worse. Not all drugs produce highly aversive withdrawal symptoms when an addicted person abstains from them (West 2006). Further, lapses may occur months, even years, after the last episode of consumption – long after the person has put withdrawal and its symptoms behind them. In fact, there is some evidence that addicts may deliberately go through withdrawal in order to lower their tolerance to the drug, and thereby return to drug taking requiring a smaller dose for the same effect (Ainslie 2000).

While there is extensive experimental evidence for reduced control by addicts (see, for instance, Hester and Garavan 2004), they retain a sufficient degree of control, measured by their responsiveness to ordinary incentives, for it to be very mysterious how this reduction of control could explain lapses except in a small minority of cases. Problems with executive function can explain impulsive behaviour: they can explain lapses that resemble the reactivation of a habit (for instance, they might be able to explain why an abstinent person with an addiction might find herself impulsively accepting the offer of a drink). But many lapses involve some degree of planning and preparation: rather than simply accepting an offer, the person resolves to seek out an opportunity for consumption and then constructs a plan to procure drugs (a plan that may be quite complex: it may be necessary to concoct a cover story for work or family, to secure money to score, and so on). Simple executive dysfunction can't explain this kind of activity. Or rather, it can't explain it *directly*: executive dysfunction may play a role in explaining lapses in these cases by playing a role in explaining how it can come about that addicts come to judge that (all things considered) they ought to procure and consume a drug.

Towards a judgment-centred account

Most of the drug procuring and consuming behaviour of addicts is entirely normal, in the following sense: it gives every appearance of being what psychologists call controlled behaviour, or of being explicable by what philosophers call belief/desire or folk psychology. Given the reasons-responsiveness of the behaviour – the way in which it is sensitive to perturbations in the environment, which give the agent reasons to inhibit or modulate their activity – the best explanation of why it unfolds as it does is that the agent *desires* to procure and consume drugs and that this desire has led them to form the *intention* to do so. Their behaviour is then controlled by this intention and other beliefs, particularly instrumental beliefs about how this is best brought about (most efficiently, at lowest cost in terms of resources and risk, and so on). Sometimes when abstinent addicts lapse they do so by consuming an immediately available drug. Episodes like that can often be explained by impulsivity, absence of control or reactivation of habits. But sometimes lapses involve a long series of behaviours aimed at procuring the drug first (or even first acquiring the money to procure the drug), and only subsequently consuming the drug hours or even days later. This long series of instrumental actions seems best explained by attributing to them the all-things-considered *judgment* that they ought to consume the drug.

Richard Holton (2009) introduced the helpful term 'judgment-shift' to refer to the overhasty revision of a resolution, in response to evidence or facts against which the resolution was meant to be proof. I suggest that judgment-shift underlies and explains many lapses. Addicted persons are especially vulnerable to drug-related judgment shifts, I claim. They have formed the resolution to abstain from drugs, and intend to abide by the resolution in circumstances like the ones in which they now find themselves. But in these circumstances they find themselves on what I have elsewhere called the 'garden path' to consumption (Levy 2016). Due to features of the neurobiology of addiction, finding themselves on the garden path disposes them to undergoing a cascade of judgment-shifts: first shifting toward judging that they ought to step down the garden path and subsequently to judging that they ought to consume.

Addiction causes a number of changes in the brain, such as those that cause the executive dysfunction already mentioned and those that cause anhedonia in response to withdrawal. Central to the addiction phenotype, however, is the response of the midbrain (mesolimbic) dopamine system to drugs of addiction (Hyman 2005; Kalivas and Volkow 2005). To understand the significance of this response, we need first to understand the system's role in ordinary cognition.

The midbrain dopamine system is widely held to be a reward prediction system. It has the role of signalling *unexpected* reward. In classic experiments performed on monkeys, activity in this system was measured when the animals were given a reward. In the initial condition, the reward (a squirt of fruit juice) was unexpected. There was a spike in phasic dopamine upon reward delivery. Subsequently, a signal (a light or a tone) was given prior to delivery of the juice. Once the monkey learned the signal that predicted the juice, there was no longer a spike in phasic dopamine in response to delivery of the reward (indicating that the spike was not tracking actual reward). Instead, the spike occurred in response to the (unexpected) signal. The signal was a sign that the world was better than the animal expected (Schultz *et al.* 1992; Schultz *et al.* 1997). Conversely, failure to deliver the reward subsequent to the signal led to dopamine falling below the resting baseline. Phasic dopamine thus seems to signal the state of the world relative to the organism's expectations. It thereby orients the animal toward the reward, making it more likely that they reap its benefits (dopamine is known to play a role in regulating attention: Corlett *et al.* 2007; Fletcher and Frith 2009).

In one way or another (by driving up dopamine directly or indirectly, or preventing reabsorption), however, addictive drugs 'hijack' this system (Carter and Hall 2012). Because addictive drugs drive up dopamine via their chemical action, they prevent it from accurately tracking reward value. When the addicted person is exposed to a cue that signals drug availability (the sight of a person with whom they have used in the past, say, or a sound, sight or smell they associate with using – anything that has been paired with consumption in the past often enough), they experience the spike in phasic dopamine that signals the availability of unexpected reward. But they also experience a spike in dopamine on consumption of the drug. For the relevant system, this spike is a signal of unexpected reward, despite the fact that, for the experienced user, the reward should be expected.

A predicted consequence of the fact that drugs of addiction produce a spike in phasic dopamine, thereby mimicking the signal of unexpected reward, is that the relevant system will not be able to adapt to the actual reward value of the drug. Instead, there will be a dissociation between the actual reward value of the drug for the person and the value placed on it by the midbrain dopamine system, helping to explain how addicts can be powerfully motivated to pursue drugs they no longer enjoy; the famous dissociation between 'liking' and 'wanting' (Berridge and Robinson 1995; Robinson and Berridge 2003). In fact, there is recent evidence that there is at least some accommodation of drug rewards with repeated exposure, both in animal models (e.g. Willuhn *et al*. 2014) and humans (e.g. Martinez *et al*. 2007). Further, recent work has shown that the phasic dopamine response to drug-related cues is of a similar magnitude to that of non-drug cues (Cameron *et al*. 2014). However, the mesolimbic system is composed of populations of functionally discrete neurons and this evidence is compatible with the existence of systems that fail to adapt to the value of drug rewards and therefore generate dysfunctionally large – or dysfunctionally orienting – prediction errors.

I think it is a mistake to think of the midbrain dopamine system as simply a reward prediction system. At least some components of the prediction errors it generates have a more general role: to signal the unexpected generally, not just unexpected reward. Corlett *et al*. (2004) showed that activation of rostral prefrontal cortex (rPFC) – a primary target of dopamine projections – was correlated with violations of expectations on a task unrelated to reward. If I am right in thinking that the dopamine system has the role of updating expectations generally, then it should be thought of as a representation-update system. It has the function of changing the organism's model of the world. There is, in fact, evidence directly linking dysregulation of this system to delusional belief formation. Corlett *et al*. (2007) measured rPFC activity in delusional patients and a control group in response to violations of expectations. They found attenuated response to violations of expectation and abnormally high responses to expected events in the patient group. Corlett *et al*. (2006) found that magnitude of prediction error response in controls predicted the likelihood of delusions following administration of the hallucinogen ketamine. All of this is evidence the midbrain dopamine system plays a role in updating the organism's causal model of the world, in response to violations of expectations.

The fact that the behaviour of lapsing addicts often appears to be controlled behaviour – the fact, that is, that it is incentive-sensitive in a very ordinary way – provides us with good reason to think that when these addicts engage in behaviour aimed at procuring and consuming drugs, their behaviour is controlled by the all-things-considered judgment that they ought, on this occasion, to consume the drug (in broadly the same kind of way in which my bread buying and toasting behaviour is controlled by the judgment that toast would be nice). I suggest that the dysfunction that addictive drugs provoke in the midbrain dopamine system provides a mechanism for judgment-shift.

The mechanism is best presented in the framework of prediction error minimisation models of brain function (Hohwy 2013; Clark 2016). According to these models, the brain constantly predicts the input it 'expects' to receive. It then updates its expectations on the basis of actual input, to minimise the prediction error: the gap between expected and actual input. Updating expectations is updating a representation: it involves an alteration in the model of the world. It is therefore updating a doxastic state (the state may, but need not, have all the features associated with a belief, such as serving as a basis for domain-general inference; nevertheless, it is a representational state that may be correct or incorrect, and that is sufficient to qualify it as doxastic). I suggest that the midbrain dopamine system is, or is part of, a prediction minimisation mechanism.

When the addicted person who has resolved to be abstinent is presented with cues predictive of drug availability (again, anything reliably enough associated with drug-taking to serve as a conditioned stimulus), she experiences a spike in phasic dopamine that signals a gap between the expected input and the actual input. As assessed by the relevant subpersonal mechanism, the world is (much) better than expected: it contains the promise of a large reward. The person must now update her model of the world, to minimise the prediction error. This process occurs at the subpersonal level: the person is unaware of the content of the expectation or of the prediction error. Whether she will undergo a change in personal-level attitudes or representations depends on whether the error can be minimised at lower levels of the processing hierarchy. According to the predictive coding framework, errors propagate up when there is no causal model available at the level at which the error occurs that can minimise it sufficiently (conversely, models propagate down). If the error is large enough and no such model is available at subpersonal levels of processing, model update may occur at the personal level.

This may be experienced by the person as a sense that some options are more rewarding than others. This is a sense that may be understood as a representational state. Suppose, for instance, the cue is a person with whom she has consumed in the past. She may experience a sense that being with the person will be rewarding, but she will be unaware that the reason she experiences this sense is because that person is associated with the availability of drugs, such that their presence is sufficient to cause a prediction error. She may instead just think that it would be fun to relive some old times with her friend. A judgment that going back to an old haunt with her friend would be rewarding may be the personal-level consequence of subpersonal representation update, caused by mechanisms that minimise prediction error.

The decision to spend some time with an old friend is innocuous enough, but it may place the person on the garden path all the way to consumption. Her friend may take her to somewhere that is more strongly associated with consumption, triggering another prediction error and another round of representation (and belief) update. She may thereby proceed step by step down the garden path, at each step unaware that she is heading toward consumption. It is only at the last step that she forms the judgment that (on this occasion, at least), procuring and consuming the drug is best. At the end of the garden path may be a relatively impulsive decision to consume. But the decision is the formation of a judgment: a controlling state that represents the world as being a certain way. Once this judgment is in place it is available to control behaviour generally, in just the same kind of way as her earlier resolution to be abstinent controlled her behaviour before she set out on this path. Note, too, that under the right conditions – when a prediction error is large enough and especially when competing rewards are not salient (lack of competition from competing rewards may be necessary, given the evidence that non-drug cues may trigger dopamine spikes of the same magnitude as drug cues: Cameron *et al.* 2014), the path from cue to personal level judgment may be much shorter.

Addicted people remain rational agents. They are not simply at the mercy of subpersonal mechanisms that cause representational states to change without apparent reason (any more than those of us who are not addicts are). We all make decisions and form judgments in ways that are guided by the outputs of subpersonal mechanisms, which dispose us to take some options seriously and dismiss others and may strongly incline us toward a particular judgment. Making decisions in this way is not a limitation on our rationality, but partially constitutive of it. We could not make decisions at all if we did not have mechanisms that narrowed the search space for us, ensuring that the set of options between which we deliberate are tractably few. This kind of mechanism often helps to constitute our rational processing. In this case, however, it is dysfunctional and produces a maladaptive response.[2] But because the person remains a rational agent, concerned to make sense of her own behaviour, she will justify her behaviour to herself in a way that is congruent with her former resolution to remain abstinent.

As Holton (2009) has emphasised, resolutions are never absolute. Rather, they come hedged with implicit escape clauses: conditions under which they are appropriately revised. I do not act irrationally or akratically if I revise or abandon my resolution to jog first thing in the morning should I realise that my partner is unwell, or that the road is flooded, or my neighbour needs urgent help. We do not, and cannot, specify the full range of such escape clauses: they are indefinitely many. Because this is true, we can quite easily deceive ourselves into judging that an escape clause has been triggered. I may, for instance, revise my resolution to skip dessert in the face of the thought that it's a special occasion. Of course, sometimes it *is* a special occasion, and I am right to do so, but I may multiply special occasions until I am consuming dessert more often than I am not. The addicted person, too, is likely to explain her behaviour by seeing an implicit escape clause as triggered ("today has been especially stressful"; "just this once won't hurt"; "it would be hurtful to my friend to refuse", or whatever). This may not be mere post-facto rationalisation, in the sense that the unavailability of such a confabulated explanation of her own behaviour might be sufficient to prevent her abandoning her resolution. She has expectations about her own behaviour too, and the inability to update these expectations in a way that maintains a sufficient degree of coherence among her personal-level beliefs may constitute a stronger pressure to maintain the resolution than to abandon it.

Judgment-shift and control

Most people – laypeople and specialists alike – regard addiction as causing or constituting a pathology of control over behaviour. On the picture I have sketched in this chapter, the behaviour of the addicted person is controlled behaviour: at each moment, she acts rationally, given her own all-things-considered judgments. That does not entail, however, that she possesses rational control over her own behaviour. Or rather, the rational control she possesses may not extend across time sufficiently for us to say that she is genuinely in control.

The agent possesses control over her behaviour, including her drug-seeking and consuming behaviour, because it is appropriately sensitive to her reasons. Such reasons-sensitivity is what control consists in. As Fischer and Ravizza (1998) have argued, an agent or a mechanism possesses control over an action or a state of affairs to the extent to which she or it is capable of tracking reasons (which they call reasons-receptivity) and responding to some of those reasons (which they call reasons-reactivity). The addicted person's behaviour exhibits a great deal of such sensitivity, and therefore of control. This is true for most of the time when she is abstinent and abides by her resolution, and it is also true when her behaviour is controlled by her judgment that, all-things-considered, she ought to procure and consume the drug. But, while her behaviour may exhibit a great deal of control whichever judgment controls it, the shift from abstinence to seeking and using drugs is caused by mechanisms that exhibit an impairment in control (as measured by the strength and range of the reasons to which they are responsive). Prediction error minimisation mechanisms may be said to realise exquisite control in agents, when they are appropriately set up to track genuinely reason-giving states. But, in this case, the mechanism is dysfunctional and its capacity to track reasons is impaired.[3]

We began by noting that accounts of addiction very often give central place to loss of control as an explanatory notion. If the account offered here is on the right track, loss – or at least significant impairment – of control is indeed central to explaining lapses, and not only in those cases in which executive dysfunction or habit is at its heart. Extant control-based theories misplace the locus of the diminution of control when it comes to explaining many lapses. They explain lapse by a loss of control of behaviour, whereas on my account the loss of control is over judgment. Because the mechanisms that cause judgement-shift are significantly impaired in the range of reasons to which they are capable of responding, the person's control over her judgments is itself significantly impaired. Sometimes, at least, this impairment is significant enough that we may regard the person as losing control.

The fact that loss of control is central to explaining lapses is important, because many of the normative implications, and indeed implications for treatment, that are supposed to follow from extant control-based accounts may follow from mine too. If, for instance, a significant impairment of control excuses, then we have good grounds for reducing or eliminating blame for addicts in many circumstances. The fact that a loss of control is central to lapses may also have implications for the incentives to which we can expect addicted people to respond, and therefore to the question of how we can help them moderate or eliminate drug use. Caution is needed in drawing normative and non-normative inferences, however: the fact that an agent lost control because a mechanism failed to be sufficiently sensitive to reasons, or was sensitive to cues that were not reasons, need not be an indication that the mechanism is generally reasons-insensitive. The same mechanism may realise control under some conditions while failing to do so under others.

Conclusion

People lapse for a variety of reasons, and there may be a multiplicity of mechanisms that explain lapses. Plausibly, they sometimes experience a loss of control over behaviour. But often they continue to engage in controlled behaviour even while lapsing. Judgment-shift explains how agents may lose control in one sense, while retaining it in another: they may experience an uncontrolled shift from one controlled state to another. It is possible that judgment-shift explains some cases of ordinary akrasia (thereby explaining how we may lose control "with calm and even with finesse" (Austin 1979: 198). More dramatic cases, like shifting from a judgment to which the agent is strongly committed and behind which she resolutely stands, may require dysfunction in the machinery controlling judgment.

In this chapter I have suggested that the failure of the midbrain dopamine system to adapt to the reward value of addictive drugs (and, possibly, of gambling too; Ross *et al.* 2008) provides a mechanism for such dramatic judgment-shifts. The way in which this system responds to cues predictive of the availability of the drug explains how the abstinent person may find herself on the garden path to consumption. If this picture is correct, it gives us targets for intervention: besides targeting the neuropsychological mechanisms that are dysfunctional, we can target the environment in which the person finds herself, to enable her to avoid the cues that trigger relapse.

Acknowledgements

I am grateful to Serge Ahmed and Hanna Pickard for detailed and challenging comments. Their comments led to significant improvements in almost every aspect of this chapter. This work was supported by a generous grant from the Wellcome Trust; WT104848/Z/14/Z, *Responsibility and Healthcare*.

Notes

1 In this chapter I use 'addiction' to encompass a problematic pattern of alcohol consumption as well as (other) drugs, and I use 'drugs' from now on to encompass alcohol. Making a distinction between 'alcohol' and 'drugs' is not only neurochemically unfounded, it is also politically invidious. It furthers what might be called the othering of those who consume illegal drugs and contributes to the perception that it is these alone that constitute the major social problem. Alcohol is responsible for more drug-related harm than any other drug (Nutt *et al.* 2010). Of course, many more doses of alcohol than of illegal drugs are consumed annually, so the incidence of harm is not a direct reflection of the harmfulness of the drug. Taylor *et al.* (2012) produced a consensus ranking of the harmfulness of drugs. Alcohol was ranked the fourth most harmful drug, behind heroin, crack cocaine and crystal meth, but ahead of many illegal drugs including cocaine, amphetamines and ecstasy.
2 At least, there is a strong case for saying the mechanism is dysfunctional *if* a subsystem in the nucleus accumbens fails to adapt to the reward value of drugs due to their production of a spike in phasic dopamine. Right now, we cannot be confident that this is in fact the case. If the response of the relevant

mechanisms is of a similar magnitude to non-drug rewards, then we may conclude that the system is working as designed, to orient the organism to rewards in its environment. In that case, the judgment-shift that occurs in addiction would be of the same kind we see in ordinary weakness of the will, and we would need to look elsewhere for what (if anything) makes addiction pathological. Addiction is, of course, complex, socially and neurobiologically, and we shouldn't expect the hypothesis offered here to constitute anything like a complete explanation of its power.

3 Fischer and Ravizza are, in my view, insufficiently sensitive to the factors that may diminish, without entirely removing, control. Elsewhere, I suggest ways of extending their account to accommodate the different dimensions along which control comes in degrees (Levy 2017). It is important to recognise the ways in which control is degreed, because its loss is almost never total: in the majority of those cases we (rightly) regard as involving a loss of control, the relevant mechanisms remain capable of responding to *some* reasons.

References

Ainslie, G. (2000) "A research-based theory of addictive motivation", *Law and Philosophy* 19(1): 77–115.

Austin, J. L. (1979) "A plea for excuses", in *Philosophical Papers, 3rd edition*, Oxford, UK: Oxford University Press, pp. 175–204.

Berridge, K. C. and Robinson, T. E. (1995) "The mind of an addicted brain: neural sensitization of 'wanting' versus 'liking'", *Current Directions in Psychological Science* 4(3): 71–76.

Cameron, C. M., Wightman, R. M. and Carelli, R. M. (2014) "Dynamics of rapid dopamine release in the nucleus accumbens during goal-directed behaviors for cocaine versus natural rewards", *Neuropharmacology* 86: 319–328.

Carter, A. and Hall, W. (2012) *Addiction Neuroethics: The Promises and Perils of Neuroscience Research on Addiction*, Cambridge, UK: Cambridge University Press.

Clark, A. (2016) *Surfing Uncertainty: Prediction, Action, and the Embodied Mind*, Oxford, UK: Oxford University Press.

Corlett, P. R., Aitken, M. R., Dickinson, A., Shanks, D. R., Honey, G. D., Honey, R. A., Robbins, T. W., Bullmore, E. T. and Fletcher, P. C. (2004) "Prediction error during retrospective revaluation of causal associations in humans: fMRI evidence in favor of an associative model of learning", *Neuron* 44(5): 877–888.

Corlett, P. R., Honey, G. D., Aitken, M. R., Dickinson, A., Shanks, D. R., Absalom, A. R., Lee, M., Pomarol-Clotet, E., Murray, G. K., McKenna, P. J., Robbins, T. W., Bullmore, E. T. and Fletcher, P. C. (2006) "Frontal responses during learning predict vulnerability to the psychotogenic effects of ketamine: linking cognition, brain activity, and psychosis", *Archives of General Psychiatry* 63(6): 611–621.

Corlett, P. R., Murray, G. K., Honey, G. D., Aitken, M. R., Shanks, D. R., Robbins, T. W., Bullmore, E. T., Dickinson, A. and Fletcher, P. C. (2007) "Disrupted prediction-error signal in psychosis: evidence for an associative account of delusions", *Brain* 130(9): 2387–2400.

Elster, J. (1999) *Strong Feelings: Emotion, Addiction and Human Behavior*, Cambridge, MA: The MIT Press.

Fingarette, H. (1988) *Heavy Drinking: The Myth of Alcoholism as a Disease*, Berkeley, CA: University of California Press.

Fischer, J. M. and Ravizza, M. (1998) *Responsibility and Control: An Essay on Moral Responsibility*, Cambridge, UK: Cambridge University Press.

Fletcher, P. C. and Frith, C. D. (2009) "Perceiving is believing: a Bayesian approach to explaining the positive symptoms of schizophrenia", *Nature Reviews Neuroscience* 10(1): 48–58.

Hester, R. and Garavan, H. (2004) "Executive dysfunction in cocaine addiction: evidence for discordant frontal, cingulate, and cerebellar activity", *Journal of Neuroscience* 24(49): 11017–11022.

Heyman, G. (2009) *Addiction: A Disorder of Choice*, Cambridge MA: Harvard University Press.

Hohwy, J. (2013) *The Predictive Mind*, Oxford, UK: Oxford University Press.

Holton, R. (2009) *Willing, Wanting, Waiting*, Oxford, UK: Oxford University Press.

Hyman, S. E. (2005) "Addiction: a disease of learning and memory", *American Journal of Psychiatry* 162(8): 1414–1422.

Kalivas, P. W. and Volkow, N. (2005) "The neural basis of addiction: a pathology of motivation and choice", *American Journal of Psychiatry* 162(8): 1403–1413.

Levy, N. (2016) "Addiction, autonomy and informed consent: on and off the garden path", *Journal of Medicine and Philosophy* 41(1): 56–73.

Levy, N. (2017) "Implicit bias and moral responsibility: probing the data", *Philosophy and Phenomenological Research* 94(3): 3–26.

Martinez, D., Narendran, R., Foltin, R.W., *et al.* (2007) "Amphetamine-induced dopamine release: markedly blunted in cocaine dependence and predictive of the choice to self-administer cocaine", *The American Journal of Psychiatry* 164(4): 622–629.

Neale, J. (2002) *Drug Users in Society*, New York, NY: Palgrave.

Nutt, D. J., King, L. A. and Phillips, L. D. (2010) "Drug harms in the UK: a multicriteria decision analysis", *The Lancet* 376(9752): 1558–1565.

Pickard, H. (2016) "Denial in addiction", *Mind & Language* 31(3): 277–299.

Pickard, H. and Ahmed, S. H. (2016) "How do you know you have a drug problem? The role of knowledge of negative consequences in explaining drug choice in humans and rats", in N. Heather and G. Segal (eds), *Addiction and Choice*, Oxford, UK: Oxford University Press, pp. 29–48.

Robinson, T. E. and Berridge, K. C. (2003) "Addiction", *Annual Review of Psychology* 54: 25–53.

Ross, D., Sharp, C., Vuchinich, R. E. and Spurrett, D. (2008) *Midbrain Mutiny: The Picoeconomics and Neuroeconomics of Disordered Gambling*, Cambridge, MA: MIT Press.

Schultz, W., Apicella, P., Scarnati, E. and Ljungberg, T. (1992) "Neuronal activity in monkey ventral striatum related to the expectation of reward", *Journal of Neuroscience* 12(12): 4595–4610.

Schultz, W., Dayan, P. and Montague, P. R. (1997) "A neural substrate of prediction and reward", *Science* 275: 1593–1599.

Taylor, M., Mackay, K., Murphy, J., McIntosh, A., McIntosh, C., Anderson, S. and Welch, K. (2012) "Quantifying the RR of harm to self and others from substance misuse: results from a survey of clinical experts across Scotland", *BMJ Open* 2(4): 10.1136/bmjopen-2011-000774.

West, R. (2006) *Theory of Addiction*, Oxford, UK: Blackwell Publishing.

Willuhn, I., Burgeno, L. M., Groblewski, P. A. and Phillips, P. E. M. (2014) "Excessive cocaine use results from decreased phasic dopamine signaling in the striatum", *Nature Neuroscience* 17(5): 704–709.

6
ADDICTION AND MORAL PSYCHOLOGY

Chandra Sripada and Peter Railton

"Moral psychology" designates inquiry into the psychological infrastructure required for moral thought and practice. Historically, moral psychology has played a large role in moral philosophy, and many of the great works in the history of ethics—Aristotle's *Nicomachean Ethics*, Hobbes' *Leviathan*, and Hume's *Treatise*, for example—open their accounts of ethics with extensive psychological theorizing. This makes sense, since *ought* plausibly implies *can*. So, a *normative* ethical theory should take into account fundamental *empirical* facts about what humans are capable of. A paradigmatic example is *addiction*, which has long raised questions about whether addicts' motivations or capacity for self-control are affected in ways that should mitigate our normative assessments of their conduct. Historically, the relationship between moral judgment and addiction has not always been happy. For centuries, and even today in some quarters, addiction has been seen as a moral failing, rendering the addict culpable. It has taken years of empirical research to develop accounts of addiction that offer scientific alternatives to such moralization. However, medicalized approaches to addiction carry their own risks, since they threaten to remove choice or agency from addicts by positing "irresistible" desires driven by underlying physiological compulsions. What is needed is to understand the interaction of motivation and agency to see *both* how addicts can be exercising significant choice *and* how the mental and physical changes they undergo result in special and evidently difficult-to-manage problems for choice and action. Once we understand this interaction, we will be in a better position to pose normative and therapeutic questions about the extent or nature of an addict's responsibility for her actions, and about what evaluative attitudes are appropriate for her, for those close to her, and for the wider society. If there is a *moral* problem with addiction, it seems unlikely to lie in addiction *per se*, but in the challenges addiction creates for personal choice, interpersonal relations, public health, and social policy.

Our aim here is not to engage these normative questions directly, but to discuss preliminary questions of how our frameworks for thinking about motivation and self-control might need to be enriched to accommodate the complex phenomenon of addiction. Moral psychology, at least as practiced by philosophers, has tended to approach addiction via highly stylized "ideal type" examples. This approach is akin to the physicist's use of an "ideal gas"—artificially simplified and free of many of the forces and interactions found in any actual gas—to develop a basic framework for thermodynamics. The value of such idealizations depends upon how fruitful they prove in advancing inquiry into, and understanding of, real systems. We hope to show that, by taking such philosophers' idealized examples as a starting point, we can build step by step to a clearer conceptual and explanatory framework for addiction that converges at many points with the evolving empirical literature on addiction.

Introduction: a simple model of action and agency

We will begin with the "orthodox" simplified model of action that has prevailed in much contemporary philosophy, economics, and decision theory (Davidson 1963; Luce and Raiffa 1957). In this model, decision and action arise from the combination of two elements, *beliefs* and *desires* (or, in economics and decision theory, *credences* and *preferences*). The *normative* theory of decision and action builds upon this simple model: *rationality* in action consists in making choices that, relative to one's beliefs, promote as far as possible the satisfaction of one's desires.

Several features of the simplified model are especially noteworthy from the standpoint of moral psychology. First, while belief and desire are treated as co-equals in generating action, they are not similarly situated with respect to rationality. Belief is seen as a representational or cognitive attitude, and therefore evaluable as true or false and subject to rational processes of learning, inference, and justification. Desire, by contrast, is seen as a non-representational or non-cognitive attitude, a tendency to exert effort on behalf of performing a given action or attaining a given goal. As such, desires can be stronger or weaker, but they cannot be true or false, or subject to rational processes like learning, inference, or justification (Smith 1994). Thus, even though action draws its goals from desire, desire itself lies essentially *outside* of rationality, except, perhaps, for some formal constraints, e.g., the transitivity of preferences.

Second, this model makes possible a simple account of what makes action *intentional*. The object of desire supplies the "aim" of action, and when this combines with an agent's means-end beliefs to give rise to behavior dynamically oriented toward realizing that aim (in light of evolving circumstances and information), then we can say the agent is acting with a given intent. This contrasts with a case in which the agent's behavior happens to satisfy one or more of her desires, but not via the operation of this kind of internal control process (Davidson 1963; for critical discussion see Bratman 1999).

Third, the simplified model enables us to give very straightforward answers to questions about the nature of free agency and willingness. An intentional action is free and willing when the individual is doing what she wants to do because she wants to do it, without external interference. Thus we have the key building blocks for a "compatibilist" approach to *responsibility*: other things equal, a person is responsible for an intentional action when she performed it freely and willingly. Responsibility is not a matter of *whether* her actions are caused, but *how*—by processes that translate her beliefs and desires into action, or by factors that interrupt or pre-empt this process. No mysterious, uncaused "agent causation" is needed.

Fourth, the simplified model does not have an obvious place for *emotion* or *affect*. To be sure, it does not exclude the idea that affective states or attitudes can give rise to beliefs or desires that can then cause action ("his sadness made him long to return home"), or the idea that emotion can *disrupt* normal causal control of action by one's beliefs and desires ("I'm sorry—in my anger, I lashed out reflexively"). But affect is not an integral part of action, and therefore not an integral component of our rationality, agency, or willingness.

Let us call the simple model, as extended into questions of rationality, agency, and responsibility, the *extended simple model*. Thinking about addiction has played a central role in moral psychology because it yields cases in which this extended simple model does not suffice to explain our intuitive sense that addiction can undermine rational, free action and diminish responsibility. Part 1 of this chapter examines ways in which an agent's motivations can come apart from, or fail to be regulated by, her goals, values, or ideals, even though the agent is "doing what she wants to do, because she wants to do it, without external interference". Part 2 takes up the issue of self-control and ways that an agent's abilities to actively and intentionally restrain wayward motivations can be undermined.

Part 1: motivation, values, and identity

Motivation has long held a central place in debates in moral philosophy, because moral considerations are supposed to guide *action*, and action cannot take place without motivation—on this Aristotle, Hume, and Kant all agree (Aristotle 1999; Hume 1978; Kant 1996). And if moral considerations are to guide human action in their own right—not owing to external motives, incentives, or threats—then somehow *recognition* of moral considerations must suffice to yield motivation. But recognition is a cognitive process, while on the extended simple model, motivation is non-cognitive and arational. This problem is made more daunting by the fact that moral considerations are supposed to be impartial and applicable to all rational agents, regardless of their individual differences in motivation or interest (Baier 1958). But on the extended simple model, an individual's reasons for action always trace back to *her* desires, whatever they may be. These difficulties for the extended simple model are sometimes called "the problem of moral motivation", or even "the moral problem" (Smith 1994).

Historically, rationalist philosophers had attempted to solve this problem by positing "sovereign reason" as a real part of the psyche, capable on its own of recognizing moral principles and deploying "the will" to supply or channel motivation. Thanks to "will power", any refractory desires can be controlled and overridden. On this picture, if an addict recognizes moral reasons for quitting—say, living up to his responsibilities as a parent—yet fails to be responsive, he is *ipso facto* to that extent rationally deficient. A picture like this seems to have been partly responsible for the long history of treating addiction as a *moral* failing: it is the addict's short-term, self-absorbed thinking and "weak will" that explain why he is not keeping his behavior "within reason"—that is, within the bounds of moderation, prudence, or morality.

However, such a "culpabilization" of addiction not only failed to prove effective in changing the behavior, it contained a problematic explanatory posit: a "sovereign reason" and "will" that somehow stands apart from the individual's actual motivational system, yet is still capable of *acting* or *exerting will-power*.

In the twentieth century rationalism waned, "sovereign reason" disappeared from empirical psychology, and something more like the extended simple model took hold. This made possible a non-culpabilizing explanation of addiction: it is a drug-induced *disease* in which the desire component in action hypertrophies into a pathology of "compulsive" or "irresistible" desires. This medical rather than moralistic approach encouraged less punitive attitudes towards addicts and greater research into the underlying causes of addiction, but it risked robbing the addict of agency altogether. By definition, what is irresistible cannot be resisted, and so the addict would no more be exercising agency in his behavior than a person who is tumbled over by a breaking wave.

So we are left with two problems that might seem quite different, the philosopher's "problem of moral motivation" and the addiction researcher's problem of understanding the ways in which agency may operate even when the individual is subject to the powerful effects of addiction upon the psyche. Despite their apparent difference, however, solving both problems requires something in common—getting beyond a conception of the motivational system as non-cognitive and arational.

We will consider two ways in which philosophical accounts of motivation have moved beyond this picture, inspired in part by thinking about addiction: (a) the idea that agency involves a *hierarchical structure* within motivation, and (b) the idea that the agent's *values* are a source and regulator of motivation.

Introducing structure and hierarchy into motivation

Interestingly, both of these revisions in philosophical accounts of motivation can be traced to the influence of a seminal article by Harry Frankfurt (1977), which introduced three intuitive, idealized examples, the "willing addict", the "unwilling addict", and the "wanton addict". Even if no actual addict exemplifies a pure form of these "ideal types"—instead, elements of willingness, unwillingness, and impulsivity are likely to be mixed within him—they nonetheless enable us to identify features of the motivational system that arguably do play an important role in actual addicts, and that had hitherto not been adequately distinguished.

When the "willing addict" reflects upon his various desires—for taking the drug, for sustaining close relations to family and friends, for sociability and a career, etc.—he recognizes that his desire to take the drug is stronger than all these other desires combined. He naturally regrets what he has to forgo in order to satisfy his addictive desire, but he is clear in his mind that the addictive life is the life for him. He *identifies* with this life, and so, even if the drug-related urges *would* prevent him from quitting, still, quitting is not a path he wishes to take. Therefore, when he takes the drug, he is acting freely and willingly.

The "unwilling addict", by contrast, "sincerely hates his addiction" and what it has done to his life, his friendships and family, and his work. He *disidentifies* with this addiction, and "always struggles desperately . . . against its thrust"—but the drug-taking urges it produces prove too strong for him to resist and he is "conquered by his own desires" (Frankfurt 1977: 12). Thus, even though he, like the "willing addict", is acting on his strongest desire when he takes the drug rather than spend time with his family, this is *not* the desire he wants to have dominate this decision, or his life more generally.

The extended simple model would say that both addicts are equally free and willing when they take the drug—they are, after all, doing "what they most want, because they most want it, without external interference"—but intuitively they seem not to be. They differ in their attitudes toward their strong drug-taking urges in a way that is not captured by speaking entirely in terms of competition between desires of different *strengths*. Like all of us, they have multiple desires, which cannot all be realized at the same time or, perhaps, ever. But what the "unwilling addict" experiences is not simply the disappointment of not being able to have everything he wants.

Can we understand the predicament of the "unwilling addict" without reintroducing a problematic "inner agent" standing over and above his motivational system? Frankfurt argues that it suffices to recognize a *hierarchy* among his desires. While the "willing" and "unwilling" addict have essentially similar *first-order* desires, the "willing addict" has a *second-order* desire that his drug-taking first-order desires be effective, while the "unwilling addict" instead has a *second-order* desire that his desires to maintain ties with friends and family, with his work, and with the larger community, be effective. One's higher-order desires have an especially close connection with one's *identity*: they embody not simply what one wants, but what one wants to want—the kind of person one is. Thus the "unwilling addict" will feel that *he* is "defeated" when his powerful addictive desires keep driving him to act in ways that destroy his relations with friends and family and ruin his ambitions.

To illustrate how agency and the structure of motivation are bound together in a person's identity, Frankfurt introduced a third intuitive "ideal type", the "wanton addict". At the first-order level, the "wanton", too, has both an overwhelmingly strong first-order desire to take a drug and various other, conflicting, desires. But the wanton has no second-order desires of any kind—whichever of his first-order desires is strongest at a given moment, that is what he'll do, absent interference. For example, if, on a given Saturday afternoon he desires to be with his kids but also to withdraw to his room and take the drug, he "either cannot or does not care which of his first-order desires wins" (Frankfurt 1977: 13)—he neither identifies with his addiction nor with his parenthood. Such a "wanton", Frankfurt argues, is able to act, and even to act freely—that is, to do what he wants to do because he wants to do it, without interference—but there is a fairly intuitive sense in which he lacks an identity. Equally, he lacks a key component of what it is to be a person: a capacity for "reflective self-evaluation" of his aims, such that he could ask himself what he really wants to want (Frankfurt 1977: 7). Frankfurt speculates that perhaps highly intelligent animals are like this, and that this may help explain why we tend to think that, even though animals can act freely—doing what they want because they want to do it, without interference—we do not think of them as possessing free will or as responsible for their actions in the ways persons normally are. This suggests that, while animals might be able to model some aspects of drug addiction, they cannot model the profound challenges to one's sense of freedom and identity faced by many human addicts (cf. Ahmed 2012). These challenges are surely central to the torment and uncertainty many addicts undergo, and pose special problems for the clinician—though they may also create pathways for treatment that would be absent in our animal brethren.

Even if actual addicts are unlikely to be pure cases of any one of these three types, this typology helps us to distinguish different kinds of conflict and self-alienation within a motivational system, and how these are related to free agency, willingness, and identity. In a more realistic example, an addict might "hate" what his addiction is doing to his family, but at the same time "love" his drug of choice and the experiences it affords (Pickard 2017). In such a case, he will confront certain questions—from himself or others—like, "Are you *really* the father you take yourself to be?" The answer to this question, Frankfurt argues, does not lie in his self-*conception*—if that self-conception doesn't correspond to the actual structure of his motivational system, then he may only be deceiving himself (Frankfurt 2002).

Introducing affect and value into motivation

Frankfurt added hierarchy to desire, but he did not challenge the view that the desires at each level are themselves non-cognitive and arational. As such, they do not represent an act or outcome as having any particular *value* or *desirability* (Frankfurt 1977, 2002). Critics soon pointed out that this markedly lessens the plausibility of the view (Watson 1987). Even the words Frankfurt used in his original article—words that he later withdrew (Frankfurt 2002)—tended to go beyond mere motive force. When the "unwilling addict" is described as "hating" his addiction, or the wanton as "not caring" which of his desires are effective, or the formation of second-order desires as linked to "a capacity for reflective self-evaluation" (Frankfurt 1977), we have moved into the realm of affect and value.

Such a move is only natural, given the connection between one's motivational system and one's identity. Imagine stripping the description of the "unwilling addict" of all affective or evaluative content. He might say, "When drug cravings occur, I really *want* to resist them. It isn't that I *care* more about having a life with family, friends, and a career, or see such a life as more worthwhile or attractive. I just have this strong desire that my desires for *these* goals be effective, not my addictive desires." This would be a surprisingly thin notion of identity—rather close, in fact, to the original characterization of the "wanton addict" who does not care which of his first-order desires win. If we are to explain the connection between motivation and identity, or to capture the distinctive kinds of motivational conflict found in addiction, we will need to incorporate into the motivational system affect, value, and representational content. Indeed, motivation in the absence of affect or value is rare: even taking *interest* in an outcome, or *caring* whether it happens, or *hoping* that it will, or being *pleased* when one makes progress toward it, are species of affect. The motivational system of intelligent animals appears to be designed, not as a battery of basic drives (such as the "salt appetite" when salt-deprived, Berridge 2004), but as an adaptive system that flexibly *allocates effort in accord with expected value* as represented by the animal. And such value—whether hedonic, nutritive, social, or informational—is encoded in the multidimensional affective system, which mediates the transitions from perception to thought and from thought to action (Behrens et al. 2007; Behrens et al. 2008; Grabenhorst et al. 2008; Barrett and Barr 2009; Lak et al. 2014). A foraging animal, for example, may find itself food-deprived and hungry in the early day, but still cache the food it gathers in anticipation of a time of greater future need. For such behavior to be possible, the brain must be capable of allocating effort in accord with an expectation of value or disvalue, not an occurrent motive force (Raby et al. 2007; Kolling et al. 2012). Indeed, the capacity of foraging animals to represent a wide array of values relevant to their well-being appears to be sufficiently sophisticated and information-sensitive that it can produce optimal foraging behavior in the face of the many trade-offs in a complex environment (in experimental simulations, humans can also exhibit optimal foraging, Behrens et al. 2007; Kolling et al. 2012).

We thus have a clue for how we need to enrich the hierarchical model: we need to see affect and evaluation as central rather than peripheral parts of the motivational system. One's identity can be reflected in the structure of one's motivations because they embody the agent's underlying feelings and values—what he cares about and why—not simply a hierarchy of non-cognitive "motive forces".

Indeed, we now are in a position to offer an alternative to the "motive force" view of desire: an *evaluative theory of desire* according to which a desire that *p* (e.g., *that I take the drug* or *that I spend the afternoon with my family*) is a compound state coupling affective evaluation and motive force: a degree of positive affect toward a state or action *p* regulates a degree of effort oriented toward realizing *p* (Railton 2002, 2012). In the case of the "unwilling addict", his first-order desire to spend time with his family involves a degree of positive affect ("caring", "liking", "attraction", "cherishing") toward the prospect of being with them, which can, when the occasion presents itself, elicit and orient a degree of motivation to spend time with his family. Moreover, he has a strong negative second-order affective attitude ("hating", "disliking", "revulsion") toward the idea of having addictive desires dominate his life, to the neglect of his family. This negative higher-order affective attitude elicits and orients a degree of motivation to "struggle" against domination of his life by addictive desires. However, while these attitudes would normally be sufficient to move a person to avoid the drug and spend time with his family, it seems to be a *distinctive* effect of addictive substances and practices (like gambling) to elevate motivation toward the addictive substance or practice out of proportion to the individual's affective evaluation of it (Berridge and Robinson 2011; Robinson et al. this volume). That is, addiction can ramp up motive force directly, to a degree independent of the normal regulation of motivation of one's evaluative attitudes. Thus the individual can feel "undermined" or "defeated" because his *values* are being "undermined" and "defeated" in exerting their normal effect on what he is moved to do (Railton 2012).

On the evaluative picture, desire involves *representation content*—the representation of an action or outcome as having a certain positive or negative expected value. Such representations can be more or less *accurate*, and more or less *sensitive to evidence* and *subject to inference*. That is, they show that desire can be brought within the scope of cognition and rationality after all.

A positive affective attitude toward *p* generates not only a motivating disposition to pursue *p*, but also positive *evaluative expectations* about what it will be like to obtain *p*. If the actual experience of *p* differs from expectation, this information can be used to update evaluative expectations going forward, permitting evaluative *learning* and subsequent reallocation of attention and effort (Lak *et al.* 2014). Importantly, the human affective system is capable of a very wide range of positive and negative attitudes—from *liking* or *loving* to *trusting* to *fearing* to *admiring* to *dreading* to *finding disgusting* to *feeling sympathy* to *blaming* to *feeling guilt or shame*—corresponding to the complex array of values that figure in a full human life. Each of these attitudes elicits and guides motive force in a distinctive way, and with distinctive evaluative expectations for subsequent actions or events. Thus all can collaborate in helping calibrate an individual's evaluative representations, depending upon whether or how much expectations are met. Thus a child feeling generalized anxiety when entering a new school will come with experience to form more specific and focused affective evaluations, learning whom to trust and whom to fear, finding companions and ceasing to panic over social rejection at each lunch hour, and concentrating more concern in the subjects where she was least prepared.

It appears to be another distinctive effect of addiction to interfere with such learning, whether by disrupting the shaping of action by affective evaluation or by distorting reward signals to over-represent drug-taking reward and under-represent other sources of reward (Levy this volume). In this respect, addiction is also a *cognitive* disorder (Hyman 2009). Note, however, that this is not in *contrast* with it being a *motivational* disorder—cognitive content and learning are pervasive features of the motivational system.

Moral psychology and motivation

Returning to issues in moral psychology, we can see how the enrichments of the theory of motivation made in response to challenges posed by addiction may enable us to reframe some long-standing problems in a promising way.

Free, willing action was glossed as "doing what you want, because you want to do it, without external interference", yet the "unwilling addict" makes this seem untenable. On a hierarchical, evaluative model of motivation, however, the gloss is closer to, "doing what you most value and identify with, because you value and identify with it, without external interference". This seems intuitively more plausible, and makes it clearer why free, willing action has such a central role in questions about how to live and how to evaluate actions and agents.

And recall that the "problem of moral motivation" involved the discrepancy between recognition of moral value as a *cognitive* state and motivation as a *non-cognitive* force. Once we understand that recognition of value is *encoded* affectively, and that motivation is normally *regulated* by affective evaluative representations, then we have a potential bridge between recognizing moral considerations and being moved to action by them in their own right. It all depends upon to which values our affective system can render us directly sensitive. For example, humans appear to be equipped with a sophisticated capacity for empathy, which can generate representations of others' points of view and mental states by means of simulations run on one's own affective system (Decety *et al.* 2012). Because these simulations involve affect, they encode as well the positive vs negative valence of what others are experiencing, and can generate corresponding motivational dispositions. If you come upon a lost child in the park, and affectively register her fear and the riskiness of her situation via empathic simulation, then your affective responses are encoding what is morally relevant in the situation. And because these are *affective* responses, they can move you to act directly on the child's behalf without mediation by self-interested motivation, and even at some expense to yourself (Crockett *et al.* 2014). To the extent that others' predicaments can touch us, or their behavior can arouse our admiration at its generosity or outrage at its cruelty, moral considerations can move us to action *via* the motivational system, not despite it.

The evaluative theory of desire implies that an addict's values—reflecting his complexity as a person, his commitments, and his capacity for empathy—will continue to shape his dispositions to act in countless ways, even when addictive desires are predominant. This means that changes in his circumstances, relationships, opportunities, resources, or incentives can shift the motivational balance and affect his choices. By improving the situations, relationships, resources, and options available to the addict, we can work with, not against, his motivational system and the underlying values it embodies.

Part 2: self-control and inner struggle

Weakness of will, judgment shift, and fallibility

A second main area in which addiction has been thought important for moral psychology is the domain of self-control. So-called "weakness of will" has been a preoccupation of ethics ever since philosophers gave up the Socratic idea that if one has genuine ethical knowledge, then one's behavior will simply follow suit. Sometimes one's judgment of what is best seems to lose out to temptation. Once again, the "unwilling addict" has served as something like a paradigm case. We have seen how, on the evaluative model of desire, we can explain how addictive wants can come to be experienced as "alien" and outside one's control. But what happens in weakness of will is *not* simple loss of control—after all, the individual goes on to eat the over-rich dessert or head to the bar rather than home in an intentional, goal-directed way.

The important fact that intentional agency is still active in cases we describe as "yielding" to temptation has led some theorists to hypothesize that the agent's evaluation or judgment of the situation itself changes, elevating the "decision weight" of the temptation, and she then is moved and acts accordingly (Watson 1999; Holton 2004). What could account for this change of mind?

Consider the following pair of option sets:

Option set 1	Option set 2
A. $100 delivered 30 days in the future	C. $100 delivered now
B. $110 delivered 31 days in the future	D. $110 delivered tomorrow

Most people choose B in option set 1 and C in option set 2. This cannot be because people *in general* prefer getting money earlier (even at some cost), since that happens in A as well as C. The choice of B in option set 1 suggests that it is the *immediacy* of the outcome in C that tips the balance in its favor. There are various ways of discounting value over time that result in such a special value for immediacy—most notably hyperbolic discounting (Ainslie 2001; Ainslie this volume) and the beta-delta model (Laibson 1997). For our purposes, what is important is that this seems to be a case of *evaluation* or *judgment shift*. The agent pursuing the short-term options doesn't, at the time of action, act contrary to what she judges best but rather what she judges best shifts in favor of the short-term.

The use of drugs, too, plausibly involves tradeoffs between immediate versus delayed outcomes. The desirable effects of taking a drug, e.g., euphoria or pain-deadening, are usually produced immediately, while the detrimental effects—hangover, withdrawal syndrome, disruptions to work and family life, etc.—tend to unfold later. For a binge-drinker, from the perspective of early in the week, the desirable and detrimental effects are both at some temporal distance, and she may judge it preferable to stay sober this weekend. But on Friday evening, the desirable effects are much closer than the detrimental ones, and her judgment shifts. While hyperbolic discounting describes the behavior of most people, and even many animals, studies have found that addicts have steeper-than-normal hyperbolic discounting functions for a range of potential goods and harms, and perhaps this explains their special vulnerability. It may also help to explain the seeming *irrationality* of addicts—while it is not irrational to act on one's strongest preference at a time, many contend it *is* irrational to have intransitive preferences over time (Davidson *et al.* 1955).

Other theorists have argued that much of what commonsense views as "will-power" actually consists of cognitive and imaginative strategies that reduce the salience of a proximate stimulus and enhance the appeal of gains to be received in the distant future (Mischel *et al.* 1989). On this view, succumbing to addictive urges isn't due to "weakness" of one's will or the overpowering nature of the opposed motive. Rather, it is due to a failure to exert one's will at all, which in turn is due to the fact that certain cognitive/imaginal skills that would have shifted the balance of motivation in favor of future gains failed to be employed.

Why might individuals fail to exercise cognitive and imaginal capacities to combat drug-directed inclinations? Neil Levy (this volume) conjectures that the neurochemical effects of drugs on mechanisms for belief updating play a role. Levy starts with the widely accepted idea that drugs of abuse produce changes to the brain's reward representations. The novel aspect of his model is that he proposes an interesting form of *doxastic* mediation: changes in behavior produced by alterations in reward representations are mediated by changes in one's beliefs. The idea builds on increasingly influential "prediction error minimization" models (Hohwy 2013; Clark 2016). Levy posits that when an addict is exposed to a cue that is associated with reward availability (such as spending time with a friend with whom one consumed drugs) she experiences a spike in phasic dopamine that signals the world is much better than expected. This signal, Levy suggests, can, in at least certain circumstances, produce downstream correspondent effects on her beliefs and rationales for action. That is, to explain the "better than expected" signal, the addict changes her beliefs about spending time with this person and the reasons to do it (e.g., I'm visiting her just to relive old times). In this way, one's doxastic barriers to using drugs are slowly, often stealthily, *eroded* away, setting one up for an eventual shift in judgment.

The preceding three explanations for judgment shift—hyperbolic discounting, failures of imagination, and "doxastic erosion"—certainly help to explain why some addicts will abandon their original intentions and decide to resume use of drugs. But it is possible that a key phenomenon in addiction remains unexplained: Some addicts report that they were sincerely and wholeheartedly opposed to their succumbing to drugs and that giving in was simply not under their control (indeed, the first step in Alcoholics Anonymous says "We admitted we were *powerless* over alcohol"). The models of judgment shift discussed so far don't obviously explain how a person could be *genuinely* powerless to prevent a lapse.

One model that might shed some insight into this more extreme kind of powerlessness (if it exists at all) builds on the observation that judgment processes, like all mental processes, are *fallible*. They are fallible not just in the superficial sense that people often make mistakes if they are careless or don't know how to do something, but rather in a deeper sense that in undertaking any complex multi-step process, some non-zero rate of "pure" error is inevitable. Forming a reflective judgment about what to do when facing a temptation is certainly a complex process that involves elaborate coordination between doxastic, attentional, mnemonic, imaginal, and regulatory processes, and so its intrinsic error rate—i.e., the rate of error that is not attributable to lack of motivation or lack of skill—will, even if small, surely be non-zero.

Now imagine that an "unwilling addict" has about the same rate of intrinsic error in "resisting temptation" as the ordinary person, perhaps 98 out of 100 attempts are successful while two are unsuccessful due to intrinsic errors in the psychological substrates underlying judgment formation. Still, if, due to her addiction, she is assaulted more frequently with transient thoughts of drugs or urges to take them, and living the kind of life she does confronts her more frequently with situations in which drugs or drug cues are proximate, then—even with a 0.98 success rate—the chance that she will fail at some point will be troublingly high, high enough that we should not be surprised, indeed we should *expect*, that she will lapse even after a substantial period off the drug. Yet if such a lapse is indeed due to (rare) intrinsic error—where susceptibility to such errors is plausibly an inherent part of human psychology—it would seem to be unfair to accuse such individuals of having a "weak will" or character flaw in matters of self-control (for detailed discussion of this "fallibility model", see Sripada 2017).

Architectural models of the mind

It is a platitude that addiction involves "inner conflict", but this idea might be understood in different ways. Thus far, the models we have discussed focus on conflict understood as stark disagreement in content, or failure of "mesh", between the motives that are the direct antecedents of action and "deeper" states that are more intimately connected with one's identity, such as one's second-order desires or one's values. This kind of conflict can be contrasted, at least in a rough and ready way, with a still more active form of conflict, which we might call "inner struggle", in which a person engages in ongoing *intentional* activity for the explicit purposes of suppressing or blocking expression of certain of her own motivational states. Addicts often describe themselves as engaged in an ongoing battle with parts of their own psyches, which is sometimes described in the language of interpersonal combat—e.g., "It took everything I had, but I beat it back" or "I fought hard but it got the better of me." It is likely that this aspect of the phenomenology of addiction reflects not just lack of mesh between one's values and one's action but rather active, intentional inner struggle.

There is an interesting, and sometimes underappreciated challenge, however, in trying to make sense of the phenomenon of active inner struggle. To see the issue, return to Frankfurt's unwilling addict. Though Frankfurt is not explicit about this, it is plausible that the unwilling addict's strongest desire is his first-order desire to use the drug. If this desire were not his strongest, then it is not clear he has a problem in agential control at all—for example, if his desire to be with his kids were his strongest, he already has the "will he wants to have" and simply stays with his kids. A problem in his agency arises because his desire to use drugs *is* his strongest and he also has a (strong) second-order desire that his desire to use drugs should not be effective in action. But given that, when two desires compete, the stronger wins (this is a basic background principle in belief/desire psychology), it is hard to see how the unwilling addict could ever actively and intentionally resist his first-order desire to use the drug—by stipulation, the latter is his strongest. The root of the problem consists in the fact that in the architectures we have considered so far, it is not obvious how a person can intentionally undertake an action with the (sole) purpose of thwarting her strongest desire.

This issue being discussed has been dubbed the "problem of synchronic self-control" (Mele 1987; Kennett and Smith 1996). One approach to addressing it consists of adding additional structure to the agent's mental architecture. Consider the following model proposed in Sripada (2014): The agent's motivational architecture has two major structural components. One component comprises a network of *spontaneous*, often unconscious processes that are linked closely to stimuli—various affective attitudes and thoughts are continually being elicited by experience, coming and going essentially unbidden. The second component comprises a network of more *controlled* processes in which serial, reflective, conscious thought and judgment take place.

These two networks operate in parallel, though not without dialogue between them, and each has its own distinctive ways of processing and using information. Each, moreover, is capable of producing intelligent, goal-directed action. However, in Sripada's model, the second component also has *proprietary* access to a variety of regulatory mechanisms—that is, this second compartment is the sole basis for determining whether or not regulation is undertaken. When regulatory processes are engaged, they can suppress the action-tendencies initiated in the spontaneous component (see also the "tripartite" architecture proposed in Dill and Holton 2014).

A theory along these lines *can* give psychological reality to talk of active inner struggle that includes intentional actions designed to thwart parts of one's own psyche. Suppose an addicted individual has just encountered a situation with salient drug cues and immediately experiences a yearning for the drug. If she were not vigilant, that is, not also engaged in controlled attention and deliberate, reflective processing of her situation, this strong desire would lead spontaneously to drug-taking behavior. However, she *is* vigilant, and step by step considers the consequences of taking versus not taking the drug, "making up her mind" that, all things considered, she should abstain—and thus forms the corresponding explicit intention. However, forming this judgment and intention doesn't *extinguish* the wayward desire, which seeks its spontaneous path to action, pulling her back toward taking the drug. Here the inhibition and control processes proprietary to the deliberative mind come into play, as she "stands apart from" her spontaneous desire and "blocks" it from taking action—an active, effortful process, requiring concentration and attention, and vulnerable to fatigue.

This sort of divided mental architecture draws support from a large literature in cognitive science and psychology that spans diverse subfields and methodologies. Specific lines of evidence include: neuroimaging studies identify distinct neural substrates of automatic and controlled processing (Niendam *et al.* 2012; Lieberman 2000); lesion studies find damage to regions supporting controlled processes yield predicted patterns of behavioral changes (Shallice and Burgess 1991); and studies leveraging multiple methods have mapped specific pathways that transmit regulatory signals that modulate or override spontaneous responses (Aron *et al.* 2004). An important aim for future research is to further elucidate the processes that produce deliberative "step by step" reasoning in greater mechanistic detail (some progress towards this goal has already been made in several largely distinct areas of cognitive science—for example, Newell 1994; Anderson 1982; Botvinick and Cohen 2014; Hazy *et al.* 2006—but these results await integration).

If a model involving a divided motivational architecture is correct, then humans may have extensive powers to actively and intentionally exert regulatory control over wayward appetitive desires. How then might a person lose control over consumption of addictive substances? One answer that has had widespread appeal to philosophers, clinicians, and neuroscientists is that addictive substances produce *irresistible desires*. William James, for example, wrote that statements like the following "abound in dipsomaniacs' mouths": "Were a keg of rum in one corner of a room and were a cannon constantly discharging balls between me and it, I could not refrain from passing before that cannon in order to get the rum" (James 1890: 543).

Though supporters of irresistible desires views don't often spell out the details, the basic idea is that addiction produces a mismatch between the strength of wayward desires on the one hand and the "strength" of one's regulatory control processes on the other—the latter cannot suppress the former no matter how vigorously engaged. This might occur either because addictive drugs hypertrophy the strength of wayward desires, or undercut the ability of regulatory control processes to attenuate these desires, or both.

There is a key problem for irresistible desires views: they fit poorly with observations of the regularity with which individuals with drug addictions *do* manage to successfully resist drug-directed desires—for minutes (e.g., putting away needles when a policeman is walking by), for hours (e.g., during work hours or when traveling on a plane), even for days and weeks. Indeed, most individuals with addictions make multiple attempts to quit, and often succeed for extended stretches of time (see, for example, Chaiton *et al.* 2016 and Chiappetta *et al.* 2014).

To be sure, some, perhaps even many, addicts who attempt to quit will eventually relapse (McLellan *et al.* 2000; Kirshenbaum *et al.* 2009; cf. Heyman 2013). The point being emphasized here, however, is that addicts typically exhibit far more success in exercising conscious, deliberate agency to regulate their drug seeking than a picture of "irresistible desires" would suggest. Indeed, a growing literature suggests that addicts are responsive to such incentives as changes in drug price or programs of monetary reward for reduced drug use (Hart *et al.* 2000; Zajac *et al.* this volume).

One way to explain why agents who have extensive capacities for intentional regulation of desires sometimes fail at it, with catastrophic consequences, consists in combining divided architectures of motivation with elements of the "judgment-shift" models discussed earlier. There we considered ways that one's considered judgments might inappropriately shift due to factors such as: a tendency towards excessive discounting of the future, errant prediction error signals, or fallibility of our reasoning faculties. If we add these sorts of elements to a divided motivational architecture, this potentially yields a joint explanation for why addicts are capable of actively and intentionally suppressing their drug-directed urges (and why this might lead to a chronic sense of active inner struggle) and why they succeed on many (or most) occasions, but long-term success, without experiencing even a single potentially devastating lapse in self-control, is to an important degree rarer.

Conclusion

The study of addiction has enriched our understanding of the structure of motivation and action. Moreover, it helps us to develop a less culpabilizing view of addiction and a greater appreciation of the agency of addicts. This will not only ultimately put us in a better position to investigate related issues in moral psychology, such as free will and moral responsibility, but will also help us to find more humane and effective ways of counteracting some of the devastating effects of addictive substances upon individual lives and society as a whole.

Acknowledgements

The authors wish to thank Hanna Pickard and Serge Ahmed for exceptionally helpful comments on an earlier draft.

References

Ahmed, S. H. (2012) "Une brève histoire des modèles animaux de l'addiction", *Biofutur* 31(338): 36–38.
Ainslie, G. (2001) *Breakdown of Will*, Cambridge, UK: Cambridge University Press.
Ainslie, G. (this volume, pp. 34–44) "The picoeconomics of addiction".
Anderson, J. R. (1982) "Acquisition of cognitive skill", *Psychological Review* 89(4): 369–406.
Aristotle (1999) *De Anima*, trans. D. Hamlyn, Oxford, UK: Oxford University Press.
Aron, A. R., Robbins, T. W. and Poldrack, R. A. (2004) "Inhibition and the right inferior frontal cortex", *Trends in Cognitive Sciences* 8(4): 170–177.

Baier, K. (1958) *The Moral Point of View: A Rational Basis for Ethics*, Ithaca, NY: Cornell University Press.
Barrett, L. F. and Barr, M. (2009) "See it with feeling: affective predictions during object perception", *Philosophical Transactions of the Royal Society, B*, 364(1521): 1325–1334.
Behrens, T. E. J., Hunt, L. T., Woolrich, M. W. and Rushworth, M. F. S. (2008) "Associative learning of social value", *Nature* 456(7219): 245–250.
Behrens, T. E. J., Woolrich, M. W., Walton, M. E. and Rushworth, M. F. S. (2007) "Learning the value of information in an uncertain world", *Nature Neuroscience* 10(9): 1214–1221.
Berridge, K. (2004) "Motivation concepts in behavioral neuroscience", *Physiology and Behavior* 81(2): 179–208.
Berridge, K. and Robinson, T. (2011) "Drug addiction as incentive sensitization", in J. Poland and G. Graham (eds), *Addiction and Responsibility*, Cambridge, MA: MIT Press, pp. 21–53.
Botvinick, M. M. and Cohen, J. D. (2014) "The computational and neural basis of cognitive control: charted territory and new frontiers", *Cognitive Science* 38(6): 1249–1285.
Bratman, M. (1987) *Intentions, Plans, and Practical Reason*, Stanford, CA: CSLI Publications.
Chaiton, M., et al. (2016) "Estimating the number of quit attempts it takes to quit smoking successfully in a longitudinal cohort of smokers", *BMJ Open* 6(6): p.e011045.
Chiappetta, V., et al. (2014) "Predictors of quit attempts and successful quit attempts among individuals with alcohol use disorders in a nationally representative sample", *Drug and Alcohol Dependence* 141: 138–144.
Clark, A. (2016) *Surfing Uncertainty: Prediction, Action, and the Embodied Mind*, Oxford, UK: Oxford University Press.
Crockett, M., Kurth-Nelson, Z., Siegel, J. Z., Dayan, P. and Dolan, R. J. (2014) "Harm to others outweighs harm to self in moral decision making", *Proceedings of the National Academy of Sciences* 111(48): 17320–17325.
Davidson, D. (1963) "Actions, reasons and causes", *Journal of Philosophy* 60(23): 685–700.
Davidson, D., McKinsey, J. C. C. and Suppes, P. (1955) "Outlines of a formal theory of value, I", *Philosophy of Science* 22(2): 140–160.
Decety, J., Michalska, K. J. and Kinzler, K. D. (2012) "The contribution of emotion and cognition to moral sensitivity: a neurodevelopmental study", *Cerebral Cortex* 2(1): 209–220.
Dill, B. and Holton, R. (2014) "The addict in us all", *Frontiers in Psychiatry*, 5: 1–20.
Fischer, J. M. and Ravizza, M. (1998) *Responsibility and Control: An Essay on Moral Responsibility*, Cambridge, UK: Cambridge University Press.
Frankfurt, H. (1977) "Freedom of the will and the concept of a person", *Journal of Philosophy*, 68(1): 5–20.
Frankfurt, H. (2002) "Replies", in S. Buss and L. Overton (eds), *Contours of Agency*, Cambridge, MA: MIT Press.
Grabenhorst, F., Rolls, E. T. and Parris, T. M. (2008) "From affective value to decision-making in the prefrontal cortex", *European Journal of Neuroscience* 28(9): 1930–1934.
Hart, C. L., et al. (2000) "Alternative reinforcers differentially modify cocaine self-administration by humans", *Behavioural Pharmacology* 11(1): 87–91.
Hazy, T. E., Frank, M. J. and O'Reilly, R. C. (2006) "Banishing the homunculus: making working memory work", *Neuroscience* 139(1): 105–118.
Heyman, G. M. (2013) "Quitting drugs: quantitative and qualitative features", *Annual Review of Clinical Psychology* 9: 29–59.
Hohwy, J. (2013) *The Predictive Mind*, Oxford, UK: Oxford University Press.
Holton, R. (2004) "Rational resolve", *The Philosophical Review* 113(4): 507–535.
Hume, D. (1978) *A Treatise Concerning Human Nature*, Sir L. A. Selby-Bigge and P. H. Nidditch (eds), Oxford, UK: Oxford University Press.
Hyman, S. E. (2009) "The neurobiology of addiction: implications for voluntary control of behavior", *The American Journal of Bioethics* 7(1): 8–11.
James, W. (1890) *Principles of Psychology*, New York, NY: Henry Holt and Company.
Kant, I. (1996) *Metaphysics of Morals*, trans. and ed. by M. Gregor, Cambridge, UK: Cambridge University Press.
Kennett, J. and Smith, M. (1996) "Frog and toad lose control", *Analysis* 56(2): 63–73.
Kirshenbaum, A. P., Olsen, D. M. and Bickel, W. K. (2009) "A quantitative review of the ubiquitous relapse curve", *Journal of Substance Abuse Treatment* 36(1): 8–17.
Kolling, N., Behrens, T. E. J., Mars, R. B. and Rushworth, M. F. S. (2012) "Neural mechanisms of foraging", *Science* 336(1): 95–98.

Laibson, D. (1997) "Golden eggs and hyperbolic discounting", *The Quarterly Journal of Economics* 112(2): 443–478.

Lak, A., Stauffer, W. R. and Schultz, W. (2014) "Dopamine prediction error responses integrate subjective value from different reward dimensions", *Proceedings of the National Academy of Sciences* 111(6): 2343–2348.

Levy, N. (this volume, pp. 54–62) "Addiction: the belief oscillation hypothesis".

Lieberman, M. D. (2000) "Intuition: a social cognitive neuroscience approach", *Psychological Bulletin* 126(1): 109–137.

Luce, R. D. and Raiffa, H. (1957) *Games and Decisions*, New York, NY: Wiley and Sons.

McLellan, A. T. et al. (2000) "Drug dependence, a chronic medical illness: implications for treatment, insurance, and outcomes evaluation", *Journal of the American Medical Association* 284(13): 1689–1695.

Mele, A. (1987) *Irrationality: An Essay on Akrasia, Self-Deception, and Self-Control*, New York, NY: Oxford University Press.

Mischel, W., Shoda, Y. and Rodriguez, M. I. (1989) "Delay of gratification in children", *Science* 244(4907): 933–938.

Newell, A. (1994) *Unified Theories of Cognition*, Cambridge, MA: Harvard University Press.

Niendam, T. A., et al. (2012) "Meta-analytic evidence for a superordinate cognitive control network subserving diverse executive functions", *Cognitive, Affective, and Behavioral Neuroscience* 12(2): 241–268.

Pickard, H. (2017) "Addiction", *Routledge Companion to Free Will*, New York, NY: Routledge, pp. 454–467.

Raby, C. R., Alexis, D. M., Dickinson, A. and Clayton, N. S. (2007) "Planning for the future by western scrub jays", *Nature* 445(7130): 919–921.

Railton, P. (2002) "Kant meets Aristotle where reason meets appetite", in C. U. Moulines and K.-G. Nierbergall (eds), *Argument und Analyse*, Paderborn, DE: Mentis.

Railton, P. (2012) "That obscure object: desire", *Proceedings of the American Philosophical Association* 86(2): 22–46.

Robinson, M. J. F., Robinson, T. E. and Berridge, K. C. (this volume, pp. 351–361) "The current status of the incentive sensitization theory of addiction".

Shallice, T. and Burgess, P. (1991) "Higher-order cognitive impairments and frontal lobe lesions in man", in H. S. Levin, H. M. Eisenberg and A. L. Benton (eds), *Frontal Lobe Function and Dysfunction*, New York, NY: Oxford University Press, pp. 125–138.

Smith, M. (1994) *The Moral Problem*, Oxford, UK: Blackwell.

Sripada, C. (2014) "How is willpower possible? The puzzle of synchronic self-control and the divided mind", *Noûs* 48(1): 41–74.

Sripada, C. (2017 accepted) "Addiction and fallibility", *Journal of Philosophy*.

Watson, G. (1987) "Free action and free will", *Mind* 96(382): 145–172.

Watson, G. (1999) "Disordered appetites: addiction, compulsion, and dependence", in *Addiction: Entries and Exits*, New York, NY: Russell Sage Foundation.

Zajac, K., Alessi, S. M. and Petry, N. M. (this volume, pp. 455–463) "Contingency management approaches".

7
IDENTITY AND ADDICTION

Owen Flanagan

Identity and identification

Alcoholics and drug addicts often speak about no longer being themselves, of having lost their way, and loved ones, friends, the law, and the mental health community typically agree. The adult addict is physically continuous with some particular baby born years before, and they have an autobiographical memory of that particular individual life. Metaphysically speaking, the addict is the same person they always were. But they are no longer the person they planned, hoped, or expected to be, or who others expected them to be. The kind of identity they have lost or are in danger of losing is the kind of identity that comes from executing authorial power to align, keep aligned, and then continually recalibrate, one's actual life in terms of one's vision of the good. Healthy individuals in healthy ecologies normally possess this kind of authorial control.

The addict like everyone else has some set of norms they aim to abide by, and ideals they aim to achieve, which, if they keep their eyes on the prize, would make them a success in life. But their addiction, which they themselves nurtured but not of course with the intention of becoming addicted, results in a mismatch between those norms and ideals, the vision, let us call it, and the reality their life enacts. Addicts in recovery speak of times when the quality of their behavior—out-of-control use, lying, cheating, stealing, sexual promiscuity—was plummeting faster than they could lower their standards. Morality and most other things that matter to the addict were bulldozed aside by the rapacious and relentless desire to use. Once in the grip of addiction, the addict behaves in ways that actively and consistently undermine achieving the self they aim at, that they wish to be. The life of an addict fully in the grip of some substance is self-undermining in a deep existential way. There is someone they wish to be, but their behavior undermines achieving that self. This is one of several reasons their predicament seems hopeless (Flanagan 2011; 2013a; 2013b; 2013c).

Addiction is an ironic disorder. The very community that provides the set of socially certified norms and aspirations for a good life, also often sets out norms for using alcohol and other dangerous drugs in ways that mark membership in that community. Many addicts become addicts because they become too good at following communal exemplars, role models, and scripts for using dangerous substances as modeled in and by the very communities they seek full adult membership into. They are substance-using overachievers. The same communities that offer up visions of a good person and a good life also offer up rituals for substance use that will inevitably catch out the vulnerable and cause them to ruin their own lives, sometimes the lives of people they love, as well as of innocent bystanders. One reason for the resiliency of addictive habits is that addiction often involves one's deep normative identification with the practices of communities in whose terms one seeks one's good, which include recreational and ceremonial use of the very substance the addict cannot eventually safely use. This has implications for both how to conceive of the nature of addiction, its ontology, and for what the addict needs to overcome, which at least for the kind of addiction I aim to discuss is more than a problem with brain circuitry and more than an unfortunate habit or peculiar dosing relation with some substance. And it is certainly not an allergy, in contrast to what the Big Book of Alcoholics Anonymous says. It is the opposite of an allergy. Addicts, some addicts at any rate, are drawn to their "drug of choice" because it is in certain respects identity conferring and identity constituting.

By my rough arithmetic, I have listened to at least twenty thousand testimonials of alcoholics and addicts over the years, mostly in rooms of AA but some in NA meetings. Almost everything people say in these meetings pertains to exactly three things: "what it was like, what happened, and what it is like now." This is not perfect evidence, but it counts as some data about how addicts conceive of what ails them (Flanagan 2011; 2013c). I have come to believe that many theorists of addiction don't remotely get how existentially loaded many addictions are. When a culture authorizes use of mind-altering and potentially addictive substances, it almost always does so in rituals that mark matters of human significance: birth, friendship, sexual coming of age, romance, marriage, creativity, authenticity, season's changing, plantings, harvests, in-group holidays, business success, death, celebration of our values, calm, equanimity, solidarity, as well as sacred, sacramental, transcendentally significant zones of life. Many addicts experience powerful ongoing identifications with the communities that initially authorized use and taught them how to use. Eventually the addict transgresses norms of proper and decorous use, and loses their welcome—sometimes in late stages even in subcultures of other addicts. They find themselves alone, out-of-control, and the subject of their own and their community's "incomprehensible demoralization" (AA Big Book 2001, p. 30). It follows that the project of overcoming addiction involves, in part, changing a lifestyle, not just a single habit, and finding ways to safely belong to a community without partaking of all its identity signaling, creating, and constituting practices. Sometimes this cannot be done, and the addict must find a new community to which to belong and in which to find their way.

The increasingly popular idea that addiction is a brain disease competes with the way addiction is experienced by addicts. No doubt using alcohol, cocaine, heroin, and so on, to the point of addiction changes the brain. But addiction is also part of an existentially charged lifestyle. This matters to the ontology of addiction, to inquiry into what addiction is, and it matters to the kind of therapeutic work needed to undo addiction. Let me explain.

Caveat: My analysis of identity-constitutive addiction might apply broadly. In this chapter I leave aside the question of whether and/or how it applies to prescription drug addiction, which often arises privately as a consequence of overusing pain medications or anti-anxiety drugs, but without similar permissions or practices of communal use.[1]

For the rest of the chapter, when I use the terms "alcoholism" and "addiction", I mean to refer to the large class of drinking or drug problems that involve loss of control, obsession and craving, *and* identification with a community of users that introduced the way of life that is now out-of-control. This type of addiction—identity-constitutive addiction—is comprised in part by strong identification with the social network in which the addiction took root. Recovery requires a certain amount of identity undoing, which is both difficult and possibly very costly.[2] Overcoming the aspects of addiction that come from disequilibria of the "wanting-liking" brain subsystems, or a "mid-brain mutiny," or the loss of PFC control over the economy of desire, come more easily than reorienting an entire person's deepest identifications.

Social psychology and identity

I start with some commonplaces from the human sciences:

- Humans are normative animals and masters of imitation (Hurley & Chater 2005).
- The practices of the community or communities one belongs to are identity constitutive.
- The practices of communities are contagious (Christakis et al. 2008). Drinking habits, smoking habits, eating habits, drug-taking habits cluster independently of rationale and coercion.
- There is pride in community membership, often in several communities or associations at once. One can identify, for example, with being a good husband and father, a good drinker with a band of brothers, and an upstanding citizen.
- The practices of communities can mutually reinforce each other or compete with each other. One can be a member of the church of the Immaculate Conception and the Mafia.

The key idea is that many substance addictions are grown in communities of use inside which using dangerously is approved and authorized, which communities, however, also abide by norms that full-on addicts reliably transgress. Across the earth, group membership commonly involves ritual acquisition for using specific mind-altering substances in specific ways. Catholic Mass and Jewish Seder require wine. The Achuar of Ecuador take ayahuasca to mark puberty. In Yemen, a Muslim country, where intoxicants are strictly proscribed, elder males chew the narcotic khat day-in and day-out. Knowing how to use in the right ways signals that one is a member of that community whatever it is. Two communities are particularly important in the development of many addictions, one largely involuntary, and the second more voluntary:

- Lineage communities that model particular historical, familial, religious, and class permissions, scripts, and conventions for using.
- The practices of adolescent communities that compete with lineage community practices, by some mix of lowering the conventions governing age of use, contexts of use, amounts used, and/or by offering new or different mind-altering substances to members from those on offer by the lineage community.

Social communion and licenses to use

Two tales:

- Mike O'Hara Jr. grew up just north of New York City in the 1960s in a large Irish Catholic family. The family identified as "Irish" even though the last Irish ancestors had come over a full century before during the potato famine of 1848. Mike's dad, Mike Sr, was a successful businessman in the city, the first in his family to go to college on the GI Bill after serving in Patton's Third Army in the Second World War. The O'Haras were a family of "drinkers." They said so, acted so, and were proud to be drinkers. They were also good Catholics, hard-working, reliable, and morally upstanding. Every now and then an aunt or uncle would have to "go on the wagon," but only one family member, poor Aunt Irene, had succumbed to the charms of the cup and died an early alcoholic death. When Mike Jr was 18, the drinking age in 1968, Mike Sr taught him how to make a martini. Mike Jr does not recall anything about that night after the ritual instruction conveying that sacred family knowledge, and the first sip toasting his birthday with his beloved father. He had a blackout. In college, Mike Jr, like almost all American students, especially those with familial permission, drank in ways that fit every definition of "abuse" and "misuse."[3] When he was 37, Mike Jr's wife asked for a divorce in part because he was a philanderer and a liar. Three years after the marriage ended, Mike Jr started to have trouble at work, he was depressed, and an annual physical showed unhappy liver enzymes. He went to AA, admitted he was an alcoholic, and tried to sober up.

- Leon Washington grew up in East St Louis, Missouri. His hard-working Mom and Grandma raised him and his two sisters. Leon was a good basketball player and fantasized in elementary school about playing in the pros. He did not make his middle school JV team. In 9th grade Leon started to hang out in the park where boys from each high school grade had separate turf and where they smoked weed. Leon was drafted into the army in 1967 immediately after graduation, and fought bravely in the battle of Hué during the Tết offensive in Vietnam, where he saw two close comrades die, their faces blown off. There were casualties among the Viet Cong and North Vietnamese People's Army, but he knew also that his battalion killed a lot of ordinary people in Hué. Leon returned home in 1969 after being honorably discharged. Students at Washington University in St Louis and University of Missouri, Columbia, were in the news everyday protesting against the war. Leon thought this was "fucked-up." Back home, Leon returned to his band of brothers, now mostly vets, who hung out near the basketball courts, smoked lots of pot, and started to use heroin. He worked as a janitor at the VA hospital, became a junkie, got hepatitis C, and went to jail four times in the 1970s. When he returned again to the hood in the early 80s, he promised himself, his Mom, and his baby's Mom, that he would never use junk again. He drank Colt 45 malt liquor with friends, smoked some pot, and started to smoke crack cocaine, which he thought was safe, or at least not as dangerous as junk. Leon has lost 30 lbs, doesn't work, lives with his Mom, and will do pretty much anything to score crack.

The purpose of these two stories is to fix the mind on the fact that humans are gregarious social animals who are raised in communities where there are relatively clear lines defining which groups one belongs to and doesn't belong to, what is in-group and what is out-group. Second, these groups typically provide various permissions and scripts for using dangerous substances. Books like Nick Reding's *Methland* (2009) and Jeet Thayil's *Narcopolis* (2012) tell gripping stories about the different permissions, rituals, economies, and substances, methamphetamine versus opium and heroin, which are authorized, especially among the young and among experimentalists, in Oelwein, Iowa, in the first decade of the 21st century, and in Bombay in the 1970s, respectively. Methamphetamine is advertised as excellent for hard-working people like the farmers of Iowa. It keeps one awake and alert. Opium and heroin have special virtues for associations of sexual experimentalists. Film and literature are filled with encomiums to drink as a source of artistic creativity, and to its central place in achieving the great good of romantic love.

Augusten Burroughs' *Dry* is a memoir of a specific association of users, gay professional men working in New York City in the 90s. Burroughs explains that bars in New York City (and many other American cities) were suddenly the public places gay men went for conviviality, sex, romance, possibly love, in the last decades of the twentieth century, and still into the first decades of the twenty-first century. "You'll never know who you'll meet or where you'll end up. It's like this fucking incredible vortex of possibility. Anything can happen at a bar" (Burroughs 2003, pp. 23–24).

After Burroughs' first attempt at abstinence, he goes into a bar and sees that it still loudly proclaims its romantic possibilities. The bar is not just a place that happens to sell drinks, like a grocery store sells laundry soap. The bar offers up a heavenly palette of aesthetic and existential satisfaction, love and beauty, eros, philia—the sublime.

> An expansive bar begins near the door and stretches back into blackness for what is probably miles . . . Behind the bar, colorful liquor bottles are lit from below like fine art. [T]hey look breathtakingly beautiful. Seeing them, I am filled with longing. It's not an ordinary craving. It's a romantic craving. Because I don't just drink alcohol. I actually love it. I turn away.
>
> *Burroughs 2003, p. 136*

In his memoir, *The Wet and the Dry*, Lawrence Osborne (2013) tries to apply the idea that addiction is a way of achieving social communion in reverse. Osborne is a posh English "drinker," who can't stop. He surmises that he has been unable to get his drinking under control because the "English are very indulgent to episodes of alcoholic insanity. They strike them as sympathetic, understandable, and a sign of being a real human being" (p. 51). He reasons that he needs to find a world where his lifestyle is not met with the same permissiveness. Brilliant idea: He will take his craft to Muslim countries where there are no permissions to drink, make some money, and get his drinking problem under control. Two birds, one stone. It doesn't work. First, Osborne is a "drinker." He isn't just someone who drinks. He identifies existentially with being a "drinker;" second, it is now second nature both for existential reasons and because of his physical dependency for him to drink.

The group or groups one identifies with typically provides permissions, as well as indoctrination, into authorized substances, proper doses, norms and rituals for using. *Drinking Cultures* (Wilson 2005) is a good anthropology that provides a needlepoint of detail to this picture from a wide array of cultures, from Ireland, France, Basque Country, Japan, the Czech Republic, UK, to several indigenous cultures in Mexico. Thomas Wilson, the editor, writes:

> drinking alcohol is an extremely important feature in the production and reproduction of ethnic, national, class, gender, and local community identities, not only today but also historically . . . In many societies, perhaps the majority, drinking alcohol is a key practice in the expression of identity, an element in the construction and dissemination of national and other cultures.
>
> *2005, p. 3*

Children see how adults behave on ceremonial occasions, at Christmas, and New Year's, at baptisms and birthdays, bar mitzvahs and funerals, and they build aspirations to someday enter these holy zones and behave as the adults do. They also hear how adults speak of the transgressions of adolescents and are attuned to how such a group behaves, what gets the attentions of the elders, techniques for participating in such groups below parental and police radar, and so on. In cultures or subcultures in which drug use is endorsed and signals membership in a group, a fellowship, a community of users, the same sort of dynamic applies. The Steeler fans gather at Moriarity's and drink only beer and shots; the gay men gather uptown and drink vodka and do lines of coke; the kids smoke weed at the basketball courts. Such information is readily publicly available.

Manliness

Communities that license using almost never endorse addiction. But they often endorse using in ways that are dangerous, that approach the line that addicts cross. Getting right up to the line is often advertised as worthy of admiration. In *A Drinking Life* (Hamill 1994), Pete Hamill vividly describes the association of drinking with being a man of a certain sort, a manly man, a working-class Irishman in Brooklyn during and after the Second World War. His father modeled the ways of the bar life. The neighborhood bar was a hopeful place, a place for conviviality. "This is where men go, I thought; this is what men do" (Hamill 1994, p. 17).

For Hamill, every birth, baptism, first communion, confirmation, beginning of war, end of war, death, and funeral constitutes a reason to drink. Every important ritual involved essentially male drinking. As an altar boy he discovers that wine is essential to the mysterious sacramental zone of life. Alcohol is involved in transubstantiation!

> And so the pattern had begun, the template was cut. There was a celebration and you got drunk. There was a victory and you got drunk. It didn't matter if other people saw you; they were doing the same thing. So if you were a man, there was nothing to hide. Part of being a man was to drink.
>
> *Hamill 1994, p. 57*

In the films all the Brooklyn boys of the 1950s saw, men drank:

> In those westerns, in the gangster movies, in the war movies, and even the love movies, the men were always drinking. They shot each other in saloons and nightclubs. They got drunk on leave and got into wild, hilarious fights in waterfront bars. Some of the movie drunks were comical, some mean. With the exception of a few cowboys, even the heroes drank whiskey. They never got drunk.
>
> *Hamill 1994, p. 43*

Even in the permissive community of Irish saloons in Brooklyn, there were rules: The hero "never got drunk." The model was to tow that line. "I didn't want to be a drunk. [A]nd yet drinking started to seem as natural to real life as breathing" (Hamill 1994, p. 107). Hamill, eventually in the throes of addiction, angry at himself and the world, now getting falling down drunk like his father, pub brawling like the ordinary grubby and rowdy ranch hand, still held to the "astounding illusion" that he could be an exceptional drinker, like the heroic upstanding Marshall who frequents the saloon, but is somehow above the fray, or like "Bogart in *Casablanca*, sitting at the bar in a pool of bitterness, drinking his whiskey. I would be like that. I would just drink, quietly and angrily and say nothing" (p. 114).[4]

Over half a century earlier in one of the very first alcoholic memoirs, *Jack Barleycorn* (London 1913/2009), Jack London tells us that, "Drink was the badge of manhood" (p. 28). The saloon was also the context, in some ways the only context, in the life of the docks of Oakland, in which men were social:

> A newsboy on the streets, a sailor, a miner, a wanderer in far lands, always where men came together to exchange ideas, to laugh and boast and care to exchange ideas, to laugh and boast and dare, to relax, to forget the dull toil of tiresome nights and days, always they came together over alcohol. The saloon was the place of congregation. Men gathered to it as primitive men gathered around the fire of the squatting-place or the fire at the mouth of the cave.
>
> London 1913/2009, p. 3

But then, years later after embracing the life of drink, identifying with it, wanting it, London finds himself in this degrading situation at the end of a long sea journey. The ship docks, and he is filled with anticipation of being a cultural tourist in Japan, learning about the country, its people, seeing its temples and gardens. "We lay in Yokohama harbor for two weeks, and about all we saw of Japan was its drinking places where sailors congregated" (London 1913/2009, p. 97).

The groups and the permissions for using are carved in multifarious ways along gender, age, racial, ethnic, and socioeconomic lines. Sometimes there are conflicting norms. Whereas the family of origin will typically try to model how adults from the lineage use responsibly, or on what special occasions irresponsible use is permitted, adolescents will almost always want to jump the gun, and, depending on the local ecology, will experiment with what is considered "cool," the latest, and so on, among near-age friends. The key is that everyone conceives of themselves as belonging or aspiring to belong to specific groups. Pride, respect, and esteem accrue from experiencing oneself as a bona fide member.

Women and addiction

I have been speaking mainly about addiction and men. This is principled. I am committed to using what I call "the natural method" in understanding addiction. Listen carefully to first-person phenomenology, to the addict's reports of what it is like to be an addict (Flanagan 2011; 2013a; 2013b; 2013c; 2016, 2017), and try to bring this into reflective equilibrium with what neuroscience, psychology, sociology and anthropology teach about addiction. I am a former addict and a man. I understand better how men got hooked from my own case and from listening to other men. But I have read about and listened to the reports of many women in recovery and I have many female friends and loved ones who were addicts. Here is one kind of story they tell. They had many of the same permissions to use socially and ritualistically as men, with many of the same rewards on offer: friendship, romance, glamour, fun, conviviality, relaxation, overcoming shyness and social awkwardness, and participation in many important social rituals. But they had none of the same permissions to be a jerk while using, and in addition there were rules about safety and prudence that restricted where and when and how much they used in public.[5] In public women were to drink like ladies. If use escalated it did so when they started to develop a solitary habit secretively, by themselves, and as relief from boredom, anxiety, depression, low self-esteem or some admixture. The script was not hard to find. There is plenty of modeling in American culture of the idea that girls and women, as opposed to boys and men, are supposed to have their vices privately, and furthermore that moods and body image, especially for girls, can be altered by various secret regimens of eating, drinking, and drug taking. All the ordinary differences in gender socialization pertaining to modesty, decorousness, restraint on pleasure-seeking, and sexual purity, intersect with what is promised from substance use, as well as the permissions and restrictions on substance use.

So, although the connection between femininity and substance use and abuse might be less visible in our cultural scripts than the connection between masculinity and substance use and abuse, its dynamics are no less real and important. Current gender norms for drinking in America indicate that women are drinking more like men, and thus less responsibly, rather than men learning to drink in "lady-like" ways.[6]

Being together and feeling together

The various groups to which one belongs as a member, or the ones an individual hopes to become a member of, not only model which substances are authorized, but also what they do—produce "high-spirits," hallucinations, feelings of euphoria, calm, chill, safety, sexiness, feeling bound to a "band of brothers," and so on.

In *Drinking: A Love Story*, Caroline Knapp describes both the social endorsement and then eventually her first-personal discovery of the calm and safety that her parents' pre-dinner drinks had advertised: "Growing up, I had an unsafe feeling. From early on, drinking provided the feel of a psychological safety net" (Knapp 1997, p. 69). Even before she experienced the fear lifting when she herself used, she saw that something magical, a quieting of fear or tension, occurred during cocktail hour. She writes this from the pose of a little girl, a sort of semi-participant observer in a familiar American middle-class ritual:

> I liked the ritual long before I started to drink myself. Without realizing it, I had learned to look forward to it. My parents were normally so quiet: they'd sit on the sofa, my mother knitting and my father staring out the window, and a tension would hang over the room like a fog, a preoccupied silence that always made me feel wary, as though something bad were about to happen. My mother would say something about her day—about how she'd ordered some new curtains say, or taken the dog to the vet—and even though my father didn't ignore her in any obvious way, you'd get the sense that he wasn't listening, that his thoughts were about six blocks down the street. Five minutes would pass like that, or ten. Then he'd drink his martini, perhaps pour the second one. He'd begin to loosen up, and within a few minutes it would feel as though all the molecules in the room had risen up and rearranged themselves, settling down into a more comfortable pattern.
>
> *Knapp 1997, pp. 38–39*

Later Knapp describes a dinner out as a young adult with her father, who had introduced her to "martinis, Spanish sherry and single-malt Scotch." Now she could engage, genuinely participate, in the ritual where the "molecules rise up and rearrange themselves." And she knew how to modify her own anxiety and awkwardness.

> I sat on my hands, I remember feeling that particularly acute brand of teenage awkwardness, unable to think of a word to say, and I remember a thick interminable silence. I also remember an empty feeling, a wariness, something I often felt in my father's presence—looking for some nod of encouragement or approval from him, hoping for something to fill the gap between us . . . [B]ut then the wine came, one glass and then a second glass. And somewhere during that second drink, the switch was flipped. The wine gave me a melting feeling, a warm light sensation in my head, and I felt like safety itself had arrived in that glass, poured out from the bottle and allowed to spill out between us . . . the discomfort was diminished, replaced by something that felt like a land of love.
>
> *Knapp 1997, pp. 39–40*

Learning to drink in the way Knapp did involved acquiring a habit, learning an activity, participating in a form of life, for the sake of producing effects that answer to deep psychological needs, such as feeling safe, secure, not scared, not anxious, not awkward (Flanagan 2011). Drinking, a necessary condition of becoming a drinker, possibly a heavy drinker, eventually even, although no one aspires to it, becoming an alcoholic, is presented in many cultural niches—perhaps it is even a dominant trope—as a site of adventure, love, romance, sex, possibility, creativity. These are things that matter to people, common sites of deep identification.

Such communities also convey consequential messages about the contours and limits on acceptable use. In *The Natural History of Alcoholism* (Valliant 1995), George Valliant explains that coming from a family or ethnic group that does not disapprove of adult drunkenness, especially one that jokes about or approves of adult drunkenness, makes the normal male social drinker highly vulnerable to alcoholism. Perhaps this finding generalizes to other kinds of substance addiction so that the more normalized heroin or cocaine addiction are in a community, the more vulnerable the youth are to those kinds of addiction.

The shame of addiction

The etiology of the kind of addiction I am focusing on typically involves first using as one's lineage community does and/or the way one's community of peers does. Sometimes the identifications involved are one and the same, other times they compete with each other, other times they co-exist. One drinks in the saloons like the elders and also smokes weed at the basketball courts in the hood. Adolescents are notorious for ambivalent identification with their lineage association. But they almost always know how to use as the elders do, even if they think those ways are for sissies and wimps; and, of course, they often revert to exactly these ways when they get older.

One who starts to use as one's lineage uses or as one's peers use is just using. If there is use of illegal substances or drinking and driving there is abuse or misuse. Some users use to the point that they become addicts. They reach a point where they want to moderate or stop, but can't. They experience craving and obsession. They also transgress the norms for permissible using of the community of fellow users—they get greedy and sloppy—and they also transgress moral norms.

Caroline Knapp writes about the eventual structure of intimate relationships: "Mostly we lie" then we drink "to drown the guilt and confusion . . . the slow erosion of integrity" (Knapp 1997, p. 196). This repetitive pattern enhances the self-pity and self-loathing. "I hated myself for living like this, but by that time I felt I'd lost control over the script" (p. 206).

Private suspicion that one has a serious drinking or drug problem leads to wonder and worry, and initially to a certain amount of private denial. Disapproval, even ordinary loving concern by friends and loved ones, encourage the alcoholic or addict to minimize, conceal, dissemble, and lie to others. This increases their maddening bewilderment and self-loathing, further energizing the cycle of minimizing, concealment, dissembling, and lying. Then there is transgressive and/or odd behavior while under the influence: promiscuity, odd emotional outbursts, stealing money or property to get a fix, and so on.

Some wonder whether the self who is revealed once one is in the grip of some substance is the "real self" or some alien form produced by the addiction (Arpaly 2017). For present purposes, we can say this much: An addict's self, like every self, is as Walt Whitman says "large" and "contains multitudes." But like everyone else, the addict only reflectively endorses some of their impulses and desires. Other impulses and desires they aim to tame, regulate, control, moderate, keep contained, possibly to extirpate by self-work. If what we mean by the "real self" is the set of reflectively endorsed ways of desiring and being, then the addict's misbehavior does not reveal their "real self." The addict performatively undermines that self. But if by "real self" we include the cacophony of all the impulses and desires that well-up in everyone, but that normally do not see the light of day, then the addict reveals them in ways non-addicts do not. Addictive substances disinhibit and weaken cortical control. This is substance abuse 101.

Furthermore, some, not all, desires and actions of the addict are new, produced by the alchemy of addiction and communal responses to the addiction. The alcoholic who weeps morosely about their plight, laments ever becoming this way, is having new thoughts and feeling new emotions caused by their addiction and possibly by their expulsion from associations they care about. The crack addict who sells herself did not have an antecedent desire to have sex with anyone who wants her or who wants to be serviced by her. But she has a new and overwhelming desire for crack cocaine, and is willing to sell herself to satisfy that new desire, perhaps even at the cost of great shame.

The facts are that on either interpretation of what the "real self" is and where it resides, the addict comes to a point where they performatively undermine their plan to be a certain kind of self, a good person, and enact instead the sort of identity no one would endorse. There is terrible shame in the fact that one embodies an inversion of one's own values.

Many think that it is bad to feel shame because shame, unlike guilt, which targets specific bad acts, involves an overall indictment of the person one is. But I think that shame is the right emotion for a full-on alcoholic or addict to experience. First, addiction is all-consuming; it captures all the addict's energy, as they crave, obsess, plan for the next drink or fix, strategize about skirting the disappointed survey of others, and about how to keep track of the lies and cover-ups. Guilt might be appropriate if the addict had a single bad habit, or a weakness in relation to some specific set of desires that sometimes pop up, take center stage, and then are hard to control, like a weakness for chocolate or strawberries or expensive shoes. Guilty pleasures are normally sated once one succumbs to one's weaknesses. Addiction is not like that from the addict's point of view. Addiction absorbs the bandwidth of attention, either as figure or ground, sometimes both at once. Addicts suffer obsessive thinking about their drug of choice, and typically some degree of constant craving. When the addict is not using, they will report craving, or mentally planning to use, or perseverating about their supply, or strategizing about how they will fit in things like doing their job and being a good parent. Being an addict is a way of life not just a habit or a hobby. The shame of addiction is abiding and penumbral. Second, shame is highly motivating. It expresses the verdict that one is living in a way that fails one's own survey, as well as that of the community upon whose judgment self-respect is legitimately based. It is quite a bit more serious than the judgment that one shows certain weaknesses or character flaws. The addict experiences him- or herself as a normative failure insofar as, because of their addiction, they cannot live up to the standards for a good life that they themselves accept (Flanagan 2013a). The addict, according to my analysis, needs to reconfigure certain aspects of their identity, the way they are constituted as a whole person, not simply to quell some stubborn desires and change some behaviors. Bernard Williams writes that "Shame is a wish to hide or disappear, and this is one thing that links shame as, minimally, embarrassment with shame as social or personal reduction. More positively, shame may be expressed in attempts to reconstruct or improve oneself" (Williams 1993, p. 90). The downside of shame—and this does happen—is that rather than motivate, it can mark utter and complete demoralization and despair. Still, it would be misguided for addicts themselves and for the others that wish to help them heal to deny that shame is what is experienced, that it is appropriate, and that it can be utilized as a powerful source of motivation to change the self and to recover.

Communities of recovery

The addict has trouble putting their reasons, their best thinking, and their considered ideals in reliable control of their actions, and they have trouble abiding by the moral norms upon which their sense of their own integrity and self-worth turns. Addicts reliably speak of addiction as involving twin normative failings: loss of control over the substance and inability to stay on track and abide by the moral order (Flanagan 2013a).

Burroughs, many months beyond his last drink, knows what he wants, but this time his will works:

> *A Ketel One martini please, very dry with olives,* I want to say. "Um just a selzer with lime," I say instead. I might as well have ordered warm tap water or dirt. I feel that uncool. And suddenly, it's like I can feel how depressing alcoholism really is. Basements and prayers. It lacks the swank factor.
>
> Burroughs 2003, p. 137

The bar, the place of "breathtakingly beautiful" backlit bottles, "the fucking vortex of possibility," the repository of the alcohol he does not merely drink but "actually loves," is contrasted with the life of recovery that embodies all that is "uncool" and un-"swank."

Somehow for recovery to work there needs to be this reckoning: the addict needs to come to terms with the veins of identification that partly constitute their malady. This might mean reconceiving the wisdom of the identification, separating the aspects of the communities to which one's addiction is bound that are worth identifying with and those aspects—using—that are not, and to identify with those who now spend time in church basements, a community that, despite lacking "the swank factor," is a community of genuine love.

Insofar as addiction is a social malady that involves identification with the practices of a community to which one initially wanted to belong, to be a good member, it makes sense that we ought to use communal tools to undo it and keep it at bay. If one thinks the addiction is only a disease of the brain one might be inclined to think that the brain is the right place to treat it and thus that pharmaceutical intervention is the right means. If one thinks that addiction is just an unfortunate behavioral habit, then one might recommend breaking that associative network, by avoiding "people, places, and things" associated with using. If one thinks addiction—at least a common kind—is a complex neuro-psycho-social phenomenon, a lifestyle, an existentially robust way of being that involves deep identification with aspects of the very kind of life that is now out-of-control, then one will think it wise to leverage it at all the locations where it has its tenacious hold. This might include the brain and behavior. But it will also include the identity-constitutive relations that still—perhaps will always—hold charms for the addict even in recovery. This helps explain how and why recovery of the kind of addiction I have focused on here ought to sensibly leverage the same needs for community identification and acceptance that in many cases lead to use and then to addiction in the first place; but now with the aim of undoing it by finding solidarity among people who get how powerfully most addicts were, and still are, motivated by powerful needs to belong (Pearce and Pickard 2012).

Acknowledgments

Thanks to Hanna Pickard, Serge Ahmed, Peg O'Connor, and Lariska Svirsky for immensely helpful comments.

Notes

1 Hanna Pickard makes this helpful suggestion for how the social permission analysis on offer might apply to the prescription drug case. She writes (personal correspondence) that in the case of prescription drugs "those with power and authority to give permission and the impression that the drug is 'medicine' and 'safe' do so. Indeed, on a conventional view of the doctor–patient relationship, once prescribed the patient is not only permitted, but required to take the drug, because their social role is to do what the doctor orders. Of course, sometimes people may over-medicate – they may exceed what is prescribed. But you might think that is exactly analogous to the point you are making about alcohol and other drugs in the relevant communities – a certain degree and kind of use is not only permitted, but encouraged and expected."
2 There are three interconnected aspects of identity that may require renegotiation: 1) The addict will need to revise their self-understanding of themself as an addict, to something like a former addict, an addict in recovery, or an addict in remission since understanding oneself as an addict is a doomsday script; 2) The addict will need to provide evidence to themself and to their community for a revision of the view that they are a hopeless drunk or junkie; 3) The addict may need to sever relations with the people, places, things, rituals, and practices, even the entire form of life, in which their addictive habit took root.
3 Abuse is the preferred language in DSM-5. The Surgeon General's report "Facing Addiction in America," https://addiction.surgeongeneral.gov/surgeon-generals-report.pdf, claims that 'abuse' is considered shaming and thus that 'misuse' is the new term du jour (Flanagan 2017).
4 Eventually Hamill used this tactic on himself to get sober: "Drinkers are forgetters." But another community he identified with, the community of writers, needs to remember. He worked successfully to grow his strongest identification with the association of "rememberers."
5 Hanna Pickard who does clinical work with female alcoholics in the UK tells me she sees women alcoholics who fit the profile I see in the USA primarily among men.
6 See: www.washingtonpost.com/national/for-women-heavy-drinking-has-been-normalized-thats-dangerous/2016/12/23/0e701120-c381-11e6-9578-0054287507db_story.html?utm_term=.027a0f9dc384.

References

Alcoholics Anonymous (AA) (2001) *Alcoholics Anonymous [Big Book]: The Story of How Many Thousands of Men and Women Have Recovered from Alcoholism*, 4th Edition, New York, NY: Alcoholics Anonymous World Services, Inc.
Arpaly, N. (2017) "Moral psychology's drinking problem", in I. Fileva (ed.), *Questions of Character*, Oxford, UK: Oxford University Press, pp. 121–131.
Burroughs, A. (2003) *Dry: A Memoir*, New York, NY: Picador.
Christakis, N. and Fowler, J. (2008) "The collective dynamics of smoking in a large social network", *The New England Journal of Medicine* 358: 2249–2258.
Christakis, N. and Fowler, J. (2012) "Social contagion theory: examining dynamic social networks and human behavior", *Statistics in Medicine* 32(4): 556–557.
Flanagan, O. (2011) "What is it like to be an addict?" in J. Poland and G. Graham (eds), *Addiction and Responsibility*, Cambridge MA: MIT Press, pp. 269–292.
Flanagan, O. (2013a) "The shame of addiction", *Frontiers of Psychiatry* 4(120): 1–11.
Flanagan, O. (2013b) "Identity and addiction: what alcoholic memoirs teach", in K. Fulford, M. Davies, R. Gipps, G. Graham, J. Sadler, G. Strangellini and T. Thornton (eds), *The Oxford Handbook of Philosophy and Psychiatry*, Oxford, UK: Oxford University Press, pp. 865–888.
Flanagan, O. (2013c) "Phenomenal authority: the epistemic authority of alcoholics anonymous", in N. Levy (ed.), *Addiction and Self-Control*, Oxford, UK: Oxford University Press, pp. 67–93.
Flanagan, O. (2016) "Willing addicts?" in N. Heather and G. Segal (eds), *Addiction and Choice*, Oxford, UK: Oxford University Press, pp. 66–81.
Flanagan, O. (forthcoming 2018) "Addiction doesn't exist, but it is bad for you", *Neuroethics* 91–98, doi: 10.1007/s12152-016-9298-z.
Hamill, P. (1995) *A Drinking Life: A Memoir*, Boston, MA: Back Bay Books.
Hurley, S. and Chater, N. (2005) *Perspectives on Imitation*, Vols 1 & 2, Cambridge, MA: MIT Press.
Knapp, C. (1997) *Drinking: A Love Story*, New York, NY: Dial Press.
Levy, N. (2013) "Addiction is not a brain disease (and it matters)", *Frontiers of Psychiatry* 4: 7–18.

London, J. (1913/2009) *John Barleycorn: 'Alcoholic Memoirs'*, Oxford, UK: Oxford University Press.
Osborne, L. (2013) *The Wet and the Dry*, New York, NY: Crown.
Pearce, S. and Pickard, H. (2012) "How therapeutic communities work: specific factors related to positive outcomes", *International Journal of Social Psychiatry* 59(7): 1–10.
Reding, N. (2009) *Methland: The Death and Life of an American Small Town*, New York, NY: Bloomsbury.
Thayil, J. (2012) *Narcopolis: A Novel*, New York, NY: Penguin.
Vaillant, G. E. (1995) *The Natural History of Alcoholism: Revisited*, Cambridge, MA: Harvard University Press.
Williams, B. A. O. (1993) *Shame and Necessity*, Berkeley, CA: University of California Press.
Wilson, T. M. (2005) *Drinking Cultures*, New York, NY: Berg.

8

THE HARMFUL DYSFUNCTION ANALYSIS OF ADDICTION

Normal brains and abnormal states of mind

Jerome C. Wakefield

Is addiction a disorder?

Despite a recent explosion of illuminating empirical and theoretical research on addiction, the addiction field is divided over some basic conceptual questions. The most fundamental question is whether addiction is a medical disorder. Addiction is currently classified as a mental/psychiatric medical disorder in official diagnostic manuals such as DSM-5 (American Psychiatric Association 2013) and ICD-10 (World Health Organization 1992). It is further considered a brain disorder by the National Institute of Drug Abuse (NIDA), based on observed well-documented neuronal changes in addicts that NIDA interprets as substance-induced damage to the brain's reward, inhibitory, and other systems (Leshner 1997; Volkow et al. 2016). The NIDA's position that addiction is a brain disorder (which NIDA terms a "brain disease" despite the fact that such damage does not clearly qualify for the label 'disease') is repeated frequently in public statements and shapes research grant awards, so can justifiably be considered the standard current view among psychiatrists and researchers.

However, the claim that addiction is caused by brain damage and thus is a brain disorder has been widely questioned. Although tissue damage may occur from excessive substance use, it is puzzling that many addictive substances should happen to cause damage at multiple brain loci in precisely the way that brings about intense motivation to continue using that substance, yielding intensified desire, heightened sensitivity to cues of availability, reduced interest in other rewards, willingness to undergo painful tradeoffs to obtain the substance, reduction of inhibitory functioning of the cortex that usually restrains impulsive action, increased delay discounting favoring immediate gratification, and so on. These forms of "damage" look like what happens when one develops an overriding "peremptory" desire, that is, a desire that grips one such that mental functioning reorganizes in primary pursuit of the object of the desire. When in the throes of passionate erotic love or suffering from starvation, it is normal for one's perceptions and reactions to reorganize around that central motive, and surely there are accompanying brain alterations. Simply identifying brain processes corresponding to the addicted individual's increased desire/wanting, reduced inhibitory control, and lessened interest in alternative rewards does not resolve the puzzle of whether addiction is normal or disordered. For example, drug-preferring rats have larger neuronal assemblies firing in response to the drug (Guillem and Ahmed 2018), but rather than being a brain disorder this could be the normal brain instantiation of strong preference.

Critics of NIDA's "brain disorder" claim argue that the observed changes, rather than being brain damage, are the expectable results of a normal brain's neuroplasticity, habituation, and restructuring in reaction to the intense learning process that occurs in response to substance use (Lewis 2017). Other critics, reacting to implausible claims of total lack of choice or control that seem implied by the brain-damage account, emphasize that the addicted individual is in fact making choices constrained by circumstances ranging from poverty and self-medication to sheer pursuit of pleasure and they deny that the series of such choices that constitute an addiction is a medical disorder (e.g., Ainslie 1992; Becker and Murphy 1988; Courtwright 2010; Hall *et al.* 2015; Heyman 2009, 2013; Levy 2006, 2011; Pickard 2012, 2016; Rachlin 2007; Satel and Lilienfeld 2014). One might fear that such non-disorder choice models could unintentionally revive the moral condemnatory approach to addiction as a vice that deserves blame and punishment, and to create more constructive approaches some critics are rethinking the relevant notions of responsibility and blame (Lewis 2017; Pickard 2017).

I am going to argue that even if one accepts the critics' argument that the evidence fails to demonstrate that addicts have brain damage, the conclusion that addiction is not a medical disorder does not follow. Relying on my evolution-based harmful-dysfunction analysis of the concept of medical disorder, I defend the view that substance addiction is a medical disorder even if addicts' brains are entirely physiologically normal. However, to allow conceptual space for my argument that addiction can be a medical disorder without there being any brain damage, I must also argue against the larger principle, which has become an ingrained assumption in much contemporary psychiatric theory and research under the label of "neo-Kraepelinianism," that mental disorders must always be brain disorders. I argue to the contrary that abnormal psychological conditions need not reflect abnormal brain conditions.

There are three caveats worth noting before we start. First, I am concerned to analyze the intuitive phenomenon of substance addiction. I will *not* aim my discussion at the category of people who fall under the official substance addiction diagnosis in DSM-5 (American Psychiatric Association 2013), labeled "substance use disorder." As I have argued elsewhere, although the DSM criteria are used universally in the US in addiction diagnosis and research, the DSM-5 criteria are invalidly broad and mistakenly classify many non-addicted individuals who use substances preferentially but not compulsively as addictively disordered (Wakefield 2013, 2015a, 2016a; Wakefield and Schmitz 2015), thus the criteria do not capture the intuitive concept and confuse the analysis of addiction. Animal studies have recently supported what statistics of addiction rates after drug exposure in humans suggest, namely, that of all those exposed to a drug who may engage in addiction-like behavior, only a subset may have a true addiction (Ahmed 2010; Ahmed *et al.* 2013; Cantin *et al.* 2010; Lenoir *et al.* 2007).

Second, my analysis addresses substance addictions, not behavioral addictions to such activities as gambling (see Whiting, Hoff, and Potenza in this volume), video gaming, or internet pornography. Only substance addictions are claimed by NIDA to be due to brain damage, and I address that claim. However, towards the end I will suggest that a unified harmful-dysfunction approach is possible to substance and behavioral addictions.

Third, I make no claim that the account presented here must cover all possible substance addictions. The final-pathway outcome of addiction is likely to have multiple potential etiologies (Redish *et al.* 2008). However, I believe that the harmful dysfunction account identifies a common factor in most major substance addictions of current concern.

The harmful dysfunction analysis of the concept of medical disorder

I have long argued that questions about the distinction between health and medical disorder are best approached within the framework of my harmful dysfunction analysis of the concept of medical disorder (Wakefield 1992a, 1992b, 1993, 1997, 1999a, 1999b, 2000a, 2000b, 2006; Wakefield and First 2003). According to this analysis, a medical disorder exists only when two conditions are both met. First, the condition must be caused by a dysfunction, that is, by a failure of some physiological or psychological mechanism to perform a natural function that it was biologically designed to perform. "Biological design" and "natural function" are understood here not in terms of any actual theistic or human intentionality or purpose or mystical "final causes", but simply in terms of the natural selection of mechanisms in the process of evolution because those mechanisms have certain effects that constitute their functions. Thus "biological function" and "biological dysfunction" construed as a failure of biological design are factual concepts in the sense that they are scientifically descriptive and not evaluative (Wakefield 1995, 2003, 2016b).

However, the failure of biological mechanisms to be capable of performing a natural function cannot by itself account for our concept of medical disorder. Biological dysfunctions in which there is a failure of a mechanism to perform its biologically designed function are ubiquitous and often harmless. For example, every time one goes out in the sun there are thousands of mutations in skin cells that cause specific genetic loci to be incapable of performing their functions and thus constitute dysfunctions at those genetic loci, yet are generally completely harmless due to the redundancy in the skin's various systems. The idea that all such harmless dysfunctions are medical disorders reduces the notion to absurdity (Wakefield 2014).

So, second, a medical disorder requires that the dysfunction must cause significant harm to the individual as judged by social values. Thus, the harmful dysfunction analysis is known as a "hybrid" fact/value view due to the insistence that both factual and value components are involved in disorder attributions.

The harmful dysfunction view allows a distinction to be drawn between harmless dysfunctions that are not medical disorders versus harmful dysfunctions that are medical disorders. Among addictions, a distinction can thus be drawn between "addiction" per se and "addictive disorder." Various cultures may have practices that render certain addictions harmless and not disorders. For example, in our society, a moderate addiction to caffeine can be harmless and not qualify as a medical disorder because caffeine use is accepted for performance enhancement and caffeine is freely available. However, if caffeine intake is harmful due to medical issues, caffeine addiction may then qualify as an addictive disorder. The distinction between addiction and addictive disorder might be likened to the DSM-5 distinction between paraphilias, which are sexual deviations but not necessarily disorders, and paraphilic disorders, in which the sexual deviation causes harm (Wakefield 2011, 2012) – although it is noteworthy that DSM-5 allows the harm in some categories of paraphilic disorder to include harm to others in the form of acting on sexual impulses with a nonconsenting individual, whereas the primary harm in addiction is the distress or impairment of the addicted individual. Like most writers on addiction, I consider only harmful addictions here, so I ignore the terminological niceties and use the term 'addiction' for addictive disorder.

The evolutionary view of addiction: a disorder of desire/choice motivational mechanisms

The evolutionary view of addiction is neutral on whether there is tissue damage from substance use. It holds that the evolutionarily novel nature or purity or amounts of substances of abuse, which exist in their present form due to our technological capabilities, can lead to genuine functional pathology without any tissue damage being necessary, simply as the "normal" response of biologically designed mechanisms to a novel input. As Nesse and Berridge explain:

> Pure psychoactive drugs and direct routes of administration are evolutionarily novel features of our environment. They are inherently pathogenic because they bypass adaptive information processing systems and act directly on ancient brain mechanisms that control emotion and behaviour . . . and can result in continued use of drugs that no longer bring pleasure.
>
> *1997: 63–64*

(For a competing view that suggests that substance use may be based on evolved affinity to the abused substances and thus that addiction is not sheerly an accidental "hijacking" of motivational mechanisms but rather a pathologically excessive form of a biologically designed response to a substance, see Hagen and Sullivan in this volume.)

It is not a disorder to ingest novel substances that go outside of biologically designed domains in the quest for pleasure, even if such activities entail risks of harm. However, taking substances that our brain was not biologically designed to handle can cause motivational dysfunctions in some individuals, most likely due to normal variations in individuals that were not a problem in the environments in which we evolved and thus not dysfunctions in themselves. The result can be overwhelming peremptory motivations to take the substance despite good reasons not to do so, constituting a motivational disorder. Note that some writers suggest that, given the negative social context of much drug-taking, it may be a rational action rather than a motivational dysfunction to choose to take a substance despite even extreme negative outcomes to oneself or others (Pickard 2012). If so, such rational decisions are not strictly addiction in the sense considered here, which presupposes the presence of a motivational compulsion that goes beyond sheer rational calculation. The recent and quite deadly opioid addiction epidemic in the US that has affected many intact middle-class communities suggests that even individuals who are not in distinctively impoverished or deprived circumstances are subject to addictive motivational pathology.

The usual desire/choice process is subject to disruption by biologically designed peremptory desires and emotions that address special challenges. For example, severe hunger, intense erotic desire, sexual jealousy, or extreme fear or rage can naturally narrow and distort cognition, override and disrupt usual desire/choice processes, and constrain choice without indicating dysfunction. Thus, analogies have been drawn between addiction and such normal processes as intense love or hunger (Lewis 2017). However, the desire/choice system is designed to be disrupted in these special circumstances to facilitate urgent fitness-enhancing reactions. The similarly intense peremptory desires that occur in addiction are not part of such biologically designed exceptions to the usual deliberation and rational choice functions and thus qualify as dysfunctions.

Despite addiction's motivational intensity, it is important to emphasize that even severe addiction involves some degree of choice and is not the total suspension of self-control sometimes portrayed by brain-disease proponents. Addiction distorts and pathologically narrows the desire/choice process, but does not eliminate it, and this is the basis for many treatment strategies.

Can a mental disorder exist in a normal brain?

Certainly, if addiction is due to brain damage, then it is a medical disorder. The converse claim—if addiction is a medical disorder, it must be because there is some lesion or other damage to the brain—is also widely held and fits with a doctrine currently popular within psychiatry that all mental disorders must be brain disorders. This doctrine is a tenet of an influential movement known as "neo-Kraepelinianism" (Hoff 2015; Klerman 1978), after the 19th-century psychiatrist Emil Kraepelin (1921, 1922). Influenced by the discovery that syphilitic infection of the brain was responsible for the scourge of general paresis and by the identification of brain pathology in Alzheimer's disease, Kraepelin hypothesized that distinct brain pathologies would eventually be found to correspond to each mental disorder. Modern neo-Kraepelinians embrace this doctrine and have been highly influential in psychiatry over the past half century. As Samuel Guze, one of the neo-Kraepelinian movement's founding theoreticians, put it: "The conclusion appears inescapable to me that what is called psychopathology is the manifestation of disordered processes in various brain systems that mediate psychological functions. Psychopathology thus involves biology" (Guze 1989: 317).

If the "brain damage" account of addiction is rejected, can addiction still be a medical disorder? That is, can there be a medical/mental disorder in a biologically normal brain? Many distinguished psychiatrists follow Guze's doctrine and argue that the answer is "no" because all mental processes take place in brain tissue, so if anything is wrong with mental functioning then there must be something wrong with brain functioning describable in anatomical/physiological terms. For example, Nobel prize-winning neuroscientist Eric Kandel states: "All mental processes are brain processes, and therefore all disorders of mental functioning are biological diseases. . . . The brain is the organ of the mind. Where else could [mental illness] be if not in the brain?" (as quoted in Weir, 2010: 30). Similarly, Nancy Andreasen, former Editor-in-Chief of the *American Journal of Psychiatry*, asserts that "people who suffer from mental illness suffer from a *sick or broken brain*" (1984: 8, emphasis in original). I assume that in postulating that there is always a brain disorder underlying any mental disorder, neo-Kraepelinians mean that there is always a dysfunction of brain mechanisms in which the function that is failing to be performed can be fully specified in brain-anatomical and brain-physiological terms without any essential reference to the psychological/mental level of description involving intentional contents or conscious experiences.

Given that every occurrence of a mental state is an occurrence of a physical brain state and so takes place in the brain, in a locational sense every mental disorder is a brain disorder. However, that does not imply that every mental disorder is a brain disorder in the stricter sense that there is a neurobiologically specifiable dysfunction. Functions and dysfunctions describable at the psychological level of organization might not correspond to functions and dysfunctions describable in purely non-psychological neurobiological terms. Malfunctions at the level of the interaction of psychological contents—where contents are designed to interact with each other in certain ways that make cognitive, conative, and emotional sense—may be emergent and not entail malfunctions describable at the brain circuit level. For example, perception is biologically designed to convey certain information about the environment, sequences of thoughts are biologically designed to occur in patterns that relate evidence to a conclusion in a way that approaches the truth, and emotions are biologically designed to occur at levels of intensity that are roughly proportional to certain triggering circumstances (e.g., fear is triggered by threat, sadness by loss, anger by violation, joy by triumph). It seems prima facie possible that such biologically designed psychological-level processes might go awry for purely psychological-level reasons without any underlying physiological disorder.

The invalidity of "all mental disorders are in the brain, therefore all mental disorders are brain disorders" is suggested by the invalidity of the analogous argument: All computer software runs in computer hardware, therefore all software malfunctions must be hardware malfunctions. Note that, contrary to some criticisms of the hardware/software analogy to the brain/mind distinction (Kendler 2012), it is not a Cartesian holdover to say that there are different levels of description of brain processes corresponding to neuronal physiological processes and representational mental contents and that biologically designed functions and corresponding dysfunctions can occur at those different levels of description. Even if a dysfunction at the psychological level does not correspond to a dysfunction at the physiological level, it does correspond to some physiological process and it is still the case that processes at the physiological level have causal influences on processes at the psychological level. Analogously, every step in the running of a computer program is an occurrence in the computer's hardware, but the functions of software involve the symbolic meanings, not the silicon-level descriptions, of the hardware configurations. Thus, there are software problems in which symbolic processing is malfunctioning that don't involve any hardware problems. To take a familiar example, the "Y2K" (i.e., "year two thousand") panic about the potential breakdown of our information processing infrastructure involved no feared hardware malfunctions. Rather, the problem was that, once the year 2000 arrived, the two-digit "year" register in most commercial software, designed only for inputs of twentieth-century years using just the last two digits of a year, would have provided inappropriate inputs (e.g., "01" would be interpreted as "1901" rather than "2001"), yielding software malfunction. For example, software calculating interest on a bank account would yield nonsense as output, yet nothing would be wrong with the computer hardware.

Consider again the Kandel passage quoted above. Intentional contents and conscious experiences do occur "in the brain," but the software analogy suggests that the fact that all mental occurrences take place in brain tissue does not imply that every mental dysfunction is a dysfunction specifiable in purely anatomical/physiological terms. The problem with the Kandel-type argument is the equivocation in moving from the correct premise that all mental disorders are brain disorders *in the locational sense* to the conclusion that all mental disorders are brain disorders *in the narrower sense that there is an underlying dysfunction describable sheerly in anatomical/physiological terms*. Granting Kandel's locational premise, it remains at least conceptually possible that there are mental dysfunctions that are not brain dysfunctions in the latter, narrower sense. The question is whether one can leave the domain of analogy and illustrate directly via an example why the inference from mental dysfunction to brain dysfunction describable in sheerly physiological terms is invalid. That is, can one specify a psychological function that is failing while no physiologically specifiable function is failing?

The gosling and the fox: how a mental disorder can exist in a normal brain

As we have seen, prevailing psychiatric ideology holds that every mental disorder implies brain damage, setting up the dichotomy: either addiction is a disorder because it involves brain damage, or addiction is not a disorder at all. I am arguing for a third approach that holds that addiction can be a genuine medical disorder even if there is no brain damage. To construct such a view, I have to establish that it is possible in principle for there to be a mental disorder that involves no brain damage. So, in this section, I develop an initial example in which something is wrong with the mind despite there being nothing wrong with the brain.

Goslings (baby geese) famously have neural assemblies dedicated to an "imprinting" function. The imprinting system is biologically designed to store an image of the first creature the gosling sees upon hatching, and the gosling then has an irresistible and irreversible desire to stay close to and follow the represented creature. The mechanical nature of the imprinting system was famously demonstrated when Konrad Lorenz (1935, 1952) caused goslings to imprint on him and follow him around as if he was their mother.

The evolutionary explanation for the imprinting mechanism is that, first, almost invariably the first creature the gosling sees upon hatching is its mother, given her involvement in bringing about the hatching. The mechanism's selection depended on this correlation. Second, linking the gosling specifically to its mother is the function of imprinting because the development and survival of the gosling depend on staying close to its mother after hatching. This is due to the existence of a variety of naturally selected coordinated mechanisms in mother and gosling such that the mother's inclination to protect, feed, and teach the gosling is triggered by the hatching and the subsequent interaction between gosling and mother goose.

Suppose, however, that a gosling hatches and by happenstance the first creature it sees after hatching is a passing fox, on which it immediately imprints in the same way that the goslings imprinted on Lorenz when he was the first creature they saw. The gosling thus leaves its mother and rigidly follows the fox around until it is eaten. When the gosling is insistently following the fox around to its death, something has certainly gone wrong with the gosling's naturally selected psychological functioning. A lot is also going right psychologically—it has stored an image, it has imprinted, it is following around the object on which it has imprinted, and so on—but something is also going wrong. Note that the closest human equivalent to such a condition, in which a child's attachment occurs to arbitrary strangers rather than a caregiver, is labeled "disinhibited social engagement disorder" and listed as a category of mental disorder in DSM-5.

Against this intuition that something has gone wrong, a neo-Kraepelinian must hold that nothing can be wrong with the gosling psychologically because all of its underlying brain mechanisms are doing exactly what they are biologically designed to do as describable strictly at a neurobiological anatomical/physiological level. The brain processes that correspond to the imprinting mechanism worked as designed, and consequently the gosling's brain was neuronally reconfigured so that it stored what at a psychological level is a detailed image of the first creature the gosling saw after hatching. That triggered a brain-based motivational/behavioral module that caused the gosling to follow and stay close to the creature whose image was stored in its imprinting register. The fact that a fox rather than the mother goose happened to be in the gosling's field of vision at the crucial moment says nothing about the gosling's brain being dysfunctional in a way describable in purely neurological terms, thus for the neo-Kraepelinian it can have no implication that the gosling has a psychological dysfunction. So, what justifies the intuition that something has gone wrong psychologically with the gosling?

To explain the intuition that the gosling is disordered, one first must acknowledge functional hierarchies that depend on the expectable environmental conditions in which various mechanisms evolved. Naturally selected functions almost always come in hierarchies. A common example is that the heart has the function of beating, which in turn has the function of circulating the blood, which has the function of bringing nutrients to the cells, and so on. Similarly, one function of the sexual organs is to yield pleasure, and one function of that pleasure is to motivate people to have sex. These multiple functions of the same organs are hierarchical in that each function was selected because it tends to lead to the success of the next function in the hierarchy; the heart beating wouldn't have been selected if it didn't cause the circulation of the blood, and sexual pleasure would not have been selected if it did not frequently enough motivate people to have sexual intercourse.

Each step in the functional hierarchy may depend on expectable environmental correlations that allow a lower level to reliably have the selected effect. For example, the possibility of sexual pleasure does not lead to intercourse if no appealing partners are available. When expectable environmental correlations fail so that higher-level functions are not performed, that is not considered a disorder but a mismatch between the organism and its environment. A disorder exists only if there is an internal structural alteration that renders the organism incapable of performing higher-level functions. Thus, some have a capacity for sexual pleasure but are incapable of engaging in sexual intercourse due to impotence, and that is a dysfunction.

The gosling's imprinting mechanism is associated with such a hierarchy of functions, some of which depend for their success on expectable environmental correlations. The storage of the image of the first seen creature after hatching has the function of storing the image of the mother, and the success of the latter function depends on the expectable environmental correlation of the mother being the first creature seen after hatching. Storing the image of the mother in turn has the function of activating the follow-and-stay-close program when seeing the mother that matches the image, and the evolution of the function of the staying-close program depended on a strong correlation between being the mother goose and having a variety of care-taking, protective, and teaching programs triggered around the time of the gosling's hatching.

Returning to the neo-Kraepelinian's challenge, if one concedes that nothing is wrong with the gosling's brain at the level of neurological description, how can anything be wrong with the gosling's functioning at the psychological level? The answer is that the dysfunction at the psychological level concerns intentional representational content, not neurophysiology; and it concerns not whether the psychological process of imprinting took place successfully (it did), but what the gosling imprinted on. The image in the brain refers to a passing fox, not to the mother, and thus involves the failure of a higher level of function of the imprinting mechanism, namely, that the gosling imprint on the mother. One can identify what has gone wrong only by going beyond brain descriptions and referring to meanings (i.e., what the gosling's brain-stored image in fact represents).

Even if one examined the brain of the gosling in a way that allowed one to identify the nature of the image within the imprinting register, one would still not know for sure whether the image was of a fox or of an unusual-looking mother goose facially similar to a fox. Even if the image looks like a goose, the accidental target of imprinting could be a passing goslingcidal goose that resembles the mother goose. Examining the stored image alone cannot establish the object whose image was stored, thus cannot establish whether the imprinting mechanism performed its higher-order function of imprinting on the mother goose. For that, one must go beyond the facts about the gosling's brain and understand the relationship of the image to the gosling's environment and to the biologically designed functions of the involved mechanisms.

True, the imprinting mechanism has accomplished several of its functions. However, it still manifests one dysfunction. Even though the gosling has successfully imprinted, the function of imprinting specifically on the mother has failed. This point is important because, in medical contexts, the presence of one harmful dysfunction is sufficient to support a disorder attribution even when most functions are being successfully performed. It makes no sense to say "cells evolved to reproduce, so they are working well in cancerous tumor growth and there is no disorder," or "anxiety mechanisms evolved to generate anxiety, so they are working well in generalized anxiety disorder and panic attacks and there is not disorder." The problem is that cells evolved to reproduce in a specifically regulated way and anxiety mechanisms evolved to generate anxiety specifically in response to perceived threatening triggers, and those specific functions have failed.

What makes the gosling's imprinting on a fox a dysfunction whereas lack of sexual partners is not a dysfunction? In imprinting, during a brief critical period at hatching, the imprinting mechanism enduringly and irreversibly fixes the image in the imprinting register. Once a failure of the expectable correlation causes that mechanism to fail to perform its storing-mother-image function, all higher levels of function are enduringly disrupted due to the permanent storage of the "wrong" image in the internal mechanism. No such incapacitation occurs in the lack of available partners, which does not preclude designed functioning if circumstances change. A dysfunction consists not merely of the failure to achieve a biologically designed function, but of the incapacity to perform a biologically designed function when within an expectable environment that relevantly matches the range of conditions for which it was selected. The gosling by these criteria does have a dysfunction, the individual without available partners does not.

The moral of this imprinting example is that some psychologically describable functions can go wrong without any physiologically describable functions going wrong. A brain that is internally learning normally may be suffering from a dysfunction due to what it is learning (for further discussion of mismatches and disorder, see Faucher and Forest, forthcoming).

Dysfunction without tissue damage: the example of carbon monoxide poisoning

The gosling example establishes the principle that there can be mental-level disorders without any brain-level disorder. However, it involves no ingested substance. Examples analogous to addiction of evolutionarily unexpectable substances that "hijack" natural systems occur in other areas of functioning. One much-discussed example is the hormone-mimicking chemicals that leach from plastics and trigger hormone receptors, potentially disrupting developmental programming. The hormone receptors are perfectly normal, but their activation by plastic-based hormone mimics is a dysfunction.

Another example is carbon monoxide poisoning. Carbon monoxide gas in significant environmental amounts is an evolutionarily novel product of incomplete human-caused combustion. Carbon monoxide happens to have chemical properties that allow it to perfuse from entirely normal lungs to blood cells and bind successfully with entirely normal hemoglobin in Lewis acid-base reactions of the kind that are biologically designed to occur with oxygen. The problem is that carbon monoxide binds more strongly than oxygen to hemoglobin, so does not disengage when it reaches the cells and does not allow a subsequent adequate rate of exchange for fresh oxygen. One can die of such hijacked hemoglobin transport even though the hemoglobin is normal and in many respects is doing exactly what it was designed to do. The dysfunction derives from the fact that the hemoglobin is performing its function with a novel substance with which it was not designed to interact.

Addiction is analogously a kind of overly strong "binding" of desire/choice systems to a kind of motivating reward they were not designed to handle. Because addictive substances act directly on reward-detecting systems each time they are used, they do not allow for natural processes of habituation. During such processes, the brain usually becomes increasingly sensitive to unexpected levels of potential reward, and motivational salience typically migrates from the reward itself at expectable levels to various indirect cues of unexpected reward levels (Robinson and Berridge 2001; Levy 2017). Because this biologically designed reduction of motivational salience does not occur in response to addictive substances, the anticipated reward is thus repeatedly strengthened in a way that it was not designed to be, yet there is no tissue damage underlying this dysfunction. Many people have died from the hijacking of hemoglobin in carbon monoxide poisoning, but many more have died from a hijacked desire system in which choice is reshaped by a substance for which the choice system was not designed. An individual with carbon monoxide poisoning has perfectly normal lungs and hemoglobin, an individual responding to leached hormone mimics from plastic has perfectly normal hormone receptors, and an addicted individual has a perfectly normal brain. Yet each of them has a dysfunction due to binding to the wrong thing. However, there is a crucial difference: carbon monoxide's binding to hemoglobin is universally powerful enough to cause harm at high levels despite some variations in sensitivity, whereas most addictive substances cause a harmful level of addiction in only a minority of individuals even when taken in substantial amounts over time. This implies that there are additional psychological or biological risk factors involved in addiction. I suggested above that one such additional factor is normal genetic variations that were not problematic in evolutionarily expectable environments but create vulnerability to substance-caused motivational dysfunction. One such normal variation that influences alcohol addiction susceptibility is whether one has the so-called "Asian alcohol gene" that lowers the likelihood of alcohol addiction by making drinking less enjoyable. Conceivably, in addition to such normal variations, some of the variations that create vulnerability could themselves be brain dysfunctions, introducing a role for brain dysfunction in the etiology of some substance addictions after all.

Harmful dysfunction as a unified account of substance addictions and behavioral addictions

The brain-damage theory against which I am arguing is most plausible as an account of substance addictions, but there is no credible model of how it would apply to behavioral addictions to gambling, video games, or internet porn. The existence of behavioral addictions remains a contentious issue, and one must be alert to the enormous false-positives potential of such diagnoses, but, assuming they do exist, the harmful dysfunction account can explain addictions to both substances and behaviors with equal ease. Just as certain novel substances introduced into our reward systems can cause addictive dysfunction, so if evolutionarily novel stimuli (e.g., gambling games designed to give the illusion of likely gain, or controllable internet-generated sexual images shaped to one's deepest fantasies) are inputted to circuits regulating behavior concerned with, for example, taking risks or sexual response, they may enduringly distort motivational functioning in vulnerable individuals. Thus, acceptance of the existence of behavioral addictions does not inherently conflict with the position that addictions are medical disorders, if one takes a harmful dysfunction approach to addiction rather than a brain-damage approach.

Social responsibility for the problem of addiction under the harmful dysfunction approach

The harmful dysfunction view places considerable responsibility on society at large for the problem of addiction. This is not only for the usual reason that negative social circumstances such as poverty and lack of life options increase drug use and thus addiction. From the harmful dysfunction perspective, the primary direct social factor in creating the phenomenon of addiction is the social creation and availability of evolutionarily novel addictive substances and activities that are capable of producing addiction in interaction with normal genetic and psychological variations that create addiction vulnerability. An analogy is to the misguided "blaming" of many individuals who are obese due to normal variations in genes regulating fat storage that were adaptive during our evolutionary past interacting with the unprecedented amount of high-calorie food marketed in temptation-maximizing ways to people in contemporary societies. Similarly, most addicted individuals would be fine in evolutionarily expectable environments. Thus, society bears a heavy responsibility for the problem of addiction. On this view, offering treatment is not just a matter of compassion but also a matter of justice (Wakefield 2015b).

References

Ahmed, S. H. (2010) "Validation crisis in animal models of drug addiction: beyond non-disordered drug use toward drug addiction", *Neuroscience and Biobehavioral Reviews* 35(2): 172–184.

Ahmed, S. H., Lenoir, M. and Guillem, K. (2013) "Neurobiology of addiction versus drug use driven by lack of choice", *Current Opinion in Neurobiology* 23(4): 581–587.

Ainslie, G. (1992) *Picoeconomics: The Strategic Interaction of Successive Motivational States Within the Person*, Cambridge, UK: Cambridge University Press.
American Psychiatric Association (APA) (2013) *Diagnostic and Statistical Manual of Mental Disorders, Fifth Edition (DSM-5)*, Arlington, VA: American Psychiatric Association.
Andreasen, N. (1984) *The Broken Brain: The Biological Revolution in Psychiatry*, New York, NY: HarperCollins.
Becker, G. S. and Murphy, K.M. (1988) "A theory of rational addiction", *Journal of Political Economy* 96(4): 675–700.
Cantin, L., Lenoir, M., Augier, E., Vanhille, N., Dubreucq, S., Serre, F., Vouillac, C. and Ahmed, S. H. (2010) "Cocaine is low on the value ladder of rats: possible evidence for resilience to addiction", *PLoS One* 28;5(7): e11592.
Courtwright, D. T. (2010) "The NIDA brain disease paradigm: history, resistance and spinoffs", *BioSocieties* 5(1): 137–147.
Faucher, L. and Forest, D. (forthcoming) *Defining Mental Disorder: Jerome Wakefield and His Critics*, Cambridge, MA: MIT Press.
Guillem, K. and Ahmed, S. H. (2018) "Preference for cocaine is represented in the orbitofrontal cortex by an increased proportion of cocaine use-coding neurons", *Cerebral Cortex* 28(3): 819–832.
Guze, S. B. (1989) "Biological psychiatry: is there any other kind?" *Psychological Medicine* 19(2): 315–323.
Hagen, E. H. and Sullivan, R. J. (in this volume) "The evolutionary significance of drug toxicity over reward," in H. Pickard and S. H. Ahmed (eds), *The Routledge Handbook of the Philosophy and Science of Addiction*, London, UK: Routledge.
Hall, W., Carter, A. and Forlini, C. (2015) "The brain disease model of addiction: is it supported by the evidence and has it delivered on its promises?" *Lancet Psychiatry* 2(1): 105–110.
Heyman, G. M. (2009) *Addiction: A Disorder of Choice*, Cambridge, MA: Harvard University Press.
Heyman, G. M. (2013) "Quitting drugs: quantitative and qualitative features", *Annual Review of Clinical Psychology* 9: 29–59.
Hoff, P. (2015) "The Kraepelinian tradition", *Dialogues in Clinical Neuroscience* 17(1): 31–41.
Kendler, K. S. (2012) "The dappled nature of causes of psychiatric illness: replacing the organic–functional/hardware–software dichotomy with empirically based pluralism", *Molecular Psychiatry* 17(4): 377–388.
Klerman, G. (1978) "The evolution of a scientific nosology", in J. Shershow (ed.), *Schizophrenia: Science and Practice*, Cambridge, MA: Harvard University Press, pp. 99–121.
Kraepelin, E. (1921) *Textbook of Psychiatry, Eighth Edition*, Edinburgh, UK: E. & S. Livingstone.
Kraepelin, E. (1922) "Ends and means of psychiatric research", *The Journal of Mental Science* 68(281): 115–143.
Lenoir, M., Serre, F., Cantin, L. and Ahmed, S. H. (2007) "Intense sweetness surpasses cocaine reward", *PloS One* Aug 1;2(8): e698.
Leshner, A. I. (1997) "Addiction is a brain disease, and it matters", *Science* 278(5335): 45–7.
Levy, N. (2006) "Autonomy and addiction", *Canadian Journal of Philosophy* 36(3): 427–447.
Levy, N. (2011) "Addiction and compulsion", in T. O'Connor and C. Sandis (eds), *A Companion to the Philosophy of Action*, Oxford, UK: Blackwell, pp. 267–273.
Levy, N. (2017) "Hijacking addiction", *Philosophy, Psychiatry, & Psychology* 24(1): 97–99.
Lewis, M. D. (2017) "Addiction and the brain: development, not disease", *Neuroethics* 10(1): 7–18.
Lorenz, K. (1935) "Der kumpan in der umwelt des vogels", *Journal of Ornithology* 83: 137–213, 289–413.
Lorenz, K. (1952) "The past twelve years in the comparative study of behavior", in C. H. Schiller (ed.), *Instinctive Behavior*, New York, NY: International Universities Press, pp. 288–317.
Nesse, R. M. and Berridge, K. C. (1997) "Psychoactive drug use in evolutionary perspective", *Science* 278(5335): 63–6.
Pickard, H. (2012) "The purpose in chronic addiction", *American Journal of Bioethics Neuroscience* 3(2): 30–39.
Pickard, H. (2016) "Denial in addiction", *Mind & Language* 31(3): 277–299.
Pickard, H. (2017) "Responsibility without blame for addiction", *Neuroethics* 10(1):169–180.
Rachlin, H. (2007) "In what sense are addicts irrational?" *Drug and Alcohol Dependence* 90(Suppl): S92–S99.
Redish, A. D., Jensen, S. and Johnson, A. (2008) "A unified framework for addiction: vulnerabilities in the decision process", *Behavioral and Brain Sciences* 31(4): 415–437.
Robinson, T. E. and Berridge, K. C. (2001) "Incentive-sensitization and addiction", *Addiction* 96(1): 103–114.
Satel, S. and Lilienfeld, S. O. (2014) "Addiction and the brain-disease fallacy", *Frontiers of Psychiatry* 4(141): 1–11.
Volkow, N. D., Koob, G. F. and McLellan, A. T. (2016) "Neurobiologic advances from the brain disease model of addiction", *New England Journal of Medicine* 374: 363–371.

Wakefield, J. C. (1992a) "The concept of mental disorder: on the boundary between biological facts and social values", *American Psychologist* 47(3): 373–388.
Wakefield, J. C. (1992b) "Disorder as harmful dysfunction: a conceptual critique of DSM-III-R's definition of mental disorder", *Psychological Review* 99(2): 232–247.
Wakefield, J. C. (1993) "Limits of operationalization: a critique of Spitzer and Endicott's (1978) proposed operational criteria for mental disorder", *Journal of Abnormal Psychology* 102(1): 160–172.
Wakefield, J. C. (1995) "Dysfunction as a value-free concept: reply to Sadler and Agich", *Philosophy, Psychiatry, and Psychology* 2(3): 233–246.
Wakefield, J. C. (1997) "Normal inability versus pathological disability: why Ossorio's (1985) definition of mental disorder is not sufficient", *Clinical Psychology: Science and Practice* 4(3): 249–258.
Wakefield, J. C. (1999a) "Evolutionary versus prototype analyses of the concept of disorder", *Journal of Abnormal Psychology* 108(3): 374–399.
Wakefield, J. C. (1999b) "Disorder as a black box essentialist concept", *Journal of Abnormal Psychology* 108(3): 465–472.
Wakefield, J. C. (2000a) "Spandrels, vestigial organs, and such: reply to Murphy and Woolfolk's 'The harmful dysfunction analysis of mental disorder'", *Philosophy, Psychiatry, and Psychology* 7(4): 253–270.
Wakefield, J. C. (2000b) "Aristotle as sociobiologist: the 'function of a human being' argument, black box essentialism, and the concept of mental disorder", *Philosophy, Psychiatry, and Psychology* 7(1): 17–44.
Wakefield, J. C. (2003) "Dysfunction as a factual component of disorder: reply to Houts, Part 2", *Behavior Research and Therapy* 41(8): 969–990.
Wakefield, J. C. (2006) "The concept of mental disorder: diagnostic implications of the harmful dysfunction analysis", *World Psychiatry* 6(3): 149–156.
Wakefield, J. C. (2009) "Mental disorder and moral responsibility: disorders of personhood as harmful dysfunctions, with special reference to alcoholism", *Philosophy, Psychiatry and Psychology* 16(1): 91–99.
Wakefield, J. C. (2011) "DSM-5 proposed diagnostic criteria for sexual paraphilias: tensions between diagnostic validity and forensic utility", *International Journal of Law and Psychiatry* 34(3): 195–209.
Wakefield, J. C. (2012) "The DSM-5's proposed new categories of sexual disorder: the problem of false positives in sexual diagnosis", *Clinical Social Work Journal* 40(2): 213–223.
Wakefield, J. C. (2013) "DSM-5: an overview of changes and controversies", *Clinical Social Work Journal* 41(2): 139–154.
Wakefield, J. C. (2014) "The biostatistical theory versus the harmful dysfunction analysis, part 1: is part-dysfunction sufficient for medical disorder?" *Journal of Medicine and Philosophy* 39(6): 648–682.
Wakefield, J. C. (2015a) "DSM-5 substance use disorder: how conceptual missteps weakened the foundations of the addictive disorders field", *Acta Psychiatrica Scandinavica* 132(5): 327–334.
Wakefield, J. C. (2015b) "Psychological justice: DSM-5, false positive diagnosis, and fair equality of opportunity", *Public Affairs Quarterly* 29(1): 32–75.
Wakefield, J. C. (2016a) "Diagnostic issues and controversies in DSM-5: return of the false positives problem", *Annual Review of Clinical Psychology* 12: 105–132.
Wakefield, J. C. (2016b) "The concepts of biological function and dysfunction: toward a conceptual foundation for evolutionary psychopathology", in D. Buss (ed.), *Handbook of Evolutionary Psychology, Second Edition*, New York, NY: Oxford University Press, pp. 988–1006.
Wakefield, J. C. and First, M. B. (2003) "Clarifying the distinction between disorder and non-disorder: confronting the overdiagnosis ("false positives") problem in DSM-V", in K. A. Phillips, M. B. First, and H. A. Pincus (eds), *Advancing DSM: Dilemmas in Psychiatric Diagnosis*, Washington, DC: American Psychiatric Press, pp. 23–56.
Wakefield, J. C. and Schmitz, M. F. (2015) "The harmful dysfunction model of alcohol use disorder: revised criteria to improve the validity of diagnosis and prevalence estimates", *Addiction* 110(6): 931–942.
Weir, K. (2012) "The roots of mental illness: how much of mental illness can the biology of the brain explain?" *APA Monitor* 43(6): 30.
Whiting, S. W., Hoff, R. A. and Potenza, M. N. (in this volume) "Gambling disorder", in H. Pickard and S. H. Ahmed (eds), *The Routledge Handbook of the Philosophy and Science of Addiction*, London, UK: Routledge.
World Health Organization (WHO) (1992) *The ICD-10 Classification of Mental and Behavioral Disorders: Clinical Descriptions and Diagnostic Guidelines*, Geneva: World Health Organization.

9

THE EVOLUTIONARY SIGNIFICANCE OF DRUG TOXICITY OVER REWARD

Edward H. Hagen and Roger J. Sullivan

Drug reward is an evolutionary conundrum. It is not surprising that neural circuitry evolved to reward or reinforce behaviors leading to the essentials of survival and reproduction, like food, water, and sex. Why, though, would these same circuits reward and reinforce the consumption of drugs of abuse, which is often harmful? Here we briefly review the history of reward-based learning, which resulted in a widely accepted evolutionary account of drug reward that we term the *hijack hypothesis*. We then critique the evolutionary bases of the hijack hypothesis. We conclude by sketching an alternative evolutionary model of human drug use grounded in drug toxicity. Specifically, avoidance of toxic drugs is a compelling hypothesis for the low use of drugs by children and women relative to men. In addition, the regulated ingestion of small quantities of toxins might have provided important medicinal and other benefits to humans and non-human animals over the course of their evolution.

Neurobiological theories of drug use are deeply intertwined with those of reward-based learning. The main idea was captured in Thorndike's *law of effect*: "Of several responses made to the same situation, those which are accompanied or closely followed by satisfaction to the animal will, other things being equal, be more firmly connected with the situation, so that, when it recurs, they will be more likely to recur" (Thorndike 1911). The related concept of *reinforcement* refers to the ability of certain stimuli, such as food, to strengthen learned stimulus–response associations. Relief from aversive stimuli could similarly "negatively reinforce" stimulus–response associations.

Early animal studies of drug addiction found evidence for both negative and positive reinforcement. In animals addicted to, e.g., morphine, relief from aversive withdrawal symptoms reinforced stimuli associated with obtaining the drug, but morphine's hedonic or euphoric effects also positively reinforced drug use (Spragg 1940; Beach 1957).

The neurobiological story of reward-based learning began with Olds and Milner's observation that rats will self-administer an electrical current to the septal region of the brain. They concluded that such intracranial stimulation was possibly the most potent reward ever used in animal experimentation to that date (Olds and Milner 1954). Its resemblance to drug addiction was immediately evident (Milner 1991).

More than two decades of experiments ensued to identify the precise neurons and neurotransmitters mediating the reinforcing effects of self-stimulation, in conjunction with work invested in understanding drug reward in its own right. The neurons critical for both septal self-stimulation and the reinforcing properties of at least some drugs turned out to be dopamine neurons in the midbrain, commonly referred to as the *mesolimbic dopamine system* (MDS). Thus were born two intimately intertwined theories, the dopamine theory of reinforcement learning and the dopamine theory of substance use and addiction, each deeply rooted in the stimulus-response paradigm at the core of behaviorism.

The hijack hypothesis

Drug reward requires an evolutionary explanation: unlike food, sex, and other natural rewards, drugs, at first glance, do not make an obvious contribution to an animal's survival or reproduction. In fact, chronic drug use is often harmful. Hence, it seems the brain should have evolved circuits to prevent drug use, rather than to reinforce it. One possibility is that drugs, like wires in the brain, are evolutionarily novel and their rewarding properties are artificial. Indeed, neurobiologists came to view drugs in the same way as intracranial electrodes, that is, as evolutionarily novel laboratory instruments to selectively activate or deactivate specific neural circuits (Wise 1996). After noting that "intravenous drug rewards establish and maintain response habits similar to those established and maintained by natural rewards," Wise (1996, 320) goes on to say that: "This should not be surprising; the brain mechanisms that make animals susceptible to brain stimulation reward evolved long before the human inventions that made intracranial self-stimulation or drug addiction possible." These human inventions include "e.g. the use of fire, pipes, and cigarette papers; the use of the hypodermic syringe and needle; agricultural skills for the harvesting and curing of tobacco; the ability to synthesize or purify drugs; the ability to concentrate, store, and transport alcoholic beverages" (Wise 1996, 320).

Subsequent highly cited review articles on the neurobiology of drug use endorsed the notion that brains are susceptible to drugs of abuse because they are evolutionarily novel and are consumed in a novel fashion. Natural rewards such as food and sex "activate" the reward system, whereas drug rewards "hijack," "usurp," "co-opt," or artificially stimulate it (for references, see Hagen, Roulette, and Sullivan 2013). Kelley and Berridge (2002, 3306), for instance, open their review with:

> Addictive drugs act on brain reward systems, although the brain evolved to respond not to drugs but to natural rewards, such as food and sex. Appropriate responses to natural rewards were evolutionarily important for survival, reproduction, and fitness. In a quirk of evolutionary fate, humans discovered how to stimulate this system artificially with drugs.

In another example, Hyman (2005, 1414) leads into a section titled "A Hijacking of Neural Systems Related to the Pursuit of Rewards" with:

> [A]ddiction represents a pathological usurpation of the neural mechanisms of learning and memory that under normal circumstances serve to shape survival behaviors related to the pursuit of rewards and the cues that predict them.

According to the hijack hypothesis, then, drugs of abuse, like intracranial electrodes, (1) are evolutionarily novel, especially in their purity or concentration, (2) are consumed in a novel fashion, and (3) provided no evolutionary fitness benefit. There are reasons to be skeptical about each proposition. (Skepticism of the hijack hypothesis is also increasing within neurobiology because most laboratory animals, when given a choice between an intravenous drug dose and a non-drug reward, choose the non-drug reward; see, e.g., Ahmed, Lenoir, and Guillem 2013).

The evolution of drugs of abuse and other pesticides

All living organisms, including all humans, are the latest members of unbroken lineages of organisms extending back to the origin of life, over 3 billion years ago. Today, almost all organisms acquire their energy directly or indirectly from oxygenic photosynthesis, which uses sunlight to reduce carbon dioxide to organic carbon, and stores chemical energy in the form of sugars and other carbohydrates. These then provide the building blocks and fuel for the growth and reproduction of the photosynthetic organisms, termed *autotrophs*. The first single-celled oxygenic photosynthetic autotrophs evolved about 2.4 billion years ago (Hohmann-Marriott and Blankenship 2011).

Unfortunately for these autotrophs, *heterotrophs* evolved that feed on them, sparking an evolutionary arms race (Dawkins and Krebs 1979) that continues to this day: heterotrophs evolved to exploit autotroph tissues and energy stores; autotrophs, in turn, evolved numerous defenses; heterotrophs then co-evolved countermeasures, and so forth.[1] Key events in this arms race include the evolution of marine animals more than 600 million years ago (Knoll and Carroll 1999), and the evolution of terrestrial plants ~400 million years ago, along with the terrestrial bacterial, fungal, nematode, invertebrate and vertebrate herbivores that feed on them (Herrera and Pellmyr 2009).

Central to our account of human drug use are the chemical defenses that evolved in marine and, later, in terrestrial autotrophs (plants). Some chemical defenses, such as tannins and other phenolics, have relatively non-specific effects on a wide range of molecular targets in the herbivore, for example binding to proteins and changing their conformation, thereby impairing their function. Other chemical defenses—neurotoxins—evolved to interfere specifically with signaling in the central nervous system (CNS) and peripheral nervous system. Various plant neurotoxins interfere with nearly every step in neuronal signaling, including (1) neurotransmitter synthesis, storage, release, binding, and re-uptake; (2) receptor activation and function; and (3) key enzymes involved in signal transduction. Plant neurotoxins have these effects, in many cases, because they have evolved to resemble endogenous neurotransmitters. Disruption of nervous system function by such toxins serves as a potent deterrent to herbivores (Wink 2011).

Because plant drugs, almost by definition, interfere with signaling in the CNS and elsewhere, they are widely believed to have evolved as plant defenses (Wink 2011). Nevertheless, among the popular plant drugs, only nicotine, which we discuss next, has been conclusively shown to serve plant defense.

Nicotine

In what follows we will often rely on studies of tobacco and nicotine because, for our purposes, tobacco is an ideal model drug. First, it is globally popular and highly addictive. Second, nicotine's role in plant defense is well documented: numerous studies of tobacco demonstrate that nicotine reduces leaf loss and plant mortality and increases production of viable seed by deterring, harming and killing herbivores (Baldwin 2001; Steppuhn et al. 2004). Third, we can draw on the extensive research on the nervous system effects of nicotine. Fourth, although the two domesticated tobacco species were probably artificially selected to increase their nicotine content, several of the more than 60 wild tobacco species have nicotine content comparable to or exceeding the domesticated species (Sisson and Severson 1990), and both wild and domesticated species were widely used by pre-Columbian Native Americans (Tushingham and Eerkens 2016).

This indicates that consumption of nicotine-rich tobacco is not simply a modern phenomenon. Finally, tobacco is usually consumed by chewing or smoking, and it is conceivable that humans chewed, or even smoked, various toxic and psychoactive plants for much of our recent evolution (Sullivan and Hagen 2002; Hardy et al. 2012).

Nicotine is a dangerous neurotoxin. In humans, oral ingestion of 4–8 mg of nicotine causes burning of the mouth and throat, nausea, vomiting, and diarrhea. Higher doses result in dizziness, weakness, and confusion, progressing to convulsions, hypertension and coma. Ingestion of concentrated nicotine pesticides can cause death within five minutes, usually from respiratory failure (Landoni 1991).

A single cigarette typically contains 10–20 mg of nicotine, enough to seriously endanger a young child and cause acute toxic symptoms in an adult. When a cigarette is smoked, much of its nicotine is burned, however, and smokers ultimately absorb 0.5–2 mg per cigarette. Tobacco chewers absorb up to 4.5 mg per "wad" (Hukkanen, Jacob, and Benowitz 2005), a dose that is often sufficient to cause severe acute toxicity in naive users.

Despite its acute toxicity, nicotine is not thought to be directly responsible for the chronic diseases caused by smoking (but see Grando 2014). Thus its toxicity, which explains why it is present in tobacco leaves in the first place, plays little role in research on tobacco use and addiction. In the framework we develop here, however, drug toxicity plays a central role.

Although the data are not yet as conclusive as they are for nicotine, a defensive role will probably be established for most other plant drugs, such as cocaine, morphine, codeine, THC, and caffeine (reviewed in Hagen et al. 2009). Like nicotine, most plant drugs are acutely toxic for humans, and the typical quantities consumed by drug abusers are often surprisingly close to the lethal dose (Gable 2004; Lachenmeier and Rehm 2015). For these reasons, in the remainder of the chapter we will often refer to recreational drugs as neurotoxic plant pesticides, which better describes their evolved function.

We will also draw on the extensive research on pharmaceuticals because these are frequently derived from plant toxins (e.g., nicotine, which has therapeutic applications and is also widely used as a pesticide), chemically resemble plant toxins, or have neurophysiological effects analogous to those of plant toxins.

Human toxin defense mechanisms

Human capabilities to detect, avoid, and neutralize plant toxins evolved over the course of our billion-year evolutionary arms race with autotroph defenses. During the final phase of this arms race, the human lineage was a lineage of primates, which diverged from other mammals roughly 65 million years ago, and which subsisted mostly on plants and insects (Fleagle 2013). As many insect species sequester plant toxins to deter predators, both elements of the primate diet required effective defenses against plant toxins. Primate toxin defense mechanisms, inherited from mammalian and vertebrate ancestors, would therefore have been continuously maintained and 'tuned' by natural selection. When human and chimpanzee ancestors diverged, probably more than 6 million years ago, the human lineage inherited a robust suite of toxin defense mechanisms. There is substantial evidence that these defense mechanisms correctly recognize all drugs of abuse as toxic (Sullivan, Hagen, and Hammerstein 2008; Hagen et al. 2009).

Taste receptors

Basic human anatomy prioritizes toxin defense, and taste buds are on the front line. Taste is responsible for evaluating the nutritious content of food and preventing the ingestion of toxic substances. The sweet and umami taste receptors, which identify two key nutrients—sugars and amino acids—belong to a small, three-member family of genes, the T1Rs, that are expressed in taste receptor cells in the tongue (Chandrashekar et al. 2006).

Bitter taste, in contrast, must prevent the ingestion of tens of thousands of structurally diverse toxins. Not surprisingly, bitter taste is mediated by a large repertoire of about 25 receptor genes, the T2Rs (Chandrashekar et al. 2006). All common recreational plant drugs taste bitter. Thus, these receptors properly recognize common psychoactive plant drugs as toxic.

In addition to their expression in tongue and palate epithelium, the sweet, umami, and bitter taste receptors are also expressed in other tissues exposed to nutrients and toxins, such as the respiratory system, gastrointestinal tract, testes, and brain (Behrens and Meyerhof 2010).

Barrier defenses

If a toxic plant substance is ingested it then encounters a "barrier defense." The body can be conceptualized as a set of compartments, such as the intestines and lungs, that are typically separated by tissue barriers comprising epithelial or endothelial cells linked together with special proteins forming "tight junctions." These tissue barriers include our skin, gastrointestinal (GI) tract, respiratory tract, and the blood–brain barrier (BBB). The barriers have several functions, such as allowing an influx of essential chemicals like sugar and oxygen into a compartment, and simultaneously preventing an influx of microorganisms and toxins (Mullin et al. 2005). The barriers achieve these effects by limiting or enhancing passive diffusion across the cells and tight junctions, and also by active mechanisms that transport essential chemicals into a compartment, and that neutralize and transport toxins out of a compartment.

Figure 9.1 illustrates the basic anatomical and cellular components of the barrier defenses against toxins and other xenobiotics. A plant toxin (represented as a pharmaceutical) comes into contact with a barrier, such as the skin, airways, lung, or intestine. If the toxin manages to enter a cell, such as an enterocyte, it then activates a complex network of proteins that neutralize and remove it in a four-phase process.

Phase 0 involves transporters—special proteins that span cell membranes and move chemicals into and out of cells using passive and active mechanisms. Efflux transporters generally remove toxins and waste products from the cell, and are typically members of the ATP binding cassette (ABC) super family. Humans have 48 ABC transporter genes, about 20 of which are efflux transporters. In Phase 0, a xenobiotic enters the cell, and an ABC transporter pumps it back out.

In Phase I, any xenobiotic remaining in the cell is chemically altered by enzymes to reduce toxicity and increase water solubility to facilitate excretion. Typically, this involves oxidation by one or more cytochrome P450 (CYP) enzymes. Humans have 57 CYP genes, about 25 of which are involved in xenobiotic metabolism (Sullivan, Hagen, and Hammerstein 2008).

In Phase II, diverse families of enzymes conjugate charged species with xenobiotic metabolites, further reducing toxicity[2] and increasing water solubility. Phase I and II xenobiotic metabolizing enzymes are most highly expressed in the liver, but are also expressed in most other tissue barriers, including skin, intestine, lung, placenta, and brain (Gundert-Remy et al. 2014). Xenobiotics can also bind to xenosensing nuclear receptors that then up-regulate expression of metabolic and transport proteins, accelerating elimination of the xenobiotic.

In Phase III, metabolites are pumped out of the cell by a transporter for renal or biliary elimination. For details on Phase 0–III, see Tóth et al. (2015), and references therein.

Significance of drug toxicity over reward

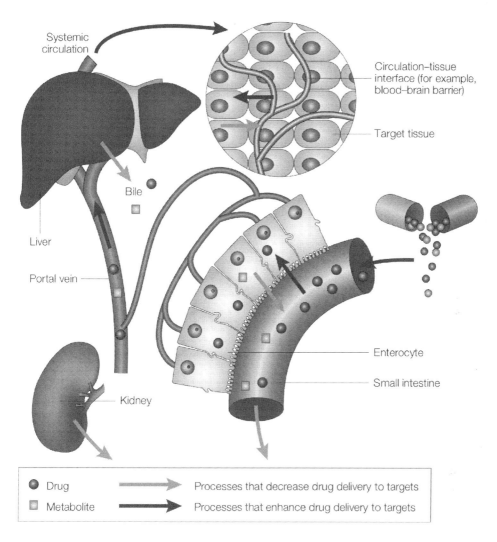

Figure 9.1 Toxin defense mechanisms of the gut "barrier," and first pass elimination. Many toxic substances (represented here as a pharmaceutical drug) entering the gastrointestinal tract are first metabolized by enzymes in the gut wall, or are transported back into the intestine. The remaining absorbed fraction enters the portal vein and is immediately routed through the liver, the principal organ of detoxification, before entering systemic circulation where it encounters other barrier defenses, such as the blood–brain barrier.

Source: Roden and George Jr (2002).

Nausea, vomiting, and conditioned taste aversion

Most nutrients, toxins, and other xenobiotics are processed in the gut, where many toxins are partially or completely neutralized and eliminated as just described. In addition, the gut is richly innervated, with large quantities of information conveyed from the GI tract to the CNS via the afferent vagal nerve. The *area postrema*, in particular, is a chemosensitive part of the brain that is outside the BBB, and therefore is also exposed to chemicals in systemic circulation. Together, these circuits can respond to toxins in the gut and blood with nausea, vomiting, and learned aversions and avoidances (Babic and Browning 2014).

Toxins that are not expelled from, or metabolized by, the gut enter the bloodstream and are then immediately routed through the liver, the principal detoxification organ, which further metabolizes and eliminates them. Any remaining toxin enters systemic circulation where it encounters other barrier tissues, such as the BBB, and further metabolism (Figure 9.1).

Psychoactive drugs like nicotine are neurotoxins, some portions of which evade all such defenses, successfully entering the brain and interfering with CNS function.

Evaluating the hijack hypothesis in light of evolved xenobiotic defenses

The multiple layers of toxin defenses involving scores of receptor, metabolic and transporter genes, numerous distinct tissue barriers, complex neural circuits, organs like the liver and kidneys, and the basic organization of the circulatory system, all demonstrate that human ancestors were regularly exposed to a large variety of dangerous xenobiotic toxins that entered the body via the skin, GI tract, and respiratory tract. These toxins often gained access to systemic circulation and, as evidenced by an extremely robust BBB that prevents most drugs from entering the brain (Pardridge 2012), many posed a substantial threat to CNS function, i.e., were psychoactive. Exposure to psychoactive plant compounds is not evolutionarily novel.

Evolutionarily novel levels of drug purity do not appear to be a general explanation for drug use and addiction either. Pure nicotine is not abused by humans, and most smokers do not prefer nicotine spray to placebo; nicotine and nicotinic receptor agonists only slightly improve smoking cessation rates; and other constituents of tobacco smoke, such as acetaldehyde, norharman and harman (MAO inhibitors), appear to potentiate the addictive properties of nicotine (Small *et al.* 2010). E-cigarettes, which deliver nicotine and flavorants, may be as or less addictive than nicotine gums, which themselves are not very addictive (Etter and Eissenberg 2015).

Evolutionarily novel methods of administration are also unlikely to explain recreational drug use and dependence, as chewing tobacco is addictive (US Department of Health and Human Services 1986) but chewing plants is not evolutionarily novel. Inhalation of toxic smoke was also probably common during human evolution. Our hominoid and hominin ancestors evolved in forest and savannah environments that regularly experienced wildfires, and our lineage might have achieved control of fire by 1 million years ago (Parker *et al.* 2016). It is therefore likely that human ancestors were frequently exposed to vaporized plant toxins, which probably helps explain the presence of robust xenobiotic defenses in our respiratory tract.[3] Furthermore, indigenous drug use often incorporates cultural techniques to "free base" psychoactive neurotoxins and to utilize physiology to avoid first-pass metabolism. For example, both betel nut (SE Asia) and coca (American Andes) are commonly mixed with a base (e.g. lime) and are chewed in the buccal cavity where the free alkaloids can cross directly into the bloodstream and into the CNS (Sullivan and Hagen 2002).

Thus, psychoactive compounds are not evolutionarily novel; they can be found in plants in concentrations similar to globally popular drugs (e.g., several wild tobacco species); evolutionarily novel purity does not explain their addictiveness (at least for nicotine); and they regularly entered systemic circulation via ingestion, contact with the skin, and inhalation, just as recreational drugs do today. Exposure to psychoactive compounds is as "natural" as exposure to sugars and starches.

Moreover, nicotine and other popular recreational plant drugs activate most known toxin defense mechanisms, including bitter taste (Wiener *et al.* 2012), xenobiotic nuclear receptors (Lamba 2004), xenobiotic metabolism (Sullivan, Hagen, and Hammerstein 2008), nausea and vomiting (Wishart *et al.* 2015), and conditioned avoidances and aversions (Lin, Arthurs, and Reilly 2017 and references therein). Human neurophysiology correctly recognizes drugs of abuse as the toxic pesticides that they are.

In summary, drug researchers correctly realized that the rewarding and reinforcing properties of toxic and harmful substances required an evolutionary explanation, but, on very scant evidence, wrongly concluded that drugs and their routes of administration were evolutionarily novel, and that this provided an adequate evolutionary account of human drug use. Plants are under strong selection to evolve compounds that "hijack" herbivore nervous systems, but for precisely the opposite effects: to punish and deter plant consumption, not reward or reinforce it.

Without considerable further evidence, it is not possible to accept that neurotoxic pesticides like nicotine are able to "hijack" reward circuits because they are evolutionarily novel, or are consumed in a novel fashion. The hijack hypothesis can only be rescued with more convincing evolutionary arguments and much stronger empirical evidence. As we explain next, the correct evolutionary account of human drug use is not yet clear.

The paradox of drug reward

The widespread recreational use of, and addiction to, several neurotoxic plant pesticides is extremely puzzling, to say the least. The reigning neurobiological paradigm of drug use, grounded in the rewarding or reinforcing effects of drugs in humans and other laboratory animals, is obviously in conflict with the reigning evolutionary biological paradigm of drug origins, grounded in the punishing effects of nicotine and other plant-based drugs on herbivores. Specifically, plants should not have evolved compounds that reward or reinforce plant consumption by herbivores, nor should herbivores have evolved neurological systems that reward or reinforce ingestion of potent plant neurotoxins. This contradiction has been termed the paradox of drug reward (Sullivan, Hagen, and Hammerstein 2008; Hagen *et al.* 2009; see also Sullivan and Hagen 2002).

Drug researchers have long recognized that drugs are toxins and have aversive effects, and that drug toxicity and aversiveness is at odds with drug reward (for review, see Verendeev and Riley 2012). Pavlov himself related experiments in which dogs learned to associate the toxic effects of morphine injections with stimuli. In the most striking cases, vomiting and other symptoms could be caused simply by the dog seeing the experimenter (Pavlov 1927). Unfortunately, drug aversion has had little influence on drug use theory (Verendeev and Riley 2012).

In the remainder of this chapter, we propose that drug toxicity explains dramatic age and sex differences in drug use. We also explore possible resolutions of the paradox of drug reward that are grounded in the neurotoxic properties of common recreational drugs.

Explaining the dramatic age difference in drug use

Users of popular neurotoxic pesticides report little-to-no use prior to the age of 10 (Figure 9.2). This is remarkable. Why are children so resistant to drug use? Although many researchers focus on the rapid adolescent transition to neurotoxic pesticide use, so far as we can tell there is essentially no investigation of the striking *lack* of child neurotoxic pesticide use. Perhaps drug researchers simply assume that parental and societal restrictions prevent child use.

This assumption seems reasonable for tobacco, as the US spends about $500 million each year on tobacco control efforts (WHO 2013). It is much less reasonable for caffeine, a bitter-tasting defensive neurotoxin that is found in 13 orders of the plant kingdom (Ashihara and Suzuki 2004), and that shows promise as a pesticide and repellant for slugs, snails, birds and insects (e.g., Hollingsworth, Armstrong, and Campbell 2002; Avery *et al.* 2005). Like nicotine, caffeine is a rewarding psychostimulant that strongly interacts with the central dopaminergic systems (Ferré 2008). Unlike nicotine, caffeine faces few social restrictions against use—it is listed by the US Food and Drug Administration as "GRAS [generally recognized as safe] for use in cola-type beverages at levels not to exceed 200 parts per million (ppm) (0.02%)" (Rosenfeld *et al.* 2014, 26). This level corresponds to 71 mg of caffeine in a 12-oz serving (although most colas contain about half that amount). For comparison, a 1 oz shot of espresso contains about 64 mg of caffeine, an 8 oz cup of coffee might contain 145 mg of caffeine, energy drinks typically contain 17–224 mg of caffeine per serving, and chocolate candy contains 11–115 mg caffeine per oz (Rosenfeld *et al.* 2014).

Despite the light regulation of caffeine compared with tobacco and nicotine, and its ready availability in colas, chocolate candies, and other child food products, child consumption of caffeine is low (Figure 9.3a, b, c), suggesting that low child use of putatively "rewarding" neurotoxic pesticides is not explained solely by parental or societal controls. What, then, does explain the dramatic lack of child neurotoxic pesticide use, and the equally dramatic "switch-like" transition to neurotoxic pesticide use during adolescence?

Plant defensive pesticides are often teratogenic, disrupting development and permanently impairing functionality. Nicotine is a teratogen that interferes with acetylcholine signaling, which has a unique trophic role in brain development. Nicotine exposure can disrupt all phases of brain assembly (Dwyer, Broide, and Leslie 2008).

Consistent with the risk that plant toxins pose to child development, there is considerable evidence for heightened toxin defenses during infancy and childhood. Although infants recognize that plants are sources of food, they are more reluctant to touch novel plants compared with other types of novel artifacts and natural objects of similar appearance, which might reflect an evolved psychological defense against plant toxins (Wertz and Wynn 2014). Neophobic food rejection occurs primarily due to visual cues. Foods that do not "look right"—green vegetables for example, or foods that resemble known bitter foods—are rejected without being placed in the mouth. Food neophobia peaks between 2 and 6, and then decreases with age, becoming relatively stable in adulthood, a developmental trajectory widely interpreted to reflect an evolved defense against plant teratogens. Children also have a higher density of taste buds on the tip of the tongue than adults and are more sensitive to bitter tastes. High bitter taste sensitivity leads to reduced consumption of bitter vegetables, especially in children. For review, see Hagen, Roulette, and Sullivan (2013).

As a starting point for future research on low child drug use, we propose a model with three elements. First, to prevent ingestion of teratogens, children are innately neophobic, picky, and have heightened bitter sensitivity. Consequently, they find most neurotoxic pesticides to be especially unpalatable.

Second, social learning plays an especially important role in toxin avoidance. Whereas learning about toxic substances via individual trial-and-error comes with the potentially high cost of ingesting a toxin, one can socially learn to avoid toxic substances from knowledgeable others "for free" (Boyd and Richerson 1985; Rogers 1988). Children should therefore be particularly attentive to information from parents and other adults that certain substances are dangerous, poisonous, or do not taste good, and assiduously avoid those substances (Cashdan 1994).

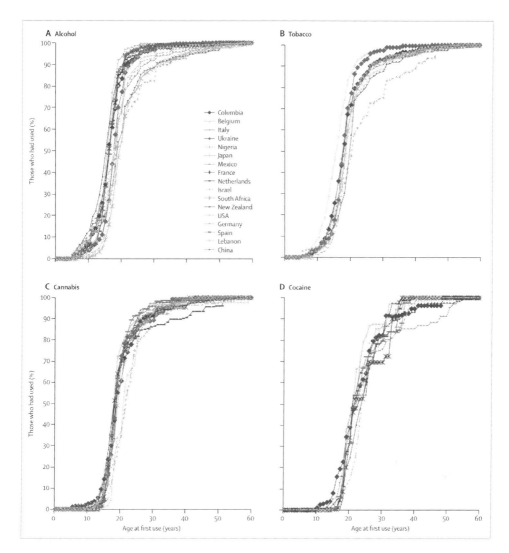

Figure 9.2 Cumulative distribution of self-reported age of first use of alcohol, tobacco, cannabis, and cocaine in a large (N = 85,052) cross-national sample of users of these substances. These patterns suggest the existence of a developmental "switch" to drug use during adolescence.

Source: Degenhardt et al. (2016).

In contrast, we expect considerable child resistance to parents' efforts to restrict access to candy and other sugary foods, which, from an evolutionary perspective, are nearly pure beneficial nutrients.

Third, in adolescence brain and other organ development is nearing completion. We propose that adolescent onset of neurotoxic pesticide use is partly related to the reduced risk of developmental disruption and consequent reduced aversion to plant toxins, which also serves to broaden diet. See Figure 9.4.

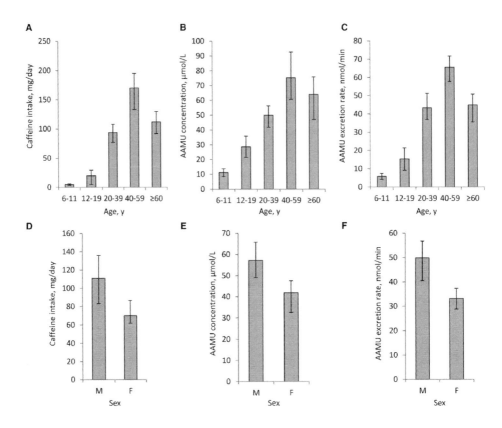

Figure 9.3 Age and sex vs. caffeine intake from dietary recall and urinary caffeine metabolite 5-acetylamino-6-amino-3-methyluracil (AAMU) in a representative sample of the US population (n = 2466).

Source: Rybak et al. (2015).

Explaining the large sex difference in drug use

More men regularly use neurotoxic pesticides (and alcohol) than women, though the extent of the male bias varies by nation, substance, age, birth cohort, and other factors. Male prevalence of smoking is almost always greater than female prevalence, for instance, albeit with considerable variation across nations (Figure 9.5). In the US there is even a male bias in caffeine intake (Figure 9.3).

The global male bias is narrower in younger cohorts, especially for the legal drugs tobacco and alcohol, and in recent years US adolescent girls (12–17) were more likely than adolescent boys to use alcohol and be non-medical users of psychotherapeutic drugs. In the US population as a whole, however, men were more likely than women to be users of all categories of drugs, including psychotherapeutic drugs and alcohol. For review, see Hagen, Roulette, and Sullivan (2013).

Over human evolution, ingestion of neurotoxic pesticides would probably have posed similar threats to men and women, but women of childbearing age faced the additional risk of disrupted fetal and infant development. Ancestral women were pregnant or lactating for much

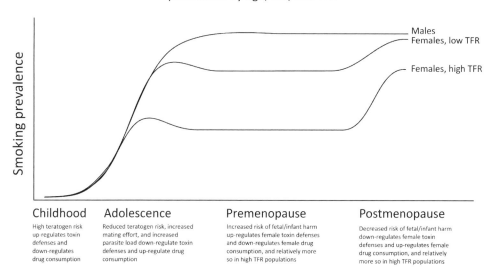

Figure 9.4 Theoretical model of age and sex differences in use of tobacco and other plant drugs. Social learning of toxic and teratogenic substances also plays an important role in each phase. TFR: Total fertility rate.

Source: Hagen, Garfield, and Sullivan (2016).

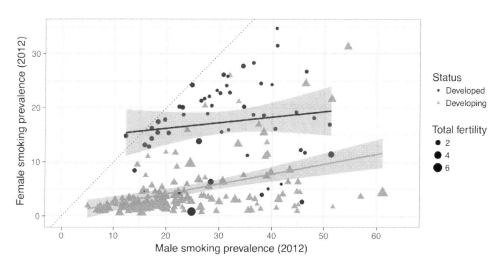

Figure 9.5 Female vs male smoking prevalence across nations. Each dot is one country. The dotted diagonal line represents equal prevalence.

Source: Hagen, Garfield, and Sullivan (2016).

of their late teens to their late thirties. At the age that young women in Western societies might begin regular use of plant drugs (and regular use of birth control), with their first pregnancy often years in the future, most women in ancestral environments were beginning about two decades of pregnancy and lactation. This could have selected for an increased ability to detect and avoid plant teratogens that would be harmful to fetuses and nursing infants, resulting in lower female use of neurotoxic pesticides that is evident even today.

There is considerable evidence for sex differences in toxin detection and disposition. Less clear is whether these differences are a consequence of greater toxin defenses in women, particularly pregnant women, or instead are byproducts of, e.g., sex differences in body size and composition.

Women have more taste buds than men and, according to most studies, are able to detect lower concentrations of bitter substances. High bitter sensitivity, in turn, generally predicts reduced vegetable intake in both women and men. Most studies indicate that women also have higher toxin metabolism rates. During pregnancy, heightened food aversions appear to help prevent ingestion of toxic plants, including coffee and tobacco, that might pose a risk to the developing fetus, especially during organogenesis. Women smokers, for example, commonly report new olfactory and gustatory aversions to tobacco during pregnancy, and the olfactory aversions are associated with women smoking less (Pletsch et al. 2008). Nicotine metabolism is accelerated in pregnancy, and activities of many xenobiotic-metabolizing enzymes are increased several-fold. For references and further discussion, see Hagen, Roulette, and Sullivan (2013).

In countries where women are pregnant and lactating more often (i.e., those with higher total fertility rates), there are fewer women smokers, even after accounting for gender inequality (Hagen, Garfield, and Sullivan 2016). The diminishing sex differences in use of some substances in younger cohorts might therefore partially reflect the global fertility transition over the last several decades that involves increased use of birth control, later age at marriage, delay of first birth, and lower total fertility, all of which would allow women, especially younger women, to increase drug intake while limiting fetal and infant exposure (Hagen, Roulette, and Sullivan 2013). We propose that social learning also plays an important role in women's decisions to avoid teratogenic plant substances (Placek and Hagen 2015; Placek, Madhivanan, and Hagen 2017). See Figure 9.4.

Possible explanations for regulated neurotoxic pesticide intake: an evolved 'taste' for drugs?

Drug toxicity would seem to predict no use whatsoever by individuals of any age or sex, contrary to the global popularity of smoking and other drug use.

Drug reward might be an accident. Over 100,000 plant defensive compounds have been identified (Wink 2011), and perhaps humans simply discovered a very few that, despite their toxicity to insects and other herbivores, accidentally trigger reward or reinforcement mechanisms (Hagen et al. 2009). This hypothesis faces a "Goldilocks" problem, though: recreational plant pesticides must be accidentally rewarding or reinforcing enough to overcome their aversive properties, yet because they are often highly toxic, they cannot be so rewarding or reinforcing for most users that they lead to immediate overdoses and death. The accidental effects of these compounds must be "just right." Hence, for each drug, the accident hypothesis involves not just one rare accident, but two.

Alternatively, *regulated* toxin intake might have produced fitness benefits in certain circumstances. Because wild plant foods are infused with defensive chemicals, plant consumers, including human ancestors, should have evolved some type of regulatory mechanism that balances intake of nutrients vs toxins so as to avoid poisoning (Torregrossa and Dearing 2009). See Figure 9.6. But regulated toxin intake occurs even in the absence of a nutrient signal. Laboratory animals regulate their self-administration of drugs at a fairly constant and stable level regardless of the dose per injection or number of lever presses required (Yokel and Wise 1976). Human cigarette smokers similarly alter their smoking behavior in response to changes in nicotine content so as to maintain a relatively constant blood concentration of nicotine (Scherer and Lee 2014). These are clues that special mechanisms might have evolved to carefully regulate plant toxin intake (Hagen *et al.* 2009; Hagen, Roulette, and Sullivan 2013).

There are many possible fitness benefits of regulated ingestion of neurotoxic plant pesticides, most of which reconceptualize these compounds as valuable medicines rather than hijackers (Sullivan and Hagen 2002; Sullivan, Hagen, and Hammerstein 2008; Hagen *et al.* 2009; Hagen, Roulette, and Sullivan 2013). Neurotoxic pesticides achieve their effects because they evolved to manipulate cellular signaling. Nicotine, for instance, mimics acetylcholine, a neurotransmitter involved in neuromuscular communication and many other important functions. In large doses, nicotine kills. In small, highly regulated doses, though, such as the ~1 mg delivered by smoking a cigarette, it might provide a number of immediate benefits. (The long-term health costs of smoking are indisputable, however.)

One possible benefit is defense against parasites. Many heterotrophic species evolved to co-opt plant toxins for prophylactic or therapeutic effects against pathogens, i.e., self-medication, also known as pharmacophagy or zoopharmacognosy. All popular recreational drugs are toxic to parasitic worms (helminths). It is not out of the question that humans and other animals evolved to seek out and ingest small quantities of neurotoxic pesticides to help combat helminths and other parasites (Sullivan, Hagen, and Hammerstein 2008; Hagen *et al.* 2009; Hagen, Roulette, and Sullivan 2013; Roulette *et al.* 2014). If the pharmacophagy hypothesis is correct, then the toxicity of drugs to parasites provides the ultimate explanation for their use by humans. See Figure 9.6.

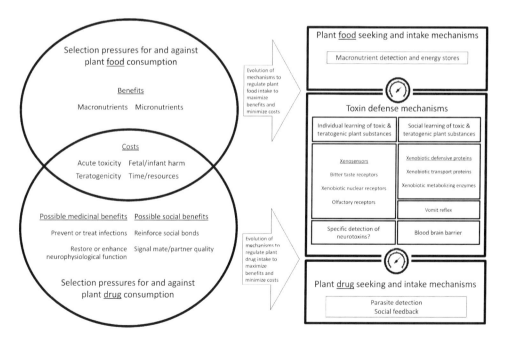

Figure 9.6 The neurotoxin regulation model of the evolution of "recreational" drug use.

Source: Modified from Hagen, Garfield, and Sullivan (2016).

Salt intake provides a useful analogy: there are complex neuronal and endocrine mechanisms, including special salty taste receptors on the tongue, that regulate intake of milligrams of this valuable environmental chemical to maintain sodium homeostasis (Geerling and Loewy 2008), even though there is no conscious awareness of its biological benefits. Similarly, bitter taste receptors and other xenosensors, in conjunction with neuronal, immunological, social learning and other mechanisms, might regulate intake of milligrams of neurotoxins for their medicinal or social benefits without any conscious awareness of these benefits.

In conclusion, popular recreational drugs are neurotoxic pesticides, varieties of which have infused the diets of human ancestors for hundreds of millions of years. These and other xenobiotics selected for a sophisticated, multilayered toxin defense system that correctly identifies all drugs of abuse as toxins. In this light, it is doubtful that recreational drugs are best characterized as evolutionarily novel hijackers of reward circuitry. Although the correct evolutionary account of recreational drug use is not yet clear, our neurotoxin regulation hypothesis (Figure 9.6) provides a compelling hypothesis for the very low use of recreational drugs by children, the low use by women of reproductive age relative to men, and the careful titration of drug intake by humans and non-human animals. The increasing evidence that non-human animals ingest plant toxins to help defend against pathogens and provide other fitness benefits should inspire similar hypotheses for human drug use.

Notes

1 Autotrophs and heterotrophs have also undergone mutually beneficial coevolution. See discussion in Hagen et al. (2009).
2 Occasionally, Phase I and Phase II metabolites are more toxic than the parent compound.
3 Even hypodermic injection is not an evolutionarily novel mode of exposure to psychoactive toxins. Although we have emphasized plant neurotoxins, numerous vertebrates and invertebrates produce potent neurotoxins that they inject into predators and prey with stingers and fangs. There is increasing evidence that humans have an innate fear of spiders and snakes, probably because many of these species are venomous and frequently attacked human ancestors with fangs (Öhman and Mineka 2001). Even today, snakebite is a major cause of morbidity and mortality in much of the world (Gutiérrez et al. 2013). The dangers of bites and stings might also explain an apparent innate fear of needles (Hamilton 1995).

References

Ahmed, S. H., Lenoir, M. and Guillem, K. (2013) "Neurobiology of addiction versus drug use driven by lack of choice", *Current Opinion in Neurobiology* 23(4): 581–587.
Ashihara, H. and Suzuki, T. (2004) "Distribution and biosynthesis of caffeine in plants", *Frontiers in Bioscience* 9(2): 1864–1876.
Avery, M. L., Werner, S. J., Cummings, J. L., Humphrey, J. S., Milleson, M. P., Carlson, J. C., Primus, T. M. and Goodall, M. J. (2005) "Caffeine for reducing bird damage to newly seeded rice", *Crop Protection* 24(7): 651–657.
Babic, T. and Browning, K. N. (2014) "The role of vagal neurocircuits in the regulation of nausea and vomiting", *European Journal of Pharmacology* 722: 38–47, doi:10.1016/j.ejphar.2013.08.047.
Baldwin, I. T. (2001) "An ecologically motivated analysis of plant-herbivore interactions in native tobacco", *Plant Physiology* 127(4): 1449–1458.
Beach, H. D. (1957) "Morphine addiction in rats", *Canadian Journal of Psychology/Revue Canadienne de Psychologie* 11(2): 104–112.
Behrens, M. and Meyerhof, W. (2010) "Oral and extraoral bitter taste receptors", in W. Meyerhof, U. Beisiegel, and H.-G. Joost (eds), *Sensory and Metabolic Control of Energy Balance*, Berlin: Springer, pp. 87–99.
Boyd, R. and Richerson, P. J. (1985) *Culture and the Evolutionary Process*, Chicago, IL: University of Chicago.

Cashdan, E. (1994) "A sensitive period for learning about food", *Human Nature* 5(3): 279–291.
Chandrashekar, J., Hoon, M. A., Ryba, N. J. P. and Zuker, C. S. (2006) "The receptors and cells for mammalian taste", *Nature* 444(7117): 288–94.
Dawkins, R. and Krebs, J. R. (1979) "Arms races between and within species", *Proceedings of the Royal Society of London. Series B. Biological Sciences* 205(1161): 489–511.
Degenhardt, L., Stockings, E., Patton, G., Hall, W. D. and Lynskey, M. (2016) "The increasing global health priority of substance use in young people", *The Lancet Psychiatry* 3(3): 251–264.
Dwyer, J. B., Broide, R. S. and Leslie, F. M. (2008) "Nicotine and brain development", *Birth Defects Research Part C: Embryo Today: Reviews* 84(1): 30–44, doi:10.1002/bdrc.20118.
Etter, J. F. and Eissenberg, T. (2015) "Dependence levels in users of electronic cigarettes, nicotine gums and tobacco cigarettes", *Drug and Alcohol Dependence* 147: 68–75.
Ferré, S. (2008) "An update on the mechanisms of the psychostimulant effects of caffeine", *Journal of Neurochemistry* 105(4):1067–1079, doi:10.1111/j.1471-4159.2007.05196.x.
Fleagle, J. G. (2013) *Primate Adaptation and Evolution*, San Diego, CA: Academic Press.
Gable, R. S. (2004) "Comparison of acute lethal toxicity of commonly abused psychoactive substances", *Addiction* 99(6): 686–696.
Geerling, J. C. and Loewy, A. D. (2008) "Central regulation of sodium appetite", *Experimental Physiology* 93(2): 177–209.
Grando, S. A. (2014) "Connections of nicotine to cancer", *Nature Reviews Cancer* 14(6): 419–429.
Gundert-Remy, U., Bernauer, U., Blömeke, B., Döring, B., Fabian, E., Goebel, C., Hessel, S., et al. (2014) "Extrahepatic metabolism at the body's internal–external interfaces", *Drug Metabolism Reviews* 46(3): 291–324.
Gutiérrez, J. M., Warrell, D. A., Williams, D. J., Jensen, S., Brown, N., Calvete, J. J., Harrison, R. A. and Global Snakebite Initiative (2013) "The need for full integration of snakebite envenoming within a global strategy to combat the neglected tropical diseases: the way forward", *PLOS Neglected Tropical Diseases* 7(6): e2162.
Hagen, E. H., Sullivan, R. J., Schmidt, R., Morris, G., Kempter, R. and Hammerstein, P. (2009) "Ecology and neurobiology of toxin avoidance and the paradox of drug reward", *Neuroscience* 160(1): 69–84.
Hagen, E. H., Roulette, C. J. and Sullivan, R. J. (2013) "Explaining human recreational use of 'pesticides': the neurotoxin regulation model of substance use vs. the hijack model and implications for age and sex differences in drug consumption", *Frontiers in Psychiatry* 4(142), doi:10.3389/fpsyt.2013.00142.
Hagen, E. H., Garfield, M. J. and Sullivan, R. J. (2016) "The low prevalence of female smoking in the developing world: gender inequality or maternal adaptations for fetal protection?" *Evolution, Medicine, and Public Health* 1(1): 195–211, doi:10.1093/emph/eow013.
Hamilton, J. G. (1995) "Needle phobia: a neglected diagnosis", *Journal of Family Practice*, 41(2): 169–176.
Hardy, K., Buckley, S., Collins, M. J., Estalrrich, A., Brothwell, D., Copeland, L. and Huguet, R. (2012) "Neanderthal medics? Evidence for food, cooking, and medicinal plants entrapped in dental calculus", *Naturwissenschaften* 99(8): 617–626.
Herrera, C. M. and Pellmyr, O. (2009) *Plant-Animal Interactions: An Evolutionary Approach*. Malden, MA: John Wiley & Sons.
Hohmann-Marriott, M. F. and Blankenship, R. E. (2011) "Evolution of photosynthesis", *Annual Review of Plant Biology* 62: 515–548.
Hollingsworth, R. G., Armstrong, J. W. and Campbell, E. (2002) "Pest control: caffeine as a repellent for slugs and snails", *Nature* 417(6892): 915–916.
Hukkanen, J., Jacob, P. and Benowitz, N. L. (2005) "Metabolism and disposition kinetics of nicotine", *Pharmacological Reviews* 57(1): 79–115.
Hyman, S. E. (2005) "Addiction: a disease of learning and memory", *American Journal of Psychiatry* 162(8): 1414–1422.
Kelley, A. E. and Berridge, K. C. (2002) "The neuroscience of natural rewards: relevance to addictive drugs", *Journal of Neuroscience* 22(9): 3306–3311.
Knoll, A. H. and Carroll, S. B. (1999) "Early animal evolution: emerging views from comparative biology and geology", *Science* 284(5423): 2129–2137.
Lachenmeier, D. W. and Rehm, J. (2015) "Comparative risk assessment of alcohol, tobacco, cannabis and other illicit drugs using the margin of exposure approach", *Scientific Reports* 5(8126), doi:10.1038/srep08126.

Lamba, V. (2004) "PXR (NR1I2): splice variants in human tissues, including brain, and identification of neurosteroids and nicotine as PXR activators", *Toxicology and Applied Pharmacology* 199(3): 251–265, doi:10.1016/j.taap.2003.12.027.

Landoni, J. H. (1991) "Nicotine", *Poisons Information Monographs, International Programme on Chemical Safety*, www.inchem.org/documents/pims/chemical/nicotine.htm.

Lin, J.-Y., Arthurs, J. and Reilly, S. (2017) "Conditioned taste aversions: from poisons to pain to drugs of abuse", *Psychonomic Bulletin & Review* 24(2), 335-351.

Milner, P. M. (1991) "Brain-stimulation reward: a review", *Canadian Journal of Psychology/Revue Canadienne de Psychologie* 45(1): 1–36, doi:10.1037/h0084275.

Mullin, J. M., Agostino, N., Rendon-Huerta, E. and Thornton, J. J. (2005) "Keynote review: epithelial and endothelial barriers in human disease", *Drug Discovery Today* 10(6): 395–408.

Öhman, A. and Mineka, S. (2001) "Fears, phobias, and preparedness: toward an evolved module of fear and fear learning", *Psychological Review* 108(3): 483–522.

Olds, J. and Milner, P. (1954) "Positive reinforcement produced by electrical stimulation of septal area and other regions of rat brain", *Journal of Comparative and Physiological Psychology* 47(6): 419–427.

Pardridge, W. M. (2012) "Drug transport across the blood–brain barrier", *Journal of Cerebral Blood Flow & Metabolism* 32(11): 1959–1972, doi:10.1038/jcbfm.2012.126.

Parker, C. H., Keefe, E. R., Herzog, N. M., O'Connell, J. F. and Hawkes, K. (2016) "The pyrophilic primate hypothesis", *Evolutionary Anthropology* 25(2): 54–63.

Pavlov, I. P. (1927) *Conditioned Reflexes: An Investigation of the Physiological Activity of the Cerebral Cortex*, Oxford, UK: Oxford University Press.

Placek, C. D. and Hagen, E. H. (2015) "Fetal protection: the roles of social learning and innate food aversions in South India", *Human Nature* 26(3): 255–276.

Placek, C. D., Madhivanan, P. and Hagen, E. H. (2017). "Innate food aversions and culturally transmitted food taboos in pregnant women in rural southwest India: separate systems to protect the fetus?" *Evolution and Human Behavior*, 38(6), 714–728.

Pletsch, P. K., Pollak, K. I., Peterson, B. L., Park, J., Oncken, C. A., Swamy, G. K., et al. (2008) "Olfactory and gustatory sensory changes to tobacco smoke in pregnant smokers", *Research in Nursing & Health* 31(1): 31–41.

Roden, D. M. and George, A. L. Jr (2002) "The genetic basis of variability in drug responses", *Nature Reviews Drug Discovery* 1(1): 37–44.

Rogers, A. R. (1988) "Does biology constrain culture?" *American Anthropologist* 90(4): 819–831.

Roulette, C. J., Mann, H., Kemp, B. M., Remiker, M., Roulette, J. W., Hewlett, B. S., Kazanji, M., et al. (2014) "Tobacco use vs. helminths in Congo basin hunter-gatherers: self-medication in humans?" *Evolution and Human Behavior* 35(5): 397–407.

Rosenfeld, L. S., Mihalov J. J., Carlson S. J. and Mattia A. (2014) "Regulatory status of caffeine in the United States", *Nutrition Reviews* 72 (no. suppl 1): 23–33, https://doi.org/10.1111/nure.12136.

Rybak, M. E., Sternberg, M. R., Pao, C.-I., Ahluwalia, N. and Pfeiffer, C. M. (2015) "Urine excretion of caffeine and select caffeine metabolites is common in the US population and associated with caffeine intake", *The Journal of Nutrition* 145(4): 766–774.

Scherer, G. and Lee, P. N. (2014) "Smoking behaviour and compensation: a review of the literature with meta-analysis", *Regulatory Toxicology and Pharmacology* 70(3): 615–628, doi:10.1016/j.yrtph.2014.09.008.

Sisson, V. A. and Severson, R. F. (1990) "Alkaloid composition of the nicotiana species", *Beiträge Zur Tabakforschung International* 14(6): 327–339.

Small, E., Shah, H. P., Davenport, J. J., Geier, J. E., Yavarovich, K. R., Yamada, H. and Bruijnzeel, A. W. (2010) "Tobacco smoke exposure induces nicotine dependence in rats", *Psychopharmacology* 208(1): 143–158.

Spragg, S. D. S. (1940) "Morphine addiction in chimpanzees", *Comparative Psychology Monographs* 15(7): 1–132.

Steppuhn, A., Gase, K., Krock, B., Halitschke, R. and Baldwin, I. T. (2004) "Nicotine's defensive function in nature", *PLOS Biology* 2(10): 1074–1080.

Sullivan, R. J. and Hagen, E. H. (2002) "Psychotropic substance-seeking: evolutionary pathology or adaptation?" *Addiction* 97(4): 389–400.

Sullivan, R. J., Hagen, E. H. and Hammerstein, P. (2008) "Revealing the paradox of drug reward in human evolution", *Proceedings of the Royal Society B* 275(1640): 1231–1241.

Thorndike, E. L. (1911) *Animal Intelligence: Experimental Studies*, New York, NY: Macmillan.

Torregrossa, A-M. and Dearing, M. D. (2009) "Nutritional toxicology of mammals: regulated intake of plant secondary compounds", *Functional Ecology* 23(1): 48–56.

Tóth, A., Brózik, A., Szakács, G., Sarkadi, B. and Hegedüs, T. (2015) "A novel mathematical model describing adaptive cellular drug metabolism and toxicity in the chemoimmune system", *PLOS ONE* 10(2): e0115533.

Tushingham, S. and Eerkens, J. W. (2016) "Hunter-gatherer tobacco smoking in ancient North America: current chemical evidence and a framework for future studies", in E. A. Bollwerk and S. Tushingham (eds), *Perspectives on the Archaeology of Pipes, Tobacco and Other Smoke Plants in the Ancient Americas*, Berlin: Springer, pp. 211–230.

US Department of Health and Human Services (1986) *The Health Consequences of Using Smokeless Tobacco: A Report of the Advisory Committee to the Surgeon General*, Bethesda, MD: US Department of Health and Human Services, Public Health Service.

Verendeev, A. and Riley, A. L. (2012) "Conditioned taste aversion and drugs of abuse: history and interpretation", *Neuroscience & Biobehavioral Reviews* 36(10): 2193–2205.

Wertz, A. E. and Wynn, K. (2014) "Thyme to touch: infants possess strategies that protect them from dangers posed by plants", *Cognition* 130(1): 44–49, doi:10.1016/j.cognition.2013.09.002.

Wiener, A., Shudler, M., Levit, A. and Niv, M. Y. (2012) "BitterDB: a database of bitter compounds", *Nucleic Acids Research* 40(1): D413–419.

Wink, M. (2011) *Annual Plant Reviews, Biochemistry of Plant Secondary Metabolism*, Vol. 40. Malden, MA: John Wiley & Sons.

Wise, R. A. (1996) "Addictive drugs and brain stimulation reward", *Annual Review of Neuroscience* 19: 319–40.

Wishart, D., Arndt, D., Pon, A., Sajed, T., Guo, A. C., Djoumbou, Y., Knox, C., et al. (2015) "T3DB: the toxic exposome database", *Nucleic Acids Research* 43(D1): D928–D934.

World Health Organization (WHO) (2013) *WHO Report on the Global Tobacco Epidemic*, available from: www.who.int/tobacco/global_report/2013/en.

Yokel, R. A. and Wise, R. A. (1976) "Attenuation of intravenous amphetamine reinforcement by central dopamine blockade in rats", *Psychopharmacology* 48(3): 311–318, doi:10.1007/BF00496868.

SECTION B

Varieties, taxonomies, and models of addiction

10
DEFINING ADDICTION
A pragmatic perspective

Walter Sinnott-Armstrong and Jesse S. Summers

Many people falsely assume that all definitions are either arbitrary stipulations or mere descriptions of common usage. However, some definitions serve concrete practical purposes. Consider cities. The United States once passed a law that provided federal funds for public transportation systems in any city in the United States. Many communities applied for the funds, but which of these communities were cities and, hence, eligible for funds? The government could not answer this question by citing common usage, because common usage is too vague. The government also did not want to decide arbitrarily, since that would be unfair and impossible to justify to communities that were excluded. What they did was to estimate how many communities would apply and for how much funding. Then they could calculate how far the available funds could go. They defined "city" so that their available funds could help as many people as possible. They ended up defining cities as communities with at least 50,000 inhabitants within a certain area. This definition did not describe common usage, and it was not arbitrary stipulation. It was based on practical concerns. They would have been subject to criticisms if they had defined cities too broadly (so as to include a town of 5,000) or too narrowly (so as to include only areas with over 5,000,000 inhabitants). The criticisms would have been that they exceeded the limits of common usage and also that their definition didn't serve their practical purpose of funding as many cities as they could afford to fund. Their solution was within these limits and served a practical purpose, so their answer was reasonable.

Other definitions still might be reasonable for other purposes. Imagine that a friend says that she would like to move to a city to live around more people. Will you correct her if she wants to move somewhere not on the government list? When an urban planning study looks at mass transportation "in cities," but they consider as cities only the very largest cities, is that a methodological problem with the study? Or are these simply cases in which a perfectly good definition of "city" fails to define what cities are for all possible purposes?

Practical or precising definitions like these occur not only in government but also in science, where the goal is often more theoretical than practical. What is sugar? Most people answer that sugar is the sweet substance that comes from sugarcane. Chemically, however, sugars are short-chain soluble carbohydrates that can come from any number of sources, including sugar beet and dried fruit. A biologist studying animal metabolism may be indifferent to the taste of such compounds, while a flavor scientist may care *only* about the taste. A nutritionist may mean only refined sugar by "sugar," and a dieter may include a wide range of high-calorie sweet things. Someone who dislikes "sugary foods" may even include sugar substitutes in his definition.

Why are there such diverse definitions? Are they all arbitrary or merely descriptive of common usage? No. These definitions reflect practical or theoretical purposes, even among chemists. Chemists have different interests in studying sugar: some may care about differences between various naturally occurring sugars, while others only study their metabolic or gustatory similarities, so they will all use definitions that fit what they find most important. In contrast, most non-chemists are interested in sugar either because they care about taste or because they care about calories, and their definitions of "sugar" will be shaped by this interest. Each group uses a definition that serves its particular purposes.

The same point applies to definitions of addiction. The DSM-5 (APA 2013) uses the word "addiction" without giving it an explicit definition. In the older DSM-IV, the two categories of "Substance Abuse" and "Substance Dependence" were taken to be more or less the clinical definition of addiction. In the DSM-5, these are replaced with a single category of "Substance Use Disorder" that comes in mild, moderate, and severe degrees. When a clinician refers to an "addiction," she likely has this category in mind.

Is the clinician right if she thinks of this diagnostic category as a definition of addiction? For reasons that will become obvious here, we provide no definition of "definition": what some people call a "definition" others will call a "characterization," a "paradigm," a "sense," etc. We'll leave further discussion of how to define "definition" for another occasion; for the present discussion, in order to make any progress in understanding definitions of addiction, we'll use the word "definition" as broadly as it's commonly used. Otherwise, we might rule out illuminating examples. For example, much clinical and popular discussion about what addiction is turns on the DSM clinical criteria of Substance Use Disorder, so we don't want to rule that out just because some would object that clinical criteria aren't a definition.

Diagnoses of a Substance Use Disorder are based on eleven symptoms: overuse, failure to cut down, time spent seeking and using, cravings, interference with other duties, relationship problems, giving up other activities, use when physically hazardous, related physical and psychological problems together with knowledge of such problems, tolerance, and withdrawal. A diagnosis of mild substance use disorder requires two to three symptoms from this list, whereas moderate substance use disorder requires four to five symptoms from this list, and severe substance use disorder requires six or more symptoms from this list. Clinicians can also specify that the substance use disorder is in early or sustained remission or on maintenance therapy.

This definition (like much of DSM-5) has been heavily criticized. One common objection is that very many people would be diagnosed as having at least a mild substance use disorder. That result seems clear, because many beer and wine drinkers build tolerance and occasionally have at least one drink more than they intended to have. However, whether this result is a problem depends not on whether the definition includes *very* many people but instead on whether it includes *too* many people. How many is too many? That depends on one's purposes. The DSM-5 was written for clinicians to use as part of a general assessment and ultimately treatment, and its target is those for whom their disorder "causes clinically significant distress or impairment in social, occupational, or other important areas of functioning" (APA 2013: 20). The DSM-5's main purpose is therefore to include people who could benefit from clinical help. That can be very many people. The broadness of the definition, in effect, allows clinicians to help people who need help but whom they could not help if the definition were narrower. The definition would be too broad if it included people who do not really need any help from clinicians, even if they ask for help, but it is hard to see why the definition goes that far. Given these purposes, then, it is not at all clear that the definition includes too many people.

What makes the DSM-5 definition useful for clinical purposes might make it useless for the purpose of scientific research. Since mild and moderate substance use disorder require only two to five of the eleven symptoms, individuals with these diagnoses need not share any symptoms at all. One patient could have five symptoms (such as overuse, failure to cut down, cravings, tolerance, and withdrawal), whereas another patient has a non-overlapping set of symptoms (such as time spent, interference with other duties, relationship problems, giving up other activities, and knowledge of their own related physical and psychological problems). Without any symptoms in common, there is little or no reason to expect scientific research to find regularities that cover both cases (though there might be a common neural basis for very different symptoms in some cases). The same point applies to severe substance use disorder because two patients with six of the eleven symptoms still might share only one symptom. Thus, the DSM-5 definition fails the purposes of research scientists even if it serves the purposes of clinicians.

Our conclusion is not that the DSM-5 is either good or bad in the way it characterizes—defines, even—addiction. Instead, like the case of sugar discussed above, different definitions serve different purposes. Each definition can be good in its own way for its own purposes. Problems arise only when someone tries to use a definition for a purpose that it was never intended or shaped for. Researchers will fail if they use definitions intended for clinicians. Clinicians will fail if they use definitions intended for researchers. However, that is not a problem with the definitions. It is only a problem with the way that people use the definitions. The definitions are still fine when used properly.

This vagueness and relativity to purposes is anathema to some. Some scientists and philosophers want to know what addiction "really is," with no qualifications about purposes or contexts. One path towards that goal might seem to follow neuroscience, which is supposed to be "harder"—more objective and precise—than psychology, especially clinical psychology. This approach might seem to be suggested by leading researchers who define addiction as a chronic, relapsing brain disease (see Leshner 1997; Hyman 2005; NIDA 2016). If addiction can be defined as a certain brain condition, then clinicians' and psychologists' behavioral tests of addiction are really just proxies for the underlying problem, which is the brain condition.

Unfortunately, however, neuroscientists cannot define addiction any more objectively than clinicians and other scientists. First, consider how a neural marker of addiction would have to be discovered. Neuroscientists would need to identify which participants in the study are addicted. Only then could they look for a neural condition that is shared by all of the addicts and by none of the non-addicts. The results would be very different if they started by assuming that everyone with a mild substance use disorder is an addict than if they started by assuming that only those with severe substance use disorders are addicts. The results would also be very different if they started by assuming that addicts are still addicts after years of "sustained remission" than if they instead started by assuming that addicts are not still addicts once they've quit. The results would also be very different if they started by assuming that there are no behavioral addictions (such as addictions to gambling, sex, or the internet) than if they instead started by assuming that behavioral addictions are addictions just like substance addictions. The neural condition that underlies "addiction" cannot be identified without controversial assumptions about the group of cases that it is supposed to underlie.

Moreover, if the class of addicts is large enough to satisfy clinicians who want to help all of those who need help with their substance use problems, then it is very doubtful that neuroscientists will find any neural condition that is shared by all addicts that is not also shared by some non-addicts. The class is simply too diverse and vague. There is perhaps more hope of finding a neural condition that underlies a narrower class of addictions, such as opioid addictions that are not in remission. However, clinicians can then reply that the discovered brain condition underlies only some but not all addictions. They do not have to admit that other cases (such as addicts in remission or behavioral addicts) are not addictions simply because they do not share the underlying brain condition that is shared by the cases that interest those neuroscientists. As a result, no neural marker can provide a definition independent of purposes.

Another approach to definitions is adopted by philosophers. Many philosophers are interested in what makes addiction a disease and in whether addicts are responsible, so they tend to define addiction in terms of harm and control. For example, one of the current authors has argued that "Addiction is a strong and habitual want that significantly reduces control and leads to significant harm" (Sinnott-Armstrong and Pickard 2013). This definition does not pretend to avoid depending on the purposes of the speaker who ascribes an addiction. Its vague terms are intended to show where purposes matter. When is a want strong enough to count as an addiction? When is the harm significant enough? When does the person lack enough control to be classified as an addict? Answers to these questions will differ depending on the purposes of the speaker. For example, a patient who lacks a moderate degree of control could justifiably be said to have significantly reduced control and hence be classified as an addict by a clinician whose purpose is to help that patient regain control so as to avoid more harm, but a judge or jury still might justifiably deny that the person lacks the degree of control that is required for legal or moral responsibility, so they might deny that this patient's reduction in control is significant for their purposes.

The point of the philosophical definition is to be vague in the right ways to allow variation in application so as to illuminate disagreements. When one observer asserts and another observer denies that the patient is addicted, the definition suggests that the asserter thinks that the want is strong enough, the harm is significant enough, and the patient's degree of control is low enough, whereas the denier believes that at least one of these conditions is not met. They can agree on how much control the patient has but still disagree about whether that degree of control is low enough to count as significantly reduced and, hence, as addiction. That disagreement may itself be determined by the context and purposes of the speakers. Similarly, they can agree on how strong the patient's want is but still disagree about whether that want is strong enough, and they can agree on how much harm is suffered but disagree on whether that harm is significant enough to warrant classifying that patient as addicted. Again, this disagreement may all be explicable by looking at the speakers' purposes and the contexts in which they're speaking. The speakers may still genuinely disagree over whether a person is addicted, but this makes it clearer where the disagreement arises.

Thus, definitions by clinicians, neuroscientists, and philosophers all eventually must depend at some point on the purposes of the speakers who assert or deny that someone is addicted. No definition of addiction is independent of such purposes.

Is this relativity to purpose a problem? Some say so. Imagine a patient who is diagnosed with an alcohol use disorder in the United States because he has two glasses of wine with lunch every day, which reduces his afternoon productivity in his job and interferes with his relationships to friends who condemn this practice. Then he quits his job and moves to an artists' colony in France where this practice is common, even encouraged. He might not meet the criteria for even mild substance use disorder in France even though he did meet them in the United States. However, if he is an addict in the United States but not in France, then his alcoholism might seem to be cured by a plane ticket.

Indeed, the patient need not travel so far as France. Imagine that a patient in a clinic volunteers for a neuroscientific study. His clinician tells him that he has a substance use disorder, but the scientist in charge tells him that he does not meet the criteria for inclusion in the study of substance use disorders. He seems to be cured of his substance use disorder or addiction simply by moving from the clinic to the lab next door!

Is that too much relativism? We think not. To avoid any contradiction, we simply have to remember that different people use different definitions, but these are not incommensurable idiolects: these definitions are all perfectly justified by their different purposes, and everyone involved can easily understand the competing definitions, even if they all choose to use their own. The clinician is justified in diagnosing the patient with a mild to moderate substance use disorder (depending on which other criteria are fulfilled) because he wants to help the patient get better, whereas the neuroscientist is justified in excluding this patient from his study of substance use disorders because this patient's symptoms locate him outside of the group that this scientist wants to study because he hypothesizes that this group shares a common and distinctive neural condition. Both the clinician and the neuroscientist are justified in reaching diagnoses that only appear to be contradictory, and they would have no problem in understanding why the other chooses an alternative definition.

A similar point applies to the trip to France. If a certain degree of lack of control over having two glasses of wine with lunch is common and accepted in France, particularly in the creative conditions of the artists' colony, but is neither common nor accepted in the business world in the United States, and if this condition causes more harm in the United States than in France, then there is nothing wrong with saying that the patient has an alcohol use disorder in the United States but ceases to have an alcohol use disorder after he flies to France. But are we saying that the therapist could cure addiction by prescribing a one-way plane ticket? If all the patient wants is not to be clinically diagnosed, then perhaps that would work. The patient, however, presumably cares less about the particular diagnosis than he cares about the underlying problems that are causing him to seek treatment. What he wants is to maintain the valuable features of his life that his alcohol use threatens, and leaving those valuable things behind by moving likely won't help. His condition does not change, but what does change is whether he overuses (beyond the norm), whether his use interferes with his (norm-relative) duties and relationships, and so on for other symptoms in the DSM-5 definition. Relevant to the philosophical definition above, what also changes is whether his want is significantly stronger than normal and whether any resulting harm or reduction in control is significant. A definition of addiction does not need to yield a bright line that classifies every case as addiction or not regardless of circumstances, effects, and purposes.

There can be extreme cases that count as addictions for any purpose because they clearly fulfill all of the relevant criteria. Any plausible definition of addiction must accommodate these paradigm cases. In order to remove relativity, one might be tempted to define addiction as only these paradigm cases or, less restrictively, as severe substance use disorder under DSM-5 criteria.[1] Then one could claim that all other cases are not really addictions but are only loosely called addiction because they are similar in some ways to the paradigm or severe cases. Such narrow definitions will serve some purposes, and nobody will deny the cases that they include as addictions. Nonetheless, they are still not adequate for other purposes, because they exclude many patients that clinicians want to help as well as many kinds of addiction that scientists want to study. In order to fulfill those other purposes, we need to move beyond the paradigm and severe cases into the borderline territory where various definitions come apart. A choice to avoid those borderline cases reflects one's purposes just as much as a choice to include them.

Hence, there is nothing wrong with a definition that allows diagnoses and classifications to vary with purposes of speakers. Indeed, allowing definitions to be relative to purposes has benefits for understanding some tricky cases. These benefits can be illustrated by three examples.

Abstinent addicts

Alcoholics Anonymous notoriously requires its members to announce, "I am an alcoholic," and related groups have similar requirements. This announcement strikes many people as odd when said by a member of Alcoholics Anonymous who has not touched alcohol for several years. Similarly, DSM-5 allows clinicians to say that a patient has a substance use disorder even when that patient is in "sustained remission" and has no cravings any more. How can someone still be an alcoholic or addict after years of abstaining?

One interpretation is that this practice is a useful fiction. It might make sense for the long-abstinent member of Alcoholics Anonymous to think of herself as still an addict, because this announcement helps her avoid relapse. Instead of thinking of herself as able to have an occasional drink like a non-addict, she thinks of herself as an addict for whom every drink creates a serious danger of relapse and contributes to her alcoholic pattern of drinking. If the goal is to prevent relapse (which is the goal of Alcoholics Anonymous), and if this definition serves as a useful fiction to help prevent relapse, then it makes practical sense to define alcoholism broadly enough to include members who have not touched alcohol in years.

Similarly, suppose that a neuroscientist discovered that the neural condition that is common to alcoholics (and perhaps other addicts) continues in an alcoholic's brain for years after remission. Then it could make theoretical sense to continue to classify a patient as an alcoholic for years after sustained remission. After all, what interests the neuroscientist is that lasting brain condition.

In contrast, imagine that a person who used to drink heavily but has not used alcohol in years cites his alcoholism as grounds for taking a week off from work to go to a luxurious alcoholism treatment center paid by his insurance. Or suppose he cites his alcoholism as an excuse for neglecting his children and spouse. Claims like these would and should be rejected. Even if it makes sense for the purpose of preventing relapse or neuroscience to classify this person as an alcoholic, that diagnosis cannot properly be used for the distinct purposes of insurance or responsibility. Different people can use the same words for different purposes, but we still need to distinguish those different purposes in order to avoid confusion and absurdity.

Pain medication

Consider next a patient who takes morphine to reduce pain from a chronic illness. She tried to cut down, but her chronic pain became too intense to bear. However, that pain is due to her chronic condition and not—at least not primarily—due to withdrawal from the pain medication. If she quit taking morphine, then she might have some withdrawal symptoms, but soon she would feel pretty much as she would have felt if she had never taken morphine to begin with. Still, feeling that way would be excruciating. Let's assume also that she develops tolerance and spends a lot of time seeking and using morphine in order to regulate her pain.

Is she addicted? Here our judgment is more complicated than in the first case, but the lesson is basically the same. She fulfills the diagnostic criteria of addiction, so she could be classified as an addict for some purposes. She certainly has a hard time quitting and her life is ruled by her use of morphine. If we're trying to excuse her when she misses a family event because, despite trying not to use morphine, she succumbs to her pain and uses, then that excuse can be valid. But the excuse looks different if we think of the excuse as being an addiction caused by regular morphine use than if we think of it as an excuse that comes from regular morphine use in response to pain. The full excuse includes both, and we suspect as a result that people may agree on the relevant facts but disagree over just how responsible the person is.

In contrast, if a neuroscientist is studying the brain condition that underlies addiction to morphine, then the neuroscientist's interests will determine whether this patient is a good candidate for a research study. If the neuroscientist is interested in the effects of long-term morphine use on tolerance, then this patient might be a good candidate, since she did develop tolerance. In contrast, if the neuroscientist is interested in how long-term use changes addicts' decision-making abilities, then this patient would seem to be a poor candidate for research, because her decision to take morphine is determined by her desire to avoid pain and not only by the effects of morphine use. Thus, whether or not this patient should be classified an addict depends on the interests of the person asking the question.

Can our earlier philosophical account settle whether or not she is an addict? This patient shows a strong want that leads to some harm and some lack of control, but how much and are they significant? Normally we would measure her level of control by asking whether she would quit if she had more reason to quit than to continue. In this case, however, it is not clear how she could have better reason to quit as long as her pain would be excruciating and debilitating if she quit taking morphine. We could still ask whether she would quit if she had strong reasons and also could quit without ensuing pain, but it is not clear why that counterfactual shows that she has control in her current circumstances where she needs the drug to stop the pain. Thus, it is not clear how much control she lacks. And what about harm? Even if she is harmed by her regular use of morphine, she still might be harmed even more if she were to quit. Of course, the harms are different—health effects instead of pain—but it is not clear whether she is harmed overall or whether overall harm instead of harm to health is what determines whether she is addicted. The general problem is that we need to determine whether a person is harmed by use when there are also harms in not using.[2] What this case highlights is that an abstract, philosophical account that applies to addiction generally will not by itself settle such difficult cases.

Erdős

Finally, consider a real case (Hoffman 1998). Paul Erdős was a famously prolific mathematician whose love for math seemed to surpass his interest in everything else. He co-authored with over 500 other mathematicians and scientists, usually after showing up at their houses (perhaps unannounced) with all of his few possessions, collaborating for a few days, then leaving for his next destination with new papers in his wake. He never married or had children, had no siblings, and most of his family died in the Holocaust. During his life, he published over 1,500 papers and made many other contributions to mathematics.

Erdős drank a tremendous amount of coffee, and, when he was almost 60, he began using amphetamines. His friends were bothered by his amphetamine use, and one of them bet Erdős $500 that he couldn't go a month without drugs. Erdős won the bet, but he claimed "You've showed me I'm not an addict. But I didn't get any work done. I'd get up in the morning and stare at a blank piece of paper. I'd have no ideas, just like an ordinary person. You've set mathematics back a month" (Hoffman 1998: 16). He began using again.

Was Erdős an addict? He was able to quit, but so are many addicts, given the right incentives—sometimes quite minimal incentives, as Contingency Management therapy exploits to good effect (Heyman 2009; Zajac *et al.* this volume). Clearly, Erdős was dependent on amphetamines insofar as he could not work as productively without them once he'd started. Just as clearly, he was not dependent on amphetamines in that he did quit for $500 and had 40 years of prolific work behind him at the time when he started using them. Did he have control over his use? Like the previous case, he demonstrated some control, though he probably felt that he had no reason ever to try to quit again, so he likely wouldn't test it again. Was his use causing harm? Perhaps, since amphetamines do have some health consequences, such as increased risk of heart attacks, and Erdős died of a heart attack at age 83. Still, again as in the previous case, any discussion of harm needs to balance the harms of use with the harms of not using. Not using clearly created some serious harms for someone who loved mathematics as much as he did.

If a neuroscientist wanted to study addiction, though, would Erdős have been a good research subject? That depends on the particular interest of the neuroscientist. Would Erdős have been a good patient for a clinical dependence treatment program? Perhaps. He clearly would have resisted treatment, and, given his priorities, he may have been right to do so. But, if amphetamines were, on balance, interfering with his life—imagine that he began having legal or health problems as a result of use that interfered even more with his productivity than his withdrawal did—then clinical dependence treatment may have been useful.

The ways in which Erdős differs from many other homeless, itinerant amphetamine users are significant, but do they show that he is any more or less an addict? Or do they show, as we here propose, that there are many ways to define "addiction" that are all useful for their own purposes, and that we are misguided if, once we're clear about these various practical purposes, we continue to insist that there must be *a* single correct definition for all, regardless of purpose.

Conclusion

We have argued that a pragmatic view of addiction is defensible, and that it solves some difficulties in understanding addiction by highlighting that various definitions of addiction serve various practical purposes. Those various definitions still might not differ in most cases, but they will differ in some cases by classifying individuals as addicts for some purposes and not as addicts for other purposes. When definitions seem to conflict in this way, these apparent conflicts need not show that these definitions are incorrect or inadequate. These conflicts might instead show only that different definitions are equally useful, though for different purposes. That is enough for pragmatists.

We have focused our attention on those definitions that are most likely to come into conflict for serious philosophical and scientific work on addiction. But there are of course many other purposes and definitions of addiction. Abstemious religious and political groups may define addiction in a broad way so as to include any regular use. By contrast, the alcohol and tobacco industries (and other commercial interests) prefer to define addiction quite narrowly, classifying regular smoking, for example, as a habit, not as an addiction.[3] What we've said here hasn't taken a position on the definitions we've considered or on how valuable the underlying purposes are. We can, however, ask whether the tobacco industry's purposes and corresponding definition are worth taking seriously, and, if not, why not. In fact, by highlighting the way in which such purposes inform *all* definitions of addiction, we have made explicit how we can engage those definitions that we might find disagreeable.

Given our pragmatic view, what is the point of philosophical reflection on addiction? One point is to illuminate related concepts and relations to those other concepts. We illustrated this above in discussing "control" and "harm," which are philosophically interesting and important concepts that reflection on the various cases of addiction help to illuminate. Philosophical reflection can also help us refine various pragmatic concepts of addiction. We saw in some of the above cases that diagnostic criteria leave a lot of grey areas in which clinicians must deploy philosophical reflection about what they are trying to accomplish. More abstractly, philosophical reflection—particularly when that reflection is informed by clinical practice and neurobiological research—helps us determine which features should be grouped together as relevant for treatment and what about those features can justify treatment even over objection. Finally, philosophical reflection on the nature of addiction can reveal what unites a disparate set of definitions. By reflecting abstractly on what pragmatic definitions have in common, we may arrive at a better understanding of them all.

Notes

1 Thanks to Hanna Pickard for suggesting that we reply to this possibility.
2 Cases like this led one of the authors to focus on the comparative element in characterizing addiction, asking not whether the drug use caused harm, but whether a person has *over*valued her own drug use, relative to the alternatives (Summers 2015). Significant harm is a cost that entails that one overvalues their drug use, so this account will largely overlap with the account in Sinnott-Armstrong and Pickard (2013). However, the comparative account directly addresses cases in which one *should* bring about significant harm because the alternative harm is worse.
3 Thanks to Serge Ahmed for highlighting this issue.

References

American Psychiatric Association (APA) (2013) *Diagnostic and Statistical Manual of Mental Disorders, Fifth Edition*, Arlington, VA: American Psychiatric Association.
Heyman, G. (2009) *Addiction: A Disorder of Choice*, Cambridge, MA: Harvard University Press.
Hoffman, P. (1998) *The Man Who Loved Only Numbers: The Story of Paul Erdos and the Search for Mathematical Truth*, New York, NY: Hachette.
Hyman, S. (2005) "Addiction: a disease of learning and memory", *American Journal of Psychiatry* 162(8): 1414–1422.
Leshner, A. I. (1997) "Addiction is a brain disease, and it matters", *Science* 278(5335): 45–47.
National Institute on Drug Abuse (NIDA) (2016) available at: www.drugabuse.gov/publications/media-guide/science-drug-abuse-addiction-basics, accessed June 12, 2016.
Sinnott-Armstrong, W. and Pickard, H. (2013) "What is addiction?" in K. W. M. Fulford, M. Davies, R. G. T. Gipps, G. Graham, J. Z. Sadler, G. Stranghellini and T. Thornton (eds), *The Oxford Handbook of Philosophy and Psychiatry*, Oxford, UK: Oxford University Press, pp. 851–864.
Summers, J. S. (2015) "What is wrong with addiction", *Philosophy, Psychiatry, & Psychology* 22(1): 25–40.
Zajac, K., Alessi, S. M. and Petry, N. M. (this volume, pp. 455–463) "Contingency management approaches".

11
DIAGNOSIS OF ADDICTIONS

*Marc Auriacombe, Fuschia Serre, Cécile Denis,
and Mélina Fatséas*

What is addiction? Core and constellation: a clarification challenge

Both substances and behaviors can be considered to be forms of addictions, understood as an abnormal long-lasting pattern of use or practice that is reinforcing and may be repeated to excess, to the point that it endangers the individual. This excess of use or practice is typically visible to an observer, but, equally, users themselves may also report it as disturbing. The dangerous consequences may sometimes spread from the individual to his environment, making addiction an individual characteristic (a disease or disorder), with environmental and social consequences (public health impact, political and societal implications).

Although excess is a common characteristic of addictions, its definition is difficult. Excess may be defined by use over a pre-determined threshold that may be defined by quantity or frequency. Excess may also be defined by any quantity/frequency as long as it has negative consequences, acute or chronic. Nevertheless, substances that are commonly taken to excess have been shown to directly activate the brain reward system, which is involved in the reinforcement of behaviors and the production of memories (Volkow et al. 2016). Similarly, behaviors that are practiced to excess have been shown to activate most of the same reward pathways activated by substances (Noori et al. 2016). The pharmacological mechanisms by which each class of substances activates the reward pathways are different, but a common outcome is the production of pleasure (an experience that motivates repetition). This is also reported for behaviors that can be practiced to excess. Although the pleasure produced by most of the substances is more intense and reliable than that produced by behaviors, inter-individual variability is important to recognize. Gambling and gaming, physical exercise, sex, and use of the Internet are all examples of behaviors for which the activation of the reward system has been documented and for which there are reports of a pleasurable effect. It is an open question whether excessive food consumption is more like a substance addiction or a behavioral addiction (Gearhardt et al. in this volume).

Loss of control over substance use or practice is considered to be the core of addiction, which must be differentiated from the surrounding constellation of preexisting risk factors and consequences, whether toxicological, physical or environmental. Once settled all these characteristics coexist making distinguishing them difficult (Figure 11.1). Loss of control is typically expressed through observer or subject reports of excessive use or behavior. This is a major dilemma for addiction modeling and research as excessive use or behavior is not easy to define.

Diagnosis of addictions

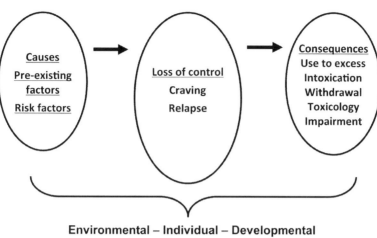

Figure 11.1 A model for addiction diagnosis criteria.

Note: Core criteria of addiction and surrounding constellation of signs and symptoms consequential to addiction need to be better assessed and controlled. The challenge is to tease apart core criteria of addiction from preexisting risk factors or individual causal characteristics as well as the consequences of addiction per se. When assessing individuals with addiction, expression of core and constellation symptoms may be mixed at any given time. Statistically, all these signs and symptoms aggregate together (Auriacombe *et al.*, 2016b, Fatséas *et al.*, 2015b, Serre *et al.*, 2015).

If defined as use over a pre-determined threshold, how do we determine the threshold? And if defined as use with consequences, how do we differentiate loss of control of use or behavior from any use in a toxic range? The dilemma is that any use, excessive or not, has consequences that are independent of the how and the why of use (i.e., independent of addiction). Be that as it may, excessive use or behavior nonetheless has short-, medium- and long-term toxicological consequences for the brain and many body parts, as well as social consequences, both direct and indirect (brain-induced social impairment). If excessive use is common in addiction, it is not enough by itself to characterize addiction, as excessive use may simply reflect a voluntary pattern of use or behavior among individuals without addiction. Most models of addiction, animal and human, have failed to introduce an appropriate distinction between addiction and its consequences. In other words, addiction is not sufficiently defined by use or behavior with consequences. Moreover, the diagnostic criteria that are driven by population-based epidemiological studies cannot make the needed distinction between causes and consequences, because these necessarily aggregate statistically.

Another important characteristic, and potential controversy, surrounding the question of what is addiction, is that it is a fairly stable condition: it persists beyond detoxification and substance/behavior abstinence. This is expressed in the repeated relapses and intense cravings that may occur when individuals with addictions attempt to control excessive use through abstinence. This craving, defined as a repeated unwanted intrusive psychological state that is characterized by an intense and compulsive desire to use a substance or to engage in hedonic behaviors, might be the core of addiction (see the section that begins on page 137 of this chapter). Contrary to withdrawal and tolerance that reverse rapidly, craving persists years after substance or behavior discontinuation and is highly predictive of relapse. From this perspective, addiction may be considered as a chronic disease, and may benefit from long-term approaches to treatment, in line with other areas of health (McLellan et al. 2000).

What are the diagnostic criteria?

The *Diagnostic and Statistical Manual of Mental Disorders* (DSM) and the *International Statistical Classification of Diseases and Related Health Problems* (ICD) have approached mental disorders as largely discrete entities that are characterized by distinctive signs, symptoms, and natural histories. Addiction diagnosis was included in the DSM and the Mental Disorders section of ICD at its implementation. Since the mid-1980s the DSM and ICD have brought considerable diagnostic reliability (Kraemer 2014). Properly assessed using the DSM (which in this respect is in advance of the ICD (Hasin et al. 2013)), addictions have high inter-rater reliabilities (Lobbestael et al. 2011), and are internally coherent and valid statistically (Kraemer et al. 2012; Nelson et al. 1999). However, no reliable biological markers have emerged yet from this approach, to the disappointment of many who consider important the progress of knowledge in neurosciences in general and very specifically in the addiction area (Kwako et al. 2016; Noori et al. 2016; Volkow et al. 2016). Some have considered this a flaw in the DSM and ICD process, which relies too much on epidemiology and statistics in its approach and does not take sufficiently into account biological knowledge in neurosciences, genetics and imaging. In response, in 2009, the Research Domain Criteria (RDoC) initiative was launched by the NIMH. Its objective was to ground mental disorders in biology by going from knowledge of neurocircuitry activity up to meaningful clinical entities (Lilienfeld and Treadway 2016). The DSM/ICD and RDoC approaches have been considered to be opposed to one another. The DSM/ICD is considered too categorical, making too many distinctions among mental disorders; and the RDoC too dimensional in its perspective, leading to confusion between biological continuity and behavioral and emotional expressions. However, to date, the RDoC approach has yet to make good on its promises. Further, categorical and dimensional perspectives do not need to be opposed, indeed both aspects are needed to characterize disorders from the perspective of interventions, whether these are in the form of prevention or treatment (Kraemer 2015), as some degree of dimensionality is needed for outcome monitoring. Recently, Kwako and colleagues have suggested a neuroscience-based framework for diagnosis of addictive disorders (Kwako et al. 2016). Nonetheless, at this time, the DSM approach is the most prevalent state-of-the-art diagnostic system for the addictions.

Current state of the art: a focus on the recent edition of the DSM-5 diagnostic criteria

Diagnosis of addiction is based on a set of criteria established by an international and multidisciplinary team of experts who consider the latest advances in research and clinical knowledge (American Psychiatric Association 2013). The DSM-5 has lumped substance-use disorders and gambling disorder into one new diagnostic category based on the many commonalities they share; and has made suggestions for a possible internet gaming disorder to be studied for inclusion in future revisions (Hasin et al. 2013a; Petry et al. 2014a; Petry et al. 2014b; Denis et al. 2012a). A new eating behavior has been characterized that very much overlaps with addictions, "Binge Eating Disorder", although it has been placed in the Eating Disorders chapter and not, in contrast with Gambling, in the Substance-related and Addictive Disorders chapter. This opens the door for food addiction and other non-substance addictions to be included as disorders in future editions of the DSM.

Features of the DSM diagnostic criteria: core and constellation

The essential feature of the DSM Substance Use Disorder is a cluster of cognitive, behavioral, and physiological symptoms indicating that the individual continues using despite significant substance-related problems. The diagnosis of a Substance Use Disorder can be applied to any reinforcing substance that may be used to excess. For certain substances, some criteria are less salient, and in a few instances may not apply. These set of criteria have also been thought to apply to reinforcing behaviors such as food, sex, gambling and gaming. As noted above, in DSM-5, of these only gambling is treated as an addictive-type disorder, retained from earlier editions with a modified set of criteria.

Overall, the diagnosis of a Substance Use Disorder is based on a pathological pattern of behaviors related to use of the substance. Criteria can be considered to fit within four groupings that correspond to impaired control, social impairment, risky use, and pharmacological adaptation. We argue that the first grouping describes the core of addiction, while the other three groupings are a constellation of pre-existing risk factors and consequences.

Core criteria of addiction

IMPAIRED CONTROL OVER SUBSTANCE USE (CRITERIA 1–4)

The individual may take the substance in larger amounts or over a longer period than was originally intended (Criterion 1). The individual may express a persistent desire to cut down or regulate substance use and may report multiple unsuccessful efforts to decrease or discontinue use (Criterion 2). The individual may spend a great deal of time obtaining the substance, using the substance, or recovering from its effects (Criterion 3). In some instances of more severe Substance Use Disorders, virtually all of the person's daily activities revolve around the substance. Craving (Criterion 4) is manifested by an intense desire or urge for use which may occur at any time.

Pre-existing risk factors and consequences

SOCIAL IMPAIRMENT (CRITERIA 5–7)

Recurrent substance use may result in a failure to fulfill major role obligations at work, school, or home (Criterion 5). The individual may continue substance use despite having persistent or recurrent social or interpersonal problems caused or exacerbated by the effects of the substance (Criterion 6). Important social, occupational, or recreational activities may be given up or reduced because of substance use (Criterion 7). The individual may withdraw from family activities and hobbies in order to use the substance.

RISKY USE OF THE SUBSTANCE (CRITERIA 8–9)

This may take the form of recurrent substance use in situations in which it is physically hazardous (Criterion 8). The individual may continue substance use despite knowledge of having a persistent or recurrent physical or psychological problem that is likely to have been caused or exacerbated by the substance (Criterion 9). The key issue in evaluating this criterion is not the existence of the problem, but rather the individual's failure to abstain from using the substance despite knowledge or evidence of the difficulty it is causing.

PHARMACOLOGICAL ADAPTATION (CRITERIA 10–11)

Tolerance (Criterion 10) is signalled by requiring a markedly increased dose of the substance to achieve the desired effect or a markedly reduced effect when the usual dose is consumed. Tolerance must be distinguished from individual variability in the initial sensitivity to the effects of particular substances. Withdrawal (Criterion 11) is a syndrome that occurs when blood or tissue concentrations of a substance decline in an individual who had maintained prolonged heavy use of the substance. After developing withdrawal symptoms, the individual is likely to consume the substance to relieve the symptoms if they have the opportunity. Withdrawal symptoms vary greatly across the classes of substances. Neither tolerance nor withdrawal is necessary for a diagnosis of a Substance Use Disorder. Furthermore, symptoms of tolerance and withdrawal occurring during appropriate medical treatment with prescribed medications (e.g., opioid analgesics, sedatives, stimulants, etc.) are specifically not counted when diagnosing a Substance Use Disorder. The appearance of normal, expected pharmacological tolerance and withdrawal during the course of medical treatment has been known to lead to an erroneous diagnosis of "addiction" even when these were the only symptoms present. Persons whose only symptoms are those that occur as a result of medical treatment (i.e., tolerance and withdrawal as part of medical care) do not qualify for the diagnosis of a Substance Use Disorder.

Diagnostic criteria change overtime. Example of changes introduced by the DSM-5

Changes in the concept of addiction have led to the evolution of its definition and its diagnostic criteria, as reflected by the successive and revised editions of the DSM since its first publication in 1952. The last edition was published in May 2013, nearly 20 years after the previous edition, the DSM-IV, published in 1994.

A major change of the last edition was the introduction of a dimensional approach: individuals exhibit a more or less severe addiction depending on number of criteria met. This dimensional perspective is a change from the previous purely categorical approach, and the DSM-5 diagnosis of Substance Use Disorder combines the DSM-IV criteria for substance abuse and for substance dependence. Indeed, item response theory (IRT) analysis conducted in many studies, of more than 200,000 subjects in total, revealed the uni-dimensionality of all DSM-IV criteria for abuse and dependence, except for one, namely, legal problems (Hasin *et al.* 2013).

In addition to the introduction of a dimensional perspective, the revision process between DSM-IV and 5 also considered whether some criteria could be dropped. The legal problems criterion was removed, based on its low prevalence and its low discrimination power in IRT analysis. This criterion was also dependent on the legislation, and therefore introduced variability of diagnosis by country.

Gambling disorder, previously integrated as pathological gambling in the DSM-IV section of Impulse-Control Disorders, is now joined to substance use disorders in the DSM-5 diagnostic category of Substance-Related and Addictive Disorders. This evolution reflects the frequent comorbidity between gambling disorder and substance use disorders (Grant and Chamberlain 2015), and their many behavioral and biological similarities (Petry *et al.* 2014a; Bosc *et al.* 2012; Rennert *et al.* 2014). The criterion "illegal acts" was removed for the same reasons that legal problems were removed from substance use disorders; and the diagnostic threshold was reduced from five to four or more criteria to improve classification accuracy. The possibility of using criteria for substance use disorders to assess gambling was shown to be feasible and reliable with the DSM-IV dependence criteria (Denis *et al.* 2012a). Future work should explore if gambling disorder might be assessed using the same criteria as those used for Substance Use Disorders in DSM-5, which include a specific focus on craving.

This is indeed a major revision between DSM-IV and 5, namely, the addition of a new criterion: craving. It should be acknowledged that this criterion was implicated by the dependence criteria in ICD-10, although the word "craving" itself is not used (World Health Organization 1993). Although this new criterion does not seem to provide any additional information statistically, the IRT analysis revealed that it fits well with the other criteria and does not perturb their factor loadings, severity and discrimination. Support for adding craving comes from human research studies (Auriacombe *et al.* 2016b; Hasin *et al.* 2013; Sayette 2016; Serre *et al.* 2015).

The core of addiction: is craving the link between behavior, brain and environment?

Craving is often cited as intrinsically linked to relapse, making it an interesting and useful criterion for research and clinical purposes (Auriacombe *et al.* 2016b; Sayette 2016). However, how to define craving represents a challenge for patients, clinicians and researchers. In the addiction field, standard definitions of craving refer to an irrepressible and intense urge to use a substance or to perform a rewarding behavior. The distinction between craving and urge is then based on intensity. However, craving is often described by individuals as an unwanted experience – an unwanted urge to use. Many definitions of craving do not make explicit this involuntary aspect. More than the intensity of the urge, it is also that it occurs at an inappropriate moment (time/place) that contributes to the associated distress (Auriacombe *et al.* 2016b). This is an ego-dystonic experience, which causes distress and discomfort for those who experience it. A further challenge to defining craving is that when experiencing an unwanted craving, individuals may lack verbal means to adequately describe and communicate their distress. As a result, although clinicians and researchers have been interested in craving for a long time (Childress *et al.* 1986; O'Brien *et al.* 1998), it has not been fully investigated because of the difficulty of pinpointing the experience within clinical and research contexts. A final challenge is to distinguish craving from the acute phenomenon of withdrawal, both in its clinical expression and in the underlying pathophysiological mechanisms. For many years, craving has been listed among symptoms of withdrawal, even though it can occur long after withdrawal symptoms have abated.

The World Health Organization (WHO) emphasized in 1954 that the term "craving" was confusing and had the disadvantage of implying negative connotations in popular English usage (World Health Organization 1955). This is not a problem for other languages in which the term "craving" does not exist. As a consequence, the use of the word "craving" introduces a need for clarification and thus can serve to contribute to a better understanding of what is meant, requiring us to characterize craving as a unique experience of individuals with addiction, thereby facilitating individuals' report of this experience. Unfortunately, WHO suggested avoiding the term "craving." The term "pathological desire" was recommended instead for describing "symbolic craving" as opposed to a form of "physical craving" more related to withdrawal. This contributed to many misunderstandings about craving vis-à-vis withdrawal. On a conceptual level, the term "craving," is sometimes defined as a subjective desire to feel the effects of the substance, and is differentiated from the term "urge," which is characterized as a consequence of craving, and represents the intention to use (Marlatt 1987).

The lack of consensus and clarity on the definition of "craving" led to the development of a multitude of very heterogeneous measurement tools (Rosenberg 2009; Sayette et al. 2000). Many studies use a visual analogue scale, allowing a simple and rapid measurement of the intensity of craving. Other studies use multi-item tools to examine different dimensions of craving (Flannery et al. 1999). Difficulty in defining and measuring craving can be explained by the complexity of the phenomenon and its multidimensional aspect (Shadel et al. 2001). A recent meta-analysis (Noori et al. 2016) has suggested that cue-mediated craving involves mechanisms that are not exclusive for addictive disorders but rather resemble the intersection of information pathways for processing reward, emotional responses, non-declarative memory and obsessive-compulsive behavior. According to the theoretical models chosen, the concept of craving can integrate cognitive, affective, motivational or physiological components. Thus, tools have been developed to better capture some of these aspects.

From a prognostic perspective, craving could be the ideal candidate to predict relapse (Miller et al. 1996; Fatséas et al. 2011; Tiffany and Wray 2012). It is of great importance, both for research and clinical purposes, to discover a measurable criterion to identify risk of relapse. Craving is often reported as a conscious precipitating factor for relapse by individuals with addiction. Although it is generally accepted that craving is a core symptom of addiction, controversy still exists concerning its role in substance use and relapse. Many theoretical models of addiction place craving as the major motivational substrate of substance use and relapse during abstinence attempts (Baker et al. 1986; Ludwig et al. 1974; Marlatt and Gordon 1980; Robinson and Berridge 1993; Wise 1988), but some others suggest that craving is not necessarily involved in substance use (Baker et al. 2004; Tiffany 1990). Two recent systematic literature reviews have tried to distinguish the predictive value of craving in relapse, treatment outcomes, and substance use in general (Serre et al. 2015; Wray et al. 2013). The first review was restricted to tobacco studies and concluded that, although craving was frequently associated with relapse, this association was not systematic (Wray et al. 2013). The association between craving and relapse seemed to be highly dependent on the time of measurement of craving (post-quit craving more predictive than pre-quit craving) and the context in which craving was measured (cue-induced craving in the laboratory is weakly associated with relapse). The second review was restricted to studies evaluating the relationship between craving and substance use in ecological conditions of daily life, through the EMA (Ecological Momentary Assessment) method (Serre et al. 2015). This method uses smartphones to collect real-time data, several times a day, in the natural environment of study participants (Stone and Shiffman 1994). The EMA offers the possibility to study prospective links between events, integrating the influence of environmental factors. This systematic review collected studies concerning all substances, and concluded that craving was associated with substance use and relapse in 92% of studies. This finding was most pronounced when craving occurred shortly (minutes or hours) before substance use. In a recent study conducted in the context of daily life using the EMA method, the role of environmental stimuli in the induction of craving and relapse was examined among patients treated for addiction to a variety of substances (Fatséas et al. 2015b). The results of this study showed that the intensity of craving was a powerful predictor of substance use in the following hours. Furthermore, exposure to factors previously associated with use, and specific to each individual, were potent inducers of craving followed by relapse, within hours of exposure to these person-specific stimuli. These person-specific factors, such as places, contexts, and emotions are very specific to each individual, linked to personal history and provide stronger inducers of craving than more universal substance-specific cues.

Craving is also reported as an important symptom among individuals with gambling disorder, persisting months after gambling abstinence (Ladouceur et al. 2007) and a key determinant of relapse in gambling disorder (Smith *et al.* 2015; Tavares *et al.* 2005). A recent study showed that craving ratings in participants with gambling disorder increased following gambling cues compared with non-gambling cues; that gambling cues in individuals with gambling disorder increased brain responses in reward-related circuitry; and that this response co-varied with craving intensity (Limbrick-Oldfield *et al.* 2017). Animal studies have suggested the addictive liability of sugar (Ahmed *et al.* 2013). Obese subjects with possible food addiction have been shown to report more severe food craving than their non-addicted counterparts (Davis *et al.* 2011; Meule and Kubler 2012; Fatséas *et al.* 2015a). Food craving has been suggested to contribute to unsuccessful attempts to reduce calorie intake, and early dropout from obesity treatment programs (Batra *et al.* 2013). A prospective link between the intensity of food craving and the decrease in dieting success and meeting other criteria for food addiction has been shown (Fatséas *et al.* 2015a; Meule *et al.* 2016).

These results support consideration of craving as a common and important criterion for all addictions. Hence, it is possible to hypothesize a simplified universal model for addiction, with craving as its specific marker (Figure 11.2).

From a treatment perspective, craving appears as a prime target for the treatment of addiction, both in psychotherapy and pharmacotherapy (Auriacombe *et al.* 2016a). Several medications aiming to reduce craving have been developed over the past 30 years (O'Brien 2005). These include naltrexone (O'Malley *et al.* 1992; Volpicelli *et al.* 1992) and acamprosate (Kranzler 2000; Mason 2001) for alcohol addiction, methadone and buprenorphine for opiate addiction (Fatséas *et al.* 2016; Auriacombe *et al.* 2003; Fareed *et al.* 2011) and nicotine patches for tobacco addiction (Shiffman and Ferguson 2008; Auriacombe *et al.* 2003). Many psychotherapeutic interventions target the management of craving. This is often done through identification of cues/triggers so as to avoid them and/or develop strategies to cope with them and thus reduce occurrences of craving. In case craving occurs a plan is anticipated to avoid use and relapse through distraction and/or getting external support or more cognitive-based interventions (Beck *et al.* 1993; Marlatt and Gordon 1985; Witkiewitz *et al.* 2013). Craving can also be useful as an indicator of treatment efficacy, and evolution of craving during treatment could be used by therapists as a marker of the impact of the implemented treatment, whether psychotherapy or pharmacotherapy (Tiffany *et al.* 2012).

Figure 11.2 A simplified universal model for addiction diagnosis, inclusive of substance and behavioral addictions.

Note: Based on human research documenting that the intensity of craving is a powerful predictor of substance use and behavior practices, it is possible to suggest a simplified model with craving as the specific mediator to use. Furthermore, exposure to factors previously associated with use, and specific to each individual, are potent inducers of craving in the hours following exposure to these stimuli. These person-specific factors, such as places, contexts, and emotions, are unique to each individual, linked to personal history, and are stronger inducers of craving than more generic cues (Auriacombe *et al.*, 2016b, Fatséas *et al.*, 2015b, Serre *et al.*, 2015).

Broadening diagnosis to severity assessment and comprehensive treatment planning

Besides measuring the severity of addiction by counting the number of endorsed DSM-5 criteria (Hasin *et al.* 2013), a more comprehensive evaluation of the disorder and its consequences are needed for clinicians and therapists. Several tools have been developed for that purpose over the past 30 years. Among them, the most widely used instrument to assess the severity of addiction in different settings and among different populations is the Addiction Severity Index (ASI) (Cacciola *et al.* 2011; McLellan *et al.* 2006). Introduced in 1980, it has since been translated into many languages (McLellan *et al.* 1980). The ASI aims to assess impairments that commonly occur in individuals with addictions and to help clinicians design better comprehensive and integrated treatments (McLellan *et al.* 2006). Although the initial ASI focused on alcohol and drugs, it was modified (mASI) by adding specific items to systematically gather data on tobacco use, gambling, eating disorders and other putative non-substance addictions (Auriacombe *et al.* 2004). The ASI and mASI produce relevant, reliable and valid data for both clinical and research evaluation (Makela 2004; Denis *et al.* 2016; Denis *et al.* 2012b). For instance, higher ASI scores have been shown to be concordant with substance use disorder diagnoses and gambling disorder; they also reasonably approximate DSM dependence diagnosis (Denis *et al.* 2016; Rikoon *et al.* 2006). The ASI interview is not designed as a self-standing diagnostic tool, but for use with individuals who have been antecedently screened and determined to have an addictive disorder. The standardized properties of the mASI permit a comprehensive and systematic assessment of all addictions independently of individuals' perceived problems and treatment settings, hence facilitating better-personalized treatment planning. The mASI may be helpful for clinicians to design the best treatment plans for a patient; for policy makers to objectively understand the needs of patients in treatment; and for care centers, other institutions, and also researchers to measure progress and outcomes in addiction treatment. For research purposes, the use of a unique non-substance-specific instrument allows researchers to better address the similarities and differences between addictions by avoiding potential confusion due to a multiplication of tools. In addition, a multifactorial assessment tool allows research to control for the impact of co-addictive disorders on treatment progress and outcome of another addictive disorder.

Looking to the future. What to anticipate for DSM6+ and ICD12+?

The most important challenge for the future of addiction diagnosis is arguably to clarify whether craving is or is not a reliable marker of addiction. This would require a clear and agreed definition of craving, e.g. as an unwanted phenomenon, to better determine how it can be distinguished from related phenomena such as urges and desires. There is also a need to better distinguish addiction from co-occurring mental disorders, and the latter from addiction-induced pseudo-mental disorders. This is a significant challenge, as mental disorders and addictive disorders both produce similar symptoms, such as anxiety, depression and thought distortion. In the case of addiction, these are consequences of intoxication, withdrawal and craving, whereas they may also be the direct expression of a mental disorder. In this respect, the current set of criteria would benefit from being organized according to what is a core expression of addiction (loss of control and craving) *versus* what is consequential or pre-existing and/or more of a severity measure.

The lumping together by DSM-5 of some non-substance addictions with the usual substance addictions should be further explored, no doubt cautiously but also with some focus and determination. If this is valid, behavioral addictions should be based on the same set of criteria as those used for substance use disorders. This has already started to occur with respect to gambling and food addiction (Denis *et al.* 2012a; Gearhardt *et al.* 2009), with some success.

For diagnoses that are noted in the DSM Appendix because of lack of evidence at the time of the DSM-5 publication (i.e., caffeine use disorder, gaming use disorder) further studies might provide sufficient evidence for eventually including these diagnoses in the Substance-related and Addictive Disorder chapter.

In addition to further study of craving, the identification of reliable biomarkers is a valuable goal to pursue, notwithstanding the many disappointments and controversies to date. Neuroimaging data have allowed for a better understanding of the dimensions of cue-reactivity, impulsivity, and cognitive control, associated with mediators and moderators of treatment outcomes in addictive disorders. However, biomarkers of treatment response have yet to be identified to date (Garrison and Potenza 2014). The combination of the neuroimaging and the findings of genetic and epigenetic studies might identify both reproducible and predictable biomarkers of addictive disorders (Volkow *et al.* 2015) that eventually could be integrated in future classification of addictive disorder.

References

Ahmed, S. H., Guillem, K. and Vandaele, Y. (2013) "Sugar addiction: pushing the drug–sugar analogy to the limit", *Current Opinion in Clinical Nutrition & Metabolic Care* 16(4): 434–439.

American Psychiatric Association (APA) (2013) *Diagnostic and Statistical Manual of Mental Disorders, Fifth Edition (DSM-5)*, Washington, DC: American Psychiatric Association.

Auriacombe, M., Denis, C., Lavie, E., Fatséas, M., Franques-Rénéric, P., Daulouède, J. P. and Tignol, J. (2004) "Experience with the addiction severity index in France. A descriptive report of training and adaptation to tobacco and non-substance-addictive behaviors", *66th College on Problems of Drug Dependence*, San Juan, Puerto Rico.

Auriacombe, M., Dubernet, J., Sarram, S., Daulouède, J. P. and Fatséas, M. (2016a) "Traitements pharmacologiques dans les addictions: pour une approche transversale et simplifiée" in M. Reynaud, A. Benyamina, L. Karila and H. J. Aubin (eds), *Traité d'addictologie (2e édition)*, Paris: Lavoisier.

Auriacombe, M., Fatséas, M., Franques-Rénéric, P., Daulouède, J. P. and Tignol, J. (2003) "Therapeutiques de substitution dans les addictions", *La Revue du Praticien* 53: 1327–1334.

Auriacombe, M., Serre, F. and Fatséas, M. (2016b) "Le craving: marqueur diagnostic et pronostic des addictions?" in M. Reynaud, A. Benyamina, L. Karila and H. J. Aubin (eds), *Traité d'addictologie (2e édition)*, Paris: Lavoisier.

Baker, T. B., Morse, E. and Sherman, J. E. (1986) "The motivation to use drugs: a psychobiological analysis of urges", *Nebraska Symposium on Motivation* 34: 257–323.

Baker, T. B., Piper, M. E., McCarthy, D. E., Majeskie, M. R. and Fiore, M. C. (2004) "Addiction motivation reformulated: an affective processing model of negative reinforcement", *Psychological Review* 111(1): 33–51.

Batra, P., Das, S. K., Salinardi, T., Robinson, L., Saltzman, E., Scott, T., Pittas, A. G. and Roberts, S. B. (2013) "Relationship of cravings with weight loss and hunger. Results from a 6 month worksite weight loss intervention", *Appetite* 69: 1–7.

Beck, A., Wright, F., Newman, C. and Liese, B. (1993) *Cognitive Therapy of Substance Abuse*, New York, NY: Guilford Press.

Bosc, E., Fatséas, M., Alexandre, J. M. and Auriacombe, M. (2012) "Similitudes et différences entre le jeu pathologique et la dépendance aux substances: qu'en est-il?" *L'Encéphale* 38(5): 433–439.

Cacciola, J. S., Alterman, A. I., Habing, B. and McLellan, A. T. (2011) "Recent status scores for version 6 of the Addiction Severity Index (ASI-6)", *Addiction* 106(9): 1588–1602.

Childress, A. R., McLellan, A. T. and O'Brien, C. P. (1986) "Abstinent opiate abusers exhibit conditioned craving, conditioned withdrawal and reductions in both through extinction", *British Journal of Addiction* 81(5): 655–660.

Davis, C., Curtis, C., Levitan, R. D., Carter, J. C., Kaplan, A. S. and Kennedy, J. L. (2011) "Evidence that 'food addiction' is a valid phenotype of obesity", *Appetite* 57(3): 711–717.

Denis, C., Fatséas, M. and Auriacombe, M. (2012a) "Analyses related to the development of DSM-5 criteria for substance use related disorders: 3. an assessment of Pathological Gambling criteria", *Drug and Alcohol Dependence* 122: 22–27.

Denis, C., Fatséas, M., Beltran, V., Bonnet, C., Picard, S., Combourieu, I., Daulouède, J. P. and Auriacombe, M. (2012b) "Validity of the self-reported drug use section of the Addiction Severity Index and associated factors used under naturalistic conditions", *Substance Use & Misuse* 47(4): 356–63.

Denis, C., Fatséas, M., Beltran, V., Serre, F., Alexandre, J. M., Debrabant, R., Daulouède, J. P. and Auriacombe, M. (2016) "Usefulness and validity of the modified Addiction Severity Index: a focus on alcohol, drugs, tobacco, and gambling", *Substance Abuse* 37(1): 168–175.

Denis, C. M., Gelernter, J., Hart, A. B. and Kranzler, H. R. (2015) "Inter-observer reliability of DSM-5 substance use disorders", *Drug and Alcohol Dependence* 153: 229–235.

Fareed, A., Vayalapalli, S., Stout, S., Casarella, J., Drexler, K. and Bailey, S. P. (2011) "Effect of methadone maintenance treatment on heroin craving, a literature review", *Journal of Addictive Diseases* 30(1): 27–38.

Fatséas, M., Collombat, J., Rosa, M. A. C., Denis, C. M., Alexandre, J.-M., Debrabant, R., Serre, F. and Auriacombe, M. (2015a) "Measuring 'craving' in food addiction: type, frequency, intensity", *Drug and Alcohol Dependence* 156: e66.

Fatséas, M., Daulouède, J.-P. and Auriacombe, M. (2016) "Médicaments de l'addiction aux opiacés: méthadone orale et buprénorphine sublinguale", in M. Reynaud, A. Benyamina, L. Karila and H. J. Aubin (eds), *Traité d'addictologie (2e édition)*, Paris: Lavoisier.

Fatséas, M., Denis, C., Massida, Z., Verger, M., Franques-Rénéric, P. and Auriacombe, M. (2011) "Cue-induced reactivity, cortisol response and substance use outcome in treated heroin dependent individuals", *Biological Psychiatry* 70(8): 720–727.

Fatséas, M., Serre, F., Alexandre, J. M., Debrabant, R., Auriacombe, M. and Swendsen, J. (2015b) "Craving and substance use among patients with alcohol, tobacco, cannabis or heroin addiction: a comparison of substance-specific and person-specific cues", *Addiction* 110: 1035–1042.

Flannery, B. A., Volpicelli, J. R. and Pettinati, H. M. (1999) "Psychometric properties of the Penn Alcohol Craving Scale", *Alcoholism: Clinical and Experimental Research* 23(8): 1289–1295.

Garrison, K. A. and Potenza, M. N. (2014) "Neuroimaging and biomarkers in addiction treatment", *Current Psychiatry Reports* 16(12): 513.

Gearhardt, A. N., Corbin, W. R. and Brownell, K. D. (2009) "Food addiction: an examination of the diagnostic criteria for dependence", *Journal of Addiction Medicine* 3(1): 1–7.

Gearhardt, A. N., Joyner, M. and Schulte, E. (this volume, pp. 182–191) "Food addiction".

Grant, J. E. and Chamberlain, S. R. (2015) "Gambling disorder and its relationship with substance use disorders: implications for nosological revisions and treatment", *The American Journal on Addictions* 24(2): 126–131.

Hasin, D. S., O'Brien, C. P., Auriacombe, M., Borges, G., Bucholz, K., Budney, A., Compton, W. M., Crowley, T., Ling, W., Petry, N. M., Schuckit, M. and Grant, B. F. (2013) "DSM-5 criteria for substance use disorders: recommendations and rationale", *American Journal of Psychiatry* 170(8): 834–851.

Kraemer, H. C. (2014) "The reliability of clinical diagnoses: state of the art", *Annual Review of Clinical Psychology* 10: 111–130.

Kraemer, H. C. (2015) "Research Domain Criteria (RDoC) and the DSM—two methodological approaches to mental health diagnosis", *JAMA Psychiatry* 72: 1163–1164.

Kraemer, H. C., Kupfer, D. J., Clarke, D. E., Narrow, W. E. and Regier, D. A. (2012) "DSM-5: how reliable is reliable enough?" *American Journal of Psychiatry* 169: 13–15.

Kranzler, H. R. (2000) "Pharmacotherapy of alcoholism: gaps in knowledge and opportunities for research", *Alcohol and Alcoholism* 35(6): 537–547.

Kwako, L. E., Momenian, R., Litten, R. Z., Koob, G. F. and Goldman, D. (2016): Addictions neuroclinical assessment: a neuroscience-based framework for addictive disorders", *Biological Psychiatry* 80(3): 179–189.

Ladouceur, R., Sylvain, C. and Gosselin, P. (2007) "Self-exclusion program: a longitudinal evaluation study", *Journal of Gambling Studies* 23(1): 85–94.

Lilienfeld, S. O. and Treadway, M. T. (2016) "Clashing diagnostic approaches: DSM-ICD versus RDoC", *Annual Review of Clinical Psychology* 12: 435–463.

Limbrick-Oldfield, E. H., Mick, I., Cocks, R. E., McGonigle, J., Sharman, S. P., Goldstone, A. P., Stokes, P. R., Waldman, A., Erritzoe, D., Bowden-Jones, H., Nutt, D., Lingford-Hughes, A. and Clark, L. (2017) "Neural substrates of cue reactivity and craving in gambling disorder", *Translational Psychiatry* 7(1): e992.

Lobbestael, J., Leurgans, M. and Arntz, A. (2011) "Inter-rater reliability of the Structured Clinical Interview for DSM-IV Axis I Disorders (SCID I) and Axis II Disorders (SCID II)", *Clinical Psychology & Psychotherapy* 18(1): 75–79.

Ludwig, A. M., Wikler, A. and Stark, L. H. (1974) "The first drink: psychobiological aspects of craving", *Archives of General Psychiatry* 30(4): 539–547.

Makela, K. (2004) "Studies of the reliability and validity of the Addiction Severity Index", *Addiction* 99(4): 398–410; discussion 411–418.

Marlatt, G. A. (1987) "Craving notes (comments on Kozlowski & Wilkinson's paper 'Use and Misuse of the Concept of Craving by Alcohol, Tobacco and Drug Researchers')", *British Journal of Addiction* 82: 42–44.

Marlatt, G. A. and Gordon, J. R. (1980) "Determinants of relapse: implications for the maintenance of behavior change", in P. O. Davidson and S. M. Davidson (eds), *Behavioral Medicine: Changing Health Lifestyles*, New York, NY: Brunner/Mazel.

Marlatt, G. A. and Gordon, J. R. (1985) *Relapse Prevention: Maintenance Strategies in the Treatment of Addictive Behaviors*, New York, NY: Guilford.

Mason, B. J. (2001) "Treatment of alcohol-dependent outpatients with acamprosate: a clinical review", *The Journal of Clinical Psychiatry* 62: Suppl 20, 42–8.

McLellan, A. T., Cacciola, J. C., Alterman, A. I., Rikoon, S. H. and Carise, D. (2006) "The Addiction Severity Index at 25: origins, contributions and transitions", *The American Journal on Addictions* 15(2): 113–24.

McLellan, A. T., Lewis, D. C., O'Brien, C. P. and Kleber, H. D. (2000) "Drug dependence, a chronic medical illness: implications for treatment, insurance, and outcomes evaluation", *JAMA: The Journal of the American Medical Association* 284(13): 1689–1695.

McLellan, A. T., Luborsky, L., Woody, G. E. and O'Brien, C. P. (1980) "An improved diagnostic evaluation instrument for substance abuse patients, the Addiction Severity Index", *Journal of Nervous and Mental Disease* 168(1): 26–33.

Meule, A. and Kubler, A. (2012) "Food cravings in food addiction: the distinct role of positive reinforcement", *Eating Behaviors* 13(3): 252–255.

Meule, A., Richard, A. and Platte, P. (2016) "Food cravings prospectively predict decreases in perceived self-regulatory success in dieting", *Eating Behaviors* 24: 34–38.

Miller, W. R., Westerberg, V. S., Harris, R. J. and Tonigan, J. S. (1996) "What predicts relapse? Prospective testing of antecedent models", *Addiction* 91: S155–172.

Nelson, C. B., Rehm, J., Ustun, T. B., Grant, B. and Chatterji, S. (1999) "Factor structures for DSM-IV substance disorder criteria endorsed by alcohol, cannabis, cocaine and opiate users: results from the WHO reliability and validity study", *Addiction* 94(6): 843–855.

Noori, H. R., Cosa Linan, A. and Spanagel, R. (2016) "Largely overlapping neuronal substrates of reactivity to drug, gambling, food and sexual cues: a comprehensive meta-analysis", *European Neuropsychopharmacology* 26(9): 1419–1430.

O'Brien, C. P. (2005) "Anticraving medications for relapse prevention: a possible new class of psychoactive medications", *American Journal of Psychiatry* 162(8): 1423–1431.

O'Brien, C. P., Childress, A. R., Ehrman, R. and Robbins, S. J. (1998) "Conditioning factors in drug abuse: can they explain compulsion?", *Journal of Psychopharmacology* 12(1): 15–22.

O'Malley, S. S., Jaffe, A. J., Chang, G., Schottenfeld, R. S., Meyer, R. E. and Rounsaville, B. (1992) "Naltrexone and coping skills therapy for alcohol dependence. A controlled study", *Archives of General Psychiatry* 49(11): 881–877.

Petry, N., Blanco, C., Auriacombe, M., Borges, G., Bucholz, K., Crowley, T., Grant, B., Hasin, D. and O'Brien, C. (2014a) "An overview of and rationale for changes proposed for pathological gambling in DSM-5", *Journal of Gambling Studies* 30: 493–502.

Petry, N. M., Rehbein, F., Gentile, D. A., Lemmens, J. S., Rumpf, H. J., Mossle, T., Bischof, G., Tao, R., Fung, D. S., Borges, G., Auriacombe, M., Gonzalez Ibanez, A., Tam, P. and O'Brien, C. P. (2014b) "An international consensus for assessing internet gaming disorder using the new DSM-5 approach", *Addiction* 109(9): 1399–1406.

Rennert, L., Denis, C., Peer, K., Lynch, K. G., Gelernter, J. and Kranzler, H. R. (2014) "DSM-5 gambling disorder: prevalence and characteristics in a substance use disorder sample", *Experimental and Clinical Psychopharmacology* 22: 50–56.

Rikoon, S. H., Cacciola, J. S., Carise, D., Alterman, A. I. and McLellan, A. T. (2006) "Predicting DSM-IV dependence diagnoses from Addiction Severity Index composite scores", *Journal of Substance Abuse Treatment* 31(1): 17–24.

Robinson, T. E. and Berridge, K. C. (1993) "The neural basis of drug craving: an incentive-sensitization theory of addiction", *Brain Research. Brain Research Reviews* 18(3): 247–291.

Rosenberg, H. (2009) "Clinical and laboratory assessment of the subjective experience of drug craving", *Clinical Psychology Review* 29(6): 519–534.

Sayette, M. A. (2016) "The role of craving in substance use disorders: theoretical and methodological issues", *Annual Review of Clinical Psychology* 12: 407–433.

Sayette, M. A., Shiffman, S., Tiffany, S. T., Niaura, R. S., Martin, C. S. and Shadel, W. G. (2000) "The measurement of drug craving", *Addiction* 95(2): S189–210.

Serre, F., Fatséas, M., Swendsen, J. and Auriacombe, M. (2015) "Ecological momentary assessment in the investigation of craving and substance use in daily life: a systematic review", *Drug and Alcohol Dependence* 148C: 1–20.

Shadel, W. G., Niaura, R., Brown, R. A., Huchison, K. E. and Abrams, D. B. (2001) "A content analysis of smoking craving", *Journal of Clinical Psychology* 57(1): 145–150.

Shiffman, S. and Ferguson, S. G. (2008) "The effect of a nicotine patch on cigarette craving over the course of the day: results from two randomized clinical trials", *Current Medical Research and Opinion* 24(2): 2795–2804.

Smith, D., Battersby, M., Pols, R., Harvey, P., Oakes, J. and Baigent, M. (2015) "Predictors of relapse in problem gambling: a prospective cohort study", *Journal of Gambling Studies* 31(1): 299–313.

Stone, A. and Shiffman, S. (1994) "Ecological momentary assessment in behavioral medicine", *Annals of Behavioral Medicine* 16: 199–202.

Tavares, H., Zilberman, M., Hodgins, D. and El-Guebaly, N. (2005) "Comparison of craving between pathological gamblers and alcoholics", *Alcoholism: Clinical and Experimental Research* 29: 1427–1431.

Tiffany, S. T. (1990) "A cognitive model of drug urges and drug-use behavior: role of automatic and non-automatic processes", *Psychological Review* 97(2): 147–168.

Tiffany, S. T., Friedman, L., Greenfield, S. F., Hasin, D. S. and Jackson, R. (2012) "Beyond drug use: a systematic consideration of other outcomes in evaluations of treatments for substance use disorders", *Addiction* 107(4): 709–718.

Tiffany, S. T. and Wray, J. M. (2012) "The clinical significance of drug craving", *Annals of the New York Academy of Sciences* 1248: 1–17.

Volkow, N. D., Koob, G. F. and Baler, R. (2015) "Biomarkers in substance use disorders", *ACS Chemical Neuroscience* 6(4): 522–525.

Volkow, N. D., Koob, G. F. and McLellan, A. T. (2016) "Neurobiologic advances from the brain disease model of addiction", *The New England Journal of Medicine* 374(4): 363–71.

Volpicelli, J. R., Alterman, A. I., Hayashida, M. and O'Brien, C. P. (1992) "Naltrexone in the treatment of alcohol dependence", *Archives of General Psychiatry* 49(11): 876–880.

Wise, R. A. (1988) "The neurobiology of craving: implications for the understanding and treatment of addiction", *Journal of Abnormal Psychology* 97(2): 118–132.

Witkiewitz, K., Bowen, S., Douglas, H. and Hsu, S. H. (2013) "Mindfulness-based relapse prevention for substance craving", *Addictive Behaviors* 38(2): 1563–1571.

World Health Organization (WHO) (1955) "The craving for alcohol; a symposium by members of the WHO expert committee on mental health and on alcohol", *Quarterly Journal of Studies on Alcohol* 16(1): 34–66.

World Health Organization (1993) *International Classification of Diseases, 10th edition*, Geneva: CH, WHO.

Wray, J. M., Gass, J. C. and Tiffany, S. T. (2013) "A systematic review of the relationships between craving and smoking cessation", *Nicotine & Tobacco Research* 15(7): 1167–1182.

12
RECONSIDERING ADDICTION AS A SYNDROME
One disorder with multiple expressions

Paige M. Shaffer and Howard J. Shaffer

> The important thing in science is not so much to obtain new facts as to discover new ways of thinking about them.
>
> *Sir William Bragg (1862–1942)*

Introduction

In this chapter we will discuss addiction as a construct, its history, and describe scientific evidence that influenced the development of the Addiction Syndrome Model. After introducing the model, we will conclude by considering some of the implications of this etiologic model for the field of addiction studies and treatment.

Addiction is a chronic relapsing disorder of brain reward, memory and related neurocircuitry, and learning. In the United States, addiction impacts nearly 1 in 10 people over the age of 12. Among brain disorders, addiction incurs more economic costs than Alzheimer's disease, stroke, Parkinson's disease, or head and neck injury. Addiction is the most costly neuropsychiatric disorder (Uhl and Grow 2004). According to the National Institute on Drug Abuse, the overall annual cost of addiction exceeds half a trillion dollars, including health- and crime-related expenses, as well as losses in productivity. Nevertheless, the consequences of addiction extend far beyond economic cost. Addiction damages individuals and their families, friends, and communities. Substance misuse and excessive behavior patterns put people at risk for developing adverse health and other social problems, such as HIV, hepatitis, overdose and death, unplanned pregnancy, family disintegration, domestic violence, criminal behavior, financial instability, housing instability, and child abuse.

Addiction can develop after exposure and interaction with a drug or activity when the relationship between the person and that drug or activity yields a desirable shift in subjective experience. Due to a set of psychosocial and neurobiological factors, some people are at greater risk for developing addiction. Currently, in the DSM-5, there are a variety of disorders related to substances and behaviors (e.g., Alcohol, Cannabis, Hallucinogens, Inhalants, Opioids, Sedatives/Hypnotics/Anxiolytics, Tobacco, Caffeine, and Gambling) (American Psychiatric Association 2013). Many people with an addiction-related diagnosis often have more than one psychiatric diagnosis (Kessler *et al.* 2005). Recent scientific research is encouraging a reconsideration of addiction (Miller and Carroll 2006). Specifically, research suggests that addiction is not a collection of distinct disorders (i.e., "addictions"), but rather varied opportunistic expressions that fall under one umbrella – addiction.

Addiction and the emergence of the addiction syndrome model: the relationship between addiction and objects of addiction

Conventional wisdom evidences the tendency to view addiction as if it could exist only if psychoactive drug use was involved. However, addiction is not the product of a substance or activity, though each of these things has the capacity to influence human experience. To illustrate, if psychoactive drug using was a necessary and sufficient cause for the development of addiction, then addiction would occur every time this pattern of drug using was present, and addictive behavior would be absent every time drug using was absent. However, due to neuroadaptation (i.e., tolerance and withdrawal) adverse consequences are often present when drug using is absent. For example, upon stopping, disordered gamblers who do not use alcohol or other psychoactive drugs can experience physical symptoms that appear to be very similar to either opioid, stimulant, or poly-substance withdrawal (Wray and Dickerson 1981; Potenza 2001). In addition, people often exceed their drug dependence threshold without experiencing addiction (e.g., post-surgical opioid use). Therefore, drug using is neither a necessary nor a sufficient cause for the development of addiction. This observation provides insight into the necessity of considering a more complex relationship for individuals who might develop addiction.

When a particular pattern of behavior or drug use reliably and robustly change emotional experience in a desirable way, the potential for addiction emerges. It is the *relationship* of a person with the object of their excessive behavior – not just the attributes of the object – that defines addiction. Consequently, the causes of addiction are multi-factorial (Zinberg 1984). It is the confluence of psychological, social and biological forces that determines addiction.

Reconsidering addiction: the emergence of the addiction syndrome model

Reconsidering the concept of addiction has profound implications for science, health care, public policy and public health, and promulgating guidelines for healthcare reimbursement. Without a clear understanding of addiction, researchers will find it very difficult to reach consensus regarding "addiction" prevalence rates, etiology, and the necessary treatment course that stimulates recovery (Shaffer 1997). Also, without a precise understanding of addiction, clinicians will encounter diagnostic and treatment matching difficulties, and satisfactory treatment outcome measures will remain lacking.

Scientific evidence supports the notion that addiction is a complex syndrome rather than various unique "addictions." This paradigm shift holds the potential to influence how we understand specific expressions of addiction. Before presenting the addiction syndrome, it is very useful to consider how we think about AIDS as a historical analogue to addiction.

On syndromes

A syndrome is a cluster of signs and symptoms related to an abnormal condition; not all signs or symptoms are present at the same time. Manifestations of a syndrome have some unique and some shared signs and symptoms. Unique components are associated with a syndrome and a specific expression of that syndrome; shared components overlap among expressions of the syndrome and with other disorders. Shared components often account for observations of comorbidity between apparently different disorders. Finally, making diagnosis very difficult, not all symptoms, signs, or disorders that comprise a syndrome are present in every expression of the syndrome.

The history of our understanding of Acquired Immune Deficiency Syndrome (AIDS) provides a useful historical analogy for understanding addiction as a syndrome. During the early 1980s, health researchers noticed a surge in a variety of what had previously been rare diseases. Clinicians viewed these diseases as distinct. Gradually, however, scientists were able to link these cases to a common immuno-suppression factor: Human Immunodeficiency Virus (HIV). We cannot overstate the importance of understanding opportunistic diseases and their relationship with a syndrome. With AIDS for example, scientists and clinicians recognize that, although each opportunistic disease is important to consider and treat, identification of common etiology is critical.

Similar to the early days of AIDS research and treatment, addiction researchers and clinicians seem to be working in silos, focusing on seemingly independent expressions of addiction (i.e., opportunistic infections). The conventional wisdom among addiction researchers and clinicians has been that people develop distinct addictions and that commonalities between them are likely spurious or a consequence of the addiction experience. These researchers and practitioners might be so focused on the individual "trees" (e.g., alcohol use disorder, gambling use disorder, cocaine use disorder) that they fail to see the larger "forest."

Identifying an addiction syndrome requires recognizing the presence of pre-morbid, comorbid, and post-morbid commonalities: phenomena shared by seemingly disparate disorders, across both *chemical and behavioral* expressions of addiction. Focusing on the unique aspects of addiction expressions tends to limit our view of the common influences shared by different expressions.

Evidence for an addiction syndrome: shared neurobiological and psychosocial antecedents

Neurobiological system non-specificity: common pathways of addiction

Scientists have identified reward neurocircuitry as the area of the brain related to the development and maintenance of addictive behaviors. Interactions with psychoactive substances and activities (e.g., sex) alike have the capacity to stimulate neurobiological reward systems in general and the brain's dopamine reward system in particular (Wise 1995; Lawrence *et al.* 2009). Functional neuroimaging studies that display images of the brain responding to a task reveal that anticipation of cocaine, money, and beauty similarly energizes the reward system (Breiter *et al.* 2001; e.g., Aharon *et al.* 2001). The observation of disparate objects stimulating similar neurobiological pathways indicates that, regardless of the object of addiction, the brain is the final common pathway for addictive behaviors.

Blum (2000) implicated a "reward deficiency syndrome" as a vulnerability to addiction. People with such dysfunction fail to experience sufficient pleasure; they adapt to this circumstance by engaging in behaviors that stimulate the faulty reward circuitry. Further,

> epidemiological, genetic, and neurobiological evidence support the notion that vulnerability to addiction (as well as impulsive and compulsive behaviors) is genetically transmitted. It is not necessary to establish that all addiction is caused by genetic vulnerability. Heavy exposure to alcohol and other drugs may set in motion brain circuitry which may have similar end results.
>
> *Blum* et al. 2000, p. 2

Clinicians and researchers similarly have implicated the dopamine system as a powerful influence on the development and maintenance of addiction. In addition to reinforcing effects, dopamine provides a sentinel warning system for unexpected or adverse experiences (e.g., punishment) (Hyman et al. 2006). Dopamine does not function in isolation. There is a complex neurobiological context that influences the effects of dopamine. In addition to the variety of brain receptor systems, the assortment of psychoactive substances also can affect these other transmitter systems (e.g., norepinephrine, GABA, glutamate, serotonin) that, in turn, interact with the dopamine system. Ingested psychoactive substances—and human activities in general—exert their influence on subjective experience through the brain's many endogenous neurotransmitter systems. For example, opioids function through the endogenous opioid system and alcohol through the GABA system. These neurotransmitter systems, as well as others, influence the dopamine system. In addition, although neurobiological reward activity represents the most well-known system that supports the addiction syndrome, other systems deserve consideration. Other parts of the brain also play a role in the development and maintenance of addiction, for example, learning and memory in the hippocampus and emotional regulation in the amygdala. Overall, neurobiological findings provide evidence that disparate objects and behaviors can stimulate similar neurobiological pathways (e.g., Potenza et al. 2011).

Genetic overlap

There is evidence suggesting substantial genetic and environmental non-specificity across different expressions of addiction. For example, genetic studies reveal common molecular mechanisms for drug addiction and compulsive running (Werme et al. 2002; Werme et al. 2000). Similarly, pathological gambling shares a common genetic vulnerability with alcohol dependence (Slutske et al. 2000). A study of male twins showed that shared genetic and environmental risk factors for substance abuse are largely substance non-specific (Kendler et al. 2003a). Kendler et al. note, "We could not find evidence for genetic factors that increase risk for individuals to abuse substance A and not also to abuse substances B, C, and D" (p. 692).

Other evidence also supports non-specific genetic vulnerability to addictive behavior. For example, Merikangas et al. (1998b) found that similar direct (e.g., exposure to drugs) and indirect (e.g., resultant family discord) factors augment genetic risk for both drug and alcohol abuse. In their study of female twins, Karkowski et al. (2000) found (1) genetic and environmental factors significantly influenced substance use in general, and (2) no evidence of a heritability or familial environmental effect for specific substances. Similar results were found in a study of Vietnam-era drug users: with the exception of heroin—which exhibited unique substance-specific genetic risk—investigators observed a common vulnerability to multi-class drug use among study participants (Tsuang et al. 1998). Finally, Bierut et al. (1998) observed that, "Although studies support the familial transmission of alcohol and substance dependence, individuals are frequently dependent on multiple substances, raising the possibility of a general addictive tendency" (p. 987). These findings provide evidence that genetics account for a general and increased risk for addiction.

Shared psychosocial antecedents

Among the participants in the National Comorbidity Survey Replication Study with a past-year diagnosis, 55% carry only a single diagnosis, 22% evidence comorbid mental disorders, and 23% report experiences that reflect three or more mental disorders (Kessler et al. 2005). The prevalence of psychopathology is increased among individuals who are dependent on multiple psychoactive substances (e.g., heroin, alcohol, cocaine; Conway et al. 2003); this is perhaps another indication of a shared vulnerability. Many substance abuse treatment seekers have increased rates of anxiety and depressive disorders (Silk and Shaffer 1996; Lapham et al. 2006; Shaffer et al. 2007). Likewise, populations with psychopathology (e.g., major depression, generalized anxiety disorder, post-traumatic stress disorder, etc.) often exhibit increased prevalence of drug use disorders (Merikangas et al. 1998a; Regier et al. 1990). New multinational research suggests that earlier life disorders are associated with the onset of similar kinds of disorders at some point later in life. Similarly, childhood adversities (e.g., parental death, divorce, substance abuse, mental illness, sexual abuse, etc.) are associated with substance use and other mental disorders during adulthood, though this association is general rather than disorder specific (Kessler et al. 2010). Finally, several studies show that comorbid psychiatric conditions typically precede alcohol abuse, cocaine use, and gambling problems (Kessler et al. 1996; Nelson et al. 1998; Kessler et al. 2008; Shaffer and Eber 2002).

Sub-clinical and social risk factors (e.g., impulsivity, poor parental supervision, delinquency) also are common across expressions of addiction (Brenner and Collins 1998; Welte et al. 2004; Whalen et al. 2001). Research shows that individuals who engage in one problem behavior also are likely to engage in others (Caetano et al. 2001; Shaffer and Hall 2002; Vitaro et al. 2001). Various socio-demographic risk factors (e.g., poverty, geography, family, peers) can influence the onset and course of drug use and other behavioral activities that increase the risk of addiction (Christiansen et al. 2002; Evans and Kantrowitz 2002; Gambino et al. 1993; Lopes 1987; Robins 1993; Shaffer et al. 1999; Wechsler et al. 1997).

Consequences of addiction: unique and shared manifestations

Unique experiences and sequelae

Addiction can manifest in many different ways, though its premorbid characteristics and some sequelae are dependent upon the object with which people interact. To illustrate, if one repeatedly interacts with alcohol or gambling and the addiction syndrome emerges, then the manifestation of this syndrome and its sequelae will have some characteristics that uniquely reflect each of these objects. For the individual using alcohol it might be liver cirrhosis, for the gambler, debt, for the smoker, pulmonary carcinoma, for an intravenous drug user, sepsis.

Shared experiences and sequelae

Despite different expressions of addiction having unique manifestations and sequelae, each expression also has shared manifestations and sequelae. Zinberg suggested that, "the experience of addiction diminishes personality differences and makes all compulsive users seem very much alike" (1984, p. 111). Various and distinct expressions of addiction also stimulate similar biopsychosocial sequelae. People who engage in substance abuse, pathological gambling or excessive shopping commonly have recognizable biopsychosocial sequelae (Shaffer and Hall 2002; Zinberg 1984; Christenson et al. 1994). Chemical and behavioral expressions of addiction also evidence similar neurobiological consequences, including the emergence of neuroadaptation (i.e., tolerance and withdrawal). Neurocognitive manifestations between behavioral and chemical expressions of addiction also are similar. For example, alcohol dependent and disordered gambling individuals share increased impulsivity and risky decision-making compared with matched controls; both of these deficits are associated with ventral prefrontal cortical dysfunction.

Parallel natural histories

There is a natural history to the course of addiction that begins with risk factors and always includes exposure to potential objects of addiction (Slutske *et al.* 2003; Vaillant 1983). Once addictive behavior patterns emerge, there is a similar natural history across various expressions. For example, Hunt *et al.* (1971) presented a seminal research synthesis demonstrating remarkably similar relapse patterns for heroin, smoking, and alcohol. The observation that drugs with important biochemical differences follow the same course suggests that the object of addiction is less relevant to the course of addiction than previously thought. These patterns reflect the dynamics of a common underlying addiction process and challenge the conventional understanding of addiction (Marlatt and Gordon 1985; Marlatt *et al.* 1988; Shaffer 1999).

There is emerging scientific evidence suggesting that, like substance use disorders, gambling disorders are dynamic: over time, individuals with these disorders might experience worsening symptoms but also might recover. People who recover do so through self-directed change and/or treatment. Many individuals who remit from addiction accomplish this on their own and without formal treatment (Schachter 1982; Shaffer and Jones 1989; Slutske 2006; Sobell *et al.* 1996; Klingemann and Sobell 2007). However, though the majority of people stop addiction on their own, it is likely that the success of this change strategy is similar to that of treatment seekers (Cohen *et al.* 1989) and the rate of success varies with the severity of the disorder.

Object non-specificity

Research suggests that addiction is not necessarily inextricably linked to a particular substance or behavior. For example, circumstantial opportunity plays a more influential role in the development of addictive behavior than individuals' preferences for certain drugs (Harford 1978). It is very common for recovering people to "hop" from one expression of addiction (e.g., opioids) to another (e.g., cocaine, alcohol, gambling, etc.) before successfully recovering from addiction. Hser *et al.* (1990) examined longitudinal patterns of alcohol and opioid use and observed a decrease in alcohol consumption at the time that opioid addiction began; likewise, during periods of decreased opioid use, alcohol consumption increased. Similar patterns have been demonstrated for illicit drugs and nicotine (Conner *et al.* 1999), alcohol abuse and bulimia (Cepik *et al.* 1995), and substance abuse and pathological gambling (Blume 1994). Finally, clinical research has shown that during early treatment for opioid dependence, as both opioid and cocaine use decreased, sedative use increased (Shaffer and LaSalvia 1992).

Concurrent manifestations of addiction

The prevalence of poly-substance abuse is well documented (Kessler *et al.* 2005; Kessler *et al.* 1997; Grant *et al.* 2004); the co-occurrence of chemical and behavioral expressions of addiction also is common (e.g., Kessler *et al.* 2008; Baker 2000; Lejoyeux *et al.* 1996; Feigelman *et al.* 1998; Christenson *et al.* 1994). For example, individuals who are dependent on psychoactive substances are more likely to be disordered gamblers (Lesieur and Heineman 1988; Feigelman *et al.* 1998; Black and Moyer 1998). Research demonstrating the frequent co-occurrence of different expressions of addiction signals the potential presence of an underlying root cause of addiction.

In addition to the shared psychosocial antecedents and co-occurring manifestations of addiction, there is evidence suggesting that expressions of addiction are associated with identifiable subclinical "shadow" syndromes that can influence the experience, manifestation and treatment outcomes associated with addiction (Boudreau et al. 2009). Shadow syndromes are subclinical clusters of signs and symptoms. Everyone has characteristic shadows; these are the features of human personality and character (Ratey and Johnson 1997). However, shadows typically go unnoticed or undetected by observers because they think of these shadows as a person's personality style (Shapiro 1965). To illustrate, in the case of shadow syndromes related to gambling disorders, research indicates that individuals with a diagnosis of pathological gambling experience characteristic reliable and identifiable subclinical patterns of symptom clusters associated with psychiatric disorders other than the diagnostic criteria for a gambling disorder (e.g., dysthymia, anxiety, and specific phobias; Boudreau et al. 2009). Although it remains to be demonstrated, it is likely that every expression of addiction has shared and unique characteristic shadows that influence how each expression manifests and how it is sustained.

Treatment non-specificity

Pharmacological treatment non-specificity (i.e., drug-specific treatment reducing the immoderate use of another drug or activity) also provides support for a syndrome model of addiction. For example, naltrexone, an opioid antagonist used for the treatment of opioid use disorders, has shown efficacy for the treatment of pathological gambling (Kim et al. 2001). In a double blind, placebo controlled study, varenicline, a partial nicotine agonist used for the treatment of nicotine dependence, has shown efficacy for alcohol use disorders (McKee et al. 2009). Methadone (i.e., opioid agonist) treatment has shown efficacy in reducing cocaine abuse among opioid dependent patients (Shaffer and LaSalvia 1992). Other research shows that topiramate (i.e., treatment for seizure disorders that acts on the brain's dopamine pathways) has efficacy in treating alcohol use disorders (Johnson et al. 2003). Similarly, researchers speculate that bupropion, an anti-depressant used in smoking cessation protocols to treat nicotine dependence, might be efficacious because of its dopaminergic and noradrenergic activities: "with the dopaminergic activity affecting areas of the brain having to do with the reinforcement properties of addictive drugs and the noradrenergic activity affecting nicotine withdrawal," rather than its anti-depressant properties (Hurt et al. 1997, p. 1201). Additional spillover treatment effects to other dopamine-mediated addictions would provide support for a syndromal theory of addiction. Finally, several non-pharmacological treatments (e.g., cognitive behavior therapy psychodynamic therapy, and behavior therapy) are efficacious in treating expressions of addiction (e.g., Potenza et al. 2011).

Modeling an addiction syndrome

The extant evidence suggests that (1) many commonalities occur across different expressions of addiction, and (2) these commonalities reflect shared etiology: a syndrome. As we mentioned earlier, a syndrome is a cluster of symptoms and signs related to an abnormal condition; not all symptoms or signs must be present in every instance of the syndrome and some expressions of a syndrome have unique signs and symptoms.

As Figure 12.1 shows, antecedents of the addiction syndrome include personal vulnerability levels, object exposure, and object interaction. Throughout the course of human development, people encounter and accumulate specific combinations of neurobiological and psychosocial elements that can influence their behavior. Some elements increase the likelihood of addiction; other factors are protective and reduce the chance of addiction (e.g., social support networks or dimensions of religiosity, Kendler et al. 2003b; Vander Bilt et al. 2004). Similarly, during their lifetime, individuals are exposed to and have access to different objects of addiction. Exposure and access to an object of addiction increases an individual's likelihood

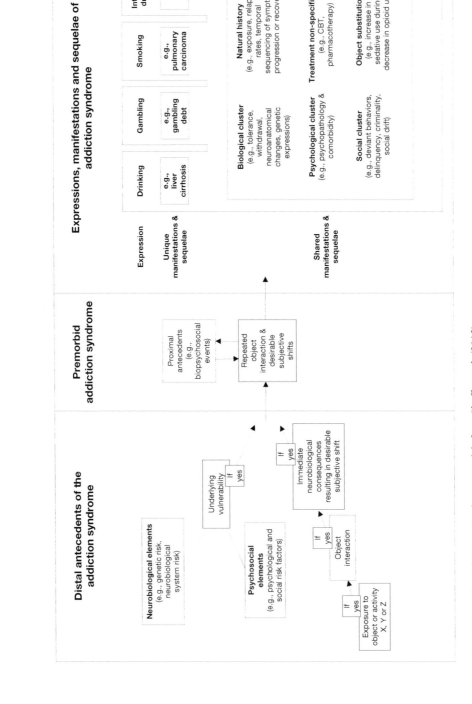

Figure 12.1 The Addiction Syndrome Model, from Shaffer *et al.* (2012).

of interacting with that object. Interacting with an object of addiction can expose at-risk individuals to neurobiological consequences that are both common to all objects of addiction (e.g., activation of reward circuitry) and unique to specific objects of addiction (e.g., type of psychoactivity).

When (1) individuals engage in repeated interactions with a specific object(s) of addiction, and (2) the neurobiological or social consequences of these interactions produce a desirable subjective shift that is reliable and robust, the premorbid stage of the addiction syndrome emerges. During this stage of the syndrome, people teeter on a delicate balance that can shift them toward either more or less healthy behavior. Although distal antecedents of addiction (see Figure 12.1) are well documented, the proximal antecedents that influence the likelihood of further syndrome development remain poorly identified. Despite the dearth of research in this area, we suggest that these proximal influences are likely to be biopsychosocial factors similar to those associated with distal influences. For example, despite being distal, genetic dispositions directly and indirectly influence current patterns of behavior; longstanding dispositions affect psychological characteristics that impact how people manage current social situations. Furthermore, distal antecedents have contemporary consequences that serve as more proximal risk factors for developing or maintaining addiction.

The addiction syndrome can manifest in many different ways; its premorbid characteristics and some sequelae are dependent upon the object with which people interact. In addition, different expressions of the addiction syndrome will share common manifestations (e.g., shifts in schedule, changing personal priorities) and sequelae (e.g., depression, neuroadaptation, deception, personal debt). Researchers and clinicians can identify the presence of the addiction syndrome when premorbid characteristics are accompanied by at least one of the shared manifestations and sequelae summarized in Figure 12.1. As we noted before, the addiction syndrome is recursive and its sequelae can generate an entirely new vulnerability profile (e.g., provoke reward system malfunction in a previously normal system). People with the addiction syndrome are at increased risk for continuing addictive behavior by developing new expressions of addiction. This chain of events is evident in many ways, but most specifically in the parallel natural histories of different manifestations of addiction, including relapse patterns, addiction hopping, treatment non-specificity, and addiction comorbidity.

The Addiction Syndrome Model: advancing the field

The Addiction Syndrome Model is broad in scope, etiological in perspective, and prescriptive in its treatment influence. Previous theorists considering addiction have limited their analysis almost exclusively to substance using behavior patterns that evidence adverse consequences. These views typically include the emergence of neuroadaptation as central to the development of addiction. However, tolerance and withdrawal are no longer central theoretical tenets— necessary and sufficient—to explain addiction (Hyman *et al.* 2006). Tolerance emerges when it becomes necessary to use an increasing dose of a substance or activity to achieve the same subjective effect as previously produced by a lower dose; withdrawal is the stereotypical set of signs and symptoms associated with stopping the use of a drug. Gradually, epidemiology, neurobiology, and other fields of inquiry relevant to addiction advanced our understanding of this phenomenon revealing that some gamblers, shoppers and others—who did not ingest psychoactive substances—evidenced neuroadaptation (i.e., tolerance and withdrawal) (Linden 2011). This observation stimulated a paradigm shift that culminated in a new diagnostic system for addiction.

For the first time, the DSM-5 groups substance use disorders and behavioral addiction disorders (i.e., gambling) within the same diagnostic class (American Psychiatric Association 2013). By including both chemical and behavioral expressions of addiction within the same diagnostic category, the American Psychiatric Association implies that these disorders are connected, thereby supporting an underlying phenomenon.

Unlike other approaches that focus almost exclusively on the objects of addiction—in addition to understanding distal influences—the syndrome model of addiction requires methods that can identify both the shared and unique proximal antecedent elements (e.g., trauma, depression, anxiety, financial stress, etc.) of addiction at the premorbid stage. Such a perspective offers the opportunity to develop a more precise understanding about the precursors of addiction expressions, yielding a diagnostic gold standard. With an improved understanding of both proximal and distal influences and a diagnostic gold standard that is not dependent upon the assessment of sequelae, clinicians and public health workers will be able to advance primary and secondary prevention programs. As Figure 12.1 summarizes, the syndrome model encourages a public health approach to addiction. Primary prevention efforts should target the left pane of the model; secondary prevention should target the middle pane of the model, which depicts the pre-morbid stage of addiction; and, finally, tertiary prevention should focus on the right side pane of the model (i.e., expressions of addiction).

At the individual level of analysis, about 80–90% of people entering recovery from addiction will relapse during the first year after treatment (Marlatt and Gordon 1985). This circumstance might be due in part to object-specific treatment approaches despite the evidence showing that objects of addiction cannot sufficiently account for the central underpinnings of addiction. From the syndromal perspective, the most effective addiction treatments are multimodal "cocktail" approaches (Marlatt 1988) that include both object-specific and addiction-general treatments. Addiction can exacerbate personal vulnerabilities and neurobiological changes. This circumstance might, in part, explain high relapse rates and new manifestations of the syndrome that often appear during the course of addiction and recovery. The syndrome model of addiction encourages clinicians to recognize that patients develop new risk factors during treatment that can interfere with recovery efforts. This model requires clinicians to develop multidimensional treatment plans that account for the many relationships among the multiple influences and consequences of addiction. The syndrome model also encourages providers to assess repeatedly the impact of these relationships on relapse, addiction "hopping," the course of the illness, and many other treatment-related outcomes.

Conclusions

Although distinct expressions of addiction have unique elements, they also share many neurobiological and psychosocial antecedents and consequents. Coupled with repeated premorbid shifts toward a desirable subjective state, neurobiological and psychosocial characteristics both define and result from the addiction syndrome. These findings impose important treatment challenges. The recursive nature of sequelae can exacerbate the difficulties associated with addiction treatment. For example, when clinicians pay insufficient attention to the etiological causes of addiction, patients can experience a cycle of remitting and exacerbating expressions of addiction. Further, the emerging behavior often serves as a risk factor for other expressions of addiction and comorbidity, increasing the likelihood of developing new manifestations of the syndrome. The addiction syndrome model encourages clinicians to develop an independent diagnostic gold standard—free from the problems of impression management that can bias self-report. This model discourages clinicians from using the manifestations and sequelae of addiction to tautologically make diagnostic decisions and inferences about the presence of addiction. Understanding distinct addictive behaviors as opportunistic rather than separate encourages the development of new, more objective diagnostic tools (e.g., fMRI, event related brain potentials [ERP] and implicit behavioral tools) to identify the shared elements of the addiction syndrome. Objective diagnostic tests will help limit the use of unnecessary clinical resources and reduce the application of treatments inappropriate for individuals at certain stages of the addiction syndrome. Objective diagnostic methods will limit demand characteristics and reduce socially desirable responses, ultimately moving the field toward increasingly reliable, valid, and clinically meaningful diagnoses. The syndrome model of addiction encourages clinicians to develop treatment "cocktails"—integrated multipart treatment programs that care for the variety of etiological and consequential problems associated with addiction. The model encourages treatment providers to develop the assessment and treatment skills necessary to care for the comorbidities associated with addiction. Finally, the model also encourages clinicians to sustain their relationship with clients, continuously assessing and caring for them, as they tend to shift expressions of addiction.

Acknowledgements

This chapter represents an adaptation of our previous work: Shaffer, H. J., Laplante, D. A., Labrie, R. A., Kidman, R. C., Donato, A. N. and Stanton, M. V. (2004) "Toward a syndrome model of addiction: multiple expressions, common etiology", *Harvard Review of Psychiatry* 12, 367–374; Shaffer, H. J., Laplante, D. A. and Nelson, S. E. (eds) (2012b).

References

Aharon, I., Etcoff, N., Ariely, D., Chabris, C. F., O'Connor, E. and Breiter, H. C. (2001) "Beautiful faces have variable reward value: fMRI and behavioral evidence", *Neuron* 32(3): 537–551.
American Psychiatric Association (2013) *Diagnostic and Statistical Manual of Mental Disorders: DSM-5*, Arlington, VA: American Psychiatric Association.
Baker, A. (2000) *Serious Shopping*, London, UK: Free Association Books.
Bierut, L. J., Dinwiddie, S. H., Begleiter, H., Crowe, R. R., Helsselbrock, V., Nurnberger, J. I., Porjesz, B., Schuckit, M. A. and Reich, T. (1998) "Familial transmission of substance dependence: alcohol, marijuana, cocaine, and habitual smoking: a report from the Collaborative Study on the Genetics of Alcoholism", *Archives of General Psychiatry* 55: 982–988.
Black, D. W. and Moyer, T. (1998) "Clinical features and psychiatric comorbidity of subjects with pathological gambling behavior", *Psychiatric Services* 49(11): 1434–1439.
Blum, K., Braverman, E. R., Holder, M. M., Lubar, J. F., Monastra, V. J., Miller, D., Lubar, J. O., Chen, T. J. H. and Comings, D. E. (2000) "Reward deficiency syndrome: a biogenetic model for the diagnosis and treatment of impulsive, addictive, and compulsive behaviors", *Journal of Psychoactive Drugs* 32: 1–112.
Blume, S. B. (1994) "Pathological gambling and switching addictions: report of a case", *Journal of Gambling Studies* 10(1): 87–96.
Boudreau, A., Labrie, R. and Shaffer, H. J. (2009) "Towards DSM-V: 'shadow syndrome' symptom patterns among pathological gamblers", *Addiction Research and Theory* 17(4): 406–419.
Breiter, H. C., Aharon, I., Kahneman, D., Dale, A. and Shizgal, P. (2001) "Functional imaging of neural responses to expectancy and experience of monetary gains and losses", *Neuron* 30(2): 619–639.
Brenner, N. and Collins, J. (1998) "Co-occurrence of health-risk behaviors among adolescents in the United States", *Journal of Adolescent Health* 22: 209–213.
Caetano, R., Schafer, J. and Cunradi, C. B. (2001) "Alcohol-related intimate partner violence among White, Black, and Hispanic couples in the United States", *Alcohol Research and Health* 25(1): 58–65.
Cepik, A., Arikan, Z., Boratav, C. and Isik, E. (1995) "Bulimia in a male alcoholic: a symptom substitution in alcoholism", *International Journal of Eating Disorders* 17(2): 201–204.

Christenson, G. A., Faber, R. J., De Zwaan, M., Raymond, N. C., Specker, S. M., Edern, M. D., Mackenzie, T. B., Crosby, R. D., Crow, S. J., Eckert, E. D., Ekern, M. D., Mussell, M. P. and Mitchell, J. E. (1994) "Compulsive buying: descriptive characteristics and psychiatric comorbidity", *Journal of Clinical Psychiatry* 55: 5–11.

Christiansen, M., Vik, P. W. and Jarchow, A. (2002) "College student heavy drinking in social contexts versus alone", *Addictive Behaviors* 27(3): 393–404.

Cohen, S., Lichtenstein, E., Prochaska, J. O., Rossi, J. S., Gritz, E. R., Carr, C. R., Orleans, C. T., Schoenbach, V. J., Biener, L., Abrams, D., Diclemente, C., Curry, S., Marlatt, G. A., Cummings, K. M., Emont, S. L., Giovino, G. and Ossip-Klein, D. (1989) "Debunking myths about self-quitting. Evidence from 10 prospective studies of persons who attempt to quit smoking by themselves", *American Psychologist* 44: 1355–1365.

Conner, B., Stein, J., Longshore, D. and Stacy, A. (1999) "Associations between drug abuse treatment and cigarette use: evidence of substance replacement", *Experimental and Clinical Psychopharmacology* 7: 64–71.

Conway, K., P., Kane, R. J., Ball, S. A., Poling, J. C. and Rounsaville, B. J. (2003) "Personality, substance of choice, and polysubstance involvement among substance dependent patients", *Drug and Alcohol Dependence* 71(1): 65–75.

Evans, G. W. and Kantrowitz, E. (2002) "Socioeconomic status and health: the potential role of environmental risk exposure", *Annual Review of Public Health* 23: 303–331.

Feigelman, W., Wallisch, L. S. and Lesieur, H. R. (1998) "Problem gamblers, problem substance users, and dual-problem individuals: an epidemiological study", *American Journal of Public Health* 88(3): 467–70.

Gambino, B., Fitzgerald, R., Shaffer, H. J., Renner, J. A. and Courtnage, P. (1993) "Perceived family history of problem gambling and scores on SOGS", *Journal of Gambling Studies* 9: 169–184.

Grant, B. F., Stinson, F. S., Dawson, D. A., Chou, S. P., Dufour, M. C., Compton, W., Pickering, R. P. and Kaplan, K. (2004) "Prevalence and co-occurrence of substance use disorders and independent mood and anxiety disorders", *Archives of General Psychiatry* 61(8): 807–816.

Harford, R. J. (1978) "Drug preferences of multiple drug abusers", *Journal of Consulting and Clinical Psychology* 46(5): 908–912.

Hser, Y., Anglin, M. and Powers, K. (1990) "Longitudinal patterns of alcohol use by narcotics addicts", *Recent Developments in Alcoholism* 8: 145–171.

Hunt, W. A., Barnett, L. W. and Branch, L. G. (1971) "Relapse rates in addiction programs", *Journal of Clinical Psychology* 27(4): 455–456.

Hurt, R. D., Sachs, D. P., Glover, E. D., Offord, K. P., Johnston, J. A., Dale, L. C., *et al.* (1997) "A comparison of sustained-release bupropion and placebo for smoking cessation", *New England Journal of Medicine* 337(17): 1195–1202.

Hyman, S. E., Malenka, R. and Nestler, E. (2006) "Neural mechanisms of addiction: the role of reward-related learning and memory", *Annual Review of Neuroscience* 29: 565–598.

Johnson, B. A., Ait-Daoud, N., Bowden, C. L., Diclemente, C. C., Roache, J. D., Lawson, K., Javors, M. A. and Ma, J. Z. (2003) "Oral topiramate for treatment of alcohol dependence: a randomised controlled trial", *The Lancet* 361(9370): 1677–1685.

Karkowski, L. M., Prescott, C. A. and Kendler, K. S. (2000) "Multivariate assessment of factors influencing illicit substance use in twins from female-female pairs", *American Journal of Medical Genetics* 96: 665–670.

Kendler, K. S., Jacobson, K. C., Prescott, C. A. and Neale, M. C. (2003a) "Specificity of genetic and environmental risk factors for use and abuse/dependence of cannabis, cocaine, hallucinogens, sedatives, stimulants, and opiates in male twins", *American Journal of Psychiatry* 160(4): 687–695.

Kendler, K. S., Liu, X.-Q., Gardner, C. O., McCullough, M. E., Larson, D. and Prescott, C. A. (2003b) "Dimensions of religiosity and their relationship to lifetime psychiatric and substance use disorders", *American Journal of Psychiatry* 160(3): 496–503.

Kessler, R. C., Chiu, W. T., Demler, O. and Walters, E. E. (2005) "Prevalence, severity, and comorbidity of 12-month DSM-IV disorders in the National Comorbidity Survey Replication", *Archives of General Psychiatry* 62(6): 617–27.

Kessler, R. C., Crum, R. M., Warner, L. A., Nelson, C. B., Schulenberg, J. and Anthony, J. C. (1997) "Lifetime co-occurrence of DSM-III-R alcohol abuse and dependence with other psychiatric disorders in the National Comorbidity Survey", *Archives of General Psychiatry* 54(4): 313–321.

Kessler, R. C., Hwang, I., Labrie, R. A., Petukhova, M., Sampson, N. A., Winters, K. C. and Shaffer, H. J. (2008) "DSM-IV pathological gambling in the National Comorbidity Survey Replication", *Psychological Medicine* 38(9): 1351–1360.

Kessler, R. C., McLaughlin, K. A., Green, J. G., Gruber, M. J., Sampson, N. A., Zaslavsky, A. M., Aguilar-Gaxiola, S., Alhamzawi, A. O., Alonso, J., Angermeyer, M., Benjet, C., Bromet, E., Chatterji, S., De Girolamo, G., Demyttenaere, K., Fayyad, J., Florescu, S., Gal, G., Gureje, O., Haro, J. M., Hu, C.-Y., Karam, E. G., Kawakami, N., Lee, S., Lépine, J.-P., Ormel, J., Posada-Villa, J., Sagar, R., Tsang, A., Üstün, T. B., Vassilev, S., Viana, M. C. and Williams, D. R. (2010) "Childhood adversities and adult psychopathology in the WHO World Mental Health Surveys", *British Journal of Psychiatry* 197(5): 378–385.

Kessler, R. C., Nelson, C. B., McGonagle, K. A., Edlund, M. J., Frank, R. G. and Leaf, P. J. (1996) "The epidemiology of co-occurring addictive and mental disorders: implications for prevention and service utilization", *American Journal of Orthopsychiatry* 66(1): 17–31.

Kim, S. W., Grant, J. E., Adson, D. E. and Shin, Y. C. (2001) "Double-blind naltrexone and placebo comparison study in the treatment of pathological gambling", *Biological Psychiatry* 49(11): 914–921.

Klingemann, H. and Sobell, L. C. (2007) *Promoting Self-change from Addictive Behaviors: Practical Implications for Policy, Prevention, and Treatment*, New York, NY: Springer Science and Business Media.

Lapham, S. C., Baca, J. C., McMillan, G. P. and Lapidus, J. (2006) "Psychiatric disorders in a sample of repeat impaired-driving offenders", *Journal of Studies on Alcohol* 67(5): 707–13.

Lawrence, A. J., Luty, J., Bogdan, N. A., Sahakian, B. J. and Clark, L. (2009) "Problem gamblers share deficits in impulsive decision-making with alcohol-dependent individuals", *Addiction* 104(6): 1006–1015.

Lejoyeux, M., Ades, J., Tassain, V. and Solomon, J. (1996) "Phenomenology and psychopathology of uncontrolled buying", *American Journal of Psychiatry* 153(12): 1524–1529.

Lesieur, H. R. and Heineman, M. (1988) "Pathological gambling among youthful multiple substance abusers in a therapeutic community", *British Journal of Addiction* 83(7): 765–771.

Linden, D. J. (2011) *The Compass of Pleasure: How our Brains make Fatty Foods, Orgasm, Exercise, Marijuana, Generosity, Vodka, Learning, and Gambling Feel so Good*, New York, NY: Viking.

Lopes, L. L. (1987) "Between hope and fear: the psychology of risk", in L. Berkowitz (ed.), *Advances in Experimental Social Psychology*, San Diego, CA: Academic Press.

Marlatt, G. A. (1988) "Matching clients to treatment: treatment models and stages of change", in D. M. Donovan and G. A. Marlatt (eds), *Assessment of Addictive Behaviors*, New York, NY: Guilford Press.

Marlatt, G. A., Baer, J. S., Donovan, D. M. and Kivlahan, D. R. (1988) "Addictive behaviors: etiology and treatment", *Annual Review of Psychology* 39: 223–252.

Marlatt, G. A. and Gordon, J. (1985) *Relapse Prevention*, New York, NY: Guilford Press.

Merikangas, K. R., Mehta, R. L., Molnar, B. E., Walters, E. E., Swendsen, J. D., Aguilar-Gaziola, S., Bijl, R., Borges, G., Caraveo-Anduaga, J. J., Dewit, D. J., Kolody, B., Vega, W. A., Wittchen, H.-U. and Kessler, R. C. (1998a) "Comorbidity of substance use disorders with mood and anxiety disorders: results of the international consortium in psychiatric epidemiology", *Addictive Behaviors* 23: 893–907.

Merikangas, K. R., Stolar, M., Stevens, D. E., Goulet, J., Preisig, M. A., Fenton, B., Zhang, H., O'Malley, S. S. and Rounsaville, B. J. (1998b) "Familial transmission of substance use disorders", *Archives of General Psychiatry* 55(11): 973–979.

Miller, W. R. and Carroll, K. (2006) *Rethinking Substance Abuse: What the Science Shows, and What We Should do About It*, New York, NY: Guilford Press.

Nelson, C. B., Heath, A. C. and Kessler, R. C. (1998) "Temporal progression of alcohol dependence symptoms in the US household population: results from the National Comorbidity Survey", *Journal of Consulting and Clinical Psychology* 66(3): 474–483.

Odegaard, S., Peller, A. and Shaffer, H. J. (2005) "Addiction as syndrome", *Paradigm* 9(3): 12–13, 22.

Potenza, M. N. (2001) "The neurobiology of pathological gambling", *Seminars in Clinical Neuropsychiatry* 6(3): 217–226.

Potenza, M. N., Sofuoglu, M., Carroll, K. M. and Rounsaville, B. J. (2011) "Neuroscience of behavioral and pharmacological treatments for addictions", *Neuron* 69(4): 695–712.

Ratey, J. J. and Johnson, C. (1997) *Shadow Syndromes*, New York, NY: Pantheon Books.

Regier, D. A., Farmer, M. E., Rae, D. S., Locke, B. Z., Keith, S. J., Judd, L. L. and Goodwin, F. K. (1990) "Comorbidity of mental disorders with alcohol and other drug abuse. Results from the Epidemiologic Catchment Area (ECA) Study", *Journal of the American Medical Association* 264(19): 2511–2518.

Robins, L. N. (1993) "Vietnam veterans' rapid recovery from heroin addiction: a fluke or normal expectation?" *Addiction* 88(8): 1041–1054.

Schachter, S. (1982) "Recidivism and self-cure of smoking and obesity", *American Psychologist* 37(4): 436–444.

Shaffer, H. J. (1997) "The most important unresolved issue in the addictions: conceptual chaos", *Substance Use and Misuse* 32(11): 1573–1580.

Shaffer, H. J. (1999) "Strange bedfellows: a critical view of pathological gambling and addiction", *Addiction* 94(10): 1445–1448.

Shaffer, H. J. and Eber, G. B. (2002) "Temporal progression of cocaine dependence symptoms in the national comorbidity survey", *Addiction* 97: 543–554.

Shaffer, H. J. and Hall, M. N. (2002) "The natural history of gambling and drinking problems among casino employees", *Journal of Social Psychology* 142(4): 405–424.

Shaffer, H. J. and Jones, S. B. (1989) *Quitting Cocaine: The Struggle Against Impulse*, Lexington, MA: Lexington Books.

Shaffer, H. J., Laplante, D. A., Labrie, R. A., Kidman, R. C., Donato, A. N. and Stanton, M. V. (2004) "Toward a syndrome model of addiction: multiple expressions, common etiology", *Harvard Review of Psychiatry* 12(6): 367–374.

Shaffer, H. J., Laplante, D. A. and Nelson, S. E. (2012a) *The APA Addiction Syndrome Handbook, Vol. 1*, Washington, DC: American Psychological Association Press.

Shaffer, H. J., Laplante, D. A. and Nelson, S. E. (2012b) *The APA Addiction Syndrome Handbook, Vol. 2*, Washington, DC: American Psychological Association Press.

Shaffer, H. J. and LaSalvia, T. A. (1992) "Patterns of substance use among methadone maintenance patients: indicators of outcome", *Journal of Substance Abuse Treatment* 9(2): 143–147.

Shaffer, H. J., Nelson, S. E., Laplante, D. A., Labrie, R. A., Albanese, M. J. and Caro, G. (2007) "The epidemiology of psychiatric disorders among repeat DUI offenders accepting a treatment sentencing option", *Journal of Consulting and Clinical Psychology* 75(5): 795–804.

Shaffer, H. J., Vander Bilt, J. and Hall, M. N. (1999) "Gambling, drinking, smoking, and other health risk activities among casino employees", *American Journal of Industrial Medicine* 36(3): 365–378.

Shapiro, D. (1965) *Neurotic Styles*, New York, NY: Basic Books.

Silk, A. and Shaffer, H. J. (1996) "Dysthymia, depression, and a treatment dilemma in a patient with polysubstance abuse", *Harvard Review of Psychiatry* 3: 279–284.

Slutske, W. S. (2006) "Natural recovery and treatment-seeking in pathological gambling: results of two US national surveys", *American Journal of Psychiatry* 163(2): 297–302.

Slutske, W. S., Eisen, S., True, W. R., Lyons, M. J., Goldberg, J. and Tsuang, M. (2000) "Common genetic vulnerability for pathological gambling and alcohol dependence in men", *Archives of General Psychiatry* 57(7): 666–673.

Slutske, W. S., Jackson, K. M. and Sher, K. J. (2003) "The natural history of problem gambling from age 18 to 29", *Journal of Abnormal Psychology* 112(2): 263–274.

Sobell, L. C., Cunningham, J. A. and Sobell, M. B. (1996) "Recovery from alcohol problems with and without treatment: prevalence in two population surveys", *American Journal of Public Health* 86(7): 966–972.

Tsuang, M. T., Lyons, M. J., Meyer, J. M., Doyle, T., Eisen, S., Goldberg, J., True, W., Lin, N., Toomey, R. and Eaves, L. (1998) "Co-occurrence of abuse of different drugs in men: the role of drug-specific and shared vulnerabilities", *Archives of General Psychiatry* 55(11): 967–972.

Uhl, G. R. and Grow, R. W. (2004) "The burden of complex genetics in brain disorders", *Archives of General Psychiatry* 61: 223–229.

Vaillant, G. E. (1983) *The Natural History of Alcoholism: Causes, Patterns, and Paths to Recovery*, Cambridge, MA: Harvard University Press.

Vander Bilt, J., Dodge, H., Pandav, R., Shaffer, H. J. and Ganguli, M. (2004) "Gambling participation and social support among older adults: a longitudinal community study", *Journal of Gambling Studies* 20(4): 373–390.

Vitaro, F., Brendgen, M., Ladouceur, R. and Tremblay, R. E. (2001) "Gambling, delinquency, and drug use during adolescence: mutual influences and common risk factors", *Journal of Gambling Studies* 17(3): 171–190.

Wechsler, H., Davenport, A. E., Dowdall, G. W., Grossman, S. J. and Zanakos, S. I. (1997) "Binge drinking, tobacco, and illicit drug use and involvement in college athletics. A survey of students at 140 American colleges", *Journal of American College Health* 45(5): 195–200.

Welte, J. W., Barnes, G. M. and Hoffman, J. H. (2004) "Gambling, substance use, and other problem behaviors among youth: a test of general deviance models", *Journal of Criminal Justice* 32(4): 297–306.

Werme, M., Lindholm, S., Thorén, P., Franck, J. and Brené, S. (2002) "Running increases ethanol preference", *Behavioural Brain Research* 133: 301–308.

Werme, M., Thorén, P., Olson, L. and Brené, S. (2000) "Running and cocaine both upregulate dynorphin mRNA in medial caudate putamen", *European Journal of Neuroscience* 12(8): 2967–2974.

Whalen, C. K., Jamner, L. D., Henker, B. and Delfino, R. J. (2001) "Smoking and moods in adolescents with depressive and aggressive dispositions: evidence from surveys and electronic diaries", *Health Psychology* 20(2): 99–111.

Wise, R. A. (1995) "Addictive drugs and brain stimulation reward", *Annual Review of Neuroscience* 18: 319–340.

Wray, I. and Dickerson, M. (1981) "Cessation of high frequency gambling and 'withdrawal' symptoms", *British Journal of Addiction* 76: 401–405.

Zinberg, N. E. (1984) *Drug, Set, and Setting*, New Haven, CT: Yale University Press.

13
DEVELOPING GENERAL MODELS AND THEORIES OF ADDICTION

Robert West, Simon Christmas, Janna Hastings, and Susan Michie

Introduction

The science of addiction is being hampered by confusion in concepts and terms, and a multiplicity of models and theoretical approaches that make little reference to each other. In this respect it has much in common with other areas of social, clinical and behavioural sciences. Technologies now exist and are being rapidly advanced that can address this problem, and other sciences are already making use of them. In particular, what are known as 'ontologies' (as used in computer science) and the 'Semantic Web' could revolutionise our ability to formulate models and theories in addiction that can then provide much-needed direction to the scientific endeavour.

The field of biology suffered from a similar problem until the development of what is known as the Gene Ontology (Ashburner *et al.* 2000). The gene ontology is not just about genes, but is a representational system for the whole of biology, unifying terms, definitions and models across species and research groups in a way that has revolutionised the field (Lewis 2017).

This chapter introduces readers to ontologies and the Semantic Web, and explores their potential use in developing and expressing models and theories of addiction in ways that allow relationships to be examined between them, and between these and more general models and theories in clinical, population and behavioural sciences. These technologies also allow investigation of construct relationships *within* models, necessary for testing and hence refining and advancing them.

We begin by describing a central challenge facing the study of addiction: the need to achieve clarity of constructs and develop consensus while at the same time recognising that divergent views have utility. We then move on to describe some key characteristics of the Semantic Web, and the ways in which these provide a pragmatic way of responding to this challenge. We conclude by looking at the potential value of existing ontologies in developing a general theory of addiction.

The challenge: clarity and diversity

Models of addiction are necessary for building addiction science and developing effective interventions to combat this problem. If they go beyond describing observed relationships (descriptive models such as tobacco price elasticity (Gallus *et al.* 2006)) and attempt to explain phenomena we refer to them as 'theories' (e.g., the dopamine theory of drug reward (Blum *et al.* 2015)). There are a plethora of models and theories of addiction differing in scope, emphasis, constructs and propositions but they have not been expressed in ways that allow them to be compared, tested or integrated.

The term 'addiction' is itself an example of this lack of clarity in relation to key constructs. Many cases of psychoactive drug use have features that lead to the users being labelled as suffering from a condition called 'addiction'. Sets of such features have been listed in 'diagnostic criteria', such as DSM-5 (American Psychiatric Association 2013) and ICD-10 (World Health Organisation 2016). These include continued use despite harmful consequences, experience of adverse withdrawal symptoms during periods of non-use, difficulties controlling use, high levels of use, and repeated strong motivation to use. Features can be present to different degrees, which means that thresholds are required for deciding whether or not a pattern of drug use is addictive. These thresholds are to a large extent arbitrary and context dependent. An alternative, rather than considering addiction to be present or absent depending on whether some threshold is exceeded, is to assess the degree to which the features are evident to specify a 'degree' of addictedness or 'severity' of addiction (Gossop *et al.* 1995).

With multiple features potentially being involved, different ways of characterising these features, and different thresholds potentially being applicable, there can be substantial differences of view in whether an individual's pattern of drug use is considered addictive, or the degree of addictedness. These differences mean that issues such as the prevalence of addictions and theories concerning the causes of addiction are subject to differing viewpoints that cannot be reconciled solely by reference to objectively determined facts.

The picture is further complicated by the fact that different psychoactive drugs have different patterns of use and show different degrees and patterns of addictive features, as do different ways of using the same drug. An additional complication is that behaviours that do not involve psychoactive drugs, such as gambling, can show similar characteristics to drug addiction and so the concept of addiction appears to apply to those behaviours as well.

A similar lack of clarity exists for many other constructs that are important to a scientific understanding of addiction. Any general model or theory of addiction would need to capture processes that have been identified as important in its development and maintenance (Orford 2001; West and Brown 2013) (Table 13.1). Although conceptually different, many of these processes have features in common, or else the distinction between them is nuanced. For example, incentive sensitisation and drive theory both involve craving. In the case of incentive sensitisation, cravings are generated in response to cues through repeated exposure enhancing a direct link between those cues and the experience of 'wanting' whereas drive theory proposes that cravings are generated through exposure to a drug leading to a state that is relieved by taking the drug. Some models and theories are fundamentally neurophysiological whereas others focus on social processes.

None of the processes believed to underlie addiction are limited to addictive behaviours; they are all involved in the development of other motivations (West and Brown 2013). What makes a behaviour pattern addictive is the strength of the motivational forces generated and/or the way that particular drugs or behaviours interact with the motivational system to create a positive feedback cycle rather than the self-correcting systems that normally operate to ensure that no one behaviour receives an unwarranted priority at the expense of others.

For example, with palatable food, our natural processes of satiation and habituation reduce its rewarding value as we eat more of it during a meal. By contrast, with cocaine and amphetamine, reward mechanisms become sensitised to the impulse-generating effects

Table 13.1 Processes commonly included in models of addiction

Processes	Description	Propositions in existing models
Cost-benefit analysis	The benefits of the addictive behaviour are judged by the addict to outweigh the costs	At least some people addicted to alcohol believe that the benefits in terms of anxiety relief and mental escape are worth the financial, social and health costs
Incentive sensitisation	Repeated exposure to addictive drugs leads to sensitisation of brain pathways that generate feelings of 'wanting' in response to drug cues independent of feelings of 'liking'	Smoking crack cocaine leads to feelings of craving in situations similar to those where this has occurred, independent of feelings of euphoria produced by the drug
Reward seeking	Addicts learn that addictive behaviours provide positive feelings of enjoyment and euphoria	Methamphetamine users seek the 'rush' provided by the drug
Attachment	Addicts become emotionally attached to drugs or addictive behaviours because these have been reliable sources of comfort or gratification	Smokers often report feeling a sense of bereavement during the early stages of stopping smoking: like they have lost a cherished friend or family member
Drive reduction	Repeated engagement in an addictive behaviour results in development of an acquired drive, which is experienced as craving, after a period of abstinence	Repeated use of nicotine alters brain physiology so that abstinence results in an acquired drive state, experienced as craving
Distress avoidance	Addicts learn that addictive behaviours relieve mental and physical distress caused by mental health problems, life circumstances, and/or withdrawal symptoms	Repeated use of heroin results in changes in brain chemistry leading to adverse mood and physical symptoms when concentrations of the drug fall below certain concentrations in the brain
Social influence	Cultural, sub-cultural, peer group and/or family norms promote or are permissive of addictive behaviour	Family and peer group are important factors influencing the development of smoking and alcohol consumption
Impaired control	Addictive behaviours develop, and are maintained by, pre-existing or acquired inefficiencies in brain systems required for impulse control	Use of stimulant drugs leads to impairment in frontal lobe functioning required to inhibit impulses to continually repeat the use leading to bingeing
Classical conditioning	Repeated pairing of stimuli (cues) associated with effects of addictive behaviours leads those stimuli to generate anticipatory reactions to those effects	Lights, images and sounds are used by gaming machine manufacturers to promote high rates of use of those machines
Operant conditioning	Addictive behaviours are followed by powerful positive or negative reinforcers (rewards or offset of aversive stimuli) in the presence of discriminative stimuli (cues), so that those stimuli come to provoke a strong impulse to engage in the behaviour	Use of psycho-stimulants is maintained by the positive reinforcing properties of these drugs

Source: West 2017.

of the drug (Berridge and Robinson 2016). In the case of alcohol dependence, addicts develop craving, rebound anxiety and adverse physical symptoms that provide a very powerful motivation to resume drinking (Seo and Sinha 2014). In the case of tobacco smoking, rapid ingestion of nicotine leads to strong cravings through multiple mechanisms, including creation of an acquired drive state, similar to hunger, when CNS nicotine concentrations become depleted (West 2009). So development of general models and theories of addiction will involve bringing together a wide range of constructs and processes relating to canonical motivational theory (i.e., a theory of what normally happens) and how abnormalities in motivation occur. These processes will be both internal and external to individuals, groups and populations.

Diversity in the study of addiction

While there is a need for greater clarity of constructs in the study of addiction, no investigator or organisation has the authority, or expertise, to propose a single unifying conceptual framework.

A general theory of addiction has yet to be developed, but a key requirement for such a theory is that it should recognise and accommodate multiple viewpoints on addiction, and not be limited to a single viewpoint such as the 'medical model' (construing addiction in term of a mental disorder, disease or disease process). Figure 13.1 shows du Plessis' classification of the types of theory that would need to be recognised and accommodated by a general theory of addiction according to eight major methodologies (du Plessis 2014). Some specific theories and several theoretical approaches have attempted to span a number of the zones in this classification. Of the specific theories, the theory of Excessive Appetites and PRIME Theory have been elaborated in some detail (Orford 2001; West and Brown 2013), respectively).

It is important not to overstate this diversity. Despite the complexities discussed earlier, the observed phenomena that are captured by a term like 'addiction' are important, and it is useful to give a label to a construct that captures these. Moreover, there are shared understandings of many of the features that characterise addictive drug use, even where there may be disagreement around details and where emphasis should be placed.

On the other hand, it should go without saying that a general theory that achieved greater clarity by simply ignoring this diversity of viewpoints would not in fact be a general theory at all. To move forward, we need a way to achieve clarity of constructs and develop consensus, while at the same time recognising that divergent views have utility.

Addressing the challenge in the Semantic Web

The technology associated with the Semantic Web provides the basis for a response to the challenge we have outlined: a way for the science of addiction to move from its current state of confused, confusing and imprecise terminology to a way of representing addiction and models of addiction that will promote collaborative working, respect and preserve different viewpoints on addiction, but nevertheless allow all research results to be integrated, and theoretical predictions to be tested, and so advance the field.

What is the Semantic Web? One way to answer this question is to chart the development of the Semantic Web from the Worldwide Web (www). The Worldwide Web has revolutionised our lives by making information available from a vast range of sources. It defines a technological

Eight major methodologies   *Types of addiction theory classified according to the eight methodologies*

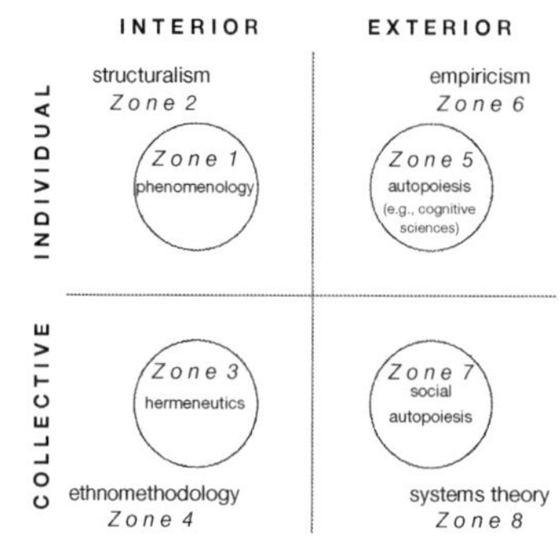

	Zone 1	Zone 2	Zone 3	Zone 4
	Phenomenology	**Structuralism**	**Hermeneutics**	**Ethnomethodology**
	• Conditioning/Reinforcement Behavioral models • Compulsion and Excessive Behavior models • Spiritual/Altered State of Consciousness models • Personality/Intrapsychic models • Coping/Social learning models • Biopsychosocial model	• Transtheoretical model • Personality/intrapsychic models	• Coping/Social Learning models • Biopsychosocial model	• Social/Environment models • Coping/Social Learning models • Biopsychosocial model • Spiritual/Altered State of Consiousness models
	Zone 5	Zone 6	Zone 7	Zone 8
	Autopoiesis Theory	**Empiricism**	**Social Autopoiesis Theory**	**Systems Theory**
	• Conditioning/Reinforcement Behavioral models • Coping/Social Learning models • Biopsychosocial model	• Genetic/Physiological models • Conditioning/Reinforcement Behavioral models • Compulsion and Excessive Behavior models • Biopsychosocial model	• Social/Environment models	• Social/Environment models • Biopsychosocial model

INDIVIDUAL — INTERIOR: structuralism Zone 2; Zone 1 phenomenology. EXTERIOR: empiricism Zone 6; Zone 5 autopoiesis (e.g., cognitive sciences).

COLLECTIVE — Zone 3 hermeneutics; ethnomethodology Zone 4. Zone 7 social autopoiesis; systems theory Zone 8.

Figure 13.1 Du Plessis' classification of addiction theories.
Source: du Plessis 2014.

framework for locating and exchanging diverse content types – for example, text, images, and films. At the heart of this framework is the Uniform Resource Locator (URL), a unique 'address' for each of the billions of different web pages. These URLs are stored as a code that means nothing to human readers, but a system of 'domain names' has been developed to link them to a name that humans can read and understand. These also have to be unique. Thus www.addictionjournal.org is a domain name for the journal *Addiction* that uniquely points to the journal's home page. Web pages use URLs to link to other pages to create the Worldwide Web.

The www has a huge limitation, however. The information contained in web pages is designed for humans to consume, and is difficult for machines to interpret, which hinders the ability of software to help in the discovery, synthesis, and summary of online content. Google, Yahoo and other search engines make their developers billions of dollars by trying to address this limitation, but for the most part all that search engines are able to do is to point to web pages that might contain the information you want; finding the right pages can often require multiple attempts to formulate a search in the right way. For example, a search for "Addiction conferences" will not return results from pages entitled "Addiction meetings"; a search for "addiction" will not return results from pages entitled "dependence". There is no representation of the underlying meaning of content that is sensitive to synonyms or closely related constructs, only the superficial textual representation of the content encoded directly in the web pages.

The Semantic Web is a radical advance on this. Instead of only making content available in text on web pages, the Semantic Web allows individual pieces of information within those pages to be assigned a meaning and a specific location. Rather than only having a string of text to interpret ("addiction"/"dependence"), within the Semantic Web units of information are explicitly defined in shared vocabularies or "ontologies" – computational representations of knowledge in a particular domain. Ontologies link together the words, expressions, and language that humans use to refer to things, with computable formal definitions of those things that allow computers to distinguish one type of thing from another.

The Semantic Web is implemented in the most general sense as a collection of statements that take the form subject-predicate-object, known as 'triples'. This very generic form of representation is then populated with entities (subjects and objects) and relationships drawn from shared, well-defined vocabularies in order to be clear on their interpretation and meaning.

Units of information are assigned Uniform Resource Identifiers (URIs), each of which points to a unique entity, concept or resource that has an address on the web, and each URI is then used in triples represented in what is known as the 'Resource Description Framework' (RDF). Computers are thus given the ability to 'understand' terms such as 'psychological condition' by how such a term is related to other terms set out in a network of relationships in RDF triples and definitions within ontologies.

This forces the definitions of all terms used within the Semantic Web to be explicit and distinguishable. Humans, by contrast, understand terms in ways that combine formal definitions with natural language meanings, which creates ambiguity. Although nuances of natural language provide a richness that is helpful in certain circumstances, ambiguities impede the scientific testing and advance of ideas. To make content computationally interpretable within the Semantic Web forces the use of clear and distinguishable representations for each different sort of thing, which explicitly addresses the sorts of confusion that are hindering progress in developing the science of addiction.

For example, Table 13.2 shows examples of propositions about addiction expressed in the form of triples. These provide a way of explicitly expressing models of addiction that allow different models to be compared and related with more general models of behaviour. Setting out each proposition in a model of addiction in this way allows that proposition to be examined and tested.

A considerable amount of unproductive debate in the field of addiction, as in many other areas of social, clinical and behavioural science, arises because the individual propositions within theories cannot be separated out for scrutiny. Thus different researchers may take differing views on any or all of the propositions in Table 13.2, but the fact that they are set out in this way allows for discussion about the merits and demerits of each one to be analysed systematically, whether by reference to evidence or inference.

Table 13.2 Examples of propositions about addiction in the form of 'triples' expressed as an approximation to natural language

addiction	is a subclass of	mental disorder
	has subclass	substance addiction
	has subclass	behavioural addiction
	inheres in	people
	has manifestation	addictive behaviour
	has attribute	addiction duration
	has attribute	addiction strength
	has attribute	addiction harmfulness
	has attribute	addiction severity
	has part	strong motivation
	has part	significant harm potential
	has manifestation	repeated motivation episodes

Diversity of perspectives in the Semantic Web

The way in which the Semantic Web works, and in particular the fact that the elements from which RDF triples are composed are explicitly assigned meaning in ontologies and by reference to their links to other elements as stated in other triples, ensures clarity of constructs. At the same time, the Semantic Web also provides a way of accommodating diversity. Indeed, a founding principle of the Semantic Web, just like the Worldwide Web, is that *anyone can say anything about anything*.

For example, different people can continue to use terms such as 'addiction' differently, since these terms will be assigned separate URIs and definitions despite sharing the same superficial label. In fact, there is a built-in way to represent different "perspectives" on the Semantic Web through use of different 'namespaces'. Stated simply, if I want to define addiction in a new way I can simply create my new meaning of addiction in my own unique namespace, and explicitly refer to that version of addiction in my content. This makes it completely clear what the term means as I am using it. Importantly it also allows for systematic comparisons between my usage and someone else's.

In principle the use of namespaces could result in anarchy if no-one agreed about the definitions of anything. But this is no different from what happens with natural language. The difference in the case of the Semantic Web and ontologies is that disagreements in how terms are used are explicit and defined. They can then be subject to scrutiny, and differences potentially resolved where such resolution is useful. In practice, many researchers *do* agree on key definitions and can therefore use a single shared namespace to reflect this consensus. This is directly analogous to referencing the definition of a construct in another paper, with the important practical difference being that such links become computable.

It is also possible that namespaces could be subject to arbitrary (in the scientific sense) bias. For example, a powerful research group might insist on use of a particular namespace for anyone wanting to receive research funding or to publish their work in a prestigious journal. Again, this is no different from what already pertains with models and theories that are expressed in natural language. The key difference is that everything is made explicit so there is a greater opportunity to expose this tendency through automated information searches and to challenge it with analysis and evidence.

The concept of a 'namespace' allows for a diversity of perspectives on the Semantic Web, and ensures that anyone can say anything about anything. There are, however, practical limits to diversity. In order to have debates about the construct of 'addiction', for example, we do need to agree on the meaning of a set of more fundamental terms and predicates we use in that debate: terms such as 'is a subclass of' and 'has attribute' (see Table 13.2 above).

The Semantic Web as a whole is underpinned by the existence of shared namespaces in which fundamental terms such as these are defined. The namespaces associated with RDF and OWL, the Web Ontology Language, provide a suite of general-purpose computable vocabulary elements that are needed to say anything at all. In the next section we will explore the potential to use other pre-existing ontologies in the development of a general theory of addiction.

Building on existing ontologies

Ontologies have been developed covering a range of topics in science, humanities and computing, and there is a very strong case for using existing, broader ontologies such as these wherever possible. Of most direct relevance to the study of addiction are the Mental Functioning Ontology (MF) (Hastings *et al.* 2012b) and the WHO's International Classification of Functioning

Disability and Health (Üstün *et al.* 2003). Concepts in the field of addiction can be represented by linking constructs from relevant ontologies (Hastings *et al.* 2012a); and addiction ontologies can be developed using principles established by the Basic Formal Ontology (BFO) (Ceusters and Smith 2010) and implemented by the Open Biological and Biomedical Ontologies (OBO) Foundry (Smith *et al.* 2007).

We now take a closer look at the potential value of two existing ontologies in developing a general theory of addiction: the Basic Formal Ontology (BFO) and the Ontology for General Medical Science (OGMS). It should be noted that the BFO represents only one of a number of top level ontologies. It is not static, and operates under the philosophy that if changes are required in order to be able to create better models, these can be implemented.

Basic Formal Ontology

The BFO has been developed to provide a top level ontology for use in domain-specific ontologies. The purpose of the BFO is to provide a categorisation of the sorts of things that exist that is generic enough to be agreed on across multiple domains, thereby providing a common vocabulary that ensures computers will be able to interpret relevant general distinctions in the same way across the different domain implementations. The Mental Functioning Ontology has been developed as one such domain-specific ontology and has been implemented in OWL. These make a suitable starting point for addiction ontologies.

There is not the space to describe them in detail but it is worth outlining key features of the BFO to illustrate how an addiction ontology may be constructed. Figure 13.2 shows the key terms in this ontology. The terms will look unfamiliar but they represent fundamental ways of representing the world using a language that makes sense once one understands the definitions.

We provide a short outline of how the BFO understands various ontological categories. It is important to note that the following descriptions are specific to the BFO and may not therefore correspond with the way in which ontological categories have been analysed within other literatures, such as the literature on metaphysics within analytic philosophy. The BFO

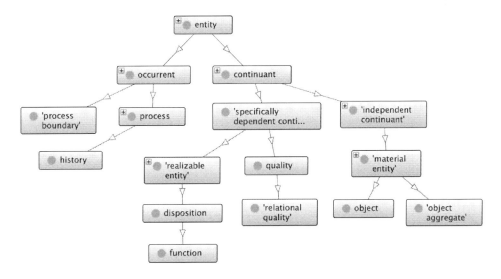

Figure 13.2 Main classes in the Basic Formal Ontology. Arrows represent the 'subclass of' relationship.

starts with the concept of 'entity'. An entity is anything that exists. Entities come in two forms: 'continuants' and 'occurrents'. Continuants exist over a period of time and come in the form of 'independent continuants' such as 'objects', and 'dependent continuants' such as 'dispositions' and 'qualities'. Occurrents inherently involve change over time, and include 'processes' and 'process boundaries'. The BFO does not distinguish between processes and events, on the basis that events necessarily involve changes over time (e.g., onsets and offsets).

The dichotomy between continuants and occurrents represents two different mutually interdependent perspectives on the world: physical and abstract objects or collections of objects and their features, and processes and events that transform these over time. Any theory or model of the physical world, including addiction, needs to be clear about where each of its constituent constructs lies in this world view.

Thus 'craving' can be construed as a 'relational quality' that has an existence over time, is necessarily attached to a person who is an 'object' and only exists in relation to something that is being craved. As such it can be measured and can be statistically and causally associated with other qualities such as 'stress' and 'processes boundaries' such as 'relapse'. But the term 'craving' can also refer to a 'process' involving neural activity within, say, the nucleus accumbens. It can also refer to a 'disposition', which is a latent characteristic that becomes expressed under certain conditions. All three of these definitions are meaningful and potentially useful, but failing to be clear about which of them one is referring to on a given occasion causes confusion.

The other top-level dichotomy in the BFO is between 'universals' and 'instances' (noting that this use of terms is different from that made in metaphysics). Universals are types or classes of entity that can exist or occur, while instances are specific cases. Thus 'addiction' is a class of 'disposition' whereas the addiction of John Smith to cocaine is an 'instance'. This distinction between universals and specifics is fundamental in any model of the world, but needs to be recognised in every case where a model of addiction is used.

Ambiguity in common language, and therefore in informal models and theories of addiction, can lead to confusion and unhelpful disagreement. Thus the 'universal' statement 'Addiction to cocaine is caused by feelings of euphoria experienced after using the drug' is ambiguous as to whether this refers to all instances of addiction or some instances, or all instances of cocaine-induced euphoria or some cases.

Most, if not all, of the subtypes of entities in the BFO are potentially relevant to building models of addiction. The entity type 'history' is particularly relevant as a subtype of process that is the sum of processes taking place in a continuant entity over time. Thus one may heuristically define the history of someone who for a period in his or her life suffers from addiction in terms of a period when he or she had not been engaged in the addictive behaviour (e.g., never used a substance), when he or she had engaged in the behaviour but did not show signs of addiction (e.g., recreational use of a substance), when he or she experienced addiction to the behaviour (e.g., diagnosed as suffering from substance use disorder), when he or she was voluntarily abstinent but subject to repeated cravings (e.g., attempting recovery from substance use disorder), and when he or she was abstinent and not repeatedly experiencing cravings (e.g., in recovery from substance use disorder).

A common confusion in models of mental entities is between 'functions' and 'functioning'. It is apparent from Figure 13.2 that functions are classified as types of disposition; they are latent enduring attributes of continuants that result in particular processes in specific circumstances. They may change over time but it is not inherent in their nature for them to change. Functions are dispositions that have been designed or evolved. They are extremely important in any model of human behaviour or psychology. For example, motivation can be regarded as a function of animals that has evolved to get animals to behave in ways that promote survival and reproduction. Motivational functioning (in the form of processes) is a realisation of variants of this function in specific circumstances, but it needs to be treated differently in the model. Failure to do so leads to confusion in the specification of measures and lack of clarity in predictions.

A major feature of the BFO needs to be considered when applying it to developing models and theories of addiction. Models of complex systems such as animals need to be able to handle the fact that there will be many instances that do not conform to a standard version. Thus we can define an elephant as a pachyderm with a trunk and large ears, but if an elephant loses its trunk in an accident it is still an elephant. This is handled in the BFO by creating 'canonical' models and allowing for deviations that can be represented as needed, without having to specify every possible way in which deviations may occur in advance. This can be used in models of addiction by creating canonical models involving the entities and their relationships that are typically observed in addiction and only extending the model as needed to represent deviations from this canonical representation.

From the preceding paragraphs it should be apparent that using the BFO as the basis for a model or theory of addiction allows this to be anchored in a set of concepts that have been defined, promotes clarity in the use of terms, and enables different models or theories to be made explicit and compared with each other. The BFO does not impose any domain-specific propositions on to models or theories made using it, beyond its very broad world view.

Ontology for General Medical Science

Many, if not most, people who work in the field of addiction find it useful to think of it as a form of clinical abnormality. The fact that it is included in the DSM and ICD diagnostic criteria illustrates this. Others may take a different view, but if one does want to model addiction as a form of clinical abnormality, it is helpful to base that model on the Ontology for General Medical Science (OGMS), which is itself based on the BFO.

Basing a model of addiction on the OGMS helps to improve the clarity of the model and can make explicit differences between one model and another. The OGMS distinguishes between 'clinical abnormality', 'disorder' and 'disease'. This distinction turns out to be extremely important for physical health as well as for mental health problems.

The OGMS introduces the entity 'life plan' of an organism. Life plan is a canonical history of the organism and includes gestation, maturation, and ageing. It then defines a 'clinical abnormality' in terms of three things being present simultaneously: 1) it is not part of the life plan of an organism of a relevant type, 2) it is causally related to an elevated risk of pain or other feelings of illness, dysfunction or death, such that 3) the elevated risk exceeds a threshold level.

This definition of clinical abnormality should look quite familiar to someone working in the field of addiction. There is nothing unique to addiction about it involving a deviation from an expected set of characteristics, this being linked to risk of harm, and there needing to be a threshold to determine whether this qualifies or is 'sub-clinical'. In the case of physical clinical abnormalities, one may set the threshold as the level at which the person affected or someone who cares about him or her feels a need to see a health professional. In the case of mental abnormalities, we need to allow for the possibility that neither the person concerned nor his or her friends or family sees a problem but there is sufficient risk of harm that others can make the determination that something needs to be done.

The OGMS then defines a disorder as an independent continuant that is a causally linked combination of components that is clinically abnormal. The disorder is the clinical abnormality but it is not the disease itself. In the case of influenza, the disorder is the presence in the body of viable cells containing the influenza virus. The disease is a disposition that is realised in a pathological process (acute inflammation) that produces abnormal bodily features that are recognised as signs (e.g., fever) and symptoms (e.g., headache).

Thus a disease is construed as a disposition of an organism to undergo pathological processes because of one or more disorders. The concept of 'disease course' is introduced as the aggregate of pathological processes in which a disease is realised. Note that a disease is a continuant while a disease course is an occurrent – the two differ at the very top level of the BFO. While this may seem an artificial distinction, few would disagree that it is useful to be able to refer to an instance of addiction as a single entity that exists in a given person over time. It is useful to be able to refer to John Smith's cocaine addiction as such, despite the varied course that the addiction may take, through initial development, subsequent abstinence, relapse and recovery.

In this formulation, addiction can be regarded as a mental disorder or set of mental disorders that are realised through pathological processes as a set of signs and symptoms. In that model addiction is not the pathological processes or the signs and symptoms – which may vary in both quality and intensity as a function of a range of environmental, physiological and psychological factors. Addiction can be viewed alternatively as the whole disease process. This is a matter of perspective. What is important are the propositions about the disorders, processes, signs and symptoms – that is where, as long as the propositions are properly formulated, there can be fruitful discussion about how they relate to the evidence.

The BFO and the OGMS are examples of ontologies that can be expressed in a language such as OWL and used to build domain-specific ontologies designed to model phenomena such as addiction. Some people working in the field of addiction may consider them too restrictive in terms of the structure they impose on thinking in the area. On the other hand, there are advantages to building on them in an addiction ontology precisely because of the clarity and structure that they bring to the process of model and theory building.

The key point is that it is up to people working in the field what they want to do. If they can come up with an ontology that better captures the phenomena observed in addiction or is more parsimonious or coherent, this will represent an advance. But, in attempting to do so, it is wasteful and confusing to ignore ontologies that already exist – rather their limitations need to be described and the ways that the novel ontology improve on them need to be explicitly stated.

Conclusion: towards a general theory of addiction

We need to move beyond current approaches to modelling addiction to one that involves entities and relationships that are clearly and explicitly defined. Given the nature of addiction, it is neither realistic nor desirable to attempt to impose a set of definitions to make this happen. The Semantic Web, however, offers a pragmatic way to achieve clarity of constructs and develop consensus, while at the same time recognising that divergent views have utility. Models of addiction can be expressed in terms of subject-predicate-object triples using Uniform Resources Identifiers (URIs) for each component. These URIs may link to a common term such as 'craving' but defined differently in different namespaces. In this way differences between use of terms will be explicit.

A considerable amount of work has already been done in other areas of science to specify concepts and relationships that can form a framework for models of addiction. In particular, the Basic Formal Ontology (BFO) provides top level constructs relating to entities that have continuing existence such as objects and their characteristics (called 'continuants'), and those that involve change over time such as processes (called 'occurrents'). Other top level frameworks may be appropriate, but developing models of addiction within the BFO will make use of the work that has been done and allow these models to be linked to others in related fields of study.

What might this look like in practice? One approach to creating a general theory of addiction would be to establish an ontology of addiction with the option for those contributing to the ontology to create their own sub-ontologies with specified namespaces. Thus one might start with a medically based ontology elaborating the Mental Functioning and Mental Disease ontologies within one namespace and build others with other namespaces, adopting the principles of the OBO Foundry to maintain discipline and efficient working within the enterprise. This ontology would define entities relevant to addiction and their relationships, relate these to entities in other fields of study in multiple disciplines, and link these with empirical findings and measures.

This would be a major undertaking but could be facilitated by the development of a web portal in which users could browse, contribute to, and query the developing ontology. The resulting ontology could then form the basis for one or more general theories of addiction. Many of the core propositions within the models and theories would be embodied within the ontology but it is likely that the representation of the models and theories would need to extend to forms of representation that are not well captured with this kind of data structure, such as equations, probabilistic formulations and conditionals.

In the end, however, use of the Semantic Web, URIs, ontologies, RDF and other ways of expressing ontologies will advance the science of addiction only to the extent that people who work in the field engage with these tools in a constructive way. These are tools that can serve a useful purpose or be misused, either through ignorance, lack of skill or on purpose. To date, these tools have proved extremely valuable in other fields of social and behavioural science (Larsen et al. 2017) and clinical science (Lewis 2017), and there is reason to believe they could lead to more rapid advances in our understanding of addiction as well.

References

American Psychiatric Association (APA) (2013) *Diagnostic and Statistical Manual of Mental Disorders 5th Edition*, Arlington, VA: American Psychiatric Association.

Ashburner, M., Ball, C. A., Blake, J. A., Botstein, D., Butler, H., Cherry, J. M. and Sherlock, G. (2000) "Gene ontology: tool for the unification of biology", *Nature Genetics* 25(1): 25–29.

Berridge, K. C. and Robinson, T. E. (2016) "Liking, wanting, and the incentive-sensitization theory of addiction", *American Psychologist* 71(8): 670–679.

Blum, K., Thanos, P. K., Oscar-Berman, M., Febo, M., Baron, D., Badgaiyan, R. D. and Gold, M. S. (2015) "Dopamine in the brain: hypothesizing surfeit or deficit links to reward and addiction", *Journal of Reward Deficiency Syndrome and Addiction Science* 1(3): 95–104.

Ceusters, W. and Smith, B. (2010) "Foundations for a realist ontology of mental disease", *Journal of Biomedical Semantics* 1: 10.

du Plessis, G. P. (2014) "An integral ontology of addiction: a multiple object as a continuum of ontological complexity", *Journal of Integral Theory and Practice* 9(1): 38.

Gallus, S., Schiaffino, A., La Vecchia, C., Townsend, J. and Fernandez, E. (2006) "Price and cigarette consumption in Europe", *Tobacco Control* 15(2): 114–119.

Gossop, M., Darke, S., Griffiths, P., Hando, J., Powis, B., Hall, W. and Strang, J. (1995) "The Severity of Dependence Scale (SDS): psychometric properties of the SDS in English and Australian samples of heroin, cocaine and amphetamine users", *Addiction* 90(5): 607–614.

Hastings, J., Le Novère, N., Ceusters, W., Mulligan, K. and Smith, B. (2012a) "Wanting what we don't want to want: representing addiction in interoperable bio-ontologies", published in the Proceedings of ICBO 2012, Graz, Austria: CEUR-WS volume 897, paper 12, http://ceur-ws.org/Vol-897/session3-paper12.pdf.

Hastings, J., Smith, B., Ceusters, W., Jensen, M. and Mulligan, K. (2012b) "*Representing mental functioning: ontologies for mental health and disease*", paper presented at the *ICBO 2012: 3rd International Conference on Biomedical Ontology*.

Larsen, K. R., Michie, S., Hekler, E. B., Gibson, B., Spruijt-Metz, D., Ahern, D. and Yi, J. (2017) "Behavior change interventions: the potential of ontologies for advancing science and practice", *Journal of Behavioral Medicine* 40(1): 6–22.

Lewis, S. E. (2017) "The vision and challenges of the gene ontology", *Methods in Molecular Biology* 1446: 291–302.

Orford, J. (2001) *Excessive Appetites: A Psychological View of Addictions*, London, UK: John Wiley and Sons Ltd.

Seo, D. and Sinha, R. (2014) "The neurobiology of alcohol craving and relapse", *Handbook of Clinical Neurology* 125: 355–368.

Smith, B., Ashburner, M., Rosse, C., Bard, J., Bug, W., Ceusters, W. and Lewis, S. (2007) "The OBO Foundry: coordinated evolution of ontologies to support biomedical data integration", *Nature Biotechnology* 25(11): 1251–1255.

Üstün, T. B., Chatterji, S., Bickenbach, J., Kostanjsek, N. and Schneider, M. (2003) "The International Classification of Functioning, Disability and Health: a new tool for understanding disability and health", *Disability and Rehabilitation* 25(11–12): 565–571.

West, R. (2009) "The multiple facets of cigarette addiction and what they mean for encouraging and helping smokers to stop", *COPD* 6(4), 277–283.

West, R. (2017) *100 Key Facts about Addiction*, London, UK: Silverback Publishing.

West, R. and Brown, J. (2013) *Theory of Addiction*, London, UK: John Wiley and Sons.

World Health Organisation (2016) *International Classification of Diseases and Related Health Problems 10th Revision*, http://apps.who.int/classifications/icd10/browse/2010/en#, Geneva: WHO.

14
GAMBLING DISORDER

Seth W. Whiting, Rani A. Hoff, and Marc N. Potenza

Gambling, or behavior involving risk of some valued commodity on the outcome of a game or event with a chance outcome (Whelan *et al.* 2007), is a ubiquitous leisure activity. Slot machines, roulette, craps, other casino gambling, animal racing, online gambling, and wagering on the outcomes of skill games such as billiards or betting on sports with friends all included, more adults gamble in a given year than those who do not (Gerstein *et al.* 1999). Despite the frequency of gambling behavior and the financial impact of the industry, only a small percentage of the population develops gambling problems. Formerly termed pathological gambling (APA 2000), gambling disorder is characterized by persistent, recurrent, and distressful problematic gambling behavior (APA 2013). Prevalence estimates of pathological gambling approximate 0.5–1%, and up to about 5% of adults may experience at least some problems with gambling (Petry *et al.* 2005; Wardle *et al.* 2007). Although the prevalence of pathological gambling is relatively low, the impact on the individual may be severe, particularly with respect to finances (Boardman and Perry 2007; Grant *et al.* 2010) and relationships in and beyond the family (Jacobs 1989).

Gambling disorder is diagnosed according to nine criteria, of which four must be present in order to qualify. The criteria cover aspects of tolerance, withdrawal, preoccupation, repeated and unsuccessful attempts to stop, gambling to escape distress, gambling to recoup losses, lying about gambling frequency, borrowing money to cover gambling losses, and suffering in other areas of life due to gambling (APA 2013). As suggested by Fantino (2008), gambling is a problem of social importance in which behavioral, cognitive, and biological components all contribute, and such a suggestion is supported by these various criteria. A discussion of each of these domains and their importance to the understanding of gambling disorder follows.

Behavioral aspects of gambling

Gambling necessarily involves interactions of the environment and the gambler, and thus a study of contextual factors and how the individual responds under gambling conditions is germane to a comprehensive analysis of the disorder. From the results of experimentation across several decades, Weatherly and Dixon (2007) proposed a behavioral model of gambling consisting of three components. First, contingency control and structural features of gambling might contribute to gambling disorder. Stimuli that provoke or prolong gambling and the arrangements of reinforcing events following behavior are included. Skinner (1953: 104) noted that a variety of forms of gambling such as roulette and slot machine betting pay the gambler on an intermittent schedule of reinforcement such that winning occurs after an unpredictable number of plays and losses. Such contingencies characteristically maintain high rates of behavioral engagement over extended time durations, even when the payouts are disadvantageous for the player. Along with the unpredictable schedules of reinforcement, factors such as high-magnitude wins or jackpots and the high rate of gambles may further increase the response-maintaining characteristics of gambling.

In addition, special outcomes such as "near misses" and "losses disguised as wins" may perpetuate gambling. Near misses are those in which a losing outcome is perceived as "close" to a win. When playing a slot machine, for example, a near miss occurs when the pay line shows two matching symbols, but the third matching symbol is just above or below the pay line (e.g., Dillen and Dixon 2008). When these outcomes are encountered, gamblers have subjectively rated these outcomes as better than losses (Dixon and Schreiber 2004), and individuals with pathological gambling have demonstrated similar reward responses in the brain following near miss outcomes as when following a winning outcome on a slot machine (Habib and Dixon 2010), although findings that show differences in near-miss and win processing have also been reported in pathological gambling (Worhunsky et al. 2014). Losses disguised as wins, like near misses, may create similar effects and may function to reinforce gambling behavior despite actually losing the player money. Here, a gambler bets some amount of money and encounters a winning outcome that pays back less money than originally wagered. Gamblers frequently and erroneously categorize these outcomes as wins and demonstrate greater arousal than when strictly losing (Dixon et al. 2015).

The authors of the behavioral model of gambling noted that many who gamble experience these outcomes similarly, but few experience gambling disorder. The second factor, verbal behavior, may explain experiential differences that discriminate these groups. Directly stated rules, whether accurate (e.g., "Slot machines are a losing game") or inaccurate (e.g., "Your luck will change if you keep gambling"), relate to gambling behavior regardless of gambling expertise (Weatherly et al. 2007). Further, derived rules, or verbal rules not directly stated that may influence behavior through relations, may bias gambling behavior in a disadvantageous manner. For example, after black was paired with "greater" stimuli and red was paired with "lesser" stimuli, roulette gamblers tended to bet more frequently on black outcomes, and vice-versa, indicating that they formed a self-rule about which alternative was better (Whiting and Dixon 2015). Although the odds of winning on each of these outcomes were matched, betting was biased as though one was less risky or more advantageous.

Third, the behavioral model incorporates motivational and individual factors to further explain differences between individuals with gambling disorder and those who gamble only recreationally. Here, additional factors warrant consideration. For example, the individual's financial state or need for money may increase or decrease the likelihood of gambling. In the same way, proximity to gambling opportunities and the local legality of gambling may affect behavior. Another individual variable, impulsivity, has garnered attention with respect to gambling disorder. Individuals with pathological gambling have demonstrated greater impulsive decision-making than individuals without pathological gambling, preferring smaller amounts of money available immediately in lieu of larger payouts following a delay (Dixon et al. 2003), and more impulsive individuals may experience a greater subjective value of gambling wins than the objective monetary loss (Rachlin et al. 2015).

Beyond these three content areas, behavioral research has noted that escape-based functions may be important in the determination of gambling disorder. In accordance with the diagnostic criteria, gambling to escape or avoid distress, or negative reinforcement, has been significantly correlated with disordered gambling and showed higher correlations with disordered gambling than betting for attention, for sensory stimulation, or for money and prizes (Miller et al. 2010). Additionally, as escape tendencies increase, so may the amount of risk taken when gambling (Weatherly et al. 2010).

Cognitive aspects of gambling

Cognitive theories of gambling assume a flawed set of beliefs, such as that gambling can be profitable or that one has a particular edge over a casino that will allow one to win in the long run (Rodgers 1998). Flawed beliefs or faulty heuristics and biases have been related to increased gambling and sub-optimal decision-making. For example, in one study, participants engaged in a talk-aloud procedure and spoke thoughts out loud while gambling, and participants who engaged in regular gambling spoke significantly more frequent irrational verbalizations than non-regular gamblers (Griffiths 1994). However, cognitive biases are seen both in people with gambling problems and those without and thus may not represent a key distinguishing feature.

Cognitive biases that may be directly relevant to gambling behavior are the gambler's fallacy, the hot-hand phenomenon, and the illusion of control. First, the gambler's fallacy is a phenomenon in which independent events are perceived as somehow related (Sundali and Croson 2006). In games of skill such as darts, for example, the results of a throw such as hitting below the bull's eye provide helpful feedback for the player; he or she can make adjustments on the next shot and (perhaps) have a greater chance at success. However, in gambling such as on slot machines, the outcome of one spin provides no relevant information in the determination of the next. Gamblers playing under this fallacy may be more likely to bet red in roulette after observing several black spins in a row, or may perceive that after a streak of losses that a win is "due" and continue playing. In contrast to the gambler's fallacy, the hot-hand phenomenon is a faulty belief in which one perceives that a streak in random events will continue (Ayton and Fischer 2004). That is, if an individual has won on several consecutive gambles, a hot-hand fallacy is evident when the gambler believes that she or he will continue to win, which may result in continued risk-taking. Both types of biases have been observed in live casino settings (Sundali and Croson 2006).

The illusion of control is a bias in which an individual believes he or she has some control in a random outcome, and thus the individual believes the likelihood of personal success is higher than the actual given odds (Langer 1975). It has been argued that gambling activities possess characteristics that may promote such a bias: slot machines allow the gambler to stop the spinning reels immediately following a button press, lottery gamblers are allowed to choose their own numbers, and craps players take turns rolling the dice themselves. Similar to the previously discussed cognitive biases, the illusion of control may be dangerous for gamblers because many gambling events are independent, random, and out of the control of the gambler. Adjusting decisions based on perceived control when none exists may contribute to negative results for gamblers. For example, when gamblers were allowed to voluntarily stop slot machine reels from spinning, they have reported that their actions may have influenced the results in a skill-related fashion and have demonstrated greater gambling persistence (Ladouceur and Sévigny 2005).

Other cognitive biases may influence gambling behaviors (see Rodgers 1998 for a review of cognitive biases in lottery gambling). Misunderstandings of odds and probabilities may perpetuate gambling when one believes the odds of winning are greater than in reality; entrapment or sunk-cost effects may increase the likelihood that gambling will persist if individuals have lost money and refuse to cut losses; perceived luckiness may draw individuals to games of chance based on an unfounded confidence (Rodgers 1998).

Biological aspects of gambling

Pathological gambling has been found to be heritable; approximately 50% of the variance in the etiology may be genetic in nature, with increasingly large genetic contributions with increasing problem-gambling severity (Eisen *et al.* 1998). Genetic factors contribute to the co-occurrence of pathological gambling and mood (Potenza *et al.* 2005), anxiety (Giddens *et al.* 2011), and substance-use (Slutske *et al.* 2000; Xian *et al.* 2014) disorders, with environmental contributions being more variable.

Specific neurotransmitter systems have been proposed to contribute to pathological gambling. First, dopamine, a neurotransmitter involved in reward processing and reward-based learning, has been reported to be elevated in individuals with pathological gambling compared with those without (Bullock and Potenza 2012), and dopamine levels have been observed to increase during exposure to gambling (Joutsa *et al.* 2012). However, significant questions remain about how central dopamine is to gambling disorder, particularly given data finding no between-group differences in dopamine receptor availability and the absence of therapeutic efficacy in randomized clinical trials of dopamine receptor antagonists (Potenza 2013a). Second, opioids, involved in feelings of pleasure and urges, have been implicated with opioid receptor antagonists demonstrating superiority over placebo controls in several placebo-controlled randomized clinical trials in the treatment of pathological gambling (Bullock and Potenza 2012; Yip and Potenza 2014). Third, serotonin has also been implicated, particularly with respect to aspects of impulse control (Crockett *et al.* 2009). Low levels of a serotonin metabolite (5-hydroxy-indole acetic acid) have been found in spinal fluid samples of individuals with pathological gambling (Evers *et al.* 2005). Last, norepinephrine, which may influence feelings of excitement, has been found in elevated levels in bodily fluids of men with pathological gambling relative to those without (Bullock and Potenza 2012). While specific neurotransmitter systems have been proposed to link to specific aspects of pathological gambling (Potenza 2001), this attribution is likely an oversimplification and, while these neurotransmitter systems have been implicated in the pathophysiology of pathological gambling (Potenza 2013b; Balodis and Potenza 2016) and may contribute to cognitive processes in pathological gambling (Potenza 2014), there is still much that we do not understand regarding the pathophysiology of pathological gambling or gambling disorder.

Specific neural structures have been implicated in pathological gambling. Several brain regions repeatedly identified include the ventral striatum and ventromedial prefrontal cortex. For example, ventral striatal activation has shown an inverse correlation with impulsive tendencies; that is, lower activation in this region has been linked to greater impulsivity, including in pathological gambling (Beck *et al.* 2009; Balodis *et al.* 2012), and such decision-making traits have been linked to development of addictions (Beck *et al.* 2009; Dalley *et al.* 2011). Relatively blunted ventral striatal activation has been observed in individuals with (as compared to without) pathological gambling during exposure to videotaped cues in a manner similar to findings in individuals with and without cocaine dependence (Potenza 2008). Blunted ventral striatal activation has also been observed in individuals with pathological gambling during reward anticipation (Balodis *et al.* 2012; Choi *et al.* 2012), with similar findings observed in alcohol dependence and nicotine dependence (reviewed in Balodis and Potenza 2015). Relatively diminished activation of the ventromedial prefrontal cortex, a brain structure implicated in decision-making and impulse control, has been observed in pathological gambling during a cognitive control task (Potenza *et al.* 2003a), cue exposure (Potenza *et al.* 2003b), simulated gambling (Reuter *et al.* 2005) and processing receipt of monetary rewards (Balodis *et al.* 2012), as well as in decision-making in individuals with pathological gambling and substance-use disorders (Tanabe *et al.* 2007). Other structures (e.g., the insula) have also been implicated in pathological gambling (Clark 2014), and how brain circuits operate in people with and without gambling problems warrants continued study (van Holst *et al.* 2014).

Differences in white matter and gray matter have been observed in individuals with and without pathological gambling. Early studies suggested both widespread (Joutsa *et al.* 2011) and more focal (e.g., in the genu of the corpus collosum; Yip *et al.* 2013) white-matter differences in individuals with and without pathological gambling, with findings suggestive of poorer white matter integrity that appears linked to greater problem-gambling severity (Chamberlain *et al.* 2016). Recently, abnormalities in crossing fibers have been reported in association with both pathological gambling and cocaine dependence, suggesting commonalities across these disorders (Yip *et al.* in press). While early studies did not identify substantial volumetric differences in gray matter related to pathological gambling, differences in subcortical volume have been reported more recently, with reduced amygdalar and hippocampal volumes observed (Rahman *et al.* 2014). The extent to which these neurobiological differences relate to specific features of pathological gambling warrants additional investigation.

Co-occurring disorders

Further adding to impairment and distress, gambling disorder frequently co-occurs with other psychiatric conditions. Gambling disorder is recognized as a substance-related disorder in the *Diagnostic and Statistical Manual of Mental Disorders-5* (APA 2013) due to the acknowledged similarities with substance abuse in symptoms, course, and neurological functioning and activation, and with such similarities gambling disorder is seldom observed in isolation.

Clinical disorders are relatively common in individuals with gambling disorder. In those with gambling disorder, the odds of a diagnosable mental health disorder are often several times more likely than in the general population and show a dose-response effect such that greater problem-gambling severity is associated with higher likelihoods of psychiatric disorders (Cunningham-Williams *et al.* 1998; Desai and Potenza 2008). Conversely, approximately 6–12% of psychiatric patients qualify for gambling disorder (George and Murali 2005; Grant *et al.* 2005), a range notably higher than that of the general population (Petry *et al.* 2005).

With the similarities between gambling disorder and other addictive disorders, and given the similar settings in which each occur, substance-use disorders are among the most commonly observed comorbid clinical conditions. In an analysis of the data collected from the National Epidemiological Survey on Alcohol and Related Conditions (NESARC), alcohol-use disorder (AUD) was the most common co-occurring disorder in individuals with pathological gambling at 73% comorbidity (Petry *et al.* 2005). It has been proposed that these conditions may be biologically linked via the presence of an allelic variant of the gene encoding the dopamine D2 receptor (Blum *et al.* 2000), although this variant is in linkage disequilibrium with another variant (*ANKK1*), which has mapped more closely to addictions (Yang *et al.* 2008). More recently, a polygenic risk score for alcohol dependence was associated with gambling disorder independent of alcohol-use disorder status (Lang *et al.* 2016). Analyses of NESARC data further revealed drug-use disorders and nicotine dependence were comorbid in 38% and 60% of individuals with pathological gambling, respectively. Mood (50%) and anxiety (41%) disorders were also relatively common (Petry *et al.* 2005). Co-occurring disorders may also be used to select pharmacotherapies for individuals with pathological gambling (Bullock and Potenza 2012), although no medications have an indication from the US Food and Drug Administration for pathological gambling or gambling disorder. Irrespective of co-occurring disorders, behavioral therapies (e.g., cognitive behavioral therapy) are arguably the mainstay of treatment for people with pathological gambling, with self-help groups (e.g., the 12-step-based Gamblers Anonymous) also widely utilized (Yip and Potenza 2014).

Along with clinical disorders, personality disorders occur at much greater rates in those with gambling disorder than those without. The most prevalent personality disorders in individuals with gambling disorder are narcissistic, antisocial, avoidant, obsessive-compulsive, and borderline personality disorders, all with approximately 13–16% rates of comorbidity in treatment-seeking populations (Dowling et al. 2015). Similar strong associations have been observed in community samples (Petry et al. 2005; Desai and Potenza 2008).

Conclusions

Pathological gambling, now termed gambling disorder in DSM-5, is a condition that can have a significant negative impact on individuals and those close to them. An improved understanding of the biological, psychological and social factors that lead to the development of gambling problems and that might be targeted in prevention, treatment and policy approaches should lead to significant public and personal health gains.

References

American Psychiatric Association (APA) (2000) *Diagnostic and Statistical Manual of Mental Disorders. Text Revision, 4th edition*, Washington, DC: American Psychiatric Press.
American Psychiatric Association (APA) (2013) *Diagnostic and Statistical Manual of Mental Disorders, 5th edition*, Washington, DC: American Psychiatric Press.
Ayton, P. and Fischer, I. (2004) "The hot hand fallacy and the gambler's fallacy: two faces of subjective randomness?" *Memory and Cognition* 32(8): 1369–1378.
Balodis, I. M., Kober, H., Worhunsky, P. D., Stevens, M. C., Pearlson, G. D. and Potenza, M. N. (2012) "Diminished fronto-striatal activity during processing of monetary rewards and losses in pathological gambling", *Biological Psychiatry* 71(8): 749–757.
Balodis, I. M. and Potenza, M. N. (2015) "Anticipatory reward processing in addicted populations: a focus on the monetary incentive delay task", *Biological Psychiatry* 77(5): 434–444.
Balodis, I. M. and Potenza, M. N. (2016) "Imaging the gambling brain", in N. M. Zahr and E. Peterson (eds), *International Review of Neurobiology, Volume 129*, Cambridge, MA: Academic Press, pp. 111–124.
Beck, A., Schlagenhauf, F., Wustenberg, T., Hein, J., Kienast, T., Kahnt, T., Schmack, K., Hägele, C., Knutson, B., Heinz, A. and Wrase, J. (2009) "Ventral striatal activation during reward anticipation correlates with impulsivity in alcoholics", *Biological Psychiatry* 66(8): 734–742.
Blum, K., Braverman, E. R., Holder, J. M., Lubar, J. F., Monastra, V. J., Miller, D., Lubar, J. O., Chen, T. J. and Comings, D. E. (2000) "Reward deficiency syndrome: a biogenetic model for the diagnosis and treatment of impulsive, addictive, and compulsive behaviors", *Journal of Psychoactive Drugs* 32(i–iv): 1–112.
Boardman, P. and Perry, J. J. (2007) "Access to gambling and declaring personal bankruptcy", *Journal of Socio-Economics* 36(5): 789–801.
Bullock, S. A. and Potenza, M. N. (2012) "Pathological gambling: neuropsychopharmacology and treatment", *Current Psychopharmacology* 1: 67–85.
Chamberlain, S. R., Derbyshire, K., Daws, R. E., Odlaug, B. L., Leppink, E. W. and Grant, J. E. (2016) "White matter tract integrity in treatment-resistant gambling disorder", *British Journal of Psychiatry* 208(6): 579–584.
Choi, J. S., Shin, Y. C., Jung, W. H., Jang, J. H., Kang, D. H., Choi, C. H., Choi, S. W., Lee, J. Y., Hwang, J. Y. and Kwon, J. S. (2102) "Altered brain activity during reward anticipation in pathological gambling and obsessive-compulsive disorder", *PLOS One* 7(9): e45938.

Clark, L. (2014) "Disordered gambling: the evolving concept of behavioral addictions", *Annals of the New York Academy of Science* 1327: 46–61.

Crockett, M. J., Clark, L. and Robbins, T. W. (2009) "Reconciling the role of serotonin in behavioral inhibition and aversion: acute tryptophan depletion abolishes punishment-induced inhibition in humans", *Journal of Neuroscience* 29(38): 11993–11999.

Cunningham-Williams, R. M., Cottler, L. B., Compton III, W. M. and Spitznagel, E. L. (1998) "Taking chances: problem gamblers and mental health disorders: results from the St. Louis Epidemiological Catchment Area Study", *American Journal of Public Health* 88(7): 1093–1096.

Dalley, J. W., Everitt, B. J. and Robbins, T. W. (2011) "Impulsivity, compulsivity, and top-down cognitive control", *Neuron* 69(4): 680–694.

Desai, R. A. and Potenza, M. N. (2008) "Gender differences in the association between gambling problems and psychiatric disorders", *Social Psychiatry and Psychiatric Epidemiology* 43(3): 173–183.

Dillen, J. and Dixon, M. R. (2008) "The impact of jackpot and near-miss magnitude on rate and subjective probability of slot machine gamblers", *Analysis of Gambling Behavior* 2(2): 121–134.

Dixon, M. J., Collins, K., Harrigan, K. A., Graydon, C. and Fugelsang, J. A. (2015) "Using sound to unmask losses disguised as wins in multiline slot machines", *Journal of Gambling Studies* 31(1): 183–196.

Dixon, M. R., Marley, J. and Jacobs, E. A. (2003) "Delay discounting by pathological gamblers", *Journal of Applied Behavior Analysis* 36(4): 449–458.

Dixon, M. R. and Schreiber, J. E. (2004) "Near-miss effects on response latencies and win estimations of slot machine players", *The Psychological Record* 53(3): 335–348.

Dixon, M. R., Whiting, S. W., Gunnarsson, K. F., Daar, J. H. and Rowsey, K. E. (2015) "Trends in behavior-analytic gambling research and treatment", *The Behavior Analyst* 38(2): 179–202.

Dowling, N. A., Cowlishaw, S., Jackson, A. C., Merkouris, S. S., Francis, K. L. and Christensen, D. R. (2015) "The prevalence of comorbid personality disorders in treatment-seeking problem gamblers: a systematic review and meta-analysis", *Journal of Personality Disorders* 29(6): 735–754.

Eisen, S. A., Lin, N., Lyons, M. J., Scherrer, J. F., Griffith, K., True, W. R., Goldberg, J. and Tsuang, M. T. (1998) "Familial influences on gambling behavior: an analysis of 3359 twin pairs", *Addiction* 93(9): 1375–1384.

Evers, E. A., Cools, R., Clark, L., van der Veen, F. M., Jolles, J., Sahakian, B. J. and Robbins, T. W. (2005) "Serotonergic modulation of prefrontal cortex during negative feedback in probabilistic reversal learning", *Neuropsychopharmacology* 60(6): 1138–1147.

Fantino, E. (2008) "Behavior analysis: thriving, but how about its future?" *Journal of the Experimental Analysis of Behavior* 89(1): 125–127.

George, S. and Murali, V. (2005) "Pathological gambling: an overview of assessment and treatment", *Advances in Psychiatric Treatment* 11(6): 450–456.

Gerstein, D., Murphy, S., Toce, M., Hoffmann, J., Palmer, A., Johnson, R., Larison, C., Chuchro, L., Buie, T., Engelman, L., Hill, M., Volberg, R., Harwood, H., Tucker, A., Christiansen, E., Cummings, W. and Sinclair, S. (1999) *Gambling Impact and Behavior Study: Report to the National Gambling Impact Study Commission*, Chicago: National Opinion Research Center.

Giddens, J. L., Xian, H., Scherrer, J. F., Eisen, S. A. and Potenza, M. N. (2011) "Shared genetic contributions to anxiety disorders and pathological gambling in a male population", *Journal of Affective Disorders* 132(3): 406–412.

Grant, J. E., Levine, L., Kim, D. and Potenza, M. N. (2005) "Prevalence of impulse control disorders in adult psychiatric inpatients", *American Journal of Psychiatry* 162(11): 2184–2188.

Grant, J. E., Schreiber, L., Odlaug, B. L. and Kim, S. W. (2010) "Pathological gambling and bankruptcy", *Comprehensive Psychiatry* 51(2): 115–120.

Griffiths, M. D. (1994) "The role of cognitive bias and skill in fruit machine gambling", *British Journal of Psychology* 85(3): 351–369.

Habib, R. and Dixon, M. R. (2010) "Neurobehavioral evidence for the 'near-miss' effect in pathological gamblers", *Journal of the Experimental Analysis of Behavior* 93(3): 313–328.

Jacobs, D. F. (1989) "Illegal and undocumented: a review of teenage gambling and the plight of children of problem gamblers in America", in H. J. Shaffer, S. A. Stein, B. Gambino and T. N. Cummings (eds), *Compulsive Gambling: Theory, Research, and Practice*, Lanham, MD: Lexington Books, pp. 249–293.

Joutsa, J., Johansson, J., Niemela, S., Ollikainen, A., Hirvonen, M. M., Piepponen, P., Arponen, E., Alho, H., Voon, V., Rinne, J. O., Hietala, J. and Kaasinen, V. (2012) "Mesolimbic dopamine release is linked to symptom severity in pathological gambling", *NeuroImage* 60(4): 1992–1999.

Joutsa, J., Saunavaara, J., Parkkola, R., Niemelä, S. and Kaasinen, V. (2011) "Extensive abnormality of brain white matter integrity in pathological gambling", *Psychiatry Research* 194(3): 340–346.

Ladouceur, R. and Sévigny, S. (2005) "Structural characteristics of video lotteries: effects of a stopping device on illusion of control and gambling persistence", *Journal of Gambling Studies* 21(2): 117–131.

Lang, M., Lemenager, T., Streit, F., Fauth-Buhler, M., Frank, J., Juraeva, D., Witt, S. H., Degenhardt, F., Hofmann, A., Heilmann-Heimbach, S., Kiefer, F., Brors, B., Grabe, H. J., John, U., Bischof, A., Bischof, G., Volker, U., Homuth, G., Beutel, M., Lind, P. A., Medland, S. E., Slutske, W. S., Martin, N. G., Voltzke, H., Nothen, M. M., Meyer, C., Rumpf, H. J., Wurst, F. M., Rietschel, M. and Mann, K. F. (2016) "Genome-wide association study of pathological gambling", *European Psychiatry* 36: 38–46.

Langer, E. J. (1975) "The illusion of control", *Journal of Personality and Social Psychology* 32(2): 311–328.

Miller, J. C., Dixon, M. R., Parker, A., Kulland, A. M. and Weatherly, J. N. (2010) "Concurrent validity of the gambling functional assessment (GFA): correlations with the South Oaks Gambling Screen (SOGS) and indicators of diagnostic efficacy", *Analysis of Gambling Behavior* 4: 61–75.

Petry, N. M., Stinson, F. S. and Grant, B. F. (2005) "Comorbidity of DSM-IV pathological gambling and other psychiatric disorders: results from the National Epidemiological Survey on Alcohol and Related Conditions", *Journal of Clinical Psychiatry* 66(5): 564–574.

Potenza, M. N. (2001) "The neurobiology of pathological gambling", *Seminars in Clinical Neuropsychiatry* 6(3): 217–226.

Potenza, M. N. (2008) "The neurobiology of pathological gambling and drug addiction: an overview and new findings", *Philosophical Transactions of the Royal Society Part B* 373(1507): 3181–3189.

Potenza, M. N. (2013a) "How central is dopamine to gambling disorder?" *Frontiers in Behavioral Neuroscience* 7: 206.

Potenza, M. N. (2013b) "Neurobiology of gambling behaviors", *Current Opinion in Neurobiology* 23(4): 660–667.

Potenza, M. N. (2014) "The neural bases of cognitive processes in gambling disorder", *Trends in Cognitive Sciences* 18(8): 429–438.

Potenza, M. N., Leung, H.-C., Blumberg, H. P., Peterson, B. S., Fulbright, R. K., Lacadie, C. M., Skudlarski, P. and Gore, J. C. (2003a) "An fMRI Stroop study of ventromedial prefrontal cortical function in pathological gamblers", *American Journal of Psychiatry* 160(11): 1990–1994.

Potenza, M. N., Steinberg, M. A., Skudlarski, P., Fulbright, R. K., Lacadie, C. M., Wilber, M. K., Rounsaville, B. J., Gore, J. C. and Wexler, B. E. (2003b) "Gambling urges in pathological gamblers: an fMRI study", *Archives of General Psychiatry* 60(8): 828–836.

Potenza, M. N., Xian, H., Shah, K. R., Scherrer, J. F. and Eisen, S. A. (2005) "Shared genetic contributions to pathological gambling and major depression in men", *Archives of General Psychiatry* 62(9): 1015–1021.

Rachlin, H., Safin, V., Arfer, K. B. and Yen, M. (2015) "The attraction of gambling", *Journal of the Experimental Analysis of Behavior* 103(1): 260–266.

Rahman, A. S., Xu, J. and Potenza, M. N. (2014) "Hippocampal and amygdalar volumetric differences in pathological gambling: associations with the behavioral inhibition system", *Neuropsychopharmacology* 39(3): 738–745.

Reuter, J., Raedler, T., Rose, M., Hand, I., Gläscher, J. and Büchel, C. (2005) "Pathological gambling is linked to reduced activation of the mesolimbic reward system", *Nature Neuroscience* 8(2): 147–148.

Rodgers, P. (1998) "The cognitive psychology of lottery gambling: a theoretical review", *Journal of Gambling Studies* 14(2): 111–143.

Skinner, B. F. (1953) *Science and Human Behavior*, New York, NY: Appleton-Century-Crofts.

Slutske, W. S., Eisen, S., True, W. R., Lyons, M. J. and Tsuang, M. (2000) "Common genetic vulnerability for pathological gambling and alcohol dependence in men", *Archives of General Psychiatry* 57(7): 666–673.

Sundali, J. and Croson, R. (2006) "Biases in casino betting: the hot hand and the gambler's fallacy", *Judgement and Decision Making* 1(1): 1–12.

Tanabe, J., Thompson, L., Claus, E., Dalwani, M., Hutchison, K. and Banich, M. T. (2007) "Prefrontal cortex activity is reduced in gambling and nongambling substance users during decision-making", *Human Brain Mapping* 28(12): 1276–1286.

van Holst, R. J., Chase, H. W. and Clark, L. (2014) "Striatal connectivity changes following gambling wins and near-misses: associations with gambling severity", *Neuroimage Clinical* 5: 232–239.

Wardle, H., Sproston, K., Orford, J., Erens, B., Griffiths, M., Constantine, R. and Pigott, S. (2007) *British Gambling Prevalence Survey 2007*, London, UK: National Center for Social Research.

Weatherly, J. N., Austin, D. P. and Farwell, K. (2007) "The role of 'experience' when people gamble on three different video-poker games", *Analysis of Gambling Behavior* 1: 34–43.

Weatherly, J. N. and Dixon, M. R. (2007) "Toward an integrative behavioral model of gambling", *Analysis of Gambling Behavior* 1: 4–18.

Weatherly, J. N., Montes, K. S. and Christopher, D. M. (2010) "Investigating the relationship between escape and gambling behavior", *Analysis of Gambling Behavior* 4(2): 79–87.

Whelan, L. P., Steenbergh, T. A. and Myers, A. W. (2007) *Problem and Pathological Gambling*, Cambridge, UK: Hogrefe and Huber.

Whiting, S. W. and Dixon, M. R. (2015) "Examining contextual control in roulette gambling", *Journal of Applied Behavior Analysis* 48: 204–208.

Worhunsky, P. D., Malison, R. T., Rogers, R. D. and Potenza, M. N. (2014) "Altered neural correlates of reward and loss processing during simulated slot-machine fMRI in pathological gambling and cocaine dependence", *Drug and Alcohol Dependence* 145: 77–86.

Xian, H., Giddens, J., Scherrer, J., Eisen, S. A. and Potenza, M. N. (2014) "Environmental factors selectively impact co-occurrence of problem/pathological gambling with specific drug-use disorders in male twins", *Addiction* 109(4): 635–644.

Yang, B. Z., Kranzler, H. R., Zhao, H., Gruen, J. R., Luo, X. and Gelernter, J. (2008) "Haplotypic variants in *DRD2, ANKK1, TTC12* and *NCAM1* are associated with comorbid alcohol and drug dependence", *Alcohol Clinical and Experimental Research* 32(12): 2117–2127.

Yip, S. W., Lacadie, C., Xu, J., Worhunsky, P. D., Fulbright, R. K., Constable, R. T. and Potenza, M. N. (2013) "Reduced genual corpus callosal white matter integrity in pathological gambling and its relationship to alcohol abuse or dependence", *World Journal of Biological Psychiatry* 14(2): 129–138.

Yip, S. W., Morie, K. P., Xu, J., Constable, R. T., Malison, R. T., Carroll, K. M. and Potenza, M. N. (2017) "Shared microstructural features of behavioral and substance addictions revealed in areas of crossing fibers", *Biological Psychiatry: Cognitive Neuroscience and Neuroimaging* 2(2): 188–195.

Yip, S. W. and Potenza, M. N. (2014) "Treatment of gambling disorders", *Current Treatment Options in Psychiatry* 1(2): 189–203.

15
FOOD ADDICTION

Ashley Gearhardt, Michelle Joyner, and Erica Schulte

Addicted to something we need to survive?

Food is a necessity for our survival. Food consumption is so essential to our basic ability to function that a number of systems have evolved to ensure that we can find and consume food. Systems associated with attention, memory, reward and motivation are integral to feeding behavior. External cues (e.g., the sight and smell of food) and internal cues (e.g., feelings of hunger) are important drivers of food consumption. When gut peptides associated with hunger become elevated, we find food-related cues in our environment more salient (Tapper et al., 2010) and foods taste more pleasant (Siep et al., 2009). Cues that become coupled with food consumption are encoded in our memory to assist us in identifying food sources in the future (Morris and Dolan, 2001). Exposure to these food cues activates neural systems involved in motivation (e.g., mesolimbic dopaminergic system), thus increasing our willingness to seek out these foods (Robinson and Berridge, 2000). Not all foods are equally capable of engaging these systems. Historically, food could often be scarce and the consumption of high-calorie foods provided a greater likelihood of surviving famine. Higher calorie foods are generally more effective in engaging attention, reward and motivation systems (Stoeckel et al., 2008), perhaps to provide an incentive to seek out these foods and increase the odds of survival. Further, inhibitory control systems can be effective in controlling behavior and motivation (Logan et al., 1997). However, given that food has historically been scarce, the ability of these inhibitory control systems to successfully reduce the drive for food may be underdeveloped and contribute to a greater vulnerability to the current obesogenic environment.

Addictive substances are capable of activating systems designed to ensure human survival (e.g., eating, mating) and instead increase vulnerability to excessive use. For example, cues associated with drugs of abuse can become increasingly more salient after repeated use and addiction-prone individuals will attend more to these cues and find them more appealing (Robinson and Berridge, 2000). These cues also become powerful triggers for biological and behavioral systems that enhance desire and motivation for the drug (Robinson and Berridge, 2008). Memories of prior drug use can also lead to craving and drug-seeking behavior (Volkow et al., 2013). Dysfunction in these systems (e.g., attention, memory, motivation) contributes to compulsive patterns of drug use despite negative consequences or repeated efforts to stop use

(Volkow et al., 2008). Thus, the systems involved in seeking out foods necessary for survival can become dysregulated in response to an addictive substance and begin to drive behaviors that are detrimental for health and well-being.

In the modern world, our food environment has changed drastically on a global level. Eating behavior is becoming increasingly driven by desire for hedonic pleasure, rather than to meet a homeostatic need. For much of human history, the higher calorie foods available were items like fruit (which have relatively high sugar levels) and meat (which have relatively high fat levels). These naturally occurring foods are typically high in either carbohydrates (e.g., fruit) or fat (e.g., nuts), but it is rare to find a food that is naturally high in both these nutrients. In contrast, our current environment is flooded with highly rewarding foods with artificially high levels of nutrient dense ingredients (e.g., sugar, fat). In the modern Western food environment, we are able to cheaply manufacture on a mass-scale processed foods that are high in both refined carbohydrates (e.g., sugar, white flour) and fat, such as ice cream, candy, pizza, and French fries. Salt and flavor enhancers further increase the rewarding nature of these foods. These processed foods are inexpensive, easily accessible and heavily advertised. The spread of this food environment across the globe has been accompanied by rising rates of obesity and diet-related disease (e.g., diabetes). The treatments designed to turn back this tide of global weight gain are generally effective in helping people lose weight initially, but most people relapse to prior eating habits and regain their lost weight (Jeffery et al., 2000). Although there are a number of factors (e.g., physical inactivity, metabolic dysfunction) that contribute to the rise of obesity, one possibility is that these highly processed foods have become potent enough to trigger an addictive response. In other words, these highly rewarding foods, like drugs of abuse, may be capable of powerfully activating systems that evolved to encourage survival and instead contribute to compulsive patterns of consumption.

In the current chapter, we will further investigate the possibility that certain foods may be addictive. First, we will discuss what foods (or attributes in food) may increase their addictive potential. Second, we will discuss individual factors associated with addictive-like eating. Third, we will review the importance of food cues and the environment in triggering eating behavior. Finally, we will discuss the treatment and policy implication of food addiction.

Which foods or food characteristics may be addictive?

The food addiction hypothesis suggests that certain foods may be capable of triggering an addictive-like response in vulnerable individuals in a similar manner to drugs of abuse (Ahmed et al., 2013; Davis and Carter, 2009). This framework theoretically parallels substance-use disorders by positing that certain foods directly contribute to problematic eating behavior for some individuals (Gearhardt et al., 2011a). An opposing theory has also been proposed to conceptualize addictive-like eating as a behavioral addiction to the act of eating, which would suggest that the type of food does not play a role in the development of an addictive-like response (Hebebrand et al., 2014). However, existing research in animal models and humans suggests that not all foods are equally implicated in patterns of addictive-like eating, providing support for the substance-based, food addiction perspective. Specifically, foods high in added fats and/or refined carbohydrates (e.g., chocolate, chips, pizza) appear to be most capable of activating addictive-like responses.

Animal studies reveal that rats exhibit behavioral indicators of an addictive-like process uniquely to foods high in added fats and/or refined carbohydrates (e.g., white flour, sugar), such as binge consumption, use despite negative consequences, and cross-sensitization to other addictive substances (e.g., amphetamine) (Avena *et al.*, 2008b; Johnson and Kenny, 2010; Oswald *et al.*, 2011; Robinson *et al.*, 2015). For instance, rats are willing to endure a foot shock in order to obtain M&Ms, a chocolate candy high in both fat and sugar (Oswald *et al.*, 2011). Animal studies have also begun to prise apart the differential roles of fat and refined carbohydrates like sugar. When sucrose (sugar) is removed from the rats' diets following intermittent access, symptoms of withdrawal have been observed, such as teeth chattering and anxiety (Avena *et al.*, 2008a). While withdrawal does not occur when fat is removed from the rats' diets, bingeing on high-fat foods seems to be more closely associated with weight gain than consumption of sucrose (Avena *et al.*, 2009). Finally, animal models have found that bingeing on high-fat, high-sugar foods (e.g., cheesecake, chips, cookies) may downregulate dopaminergic receptors in a similar manner to prolonged exposure to drugs of abuse (Johnson and Kenny, 2010; Robinson *et al.*, 2015).

In parallel with animal model research, foods high in added fats and/or refined carbohydrates (e.g., cake, French fries) also appear to be most related to indicators of addictive-like eating. Individuals frequently report greater cravings for foods high in added fats and/or refined carbohydrates (White and Grilo, 2005) and consumption of large quantities of these foods while feeling unable to control the amount consumed (Vanderlinden *et al.*, 2001), which relate to two diagnostic indicators of substance-related and addictive disorders in the *Diagnostic and Statistical Manual of Mental Disorders* (DSM) (American Psychiatric Association, 2013). Additionally, a recent study asked individuals to report how likely they were to experience the DSM symptoms of substance-use disorders (e.g., consumption in larger amounts or longer periods than intended, use despite negative consequences) in response to nutritionally diverse food items. Notably, the foods most implicated in addictive-like responses were high in both added fats and refined carbohydrates (e.g., chocolate, pizza, ice cream, muffin) (Schulte *et al.*, 2015). Further, foods with a high glycemic load, which measures the body's blood sugar spike after consumption, appeared to be particularly problematic for individuals who reported experiencing indicators of addictive-like eating (Schulte *et al.*, 2015). In contrast, foods without added fats or refined carbohydrates (e.g., nuts, fruits, vegetables) have limited relation to addictive-like eating behavior. High-fat, high-refined carbohydrate foods are also reported to be consumed in problematic patterns seen with drugs of abuse, such as bingeing (Yanovski *et al.*, 1992; Rosen *et al.*, 1986; Vanderlinden *et al.*, 2001) and consumption to cope with negative emotions (Zellner *et al.*, 2006; Epel *et al.*, 2001; Oliver *et al.*, 2000).

In summary, evidence is growing for the idea that foods high in added fats and/or refined carbohydrates are most closely implicated in addictive-like eating behavior. Animal and human studies have found that high-fat, high-refined carbohydrate foods may be associated with behavioral characteristics of substance-use disorders (e.g., consumption in larger amounts or longer periods than intended, inability to cut down despite desire to do so, craving). While animal models have demonstrated that prolonged consumption of these foods lead to changes in reward-related networks, research in humans is warranted to elucidate the neurobiological similarities (e.g., downregulation of dopamine) and differences (e.g., influence of appetitive response for food) between addictive-like consumption of food and drugs of abuse. This is especially important as there are drugs of abuse, like cocaine, that act directly on dopamine neurons, whereas highly palatable foods have more indirect effects on the brain. Future studies should examine whether repeated exposure to high-fat, high-refined carbohydrate foods may activate reward and executive control networks, akin to drugs of abuse, to increase the rewarding salience of these foods and motivate compulsive, addictive-like consumption in vulnerable individuals.

Who might be addicted?

Addiction is the result of not only the attributes of the substance, but also individual risk factors. There is no drug of abuse that is addicting to all or even most people who consume it. For example, alcohol is consumed by approximately 90% of people, but the prevalence of an alcohol use disorder is around 10% (American Psychiatric Association, 2000). Similarly, with cocaine, only 15–16% of individuals who use this drug become addicted to it within ten years of use (Wagner and Anthony, 2002). Thus, when investigating addiction, it is important to evaluate which individuals are most prone to developing problematic use and what factors may be placing them at increased risk.

Although there is no officially recognized diagnosis of food addiction, the Yale Food Addiction Scale (YFAS) is a valid and reliable tool designed to assess addictive-like eating. The YFAS applies the diagnostic criteria for substance-related and addictive disorders (SRAD) from the DSM to the consumption of highly rewarding foods (Gearhardt *et al.*, 2009). The original YFAS was published in 2009 and was based on the DSM-IV substance dependence criteria (Gearhardt *et al.*, 2009). The YFAS assesses seven symptoms of substance dependence (e.g., continued use despite negative consequences, withdrawal, tolerance) and clinically significant impairment or distress. To meet the threshold for YFAS food addiction, individuals need to endorse three or more symptoms in a 12-month period and impairment/distress. Recently, a new version of the YFAS (YFAS 2.0) was released to reflect changes to the SRAD section in DSM-5, including the addition of new symptoms (e.g., craving, interpersonal problems) and a spectrum of diagnostic options that ranges from mild to severe (Gearhardt *et al.*, 2016). The YFAS is available in both adult and children versions and has been translated into a number of different languages (e.g., German, French, Spanish, Turkish) (Meule and Gearhardt, 2014).

Elevated scores on the YFAS are associated with a number of eating-related problems. The prevalence of FA is about 19.9% based on a recent meta-analysis (Pursey *et al.*, 2014). Women and individuals with obesity exhibit a higher prevalence of food addiction relative to men and normal weight participants (Pursey *et al.*, 2014). In adults, higher YFAS scores are associated with an increased likelihood of obesity, binge eating, emotional eating, and diet-related health outcomes (e.g., elevated cholesterol) (Meule and Gearhardt, 2014). In children, more symptoms of food addiction are related to increased BMI, more emotional eating, and less satiety responsiveness (Gearhardt *et al.*, 2013). Factors associated with substance addictions are also associated with YFAS food addiction. For example, increased craving, higher impulsivity and a greater motivation to use in response to negative affect are central aspects of addiction (Li and Sinha, 2008). Similarly, YFAS food addiction is associated with higher craving (particularly for foods high in refined carbohydrates and fat), increased impulsivity, and elevated motivation to eat palatable foods in response to negative affect (Meule and Gearhardt, 2014). Individuals who endorse more addictive-like eating also exhibit biological responses that are implicated in addictive disorders, such as greater neural activation of reward-related regions in response to food cues (Gearhardt *et al.*, 2011c). One limitation of this research is that the majority of these studies have been cross-sectional, thus the causal relationship between food addiction and these factors is unclear. Thus, more research is needed to evaluate what risk factors may make an individual more prone to develop food addiction symptoms.

The role of cues

The incentive-sensitization theory of addiction proposes one explanation for why some people display greater susceptibility to addictive behaviors (Robinson and Berridge, 2000).

This theory, which has been applied to eating behavior as well as substance use (Berridge, 2009), distinguishes between two drivers of consummatory behavior – "liking" and "wanting." Under the incentive-sensitization theory, "liking" refers to the hedonic pleasure one derives from a reward, for example enjoying the taste and experience of eating a certain food. "Wanting," on the other hand, refers to a strong motivation to obtain and consume a reward such as food. While, initially, a reward tends to be both wanted and liked, with repeated consumption an individual becomes sensitized to cues associated with the reward, triggering excessive "wanting" even in the absence of liking (Robinson and Berridge, 2008). "Wanting" is thus thought to be the driving force motivating compulsive consummatory behavior, providing a hypothesis for why such behavior continues despite individuals deriving lower enjoyment from the reward (Robinson and Berridge, 2008). Importantly, according to the incentive-sensitization theory, "wanting" only occurs in the presence of relevant cues (Robinson and Berridge, 2000). These cues begin to develop incentive salience on their own, becoming especially attractive and prompting increased "wanting" and reward-seeking behavior (Robinson and Berridge, 2008; Robinson and Flagel, 2009). In humans, obese compared with non-obese individuals show sensitization to snack foods, finding them increasingly reinforcing, or "wanted," after daily consumption (Temple et al., 2009). The ability of cues to trigger increased "wanting" could explain why individuals with substance use disorders often relapse even following a period of sobriety, or why a diet becomes difficult to adhere to when one smells a tray of freshly baked cookies.

Even under the incentive-sensitization theory, not everyone shows an addictive response to foods. There appear to be individual differences in one's sensitivity to cues, such as the classification of individuals as sign-trackers versus goal-trackers (Robinson et al., 2014; Meyer et al., 2012). Sign-tracking and goal-tracking refer to different responses that occur as a result of Pavlovian conditioning. When conditioned to associate the presentation of a cue (e.g., a lever or light) with the delivery of a reward (e.g., food or cocaine), sign-trackers have a tendency to become attracted to and interact with the cue itself, while goal-trackers will instead focus on the area where they expect the reward to be delivered. In animal studies, sign-tracking behavior may include the animal touching or even attempting to eat the cue (Flagel et al., 2008). Sign-trackers and goal-trackers do not show any difference in their ability to learn the association, but rather show a bias toward a certain conditioned response (i.e., approaching either the cue or the reward). Thus, there appear to be individual differences in the attribution of incentive salience to the Pavlovian conditioned stimulus (Robinson and Flagel, 2009). Sign-tracking is associated with greater impulsivity and risk for addictive behaviors (Flagel et al., 2009). For example, rats classified as sign-trackers are more likely than goal-trackers to show reinstated food-seeking behavior when presented with a previously associated cue (Yager and Robinson, 2010), suggesting that this cue holds increased incentive salience and is triggering wanting. Thus, our cue-rich food environment is likely particularly problematic for those prone to a sign-tracking response, and in fact obese individuals show an attentional bias toward food-related cues (Yokum et al., 2011). A recent study has shown sign- and goal-tracking to be observable in humans, finding that sign-trackers were more responsive to food cues and more likely to be obese than goal-trackers (Versace et al., 2016). While research on sign- and goal-tracking in humans is still in the early stages, this concept provides a promising theory for why certain foods may prompt an addictive response in some individuals but not in others.

Treatment and policy implication

If certain foods or food attributes are found to be addictive, this may lead to a paradigm shift in the way we treat eating-related problems and the policies that shape our current food environment. Our current treatments for obesity have limited long-term effectiveness and while treatments for binge-eating disorder (e.g., Cognitive Behavioral Treatment [CBT] for eating disorders) reduce binge eating they are not effective at reducing body weight (Grilo et al., 2011). One possibility for the limited effectiveness of these treatments is that some of the mechanisms driving the overeating are not being addressed. If highly rewarding foods are capable of triggering an addictive response in some people, integrating addiction intervention approaches may be beneficial in some cases. Although there are a number of similarities in empirically supported treatments for addiction and overeating (e.g., monitoring consumption, developing alternative coping strategies), there are also distinct factors (Gearhardt et al., 2011b). For example, addiction perspectives highlight the importance of cues in triggering consumption and relapse prevention. An increased focus on these factors for individuals with more food addiction symptoms may be important. There are also pharmacological approaches in addiction that target reward and motivation systems (e.g., Contrave). If addictive mechanisms are found to play a role in overeating, addiction pharmacology approaches may prove to be useful for eating-related problems.

One aspect of an addiction perspective that may not be easily integrated into existing eating treatments is the role the food plays in excess food consumption. The CBT for binge eating take the approach that there are no good foods or bad foods (Fairburn and Harrison, 2003). One of the aims of treatment is to reduce these beliefs about these foods and to encourage consumption of all foods in moderation. However, an addiction perspective may challenge this perspective. Foods may differ in their addictive potential and their ability to trigger a cascade of biological and psychological effects that may increase the likelihood of overeating. Thus, there is likely an interaction between the risk factors of the person, but also the risk potential of the food. If certain foods have an addictive potential, this would not require that all individuals would need to strictly abstain from eating them. Instead, empirically supported treatments for addiction that aim to reduce harm by helping individuals learn to consume in a moderate way may be beneficial (Marlatt and Witkiewitz, 2002). For example, harm reduction approaches for alcohol acknowledge that alcoholic beverages vary in their risk potential with low-alcohol beer typically being less risky than hard liquor. Also, situations differ in their risk potential, such that trying to consume higher-risk alcoholic beverages when emotions are intense or around certain people may increase the likelihood of harm. However, certain higher-risk beverages (e.g., grain alcohol) may be difficult to consume moderately even in safer situations and therefore abstinence from this beverage may be more successful. In the context of eating, this approach may suggest that abstinence from the highest-risk foods in situations that are particularly triggering (e.g., times of extreme hunger, sadness, loneliness or isolation) would be beneficial. Further, some (but not all) individuals might find that specific foods are too triggering to eat regardless of the context and removing this food from the diet may be more successful.

Contingency management (CM) techniques may be adapted to emphasize abstinence, and this approach is of growing interest for the treatment of substance-related addictions. Contingency management is based on operant conditioning principles and provides individuals with rewards (e.g., gift cards) for treatment participation or achieving goals (e.g., abstaining from use) (Prendergast et al., 2006). Contingency management aims to reduce the rewarding value of addictive substances by reinforcing behavior inconsistent with drug use. Previous studies have observed that adding CM to standard treatment (e.g., psychoeducation, coping skills), relative to standard treatment alone, resulted in greater likelihood and longer duration of abstinence, reduced drop-out rates, and increased attendance of treatment sessions (for a review, see Zajac et al. in this volume). Relatedly, Kong and colleagues (2016) observed that alternative reinforcers may be useful in reducing the reinforcing effects of food. Thus, CM may also be effective in helping some individuals abstain or reduce their consumption of potentially addictive foods.

For example, individuals with addictive-like eating behavior may be rewarded with incentives for abstaining from highly triggering foods. Overall, future research is needed to evaluate whether applying these addiction-related intervention approaches would improve clinical outcomes related to obesity and binge eating.

However, it is important to note that addiction does not just impact people who have a clinical disorder. Generally, addictive responses to a substance occur on a spectrum and widespread sub-clinical response can lead to major public health consequences. Again, returning to alcohol, only a relatively small subsection of people meet for an alcohol use disorder (American Psychiatric Association, 2000). Yet, alcohol is the third leading cause of preventable death due in large part to risky use (e.g., driving drunk) even in those without a clinical disorder (Mokdad et al., 2004). The same is likely true with potentially addictive foods. The majority of people will not develop a clinical level of food addiction, but for many people these foods may elicit enough of an addictive response to encourage overconsumption. Given that even a couple extra hundred calories a day can result in excess weight gain, even mild addictive responses to these foods may lead to negative health outcomes like obesity. The most effective strategies to deal with the widespread negative public health consequences of an addictive substance are through policy (Strang et al., 2012). Although there is generally public support for educational approaches to addressing addiction, educational initiatives have limited effectiveness. Rather, environmental and economic policies are more potent weapons to turn back the tide of addiction. For example, in the case of tobacco, raising prices through taxes, restricting access through clean-air laws and reducing cues by limiting advertising have been effective in lowering smoking rates in the United States (Lantz et al., 2000). Similar food-related policies have been proposed and, while momentum may be building for certain initiatives like soda taxes, there has generally been limited public support. Given that calorie-dense, nutrient-poor foods are implicated in many chronic diseases (e.g., diabetes, heart disease), the role of policies to encourage healthier eating practices (e.g., taxation, marketing restrictions) may be important for improving public health regardless of whether these foods have an addictive potential. However, evaluating the addictive nature of these foods may be important in shifting public opinion. Individuals who believe that certain foods are addictive indicate a greater support for policy initiatives to reduce obesity, regardless of their political orientation, body mass index, or personal food addiction symptoms (Schulte et al., 2016). Although the science in this area is just beginning to build, if highly rewarding foods are found to be addictive this may be a game changer both for treatment and policy approaches to reduce compulsive overeating.

References

Ahmed, S. H., Avena, N. M., Berridge, K. C., Gearhardt, A. N. and Guillem, K. (2013) "Food addiction" in *Neuroscience in the 21st Century: From Basic to Clinical*, New York, NY: Springer.

American Psychiatric Association (APA) (2000) *Diagnostic and Statistical Manual of Mental Disorders, 4th edition (text revision)*, Washington, DC: American Psychiatric Association.

American Psychiatric Association (2013) *Diagnostic and Statistical Manual of Mental Disorders, 5th edition: DSM-5*, available online at: http://dsm.psychiatryonline.org/book.aspx?bookid=556.

Avena, N. M., Bocarsly, M. E., Rada, P., Kim, A. and Hoebel, B. G. (2008a) "After daily bingeing on a sucrose solution, food deprivation induces anxiety and accumbens dopamine/acetylcholine imbalance", *Physiology and Behavior* 94(3): 309–315.

Avena, N. M., Rada, P. and Hoebel, B. G. (2008b) "Evidence for sugar addiction: behavioral and neurochemical effects of intermittent, excessive sugar intake", *Neuroscience & Biobehavioral Reviews* 32(1): 20–39.

Avena, N. M., Rada, P. and Hoebel, B. G. (2009) "Sugar and fat bingeing have notable differences in addictive-like behavior", *The Journal of Nutrition* 139(3): 623–628.

Berridge, K. C. (2009) "'Liking' and 'wanting' food rewards: brain substrates and roles in eating disorders", *Physiology & Behavior* 97(5): 537–550.

Davis, C. and Carter, J. C. (2009) "Compulsive overeating as an addiction disorder. A review of theory and evidence", *Appetite* 53(1): 1–8.

Epel, E., Lapidus, R., McEwen, B. and Brownell, K. (2001) "Stress may add bite to appetite in women: a laboratory study of stress-induced cortisol and eating behavior", *Psychoneuroendocrinology* 26(1): 37–49.

Fairburn, C. G. and Harrison, P. J. (2003) "Eating disorders", *The Lancet* 361: 407–416.

Flagel, S. B., Akil, H. and Robinson, T. E. (2009) "Individual differences in the attribution of incentive salience to reward-related cues: implications for addiction", *Neuropharmacology* 56: 139–148.

Flagel, S. B., Watson, S. J., Akil, H. and Robinson, T. E. (2008) "Individual differences in the attribution of incentive salience to a reward-related cue: influence on cocaine sensitization", *Behavioural Brain Research* 186(1): 48–56.

Gearhardt, A. N., Corbin, W. R. and Brownell, K. D. (2009) "Preliminary validation of the Yale Food Addiction Scale", *Appetite* 52(2): 430–436.

Gearhardt, A. N., Corbin, W. R. and Brownell, K. D. (2016) "Development of the Yale Food Addiction Scale Version 2.0", *Psychology of Addictive Behaviors* 30(1): 113–121.

Gearhardt, A. N., Davis, C., Kuschner, R. and Brownell, K. D. (2011a) "The addiction potential of hyper-palatable foods", *Current Drug Abuse Reviews* 4(3): 140–145.

Gearhardt, A. N., Roberto, C. A., Seamans, M. J., Corbin, W. R. and Brownell, K. D. (2013) "Preliminary validation of the Yale Food Addiction Scale for children", *Eating Behaviors* 14(4): 508–512.

Gearhardt, A. N., White, M. and Potenza, M. (2011b) "Binge eating disorder and food addiction", *Current Drug Abuse Reviews* 4(3): 201–207.

Gearhardt, A. N., Yokum, S., Orr, P. T., Stice, E., Corbin, W. R. and Brownell, K. D. (2011c) "Neural correlates of food addiction", *Archives of General Psychiatry* 68(8): 808–816.

Grilo, C. M., Masheb, R. M., Wilson, G. T., Gueorguieva, R. and White, M. A. (2011) "Cognitive–behavioral therapy, behavioral weight loss, and sequential treatment for obese patients with binge-eating disorder: a randomized controlled trial", *Journal of Consulting and Clinical Psychology* 79(5): 675.

Hebebrand, J., Albayrak, Ö., Adan, R., Antel, J., Dieguez, C., De Jong, J., Leng, G., Menzies, J., Mercer, J. G., Murphy, M., van der Plasse, G. and Dickson, S. L. (2014) "'Eating addiction', rather than 'food addiction', better captures addictive-like eating behavior", *Neuroscience & Biobehavioral Reviews* 47: 295–306.

Jeffery, R. W., Epstein, L. H., Wilson, G. T., Drewnowski, A., Stunkard, A. J. and Wing, R. R. (2000) "Long-term maintenance of weight loss: current status", *Health Psychology* 19(1): 5.

Johnson, P. M. and Kenny, P. J. (2010) "Dopamine D2 receptors in addiction-like reward dysfunction and compulsive eating in obese rats", *Nature Neuroscience* 13(5): 635–641.

Kong, K. L., Eiden, R. D., Feda, D. M., Stier, C. L., Fletcher, K. D., Woodworth, E. M., Paluch, R. A. and Epstein, L. H. (2016) "Reducing relative food reinforcement in infants by an enriched music experience", *Obesity* 24(4): 917–923.

Lantz, P. M., Jacobson, P. D., Warner, K. E., Wasserman, J., Pollack, H. A., Berson, J. and Ahlstrom, A. (2000) "Investing in youth tobacco control: a review of smoking prevention and control strategies" *Tobacco Control* 9(1): 47–63.

Li, C.-S. R. and Sinha, R. (2008) "Inhibitory control and emotional stress regulation: neuroimaging evidence for frontal–limbic dysfunction in psycho-stimulant addiction", *Neuroscience & Biobehavioral Reviews* 32(3): 581–597.

Logan, G. D., Schachar, R. J. and Tannock, R. (1997) "Impulsivity and inhibitory control", *Psychological Science* 8(1): 60–64.

Marlatt, G. A. and Witkiewitz, K. (2002) "Harm reduction approaches to alcohol use: health promotion, prevention, and treatment", *Addictive Behaviors* 27(6): 867–886.

Meule, A. and Gearhardt, A. N. (2014) "Five years of the Yale Food Addiction Scale: taking stock and moving forward", *Current Addiction Reports* 1(3): 193–205.

Meyer, P. J., Lovic, V., Saunders, B. T., Yager, L. M., Flagel, S. B., Morrow, J. D. and Robinson, T. E. (2012) "Quantifying individual variation in the propensity to attribute incentive salience to reward cues", *PLOS One* 7, e38987.

Mokdad, A. H., Marks, J. S., Stroup, D. F. and Gerberding, J. L. (2004) "Actual causes of death in the United States, 2000", *JAMA: The Journal of the American Medical Association* 291(10): 1238–1245.

Morris, J. and Dolan, R. (2001) "Involvement of human amygdala and orbitofrontal cortex in hunger-enhanced memory for food stimuli", *The Journal of Neuroscience* 21(14): 5304–5310.

Oliver, G., Wardle, J. and Gibson, E. L. (2000) "Stress and food choice: a laboratory study", *Psychosomatic Medicine* 62(6): 853–865.

Oswald, K. D., Murdaugh, D. L., King, V. L. and Boggiano, M. M. (2011) "Motivation for palatable food despite consequences in an animal model of binge eating", *International Journal of Eating Disorders* 44(3): 203–211.

Prendergast, M., Podus, D., Finney, J., Greenwell, L. and Roll, J. (2006) "Contingency management for treatment of substance use disorders: a meta-analysis", *Addiction* 101(11): 1546–1560.

Pursey, K. M., Stanwell, P., Gearhardt, A. N., Collins, C. E. and Burrows, T. L. (2014) "The prevalence of food addiction as assessed by the Yale Food Addiction Scale: a systematic review", *Nutrients* 6: 4552–4590.

Robinson, M. J., Burghardt, P. R., Patterson, C. M., Nobile, C. W., Akil, H., Watson, S. J., Berridge, K. C. and Ferrario, C. R. (2015) "Individual differences in cue-induced motivation and striatal systems in rats susceptible to diet-induced obesity", *Neuropsychopharmacology* 40(9): 2113–2123.

Robinson, T. E. and Berridge, K. C. (2000) "The psychology and neurobiology of addiction: an incentive-sensitization view", *Addiction* 95: S91–S117.

Robinson, T. E. and Berridge, K. C. (2008) "Review. The incentive sensitization theory of addiction: some current issues", *Philosophical Transactions of the Royal Society London B: Biological Sciences* 363(1507): 3137–3146.

Robinson, T. E. and Flagel, S. B. (2009) "Dissociating the predictive and incentive motivational properties of reward-related cues through the study of individual differences", *Biological Psychiatry* 65(10): 869–873.

Robinson, T. E., Yager, L. M., Cogan, E. S. and Saunders, B. T. (2014) "On the motivational properties of reward cues: individual differences", *Neuropharmacology* 76: Pt B, 450–459.

Rosen, J. C., Leitenberg, H., Fisher, C. and Khazam, C. (1986) "Binge-eating episodes in bulimia nervosa: the amount and type of food consumed", *International Journal of Eating Disorders* 5(2): 255–267.

Schulte, E. M., Avena, N. M. and Gearhardt, A. N. (2015) "Which foods may be addictive? The roles of processing, fat content, and glycemic load", *PLOS One* 10(2): e0117959.

Schulte, E. M., Tuttle, H. M. and Gearhardt, A. N. (2016) "Belief in food addiction and obesity-related policy support", *PLOS One* 11(1): e0147557.

Siep, N., Roefs, A., Roebroeck, A., Havermans, R., Bonte, M. L. and Jansen, A. (2009) "Hunger is the best spice: an fMRI study of the effects of attention, hunger and calorie content on food reward processing in the amygdala and orbitofrontal cortex", *Behavioural Brain Research* 198(1): 149–158.

Stoeckel, L. E., Weller, R. E., Cook 3rd, E., Twieg, D. B., Knowlton, R. C. and Cox, J. E. (2008) "Widespread reward-system activation in obese women in response to pictures of high-calorie foods", *NeuroImage* 41(2): 636–647.

Strang, J., Babor, T., Caulkins, J., Fischer, B., Foxcroft, D. and Humphreys, K. (2012) "Drug policy and the public good: evidence for effective interventions", *The Lancet* 379(9810): 71–83.

Tapper, K., Pothos, E. M. and Lawrence, A. D. (2010) "Feast your eyes: hunger and trait reward drive predict attentional bias for food cues", *Emotion* 10(6): 949.

Temple, J. L., Bulkley, A. M., Badawy, R. L., Krause, N., McCann, S. and Epstein, L. H. (2009) "Differential effects of daily snack food intake on the reinforcing value of food in obese and nonobese women", *The American Journal of Clinical Nutrition* 90(2): 304–313.

Vanderlinden, J., Dalle Grave, R., Vandereycken, W. and Noorduin, C. (2001) "Which factors do provoke binge-eating? An exploratory study in female students", *Eating Behaviors* 2(1): 79–83.

Versace, F., Kypriotakis, G., Basen-Engquist, K. and Schembre, S. M. (2016) "Heterogeneity in brain reactivity to pleasant and food cues: evidence of sign-tracking in humans", *Social Cognitive and Affective Neuroscience* 11(4): 604–611.

Volkow, N. D., Wang, G. J., Fowler, J. S. and Telang, F. (2008) "Overlapping neuronal circuits in addiction and obesity: evidence of systems pathology", *Philosophical Transactions of the Royal Society London B: Biological Sciences* 363(1507): 3191–3200.

Volkow, N. D., Wang, G. J., Tomasi, D. and Baler, R. D. (2013) "Obesity and addiction: neurobiological overlaps", *Obesity Reviews* 14(1): 2–18.

Wagner, F. A. and Anthony, J. C. (2002) "From first drug use to drug dependence: developmental periods of risk for dependence upon marijuana, cocaine, and alcohol", *Neuropsychopharmacology* 26(4): 479–488.

White, M. A. and Grilo, C. M. (2005) "Psychometric properties of the Food Craving Inventory among obese patients with binge eating disorder", *Eating Behaviors* 6(3): 239–245.

Yager, L. M. and Robinson, T. E. (2010) "Cue-induced reinstatement of food seeking in rats that differ in their propensity to attribute incentive salience to food cues", *Behavioural Brain Research* 214: 30–34.

Yanovski, S. Z., Leet, M., Yanovski, J. A., Flood, M., Gold, P. W., Kissileff, H. R. and Walsh, B. T. (1992) "Food selection and intake of obese women with binge-eating disorder", *The American Journal of Clinical Nutrition* 56(6): 975–980.

Yokum, S., Ng, J. and Stice, E. (2011) "Attentional bias to food images associated with elevated weight and future weight gain: an fMRI study", *Obesity (Silver Spring)* 19(9): 1775–1783.

Zajac, K. Alessi, S. M. and Petry, N. M. (this volume, pp. 455–463) "Contingency management approaches".

Zellner, D. A., Loaiza, S., Gonzalez, Z., Pita, J., Morales, J., Pecora, D. and Wolf, A. (2006) "Food selection changes under stress", *Physiology & Behavior* 87(4): 789–793.

16
"A WALK ON THE WILD SIDE" OF ADDICTION

The history and significance of animal models

Serge H. Ahmed

August 1, 1936 is an infamous date in modern World history as it marked the opening of the Olympic Games held in Berlin, Nazi Germany. Almost entirely forgotten, it was also the day that an infrahuman animal – as nonhuman animals were commonly referred to at the time – was first reported to have behaved in a way indicating the presence of an addiction-like desire or striving for a drug, that is, a desire so predominant to take "precedence over practically all other needs and impulses" (Spragg 1940). This seminal observation was made by Sidney Spragg at Yale Laboratories of Primate Biology in New Haven, a laboratory that had been recently founded and headed by Robert Yerkes, a well-known pioneer in the study of primate behavior and cognition. Spragg reported this finding in a 132-page monograph published in 1940, thereby launching the field of research on nonhuman animal models of addiction. I will describe and discuss this landmark study in some detail below, and use it as the starting point for a walk on the animal side – not the wild one – of addiction research to explore how this research has evolved to this day and how it challenges, more or less successfully, the human uniqueness of addiction.

Until 1936, intense desires for drugs were believed to be unique to human addicts. In the words of Alfred Lindesmith, a then-leading expert on opiate addiction:

> addiction in human beings involves complex associations and perceptions which are achieved by means of verbal symbols. The genuine addict [. . .] desires morphine and knows it. [. . .] Genuine addiction has never been established in infrahuman organisms. [. . .] It is our contention that this gap could only be closed if the [animals] could be taught to speak.
>
> *Lindesmith 1946*

When Lindesmith expressed this then widely held view, there was already significant research on addictive substances in nonhuman animals – beginning with work on morphine in dogs by Claude Bernard in Paris. But this research focused almost exclusively on the effects of addictive substances on the body and its organs, and largely ignored their behavioral effects (Faust 1900; Tatum *et al.* 1929). As a matter of fact, drugs were administered "by force" to the animals using the recently-developed hypodermic syringe. This typically required immobilizing them through handling, a procedure that animals tried to resist, at least initially. No effort was made to allow animals to freely self-inject a drug. It was not until the early 1960s that the first systematic research on drug self-injection by freely moving animals was conducted (see below).

Before Spragg's inaugural work, some researchers had already claimed to have witnessed what they believed to be the behavioral manifestation of a desire or craving for a drug in nonhuman animals, mainly dogs, but this evidence was at best anecdotal (Faust 1900; Tatum *et al.* 1929). For instance, in their initial study on cocaine sensitization, Tatum and Seevers described in some detail the case of a female dog that behaved

> as if she desired the drug. About a month after injections were begun it was noticed that the animal became very excitable whenever the attendant came near [. . .]. This was especially true if the animal saw preparations for the injection, particularly the hypodermic syringe. During the next three or four months the excitement became more and more evident at the prospect of receiving the injection. [. . .] when allowed out of the cage, [she] would jump against the attendant, then lie down on her back to receive the injection. After receiving the drug she was apparently satisfied, and would of her own accord reenter the cage. Food offered at the time of injection, even after a 48-h fast, would be refused in favor of the injection.
>
> *Tatum and Seevers 1929*

Apparently, during chronic cocaine administration, the desire for cocaine had taken precedence over food and had abolished all initial resistance to the injection procedure. In the absence of other criteria, the passive submission of an animal to the injection procedure was frequently used at the time as an objective criterion for drug craving. In one study on chronic morphine, a dog behaved so cooperatively during the daily injection that the authors felt compelled to state "[she] seems actually to beg for a dose" (Plant and Pierce 1928).

For Sidney Spragg, however, "passive cooperation for hypodermic injection" could not in itself be considered adequate evidence for an addiction-related drug desire. More direct evidence had to be adduced to prove beyond any reasonable doubt that a nonhuman animal could experience and manifest a genuine desire for a drug. Obtaining this evidence was particularly relevant and important to test Lindesmith's theory about the uniqueness of addiction in humans. To test this hypothesis, Spragg deemed it necessary to turn to a nonhuman species that was closer "in the phyletic scale" to humans than dogs or even monkeys, to make sure that it was capable of forming a delayed association between a hypodermic drug injection and its effects which did not "become manifest for some ten or fifteen minutes" (Spragg 1940). Spragg reasoned that organisms unable to form delayed associations could not desire a drug, even during withdrawal distress, mainly because they could not expect the delayed effects of the hypodermic injection. This explains why he did not opt for monkeys (e.g., "macacus rhesus") which, at the time, were considered to be too cognitively limited to form delayed associations. In addition, there was also no convincing evidence of a desire for drugs in monkeys. Notably, in their study on cocaine sensitization that involved monkeys as well as dogs, Tatum and Seevers noted that rhesus monkeys did not show "any sign of craving for the drug; in fact the reverse was true. When approached with the syringe the animal presented a picture of extreme terror" (Tatum and Seevers 1929). However, Tatum and Seevers did not attribute this behavior to an inability to form delayed associations. According to them, "the absence of evident desire for the drug in [. . .] monkeys may be due to the fact that these animals do not accept handling as kindly as do dogs. Thus, dislike of being handled may mask or prevent the manifestations of conscious desire" (Tatum *et al.* 1929).

In the end, Spragg chose to conduct his research on four young adolescent chimpanzees, three males (Lyn, Velt and Frank) and one female (Kambi). Chimpanzees are phylogenetically closer to humans than any other animal species. They had also recently been shown, notably by Robert Yerkes himself, to be capable of resolving complex behavioral problems, involving multiple options and relatively long reward delays (Yerkes 1940). Since a chimpanzee would require two or three attendants to immobilize them, the use of a hypodermic syringe by force was not an option. All subjects were thus trained to accept "kindly" the injection procedure. This required ingenuity, food bait, and a lot of patience until a chimpanzee fully complied with the hypodermic injection procedure. Only then could chronic morphine administration begin. Each chimpanzee was administered a dose of morphine (2–4 mg/kg) twice a day during several months, one dose in the morning, the other in the afternoon. For each subject, the daily injection ritual unfolded as follows. The experimenter first collected the subject from its living cage which was located in the backyard of the research buildings. The subject was then walked on a leash toward a specific room in a nearby building to be injected with a morphine-loaded hypodermic syringe. After the injection, the subject remained in the room for 10–15 minutes and then was walked back on the leash to its living cage where it was left alone until the next injection. No injection of morphine was made outside of the injection room. This was to make sure that if the subject experienced drug withdrawal distress, it would yearn to go to the injection room to receive a dose of morphine for relief.

Over the course of chronic morphine administration, Spragg began to notice the emergence and progressive intensification of what he interpreted as behavioral signs of a "powerful desire" for morphine. These signs manifested only during drug withdrawal (e.g., in the morning, about 16 h after the afternoon morphine injection) and were otherwise absent. They included the following:

1. the chimpanzee showed "eagerness to be taken from the living cage by the experimenter";
2. it attempted to get to the injection room quickly, "tugging at the leash and leading the experimenter toward and into that room; and exhibiting frustration when led away from the injection room and back to the living cage without having been given an injection";
3. it showed "more or less definite solicitation of the injection, by eager cooperation in the injection procedure or even occasionally by initiation of the procedure itself" (e.g., opening of the box containing the morphine-loaded syringe, taking the syringe and giving it to the experimenter).

Spragg 1940

Lyn – a "lively, friendly" sexually immature adolescent male – was the first chimpanzee to exhibit these signs after five months of chronic morphine administration. This was on August 1, 1936. On the page of his notebook corresponding to this day, Spragg wrote:

> Lyn whimpered and pulled at the leash when I tried to put him back in his cage after weighing. He then grasped the leash in his hand and pulled back toward the Maternity Building (in which the injection room is located). I let him lead me and he went back, between the Maternity and Nursery Buildings, looking in each doorway. He looked in through the open doorway of the injection room, then pulled me into the room after him, jumped up at once on the box (on which injections are regularly made), turned toward me and grunted several times. Then he stretched out full length on the box. When I attempted to get him off the box, and lead him from the room, he

whimpered and grimaced, and tugged against me to remain on the box. I tried once more, without success. Then I took him in my arms and carried him back to his living cage, without any protest on his part. However, he whimpered and cried when I put him back in his living cage, and then barked angrily and threateningly at me when I closed the cage door.

Spragg 1940

Even today, more than 80 years after this observation, it is difficult not to see in Lyn's behavior, gesture, and vocalization unequivocal evidence for a genuine desire or striving for morphine. Even Lindesmith conceded this, although he remained skeptical about the comparison with addiction in humans (Lindesmith 1946). However, aware that mere observation of behavior could never be guaranteed to be free from anthropomorphic bias, Spragg devised an additional experimental test to obtain a clear and objective demonstration. This led him to design the first drug choice experiment ever realized in a laboratory to test if chimpanzees' desires for the drug "pervade the behavior of the organism and predominate over other, normally primary, desires" (Spragg 1940). This experiment involved an exclusive choice between a dose of morphine and a preferred fruit (e.g., a banana). Briefly, once a day, chimpanzees (Velt and Kambi) had to choose between two sticks, one painted in black, the other in white. The black stick could be used as a key to open a black box to get access to a fruit reward (a whole banana or half of an orange) while the white stick could be used to open a white box to get access to a morphine-loaded syringe. Choices were made under four different physiological states: hungry (i.e., not fed for 16–18 h) and needing morphine (i.e., not given morphine for 16–18 h); hungry and given morphine recently; given food recently and needing morphine; given both food and morphine recently. Overall, the chimpanzees chose morphine nearly exclusively during withdrawal distress; otherwise they chose food (Spragg 1940).

To Spragg and other animal researchers after him, this research demonstrated beyond any reasonable doubt morphine addiction in chimpanzees and, consequently, refuted Lindesmith's hypothesis about the uniqueness of human addiction. Chimpanzees did not have to be able to speak for them to learn that a morphine injection would eventually alleviate withdrawal distress, or to desire such an injection as a means to alleviating withdrawal distress when it was experienced. However, according to Spragg, although he had succeeded in showing a genuine addiction-like desire for a drug where others had previously failed, this was mainly because he had chosen chimpanzees as experimental subjects. For only chimpanzees were considered to be capable of forming the delayed associations necessary to bridge the temporal gap between the hypodermic drug injection and its subsequent relieving effect on withdrawal. This explains why Spragg was reluctant to extrapolate his conclusions to other nonhuman animal species, including monkeys. In fact, he believed that any species low in the phyletic scale, like rats, "could probably never become addicted to morphine, simply because they are not capable of forming associations of this order" (Spragg 1940). In other words, he simply moved Lindesmith's addiction dividing line down to a lower level in the "phyletic scale". Subsequent research on drug choice in opiate-dependent animals has shown that the basis of this reluctance is largely borne out in rats (Lenoir *et al.* 2013) but not in monkeys that, during heroin withdrawal, also prefer to use the drug over food (Griffiths *et al.* 1981; Negus 2006).

The fact that a nonhuman animal could express "a desire for the drug" during withdrawal distress was taken by Spragg and his fellow researchers as proof that "addiction has a firm organic basis." Withdrawal distress was thus not merely a "play of sympathy" to win an addict an additional drug dose. S/he really needed this dose to feel and function normally.

> Drug addiction, whether in the human or the chimpanzee subject, can be considered as a state of equilibrium, the departure of which creates a condition that generates powerful motivations to restore that equilibrium – motivations that pervade the behavior of the organism and predominate over other, normally primary, desires.
>
> *Spragg 1940*

Apart from the neurobiological details, one is struck by the similarity between Spragg's explanation of addiction and the current allostatic theory of addiction (Koob 2009), revealing a remarkable historical continuity in certain addiction theories.

However, an important difference between human addicts and Spragg's morphine-addicted chimpanzees is that the latter did not develop a state of addiction as a consequence of their own drug-use behavior. Such a state was literally forced on them by the experimenter, thereby leaving entirely open the question of what leads individuals to use a drug over a sufficiently long period of time that they become and remain addicted. In humans, many different motivations or purposes for chronic drug use exist, including, for instance, the search for pleasure, the self-medication of a preexisting negative psychological condition and/or the attempt to escape from or cope with a harsh reality (Pickard 2012; Pickard in this volume). Like many other researchers, Spragg presupposed that chimpanzees (and *a fortiori* other nonhuman animals) had none of these motivations. They manifested a desire for morphine only after they became physiologically dependent on the drug, not before they became dependent, thereby ruling out the search for pleasure as a significant factor in their desire for the drug. They were also deemed to be "not neurotic or unstable, and did not take the drug in order to escape from reality" (Spragg 1940). Obviously, this statement entirely obliterates the importance and significance of the life history and situation of experimental animals – an issue that I will return to later on. However, according to this presupposition, even if drug-naïve chimpanzees were given every opportunity to learn to self-inject morphine or another drug, they would never acquire this behavior, and thus could never go on to develop addiction without being forced into it by an experimenter. Addiction had to be inoculated in advance to install the precondition for a drug desire in a nonhuman animal. This presupposition will probably sound paradoxical to any current animal researcher who endorses the view that addiction is a disorder of compulsion, characterized by a loss of control over drug use. How could one reproduce this loss of control in a nonhuman animal that has no control over drug use from the outset? As I will show below, this question seemed at first to be resolved by the advance of the standard drug self-administration paradigm, only to resurface later under a different guise.

The belief that addiction had to be "inoculated" to allow for the emergence of a drug desire proved remarkably persistent and remained influential until the early 1960s, when the first studies on intravenous (IV) drug self-administration in nonhuman animals were conducted (Weeks 1962; Thompson and Schuster 1964). These early studies helped to define the drug self-administration paradigm that became standard in the field until this day. Briefly, during a typical session of drug self-administration, an animal with an intra-jugular catheter is connected to a drug infusion system, and placed in a cage where it has access to a response operandum (e.g., a lever) that controls the drug infusion system. This allows an animal to self-inject a drug dose directly and rapidly into its bloodstream, without being forced to take the drug by an experimenter. However, despite this, the first two studies on IV drug self-administration were conducted on subjects previously addicted to morphine by prior, forced drug administration. Researchers were so convinced that addiction had to be inoculated to "create" a motivation for the drug that they did not even consider testing a control group of drug-naïve animals for comparison (Weeks 1962; Thompson and Schuster 1964).

In this context, it must have been a shock when other researchers began to report that initially drug-naïve animals did in fact readily learn to self-inject intravenously "those drugs which man abuses severely," including cocaine, amphetamine, nicotine, and morphine (Pickens and Thompson 1968; Deneau et al. 1969). No drug-dependent "state of equilibrium" had to be inoculated and no withdrawal distress had to be relieved to make a drug desirable. Drugs were apparently able by themselves to generate a rewarding sensation that could positively reinforce behavior. In the words of some of the researchers who made this momentous discovery, the drug self-administration behavior of an initially drug-naïve animal is its "unhindered decision and a direct reflection of his preference [. . .] to exist under the influence of the drugs' effects" (Deneau et al. 1969). It was truly a momentous discovery, as it launched a massive research effort, still active to this day, on the role of positive reinforcement in drug self-administration, almost eclipsing entirely all previous research on negative reinforcement, including Spragg's inaugural study on morphine-dependent chimpanzees. From this moment in the history of animal research on addiction onwards, most researchers have studied how naïve animals initially learn the drug self-administration behavior, without being overly concerned with whether or not this behavior, once acquired, will eventually cause a state of physiological dependence or, for that matter, any other sign of addiction. In other words, the search for drug pleasure took precedence over the avoidance of drug withdrawal agony as the major source of motivation in animal research on addiction. Nonetheless, one thing did not change. Like Spragg before them, researchers who discovered that nonhuman animals were responsive to the positive reinforcing effects of drugs interpreted this finding as evidence that ruled out the importance of other sources of motivation in addiction, including once again a need to self-medicate a preexisting psychological condition and/or an attempt to escape a harsh reality.

The conclusion that addiction was primarily driven by the pursuit of drug reward seemed to be further strengthened by subsequent research showing that, under some conditions, animals would respond repeatedly and persistently in order to be under the influence of a drug, even when this pursuit interfered with other biologically-important activities, such as eating, sleeping and drinking, eventually culminating in health deterioration and death (Johanson et al. 1976; Bozarth and Wise 1985). Importantly, this self-destructive behavior was also observed when animals were explicitly offered a choice between an IV drug and a nondrug reward, typically a palatable food. For instance, in one early drug choice experiment, monkeys were allowed to choose between a large IV dose of cocaine and food every 15 minutes, 24h/24. All monkeys rapidly ended up choosing cocaine quasi-exclusively and the experiment had to be discontinued after eight days because of "concern for the health of the animals" (Aigner and Balster 1978). The experimenter now had to intervene, not to force addiction onto the animals, but instead to protect them from its self-destructive consequences. Overall, subsequent research on drug choices confirmed this finding in monkeys under a wide range of experimental conditions (Banks and Negus 2012). In rats, however, this outcome has been reproduced so far only under a much more restricted set of conditions, notably only in choice settings where animals can choose almost continuously between the drug and the nondrug options, and hence where their current choices can be influenced by the intoxicating effects of prior drug choices. In other settings where this drug influence is not possible – for instance, by imposing a sufficiently long interval between successive choices – most rats prefer the alternative option to drugs (Ahmed 2010; 2017). The origin of this species-specific difference in drug choice outcomes between rats and monkeys is not fully understood yet, but it may have something to do with rats' lower ability to form "delayed associations", as was suggested early on by Spragg (Ahmed 2017). Regardless, under certain specific conditions, such as when current choices can be influenced by the intoxicating effects of prior drug choices, most, if not all, animals will use drugs despite severe negative consequences, even when they have the choice to behave differently.

The early observation of this self-destructive behavior in nonhuman animals, combined with the then recently-discovered reward circuits of the brain by Olds and Milner (called "reward centers" at the time) (Olds and Milner 1954), conspired to give birth to the current dominant theory of addiction. Although there are now several variants of this theory, they are all built around the same theoretical core. Drugs cause addiction because they directly activate brain reward circuits, mainly the midbrain dopamine neurons and their projections to the ventral striatum, without involving the complex perceptual processes that are normally recruited by nondrug rewards. By directly boosting these circuits, drugs generate an abnormally high, albeit fake, reward signal that, though associated with no real benefit, nevertheless reinforces the self-administration behavior and promotes its repetition (Nesse and Berridge 1997; Wise and Koob 2014; Luscher 2016).

However, it is notable that although the resulting behavior is sometimes interpreted as a reflection of a voluntary, goal-directed pursuit of drug reward or euphoria, this need not be the case. Nothing forbids us from considering this drug reinforcement process as purely automatic. According to such a view, an animal would not self-administer a drug because it desires or wants to experience its euphoric or hedonic effects, but because the drug would push the "do it again" button (as it were) in its brain. It is important to note that, according to this alternative view, drug self-administration behavior would be automatic, purposeless, and uncontrolled since initial acquisition – and not only as the end result of a long history of repeated drug use, as currently hypothesized by some researchers (Everitt and Robbins 2016). Hence there would be no real loss of control over drug use according to this view, since at no time was the animal in control of its behavior. It is troubling to realize that, even to this day, we continue to ignore which of these two views is more valid: is drug self-administration by a nonhuman animal a goal-directed behavior or a mere behavioral automatism triggered by the drug? This uncertainty is mainly due to the fact that the standard procedures used to probe whether a nondrug behavior is goal-directed or not (e.g., outcome devaluation procedures) are poorly applicable to drug-use behavior despite several attempts.

Regardless of how this uncertainty will ultimately be resolved, the discovery that initially non-addicted animals would learn to self-inject drugs dramatically changed theories about the susceptibility of nonhuman animals to develop addiction on their own – as well as theories about the human uniqueness of addiction. As noted above, before this discovery, most researchers believed that addiction was unique to humans, only later coming to realize that animals could also develop an addiction-like state – but only when it was forced on them. After this discovery, however, researchers began to seriously consider the possibility that animals were not only vulnerable to develop addiction on their own, but may even be more vulnerable than humans. Only nonhuman animals seemed genuinely unable to stop using a drug to avoid self-destruction and death – and this even when they had the choice. Addiction was thus progressively and increasingly viewed as resulting, not from the "human in us", but from the "animal in us." In the early 1990s, Frank Logan, a pioneer in animal research on decision-making, summarized this view as follows: "drug addiction is primarily the result of the animal in us. Indeed, there are so many animal processes leading to drug use and abuse that the question is not so much why do some people become addicted as why doesn't everyone?" (Logan 1993). For Logan, despite the rhetoric in his question, the answer seemed obvious. Most people do not become addicted after initiating drug use because, unlike animals, people are not primarily driven by the pursuit of immediate reward; they also follow a "self-control drive." "[. . .] Control of drug use depends on the human in us, our unique cognitive capacity for self-control. Accordingly, motivation for change is essentially a matter of keeping the cortex over the midbrain." Thus, according to Logan, addiction involves having lost control over their midbrain – over the animal within. Though rather simplistic, this view nevertheless prefigures, without the details, some current neurobiological theories of addiction that define addiction mainly as a loss of control of the prefrontal cortex over the subcortical mechanisms underlying the drug-use behavior (Volkow *et al.* 2011). However, on the assumption that this view is correct (which does not seem to be the case (Heather 2017; Pickard in this volume)), it raises once again the question about the validity of using nonhuman animals to model this hypothetical loss of control in human addiction. If animals do not possess a self-control function or only a very rudimentary one, then how can they be used validly to reproduce a loss of this very function (Ahmed 2008; Pickard and Ahmed 2017)?

However, some researchers were skeptical from the outset about the emerging consensus that nonhuman animals were inherently vulnerable to drugs (Alexander and Hadaway 1982; Alexander *et al.* 1985). For such views place great emphasis on the internal brain reward machinery, while neglecting the life history and situation of animals used in addiction research. Animals such as rats, monkeys, and chimpanzees descend from highly social species, but in experimental settings they are typically raised and tested in asocial environments that one would consider to be extreme – abnormal and even traumatic if applied to a human individual. For instance, in Spragg's foundational study, the subjects were young adolescent chimpanzees, separated early from their parent groups and caged alone in a small enclosure. Similarly, laboratory rats used in addiction research are typically raised and tested in a state of near total isolation and/or extreme environmental poverty since weaning. When one considers this seriously, it becomes obvious that animals could use drugs for other reasons than to alleviate drug withdrawal distress, to seek pleasure and/or to follow blindly the dictate of a brain automatism. Contrary to Spragg's early contention and many other researchers after him, it is in fact very likely that animals may also use drugs to self-medicate a preexisting negative psychological condition and/or to try to escape a harsh experimental reality, as humans often do (Khantzian 1985; Pickard 2012; Pickard in this volume). Furthermore, since animals in such artificial conditions have no social life, they have nothing of social significance to lose or to damage by continuing to use drugs that bring negative social consequences – as drugs often do in humans – and, consequently, no relevant social disincentive to continue using. Inversely, using a drug while alone obviously reduces the potential social benefits to animals that drugs are known to bring to people (Müller and Schumann 2011). If a rat only has access to a drug while alone, it cannot utilize that drug for any social benefit, in contrast with a rat living in society (Wolffgramm 1991).

Bruce Alexander and colleagues were among the first in the late 1970s to begin to assess the influence of the social setting or situation on drug use by laboratory animals, in the now famous Rat Park experiment (Alexander *et al.* 1978; Hadaway *et al.* 1979; Alexander *et al.* 1981). This experiment demonstrated that the social situation did indeed matter – a factor that is currently increasingly being reconsidered in addiction research on nonhuman animals (Ahmed 2005; Heilig *et al.* 2016). When given a continuous choice between two bottles – one bottle containing water, the other morphine dissolved in water – rats living in a large park with many other peers of both sexes drank little of the morphine solution compared with rats living alone in a small metal cage. This difference was observed after rats had been forced to consume morphine orally during several weeks (i.e., by having only access to the morphine bottle) before being offered the two-bottle choice (Alexander *et al.* 1978). Making the bitter morphine solution more palatable by adding sugar increased morphine intake in socially-isolated rats but had little impact on morphine intake in rats living in Rat Park (Hadaway *et al.* 1979; Alexander *et al.* 1981). No matter how the experimenters tried, they failed to entice rats living in Rat Park to consume oral morphine despite its continuous availability. This outcome dramatically contrasts with the self-destructive drug-use behavior reported previously in lone animals offered a continuous choice between an IV drug and a nondrug food reward. According to Alexander and colleagues, rats living in Rat Park did not take oral morphine mainly to avoid those of its effects that likely interfere with the performance of self-reinforcing species-specific behaviors, mainly social and sexual, in an "otherwise competitive society" (Alexander *et al.* 1978; Hadaway *et al.* 1979; Alexander *et al.* 1981). These researchers also recognized that oral morphine was a relatively weak reinforcer, mainly because of its slow bioavailability. Rats "must drink an appreciable volume of fluid to gain a somewhat delayed effect" (Alexander *et al.* 1985). As a matter of fact, even socially-isolated rats were initially highly reluctant to drink the morphine solution when given a choice between it and plain water. They began to drink the morphine solution at appreciable levels only after many weeks of forced isolation and exposure to it (Alexander *et al.* 1978) or when it was sweetened with sugar (Hadaway *et al.* 1979; Alexander *et al.* 1981). Even then, socially-isolated rats did not develop a self-destructive pattern of drug use similar to that seen previously with IV drug self-administration in otherwise comparable drug access conditions. According to Alexander and colleagues, socially-isolated rats took more morphine than rats in Rat Park, not because they found morphine more rewarding and/or appealing, but mainly because they had less motivation to avoid it. Their solitude did not expose them to the social negative consequences that were hypothesized to motivate rats in Rat Park to avoid morphine (Alexander *et al.* 1985). Finally, these researchers also considered the possibility that socially-isolated rats could take morphine to relieve or self-medicate the affective distress associated with isolation – a factor that the Rat Park experiment was originally primarily intended to test – but they eventually ruled it out mainly because of insufficient evidence (Alexander *et al.* 1978; Alexander *et al.* 1981; Alexander *et al.* 1985).

The Rat Park experiment is often cited as evidence that rats can forego drugs and avoid self-destructive drug use if offered different choices. However, according to the researchers who conducted this seminal experiment, a more likely, albeit still unproven, hypothesis is that rats chose to forego morphine in Rat Park because it would have brought about negative social consequences without significant, if any, drug benefit. If true, this hypothesis raises a paradox. How could rats be able to refrain from using a continuously available drug to avoid its negative consequences in some experiments (i.e., the Rat Park experiment), but not be able to refrain from self-injecting a drug to avoid inanition and death in other experiments (i.e., those where lone rats are offered a continuous choice between a drug and nondrug reward, both of which can be sampled as much or as little as they choose)? This paradox is all the more troubling given that the oral route of drug self-administration used in the Rat Park experiment is associated with longer delays of effects than the IV route used in other experiments, and that rats, as anticipated long ago by Spragg, have a limited ability to form delayed associations (Ahmed 2017). Rats are biologically predisposed to learn to avoid foods that make them subsequently ill (Garcia *et al.* 1974), but this conditioned food aversion is unlikely to explain why Rat Park rats as opposed to socially-isolated rats did not take morphine.

One way to resolve this paradox would be to conduct a variant of the Rat Park experiment where, all else being equal, rats would have continuous access to a more powerful form of drug than oral morphine. For instance, how would rats behave if given access to intravenous heroin or even cocaine in a Rat Park setting? Will they avoid using the drug, like in the original Rat Park experiment, or will they use it self-destructively, thereby eventually prompting the doom of the Rat Park society? However, a variant of the Rat Park experiment involving a society of free-roaming rats each connected continuously to an IV drug infusion system is technically quite daunting, if not unfeasible. Nonetheless, it is a timely experiment to do because it would help to resolve a major, albeit largely overlooked, paradox in the field.

In conclusion, a walk on the animal side of addiction research reveals how views on the human uniqueness of addiction have changed over the past century. Before Spragg's foundational study on morphine-addicted chimpanzees, addiction was believed to be unique to humans – to be mainly the result of "the human in us." After this study, researchers came to believe that nonhuman animals could develop a desire for drugs so strong that it could even surpass other primary needs but only after they were forced into a state of addiction by an experimenter. Today, what happened on August 1, 1936 between Spragg and Lyn may look like a parody of Carl Elliott's portrayal of the current orthodox theory of addiction where an addict (played by Spragg) "must go where the addiction" (played by Lyn) "leads her, because the addiction holds the leash" (Elliott 2002: 48). What is obviously wrong with this parody, however, is that Lyn does not truly hold the leash; it is the experimenter who holds the leash, it is the experimenter that forced addiction into Lyn by passive drug administration. However, passive drug-administration experiments lead to the troubling question: how can we study addiction – as found in humans – in nonhuman animals if they are not allowed to self-administer the drug from the outset? At first, this problem seemed to be resolved by the advance of the drug self-administration paradigm, but it resurfaced later under a different guise. Indeed, a major consequence of the drug self-administration paradigm was to turn the theory of the human uniqueness of addiction upside down. Nonhuman animals were now predominantly viewed as inherently susceptible to self-destructive drug use, even when they had the choice not to self-administer drugs. The experimenter now had to intervene, not to force addiction onto the animals, but instead to protect them from its self-destructive consequences. Addiction was no longer seen as the result of the "human in us" (i.e., our cortex) but instead of the "animal in us" (i.e., our midbrain). An addict must go where the animal within leads her, because the animal within holds the leash. But, then, another troubling question emerges: How can we validly reproduce addiction in a nonhuman animal, if it has no human within? An important way out of this problem has been to look outside of the animal to the choice setting, and attribute the vulnerability of nonhuman animals to drugs, not to their "lower nature", but mainly to the artificiality of their experimental situations and living conditions. This is the main lesson of the Rat Park experiment. It is indeed possible that animals use drugs self-destructively mainly because they are not living an ecologically-relevant social life and thus have no major disincentives to use. If true, then, how and to what extent could one reliably extrapolate this self-destructive drug-use behavior to human addicts?

Acknowledgements

I would like to thank Hanna Pickard for her very helpful comments and for her precious editorial correction on the manuscript. This work is supported by the French Research Council (CNRS), the Université de Bordeaux, and the Fondation pour la Recherche Médicale (FRM, DPA20140629788).

References

Ahmed, S. H. (2005) "Imbalance between drug and non-drug reward availability: a major risk factor for addiction", *European Journal of Pharmacology* 526(1–3): 9–20.

Ahmed, S. H. (2008) "The origin of addictions by means of unnatural decision", *Behavioral and Brain Sciences* 31(4): 437–438.

Ahmed, S. H. (2010) "Validation crisis in animal models of drug addiction: beyond non-disordered drug use toward drug addiction", *Neuroscience & Biobehavioral Reviews* 35(2): 172–184.

Ahmed, S. H. (2018, in press) "Trying to make sense of rodents' drug choice behavior", *Progress in Neuropsychopharmacology and Biological Psychiatry*.

Aigner, T. G. and Balster, R. L. (1978) "Choice behavior in rhesus monkeys: cocaine versus food", *Science* 201(4355): 534–535.

Alexander, B. K., Beyerstein, B. L., Hadaway, P. F. and Coambs, R. B. (1981) "Effect of early and later colony housing on oral ingestion of morphine in rats", *Pharmacology Biochemistry and Behavior* 15(4): 571–576.

Alexander, B. K., Coambs, R. B. and Hadaway, P. F. (1978) "The effect of housing and gender on morphine self-administration in rats", *Psychopharmacology (Berl)* 58(2): 175–179.

Alexander, B. K. and Hadaway, P. F. (1982) "Opiate addiction: the case for an adaptive orientation", *Psychological Bulletin* 92(2): 367–381.

Alexander, B. K., Peele, S., Hadaway, P. F., Morse, S. J., Brodsky, A. and Beyerstein, B. (1985) "Adult, infant, and animal addiction", in S. Peele (ed.), *The Meaning of Addiction: An Unconventional View*, San Francisco, CA: Jossey-Bass Publishers, pp. 73–96.

Banks, M. L. and Negus, S. S. (2012) "Preclinical determinants of drug choice under concurrent schedules of drug self-administration", *Advances in Pharmacological Sciences 281768*.

Bozarth, M. A. and Wise, R. A. (1985) "Toxicity associated with long-term intravenous heroin and cocaine self-administration in the rat", *JAMA: The Journal of the American Medical Association* 254(1): 81–83.

Deneau, G., Yanagita, T. and Seevers, M. H. (1969) "Self-administration of psychoactive substances by the monkey", *Psychopharmacologia* 16(1): 30–48.

Elliott, C. (2002) "Who holds the leash?", *American Journal of Bioethics* 2(2): 48.

Everitt, B. J. and Robbins, T. W. (2016) "Drug addiction: updating actions to habits to compulsions ten years on", *Annual Review of Psychology* 67: 23–50.

Faust, E. S. (1900) "Ueber die ursachen der gewöhnung an morphin", *Archiv für experimentelle Pathologie und Pharmakologie* 44(3–4): 217–238.

Garcia, J., Hankins, W. G. and Rusiniak, K. W. (1974) "Behavioral regulation of the milieu interne in man and rat", *Science* 185(4154): 824–831.

Griffiths, R. R., Wurster, R. M. and Brady, J. V. (1981) "Choice between food and heroin: effects of morphine, naloxone, and secobarbital", *Journal of the Experimental Analysis of Behavior* 35(3): 335–351.

Hadaway, P. F., Alexander, B. K., Coambs, R. B. and Beyerstein, B. (1979) "The effect of housing and gender on preference for morphine-sucrose solutions in rats", *Psychopharmacology (Berl)* 66(1): 87–91.

Heather, N. (2017) "Is the concept of compulsion useful in the explanation or description of addictive behaviour and experience?" *Addictive Behaviors Reports* 6: 15–38.

Heilig, M., Epstein, D. H., Nader, M. A. and Shaham, Y. (2016) "Time to connect: bringing social context into addiction neuroscience", *Nature Reviews Neuroscience* 17(9): 592–599.

Johanson, C. E., Balster, R. L. and Bonese, K. (1976) "Self-administration of psychomotor stimulant drugs: the effects of unlimited access", *Pharmacology Biochemistry and Behavior* 4: 45–51.

Khantzian, E. J. (1985) "The self-medication hypothesis of addictive disorders: focus on heroin and cocaine dependence", *American Journal of Psychiatry* 142(11): 1259–1264.

Koob, G. F. (2009) "Neurobiological substrates for the dark side of compulsivity in addiction", *Neuropharmacology* 56: 18–31.

Lenoir, M., Cantin, L., Vanhille, N., Serre, F. and Ahmed, S. H. (2013) "Extended heroin access increases heroin choices over a potent nondrug alternative", *Neuropsychopharmacology* 38(7): 1209–1220.

Lindesmith, A. R. (1946) "Can chimpanzees become morphine addicts?" *Journal of Comparative Psychology* 39: 109–117.

Logan, F. A. (1993) "Animal learning and motivation and addictive drugs", *Psychological Reports* 73(1): 291–306.

Luscher, C. (2016) "The emergence of a circuit model for addiction", *Annual Review of Neuroscience* 39: 257–276.

Müller, C. P. and Schumann, G. (2011) "Drugs as instruments – a new framework for non-addictive psychoactive drug use", *Behavioral and Brain Sciences* 34(6): 293–310.

Negus, S. S. (2006) "Choice between heroin and food in nondependent and heroin-dependent rhesus monkeys: effects of naloxone, buprenorphine, and methadone", *Journal of Pharmacology and Experimental Therapeutics* 317(2): 711–723.

Nesse, R. M. and Berridge, K. C. (1997) "Psychoactive drug use in evolutionary perspective", *Science* 278(5335): 63–66.

Olds, J. and Milner, P. (1954) "Positive reinforcement produced by electrical stimulation of septal area and other regions of rat brain", *Journal of Comparative and Physiological Psychology* 47(6): 419–427.

Pickard, H. (2012) "The purpose in chronic addiction", *AJOB Neuroscience* 3(2): 40–49.

Pickard, H. (this volume, pp. 9–22) "The puzzle of addiction".

Pickard, H. and Ahmed, S. H. (2017) "How do you know you have a drug problem? The role of knowledge of negative consequences in explaining drug choice in humans and rats", in N. Heather and G. Segal (eds), *Addiction and Choice*, Oxford, UK: Oxford University Press, pp. 29–48.

Pickens, R. and Thompson, T. (1968) "Cocaine-reinforced behavior in rats: effects of reinforcement magnitude and fixed-ratio size", *Journal of Pharmacology and Experimental Therapeutics* 161: 122–129.

Plant, O. H. and Pierce, I. H. (1928) "Studies of chronic morphine poisoning in dogs I. General symptoms and behavior during addiction and withdrawal", *Journal of Pharmacology and Experimental Therapeutics* 33(3): 329–357.

Spragg, S. D. S. (1940) "Morphine addiction in chimpanzees", *Comparative Psychology Monographs* 15: 1–132.

Tatum, A. L. and Seevers, M. H. (1929) "Experimental cocaine addiction", *The Journal of Pharmacology and Experimental Therapeutics* 36(3): 401–410.

Tatum, A. L., Seevers, M. H. and Collins, K. H. (1929) "Morphine addiction and its physiological interpretations based on experimental evidences", *Journal of Pharmacology and Experimental Therapeutics* 36: 447–475.

Thompson, T. and Schuster, C. R. (1964) "Morphine self-administration, food-reinforced, and avoidance behaviors in rhesus monkeys", *Psychopharmacologia* 5(2): 87–94.

Volkow, N. D., Baler, R. D. and Goldstein, R. Z. (2011) "Addiction: pulling at the neural threads of social behaviors", *Neuron*, 69(4): 599–602.

Weeks, J. R. (1962) "Experimental morphine addiction: method for automatic intravenous injections in unrestrained rats", *Science* 138(3537): 143–144.

Wise, R. A. and Koob, G. F. (2014) "The development and maintenance of drug addiction", *Neuropsychopharmacology* 39(2): 254–262.

Wolffgramm, J. (1991) "An ethopharmacological approach to the development of drug addiction", *Neuroscience & Biobehavioral Reviews* 15(4): 515–519.

Yerkes, R. M. (1940) "Laboratory chimpanzees", *Science* 91(2362): 336–337.

PART II

Explaining addiction
Culture, pathways, mechanisms

SECTION A

Anthropological, historical, and socio-psychological perspectives

17
POWER AND ADDICTION

Jim Orford

When we talk about "addiction", we generally think of individual people who experience it or suffer from it. The focus is on individuals. When it comes to the science of addiction, the dominant conceptions are psychobiological and the master disciplines are psychiatry specifically or the psych- and neuro- sciences more generally. That overarching framework is reflected in our preoccupations with such things as the personal histories of addicted individuals, how they get into the state of addiction, how they make decisions, experience conflict, and what treatments work for them. My argument in this chapter will be that the addiction-as-individual-abnormality approach is partial, fragmented and ultimately ineffective, and that it is so because it does not acknowledge the importance of power and therefore leaves distortions of power in place. I try to offer a more social model of addiction in which power is the central concept. A longer version of this idea can be found in my book, *Power, Powerlessness and Addiction* (Orford 2013) although here I go further by speculating about some of the consequences of taking the broader social view of addiction that I am advocating.

Being addicted is a disempowering experience

Although I want to move us away from an exclusive focus on the individual, let me begin by focusing, as most theories of addiction do, on the experience of the 'addicted' individual. I suggest that experience can be well understood as one of disempowerment, one in which normal individual agency is restricted, sovereignty over one's behaviour curtailed. There is almost universal agreement among scientists and philosophers that addiction, whilst it does not totally undermine responsibility for one's actions, nevertheless it does significantly diminish it (Poland and Graham 2011). How exactly it does so is the subject of a number of chapters in this volume and I will not therefore linger on that intriguing question. Suffice it to say here that my own understanding of how addiction diminishes responsibility, explained in *Excessive Appetites* (Orford [1985] 2001), rests on the idea that addiction creates a conflict between the opposing motives created by a strongly reinforcing but seriously damaging consumption habit. More recent but similar ideas have drawn on the behavioural economics principle of delay discounting (Levy 2006; Ainslie 2011). They also come to the conclusion that addiction sets up a conflict of interests, disunifying and fragmenting decision-making and choice over time. This idea that addiction undermines the capacity to consistently make choices that are in keeping with other life plans, was expressed very clearly in the following quotation from a man who speaks of his loss of control over gambling:

> I crossed the line from fun to addiction 5 years or so ago, I know it's getting worse when I get loans to cover my losses i.e wages I've spent, selling my belongings, phones etc, increasing my overdraft limit . . . I don't enjoy it anymore but I can't stop, I feel like a prisoner of evil being forced to work for nothing and hand my money to the devil without any control over my actions. My social life has been affected, I don't see anyone anymore, I can't be happy around people as my thoughts are on gambling and my losses. I should be happy, I have a loving girlfriend and caring family, I have a job and have been successful.[1]

Like most people, his life plans would prioritise his closest relationships and his accumulation of assets, were it not for his addiction which is squeezing such things into a smaller part of his life space. He feels like a 'prisoner' and uses the colourful language of evil and the devil. But the feeling of being in the grip of something that seems to demand that one deviates from the normal, rational and consistent path of one's life – sometimes in a quite circumscribed area of life and sometimes in a way that seems totally destructive of the personality – is common amongst people who are addicted. Samuel Taylor Coleridge referred to his opium habit in letters to friends as, 'a Slavery more dreadful than any man who has not felt its iron fetters eating into his very soul, can possibly imagine' and 'This free-agency annihilating Poison' (Lefebure 1977, pp. 51, 57). Heather (2016) helpfully reminds us that the word 'addiction' derives from the Latin *addictio*, which referred to the state of bondage, amounting to slavery, assigned by legal process to a person who was unable, through debts or in some other way, to fulfil their obligations.

So why have modern theories of addiction, including those that recognise how addiction erodes consistency of choice, not gone one step further and been willing to use the language of disempowerment? I think this is because we have been caught between a reductionist biomedical concept of addiction that has no place for social and political science terms such as power and disempowerment, and the scepticism of social scientists who are often dismissive of the overly individualistic idea of addiction. I am in no doubt that addiction exists and that the experience of being addicted is a deeply discomforting one that harms one's capacity to be a fully autonomous self, governing one's own behaviour. But that misses great tracts of experience and writing on the subject of addiction that place the experience of the disempowered addicted individual in a wider social context. In what follows I try to summarise some of that context under four headings using the uniting concepts of power and powerlessness.

Family members subordinated by the power of addiction

The individual experience of disempowerment at the hands of a dangerously addictive substance or activity is contagious and spreads out to those who are near at hand, most particularly close family members. Nothing speaks so loudly about the limitations of the individual approach to addiction than the neglect that affected family members have experienced at the hands of addiction professionals and academics, and the pathologising theories about them, such as abnormal personality and codependency theories, which dominated for so long (Orford *et al.* 2005). It is the wives and husbands, mothers and fathers, daughters and sons, sisters and brothers and other family members who constitute the numerically largest group of people whose interests are attacked and whose ability to control their own destiny and that of their families is undermined by addiction. Their experiences, including their feelings of impotence and frustration, have been documented in the reports of qualitative research from a number of countries (Orford et al. 2005; Mathews and Volberg 2012) and detailed examples can be found in biographies and autobiographies such as those of the wives of the poets Samuel Taylor Coleridge (Lefebure 1986) and Dylan Thomas (Thomas and Tremlett 1986) and the mothers and fathers of popular modern singers (Doherty 2006; Winehouse 2012).

Affected family members share with their addicted relatives confusion about where responsibility lies and a tendency, shared with all disempowered groups, to attribute responsibility to themselves. Where their addicted relatives are confused by their own apparently irrational behaviour and experience guilt and shame about what they are doing to themselves and others, affected family members are confused about whether blame lies with their relatives, with 'addiction', with the purveyors of substances or activities to which the relatives are addicted, or to themselves. Social dominance theory (SDT: Sidanius and Pratto 1999), one of the most pertinent social psychological theories of power, is helpful here. According to SDT, one of the chief power hierarchy-maintaining 'myths' consists of beliefs that attribute responsibility for subordination to the disempowered themselves or to the groups or communities of which they are part. Just as the powerless everywhere have a tendency to put up with things whilst trying to protect themselves from the worst effects of subordination, so family members affected by addiction often cope by putting up with it rather than standing up to it or escaping from it (Orford et al. 2013).

The disempowering effects of addiction ripple out even further. The dilemma of how to cope with addiction and the confusion about responsibility that accompanies it apply to those beyond the family, including friends and colleagues, and potentially therefore to us all who witness addiction in others. The difficulties faced by colleagues of Charles Kennedy, leader of the British Liberal Democrat Party, in coping with his alcohol addiction is a very good case in point (Hurst 2006). The feeling of loss of control or sovereignty over one's own affairs associated with addiction can extend to a whole community. For example, Bourgois (2003) referred to a 'culture of terror' affecting parts of Harlem when he was carrying out his ethnographic research on the crack cocaine market there in the 1980s. A recent British example of gambling's community harm has been the powerlessness felt by local authorities over the diversity of their high streets as a result of the clustering of 'betting shops' in poorer areas as a consequence of the introduction of a new type of high powered, fast, high stake, gambling machine (Orford 2012). What is common to those apparently very different examples is the way in which addiction has the effect of interfering with priority plans and interests, rendering people such as colleagues or community residents or representatives less potent in the process. Impotence is a common feeling in the face of addiction.

Addiction strikes where power to resist is weakest

So far I have said nothing about how addiction arises in the first place. In line with the dominant individualistic approach to addiction, there has been an emphasis on the personal histories, personalities, attitudes and lifestyles of those who develop an addiction. But an approach that emphasises power shifts the focus and draws attention to the evidence that addiction flourishes best in circumstances where people or communities have fewer resources of knowledge, money or status, in other words where the power to resist is at its weakest. It can point, for example, to evidence of negative social gradients for alcohol-related hospitalisation and mortality rates (Harrison and Gardiner 1999), drug dependence rates (Swendsen et al. 2009) and rates of gambling problems (Wardle et al. 2007). That is despite the fact that indices of alcohol or drug use or engagement in gambling often show less of a social gradient or even show an opposite effect, with those of higher income or socio-economic status drinking or taking drugs more frequently or spending more money on gambling. This supports the idea that those with fewer resources or of lower status, although they may not always be the greater consumers, nevertheless consume dangerous products in riskier ways such as binge drinking (Tomkins *et al.* 2007) or using unsafe drug-taking methods, or consuming in riskier environments due to the illegality or stigmatised nature of the activity (Rhodes *et al.* 2005), or simply because for those with limited resources the activity, such as gambling, is more costly in relation to income (Orford 2011). It can also point to evidence that indigenous and aboriginal people, immigrant and ethnic minority groups, and lesbian, gay and bisexual people are more at risk, and that this heightened risk is related to disempowerment in the form of loss, trauma and deprivation (Ortiz-Hernández *et al.* 2009; Whitbeck *et al.* 2004).

An additional explanation for negative social gradients in prevalence rates is that once the dangers of particular products are better known and when help to change behaviour becomes more available, those with better access to help or with greater resources are the more likely to be able to take advantage. That has been most clearly demonstrated in the case of tobacco smoking (Chilcoat 2009) but is likely to be generally true (Cooper *et al.* 2016). Furthermore, there is evidence of a relationship between drug problems and poorer neighbourhoods going back to the work of Chein and colleagues (1964) in New York in the early 1960s, and the more recent evidence of a link between higher alcohol outlet density and area deprivation (Huckle *et al.* 2008) as well as similar findings for gambling outlet density (Marshall 2005).

Much of that work has pointed to the vulnerability of relatively disempowered individuals, groups or communities. But others such as Bourgois (2003), Singer (2001), and Nasir and Rosenthal (2009), who studied drug use in a largely Puerto Rican immigrant community in New York, the economically declined city of Hartford, Connecticut, and amongst gang members in slum areas in Makassar, Indonesia, respectively, speak of 'structural vulnerability' and are highly critical of what they see as the prevailing view that drug misuse could be understood as a problem of individuals' lifestyle choices. As one leading British drugs researcher has put it:

> Where drugs such as heroin and crack-cocaine are concerned, the most serious concentrations of human difficulty are invariably found huddled together with unemployment, poverty, housing decay and other social disadvantages.
>
> *Seddon, 2006, p. 680*

The trade in addictive products: the hidden faces of power

Those who are most vulnerable to addiction and those who are personally disempowered by it are mostly only dimly aware of the way power is being exercised within the vast networks involved in the chains of supply of addictive products. The legal alcohol and gambling industries are colossal, international, increasingly dominated by a small number of mega-corporations with headquarters in developed countries, with huge budgets that can be used to develop new products, engaging in a wide range of marketing activities including the sponsoring of sport, lobbying legislators and fighting restrictive regulations, including regulations aimed at benefitting public health (Jernigan 2009; Laranjeira *et al.* 2007; Orford 2011; Singer 2008). They put enormous efforts into 'social responsibility' activities, including the setting up of what might be called 'social aspects/public relations organisations' (SAPROs), promoting 'harm minimisation', about which many are sceptical (Barraclough and Morrow 2008; Babor 2009). Scientists are co-opted in a variety of ways, for example by engaging them directly, offering research funding or other kinds of support, encouraging them in promoting theories and research that support their interests and criticising those that do the opposite, and supporting the nomination to policy committees and boards of those most accommodating of their interests (Adams 2011; Babor 2009; Orford 2013).

Their power is such that they can exert it directly in those ways, mostly legally and mostly in a comparatively civilised way, but using bribes and threats when necessary. But, as power theorists have explained, power has a number of faces and it is the more subtle forms of power that are the more effective because they are less transparent and can easily masquerade as something much more benign (Lukes 2005). The second face of power, according to the power theorists, involves influencing the agenda by trying to make sure that matters that are up for debate comply with industry interests and those that threaten those interests are kept off the agenda. The most obvious example is the influence that the alcohol and gambling industries exert on the research agenda, either by directly funding research or by setting up organisations funded from industry sources to support research (Babor 2009; Orford 2011).

But it is the third face of power that is probably the most effective of all, exercised as it is with scarce recognition that power is involved at all. This is the establishment of an official or accepted discourse around the subject. If people, particularly those who control policy, can be influenced to think and talk about the subject in a way that supports the interests of the suppliers of dangerous products, then the job is done. I have suggested that there are five discourses, strongly supported by the drinks and gambling industries, which together offer difficult to resist support for the continued expansion of the legal provision of commodities with addiction potential (Orford 2011, 2013). These are the discourses that say: that these potentially harmful products are actually harmless forms of entertainment and leisure activity; that they are being commercially provided in a way that is just like any other form of legitimate business and should not therefore be over-regulated; that this contributes positively to a country's cultural and economic well-being; that people have a right to choose to use these products as they wish; and that consumers have a responsibility to themselves and others to consume in a way that is sensible and responsible. The first and last of those are perhaps the most important of all: the products being sold are *essentially* safe and harmless, rather than inherently dangerous, and those who experience problems, who in any case are said to be in a tiny minority, are not consuming 'responsibly', and have misused the freedom to choose which we are all said to have. They should be identified, re-educated and if necessary excluded, preferably 'self-excluded' to use the expression much favoured in the gambling policy field (Nower and Blaszczynski 2006).

The global illicit drug market is equally colossal. Although the power that is exercised along those supply chains is often naked and frequently violent, what has impressed a number of those who have studied the subject is that in many respects they are not unlike the chains of supply of legal addictive products. Their methods are often highly professional, their ability to change with circumstances is impressive, and the numbers of people operating in the grey area between the legal and illegal, or facilitating the trade as politicians or law-enforcement officers, is large (Feiling 2009; Singer 2008). A notable feature of the illegal trade in substances is the power differentials involved, with some highly placed people becoming very rich on the proceeds of substances to which other people are becoming addicted, and at the other end of the scale low-level operatives such as farmers, lab workers and 'runners' who are likely to be poor, vulnerable and exploited. The latter, like the personally addicted and their close family members, are likely to know relatively little about how the bigger system of which they are a very small part works (Chin 2009; Ruggiero and Khan 2006).

Addiction change as liberation

One of the extraordinary things about addiction is that, despite the disempowerment that addiction brings and the ceaseless attempts of powerful suppliers to maintain sales, every year people in their hundreds of thousands manage to re-establish autonomy in a sphere in which they had previously forfeited dominion over their own actions.

There are numerous examples of addiction change methods and models that suggest that the process is very far removed from one of individual cognitive change, but is rather one in which power is exercised benignly and constructively by others in a way that enables the previously disempowered to re-establish self-mastery over part of their lives where it had been lost. Examples include: the taking of oaths and pledges (Longmate 1968); company employee assistance programmes that combine 'constructive confrontation' with 'supportive confrontation' (Trice and Sonnenstul 1990); a recognition of the 'pressures to change' that family members and others can exert toward change (Webb et al. 2007), and programmes that encourage family members to increase those pressures (Barber and Crisp 1995); contingency management methods that use financial reward to encourage behaviour change (Budney et al. 2000); the benign authority exerted by family drug and alcohol courts (Harwin et al. 2011); and the subtle but admittedly directive motivational interviewing that uses the counsellor's expert power and the power differential between counsellor and client to gradually erode a client's resistance to change (Miller and Rollnick 2002).

In fact it was the power paradox inherent in Alcoholics Anonymous and other 12 Step mutual help organisations that first alerted me to the idea that the addiction change process might also be understood in power terms. A central idea is that one must admit powerlessness over the object of one's addiction and be humble enough to surrender control to some 'higher' source of power (Fowler 1995). There are parallels here with general ideas of 'conduct reorganisation' that cut across therapeutic, religious and political spheres, in which change is likened to a conversion process during which a person seeking change 'surrenders' to an individual or group that offers a model of change and provides a role model for the new self (Sarbin and Nucci 1973). There are also obvious parallels here with ideas from liberation psychology, well-known in Latin America but less known in mainstream individualistic Western medicine and psychology (Burton and Kagan 2005). Change, according to liberation psychology, is usually promoted by critical reflection about one's current position, growing appreciation of how power relations have worked to one's disadvantage, a turning away from resignation and fatalism, and contemplation of a different future. Transformative change may benefit from help from 'catalytic agents', in the same way that addiction change may require expert help and commitment to some credible, structured programme of behaviour change. It is notable that one of the most important change processes identified in the 'transtheoretical' model of addictive behaviour change (DiClemente and Prochaska 1982) was 'self-liberation'.

Failure to recognise power leads to selective attribution of responsibility

What are the effects of failing to recognise the importance of power and powerlessness for addiction and the way in which they might link fields of enquiry that are otherwise kept separate? Those who have written about power and freedom understand, in a way that addiction science appears not to, that the powerless, while being exploited for others' profit, tend to hold themselves accountable for their problems and are 'responsibilised' (Rose 1999) by the powerful. As the late David Smail (2005), a prominent British community psychologist, put it, the powers that shape our lives are mostly 'distal' ones, too far over our 'power horizon' to be appreciated. Hence what the powerless need most is not some form of cognitive therapy to develop insight but rather increased awareness, or 'outsight', into how their interests have been hijacked and their vulnerabilities exploited.

It has been suggested, further, that treatment providers may be contributing to this failure to appreciate the full picture and hence may be colluding in the attributing of responsibility to the powerless. By uncritically accepting the dominant discourse about addiction, we focus our attention on a deviant minority who need our guidance in order to re-take responsibility for their lives. In the process are we entering into an alliance with governments that have conflicted interests? The latter include, not only an interest in the protection of citizens and the promotion of public health, but also interest in such things as contributing to an international 'war on drugs', returning people to work, and supporting industries that trade in products that are dangerous for health. Do we become, in effect, part of the establishment, defenders of the status quo? Some commentators in the gambling field are amongst those who have seen clearly how the interests of the industry–government alliance are supported by the dominant focus on a deviant minority:

> By the mid-1990s, the gambling industry had already grasped . . . that a medical diagnosis linked to the excessive consumption of its product by some individuals could serve to deflect attention away from the products' potentially problematic role in promoting that consumption, and onto the biological and psychological vulnerabilities of a small minority of its customers.
>
> Schüll 2012, p. 261

Meanwhile, addiction research is fragmented, with researchers operating in disconnected silos, the largest and dominant group focused on identifying, studying and treating the deviant and powerless minority. Public health, whilst placing a greater emphasis on health promotion and prevention, has also been criticised for stressing individual responsibility for health, hence risking a victim-blaming stance and failing to take a critical position that understands the importance of power and social structures that support exploitation and oppression (Marks 2002). Can we reconcile an empowering approach to the treatment of individuals with an appreciation of the wider social contexts of power?

The subversive implications of a power and powerlessness view of addiction

What, then, might be some of the consequences of taking the broader, less fragmented view of addiction, centred on the concepts of power and powerlessness, which I am promoting?

First there are considerable implications for addiction research that, over a period of time, would start to look rather different in a number of ways. Informed by power and powerlessness theory, there would be more emphasis on the way in which power is employed and powerlessness reinforced, with more research on how those who trade in dangerous and unhealthy products operate. Such research would include how such industries promote their products to relatively powerless groups and communities, how they resist control and regulation, how they influence the public health agenda, and how governments and others, including researchers and charitable organisations, become dependent and complicit (Adams 2011; McCambridge et al. 2014). New or currently marginal types of research might come to the fore; for example, studies of groups and movements campaigning against the promotion of addictive products or campaigning for the prevention of addiction or improved treatment (Marshall and Marshall 1990; Wright 2009).

But the biggest change that might be hoped for is a more fruitful integration across currently separate research domains. Instead of research on people's personalities and backgrounds, family research, study of communities, research on legal industries, research on the illicit drug trade, and treatment research, all coming from disparate disciplines and traditions, we would hope to see research that crossed those boundaries. It would therefore not just draw attention to the fact that power is being exercised and that there are those who are being rendered powerless by addiction, but would more adequately articulate how power is exercised in a way that impinges on the lives of the powerless. One could imagine, for example, research that married the local marketing decisions of traders in addictive products with the impact of those decisions on local families; or study of the use of different addiction discourses among members of both relatively powerful and powerless groups such as policymakers and senior traders on the one hand and those vulnerable to addiction on the other.

We might also envisage a significant change in the way we try to help people who are having difficulty with addiction. One change that might be accommodated relatively easily, because there are some welcome signs of a small movement in that direction already (Copello *et al.* 2010; Orford *et al.* 2013), is an opening up of 'treatment' – although that word might not be thought the most appropriate – to those who are experiencing personal difficulties with addiction as partners, parents, children, siblings and other close family members and friends. Using a power framework and seeing them as amongst those most disempowered by addiction should enable help for affected family members to be moved from the periphery nearer to centre stage.

A more subversive change to the way addiction is treated could be envisaged. Instead of seeing addiction problems as purely personal, requiring individual behaviour and attitude change, 'treatment' – the terminology would certainly now need to change – might combine the personal and the political (Smail 2005). It might couple insight and outsight, and encourage both individual behaviour change and campaigning against exploitation, in a way that has always been antithetical to 12 Step and medical and psychological treatments alike. It has generally been assumed that any attempt to attribute responsibility for one's behaviour to others or to one's circumstances represents resistance to personal change. However, that assumption has never to my knowledge been tested, mainly because even raising the question is heretical. However, recognising that one is being exploited may not be inconsistent with acknowledging personal responsibility, as the following gambling website comment suggests:

> Of course it was all of my own doing, no-one forced me to do this. But these machines are dangerously, dangerously addictive – if I could outlaw them I would, I think they're absolute poison, but I know it's not that simple. But morality has to step in somewhere here, where the bookies are concerned. They're making their money but at what cost? I know these machines have brought many, many people to their knees and in the most desperate situations. Someone needs to be accountable for all this, because it's not as simple as blaming the gambler for just being weak-willed.[2]

The question arises of whether the addiction power and powerlessness framework that I have briefly outlined, and the consequent changes to addiction research and treatment that I have imagined, could be accommodated within the structures that currently control addiction-relevant academic and professional training and practice? At the very least there would need to be a quite radical change in the way new recruits to addiction science and professions are taught about the subject. The addiction curriculum would need to change. I hope the present chapter might make a small contribution to that change.

Notes

1 Gambling Watch UK website, gamblingwatchuk.org.
2 Gambling Watch UK website, gamblingwatchuk.org.

References

Adams, P. J. (2011) "Ways in which gambling researchers receive funding from gambling industry sources", *International Gambling Studies* 11: 145–52.

Ainslie, G. (2011) "Free will as recursive self-prediction: does a deterministic mechanism reduce responsibility?" in J. Poland and G. Graham (eds), *Addiction and Responsibility*, Cambridge, MA: The MIT Press, pp. 55–88.

Babor, T. F. (2009) "Alcohol research and the alcoholic beverage industry: issues, concerns and conflicts of interest", *Addiction* 104: 34–47.

Barber, J. G. and Crisp, B. R. (1995) "The 'pressures to change' approach to working with the partners of heavy drinkers", *Addiction* 90(2): 269–76.

Barraclough, S. and Morrow, M. (2008) "A grim contradiction: the practice and consequences of corporate social responsibility by British American Tobacco in Malaysia", *Social Science and Medicine* 66(8): 1784–96.

Bourgois, P. (2003) *In Search of Respect: Selling Crack in El Barrio*, 2nd Edition, Cambridge, UK: Cambridge University Press.

Budney, A. J., Higgins, S. T., Radonovich, K. J. and Novy, P. L. (2000) "Adding voucher-based incentives to coping skills and motivational enhancement improves outcomes during treatment for marijuana dependence", *Journal of Consulting and Clinical Psychology* 68(6): 1051–1061.

Burton, M. and Kagan, C. (2005) "Liberation social psychology: learning from Latin America", *Journal of Community and Applied Social Psychology* 15(1): 63–78.

Chein, I., Gerard, D. L., Lee, R. S. and Rosenfeld, E. (1964) *Narcotics, Delinquency and Social Policy: The Road to H*, London, UK: Tavistock.

Chilcoat, H. D. (2009) "An overview of the emergence of disparities in smoking prevalence, cessation, and adverse consequences among women", *Drug and Alcohol Dependence* 104S: S17–S23.

Chin, K. L. (2009) *The Golden Triangle: Inside Southeast Asia's Drug Trade*, London, UK: Cornell University Press.

Cooper, H., Linton, S., Kelley, M., Ross, Z., Wolfe, M., Chen, Y.-T., Zlotorzynska, M., Hunter-Jones, J., Friedman, S., Des Jarlais, D., Semann, S., Telpalski, B., DiNenno, E., Broz, D., Wejnert, C. and Paz-Bailey, G. F. (2016) "Racialized risk environments in a large sample of people who inject drugs in the United States", *International Journal of Drug Policy* 27(January): 43–55.

Copello, A., Templeton, L., Orford, J. and Velleman, R. (2010) "The 5-Step Method: evidence of gains for affected family members", *Drugs: Education, Prevention and Policy* 17(s1): 100–112.

DiClemente, C. C. and Prochaska, J. O. (1982) "Self-change and therapy change of smoking behavior: a comparison of process of change in cessation and maintenance", *Addictive Behaviors* 7: 133–142.

Doherty, J. (2006) *Pete Doherty: My Prodigal Son*, London, UK: Headline.

Feiling, T. (2009) *The Candy Machine: How Cocaine Took over the World*, London, UK: Penguin.

Fowler, J. W. (1995) "Alcoholics Anonymous and faith development", in B. S. McCrady and W. R. Miller (eds), *Research on AA: Opportunities and Alternatives*, New Brunswick, NJ: Rutgers University Press.

George, S. and Bowden-Jones, H. (2015) "Family interventions in gambling", in H. Bowden-Jones and S. George (eds), *A Clinician's Guide to Working with Problem Gamblers*, London, UK: Routledge.

Harrison, L. and Gardiner, E. (1999) "Do the rich really die young? Alcohol-related mortality and social class in Great Britain, 1988–1994", *Addiction* 94: 1871–1880.

Harwin, J., Ryan, M. and Tunnard, J. (2011) *The Family Drug and Alcohol Court (FDAC) Evaluation Project Final Report*, London, UK: Brunel University Press.

Heather, N. (2016) "On defining addiction", in N. Heather and G. Segal (eds), *Addiction and Choice: Rethinking the Relationship*, Oxford, UK: Oxford University Press.

Huckle, T., Huakau, J., Sweetsur, P., Huisman, O. and Casswell, S. (2008) "Density of alcohol outlets and teenage drinking: living in an alcogenic environment is associated with higher consumption in a metropolitan setting", *Addiction* 103(10): 1614–1621.

Hurst, G. (2006) *Charles Kennedy: A Tragic Flaw*, London, UK: Politico's Publishing.

Jahiel, R. I. and Babor, T. F. (2007) "Industrial epidemics, public health advocacy and the alcohol industry: lessons from other fields", *Addiction* 102(9): 1335–1339.

Jernigan, D. H. (2009) "The global alcohol industry: An overview", *Addiction* 104(Suppl 1): 6–12.

Laranjeira, R., Marquest, A. C., Ramos, S., Campana, A., Luz, E. and Franca, J. (2007) "Who runs alcohol policy in Brazil?" *Addiction* 102: 1502–1505.

Lefebure, M. (1986) *The Bondage of Love: a Life of Mrs Taylor Coleridge*, London, UK: Victor Gollancz.

Lefebure, M. (1977) *Samuel Taylor Coleridge: A Bondage of Opium*, London, UK: Quartet Books.

Levy, N. (2006) "Autonomy and addiction", *Canadian Journal of Philosophy* 36(3): 427–447.

Longmate, N. (1968) *The Water Drinkers: A History of Temperance*, London, UK: Hamish Hamilton.

Lukes, S. (2005) *Power: a Radical View*, 2nd Edition, Basingstoke, UK: Palgrave Macmillan.

Marks, D. F. (2002) "Editorial essay. Freedom, responsibility and power: contrasting approaches to health psychology", *Journal of Health Psychology* 7(1): 5–19.

Marshall, D. (2005) "The gambling environment and gambler behaviour: evidence from Richmond-Tweed, Australia", *International Gambling Studies* 5(1): 63–83.

Marshall, M. and Marshall, L. B. (1990) *Silent Voices Speak: Women and Prohibition in Truk*, Belmont, CA: Wadsworth Publishing Company.

Mathews, M. and Volberg, R. (2012) "Impact of problem gambling on financial, emotional and social well-being of Singaporean families", *International Gambling Studies* 13(1): 127–140.

McCambridge, J., et al. (2014) "Vested interests in addiction research and policy. The challenge corporate lobbying poses to reducing society's alcohol problems: insights from UK evidence on minimum unit pricing", *Addiction* 109(2): 199–205.

Miller, W. R. and Rollnick, S. (2002) *Motivational Interviewing: Preparing People for Change*, 2nd Edition, New York, NY: Guilford Press.

Nasir, S. and Rosenthal, D. (2009) "The social context of initiation into injecting drugs in the slums of Makassar, Indonesia", *International Journal of Drug Policy* 20(3): 237–243.

Nower, L. and Blaszczynski, A. (2006) "Impulsivity and pathological gambling: a descriptive model", *International Gambling Studies* 6(1): 61–75.

Nussbaum, M. C. (2011) *Creating Capabilities: The Human Development Approach*, Cambridge, MA: Harvard University Press.

Orford, J. (2001) *Excessive Appetites: A Psychological View of Addictions*, 2nd Edition, Chichester, UK: Wiley.

Orford, J. (2011) *An Unsafe Bet? The Dangerous Rise of Gambling and the Debate We Should Be Having*, Chichester, UK: Wiley-Blackwell.

Orford, J. (2012) "Gambling in Britain: the application of restraint erosion theory", *Addiction* 107: 2082–2086.

Orford, J. (2013) *Power, Powerlessness and Addiction*, Cambridge, UK: Cambridge University Press.

Orford, J., Natera, G., Copello, A., Atkinson, C., Mora, J., Velleman, R., Crundall, I., Tiburcio, M., Templeton, L. and Walley, G. (2005) *Coping with Alcohol and Drug Problems: The Experiences of Family Members in Three Contrasting Cultures*, London, UK: Brunner-Routledge.

Orford, J., Velleman, R., Copello, A., Templeton, L. and Ibanga, A. (2010) "The experiences of affected family members: a summary of two decades of qualitative research", *Drugs: Education, Prevention and Policy* 17(s1): 44–62.

Orford, J., Velleman, R., Natera, G., Templeton, L. and Copello, A. (2013) "Addiction in the family is a major but neglected contributor to the global burden of adult ill-health", *Social Science and Medicine* 78: 70–77.

Ortiz-Hernández, L., Tello, B. L. G. and Valdés, J. (2009) "The association of sexual orientation with self-rated health, and cigarette and alcohol use in Mexican adolescents and youths", *Social Science and Medicine* 69(1): 85–93.

Poland, J. and Graham, G. (2011) *Addiction and Responsibility*, Cambridge, MA: MIT Press.

Rhodes, T., Singer, M., Bourgois, P., Friedman, S. R. and Strathdee, S. A. (2005) "The social structural production of HIV risk among injecting drug users", *Social Science and Medicine*, 61(5): 1026–1044.

Rose, N. (1999) *Powers of Freedom: Reframing Political Thought*, Cambridge, UK: Cambridge University Press.

Ruggiero, V. and Khan, K. (2006) "British South Asian communities and drug supply networks in the UK: a qualitative study", *The International Journal of Drug Policy* 17(6): 473–483.

Sarbin, T. and Nucci, L. (1973) "Self-reconstitution processes: a proposal for reorganising the conduct of confirmed smokers", *Journal of Abnormal Psychology* 81: 182–195.

Schüll, N. D. (2012) *Addiction by Design: Machine Gambling in Las Vegas*, Princeton, NJ: Princeton University Press.

Seddon, T. (2006) "Drugs, crime and social exclusion: social context and social theory in British drugs – crime research", *British Journal of Criminology* 46(4): 680–703.

Sidanius, J. and Pratto, F. (1999) *Social Dominance: An Intergroup Theory of Social Hierarchy and Oppression*, Cambridge, UK: Cambridge University Press.

Singer, M. (2001) "Toward a bio-cultural and political economic integration of alcohol, tobacco and drug studies in the coming century", *Social Science and Medicine* 53(2): 199–213.

Singer, M. (2008) *Drugging the Poor: Legal and Illegal Drugs and Social Inequality*, Long Grove, IL: Waveland Press.

Smail, D. (2005) *Power, Interest and Psychology*, Ross-on-Wye, UK: PCCS Books.

Swendsen, J., Conway, K. P., Degenhardt, L., Dierker, L., Glantz, M., Jin, R., Merikangas, K. R., Sampson, N. and Kessler, R. C. (2009) "Socio-demographic risk factors for alcohol and drug dependence: the 10-year follow-up of the national comorbidity survey", *Addiction* 104(8): 1346–1355.

Thomas, C. and Tremlett, G. (1986) *Caitlin: Life with Dylan Thomas*, London, UK: Secker and Warburg.

Tomkins, S., Saburova, L., Kiryanov, N., Andreev, E., McKee, M., Shkolnikov, V. and Leon, D. A. (2007) "Prevalence and socio-economic distribution of hazardous patterns of alcohol drinking: study of alcohol consumption in men aged 25–54 years in Izhevsk, Russia", *Addiction* 102(4): 544–553.

Trice H. M. and Sonnenstuhl, W. J. (1990) "On the construction of drinking norms in work organizations", *Journal of Studies on Alcohol* 51(3): 201–220.

Wardle, H., Sproston, K., Orford, J., Erens, B., Griffiths, M., Constantine, R. and Pigott, S. (2007) *British Gambling Prevalence Survey 2007*, London, UK: National Centre for Social Research.

Webb, H., Rolfe, A., Orford, J., Painter, C. and Dalton, S. (2007) "Self-directed change or specialist help? Understanding the pathways to changing drinking in heavy drinkers", *Addiction Research and Theory* 15(1): 85–95.

Whitbeck, L. B., Adams, G. W., Hoyt, D. R. and Chen, X. (2004) "Conceptualizing and measuring historical trauma among American Indian people", *American Journal of Community Psychology* 33(3–4): 119–130.

Winehouse, M. (2012) *Amy, My Daughter*, London, UK: HarperCollins.

Wright, A. (2009) *Grog War*, 2nd Edition, Broome, Australia: Magabala Books.

18
SOCIOLOGY OF ADDICTION

Richard Hammersley

Inebriety's hazards have long been recognized, evolving into ideas of addiction during the 19th century. Before and since, periodically politicians have taken moral stances against things one might be addicted to: Tea and coffee; alcohol; marijuana; opiates; cocaine; various amphetamine derivatives; tobacco products; the dark net as a source of drugs. During these so-called "moral panics" a drug is linked to criminality, health problems and immorality, especially amongst "vulnerable" people such as youth, or women (Carnwath and Smith 2003). Panics are often followed by complacency about the harms intemperate use of alcohol or drugs can bring once the habit becomes established in the population. Drug concerns wax and wane over the centuries.

For example, during the popularization of cigarette smoking cigarettes were linked with moral degeneracy (especially for women), criminality, poisonous contamination of the product, inability to control use, as well as health problems. There were widespread bans (Borio 2007). Then, in the face of popular demand for cigarettes and huge profits, there followed 40 years of complacency about smoking-related mortality (Proctor 2012). Currently, cannabis is moving into complacency, with unknown results. For it seems likely that cannabis is "addictive", like most drugs, although comparatively harmless (Nutt *et al.* 2010). Other possible addictions have similar trajectories. Recent candidates include online games, social media and sugar. Diffuse this widely and there is a risk of addiction simply meaning a strong want, habit or liking for something – especially something forbidden, new or not customary.

In everyday life "addiction" can be a grand simplifying explanation for undesirable behaviors. Metaphorically, people refer to being "addicted" to things that they do, despite anticipated and real regrets; watching trashy TV for instance. Common technical usage of addiction locates it within the individual person, rather than it being a social problem, or an example of the social construction of madness. Additionally it is commonly assumed that addiction undermines both volition and cognition, because it is in large part caused biologically, evidenced by the fact that chronic uses of various drugs have measurable and durable effects on the brain. However, as shall be seen, this argument is probably tautologous. Even academic writing about "addiction" can be ambiguous about its meaning and drift from one meaning to another.

Here "addiction" will refer to the ambiguous topic and "dependence" to a technically defined condition or syndrome. Focus is on sociology's contribution to addictions to drugs and alcohol, rather than on tobacco, gambling, or more debatable "addictions." Defining sociology loosely to encompass some anthropology, social policy, politics, history, geography, criminology and psychology. The economics of addiction has developed somewhat separately and is excluded.

What are "drugs?"

Drugs, including alcohol, tobacco and caffeine products, are long-traded products of enormous economic importance that have helped shape civilizations and the patterns of globalization. Which substances "count" as "drugs" – rather than benign or beneficial food products, herbs or spices, medicines, or harmless vices – is constantly contested and redefined by consumers, the relevant industries, governments, health professionals and other interested parties. Sociological research on drug use and drug users has usually studied drugs of contemporary concern. Because concerns are driven by fear of "addiction", research is biased towards drugs of addiction, even when it contests the nature and meaning of addiction and dependence.

There is a need for a sociology of ordinary substance use that does not unquestioningly problematize it (Hammersley 2011), but beyond alcohol use the idea of normal, unproblematic drug use is controversial. Such a theory would include a description of what differentiated acceptable, normative but unacceptable, and problematic drug use, and would have to explain how most people use drugs without developing problems (Hammersley 2014), as well as considering how people use drugs, as they do other consumer goods and activities, to foster and maintain social relationships, to define and experience pleasure, and to display social and personal identity and social status (Parker 2005).

There are double standards regarding "drugs" and commonplace substances. For example, "agitation" is a "symptom" of cocaine or amphetamine use, but tolerated for caffeine products. The purported harms and benefits of drugs are not due solely to the objective action of the chemicals ingested, but also depend on the mind-set of the user and the social and physical setting of use (Zinberg 1984).

Brain disease model of addiction

It has long been proposed that addiction is a disease of some kind, although whether it is an identifiable organic disease entity, a "syndrome" or merely a mental health problem worthy of professional help has always been contentious. The "Brain Disease Model of Addiction" (Volkow *et al.* 2016) is the most recent model. The BDMA proposes that addiction problems are due solely to the short- and long-term effects of drugs on the brain and nervous system. Thus, psychological, sociological, political and cultural factors are not relevant. It may seem odd to include a reductionist model in a chapter on sociology, but BDMA matters because it is highly influential, originating in the USA National Institute on Drug Abuse. Also because, just as atheism requires faith in the belief there is definitely no God, so BDMA requires the radical sociology theory that people's thinking, everyday lives, and societies have no impact on addiction. Moreover, BDMA utilizes flawed reasoning: (1) Drugs affect brain systems in understood and durable ways (true). Therefore they are sufficient causes of addiction (false). For instance, few people given morphine as part of medical or surgical procedures, even over the long term, become dependent. (2) Considering addiction to be a disease is humane because it allows problems to be treated under medical insurance and is better than criminalizing addiction (true). Therefore BDMA is the only scientific alternative to viewing drug dependence as a moral and criminal problem (false). For instance, many conditions that are unarguably diseases, including diabetes and cardiovascular disease, are caused primarily by social and psychological factors and are highly amenable to social and psychological interventions, even once the disease state is reached. Another difficulty with BDMA is that, unless one counts substituting one addictive drug for another as a treatment, so far biomedical interventions against addiction have not succeeded, whereas psychological and social ones have (Hammersley 2014). As discussed below, more commonly drugs are theorized as deviance or as a healthcare problem.

The myth of addiction

There are unexpectedly convincing arguments that drug addiction is largely a myth (Davies 1997; Peele 1990; Szasz 1974), based on stigmatizing some drugs and types of user as inherently problematic, then seeking confirmatory evidence whilst ignoring similar behaviors and problems with other drugs, and other types of user. Yet, for instance, heroin is not intrinsically more harmful than alcohol.

Davies (1997) has shown that addiction is subject to attribution errors including the "fundamental attribution error," which is to attribute people's undesirable substance use to their stable, internal characteristics, as opposed to temporary, external causes, whereas desirable characteristics are attributed to the person's stable, internal characteristics. Also "self-serving bias," which is to attribute your own temperance to your stable, internal characteristics but your bouts of intoxication to external forces, but to do the opposite for other people. The myth diverts attention away from the serious social and economic problems underpinning addictions, which are challenging and unpopular to address. It can be politically expedient to attribute blame for crime, ill health and immorality to a toxic combination of the defective addict and the uncontrollable drug.

Opprobrium regarding addiction is systematically biased depending upon the social status of the user, and the acceptability of the drug being used in the way that it is used. For example, al fresco drinkers are judged and policed differently depending on whether they are dressed like students or unemployed working class youth (Galloway *et al.* 2007). Moreover, users with more social capital can manage heavy and unacceptable drug and alcohol habits more successfully. Management techniques include being able to be unobtrusive and even elide "addiction" entirely (Shewan and Dalgarno 2005), and being able to manage public disclosures of addictive drug use. One method is to confess to an addiction, attend some form of treatment, then to be in recovery. This can be easier than debating the niceties of recreational hard drug use, or persistent drunk driving, with your family, workplace, or the press.

The addiction legend is durable because it serves multiple functions in society (Hammersley and Reid 2002) including providing a clear answer to drug problems, and other social problems such as crime, abuse, homelessness and poverty. Indeed BDMA and other disease models can be politically expedient because they place the blame for the problems of some of the most socially disadvantaged on addiction, absolving users of responsibility for their actions, but also denying them agency for their recovery. "Lumpen abuse" (Bourgois and Schonberg 2009) is the unintended result and, arguably, crime, abuse, homelessness and poverty cause addiction more than the other way around. The legend is also socially and economically useful to a wide range of social groups including the media, politicians, the anti-drugs industry, biomedical and pharmaceutical industries, the illicit drugs industry, law enforcement, religion (as a means of reconciling evil behavior, free will and a beneficent God), and drug users and ex-users.

Some truth in the legend?

The norm in addiction research is probably to accept the biases discussed above but assume that when people use some drugs heavily dependence is largely determined by biological processes:

Anyone who smokes tobacco heavily on a daily basis, or injects heroin several times a day, is liable to become physiologically dependent. There are complex social and psychological issues determining the life course of an addiction. Many people recognize signs of dependence and moderate their use to prevent it (Granfield and Cloud 1999). How we collectively think about "addiction" is biased towards people who struggle to manage dependence, because addiction research is usually created by professionals with intimate knowledge of such clients, who may be atypical of drug users, or even of people who would fit criteria for drug dependence, if they were assessed. Additionally, dependence on some drugs is taken far less seriously than on others. There are also complex issues surrounding how, and by whom, substance use disorders are defined for medical insurance, diagnosis, and obtaining benefit (Sobell and Sobell 2006). To illustrate these points, consider for example the concept of responsible drinking and the manufacture of a recent opiate epidemic in the American Midwest.

The alcohol industry is motivated primarily by profit. Maximizing sales does not harmonize with health advice. For example, across Europe excluding teetotalers the mean weekly intake of alcohol is just above current UK recommended "safe" limits (Anderson and Baumberg 2006), suggesting that high levels of drinking are normal. The concept of responsible drinking is strongly promoted by the alcohol industry, which much prefers to locate alcohol problems as belonging to an unfortunate minority of alcoholics. Yet even their own statistics show that 3/10 drinkers drink above recommended safe limits (www.portmangroup.org.uk/research, accessed 19/04/2016). The industry opposes measures to reduce population-level drinking such as minimum pricing of alcohol and reductions in the legal blood alcohol limit for driving. It also lobbies strongly to prevent regulation and subverts existing controls over alcohol marketing when it can (Savell et al. 2016). For example, alcohol logos are prominent at music and sporting events. It is plausible, but impossible to prove, that this appeals to under-18s or suggests that alcohol enhances music or sport. In short, largely mythical "responsible" drinkers are freely drenched with alcohol marketing on the pretext that this does not promote heavy drinking or encourage young people to drink.

The USA has experienced a new opiate epidemic, particularly in the Midwest (Bourgois 2000). This was formed by the endemic overprescribing of various prescription opioids for pain, which were cynically marketed by their manufacturers to healthcare professionals as being safe and non-addictive, and hence not requiring rigorous monitoring and control. This sometimes involved quite blatant inducements to prescribe the product. Mass use led to many people becoming dependent, and to much leakage on to the black market. When prescribing was tightened up then many users switched to heroin, which is easier to obtain illegally (Volkow 2014). The new epidemic was created on the basis of a disingenuous separation between addictive street opiates and supposedly safe prescription painkillers, although the different drugs are functionally equivalent for users.

The impact of sociology

Over the past 60 years or so, sociology has made a major but largely behind the scenes contribution to addiction research, often fronted by sympathetic clinicians. Seemingly pathological addictive behaviors are often caused by a complex of psychological, economic, political, social and cultural factors, as well as the drug itself. Therefore, rather than being non-volitional, addictive behaviors can be changed by changing the conditions of use. Major examples (see Hammersley 2014) include the uptake of safer injecting practices and consequent control of HIV infection, and the ability of addicts to plan and manage their own recovery rather than being the passive recipients of set treatment programs. Also, the ability of some alcoholics to resume controlled drinking, and the existence of heroin users who consume in a controlled manner and reduce the problems they experience. Not forgetting the sometimes suppressed finding that most cocaine users quit without assistance, often because they realize that they are spending too much money.

Also, cognitive behavioral therapies, especially third-wave CBT, have the core theoretical assumption that how the person thinks about themselves and their problems, and manages their behavior and environment, affects their wellbeing and ability to function. It is not clear how the effectiveness of CBT can be reconciled with BDMA. Indeed the British Psychological Society Division of Clinical Psychology has questioned the appropriateness of a model of mental health problems that considers them to be disease-like conditions predominantly caused by biology (Division of Clinical Psychology 2013). Various sociological theories are important to the understanding of addiction.

Structural theories

Risk

There is a substantial body of sociological and psychological theory on how people construct, perceive and manage risk, mostly not considering addiction (e.g., Beck 1992; Slovic 2013; Douglas 2013). Space precludes detailed review but the following points are important. (1) Definitions of the risks that are worth managing vary from place to place and are contested. (2) The information needed to calculate what the risks of harm really are is difficult or impossible to obtain and incomplete, so people have to guess. (3) People take more risks with purportedly "safer" things (see the Midwest opiate epidemic above). (4) Developed societies are increasingly centralising risk, security and safety, and moving away from a "modernist" view that innovation, novelty and technology are good, towards fearfulness. Another point (5) is that when risks are presented they are almost always presented as frightening and accurate facts that require urgent action, with little or no mention of points 1–4 (Bauman 2013). Addiction is rife with problems presented in this manner. Recent examples include binge drinking, mephedrone, synthetic cannabinoids, Necknomination and Buckfast Tonic Wine. Risk perception can be manipulated for political, economic or personal ends.

Despite the rise of terrorism fears, drugs remain highly newsworthy and typically presented as an enormous problem (Manning 2013). Public health continues to try and inform people of the risks of drugs and encourage appropriate action. However, the objective risks of different behaviors are distorted and exaggerated, especially for the drugs considered most dangerous including heroin, cocaine and methamphetamine. Consequently, public health practitioners struggle against the attitude that anything but the most exaggerated and bleak view of the risks of drugs is tantamount to encouraging use.

Lumpen abuse

This sets the scene for using purported risks to perpetuate existing power structures, although addicts are not necessarily disadvantaged on purpose. Bourgois and Schonberg (2009) theorize that many of the problems of homeless heroin and crack users are due to their being institutionally victimized at the absolute margins of American society. Their ethnography documents that the righteous dopefiend lifestyle is about coping, and has an understandable and sensible social order, which is part of the American way, and that dopefiends often have life histories including very serious abuse and trauma. This research documents inadvertent cruelty towards the research participants by supposedly caring agencies (Bourgois and Schonberg 2009) that would constitute scandal were they another vulnerable group, such as the elderly. The cruelty is rationalized as being required because of the difficulty of drug injectors as clients and the need to enforce rules. The USA has been particularly brutal towards drug injectors, but milder difficulties occur in countries such as the UK with far more liberal practices (Hammersley and Dalgarno 2012; McKeganey 2010).

There are very large medical, social care and criminal justice enterprises with vested interests in treating addicts rather than solving complex social problems. Vested interests also include several huge industries that have been documented manipulating the relevant policy, legal and research agendas in their favor (see above). The illegal drugs industry also benefits massively from its products being illegal, unregulated and untaxed, and has the funds to penetrate and corrupt local and national government and justice (Castells 2000).

Simultaneously, alcohol related problems are trivialized, again to the detriment of the people suffering the problems. For example, alcohol related mortality is on the rise in the UK whilst other mortality falls (Weissenborn and Nutt 2012), and alcohol is related to vast numbers of accident and emergency admissions, yet it is only recently that there have been policies and practices considering excessive alcohol use as a factor in disorder, accidents and other health problems.

Deviant subculture theory

Further articulating social structure, this theorizes drug use as the product of a deviant subculture where activities such as drug use and supply serve normal social, cultural, economic and personal functions (Agar 1973; Bourgois 1995; Johnson *et al.* 1985), including power, signifying social status, earning a living, and meaningful occupation, often in conditions limiting access to conventional sources of these functions. This contrasts with theories of drug use as disease-like, compulsive, addiction and is an approach extending back to the 1950s (Becker 1953; Whyte 1955). Deviant subculture theory is informative about drug use, notably marijuana use, which involves specific dialect and artifacts (Golub *et al.* 2005) and is also informative about street level drug dealing, but it may be less helpful as an account of problematic drug use. Indeed, the ethnographic research underpinning the theory does not strongly differentiate "drug use" and "problematic use". Yet entirely functional accounts are limited in their capacity to explain why some people choose drugs widely perceived to be much more dangerous, or engage in drug supply, or why some people develop serious problems related to use whereas others do not. Probably, addiction is not a wholly functional lifestyle choice but rather an attempt to cope with very difficult life challenges (Bourgois and Schonberg 2009; Hammersley *et al.* 2016).

Deviant subculture theory was developed primarily during ethnographic research on social groups whose identity is self-defined in large part by their drug use. The world still contains such deviant groups, but increasingly many young people mix and match different substances on different occasions, with different friends, without strongly self-identifying as users of a specific drug (Aldridge *et al.* 2011). We do not know enough about this sort of drug use by non-deviant people, as opposed to habitual or addictive use by addicts.

Social identity

How and why do addiction or drug use become important parts of a person's identity? Deviant subculture theory focuses on conditions where people self-identify as being part of a social network of drug users or addictions and such conditions certainly exist sometimes, often when drugs are introduced to a friendship and kinship network (Golub *et al.* 2005). At the extreme the idea of temperance or abstinence can be perceived as absurd by the network and quit attempts treated with "derision" (Harris *et al.* 2005).

However, social identity is also formed by the perceptions of others, complicated because drug use can be a form of hidden or concealable stigma (Goffman 1990). Illicit drug use and alcoholism are subject to quite extreme stigmatization, including a focus on the most extreme cases involving death, disfigurement, criminality and social exclusion, and media fascination with the more gruesomely photogenic aspects of drug use and extreme drinking (Huggins 2010). For drug users – who are the majority of society if we include alcohol – the minimization of drug stigma is central to maintaining a functional social identity. Common mechanisms include attributing visibly inappropriate drug use to one-off situationally caused lapses (Davies 1997), defining other styles of drug use as worse (Rodner 2005), and simply concealing use (Shewan and Dalgarno 2005).

Identity, suffering and embodiment

Consequently, most drug users will present themselves differently in different social settings for socially functional reasons. To fellow users they may present themselves as taking drugs on purpose because they enjoy them, while to healthcare professionals they may present themselves as more addicted (Davies 1997). There are complex social and cultural mechanisms that make it difficult for people to receive appropriate help without presenting themselves as suffering (Fassin 2002). For example, drug injectors should be vaccinated against hepatitis C, but it may be difficult to ask for this without full engagement with whatever addiction services are on offer whether or not they wish or require these. Socially marginalized people may often be placed in a position where to receive help they need to display their suffering, yet they may have to conform to an ideology of addiction if they reveal their problem (Sobell and Sobell 2006), and then struggle to be "normal" (Nettleton *et al.* 2013). Similar problems apply for "people who use drugs" but who do not wish "drug user" to become part of their identity (Hammersley *et al.* 2001).

Drugs affect the body and the condition of the body signifies the social meaning of drug use (Weinberg 2002). The embodied drug injector tends to be represented as extreme and deviant, especially in the media, but also in the medical literature and by users themselves (Huggins 2006). Drug injecting comes to typify the embodiment of drug use, partly because it is photogenic, although it is relatively rare. Little is known about the embodiment of other forms of drug use. For example, the practice of smoking marijuana before going on shift in tedious jobs is widely known but barely studied (Dreher 1983).

Conclusions

Sociological understandings have made an important but under-appreciated contribution to research on and interventions for addiction. They have contributed to an understanding of the issues that is more nuanced than the idea of addiction as a disease, and have facilitated successful interventions that rely upon the rational volition of addicts, connecting theoretically with cognitive behavioral interventions. Addiction continues to be constructed mainly as a problem of the deviant other and there is a need for further sociological research on the nature of normal substance use, whatever that is.

References

Agar, M. (1973) *Ripping and Running: A Formal Ethnography of Urban Heroin Addicts*, New York, NY: Seminar Press.
Aldridge, J., Measham, F. and Williams, L. (2011) *Illegal Leisure Revisited: Changing Patterns of Alcohol and Drug Use in Adolescents and Young Adults*, London, UK: Routledge.
Anderson, P. and Baumberg, B. (2006) *Alcohol in Europe: A Public Health Perspective*, Brussels: European Commission.
Bauman, Z. (2013) *Liquid Fear*, London, UK: Wiley.
Beck, U. (1992) *Risk Society: Towards a New Modernity*, New York, NY: SAGE Publications.
Becker, H. (1953) "Becoming a marihuana user", *American Journal of Sociology* 59(3): 235–242.
Best, D., Dawson, W., De Leon, G., Kidd, B., Malloch, M., McSweeney, T., Gilman, M., Stevens, A., Thom, B., Zandvoort, A., et al. (2010) *Tackling Addiction: Pathways to Recovery*, London, UK: Jessica Kingsley Publishers.
Bjerg, O. (2008) "Drug addiction and capitalism too close to the body", *Body & Society* 14(2): 1–22.
Borio, G. (2007) "The tobacco timeline" Available online at: http://archive.tobacco.org/History/Tobacco_History.html [accessed 04/14, 2016].
Bourgois, P. (1995) *In Search of Respect: Selling Crack in El Barrio*, Cambridge, UK: Cambridge University Press.
Bourgois, P. (2000) "Disciplining addictions: the bio-politics of methadone and heroin in the United States", *Culture, Medicine and Psychiatry* 24(2): 165–195.
Bourgois, P. (2010) "Useless suffering: the war on homeless drug addicts", in H. Gusterson and C. Besteman (eds), *The Insecure American: How We Got Here and What We Should Do About It*, Berkeley, CA: University of California Press, pp. 238–254.
Bourgois, P. and Schonberg, J. (2009) *Righteous Dopefiend*, Berkeley, CA: University of California Press.
Brandell, J. R. (2013) *Psychodynamic Social Work*, New York, NY: Columbia University Press.
Carnwath, T. and Smith, I. (2003) *Heroin Century*, London, UK: Taylor & Francis.
Castells, M. (2000) *End of Millennium, Volume III: The Information Age: Economy, Society and Culture*, London, UK: Wiley.
Craib, I. (2001) *Psychoanalysis: A Critical Introduction*, London, UK: Wiley.
Davies, J. B. (1997) *The Myth of Addiction*, London, UK: Psychology Press.
Division of Clinical Psychology (2013) *Classification of Behaviour and Experience in Relation to Functional Psychiatric Diagnoses: Time for a Paradigm Shift*, Leicester, UK: British Psychological Society.
Douglas, M. (2013) *Risk and Acceptability*, London, UK: Taylor & Francis.
Dreher, M. (1983) "Marihuana and work—cannabis smoking on a Jamaican Sugar Estate", *Human Organization* 42(1): 1–8.
Fassin, D. (2002) "The suffering of the world. Anthropological considerations on contemporary polities of compassion", *Evolution Psychiatrique* 67(4): 676–689.
Galloway, J., Forsyth, A. J. M. and Shewan, D. (2007) *Young People's Street Drinking Behaviour: Investigating the Influence of Marketing & Subculture*. London, UK: Alcohol Education Research Council.
Goffman, E. (1990) *Stigma: Notes on the Management of Spoiled Identity*, London, UK: Penguin.
Golub, A., Johnson, B. D. and Dunlap, E. (2005) "Subcultural evolution and illicit drug use", *Addiction Research & Theory* 13(3): 217–229.
Granfield, R. and Cloud, W. (1999) *Coming Clean: Overcoming Addiction without Treatment*, New York, NY: New York University Press.
Hacking, I. (1999) *The Social Construction of What?* Cambridge, MA: Harvard University Press.
Hammersley, R. (2011) "Developing a sociology of normal substance use", *International Journal of Drug Policy* 22(6): 413–414.
Hammersley, R. (2014) "Constraint theory: a cognitive, motivational theory of dependence", *Addiction Research & Theory* 22(1): 1–14.
Hammersley, R. and Dalgarno, P. (2012) *Drugs: Policy and Practice in Health and Social Care*, Edinburgh, UK: Dunedin Academic Press.
Hammersley, R., Dalgarno, P., McCollum, S., Reid, M., Strike, Y., Smith, A., Wallace, J., Smart, A., Jack, M., Thompson, A. and Liddell, D. (2016) "Trauma in the childhood stories of people who have injected drugs", *Addiction Research & Theory* 24(2): 135–151.
Hammersley, R., Jenkins, R. and Reid, M. (2001) "Cannabis use and social identity", *Addiction Research & Theory* 9(2): 133–150.

Hammersley, R. and Reid, M. (2002) "Why the pervasive addiction myth is still believed", *Addiction Research and Theory* 10: 7–30.
Harre, R. and Moghaddam, F. M. (2012) *Psychology for the Third Millennium: Integrating Cultural and Neuroscience Perspectives*, New York, NY: SAGE Publications.
Harris, M., Fallot, R. D. and Berley, R. W. (2005) "Qualitative interviews on substance abuse relapse and prevention among female trauma survivors", *Psychiatric Services* 56(10): 1292–1296.
Hinton, S., Signal, T. and Ghea, V. (2015) "Needle fixation profile: an exploratory assessment of applicability in the Australian context", *Substance Use & Misuse* 50(11): 1449–1452.
Huggins, R. (2006) "The addict's body: embodiment, drug use and representation", in D. Waskul and P. Vaunini (eds), *Body/Embodiment: Symbolic Interaction and Sociology of the Body*, Farnham, UK: Ashgate, pp. 165–180.
Huggins, R. (2010) "Images of addiction: the representation of illicit drug use in popular media", in L. J. Moore and M. Kosut (eds), *The Body Reader: Essential Social and Cultural Readings*, New York, NY: New York University Press, pp. 384–398.
Johnson, B. D., Goldstein, P. J., Preble, E., Schmeidler, J., Liption, D. S. and Spunt, B. (1985) *Taking Care of Business. The Economics of Crime by Heroin Abusers*, Lexington, MA: Lexington Books.
Manning, P. (2013) *Drugs and Popular Culture*, London, UK: Taylor & Francis.
McKeganey, N. (2010) *Controversies in Drugs Policy and Practice*, London, UK: Palgrave Macmillan.
Nettleton, S., Neale, J. and Pickering, L. (2013) "'I just want to be normal': an analysis of discourses of normality among recovering heroin users", *Health* 17(2): 174–190.
Nutt, D. J., King, L. A., Phillips, L. D. and Independent Sci Comm Drugs (2010) "Drug harms in the UK: a multicriteria decision analysis", *The Lancet* 376(9752): 1558–1565.
Parker, H. (2005) "Normalization as a barometer: recreational drug use and the consumption of leisure by younger Britons", *Addiction Research & Theory* 13(3): 205–215.
Peele, S. (1990) "Addiction as a cultural concept", *Annals of the New York Academy of Sciences* 602: 205–220.
Proctor, R. N. (2012) "The history of the discovery of the cigarette–lung cancer link: evidentiary traditions, corporate denial, global toll", *Tobacco Control* 21(2): 87–91.
Rodner, S. (2005) "'I am not a drug abuser, I am a drug user': a discourse analysis of 44 drug users' construction of identity", *Addiction Research & Theory* 13(4): 333–346.
Savell, E., Fooks, G. and Gilmore, A. B. (2016) "How does the alcohol industry attempt to influence marketing regulations? A systematic review", *Addiction* 111(1): 18–32.
Shewan, D. and Dalgarno, P. (2005) "Evidence for controlled heroin use? Low levels of negative health and social outcomes among non-treatment heroin users in Glasgow (Scotland)", *British Journal of Health Psychology* 10(1): 33–48.
Slovic, P. (2013) *The Feeling of Risk: New Perspectives on Risk Perception*, London, UK: Taylor & Francis.
Sobell, M. B. and Sobell, L. C. (2006) "Obstacles to the adoption of low risk drinking goals in the treatment of alcohol problems in the United States: a commentary", *Addiction Research & Theory* 14(1): 19–24.
Szasz, T. S. (1974) *Ceremonial Chemistry*, New York, NY: Doubleday.
Volkow, N. D. (2014) "Prescription opioid and heroin abuse", available online at: www.drugabuse.gov/about-nida/legislative-activities/testimony-to-congress/2014/prescription-opioid-heroin-abuse.
Volkow, N. D., Koob, G. F. and McLellan, T. (2016) "Neurobiologic advances from the Brain Disease Model of Addiction", *New England Journal of Medicine* 373(4): 363–371.
Weinberg, D. (2002) "On the embodiment of addiction", *Body & Society* 8(4): 1–19.
Weissenborn, R. and Nutt, D. J. (2012) "Popular intoxicants: what lessons can be learned from the last 40 years of alcohol and cannabis regulation?" *Journal of Psychopharmacology* 26(2): 213–220.
Weppner, R. S. (1981) "Status and role among narcotic addicts: III some social and psychiatric characteristics", *Archives of General Psychiatry* 15: 599–609.
Whyte, W. F. (1955) *Street Corner Society*, Chicago, IL: University of Chicago Press.
Zinberg, N. (1984). *Drug, Set and Setting: The Basis for Controlled Intoxicant Use*, New Haven, CT: Yale University Press.

19

THE FUZZY BOUNDARIES OF ILLEGAL DRUG MARKETS AND WHY THEY MATTER

Lee D. Hoffer

Introduction

In this chapter I reflect on my ethnographic research[1] on local heroin markets and argue that the convention of separating (and isolating) the operation of illegal drug markets from drug addiction and the lifestyles of people addicted is unrealistic.[2] Further, I examine what changing this perspective might mean.

In the field of drug addiction, illegal drug markets receive far less attention than other related topics, and in some ways this makes sense. An illegal drug economy is much greater than the local community of people who repeatedly buy, use, and are addicted to a substance. Recognizing this, US drug policy is divided into supply and demand reduction approaches, the former being concerned with stopping the trafficking of illegal drugs (i.e., drug sales, transportation, and distribution) and the latter everything else (i.e., drug use interventions, prevention, and treatment). As local illegal drug markets are the mechanism through which drugs are supplied to users, it makes sense that this topic is central to supply reduction efforts. Yet, people who use and become addicted to an illegal drug cannot escape participating in the market in which the drug they use is sold.[3] For instance, people with heroin dependence disorders often *buy* (or acquire) heroin daily. This makes the illegal drug market a feature of addiction and, more importantly, the relationships fostered though acquiring drugs critical for understanding this chronic condition.

But another important reason why local illegal drug market research is not prominent in studies of addiction is that we often simplify basic operational features of illegal drug markets, obscuring local interactions and dynamics.[4] Transposing mainstream neo-classical economic notions of supply, demand, and price provides a straightforward way to understand local heroin markets. But we also know the economy of an illegal drug is hierarchical with cartels connecting to users via worldwide supply chains. At the retail level, through *a priori* reasoning, we know heroin users are connected to drug dealers. And if a dealer gets arrested, i.e., removed from the market, buyers must reconnect to access the drug. Finally, we recognize people sell and buy illegal drugs in both open-air markets/settings, as well as in private venues. Some also buy drugs over the internet.

In illegal drug markets, one feature is common: drug users often share information with and get assistance from their peers when buying or acquiring drugs. Thus, little of what occurs in heroin markets can be characterized as an anonymous transaction between strangers.[5]

I emphasize how, under close inspection, the inherently cooperative element of illegal drug markets undermines our conventional notion of them as traditional markets. Specifically, when heroin users repeatedly help their peers *buy* heroin (a common behavior I call "brokering," or what heroin users commonly referred to as "copping drug for others"). This act dissolves boundaries imagined to separate the illegal drug market from the everyday relationships people addicted to heroin have with their peers.

Instead of thinking of such behaviors as favors, assistance, survival, or an instrumental economic strategy, I suggest these acts are muddled. Cooperation here, which is part gift, part favor and part commodity exchange, supplies local markets with unique capabilities. Trading in this resource also occurs outside of any familiar drug market context, i.e., when people are *not* actually selling drugs. As a result, it becomes subsumed by the social connections people who are addicted have with their peers. Reminiscent of the notion of markets embedded in cultural systems (Polanyi 1944), but operating in reverse, here the economics of the illegal drug market become conflated with and eventually indistinguishable from the central processes through which heroin addicts *form* relationships with other addicts. In other words, in the process of addiction practical distinctions between the heroin lifestyle and the heroin market disappear. This boundary-blurring effect has important consequences for how local illegal drug markets operate, as well as the behavior of its primary participants, i.e., people addicted to heroin.

A buyers' perspective on heroin markets

Research on the importance of illegal drug markets has long emphasized drug dealing, drug dealers, and/or drug dealing organizations. As a capitalistic enterprise, *selling* drugs is what typifies drug markets. It follows then that drug sellers should be the focus of analysis. For ethnographic researchers, how sellers manage relationships, the social context of sales, and seller–buyer interactions characterize ways to understand these markets (Page and Singer 2010). Although I have conducted extensive fieldwork with heroin dealers (Hoffer 2006) and do not deny that dealers or dealing behavior is important, here I emphasize the other end of this trade: what contributions *buyers*, i.e., heroin users, make to these markets.

Overlooking the buyer in how heroin markets operate is easy. It is assumed dealers control, either directly or indirectly, much of a buyer's ability to purchase heroin, and in this way buyers have diminished agency. Economically, dealers set drug price and this is central. As a result, it might be suggested buyers do not belong in discussions of heroin markets.[6] Yet to exclude buyers implies a heroin addict's *only* market contribution is buying; this refocuses attention on selling (i.e., the importance of dealers) and reinforces the convention. In a break from this circular reasoning, I argue most heroin users, but especially those who are addicted to the drug, have a much greater role in how local heroin markets *function* than previously considered. In fact, the success of a local heroin market is equally (if not more) predicated on how people addicted to heroin *participate* in it than dealers selling drugs. Specifically, buyers offer local heroin markets an unassuming yet extremely robust resolution for two fundamental challenges all drug markets face: 1) providing users (i.e., themselves and their peers) with *continuous* access to the drug and 2) making *initial market connections* easy (and nearly risk free) for market newcomers or those seeking new dealers.

Questions, perspectives and caveats

The current US heroin epidemic has underscored the importance of understanding drug dependence as a complex and chronic disease. It also highlights familiar questions about local heroin markets, namely: How do new heroin users initially access their local heroin market?

How do people find heroin? With local dealers and sellers constantly being arrested, how are users able to maintain consistent access? And, finally, why are people addicted to heroin able to acquire the drug so easily?

It is beyond the scope of this chapter to consider *all* the complexity in answering these questions. In order to restrict discussion, I focus on a neophyte who has: 1) already used heroin, 2) is motivated to use again and 3) has means to purchase the drug. I will also assume this person is initially cautious and afraid to simply walk into what they may think is an open-air drug market and solicit a stranger to buy drugs. These assumptions focus attention on the role of buyers in heroin market operations. Nonetheless, the initial pathways people find to local heroin markets are extremely diverse. Some new heroin users go directly to the street to purchase heroin. Even in well-established open-air sales venues, however, this approach is dangerous, ill-advised, nor guaranteed to work. Other people acquire drugs by trading goods or service (e.g., trading sex for drugs). Other new users may get drugs for free or buy them over the internet. Yet despite this variation, the majority of heroin users pay cash for their drugs and do so in person, meaning they interact with someone.

Having money to buy heroin is another purposeful assumption. Obviously, some people who use or are addicted to heroin are poor and do not have resources to buy the drug; others are made poor from buying it. There is a long social science tradition that expertly describes the political economy of heroin, weaving drug use, addiction, poverty, and race into a complex tapestry useful for understanding illegal drug sales (Singer 2007). In this chapter, I do not duplicate this work. Instead, I emphasize something different; despite the challenges of poverty, most people who use illegal drugs have the means (i.e., money) to *buy* them. A fact as true for heroin as with any illegal drug.

Market access

Assuming the motive(s) and money are in place, how does a neophyte heroin user *access* their local heroin market? Here repeatedly buying heroin is of interest, as repeatedly using it foreshadows dependence. Right away, however, there are problems. How does one identify or locate a genuine heroin seller? How does one discretely communicate interest in buying heroin? How does a buyer avoid arrest or how does one know a seller is not the police? How does one trust a seller will not rob or otherwise harm them? How does one avoid being ripped-off? What is the true market price of heroin, anyway? Maybe most importantly, how does a person new to buying know they are buying heroin and not Fentanyl or another substance that will kill them?

These questions scaffold an important but overlooked distinction: accessing the market and purchasing heroin are distinct activities. Because heroin is expensive,[7] it is easy to assume money and access are the same thing. But it is misleading to assume money is the *only* relevant resource needed to make the transaction possible. None of the barriers that our potential buyer must contend with is resolved with money. Money does not guarantee, for a person buying "heroin," that they are actually buying heroin; or that they are not buying from the police; or that they are not getting ripped-off, etc. The plight of our neophyte underscores that gaining *access to* a dealer and *buying* drugs from a dealer are often distinct activities.[8] Money is required but so is market *access*. Such access is also relevant to non-neophytes. Experienced heroin addicts enviably, and often repeatedly, experience access issues when their dealer gets arrested, cannot be located, decides not to sell, or does not have product. Selling heroin is risky, and for people who are drug dependent access is a concern that *never* goes away.

Heroin users, and those who are dependent, greatly diminish these risks by leveraging peer support. This support lies at the intersection of how the heroin market intrudes into the everyday life of a heroin addict. Despite the formidable barriers that I note confront new heroin users, for most neophytes these barriers are non-existent. Simply stated, *people who start using heroin are typically doing this with another user or users*. People who initiate heroin use are almost always introduced to the drug by another heroin user, i.e., a family member, friend or other acquaintance. This means a new heroin user can easily circumvent the majority of barriers associated with connecting to the market by appropriating a connection to it through a peer. Peers often purchase heroin *for* new users. Various theories on drug addiction recognize that in the process of becoming addicted, new users are socialized by their more experienced drug-using peers (Becker 1963). This mentoring includes connecting to the market. At some later point, our new user may develop their own direct and personal connections to a dealer, but initial connections are seldom autonomous.

Here the people one is initially using heroin with are the same people supplying the drug. As this person (or people) supplying the drug is not actually a drug dealer but a peer, from the very beginning of a heroin user's career non-market and market relations comingle. As a result, from a heroin user's perspective there is no practical separation between what is and is not the market. Or that all social relations, within their new heroin life, are also relations potentially relevant to acquire the drug. But even if early lessons on this matter are difficult for new users to internalize, they are frequently reinforced as a person progresses from use to dependence.

Continuous access

A curious feature of most local heroin markets is that, despite the unstable nature of drug dealing, this instability does not translate into diminished drug access for buyers. In two decades of nearly continuous fieldwork with non-in-treatment urban heroin users, I have experienced only a handful of instances in which heroin users self-report the drug being scarce or difficult to locate[9] and these instances typically involved extenuating circumstances, i.e., when a user was new to the city or just leaving a very long prison stay. Here we might add, although the objective of heroin sellers is to make their product available to buyers, directly advertising, marketing, and selling heroin is problematic. Successful long-term dealers do not sit on the street corner selling heroin; they operate in private settings with select clienteles.

The challenge for heroin dealers realizing their motives to make money can be imagined then in the following terms: how can they *only* allow genuine buyers across the buyer–seller divide? As they are restricted by the illegal nature of their occupation, in what ways can they facilitate bringing buyers to them? Here one solution is that a dealer can hire, train, and pay agents to sell on their behalf, but this is expensive (Levitt and Venkatesh 2000), managerially challenging (Hoffer 2006), and not guaranteed to work. But in a majority of cases such efforts are entirely unnecessary, and much more trouble than they are worth. Like solutions offered in making initial market connections, the illegal drug market has devised an ingenious solution: it outsources both its seller location and deal-making services to the most highly motivated participants of its community: those addicted to the drug.

A very common but overlooked type of transaction operating in *all* heroin markets is brokered drug sales. A brokered sale is simply when a heroin user bridges a buyer–seller divide in order to purchase heroin for another heroin user (a buyer). A broker is a customer who acts as a "go-between," or intermediary, and when a heroin user brokers they take another heroin user's money to a seller and make a purchase for them. The broker then returns to the buyer with the heroin purchased and for this effort is typically rewarded by receiving some of the heroin purchased.[10] In this way, brokering offers an opportunity for someone addicted to heroin to use the drug *without spending any money*! Although having cash may be considered a sufficient condition to acquire heroin in this instance it is neither sufficient nor necessary. Having access to a heroin seller is a *necessary* condition to acquire the drug, but what many heroin users have figured out is that while access is necessary it can also be *sufficient* to acquire the drug. The proof of this is brokering. Money is not required to broker because the broker is using someone else's money, but access to a seller is necessary. Brokering therefore commodifies a necessary condition *of* a sale and trades it for a sufficient condition to acquire the drug. This makes brokering a very common form of exchange in heroin markets and data our research team are currently preparing for publication indicates this. Among a sample of 201 active heroin users recruited from our local syringe exchange program, 90% self-reported buying drug for someone else in their lifetime, and 83% reported doing this in the last 30 days.

Brokering offers a built-in and highly sustainable market mechanism for consistently bringing new (or experienced) people across the buyer–seller divide in local heroin markets. More importantly, it is a mechanism that facilitates dealers making sales but which is operated entirely by the people who have most to potentially gain from this action: people addicted to the drug. So, reflecting on the competitive challenges dealers face, most dealers do not need to employ people to find buyers, or sell drugs, because their customers are highly motivated to provide both of these services to their peers. It should be noted that a broker is not working for a dealer but rather they are working for themselves and the buyer.

Relations between dealers and brokers are symbiotic. Dealers only realize profits and stay in business because they sell product. For this reason, dealers who I have interviewed sometimes support brokering because it brings them customers. However, brokers can also be very problematic for dealers because they alter the "deal" being made, in effect making what is sold more expensive to buyers. Brokers can also diminish a dealer's reputation by telling buyers, sometimes falsely, that the bad deal a buyer gets was precisely the deal that was offered by the dealer. Often a dealer does not know the drugs they are selling to a customer are actually going to some other buyer. Brokers typically do not advertise this. But brokering is more consequential than what is actually happening in these relationships, all of which are subject to change. Instead the point here is that dealers do not control brokering, customers do. And this process, which ultimately a dealer's business and the market relies upon, is underwritten by the broker's addiction.

There is nothing particularly innovative about brokering and it has been recognized in the literature, in various forms and names, since social scientists have researched the behaviors of people addicted to drugs (Preble and Casey 1969). Brokering is a solution for markets of any illegal good or service.[11] For heroin, this occurs across all manner of relationships that users participate in. Friends may broker for friends, they may also broker for family, associates, or even strangers. It is also notable that individual characteristics such as race, sex, or socioeconomic status do not obstruct brokering. Here buying and brokering are interchangeable. A heroin user who brokers for a buyer one day may be seeking a broker to buy for them the next. Unlike drug dealers, brokers are not different from or superior to buyers; the act garners little status among heroin users.

Despite being extremely common, including brokering in discussions of how local illegal drug markets operate has been largely ignored. Although reasons for this are unclear, it does underscore that, much like some theories of addiction, copping drugs for others (i.e., brokering) is often viewed as only relevant to an *individual*, i.e., a strategy an individual addict employs to offset the cost of their addiction. Framed as something only needy (i.e., poor) heroin users do in desperation or without any other means to acquire the drug (Goldstein 1981), brokering has been considered no more significant than any number of petty hustles users employ for this purpose, i.e., shoplifting, panhandling, theft, etc. But brokering also suffers another major problem for theorizing: it does not *fit* into our conventional notion of illegal drug markets.

Including brokering

A local drug market is where buying and selling takes place. Heroin addicts facilitating purchases for other heroin addicts is clearly out of place in this context. So why include it? Simply stated, brokering can correct views about local illegal drug markets in several significant ways.

First, to include brokering we can fully appreciate why these markets are so perdurable. Just like the concept of "small world networks" (Watts and Strogatz 1998) that facilitates efficiencies in connecting with people though leveraging social network connections, brokering brings considerable efficiency to drug distribution. Brokering allows heroin users to extract resources from their social relations by converting heroin "search costs" into a commodity that they can exchange to acquire heroin. Any user who has a connection to a dealer can serve the market in precisely the same capacity as a dealer.[12] In this way, brokering diffuses drug access and minimizes the impact of removing any individual dealer from the market. To this economic system one drug dealer becomes inconsequential, which is a hard lesson we have learned in the 80-year war on drugs. It also gives the local market unprecedented flexibility to adapt, seamlessly counteracting mass incarceration tactics. But it also allows the market to easily reproduce, relocating within any and all settings heroin users inhabit.

This form of exchange contrasts with the traditional view of local illegal drug markets. The convention is that one can separate, isolate, and divorce the "market" from the everyday social relationships heroin addicts have with their peers. The market is external, i.e., buyers must "go to" the market to purchase heroin. This conceptualization also provides a geography for this space, i.e., a place buyers and police can locate. Dealers and *where* they sell the drug have historically defined heroin markets. Also, as with any commodity, the conventional belief promoted originally by Adam Smith is that the invisible hand coordinates the local heroin market via supply, demand, and price. But it remains unknown how this can take place when buyers and sellers cannot legitimately meet.

Although a broker may act much like a dealer, they are not a dealer. Instead of thinking of brokers as dealers, they are more aptly market coordinators or "invisible hand facilitators." A broker is not contributing any money or heroin to the market, they make no financial investment. But brokers facilitate transactions and the flow of both market information and goods to and from dealers. Confusing the conventional economic logic associated with price, brokers can "set" the price of heroin, yet dealers still sell heroin for the "retail price" whether or not the person buying is a broker. One way brokers profit from brokering is by charging the buyer *more* than what they can purchase heroin for. In our recent survey, 39% of heroin users who reported brokering self-reported inflating the price to buyers "often or almost always." Brokers may also take some drug out of the package before delivering it back to the buyer. Once again, 22% of heroin users who reported brokering self-reported in our survey doing this "often or almost always." This activity, known as "pinching," also raises the price of heroin, effectively resetting the "deal." In short, brokers extract payments from heroin buyers to help coordinate local supply, demand, and price issues. The irony is that brokers are also heroin buyers, which means such payment acts as a self-imposed tax used to finance the services of the market's most relevant participants, i.e., people addicted to heroin. But where this new version of the local heroin market resides, or is located, further complicates matters.

A broker may purchase heroin from a dealer on the street, in a home, or in some other setting, but actual transfers between broker and buyer occur elsewhere. This separation is critical as the broker protects the location of and subsequently access to the dealer. As a set of transportable interactions, brokering calls into question *where* the local heroin market operates and whether geography is still relevant. Heroin users cop for their peers in *all* market environments, including "open-air" markets. This means a "dealer" selling heroin on the street may be a dealer, but they may just as likely be a broker. In short, brokering complicates the target of supply reduction policy: the heroin seller. Although these distinctions may only seem semantic, they highlight a point de-emphasized by our current understanding: to acquire drugs, *relations* between customers (i.e., users) are just as consequential as relations between customers and dealers.

Finally, brokering gives us unique insight on the psychosocial dynamics associated with being heroin dependent. Compared with direct drug selling, brokering occurs between *peers*. In this way, it often is ensconced within a much broader set of social relations (i.e., friendship, family, etc.) that are important to users. It also serves to create close relations between addicts through embedding drug economics (and the drug market) within these social relations. But this co-mingling can be fraught with stress, ambiguity, and tension. When a heroin user *buys* heroin directly from a dealer, provided they know roughly what their money should purchase, they can evaluate the quality of the deal and make a judgment about the dealer. Simple. When a friend brokers for a friend it can potentially enhance bonds of affinity. But it can also, and just as easily, generate conflict because the broker is providing something the buyer does not have, and is purposefully keeping from them: access to a dealer. This is further complicated because transfers are always *indirect*. This can perpetually confuse peer relations. Is my friend, who is buying heroin for me (i.e., the broker), being selfish and ripping me off? Or are they helping me? The buyer can never fully resolve this position. But as heroin buyers become brokers, in the same fashion this issue can challenge the self-concept of people dependent on heroin, forcing them to consider the ways they are treating the people (i.e., drug-using peers) they feel closest to and who they rely upon for their wellbeing (see Hoffer 2016).

Brokering and addiction

In my research experience, people addicted to heroin do not autonomously buy the drug like buying food in a grocery store. The economy does not allow for this. To acquire heroin, brokering implies that the disorder of (heroin) addiction, while both scientifically and popularly characterized as an individual brain disease, also encompasses highly social behavior. As I present in this chapter, it demands social cooperation to access the heroin marketplace, as well as to maintain that access. This may challenge some of what we think about addiction.

Addiction portrayed as a comprehensive self-control disorder or one characterized by diminished executive function, relative to decision-making, is difficult to resolve in the version of local heroin markets I am presenting and that includes brokering. These features of addiction, however, do support the traditional imagination of this economy, i.e., the one that excludes brokering and emphasizes users directly buying drugs from dealers. This conventional narrative is simple and convincing. Dealer–buyer relations are innately adversarial, maybe even hostile. Only engaging dealers, buyers do not need to rely on others. In such a market setting, selfishness, distrust, conflict, violence, impulsivity, and aggression are fostered, even rewarded and encouraged. Such traits may protect a buyer in getting what they paid for. This, in turn, further separates and disconnects people with addictions from the rest of us, making cooperation something exceptional or rare, only maybe occurring when addicts are actually using drugs. A buyer's relevance becomes exclusively about buying, which conforms to a very antisocial or individualized perspective on addiction.

A local heroin market incorporating brokering provides a different narrative. Brokering (in the capacity of being a broker) requires addicts to delay the immediate gratification associated with using heroin. This is something we understand people with addictions are, by definition, incapable of. Brokering is, therefore, incommensurate with a core logic in theories on addiction. But, having witnessed this behavior a countless number of times, it is impossible to deny that brokering works, and is quite popular. Furthermore, it only works, and is significant, because addicts *return* to their peers with the drug, in exchange for a potentially smaller portion of the drug than they actually purchase. This implies that heroin addicts behave in a *real* economy much differently than they do in delayed-discounting experiments conducted in laboratory settings intended to measure how addicts' economic decision-making and brains are different. If people addicted to heroin lacked self-control or an ability to delay desires to use, brokers could not fulfill their exchange obligations. It just would not work. Furthermore, for addicts seeking brokers, they could hardly initiate such transactions without trusting the people (at some basic level) to whom they give their money.

Of course, brokering rip-offs do occur and this is a risk buyers accept. In my experience, however, wholesale theft is rare in these cases and, if this does occur, it promotes swift social sanctions. Such sanctioning is typically nonviolent because the consequences of violence are simply too expensive. I would also suggest violence from bad brokered deals is rare because people who are addicted to heroin understand the consequences of their actions (in the short term), once again defying long-standing beliefs about this illness. Instead, people who are drug dependent often respect the long-term nature of their condition. Consequently, short-term conflicts in brokering are often merely accepted, i.e., the inevitable cost of doing business. Also, and when compared with direct transactions with dealers, brokered transactions are a special case because they always include ambiguity, which confuses interpretation and subverts retribution. Outside of outright theft, which again is uncommon, it can be difficult to determine if a rip-off occurred in this form of exchange.

If a heroin user does believe another heroin user ripped them off in a brokered transaction the consequence is that they will likely not use the person to broker in the future *and* they will tell their friends not to use them. People addicted to heroin can, and do, become socially isolated by their peers through reputational processes associated with bad brokering. Heroin users often describe such peers (i.e., bad brokers) as "greedy," meaning in this instance not able to abide by the social contract of brokering. These pressures are significant as they can prevent people addicted to heroin from accessing the drug. I offer that this sanctioning, which is enacted at the individual level, reinforces a strong community norm that brokers should not steal from buyers. Users become aware that brokering thefts hurt *all* heroin users as this is a strategy *all* heroin users can potentially benefit from in acquiring heroin for free. However, there is also another important buffer at work here that lubricates this transaction.

Based on the interviews I have conducted, people buying heroin *accept* that some hustling (i.e., pinching, inflating price, etc.) may occur during a brokered transaction because they know brokers are just like them. Heroin users are aware that the people they are asking to buy heroin for them cannot be fully trusted because they too are heroin addicts and may, at times, be equally suspect. This implies that people who are addicted have a more advanced and practical self-awareness of their condition than they are given credit for in addiction theory. They know what addiction means. Associated with this, there is a sense that people with such disorders do not have the capacity for empathy, which allows them to freely and wholly participate in the deviance of illegal drug use, as well as other crime. This too is problematic in what I have presented. Even when some hustling does occur, a heroin addict (buyer or broker) *rarely* ignores, disregards, or utterly discounts another heroin addict's situation within the context of brokering. This would be exceptionally disrespectful as this mutuality is precisely what serves as membership criteria for the Life. Many users view brokering as helping a peer meet the needs of their addiction, as well as helping themselves. This is the social contract addicts must negotiate.

In local heroin markets people who are addicted to heroin often buy drugs directly from dealers, suggesting some of what we understand about addiction using a more traditional perspective on heroin markets is true. Nevertheless, for the very important reasons noted in this chapter, brokering is popular and it too supplies users with heroin. Blurring the distinction between what is the market and what is simply part of the lifestyle of a person who is addicted, this exchange is layered with both practical and symbolic meaning. It is how users actively participate in the distribution of drugs and, hence, the local market. It is also a transaction in which an addict, however briefly, entrust their peers with the responsibilities of their addiction. Because of the ways this exchange is accepted, it operates as a force in forming, extending, managing, and generally supporting an addict's relationships. Finally, brokering allows users to demonstrate their compliance with norms vital to the most important (and often only) social group supporting their condition.

For theories on addiction the complexities noted above are potentially problematic. Brokering suggests people addicted to heroin maintain cognitive skills, decision-making abilities, intelligence, self-awareness, and empathy for others. And in my experience observing this transaction among heroin addicts who have decades of experience using the drug, such capacities are not lost over time. As a chronic and progressive "brain disease," giving addicts credit for such skills may make non-addicts uncomfortable, and yet it is very difficult to explain brokering without recognizing the core competences in what they are doing and that make the exchange possible. Accepting brokering does not mean repudiating opiate use disorder or denying anything about this condition. In over 20 years of witnessing heroin addiction and how it devastates the lives of people I have personally gotten to know (i.e., my research participants), I would never remotely suggest this conclusion. And yet, for just as long, I have witnessed people broker heroin sales for their peers and exhibit, either directly or indirectly, the traits noted above.

To inform theories on addiction, brokering tempers, qualifies, moderates, and maybe even backstops a highly individual-oriented science of addiction, and its reductionistic yet all-encompassing slogan, "addiction is a brain disease." I have found it difficult to determine what specifically about brokering reflects our notion of brain disease. This chapter exclusively features a *social* behavior enacted by people addicted to heroin and how its relational elements support local market operations. This is complex social behavior, meaning what is relevant occurs outside the confines of an addict's head, and that point seems to require added emphasis in our continued effort to characterize this disorder.

Conclusion

I concede it seems counter-intuitive elevating anything besides the motives of people making money selling drugs, i.e., drug dealing, in relevance to why local heroin markets function. From my own ethnography of heroin dealers (Hoffer 2006), I know they are important and often very skillful at their occupation. But just because dealers *want* to be successful and can make money does not preordain that they will. Moreover, this is hardly an adequate explanation for how localized markets facilitate consistently bringing users into sales relationships or providing consistent drug access for those already buying. In most relevant ways, an illegal drug economy is completely invisible.

Although an essential component of local heroin market operations, brokering is easily overlooked. This may be viewed no differently from other strategies heroin addicts employ to meet the economic demands of their addiction, such as shoplifting. Clearly, people who use heroin have other ways besides their peers to acquire heroin, such as going into open-air markets. This is also only one of a number of acts, such as drug sharing, that facilitate communal bonds developing between people addicted to heroin. Heroin users also sometimes work for dealers, meaning brokering is clearly not the only way users are involved in the illegal drug market. But brokering is unique because of the way in which it combines *all* of the above. It is self-financed (i.e., by users), builds community, directly provides the means for easily bringing new heroin users into the market, and for those who already participate supplies uninterrupted access to the drug. It is also maybe the only way heroin users can acquire heroin without incurring any (additional) costs, economic or otherwise.

More broadly speaking, my objective in this chapter has been to demonstrate how boundaries overlap between what we label the local heroin market and the everyday lives of people addicted to heroin. Accepting this indistinct border significantly complicates our ability to apply conventional economic logic, tools or methods to make sense of this coordination. Here we are challenged by the fact that an illegal drug market is *both* a system external to an individual (and operates via the invisible hand, i.e., neoclassical economic theory) but also one that embeds itself in the lives of users (and operates via dynamics of gift exchange, reciprocity and other non-market based transfers). And these notions concerning the essence of markets have historically been incompatible. Here it is clear that people who use and become dependent on heroin shape their relationships, community, and self-concept through the cooperative (and often conflicting) processes of acquiring drugs. These dynamics, that occur in tandem with and sometimes as frequently as using the drug, should not be discounted in understanding addiction.

Finally, this chapter presents how the relevance of local heroin markets extends beyond the act of selling or buying drugs. As a set of intimate and transportable relations operating in the community of heroin buyers, brokering has complex influences on the lives of addicts. It also makes them essential to drug distribution. One might conclude from this chapter that policies to combat addiction should seek to arrest even more heroin users to overcome brokering. But this policy has already failed. A more enlightened approach is to accept this relationship between the drug market and drug addiction and recognize that by offering more demand reduction opportunities, for instance an open-door policy of drug treatment on demand, it can serve to meet supply reduction objectives. Here treatment not only removes buyers but has the compounding effect of removing brokers too. Our 80-plus year war on drugs has proven that arresting ever increasing numbers of dealers and users has no appreciable effect on heroin access. To implement lasting effects, we must spend our efforts and resources in support of people who are addicted to heroin and seek to disconnect both from this disease but also its market.

Notes

1 Ethnographic research is a qualitative methodology cultural anthropologists have been using since the turn of the nineteenth century to understand culture and behavior. It involves open-ended interviewing and observational (fieldwork) techniques conducted over time with participants. The intent of this approach is to generate rich detail about how other people live, as well as what they believe and how they behave.
2 I use the terms "addiction" and "drug dependence" interchangeably in this chapter as a way to characterize the comprehensive condition of opiate use disorder. Here I am most interested in people repeatedly purchasing heroin in their progression from use to disorder.
3 Only users who home produce a substance, *never* selling, giving, trading or otherwise distributing it, or relying on others in these capacities, operate outside the drug "market."
4 Another barrier in doing research on this topic is that it requires gaining the trust of drug users and dealers. By employing ethnographic fieldwork research methods, I have conducted research on the operations of various drug markets, interviewing and observing numerous users and dealers, since 1992.
5 Internet drug transactions may be considered anonymous and occurring between strangers. However, such sales only succeed because buyers can freely share information about sellers.

6 In a recent manuscript submission (for another publication) a reviewer commented that buyers were not part of local heroin markets.
7 Even "cheap" heroin is more expensive by weight than gold.
8 This is clearly evident in many cases in which a heroin user has access without money or vice versa.
9 In the author's experience, even asking a heroin user if it is difficult to find heroin is so unrealistic it can undermine the credibility of an ethnographer if asked.
10 If a drug reward is not given to the broker, the broker is less inclined in the future to buy for the user, which acts to reinforce this practice.
11 Brokering processes operate to provide access to all manners of quasi-legal and illegal markets. To keep activity hidden, a wide range of clandestine activities such as spying, terrorism, the trade in body parts, human trafficking, and political corruption rely on mediated exchange.
12 Most brokers do not transact large quantities of heroin. By definition, they only ever transact quantities of heroin for the personal use of their buyers.

References

Becker, H. S. (1963) *Outsiders: Studies in the Sociology of Deviance*, New York, NY: Free Press.
Goldstein, P. J. (1981) "Getting over: economic alternatives to predatory crime among street drug users", in J. A. Inciardi (ed.), *The Drug–Crime Connection*, Beverly Hills, CA: Sage Publications, pp. 67–84.
Hoffer, L. (2006) *Junkie Business: The Evolution and Operation of a Heroin Dealing Network*, Belmont, CA: Thompson Wadsworth.
Hoffer, L. (2016) "The space between community and self-interest: conflict and the experience of exchange in heroin markets", *The Economics of Ecology, Exchange, and Adaptation: Anthropological Explorations Research in Economic Anthropology* 36: 167–196.
Levitt, S. D. and Venkatesh, S. A. (2000) "An economic analysis of a drug-selling gang's finances", *The Quarterly Journal of Economics* 115(3): 755–789.
Page, B. J. and Singer, M. (2010) *Comprehending Drug Use: Ethnographic Research at the Social Margins*, London, UK: Rutgers University Press.
Polanyi, K. (1944) *The Great Transformation*, Boston, MA: Beacon Press.
Preble, E. and Casey, J. J. (1969) "Taking care of business: the heroin user's line on the street", *International Journal of the Addictions* 4(1): 1–24.
Singer, M. (2007) *Drugging the Poor: Legal and Illegal Drug Industries and the Structuring of Social Inequality*, Prospect Heights, IL: Waveland Press.
Watts, D. J. and Strogatz, S. H. (1998). "Collective dynamics of 'small-world' networks", *Nature* 393(6684): 440–442.

20
MULTIPLE COMMITMENTS
Heterogeneous histories of neuroscientific addiction research

Nancy D. Campbell

Multiple shifts in the conceptual practices of addiction research have occurred over the past century—from battles over addiction's proper name to the waxing and waning of an array of hypotheses about etiology, neural mechanisms, importance of social and environmental context, and the role of learning, memory, motivation, and reinforcement. In the 1990s, addiction was reified as a brain disease at an ontological level in which permanent and irreversible changes to brain structure and function were considered inherent in a unified object—the "addicted brain." Then-director of the National Institute on Drug Abuse (NIDA) proclaimed addiction a "brain disease, and it matters" (Leshner 1997). Yet today addiction stands as the epitome of a disorder of neuroplasticity, modelled upon brain disease but no longer equated with it (Volkow, Koob & McLellan 2016). No longer cast as stable, rigid, or closed, the "addicted brain" is portrayed as exemplary in its "exquisite[. . .] open[ness] to its milieu" (Rose & Abi-Rached 2013: 52). This chapter traces one of the multiple histories of addiction research, acknowledging how diverse commitments of addiction research indicate the complexity of addiction as an object of knowledge.

Grounded in Science and Technology Studies (STS), a field that makes historical and sociological sense of science by attending to ontological and epistemological commitments, as well as material conditions underpinning scientific theory and practice, this chapter considers the view that "no object, no body, no disease, is singular. If it is not removed from the practices that sustain it, reality is multiple" (Mol 2003: 6). This chapter presents an account of "addiction" not as a unified object within a unified "paradigm," but as a heterogeneous object of a variety of knowledges based primarily on animal models of a highly heterogeneous human condition.

"Addiction" names a non-specific disease state that has ranged historically from allergy to autoimmune to infectious to metabolic to neural. Although a congeries of unified theories of addiction has been advanced over the past century, including several represented in this volume, none has garnered settled consensus. As a historically contingent and intrinsically heterogeneous condition, addiction provides a productive site for exploring "disunity within an apparent unity" (Young & Breslau 2016). Those seeking to characterize underlying neural mechanisms of addiction and dependence encounter striking obduracy. Whereas other diseases once considered "moral problems" such as anemia or pellagra are now viewed as tractable, addiction stubbornly remains a scientific and clinical puzzle.

The emergence of addiction as a scientific puzzle

Science was brought to bear upon narcotic drug addiction in the late nineteenth century (Berridge & Edwards 1987; Hickman 2007). A volume surveying four thousand extant studies, *The Opium Problem* (Terry & Pellens 1928), was produced by the Committee on Drug Addiction (CDA), which coordinated the search for a non-addicting analgesic in the United States (Acker 2002). Terry and Pellens called for a non-reductionist science to "supersede the chaos of contradictory opinion" (1928: 928). Physicians prescribing morphine were considered primary vectors of "chronic opium intoxication," although heroin use for "purposes of dissipation" was on the rise. While iatrogenic addiction declined with changes in medical practice (Courtwright 2001), neither morphine nor heroin became obsolete. Recognizing that solving "the opium problem" required focusing scientific acumen on a social problem widely considered a vice, sin, or crime, CDA migrated to the National Research Council (NRC), housed within the National Academies of Science (NAS), in 1939 (Eddy 1973). The CDA coordinated medicinal bench chemistry, animal pharmacology, neurophysiological research, and human (clinical) research, later changing its name to the College on Problems of Drug Dependence (CPDD).

Great faith in science made the strategic value of redefining addiction as disease evident to those arguing against criminalization on humanitarian grounds. Lawrence Kolb Sr saw addiction not as crime but as psychopathology (Kolb 1925a; 1925b; 1925c), and sought to remake addiction through the lens of mental hygiene. Hygienic approaches were institutionalized when Kolb became the first Chief Medical Officer at the US Narcotic Farm, which opened in Lexington, Kentucky, in 1935. This vast clinical/penal institution provided a natural laboratory for studying addiction, housing the Addiction Research Center (ARC), so named in 1948 when it was declared a National Institute of Mental Health (NIMH) intramural laboratory (Campbell 2007). The ARC later became the intramural research program of the National Institute on Drug Abuse (NIDA).[1]

Addiction is understood to result from complex entanglements of pharmacological effects and vulnerabilities deriving from aspects of social life, socioeconomic status, and political position. Typically, scientists bracketed the latter effects. Longtime CDA chair Nathan B. Eddy wrote that even if safer analgesics were found, "We shall still have the opium-producing countries . . . We shall still have the established machinery for illicit production and distribution of heroin . . . [and] We shall still have social & psychological forces that encourage potential addicts to dose themselves with drugs" (1963: 679). Clinical researchers—and their emerging preclinical colleagues—characterized craving, dependence, tolerance, withdrawal, and relapse as systematic neurophysiological phenomena in a stable pattern of investigation that lasted until the emergence of neuropharmacology in the 1960s. The expansion of animal research in the 1960s replied directly to statutory requirements, but new technologies increasingly offered researchers methods for looking at animal and later at human brains.

Fresh to the topic when hired in 1957, ARC Research Director William R. Martin "slowly c[a]me to realize that it is a unique and yet fundamental type of neuro- and psychopathology which is among the most important, if not the most important, of all mental health problems" (Martin, Sloan & Eades 1978: 103). Finding that drugs differentially alter the excitability of the central nervous system (CNS),[2] Martin (1967) proposed the existence of multiple opiate receptors that he named *mu*, *kappa*, and *sigma*. Hypothesizing that the typical addict was a "hypophoric individual with increased needs and wants," Martin proposed a homeostatic theory of tolerance important for "opponent process theories" (Martin, Hewett, Baker & Haertzen 1977; Solomon & Corbit 1974; Koob & Bloom 1988). Martin came to view addiction as an affective disorder underpinned by deficiency states. Foreshadowing theories of neuroadaptation (Koob & LeMoal 2006: 11, 14), Martin's work helped remodel addiction research as a molecular matter.

Advocacy of "chemotherapy" for addiction was grounded in connections between affect—feelings and mood states—and endogenous neurohumors modified by drugs. Martin's careful descriptions of subjects' feeling states demonstrated that affect was key to behavior. Noting clinical similarities but also differences between depression and the "exaggerated need states," Martin dubbed these "hypophoria":

> With regard to self-image the euphoric state is polarly [sic] opposite to hypophoria except in the estimation of self worth. Hypophoric patients feel they are worthy and deserving even though they are unappreciated. In contrast a high percentage of depressed patients feel unworthy... [H]ypophoric patients feel hopeful, can experience joy, enjoy humor, and laugh readily. In contrast, hopelessness and sadness are among the most common symptoms of depressed patients. Another important difference between the hypophoric and the depressed patient is the lack of guilt feelings in the hypophoric patient.
>
> Martin, Hewett, Baker & Haertzen 1977; Martin 1984, 3–5

The ARC's social location and expertise guaranteed substantive interaction between clinic and bench from the 1930s to the 1970s (Klein 2008). However, the research unit was ill-equipped for molecular biological work on receptor complexes, peptides, and proteins. As the race to visualize the opiate receptor heated up at NIH and other locations in the 1970s, addiction research decentralized. The domain became far more heterogeneous as the social, material, and institutional conditions of possibility moved towards behavioral research.

Implications of the behavioral revolution for neuroscience

Attention to affect was important to early addiction research, whereas behavioral researchers "black-boxed" affect in favor of close observation and operationalization of behavior. The "behavioral revolution" wrung moralism and stigma, psyche and society out of the addiction equation. "The behavior is always right" became the mantra in an approach that viewed drugs as "reinforcers" (Glickman & Schiff 1967; Schuster 2003). Although rooted in studies of intracranial self-stimulation (Olds & Milner 1954; Olds, Killam & Bach-Y-Rita 1956), the concept of "brain-stimulation reward" (BSR) was not applied to addictive behaviors until the 1970s (Kornetsky 2007). Brain stimulation was rewarding in and of itself—a "reinforcer" activated by stimuli ranging from drugs to activities or other repetitive behaviors. The existence of "brain reward circuitry," in which dopaminergic projections from the Ventral Tegmental Area (VTA) to the Nucleus Accumbens (NAc) are implicated in mood disorders and addictions (Russo & Nestler 2013), was postulated.

Behavioral studies traced acute effects and evolved a useful categorization system on that basis. Yet correlates between brain and behavior remained elusive. Vast numbers of people modify their relationships to drug of abuse over time—most without treatment. Behavioral approaches were generative for thinking about "alternative reinforcers'" as motivating some individuals to move on from addictions to more healthy or "natural" pleasures. Behavioral concepts of "drug abuse" were enshrined in the 1980s, setting the stage for neuroscience.

> Drug abuse is a far more complex phenomenon than previously thought, and it is now recognized that drug abusers represent a highly heterogeneous group, and the patterns

leading to dependence are diverse.... A reasonable assertion is that the initiation of drug abuse is more associated with social and environmental factors, whereas the movement to abuse and addiction are more associated with neurobiological factors.

Glantz & Pickens 1992 quoted in Koob & LeMoal 2006: 8

One theory that won wide acceptance held that deregulation of particular brain-reward circuits moved individuals from "occasional, controlled drug use to the loss of behavioral control over drug-seeking and drug-taking that defines chronic addiction" (Koob & LeMoal 1997 cited in Koob & LeMoal 2006: 3). Also crucial for that phase transition was the so-called dark side, a "negative emotional state when access to the drug is prevented" (Wise & Koob 2014). Pessimistic personality states, "self-care deficits," and "emotionally disordered lives" were seen as drug effects, rather than etiological factors in previous thinking about "predisposition" and even "original sin" that were once central to the notion that some individuals had "addictive personalities." "Additional insult[s] to the personality produced by the direct effects of the drugs themselves perpetuate, and actually create, such character flaws" (Koob & LeMoal 2006: 7).

By the late 1980s, neuro-imaging laboratories using technologies such as PET, SPECT, and MRI entered addiction research. Neuroscientists working on addiction base their accounts of the reinforcing effects of drugs and behaviors on the concept of brain-reward circuitry (Koob & LeMoal 1997, 2001, 2005, 2006; Wise 1980, 1989, 1998, 2004). The mesolimbic dopamine reward hypothesis of addiction posited dopamine as acting on a "common dopaminergic substrate" to "stamp" habit into memory (Wise 1980 quoted in Koob & LeMoal 2006, 378). Behavioral work established that both "natural" rewards (food and water) and "unnatural" rewards (drugs and activities such as gambling) worked upon a common neural circuitry. The implications of dopamine in learning, motivation, and reinforcement become important for studying the neuro-circuitry of addiction as a form of neuroadaptation.

Molecular targets and translational ambitions

Research specialization and adoption of narrow translational goals has oriented addiction neuroscience towards medications development based on identifying molecular targets and drugs to hit them. By the 1990s, the "receptor fever" of the 1970s (Kuhar 2005) had given way: "early [receptor] research failed to find consistent changes in opioid receptors, monoamine transporters and other targets that could account for drug tolerance and dependence" (Nestler 2004: 211). Moving "beyond the receptor," investigators elucidated extremely complex neurotransmitter and receptor systems, dashing hopes that "relatively simple upregulation or downregulation of the drug target" resulted in drug dependence (Nestler 2004; Kuhar 2005). The advent of proteinomics enabled study of mechanisms of addiction at neuronal and synaptic sites, and research honed in on particular proteins such as Brain-Derived Neurotrophic Factor (BDNF), the current front-runner neuroplasticity candidate involved in drug-seeking and relapse (Kalivas & O'Brien 2008). The BDNF supports "good" (natural) neuroplasticity that is "usurped" in cases of addiction. The BDNF is implicated in cellular mechanisms of normal learning and memory, as well as in "incubation of craving" and long-term gene and protein expression involved in relapse and abstinence (Carter, Hall & Illes 2012: 36). But BDNF acts differently with different drugs.[3] Expectation of reward may be insufficient motivation for drug-seeking behavior, a complex amalgamation of associations between a "reinforcer" and the "people, places, and things" involved in "addiction" or "addiction-like behavior" (Carter, Hall & Illes 2012). It is difficult if not impossible to separate pharmacological effects from significant drug-associated social cues. Indeed, BDNF plays multiple roles in cocaine abuse depending on brain region, cell type, and phase of addiction (Li & Wolf 2015: 240). The BDNF mediates neuronal signaling by strengthening or weakening synaptic connections in relatively long-lasting ways, presenting a "daunting therapeutic target" for cocaine addiction (Li & Wolf 2015: 240).

The flexible explanatory powers of neuroplasticity are strained as addiction neuroscientists explore ontological and epistemological uncertainties. Dispensing with reductionism, addiction geneticists probe individual variation and population-wide differences. Social factors, including early-life trauma, take on importance as epigenetic understandings of stressful experiences are incorporated into mappings of addictive processes. Intense individual variation in response to opiates, long noted in the clinic, indicates existence of multiple plasticities ranging from "homeostatic" plasticity to "synaptic" or "wholesale" plasticity (the latter indicating overall change in neural excitability, rather than at specific synapses (Nestler 2013). Structural forms of neuroplasticity alter the number of synapses and expand or contract neuron size, a finding that changed the overall ecology of the addiction research field, creating new "trading zones." "More than any other commonly studied form of experience-dependent plasticity, we are beginning to understand the potential causal relationships between the neural circuit adaptations elicited by drugs of abuse and the behavioural consequences of that experience" (Kauer & Malenka 2007: 855).

Addiction has become a proving ground for researchers studying neuroplasticity at the subcellular level. Studies of the role of long-term potentiation (LTP) in addiction was a novel direction in the early 2000s (Bliss, Collingridge & Morris 2003; Li & Wolf 2015) that initially engendered skepticism. Nestler, who set out to investigate what chronic drug exposure changed in the brain by hypothesizing a role for intracellular signaling, noted that NIDA was

> overly focused (one might say addicted) on a single circuit. NIDA support went overwhelmingly to researchers focused on the VTA–NAc pathway, despite the fact that it is just one part of a series of parallel, distributed circuits that are known to control reward and motivation. . . . In a similar vein, the field overly invested in use of the self-administration paradigm to study drug reward. Innumerable studies examined the ability of dopamine related and other pharmacological agents to regulate drug self-administration behavior.
>
> *Nestler 2004: 214*

Molecular-level occurrences pointed towards multiple mechanisms. Addiction was delocalized as a product of processes occurring in various brain regions rather than confined to one brain-reward system or region of interest. Addiction came to be seen as an outcome of interaction between endogenous and exogenous impacts on the brain: extra-cognitive affect and unconscious emotional experiences (Berridge & Winkielman 2003); "salience" or incentive sensitization (Robinson & Berridge 2008); transitions in locus of control as drug-seeking and taking moves from episodic to habitual (Everitt & Robbins 2005); and experiences of trauma and abuse.

Addiction research trajectories converged on traumas ranging from war to childhood sexual abuse, stresses replicated in the animal laboratory in such tasks as the "forced swim." Claims of the co-occurrence of very high prevalence rates of addiction and abuse surfaced in PTSD research (Bergen-Cico 2012; Fullilove *et al.* 1993). Trauma's heterogeneity is obscured by emphasis on its etiology in a singular event. This complicates its universal or standardized status as a diagnosis (Young & Breslau 2016). Before the 1990s, trauma and the diagnostic category of PTSD were rarely applied to substance abuse, as illustrated by the first NIDA Research Monograph to address violence, *Drugs and Violence: Causes, Correlates, and Consequences* (De La Rosa, Lambert & Gropper 1990), which contained no reference to trauma or PTSD. Feminist trauma researchers studied the etiological role of sexual abuse in addictive disorders, conceptualizing "those exposed to the harsh conditions of the inner city" (Fullilove *et al.* 1993) as carrying increased "level of burden" that made treatment overwhelmingly difficult (Brown, Huba & Melchior 1995: 345). Allies pushed for "trauma-informed" treatment in the wake of a major national study, the Women, Co-occurring Disorders, and Violence Study (WCDVS) sponsored by the Substance Abuse and Mental Health Services Administration (SAMHSA), and other studies revealed that early childhood abuse accounted for over half of serious drug-use problems (Dube *et al.* 2003).

Trauma-focused researchers found experiences of trauma are common for women suffering co-morbid polysubstance abuse and mental illness diagnoses. Post-traumatic stress disorder is one of the most robust gender differences (Torchalla *et al.* 2012: 109); the multiple forms of trauma that women experience are understood in gendered terms (Knight 2015; SAMHSA 2014, TIP 57). Similar to addiction research, trauma research is also a molecular matter and some of the very neuro-circuitry involved in vulnerability to substance abuse disorders is involved in trauma response. Dysregulation of the hypothalamic-pituitary-adrenal (HPA) axis, which mediates response to threat by activating coping mechanisms (McEwen 1988), is hypothesized to be responsible for vulnerability to substance abuse disorders among trauma victims. Post-traumatic stress disorder gradually moved from a peripheral condition to the core of neuroscience, coming to be studied as an anxiety disorder involving impaired signaling, memory, and learning—all of which are modified by CNS-active drugs (Krystal *et al.* 1995). As a conceptual framework, neuroplasticity makes it possible to see that previously learned responses or "coping mechanisms" might be over-learned in ways that translate into addictions, but might also be unlearned.

Neuroplasticity may be a fruitful route towards recognizing heterogeneities previously subsumed into both "addiction" and "trauma." Maturation of neuroplasticity as an interpretive framework was enabled by concepts and technologies that made it possible to see inside the brain—but also made it harder to see outside the brain. If multiple circuitries are involved in the patterns of reward and reinforcement that we call addiction in response to trauma, it would be a mistake to privilege one. Neuroadaptation and plasticity researchers recognize that "drug addiction represents a break with homeostatic brain regulatory mechanisms that regulate the emotional state of the animal" (Koob & LeMoal 2006: 435). Addiction represents neither a simple break with homeostasis nor a predictable dysregulation of hedonic and executive functions, but is rather an allostatic or dynamic break, which, if activated too often, leads to dysregulation (Sterling & Eyer 1988) and chronic elevation of reward thresholds that results in the very failures of self-regulation and emotional dysfunction attributed to drug addicts and those who suffer PTSD. Importing the psychosocial concept of allostasis (Sterling & Eyer 1988; Krieger 2011: 192), Koob and LeMoal write: "The view that drug addiction and alcoholism are the pathology that results from an allostatic mechanism that usurps the circuits established for natural rewards provides a realistic approach to identifying the neurobiological factors that produce vulnerability to addiction and relapse" (2006: 17). While the degree to which natural reward systems are usurped by pathologically unnatural ones remains unclear, most current neurobiological theories of addiction buy into neuroadaptation. Controversy over the role of the mesolimbic dopamine system remains. For some, addiction is all about the dopamine and its effects on learning and motivation; for others, dopamine is a bit player in the adaptations occurring in brain reward circuitries responsive to trauma and stress, "reward" and aversion.

Understanding addiction as "traditionally underappreciated as a disease rooted in neuropathology," Charles P. O'Brien likens it to an "overlearned memory" that is part of a "pathology of staged neuroplasticity" (Kalivas & O'Brien 2008). "Sociological" vulnerabilities combine with genetic and developmental vulnerabilities and "repeated pharmacological insults," affecting how individuals interpret stimuli and evolve behavioral strategies to respond to those deemed "motivationally relevant" (Kalivas & O'Brien 2008: 166). According to Kalivas and O'Brien, "enduring drug-induced neuroplasticity establishes a maladaptive orientation to the environment" that usurps normal learning and motivation circuits by supplanting a pathological form of neuroplasticity (2008, 167–168). But the mechanisms by which usurpation and supplantation work remain unknown. Why do these processes occur in some people, rather than for all who are similarly exposed to drugs of abuse? Trauma holds but one explanatory key to individual variation, explaining vulnerability only for some.

Conclusion

This chapter charts currents of continuity and discontinuity, unity and disunity, obduracy and plasticity important in the conceptual history of addiction neuroscience. Despite attempts to unify object and approach, the social and affective complexity of addiction exceeds them. Social and cultural histories of diverse moments of translation in clinical and research-oriented sciences perhaps tell us as much about the societies in which these concepts emerge as they do about the ontological objects under study. Despite the so-called NIDA paradigm—and concomitant notions of brain reward circuitry, allostasis, and the dopamine hypothesis—shifts in acceptance of the view that addiction is a brain disease are underway (Lewis 2015; Szalavitz 2016). Explanatory frameworks for addiction have long been inadequate to account for the entwined psychological and emotional effects that comprise the addictions; differences between acute and prolonged exposure; individual variation in response to drugs; and social learning about these effects at the individual and societal levels. Recent efforts to elucidate the neurochemobiological processes and mechanisms thought to underlie the changes that move non-addicted brains into addictive states have been multiple and are not easily subsumed into the brain disease model.

Notes

1 NIDA was established in September 1973. The ARC moved to Baltimore, Maryland, in the late 1970s (Campbell 2007).
2 This finding was the basis for the Addiction Research Center Inventory (ARCI), a survey instrument still used to classify drugs by their subjective effects (Haertzen et al. 1963).
3 A 2012 Nestler lab study, Koo, J.W., et al. "BDNF Is a Negative Modulator of Morphine Action", *Science*, vol. 338, no. 124, shows that chronic morphine exposure "creates reward by inhibiting BDNF," while cocaine exposure increases BDNF.

References

Acker, C. J. (2002) *Creating the American Junkie: Addiction Research in the Classic Era of Narcotic Control*, Baltimore, MD: Johns Hopkins University Press.
Bergen-Cico, D. (2012) *War and Drugs: The Role of Military Conflict in the Development of Substance Abuse*, New York, NY: Routledge.
Berridge, K. and Winkielman, P. (2003) "What is an 'unconscious emotion'? The case for unconscious 'liking'", *Cognition and Emotion* 17(2): 181–211.
Berridge, V. and Edwards, G. (1987) *Opium and the People: Opiate Use in Nineteenth-Century England*, New Haven, CT: Yale University Press.
Bliss, T., Collingridge, G. and Morris, R. (2003) *Long-term Potentiation: Enhancing Neuroscience for 30 Years*, Oxford, UK: Oxford University Press.
Brown, V. B., Huba, G. J. and Melchior, L. A. (1995) "'Level of burden': women with more than one co-occurring disorder", *Journal of Psychoactive Drugs* 27(4) 339–346.

Campbell, N. D. (2000) *Using Women: Gender, Drug Policy, and Social Justice*, New York, NY: Routledge.

Campbell, N. D. (2007) *Discovering Addiction: The Science and Politics of Substance Abuse Research*, Ann Arbor, MI: University of Michigan Press.

Campbell, N. D. (2011) "The metapharmacology of the 'addicted brain'", *Journal of the History of the Present* 1(2): 194–218.

Carter, A., Hall, W. and Illes, J. (eds) (2012) *Addiction Neuroethics*, Cambridge, UK: Cambridge University Press.

Courtwright, D. T. (1982/2001) *Dark Paradise: A History of Opiate Addiction in America*, Cambridge, MA: Harvard University Press.

De La Rosa, M., Lambert, E. Y. and Gropper, B. (1990) *Drugs and Violence: Causes, Correlates, and Consequences*, NIDA Research Monograph, Vol. 103, Rockville, MD: US Department of Health and Human Services.

Dube, S. R., Felitti, V. J., Dong, M., Chapman, D. P., Giles, W.H. and Anda, R. F. (2003) "Child abuse, neglect, and household dysfunction and the risk of illicit drug use: the adverse childhood experiences study", *Pediatrics* 111(3): 564–572.

Eddy, N. B. (1973) *The National Research Council Involvement in the Opiate Problem*, Washington, DC: National Academies Press.

Eddy, N. B. (1963) "The chemopharmacological approach to the addiction problem", *Public Health Reports* 78(8): 673–680.

Everitt, B. J. and Robbins, T. W. (2005) "Neural systems of reinforcement for drug addiction: from actions to habits to compulsion", *Nature Neuroscience* 8(11): 1481–1489.

Fullilove, M. T., Fullilove, R. E., Smith, M., Winkler, K., Michael, C., Panzer, P. G. and Wallace, R. (1993) "Violence, trauma, and post-traumatic stress disorder among women drug users", *Journal of Traumatic Stress* 6(4): 533–543.

Glantz, M. and Pickens, R. (1992) *Vulnerability to Drug Abuse*, Washington, DC: APA Press.

Glickman, S. E. and Schiff, B. B. (1967) "A biological theory of reinforcement", *Psychological Review* 74(2): 81–109.

Haertzen, C. A., Hill, H. E. and Belleville, R. E. (1963) "Development of the Addiction Research Center Inventory (ARCI): selection of items that are sensitive to the effects of various drugs", *Psychopharmacologia* 4(3): 155–166.

Hickman, T. A. (2007) *The Secret Leprosy of Modern Days: Narcotic Addition and Cultural Crisis in the United States, 1870–1920*, Amherst, MA: University of Massachusetts Press.

Kalivas, P. W. and O'Brien, C. P. (2008) "Drug addiction as a pathology of staged neuroplasticity", *Neuropsychopharmacology Reviews* 33(1): 166–180.

Kauer, J. A. and Malenka, R. C. (2007) "Synaptic plasticity and addiction", *Nature Reviews Neuroscience* 8(11): 844–858.

Klein, D. F. (2008) "The loss of serendipity in psychopharmacology", *JAMA: The Journal of the American Medical Association* 299(9): 1063–1065.

Knight, K. R. (2015) *Addicted.pregnant.poor*, Durham, NC: Duke University Press.

Kolb, L. (1925a) "Drug addiction in its relation to crime", *Mental Hygiene* 9: 74–89.

Kolb, L. (1925b) "Types and characteristics of drug addicts", *Mental Hygiene* 9: 300–13.

Kolb, L. (1925c) "Pleasure and deterioration from narcotic addiction", *Mental Hygiene*, 9: 699–724.

Koob, G. F. and Bloom, F. E. (1988) "Cellular and molecular mechanisms of drug dependence", *Science* 242(4879): 715–723.

Koob, G. F. and LeMoal, M. (1997) "Drug abuse: hedonic homeostatic dysregulation", *Science* 278(5335): 52–58.

Koob, G. F. and LeMoal, M. (2001) "Drug addiction, dysregulation of reward, and allostasis", *Neuropharmacology* 24(2): 97–129.

Koob, G. F. and LeMoal, M. (2005) "Plasticity of reward neurocircuitry and the 'dark side' of drug addiction", *Nature Neuroscience* 11(8): 1442–1444.

Koob, G. F. and LeMoal, M. (2006) *Neurobiology of Addiction*, New York, NY: Elsevier.

Koob, G. F., Sanna, P. P. and Bloom, F. E. (1998) "Neuroscience of addiction", *Neuron* 21(3): 467–476.

Kornetsky, C. (2007) "A walk through the history of research in drug abuse trends and fads: part II", *Drug and Alcohol Dependence* 90(2–3): 312–316.

Krieger, N. (2011) *Epidemiology and the People's Health*, Oxford, UK: Oxford University Press.

Krystal, J. H., Bennett, A. L., Bremner, J. D., Southwick, S. M. and Charney, D. S. (1995) "Toward a cognitive neuroscience of dissociation and altered memory functions in post-traumatic stress disorder", in M. J. Friedman, D. S. Charney and A. Y. Deutch (eds), *Neurobiological and Clinical Consequences of Stress: From Normal Adaptation to Post-traumatic Stress Disorder*, Philadelphia, PA: Lippincott Williams & Wilkins Publishers.

Kuhar, M. J. (2005) Interview with the author, June, Orlando, FL.

Leshner, A. I. (1997) "Addiction is a brain disease, and it matters", *Science* 278(5335): 45–47.

Lewis, M. (2015) *The Biology of Desire: Why Addiction Is Not a Disease*, New York, NY: Public Affairs Publishers.

Li, X. and Wolf, M. E. (2015) "Multiple faces of BDNF in cocaine addiction", *Behavioural Brain Research* 279: 240–254.

Martin, W. R. (1967) "Opioid antagonists", *Pharmacological Review* 19(4): 463–521.

Martin, W. R. (1984) "Relationship of biological influences on the subjective states of addicts" in G. Serban (ed.), *The Social and Medical Aspects of Drug Abuse*, New York, NY: Spectrum Publications, pp. 1–6.

Martin, W. R., Hewitt, B. B., Baker, A. J. and Haertzen, C. A. (1977) "Aspects of the psychopathology and pathophysiology of addiction", *Drug and Alcohol Dependence* 2(3): 185–202.

Martin, W. R., Sloan, J. W. and Eades, C. G. (1978) "Addiction Research Center, 1963–75: neuropharmacology and neurochemistry", in W. R. Martin and H. Isbell (eds), *Drug Addiction and the US Public Health Service*, Rockville, MD: US Department of Health, Education, and Welfare, pp. 100–17.

McEwen, B. (1998) "Stress, adaptation, and disease: allostasis and allostatic load", *Annals of the New York Academy of Sciences* 840(1): 33–44.

Mol, A. (2003) *The Body Multiple: Ontology in Medical Practice*, Durham, NC: Duke University Press.

Nestler, E. J. (2004) "Historical review: Molecular and cellular mechanisms of opiate and cocaine addiction", *Trends in Pharmacological Sciences* 25(4): 210–218.

Nestler, E. J. (2013) *Cellular Biology of Addiction Course 2013, Transcriptional and Epigenetics Mechanisms of Addiction*, YouTube video, 11 August. Available from: www.youtube.com/watch?v=gDdLaC_Xtvc&ebc=ANyPxKq4kdiFRxVs5RY8MpjbDpEbCuNfkI3CNqQrsWVnQgaC4xm2RM2b9n1bK1BqYtew8oBRPhJe3IiuzspDMTSyUMBii_1vUg [accessed 22 March 2016].

Olds, J. and Milner, P. (1954) "Positive reinforcement produced by electrical stimulation of septal area and other regions of the brain", *Journal of Comparative Physiology & Psychology* 47: 419–427.

Olds, J., Killam, K. F. and Bach-Y-Rita, P. (1956) "Self-stimulation of the brain used as a screening method for tranquilizing drugs", *Science* 124(3215): 265–266.

Poole, N. (2015) "Integrating trauma with addiction research and treatment" in L. Greaves, N. Poole and E. Boyle (eds), *Transforming Addiction: Gender, Trauma, Transdisciplinarity*, New York, NY: Routledge.

Robinson, T. E. and Berridge, K. C. (2008) "The incentive sensitization theory of addiction", *Philosophical Transactions of the Royal Society B: London* 363(1507): 3137–3146.

Rose, N. and Abi-Rached, J. (2013) *Neuro: The New Brain Sciences and the Management of the Mind*, Princeton, NJ: Princeton University Press.

Russo, S. J. and Nestler, E. J. (2013) "The brain reward circuitry in mood disorders", *Nature Reviews Neuroscience* 14(9): 609–625, doi:10.1038/nrn3381.

Schuster, C. R. (2003) Author's interview with Bob Schuster, San Juan, Puerto Rico.

Solomon, R. L. and Corbit, J. D. (1974) "An opponent-process theory of motivation: temporal dynamics of affect", *Psychological Review* 81(2): 119–145.

Sterling, P. and Eyer, J. (1988) "Allostasis: a new paradigm to explain arousal pathology", in S. Fisher and J. Reason (eds), *Handbook of Life Stress, Cognition and Health*, Hoboken, NJ: John Wiley and Sons.

Substance Abuse and Mental Health Services Administration (SAMHSA) (2014a) *Trauma-Informed Care in Behavioral Health Services*, Treatment Improvement Protocol (TIP) Series 57, HHS Publication No. (SMA) 13–4801, Rockville, MD: SAMHSA.

Substance Abuse and Mental Health Services Administration (2014b) *SAMHSA's Concept of Trauma and Guidance for a Trauma-Informed Approach*, HHS Publication No. (SMA) 14–4884, Rockville, MD: SAMHSA.

Szalavitz, M. (2016) *Unbroken Brain: A Revolutionary New Way of Understanding Addiction*, New York, NY: St. Martin's Press.

Terry, C. E. and Pellens, M. (1928/1970) *The Opium Problem*, Montclair, NJ: Patterson Smith.

Torchalla, I., Nosen, L., Rostam, H. and Allen, P. (2012) "Integrated treatment programs for individuals with concurrent substance use disorders and trauma experiences: a systematic review and meta-analysis", *Journal of Substance Abuse Treatment* 42(1): 65–77.

Volkow, N. D., Koob, G. F. and McLellan, A. T. (2016) "Neurobiologic advances from the brain disease model of addiction", *New England Journal of Medicine* 374(4): 363–371.
Wise, R. A. (1980) "The dopamine synapse and the notion of 'pleasure centers' in the brain", *Trends in Neurosciences* 3(4): 91–95.
Wise R. A. (1989) "The brain and reward", in J. M. Liebman and S. J. Cooper (eds), *The Neuropharmacological Basis of Reward*, Oxford, UK: Oxford University Press, 377–424.
Wise, R. A. (1998) "Drug-activation of brain reward pathways", *Drug and Alcohol Dependence* 51(1–2): 13–22.
Wise, R. A. (2004) "Dopamine, learning and motivation", *Nature Reviews Neuroscience* 5(6): 483–494.
Wise, R. A. and Koob, G. F. (2014) "The development and maintenance of drug addiction", *Neuropsychopharmacology* 39(2): 254–262.
Young, A. and Breslau, N. (2016) "What is 'PTSD'? The heterogeneity thesis", in D. E. Hinton and B. J. Good (eds), *Culture and PTSD: Trauma in Global and Historical Perspective*, Philadelphia, PA: University of Pennsylvania Press, pp. 135–154.

SECTION B

Developmental processes, vulnerabilities, and resilience

21
THE EPIDEMIOLOGICAL APPROACH
An overview of methods and models

James C. Anthony

This handbook chapter has modest aims. It offers an overview of epidemiological methods and models used to study processes of becoming dependent upon tobacco, alcohol, cannabis, cocaine, heroin, and other internationally regulated drugs, with 'drug dependence' serving more or less as a synonym for 'drug addiction.' The central conceptual model is one of agent–host–environment interaction, and here the concept of 'agent' encompasses drug compounds in drug epidemiology just as the 'agent' might be the influenza or Zika viruses in other branches of communicable disease epidemiology. Several statistical models are mentioned, including models for Hill function parameter estimation, as can be used when a dose–response relationship has a sigmoidal shape. Examples and illustrations are provided from past and recent epidemiological studies.

Epidemiology as 'population science' studying agent–host–environment interactions

At their best, epidemiology's methods and models build from concepts of 'population' and 'community,' assemble collected facts into 'chains of inference' about health and disease, and produce evidence-based strategies and tactics intended to reduce disease rates and improve the public's health. Under these conditions, epidemiology can move as quickly as (1) John Snow's studies of London's 1854 cholera epidemic and its consequential epidemic-preventing improvements in water supply, decades before bacteriologist Robert Koch's discovery of the *Cholera vibrio* causal mechanism, and (2) Joseph Goldberger's United States Public Health Service (US) research on pellagra and its consequential outbreak-preventing dietary improvements, years before nutritional chemists discovered niacin's central role. By 1970 it had been established that epidemiology, as a population science, had capacities to identify malleable agent–host–environment interactions and thereby to reduce disease incidence rates much faster than was being accomplished via reductionistic individual-level studies of causal mechanisms at the level of cells, tissues, organs, and patients.

Under unfavorable conditions, epidemiologists' efforts to improve public health can be delayed by a half-century or more when opposing forces are confronted, such as state reliance upon 'sin tax' revenues and an unconstrained industry profit motive, coupled with residual

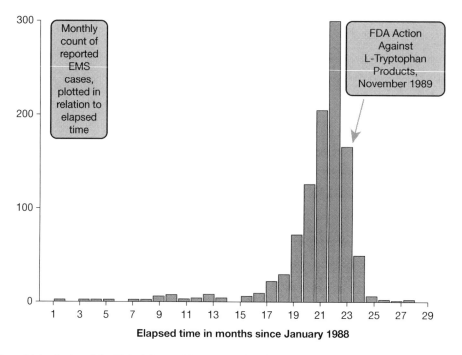

Figure 21.1 A plot of the United States 1988–1990 outbreak of dietary tryptophan-associated cases of the Eosinophilia-Myalgia Syndrome (EMS) in epidemic curve form, January 1988 through June 1990.

Source: Adapted by the author from Figure 2 in Swygert, L. A., Maes, E. F., Sewell, L. E., Miller, L., Falk, H., Kilbourne, E. M., (1990) "Eosinophilia-myalgia syndrome. Results of national surveillance," *JAMA: The Journal of the American Medical Association* 264(13):1698–1703. PubMed PMID: 2398610.

uncertainties about cause–effect inferences. Here, telling details are seen in still-blocked public health progress long after discovery of toxic effects from 'exposures' to agents such as tobacco in cigarettes and sugar in soft drinks. Nevertheless, with government and industry cooperation, rapid success can be achieved.

An instructive example can be seen in Figure 21.1, which depicts a United States (US) epidemic curve, 1988–1990, for a sometimes fatal eosinophilia-myalgia syndrome (EMS), observed as progressively worsening muscle pain, generally with fatigue, plus elevated pro-inflammatory white blood cells indicative of immune response and inflammation, often seen with allied clinical features. Initial clinician case reports about EMS launched government-initiated nationwide surveillance, plus several small scale 'controlled retrospection' case-control studies typified in the next section. Suspected exposures to L-tryptophan-containing products (e.g., LTCP dietary supplements) were identified as potential EMS causes before November 1989. That month, the US Food and Drug Administration (FDA) asked vendors to disrupt LTCP sales and encouraged users to abstain. The epidemic ended soon afterward. The FDA investigation of LTCP batches disclosed likely contamination via a manufacturing process change. Swygert *et al.* (1990) provided details about EMS-LTPC; Goodman and colleagues (2012) summarized and added details on other epidemiology victories in public health.

In these introductory examples, readers can see 'agents' and elements of 'environment' for agent–host–environment (AHE) interaction models that epidemiologists often use in research. Host characteristics, both genetic and endogenous (e.g., hormones), become prominent when behavioral choice issues surface (Breslow 1951). Illustrative of host characteristics was one female case fatality in Snow's research—namely, an out-migrant from Broad Street's neighborhood, but her choice was to drink bottled water from its pump. Similarly, when a first tobacco cigarette is offered, host susceptibility and environmental conditions (e.g., 'peer pressure') influence whether it is smoked, and afterward AHE interactions shape later choices (Wagner & Anthony 2002). Choice also governs consumption of sugary soft drinks, which in some communities cost less than cow's milk. With this note, we come full circle back to agent characteristics (e.g., relative price) and environment (e.g., supply-demand conditions governing price; available alternatives).

At this point in the chapter, the reader might already be wondering what all this has to do with starting or stopping drug use, or with whether someone who persists in drug use will become affected by a drug dependence syndrome or addictive state. Box 21.1 offers some illustrations of some specific facets of 'agent' and 'host' and 'environment' that come into play in studies of starting drug use and becoming an affected case of serious drug involvement of the type we see in dependence and addiction syndromes. Box 21.2 provides a summary overview and mentions some of the issues faced by epidemiologists who have tried, generally unsuccessfully, to conduct population studies of 'quitting' drug use, as well as 'remission' or 'recovery' processes. It seems that epidemiology, as a 'population science,' faces some inherent challenges and seriously biased samples when the goal is to study these quitting, recovery, and remission processes. These are topics better suited for 'clinical investigations' given these still-insurmountable biases that affect epidemiological samples of pre-specified community-dwelling populations.

Box 21.1 Suspected agent–host–environment determinants

When asked to explain how it happens that most people who initiate drug use do not transition to addiction, epidemiologists have to confess a good bit of ignorance. Hypotheses are abundant, and generally have been articulated in relation to the agent–host–environment (AHE) model described in this chapter. Firm definitive epidemiological evidence is scarce.

We might start by considering the drug as an 'agent' and talk about properties such as 'addiction liability' or 'potential' to serve as a reinforcer in the behavioral pharmacologist's sense of the term. An epidemiologist would tend not to think of these characteristics as properties of the agent because what the epidemiological study can disclose also is determined by host and environmental variables. The importance of modulating host characteristics was made clear in early laboratory experiments with healthy drug-naïve human subjects. For roughly 50% who received morphine or some other opioid without being told the compound's identity, there was some enthusiasm for trying the drug a second time, but the others said they never wanted to try it again. The 'mechanisms' underlying this host-related variation in response remain unknown, but clearly might include whether the healthy individuals were feeling some degree of pain or depressed mood when the opioids were administered such that symptom relief might be a determinant of the more positive response. Viable

(continued)

(continued)

alternatives include pharmacogenetic substrates of kappa opioid receptor activation and associated dysphoric responses to KOR agonist opioids. Few epidemiological studies have had the 'resolving power' to secure definitive evidence on these mechanisms and sources of variation in response to initial drug exposures.

We can add socio-cultural and environmental components to these sources of variation by paying attention to epidemiological evidence on gene variations that account for acetaldehyde dehydrogenase (ALDH) enzymes, especially the ALDH2*2 variant causally linked with the generally noxious 'facial flushing' response to drinking alcohol. There is some evidence that the ALDH2*2 gene allele is protective against progression from first drink to sustained drinking and alcoholism, but there clearly are gene–environment interactions such as positive peer pressure and other sociocultural influences on drinking. Especially for male subgroups with the ALDH2*2 allele, its otherwise protective influence is dampened by social environments conducive to drinking.

Over and above anything that might be thought of as a 'property' of the drug-agent, or a source of 'host' vulnerability, the 'E' component of the AHE model plays an especially fundamental role. A rate-limiting step for progressing from first drug use to more serious drug involvement involves whether the generating reservoir for the drug (croplands or manufacturing plants) can deliver a consistent supply of drug into the environment of a first-time user. During the Vietnam era, a soldier with a completed tour of duty who enjoyed opium smoking for the first time on the night before returning to the US might never again have experienced a chance to use opium a second time, leaving aside the importance of any hypothesized influence of the drug-agent or host vulnerability.

Box 21.2 Epidemiology and the 'maturing out' hypothesis

Box 21.1 addresses issues pertinent to how it happens that some drug users get started but never persist. It clarifies how individual case studies and epidemiology's studies of groups can teach us that the explanations for these variations in the response to initial drug exposures can be traced to conditions and processes across the scale from the macro-level to the micro-level. At the macro-level, we can see that ambient availability of the compound is important. The Vietnam vet who had smoked opium a first time before the day of departure from Vietnam, returned to the US mainland, and never again had a chance to try opium, will not progress along the continuum of opium involvement, irrespective of genetic or other susceptibilities. At the micro-level, the Korean newly incident drinker with an ALDH2*2 gene allele configuration might experience an aversive facial flushing and associated noxious EOH responses, but might be affected by peer pressure or other cultural influences to 'drink through' the noxious response until the level of intoxication ('numbness') is so great that ALDH2*2-attributable symptoms are felt negligibly, if at all. With respect to such macro-level and micro-level influences, along with meso-level influences such as peer pressure or local area 'sin' taxes on EOH purchases, past epidemiological studies have implicated all of these influences. However, epidemiology never has been in a good position to assert that the influences at one level are more important than the influences at the other levels. Here again, the concept of 'resolving power' is appropriate. Epidemiology does not have the 'resolving power' to make a proper accounting of the relative importance of these various levels of influence. Typically, in the epidemiological context, many potentially explanatory variables are at work, and there generally is no definitive experimental approach as might be used to focus on one influence while holding constant all other potential influences. The exception in both infectious disease epidemiology and drug dependence epidemiology involves the influence of the agent and environment. If the host, after an initial exposure contact, will stay within an environment where there is no contact with the agent, there can be no progression beyond first contact.

A substantially different problem is faced when the research task is to ask how many develop a dependence syndrome (or 'addiction') and then stop using altogether. In his US classic study on a 'maturing out' phenomenon, published in 1962, sociologist Charles Winick offered these thoughts:

> There have been no studies of the age at which addicts *stop* taking drugs. There has been considerable acceptance in both lay and professional circles of the thesis that many addicts never stop using drugs, but continue as addicts until they die, except for unsuccessful attempts at withdrawal or for periods of enforced abstinence in jails or hospitals. There is some feeling that there is a high state of relapse among addicts, and this had led to considerable scepticism about addict's ability to remain abstinent, and how many addicts do remain abstinent.
>
> *Winick 1962, p. 1*

Winick also wrote: "There is some reason to believe that in the 1950s there was an average period of not more than two years between the time that addicts began taking drugs and the time they were observed or reported by a community law-enforcement or medical facility as addicts (ibid.)." At present, some 60 years later, we now have good reason to contradict Winick's estimate of a two-year lag time between first drug use and official recognition of a case in the US (e.g., see Cepeda *et al.* 2016; Yeh 2017). Nonetheless, there is considerable evidence in favor of his thesis that many users do 'mature out' of their syndrome, show remission, and stop using altogether, as has been observed in studies of users of the internationally regulated drugs, as well as alcohol consumers, not only within the US, but also in other countries (Fleury *et al.* 2016; Sobell *et al.* 2000).

The just-cited review articles by the Sobell and Fleury groups provide an excellent overview of methods challenges faced in research of this type, and provide reasons to mistrust estimates from cross-sectional epidemiological field studies on this topic, even when the samples are large and nationally or regionally representative. In brief, we can have some degree of confidence when epidemiological studies are used to study all who have tried a drug and to estimate the likelihood of becoming a case of a drug dependence or addiction syndrome (e.g., Anthony *et al.* 1994), especially when there has been a focus on initial short intervals after first use (e.g., Parker & Anthony 2015). In this context, we can make a reasonable assumption that drug involvement has not yet caused incarceration (e.g., moving the user from the community into a prison or long-stay treatment institution), has not become a major impediment to study participation as might be the case for an individual with severe impairments attributable to drug use, and is a minor element in causes of death (Lopez-Quintero *et al.* 2015). In contrast, most epidemiological samples of this type will not include the full spectrum of impairments seen in clinical investigations of treated or incarcerated cases. The samples are affected by what is called a 'left-censoring' bias such that the cases are survivors, and drug-dependent cases who have died are removed from view, as well as a 'left-truncation' bias such that some cases in the community are known to exist but never would appear on the roster of dwelling-unit participants (e.g., if living under a bridge), or would not be interested in study participation, even when they are included on the sampling roster. It is for this reason that most studies

(continued)

> *(continued)*
>
> of whether and how often cases 'stop' using drugs generally are not epidemiological studies. Rather, virtually all have been clinical investigations of the type conducted by Winick, with samples limited to cases seen in either treatment or prison facilities.
>
> Gene M. Heyman, a lecturer in psychology, apparently was not aware of problems such as left-censoring and left-truncation when he wrote an interesting empirical contribution based on four large sample epidemiological studies in the US (Heyman 2013). Three of the four studies represent empirical contributions made by this chapter's author (JCA), who has declined to produce estimates of recovery or remission for the reasons just mentioned, and who has consulted on methods research that confirms suspicions that these large sample epidemiological studies in the US often do not capture the experience of the most seriously impaired cases.
>
> Concerns about this bias in large sample epidemiological research do not negate Heyman's attempt to characterize drug dependence or addiction in relation to a behavioral choice paradigm, which can be an important paradigm, particularly in the earlier stages of the syndrome. Nonetheless, the available estimates do not make a strong case for or against the hypothesis that addictive states sometimes can be characterized as 'brain diseases,' although Heyman seems to wish we would abandon that hypothesis. It may be necessary to recognize the heterogeneity of drug-induced conditions, such that some affected cases are told that their drug use is having a toxic effect (e.g., on the heart or liver) and can quit using the drug on that same day, whereas others given the same information continue to use the drug without quitting or remission of symptoms. Evidence that some can quit is not the same as evidence that none have acquired a 'brain disease' as a consequence of repeated drug exposure.

Examples from research on tobacco smoking

Epidemiologists discovered sustained tobacco smoking as a malleable causal determinant of pulmonary cancer with focus on 'pack-years' concepts, with little attention to behavioral choice issues or processes such as tobacco dependence syndromes that foster sustained smoking. Before 1900 clinician-pathologists rarely saw pulmonary tumors, but pulmonary cancer death rates climbed after mass-produced tobacco cigarettes were introduced in the late 19th century. Nevertheless, in 1951, the much-admired *Rosenau Preventive Medicine and Hygiene* textbook had no indexed 'tobacco' entry; its chapter on oral-pulmonary cancer mentioned mechanical injury from pipe stems as a possible exogenous determinant, but said little else about tobacco smoking (Breslow 1951).

Concurrently, looking within specific community populations for cancer cases and matched non-cases, epidemiologists found that the odds of agent exposure (i.e., tobacco smoking) was substantially greater for cases than corresponding 'exposure odds' for non-cases, with 'odds' calculated as 'the *proportion exposed* divided by its arithmetic complement (by which we mean the *proportion not exposed*).' By inference, this 'odds ratio' was used to estimate exposure–disease associations statistically as a 'relative risk' (RR), where a statistically null RR=1.0, an inverse RR<1.0 would signal a departure from the null in the direction of the possibility that smoking protects against cancer, and an RR>1.0 would signal a departure from the null in the direction of a tangible smoking–cancer association, possibly consistent with a cause–effect pathway.

The magnitude of the smoking–lung cancer association often was found to be quite strong (e.g., RR=16). Eventual evidence indicated a 'dose–response' relationship such that for each unit increase in tobacco smoking pack-years there was increased relative risk of disease.

Case-control approaches

In this tobacco–cancer example, readers can see basic elements of epidemiology's AHE interaction model in a 'controlled retrospection case-control study design,' distinctive because it selects or 'samples on the outcome of interest.' That is, for conditions of relatively low incidence, the design's sampling or selection probability for affected cases in study populations is large, while the corresponding probability for non-case controls is small.

The basic architecture for case-control research typically involves specifying a discrete population for sampling of affected cases, with the non-case controls coming from the same population such that sources of variation in environmental conditions, host characteristics, and at least some agent-exposures (e.g., ambient toxins) are constrained. After informed consent and recruitment, the primary suspected causal determinants are measured in an array of alternative potential determinants, often with sham hypotheses as well, and the assessment approach is 'blinded' (i.e., completed without knowledge of case and control status), although generally there are attempts to evaluate exposures pre-dating case onset. Thereafter, odds ratio estimates of relative risk (RR) are derived, currently with conditional or unconditional generalized linear modeling (GZLM) and logistic link for RR estimation and statistical adjustment for important covariates. For pragmatic reasons, controls within populations sometimes are matched to cases, which induces more balance in the moments of distributions for age, sex, self-identified ethnicity, and neighborhood of residence—that is, ensuring they are 'held constant' for RR estimation. (In this context, the 'moments' of a variable's distribution refer to some familiar features such as (1) its mean, and (2) its variance, as well as some possibly less familiar features such as (3) its skewness and (4) its kurtosis. If the age distribution of controls in an epidemiological study is correctly matched to the age distribution of its cases, then these two distributions will have more or less the same mean, variance, skewness, and kurtosis. That is, the distributions will be balanced, just as one might seek to achieve when using randomization or random assignments in controlled trials.)

A genetics-matched case-control approach is realized when RR is estimated for disease-discordant monozygotic (MZ) twin pairs, disclosing relative importance of environmental determinants. With matching, RR is estimated using conditional contingency table and logistic GZLM statistics.

Followup or 'cohort' approaches

Circumspect critics always find much to complain about in case-control studies, even when well-established research guidelines are followed (Sanderson *et al.* 2007; *Strengthening the Reporting of Observational Studies in Epidemiology*, http://strobe-statement.org/index.php?id=available-checklists, last accessed 27 February 2017). Even with 'controlled retrospection,' there can be temporal misalignment when case onsets occur before suspected causal determinants are assessed; pathological processes might distort values or measurements of the determinants under study. As an example, consider a study of monozygotic identical co-twins, one of whom has developed a schizophrenia-like psychosis whilst the other has not. Looking back, we see that the psychosis case had been a frequent cannabis smoker; the discordant co-twin had not been. In addition to noting the possibility that the frequent cannabis smoking has caused the psychosis to occur, we have to account for why it is that the case-twin became a frequent cannabis smoker, but the non-case-twin did not smoke frequently. It is possible that an unfolding pathological process leading toward onset of the psychosis in the affected case-twin altered the reinforcing function of cannabis smoking and fostered persistence and increasing frequency. That is, it is the unfolding pathological process, in the background, that accounts for both the increased frequency of cannabis smoking and the emergence of a psychosis, even if and when the cannabis smoking had nothing to do with psychosis onset. The co-twin is unaffected and did not become a frequent cannabis smoker because the co-twin had no unfolding pathological process in the background.

Some judge that longitudinal followup or cohort studies are superior for RR estimation, with temporal misalignment possibly removed by first measuring a suspected exposure, sampling on values of that exposure, and subsequent followup to ascertain newly incident cases occurring after exposure. Thereafter, disease rates and RR are estimated (e.g., via GZLM).

Several validity threats may appear. First, many pathologies develop silently and slowly. These disease processes might account for agent-exposure as observed at first assessment. Murray *et al.* (1983) discuss that sustained acetaminophen use might cause interstitial nephritis and end-stage renal disease (IN, ESRD), but headache early in IN/ESRD pathological processes can prompt acetaminopen use. This 'confounding by indication' can be hard to remove even in the best pharmacoepidemiological followup studies. Second, genetic or early-life susceptibility traits might account for sequenced occurrence of (1) suspected agent-exposures and (2) the hypothesized disease outcome, as is suspected to explain currently debated cannabis-psychosis associations (Anthony *et al.* 2017), unless the explanation is an unfolding of a pathological process that drives both cannabis-smoking frequency and occurrence of a psychosis, as outlined above. (Here, we have an example of a spurious agent–disease correlation such that there is a background condition or process that accounts for co-occurrence of two interdependent happenings: (1) agent-exposure and (2) disease outcome.) Furthermore, unlike the 'one-time-assessment' case-control approach, all contemporary followup studies face biased RR estimates from differential sample losses during followup, as participants lose interest or otherwise disappear from view and cannot be tracked down. Also, except for absorbing variables (such as dead/alive), repeat measurements can be reactive; the first measurement act can change future measurement values artifactually (Anthony 2010; Seedall & Anthony 2015). To illustrate, in many drug dependence studies with repeated measurements, the respondent first is asked about lifetime frequency of use. Individuals with no use of the drug are not asked to complete the dependence syndrome assessment for that drug, and often a threshold will be set such that dependence is assessed only when the lifetime frequency threshold has been surpassed. For tobacco cigarettes, a common lifetime frequency threshold is 100 cigarettes smoked; for heroin and other drugs, using it more than five times in the lifetime (Heyman 2013). Upon reflection, alert respondents can learn that our survey assessment sessions will become longer in relation to which drugs have been used on more than five occasions. For each drug subtype used more than five times, they will get another dependence assessment module. We have learned that when our field workers return to make the followup assessments (e.g., after one–three year intervals), some respondents recall this 'response cost' of acknowledging such drug use, and they contradict the report given at the time of the original measurement in order to shorten the second assessment session. When this happens, the first measurement is a 'reactive measurement' that changes subsequent measurement values artifactually.

A balanced appraisal of accumulated evidence from case-control and followup studies on the tobacco–cancers association can be seen in Haenszel (1966), 15 years after the Maxcy text's omissions:

The epidemiological approach

In the early days when all the evidence was derived from controlled retrospective [case-control] studies (and one could have reservations about the adequacy of the study designs and choice of controls) there was serious question as to whether the results correctly reflected a true difference in [lung cancer] risk between smokers and non-smokers (in the observational sense) or were merely artifacts arising from defects in study procedures. The confirmation by three forward [followup] studies of the retrospective results seems to have disposed of the possibility that the early findings were statistical artifacts. . . . There is now general agreement on the truth of the observation that persons smoking one pack of cigarettes or more a day have a lung cancer risk about 16 times that of non-smokers. . . . What is still in dispute is the interpretation as to whether this represents cause and effect or merely reflects an indirect association with other characteristics or attributes of smokers [as might be] responsible for the higher lung cancer risks.

pp. 426–427

Haenszel went on to summarize evidence on tobacco smoking as a lung-cancer cause, making three points: (1) the robust size of RR estimates (e.g., RR=16), (2) the 'dose-response' regularity of lung-cancer risk plotted by pack-years, and (3) observed reductions in lung cancer risk among smokers who quit smoking (Hammond & Horn 1958). In these few statements, Haenszel articulated current guidelines epidemiologists later adapted to form 'chains of inference' leading from collected observations toward community work that encourages active smokers to set quit dates and to complete medication-assisted smoking cessation programs.

Case-crossover approaches

"What source accounts for agent-exposure of exposed individuals and non-exposure of non-exposed individuals?" We already confronted this challenging question on pages 259–260 and its example of using a discordant co-twin design to study the cannabis–psychosis association. In that context, we note that an underlying susceptibility trait or pathological process might account for agent-exposures in some individuals, but not others. This challenging question also confronts anyone using between-individual observational research approaches outside the context of co-twin designs, including the case-control and followup study designs just reviewed.

Table 21.1 Illustration of analysis approach for case-crossover data using a simplified Mantel-Haenszel matched-pair odds ratio estimator

		Exposure to triggering-event during the Hazard Interval	
		No	Yes
Exposure to the triggering-event during the Control Interval	No	n11	n12
	Yes	n21	n22

Notes:
1 Odds ratio (OR) is calculated as OR = $n12/n21$.
2 Standard deviation can be calculated as $\ln(OR) = (1/n12 + 1/n21)^{1/2}$.
3 Cells n11 and n22 do not contribute to OR estimation.
4 When the hypothesized triggering-event is 'absorbing' (occurs once and only once), cell n22 is a structural zero.

In the next section we will see some examples of using random assignment to allocate agent-exposures to some individuals, but not to others. In these examples, it is randomization that accounts for agent-exposures in the exposed and for non-exposure in the non-exposed.

There is yet another way to address this challenging question, which involves the epidemiological case-crossover approach and its harnessing of 'subject-as-own-control' logic. That is, for each participant in case-crossover research, we have an 'n-of-1' experiment in which we ask this kind of question: "Was there anything atypical that the affected victim was doing just before onset of the disease?" By building up multiple replications of these 'n-of-1' experiments, we try to estimate the degree to which an exposure occurring just before the onset of the disease might be accounting for an excess risk of that disease outcome.

The case-crossover approach is most suitable for tests of 'triggering' hypotheses, as in the hypothesis that an agent-exposure is triggering abrupt disease onset, but the concept of 'abrupt' is somewhat relative and will vary, outcome by outcome, based on judgments made about the nature of the pathological process linking exposure and outcome-event. For example, if the hypothesis is that being distracted by an incoming mobile phone call might cause a driver to have a car crash when the crash otherwise would not happen, the pre-crash 'hazard' interval almost surely would be specified as a matter of minutes before the crash. The exposure question involves whether an incoming call is received during that short interval just before the crash has occurred. The 'control' interval then is specified as a randomly sampled discrete alternative interval of the same duration, chosen to be representative of the case's life-experiences in the vicinity of the crash-event, but with no overlap into the hazard interval. In order to estimate the exposure–disease association using a 'odds ratio' type statistic, we would look at each car crash driver's experience (e.g., as manifest in the mobile phone company's records) and check whether there had been an incoming call to that driver during the hazard interval, but no such call during that driver's control interval. This number of exposure-discordant drivers, with this pattern of 'exposure-positive, exposure-negative' (E+, E-), is one that tends to favor the phone–crash hypothesis. In contrast, consider that there generally will be at least one driver who has had no incoming call during the hazard interval but who receives an incoming call during the control interval. This number of 'exposure-discordant' drivers, with this pattern of 'exposure-negative, exposure-positive' (E-, E+) is one that tends to favor the alternative null (no triggering) hypothesis. The ratio of these two numbers, (E+, E-) cases divided by (E-, E+) cases, provides a test of the triggering hypothesis, with the size of the ratio used to provide an estimate of relative risk.

To re-cap, in practice, for case-crossover research, after specification of study populations, surveillance identifies newly incident cases. For each detected case, measurements assess whether agent-exposure occurred during a 'hazard interval' just before case onset, or during each case's pre-specified 'control interval' of equivalent duration. The logic follows thinking that if an agent-exposure triggers the case onset, then observed exposure odds for the hazard interval should exceed expectation based on exposure odds during the control interval.

One of the most attractive features of epidemiological case-crossover research is the use of the subject as his/her own control. This approach finesses a question that always bedevils behavioral genetics research on MZ co-twins, when the goal is to identify environmental determinants of disease outcomes. In these MZ co-twin studies, relative risk estimation is based on the number of twin pairs with an exposed case and with an non-exposed co-twin control (E+, E- discordance), divided by the number of twin pairs with an exposed co-twin control and a non-exposed case (E-, E+ discordance). The 'fly in the ointment' unanswered question in these between-individual studies is this: "Why did one twin get exposed when the other MZ co-twin did not get exposed, even though both twins shared genetic vulnerability traits and many facets of environment up to the time of the study?" The epidemiological case-crossover approach, with its within-individual approach, also matches for all genes (by virtue of a 'subject as own control' design), and in addition it matches for all environmental experiences (and for gene-by-environment interactions), up to the time of the study interval. In this sense, more than is true in behavior genetics research on MZ co-twins, the result of case-crossover research is enhanced control of genetic, epigenetic, and early-life determinants of the case outcome, as well as many gene–environment interactions. Self-matching in case-crossover research motivates conditional contingency tables, as illustrated in Table 21.1, or logistic GZLM for RR estimation.

Fairman and Anthony (2017) recently reviewed case-crossover studies pertinent to drug-use disorders. They used the approach in tobacco research to study cigar smoking as a possible trigger-event for onset of smoking the cannabis-tobacco formulation known as a 'blunt.' This work is described on page 272.

Experimental approaches

Outside epidemiology circles, there is little awareness that the epidemiology discovery process often has included formal randomized controlled trials, including randomized trials of hypothesized preventive interventions. Discussions of these 'epidemiology experiments' generally cite Jenner on smallpox, Lind on scurvy, and Goldberger on pellagra, all completed before 1920 (Jukes 1989), although many epidemiologists (including John Snow) also experimented with non-human species. In 1966, Lilienfeld's discussion of experiments was succinct, with focus on quasi-experimental contrasts of childhood rates of permanent teeth found with decay, missingness (lost after eruption) or filled (DMF) in Newburgh and Kingston (New York), chosen for comparability in the pre-fluoridation era. In May 1945 the Newburgh community started water fluoridation; Kingston did not. Evaluated in 1954–1955, estimated age-stratified Newburgh rates of DMF teeth were substantially lower than corresponding Kingston rates (Lilienfeld 1966, p. 107).

Post-dating Lilienfeld's 1966 review, many population-based tobacco-smoking prevention experiments have been completed, some with promising evidence of statistically robust reductions in cumulative incidence rates of becoming a newly incident tobacco smoker by mid-adolescence (e.g., see Kellam & Anthony 1998). Attempts at replication have not always been successful (e.g., see Kellam *et al*. 2014).

Mathematical modeling approaches

Any serious review of epidemiological methods and models cannot omit mathematical modeling of disease epidemics in human populations, given prominence of deterministic differential equation models for simulation of person-to-person disease spread introduced in the early 20th century, with later elaboration via stochastic modeling. Frauenthal (1980) provides undergraduate-level introductions to these models for those having the calculus.

One elaboration of these mathematical models in epidemiology can be seen in a relatively new area of 'in silica' research known as 'Agent-Based Models' (ABM). We can think of each ABM as a computational model of interactions between human agents within a hypothesized population that consists of heterogeneous units or sub-groups. An interaction of two individual hosts within the hypothesized population can be characterized as a micro-level interaction that gives rise to macro-level behaviors such as the sharing of drug compounds and experiences.

Although not yet a topic for epidemiological research, there is a hypothesis about persistence of drug use after a first trial use by a naïve user, with a refinement of predictions that trial use of the drug compound might be followed by drug-seeking and by a second trial with the drug if the experience is shared by two or more individuals (i.e., with potential social reinforcement involved in the experience), versus a lower likelihood of drug-seeking and persistence of use when the first drug-using experience is 'solo' (that is, the drug user in isolation of others). The ABM would not involve experimental administration of a drug to one naïve user versus two naïve users, but rather would involve setting up a computational model in order to simulate potential outcomes at the individual level and also in order to evaluate which types of individual drug experiences might be more likely to give rise to a community outbreak of use of the novel compound.

The ABM also hold promise for the study of drug markets with a mix of legal and illegal wholesale and retail exchanges. They can bring into play heterogeneities that might be predicted to follow variations in locational placement of individuals or groups of individuals, their wealth, their social connections, and other facets of human experience such as cognitive processes and motivations. The result of advanced ABM research should be a greater understanding of the 'reservoirs' out of which drug-agents emerge to come into contact with human hosts, as well as processes through which changes in shared environments can be created, sometimes via exchange of tangible goods or services (e.g., trading sex acts for drugs), and sometimes via imitation processes or diffusion of innovations through social networks. Online tutorials can provide interested readers with more background information on ABM (www.openabm.org/faq-page).

In recent progress, agent-based computational models for tobacco smoking simulation studies have been devised. This work yields evidence that e-cigarette introduction may have markedly less influence on population rates of tobacco smoking initiation, as compared with its influence on prevailing numbers of persistent smokers in a population (Cherng *et al.* 2016).

Illustrations from cross-sectional population surveys ('transversal' design)

Data from cross-sectional (transversal) population surveys of tobacco smoking and other drug use mainly have been used to estimate, for populations, how many individuals have ever used (or are current users), and to estimate correlational statistics disclosing whether drug-use rates vary across subgroups. What might surprise chapter readers is that properly designed and constructed cross-sectional surveys can yield satisfactory evidence approximating, and sometimes better than, evidence from studies designed, *de novo*, with case-control, followup, or case-crossover approaches. This section's illustrations are meant to prompt new thinking about evidence still to be harvested from cross-sectional surveys pertinent to drug dependence syndrome.

In each illustration, the data are from recent US National Surveys on Drug Use and Health (NSDUH). Populations specified for NSDUH have encompassed all non-institutionalized civilian residents aged 12 years and older living within the 50 states and District of Columbia, including non-household dwelling unit residents (e.g., college dormitories, homeless shelters). Each year, NSDUH has drawn a new multi-stage area probability sample for completed informed consent and recruitment of sampled participants, typically with 70%+ participation levels. Thereafter, measurements mainly have been via audio-enhanced computer-assisted self-interviews via portable laptop devices; quality control checks produced codebook-documented analysis-ready datasets. In the illustrations, except where noted, analysis-weighted statistical models are being used to estimate study parameters underlined below, with calculus-based 95% confidence intervals (error margins) to disclose statistical precision of estimates and statistical inferences at $p<0.05$.

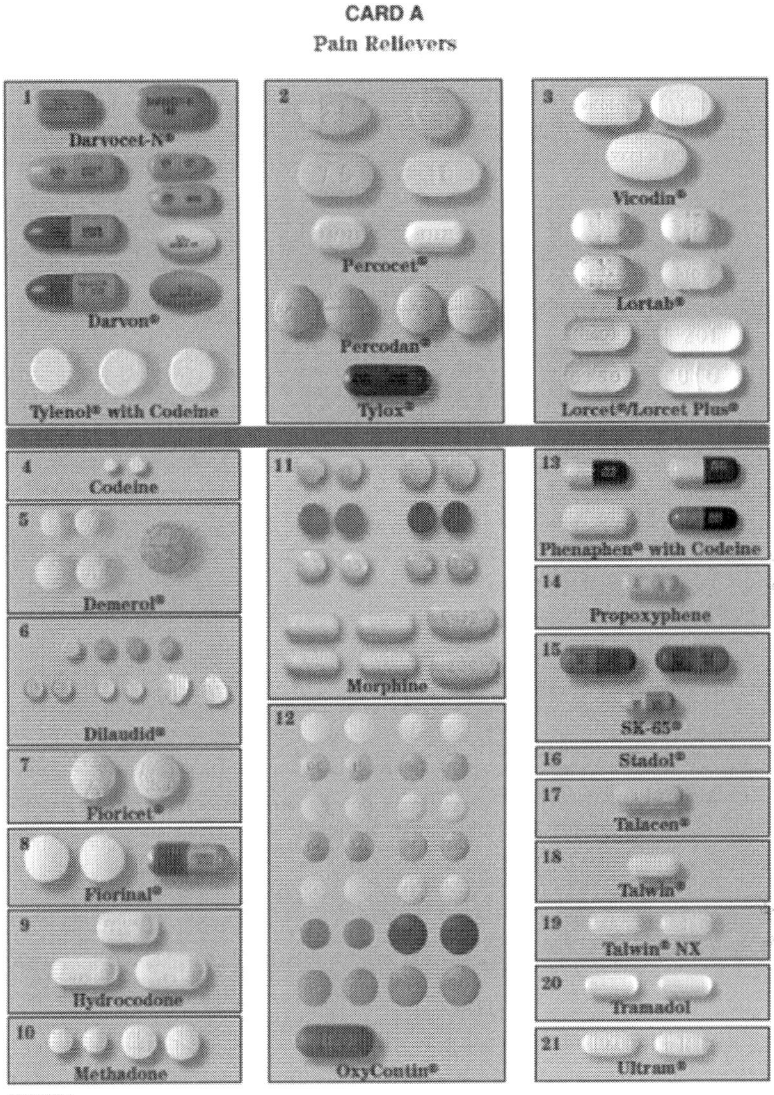

The epidemiological approach

CARD A
Pain Relievers

Figure 21.2 Example of 'color drug showcard' used in the 2014 United States National Survey on Drug Use and Health. (Public domain image of 'Card A' for prescription pain relievers. The color version may be viewed online.)

Source: http://samhda.s3-us-gov-west-1.amazonaws.com/s3fs-public/field-uploads-protected/studies/NSDUH-2014/NSDUH-2014-datasets/NSDUH-2014-DS0001/NSDUH-2014-DS0001-info/NSDUH-2014-DS0001-info-questionnaire-showcards.pdf, last accessed 27 February 2017.

Figure 21.2 is an example of an NSDUH color drug 'showcard' used to enhance response validity of drug use assessments, here pertaining to prescription pain relievers used extra-medically. Parker and Anthony (2015) provide details about NSDUH materials and methods, including the construct of using a drug 'extra-medically' (i.e., outside boundaries set by approved indications and by prescribing clinician intent, such as 'to get high' or other such feeling states). Figure 21.3

> **Pain Relievers**
>
> These questions are about the use of **pain relievers**. We are **not** interested in your use of "over-the-counter" pain relievers such as aspirin, Tylenol, or Advil that can be bought in drug stores or grocery stores without a doctor's prescription.
>
> Ask your interviewer to show you Card A.
>
> Press [ENTER] to continue.
>
> Card A shows pictures of some different kinds of prescription pain relievers and lists the names of some others. These pictures show only pills, but we are interested in your use of **any** form of prescription pain relievers that were **not** prescribed for you or that you took only for the experience or feeling they caused.
>
> Please look at Card A carefully as you answer the next questions.
>
> Press [ENTER] to continue.
>
> Please look at the pain relievers shown in Box 1 **above** the red line on Card A.
>
> Have you ever, even once, used **Darvocet, Darvon, or Tylenol with codeine** that was **not** prescribed for you or that you took only for the experience or feeling it caused?
>
> 1 Yes
> 2 No
> DK/REF
>
> Please look at the pain relievers shown in Box 2.
>
> Have you ever, even once, used **Percocet, Percodan, or Tylox** that was **not** prescribed for you or that you took only for the experience or feeling it caused?
>
> 1 Yes
> 2 No
> DK/REF

Figure 21.3 Examples of computerized survey questions used to assess extra-medical use of prescription pain relievers in the 2014 United States National Survey on Drug Use and Health.

Source: Public domain at: http://samhda.s3-us-gov-west-1.amazonaws.com/s3fs-public/field-uploads-protected/studies/NSDUH-2014/NSDUH-2014-datasets/NSDUH-2014-DS0001/NSDUH-2014-DS0001-info/NSDUH-2014-DS0001-info-questionnaire-specs.pdf, last accessed 27 February 2017.

Note: DK/REF means the response was "Don't Know" or "Refused."

displays standardized NSDUH items used in 2014 to assess lifetime history of using prescription pain relievers (PPR) extra-medically. Items not shown are used for measurements of age, month, and year of first and most recent extra-medical use, as well as details about recency and frequency of use, and routes of administration (e.g., via injection).

Case-control illustration

Illustrating how a case-control study can be nested within a cross-sectional survey design, Santiago-Rivera et al. (under review), who studied 620 newly incident heroin users identified in NSDUH cross-sectional surveys between 2003 and 2014, found 164 who had transitioned rapidly into a heroin dependence syndrome within 12 months after first heroin use and 456 non-dependent users. Based upon covariate-adjusted logistic GZLM, neither being male nor ethnic self-identification (e.g., non-Hispanic White) predicted rapid-onset dependence ($p>0.05$), but injecting heroin signaled a 2.6-fold greater odds of becoming heroin dependent quite rapidly after heroin onset ($p<0.004$).

Table 21.2 Estimated risk of transitioning and becoming an opioid dependence case no longer than 12 months after onset of starting to use prescription pain relievers extra-medically. Estimated risk of transitioning to opioid dependence (A), 95% confidence intervals (B), and age-specific meta-analysis summary estimates

Year pair	12–13 y	14–15 y	16–17 y	18–19 y	20–21 y
(A) Estimated risk of becoming an opioid dependence case (per 100 newly incident EMPPR users)					
2002–2003	4.2	6.3	4.7	2.8	1.8
2004–2005	3.2	9.8	5.1	3.3	2.4
2006–2007	5.2	7.9	3.9	1.7	3.4
2008–2009	5.4	7.5	4.8	4.2	3.2
2010–2011	8.2	5.8	7.1	6.1	2.4
2012–2013	5.1	6.4	6.1	4.8	2.1
(B) 95% confidence intervals for (A) estimates (per 100)					
2002–2003	2.3, 7.4	4.3, 9.1	3.1, 7.1	1.5, 5.1	0.9, 3.7
2004–2005	1.7, 6.0	6.9, 13.6	3.4, 7.6	1.8, 5.9	1.1, 5.0
2006–2007	2.8, 9.7	5.2, 11.7	2.3, 6.5	0.8, 3.4	1.5, 7.7
2008–2009	2.6, 11.0	5.4, 10.4	3.4, 6.7	2.4, 7.3	1.3, 7.8
2010–2011	4.4, 14.7	3.6, 9.2	4.7, 10.7	3.4, 10.7	0.9, 6.1
2012–2013	2.6, 15.1	3.8, 10.4	3.9, 9.5	2.4, 9.2	0.8, 5.2
Meta-analysis summary estimates & 95% confidence intervals (per 100)[a]	5.1 (3.9, 6.6)	7.4 (6.3, 8.7)	5.2 (4.4, 6.2)	3.6 (2.5, 5.0)[b]	2.4 (1.7, 3.4)

Source: Data from Restricted-use Data Analysis System subsamples of the National Surveys on Drug Use and Heath, United States 2002–2013.

Notes:
a The I-squared statistic quantifies heterogeneity of estimates across years.
b Here, the I-squared statistic has $0.05 < p < 0.15$ so the 95% CI are from 'random effects' estimation; the corresponding 'fixed effects' interval is 2.8, 4.6. All other meta-analytic 95% CI are from 'fixed effects' estimation (due to I-squared $p > 0.15$).

In this illustration of a case-control approach to cross-sectional survey data, there is reason to judge that a followup study might produce estimates and conclusions not appreciably different with respect to male sex and ethnicity. The heroin injection odds ratio is an approximation of what might be seen in a followup study, in that some newly incident heroin users might have started injecting after onset of heroin dependence, rather than before dependence onset.

At this point in the chapter, it might be helpful to mention some of the limitations of epidemiology, and the need to turn to a 'clinical investigation' approach without the stipulations of epidemiology's canonically required pre-specifications of a study population. Consider epidemiological sampling and assessment of any pre-specified study population. Except in countries such as Denmark, with a national identity card and its national registry of citizen-nationals and non-citizen legal residents, the sampling of a national level, state-level, or city-level study population will require some form of multi-stage area probability sampling for which a 'sampling frame' is required down to the level of each individual's dwelling unit. In a country such as the US, this 'sampling frame' will not necessarily include all of the heroin injectors or heroin dependence cases in the population. Some of them might be living under bridges and have no fixed dwelling unit address; these are 'left-truncated' heroin users, to use epidemiology's technical term for individuals who are known to exist, alive, but who are not included in the sample of the study population. Some of them might have a fixed dwelling unit address (e.g., within a girlfriend's or boyfriend's home), but would not be listed as a member of that dwelling unit when the field survey staff comes by to ask who lives in the household. That is, the adult household member responsible for saying who lives there might not think of that boyfriend or girlfriend as a member of the household. This is another example of left-truncation in any potential epidemiological sample of heroin users. Finally, even when seriously involved heroin users show up in the sampling frame and on the dwelling unit roster, are sampled for the survey, and are alive on the date of sampling, they might not be alive when the staff member returns for assessment, or perhaps they have entered an institutional facility such as a prison or long-term treatment program. This would be an example of 'left-censoring' as distinct from 'left-truncation,' either of which represent forms of bias in estimation of epidemiological parameters of interest when we study either 'starting' or 'stopping' drug use. We can constrain these biases when the specified subpopulations of interest are newly incident drug users, and when the goal is to estimate the cumulative occurrence of drug dependence or addiction syndromes. These biases are much more difficult to constrain when the goal is to estimate quitting, recovery, or remission parameters. Box 21.2 of this chapter provides some additional detail on these topics, and our reasons to encourage clinical investigations in place of epidemiological investigations of what happens after a person has become drug dependent.

Followup approximation illustrations

During the mid-1980s, while conducting followup studies, our epidemiology research group became aware that our cross-sectional survey estimates for the risk of becoming a newly incident case affected by drug problems were not appreciably distant from our risk estimates based on the followup assessments. We decided to separate the estimation problem into two parts. In the first part, we estimated a parameter known as the 'cumulative incidence proportion among survivors' (CIPAS). Box 21.3 provides an overview of the CIPAS parameter and its relationships with related epidemiological concepts such as the 'attack rate' observed after a point source agent-exposure (e.g., a luncheon where a contaminated food item is served). Where others use the term 'lifetime prevalence' we have used the term 'cumulative incidence proportion' and CIPAS (Lopez-Quintero & Anthony 2015).

Box 21.3   Attack rates, lifetime prevalence, and CIPAS

In every epidemiologist's 'first course in epidemiology' we learn about food-specific attack rates and the concepts of 'cumulative incidence' as well as the 'incidence density' function. The concept of an 'attack rate' often is introduced in the context of a public health department's food safety functions, and the task of identifying which food item might be responsible for an outbreak of gastroenteritis or a related illness after a 'pot luck' luncheon to which prepared food has been contributed from many kitchens. There is no real chance of a forward-looking study, with measurements of food consumption, item by item, before the luncheon. Instead, the sickness experience of the luncheon attendees is reconstructed after the fact, often by having all attendees fill out a questionnaire or complete an interview about which food items were eaten and which illness symptoms were experienced. In some instances, this effort is incomplete because the first case of illness might not occur until several hours or days after the luncheon. Some of the luncheon attendees might have migrated back to their home communities. Others might be too sick to complete a questionnaire or interview. Rarely, some cases might have died before there is a way to assess what foods they have consumed.

The food safety epidemiologists learn how to secure a list of luncheon attendees and food preparers, as complete as possible, and then to assess which items on a list of foods might have been consumed by each individual. This process makes it possible to create a 'denominator' for the food-specific attack rate, which counts every individual who ate each specific food item. Typically, there is a separate ascertainment of illness cases in relation to a working case definition, which might be something akin to 'vomiting and/or diarrhea' and an effort to ensure that there is an unbiased assessment of case status relative to food status.

For each food item, the denominator of exposed consumers is counted up, and, as of the time of the assessment, the epidemiologist counts up the cumulative number of illness cases who had been exposed to that food item. If any cases have died, the result is a 'cumulative incidence proportion among survivors' (CIPAS) and it typically is given the name of a 'food-specific attack rate' so as to distinguish it from the overall attack rate that describes the experience of all luncheon attendees who ate at least one food item. Most often, the number of decedents is small and can be ignored, as is the number of luncheon attendees from a distance and who could not be assessed to provide information about the food consumption and illness experience. A separate form of incidence rate, known as the 'incidence density,' can be derived by estimating the occurrence of cases as a function of time elapsed since the luncheon, as the time interval under study approaches zero (i.e., borrowing from the calculus concept of a first derivative).

The investigation of the food safety epidemiologist proceeds by estimating food-specific attack rates for each food item, and looking for differences in attack rates for subgroups defined by what they ate and what they did not eat. In a famous epidemiology laboratory exercise, the outbreak can be traced to egg salad consumption, and implicatively back to an infective agent such as Salmonella, because there is a quite large difference in the attack rate for those who did versus those who did not eat the egg salad at the luncheon. For almost all other food items, there is no difference between attack rates for those who did versus those who did not eat each food item.

Roughly 60 years ago, social scientists conducting a survey of mental illnesses in religious communities of the US decided to form similar attack rates, but apparently they were not aware of epidemiology's traditional concept of an 'attack rate' and apparently they did not know that

(continued)

(continued)

epidemiologists draw a sharp distinction between the 'incidence' of a condition ('becoming a case') versus the 'prevalence' of a condition ('being a case'). Instead, they coined the term 'lifetime prevalence' in order to label a proportion akin to an attack rate for their study sample overall and for subgroups of their study sample. Psychiatrist-epidemiologist Ernest M. Gruenberg, an expert on the differences between incidence versus prevalence, reviewed their work and critiqued the concept of 'lifetime prevalence,' labeling it as a 'gimmick' that served no useful function. Gruenberg appreciated that 'lifetime prevalence' actually is a cumulative incidence proportion among survivors to the time of the field survey, and that it had more to do with an 'attack rate' than an 'incidence density.' Lopez-Quintero and Anthony (2015), in a recent study that harnesses the food-borne illness outbreak approach for research on drug dependence as might be attributable to specific drug compounds, have described this history, as well as recent critiques of the 'lifetime prevalence' concept by experts who have urged 'retirement' of the errant concept.

Next, after formalizing a CIPAS estimate in this fashion, and after completion of the cross-sectional 1990–1992 US National Comorbidity Survey (NCS), we looked drug-by-drug and estimated CIPAS to gauge drug-specific probability of developing a dependence syndrome by age 54 years, recognizing that the cross-sectional estimates would be better for low lethality compounds such as cannabis. (Interpretation is more difficult for compounds such as heroin, and for individuals such as polydrug users, for whom death rates are larger; see Lopez-Quintero et al. 2015.)

Estimated in this fashion, we found that for US community residents aged 18 to 54 years old, the NCS-estimated probability of developing a drug dependence syndrome varied drug by drug, as measured after the first occasion of extra-medical drug use. Among those who had smoked tobacco cigarettes at least one time, an estimated 1/3rd had developed tobacco dependence. Among those who had tried heroin at least one time, irrespective of administration route, an estimated 1/4th to 1/5th had developed heroin dependence. Corresponding estimates for other drug subtypes were: cocaine (1/6th); alcohol (1/6th to 1/7th); amphetamines and stimulants other than cocaine (1/9th), and cannabis (1/9th to 1/11th), as reported in Anthony et al. (1994).

In this instance, comparison of the NCS cross-sectional approximation with corresponding estimates from the more recent National Epidemiologic Survey on Alcohol and Related Conditions (NESARC) followup study is possible only for cocaine and cannabis. Ignoring minor differences in diagnostic criteria and assessment procedures for these drug subtypes, we find overlap of the 95% confidence intervals (CI) for the cocaine and cannabis estimates, and no marked difference in the summary estimates for these proportions (Lopez-Quintero et al. 2011). Contrasts for tobacco and alcohol were not possible due to noteworthy differences in approach (e.g., a NESARC 'gating' approach for dependence on nicotine) and NESARC followup study estimates for heroin have not yet been published.

As for the second part of our estimation problem, in earlier work and in a series of more recent contributions, our research group has focused on the fairly rapid transition from first drug use until onset of specific drug dependence problems and on DSM-type drug dependence syndromes, with estimates derived from recent cross-sectional NSDUH surveys. Table 21.2 illustrates this example of an approximation of the age-specific incidence rates for making a fairly rapid transition from the first extra-medical use of an opioid prescription pain reliever (PPR) into the fully formed opioid dependence syndrome—that is, with dependence syndrome onsets occurring no longer than 12 months after starting to use PPR extra-medically. Panels A and B, respectively, show age-specific estimates of these rapid-onset transition probabilities from each of six independent replication samples, and corresponding 95% CI. At the bottom are meta-analysis summary estimates based on these six independent replication samples, showing point estimates that range from 3.6% to 7.4% for the age range from 12–13 years to 18–19 years, and a tangibly lower estimate of 2.4% among those with PPR onset delayed until 20–21 years (Parker & Anthony 2015).

Vsevolozhskaya and Anthony (2016) recently turned to a similar approximation for a very rapid transition from onset of tobacco smoking until formation of a tobacco cigarette dependence (TCD) syndrome, and used a functional analysis model to estimate Hill function parameters via cross-sectional NSDUH survey data on 1,546 newly incident tobacco cigarette smokers identified for 2004–2013. In this instance, the estimated incidence rates were for the first three months after the first tobacco cigarette was smoked. Drawing upon prior estimates pertinent to male–female differences in rapid emergence of tobacco dependence clinical features (e.g., O'Loughlin *et al.* 2003), we hypothesized a female excess risk of the TCD syndrome once smoking frequency was adjusted statistically. The evidence contradicted our hypothesis about female excess. Among those smoking fewer than five days during the 30 days before assessment, newly incident female smokers actually had lower TCD risk. No statistically robust male–female differences were found among newly incident smokers with a greater number of recent smoking days. In our self-critique and discussion of study limitations, we drew attention to some issues of statistical power and precision. We suspect that our study's relatively small numbers of very frequent users might have hampered our capacity to produce statistically precise dependence risk estimates for smokers who had progressed to near-daily or daily smoking within three months of the first cigarette (Vsevolozhskaya & Anthony 2016). Unlike some of the issues faced in design and conduct of epidemiological investigations, these are issues of research design that cannot be forecast with certainty until after the study data have been collected and analyzed. However, in this instance, we launched the study with an idea of a female excess risk and we had statistical power to detect that excess risk; we did not find a female excess risk.

Reflecting on studies of this type, we cannot claim that the cross-sectionally derived estimates are more than approximations of what a followup study might yield. Nevertheless, these approximations provide starting values that are needed to plan future followup studies, and the estimates do not suffer biases associated with differential sample loss during followup. In addition, the 'reactive measures' problem described on page 260 can be avoided in a cross-sectional research design.

As we now consider future research designs that would have to deal with reactive measurements and sample attrition, we think that a very large sample followup study with at least three quarterly followup assessments might be required to address issues of feedback loops in order to produce more definitive evidence than we were able to secure with our cross-sectional design. The fundamental problem is that any incipient dependence process might drive up the frequency of smoking days during any given 30-day interval after smoking the first cigarette. Three 'waves' of followup assessment, at minimum, would be required to estimate these feedback loops, and, unless assumptions were made, five or more followup waves of assessment might be required (Vsevolozhskaya & Anthony 2016; Paxton *et al.* 2011; Richardson 1994).

Case-crossover illustration

To illustrate case-crossover methods and statistical modeling based on cross-sectional NSDUH data from 2009–2013, we can turn to the Fairman and Anthony (2018) investigation of whether onset of tobacco cigar smoking might trigger newly incident onsets of the cannabis-tobacco formulation known colloquially as a 'blunt' (often a hollowed-out tobacco cigar then filled with cannabis herb), already mentioned on page 271, with a conditional contingency table layout shown in Table 21.1. The blunt formulation that combines cannabis herb with tobacco may be of special interest because this combination is widely used outside the US, and because blunt smokers seem to progress rapidly to higher levels of cannabis problem severity, as compared with cannabis smokers who do not smoke blunts (Fairman 2015).

For this case-crossover research, we took advantage of NSDUH assessments of month and year of onset of cannabis use (any form) and separately assessed month and year of first blunt smoking. Within the population samples, a total of 4,868 newly incident blunt smokers were identified. Almost all (n=3,493) had started blunt smoking in a month that was not the same as the month of first cannabis use (all forms). The vast majority started blunt smoking in a year other than the year of first cigar smoking, making them uninformative in case-crossover RR estimation, which is based on discordance of quite proximal hazard and control intervals, as shown in Table 21.1.

For case-crossover estimation of the cigar-blunts RR, the hazard interval was specified as the calendar month prior to the blunt onset month (i.e., t-1 month relative to blunt onset in month t), and 128 blunt smokers were found to have started cigar smoking during that hazard interval. Three alternative expectations were specified, with the first calendar month prior to the hazard interval as the primary control interval (t-2). When this primary control interval was used, 74 cigar initiates were found among newly incident blunt smokers, and the RR estimate was 1.7 (95% CI = 1.3, 2.3; $p<0.001$), such that the month after cigar-smoking onset serves as a month of excess risk for starting to smoke blunts in the US.

In the alternative control interval specifications, the strength of association increased. Using an expectation based on averaging the 1st through 11th months prior to the hazard interval (t-2 through t-12), the RR estimate was 2.7 (95% CI = 1.9, 3.9; $p<0.001$). By setting the control interval as the next prior calendar month (t-3), the RR estimate was 2.9 (95% CI = 2.1, 4.2; $p<0.001$). Hence, all three alternative control interval specifications led to rejection of the null hypothesis, which gives some standing to the Fairman-Anthony cigar-blunt triggering hypothesis. In this instance, RR are from unweighted sample data; analysis-weighted estimates do not differ appreciably.

Closing remarks

Many details and important contributions had to be left out of this relatively brief chapter on epidemiological methods and models used to study processes of becoming drug dependent. Most regrettable are (a) the cursory review of epidemiological contributions made by the many different important epidemiology research teams in this field, and (b) the emphasis upon studies completed by this author and his colleagues. Epidemics were mentioned, but the study of heroin and other drug epidemics had to be neglected completely in this handbook chapter, which also is regrettable.

Interested readers will find an especially comprehensive literature review of prior contributions by an array of international research teams in the journal article for which Lopez-Quintero and colleagues (2011) were responsible, with Anthony (1994) providing details on older studies from the 1960s through the 1980s.

Acknowledgments

The author wishes to thank his kind editors (HP, SA). Postdoctoral research associates Mahdur Chandra, PhD, and Olga Santiago-Rivera, PhD, deserve acknowledgments for help with the bibliography and heroin estimation, respectively. The preparation of this manuscript was supported in part by a National Institutes of Health, National Institute on Drug Abuse Senior Scientist and Mentorship Award to JCA (K05DA015799).

References

Anthony, J. C. (2010) "Novel phenotype issues raised in cross-national epidemiological research on drug dependence", *Annals of the New York Academy of Sciences* 1187: 353–369.

Anthony, J. C., Warner, L., and Kessler, R. (1994) "Comparative epidemiology of dependence on tobacco, alcohol, controlled substances, and inhalants: basic findings from the National Comorbidity Survey", *Experimental and Clinical Psychopharmacology* 2(3): 244–268.

Anthony, J. C., Lopez-Quintero, C., and Alshaarawy, O. A. (2017) "Cannabis epidemiology: a selective review", *Current Pharmaceutical Design* 22(42): 6340–6352.

Breslow, L. (1951) "Senescence, chronic disease, and disability in adults", in K. F. Maxcy (ed.), *Rosenau Preventive Medicine and Hygiene, Seventh Edition*, New York, NY: Appleton-Century-Crofts, Inc., pp. 72–755.

Cepeda, A., Nowotny, K. M., and Valdez, A. (2016) "Trajectories of aging long-term Mexican-American heroin injectors: the 'maturing out' paradox", *Journal of Aging and Health* 28(1): 19–39.

Cherng, S. T., Tam, J., Christine, P. J., and Meza, R. (2016) "Modeling the effects of e-cigarettes on smoking behavior: implications for future adult smoking prevalence", *Epidemiology* 27(6): 819–826.

Fairman, B. J. (2015) "Cannabis problem experiences among users of the tobacco-cannabis combination known as blunts", *Drug and Alcohol Dependence* 150: 77–84.

Fairman, B. J. and Anthony, J. C. (2018) "Does starting to smoke cigars trigger onset of cannabis blunt smoking?" *Nicotine & Tobacco Research* 20(3): 355–361.

Fleury, M. J., Djouini, A., Huynh, C., Tremblay, J., Ferland, F., Menard, J. M., and Belleville, G. (2016) "Remission from substance use disorders: a systematic review and meta-analysis", *Drug and Alcohol Dependence* 168: 293–306.

Frauenthal, J. C. (1980) *Mathematical Modeling in Epidemiology*, New York, NY: Springer-Verlag.

Goodman, R. A., Posid, J. M., and Popovic, T. (2012) "Investigations of selected historically important syndromic outbreaks: impact and lessons learned for public health preparedness and response", *American Journal of Public Health* 102(6): 1079–1090.

Haenszel, W. (1966) "Quantitative evaluation of etiologic factors in lung cancer", in A. M. Lilienfeld and A.J. Gifford (eds), *Chronic Diseases and Public Health*, Baltimore, MD: The Johns Hopkins Press, pp. 425–438.

Hammond, E. C. and Horn, D. (1958) "Smoking and death rates—report on forty-four months of followup of 187,783 men. Death rates by cause", *Journal of the American Medical Association* 166(11): 1294–1330.

Heyman, G. E. (2013) "Quitting drugs: quantitative and qualitative features", *Annual Review of Clinical Psychology* 9: 29–59.

Jukes, T. H. (1989) "The prevention and conquest of scurvy, beri-beri, and pellagra", *Preventive Medicine* 18(6): 877–883.

Kellam, S. G. and Anthony, J. C. (1998) "Targeting early antecedents to prevent tobacco smoking: findings from an epidemiologically based randomized field trial", *American Journal of Public Health* 88(10): 1490–1495.

Kellam, S. G., Wang, W., Mackenzie, A. C., Brown, C. H., Ompad, D. C., Or, F., Ialongo, N. S., Poduska, J. M., and Windham, A. (2014) "The impact of the good behavior game, a universal classroom-based preventive intervention in first and second grades, on high-risk sexual behaviors and drug abuse and dependence disorders into young adulthood", *Prevention Science* 15: S6–18.

Lilienfeld, A. M. (1966) "Epidemiologic methods and inference", in A. M. Lilienfeld and A. J. Gifford (eds), *Chronic Diseases and Public Health*, Baltimore, MD: The Johns Hopkins Press, pp. 99–112.

Lopez-Quintero, C. and Anthony, J. C. (2015) "Drug use disorders in the polydrug context: new epidemiological evidence from a foodborne outbreak approach", *Annals of the New York Academy of Sciences* 1349(1): 119–126.

Lopez-Quintero, C., Pérez de los Cobos, J., Hasin, D. S., Okuda, M., Wang, S., Grant, B. F., and Blanco, C. (2011) "Probability and predictors of transition from first use to dependence on nicotine, alcohol, cannabis, and cocaine: results of the National Epidemiologic Survey on Alcohol and Related Conditions (NESARC)", *Drug and Alcohol Dependence* 115(1–2): 120–130.

Lopez-Quintero, C., Roth, K. B., Eaton, W. W., Wu, L. T., Cottler, L. B., Bruce, M., and Anthony, J. C. (2015) "Mortality among heroin users and users of other internationally regulated drugs: a 27-year followup of users in the epidemiologic catchment area program household samples", *Drug and Alcohol Dependence* 156: 104–111.

Maxcy, K. F. (1951) *Rosenau Preventive Medicine and Hygiene, Seventh Edition*, New York, NY: Appleton-Century-Crofts, Inc.

Murray, T. G., Stolley, P. D., Anthony, J. C., Schinnar, R., Hepler-Smith, E., and Jeffreys, J. L. (1983) "Epidemiologic study of regular analgesic use and end-stage renal disease", *Archives of Internal Medicine* 143(9): 1687–1693.

O'Loughlin, J., DiFranza, J., Tyndale, R. F., Meshefedjian, G., McMillan-Davey, E., Clarke, P. B., Hanley, J., and Paradis, G. (2003) "Nicotine-dependence symptoms are associated with smoking frequency in adolescents", *American Journal of Preventive Medicine* 25(3): 219–225.

Parker, M. A. and Anthony, J. C. (2015) "Epidemiological evidence on extra-medical use of prescription pain relievers: transitions from newly incident use to dependence among 12–21 year olds in the United States using meta-analysis, 2002–13," *PeerJ* 3: e1340, doi.org/10.7717/peerj.1340.

Paxton, P., Hipp, J. R., and Marquart-Pyatt, S. (2011) *Non-Recursive Models: Endogeneity, Reciprocal Relationships, and Feedback Loops* (Vol. 168), Los Angeles, CA: Sage Publications, Inc.

Richardson, T. (1994) "Equivalence in non-recursive structural equation models", in R. Dutter, and W. Grossman (eds), *Compstat: Proceedings in Computational Statistics 11th Symposium held in Vienna, Austria*, Berlin Heidelberg: Springer-Verlag, pp. 482–487.

Sanderson, S., Tatt, I. D., and Higgins, J. P. (2007) "Tools for assessing quality and susceptibility to bias in observational studies in epidemiology: a systematic review and annotated bibliography", *International Journal of Epidemiology*, 36(3): 666–676.

Santiago-Rivera, O. J., Havens, J. R., and Anthony, J. C. (forthcoming 2018) "If we cannot prevent heroin use, can we 'prevescalate' it? Estimated transition rates for newly incident heroin users in the United States before and during heroin epidemic years 2005 through 2013", unpublished manuscript.

Seedall, R. B. and Anthony, J. C. (2015) "Monitoring by parents and hypothesized male-female differences in evidence from a nationally representative cohort re-sampled from age 12 to 17 years: an exploratory study using a 'mutoscope' approach", *Prevention Science* 16(5): 696–706.

Sobell, L. C., Ellingstad, T. P., and Sobell, M. B. (2000) "Natural recovery from alcohol and drug problems: methodological review of the research with suggestions for future directions", *Addiction*, 95(5): 749–764.

Swygert, L. A., Maes, E. F., Sewell, L. E., Miller, L., Falk, H., and Kilbourne, E. M. (1990) "Eosinophilia-myalgia syndrome. Results of national surveillance", *JAMA: The Journal of the American Medical Association* 264(13): 1698–1703.

Vsevolozhskaya, O. A. and Anthony, J. C. (2016) "Inter-relationships linking probability of becoming a case of nicotine dependence with frequency of tobacco cigarette smoking", *Nicotine & Tobacco Research* 18(12): 2278–2282.

Wagner, F. A. and Anthony, J. C. (2002) "Into the world of illegal drug use: exposure opportunity and other mechanisms linking the use of tobacco, marijuana, and cocaine", *American Journal of Epidemiology* 155(10): 918–925.

Winick, C. (1962) "Maturing out of narcotic addiction", *Bulletin on Narcotics* 14(1): 1–7.

Yeh, H. H. (2017) "Contributions to epidemiological research on the use of heroin and other opioid compounds", Michigan State University, East Lansing, MI. Unpublished dissertation.

22
A GENETIC FRAMEWORK FOR ADDICTION

Philip Gorwood, Yann Le Strat, and Nicolas Ramoz

Introduction

It is now known that addiction to a substance results from the combination of genetic factors on one hand (the inherited vulnerability) and environmental factors on the other (frequency and modality of drug exposure, social and educational background) (Ducci 2012; Wang 2012). The same model could be applied to behavioral addictions (without substance), such as eating disorders, sex addiction, gambling disorder, video game playing, physical exercise, work addiction (workaholism) or connected media use.

To get quickly to the point, there is no "gene of addiction" but genes of vulnerability that increase the risk to ultimately develop an addiction. The tools and strategies used for genetic research in addiction are the same as those developed for genetic diseases in general, with several specific difficulties. First, using genetics for the research devoted to addictive disorders heavily relies on an appropriate diagnosis, and this is not simple for many reasons. How should we assess a patient who has never had an experience with the studied addictive substance? A subject might indeed have such a high risk that he/she prevents themselves from experiencing any exposure to that substance (for example, children of alcoholics (COA)). Psychiatric comorbidity may be a factor explaining the diagnosis of addictive disorder much more than genes (for example, patients with bipolar disorder have a dramatically increased risk to dependence during lifetime). This last may be difficult to define as the outward manifestation of a specific phenotype because, in contrast to the "classic" genetic disorders due to mutations in one or several genes, some addictive cases are not due to genetic factors. Second, the definition of the addictive phenotype is not yet validated as a categorical phenomenon (yes or no), as there is a continuum as opposed to a sharp line between "no symptoms at all" to "full severity level." As all studies are presently relying on diagnoses that are therefore based on qualitative judgement, it may be that we lose the largest part of the genetic information with such a method.

However, the influence of genetic factors in addictive behaviors, as well as the ability to develop a "substance use disorder" (the older concepts of abuse and dependence were replaced by these terms in DSM-5), is now widely described and agreed (Le Strat 2008a; Bierut 2011; Ramoz 2015). The involved genetic factors may be expressed at different moments in a person's life to explain the vulnerability, sometimes leading to an addictive disorder.

Genetic component: heritability in addiction

Epidemiological studies have clarified the respective contributions of genetic and environmental components in addiction. A familial aggregation for addiction is widely reported: that is, a tendency to observe the pathology most frequently in relatives of sick patients than in healthy subjects. Furthermore, the studies on adopted children and the comparisons of the concordance rate in "identical twins" (monozygotic twins, who share the same genome) versus "fraternal" twins (dizygotic twins, who share up to 50% of the same genome) have demonstrated the presence of genetic factors involved in addiction. Thus, a genetic component is suspected because the biological parents of adopted patients are more frequently affected than the adoptive parents of healthy subjects, or when the monozygotic twins are more alike than dizygotic twins for the analyzed addictive disorder.

The heritability (h^2) is the proportion of phenotypic variance due to genetic factors and is easily estimated from concordance rates (the concordance for the disease in monozygotic versus dizygotic twins). A review reports an average heritability of around 70% for cocaine and opioids dependence, 60% for alcohol or tobacco dependence and 40% for hallucinogens (Ducci 2012). For a comparator, several predisposition genes have already been identified in diabetes, although its h^2 ranges between 30–60%, i.e., less than in many addictive disorders. This observation suggests that it is indeed possible to identify the involved genetic factors through the comparison of affected cases and healthy controls (case-control design) or in families for which at least one subject has an addictive disorder (family-based approach).

The genetics of addictive disorders is probably complex, and could be oligogenic (few genes are involved) or polygenic (many genes are significantly involved), or, in only rare exceptions, it may have a monogenic origin (one gene only). There are two human genetic models that explain how each gene or locus (a specific region of the genome) can impact the risk of a complex disorder such as addictive disorders. These models are called "common disease/common variant" (CDCV) and "multiple rare variants" (MRV) (Figure 22.1). Together they form the allelic spectrum that could explain the continuum of patients with addictive disorder. The CDCV model puts forward common genetic variants (allelic frequency >5%) that confer a modest genetic relative risk (≤1.5) toward developing a complex trait. The MRV model proposes that many different rare genetic variants (frequency ≤1%) confer a very strong genetic relative risk (≥10) toward developing a complex trait. The rare variants (and rare effects) are very difficult to identify because they require the screening of hundreds of thousands of subjects.

While the tools for sequencing of the human genome have greatly evolved and are used at a high-throughput, they are nonetheless used rarely, especially in addiction, due to their expensive cost and the length of time that is needed for data analysis. The genetics of addictive disorders therefore uses different tools and strategies, widely provided for human genetics.

Strategies and methods of genetic analysis

Two complementary approaches can be implemented to identify genetic factors in addiction. The first strategy, appropriate for when one has no information about the biological origin of the disorder, is called "reverse genetics". It consists in screening genetic markers covering the entire genome (genetic variants that are roughly present every 1.000 base pairs on the whole genome) in the goal to discover a locus (a chromosome region) that could be involved in the disorder (because at least one of these markers is more often present in patients than in controls, or is inherited by affected family members and not inherited by unaffected family members).

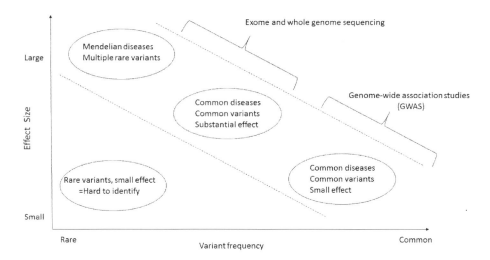

Figure 22.1 Expected distribution of genetic variants according to their effect size in complex diseases, including addictive disorders, according to the models "common disease/common variant" and "multiple rare variants." The "allelic frequency" is indicated in abscissa from rare (lower than 1%) to common (upper than 5%). The vertical axis or "effect size" indicates the relative genetic risk conferred to the disorder from small (lower than 1.5 times) to large (over 10 times). Considering the number of screened patients and the technique (GWAS or sequencing), four types of genetic disorders can be distinguished as indicated in the bubbles.

This locus (region) is then analyzed to detect the genes that it contains, as they could potentially be involved. The second strategy is called the "candidate gene" study or the "classic genetics" study. It searches for mutations or screens for polymorphic markers in one (or more) gene(s) selected for their (potential) role in an addictive disorder. These strategies are based on the study of population cohorts (cases versus controls) or of familial cohorts that may rely on the "trios" (one patient, the father and the mother), "siblings" (an affected individual and his affected sibling), or on "multiplex families" (several affected subjects in at least two generations). When the clinical cohorts are available, several genetic and statistical tools (linkage, linkage disequilibrium, associations) are used to identify the genes involved in addiction, and each method has advantages and limitations.

Genetic linkage

The genetic linkage analysis (linkage) is used to locate one or more chromosomal regions or loci that do not segregate with the disorders in families (Figure 22.2). This means that when an allele (one form of the genetic variant) is more frequently transmitted only to affected family members than it would be by chance, and less frequently transmitted only to unaffected members than it would be by chance, then this co-segregation (co-transmission) of the allele and the disorder points in favor of a significant role of this allele in the disorder. Linkage analysis is performed on a candidate region or the entire genome to identify all loci and discover the role of known genes – but also potentially to discover new ones in the addictive disorder. Genome scans were made possible thanks to the availability of different maps of the human genome based on genetic variations (especially single nucleotide polymorphisms or SNPs). There are two methods of linkage analysis: parametric and nonparametric. The parametric analysis requires an exact genetic model for various parameters such as the mode of Mendelian transmission (recessive, dominant or co-dominant), the prevalence of the disease, the frequency of the deleterious allele, the expressivity of the disease, and the allele frequencies of the markers genotyped. However, since these parameters are not known for many complex diseases, such a parametric analysis is often not feasible. In such cases, a nonparametric analysis is used instead. The nonparametric analysis does not refer to any fixed genetic model. This method is thus better suited for the study of loci involved in multigenic diseases such as addictive disorders. In return, it does not use some information that may be available and in fact very useful (family cohorts and parameters): the price to pay is that its statistical power is low compared with parametric methods. The genetic linkage analysis has been successfully used in monogenic disorders (one to one gene–disease mapping) but it is less helpful to identify genes involved in complex disorders such as addictive disorders. Indeed, with an unclear definition of what an addictive disorder is (apart from a list of unspecific diagnostic criteria), together with the imprecise divisions between the different addictive disorders (because of their high comorbidity), and strong influence of cultural background (at least for the facility to get the addictive substance), many parameters are not known. When using parametric approaches with such unknowns, it is necessary to test many models to see which one best fits the data, a procedure that increases the risk of false positive results because of chance findings.

Linkage disequilibrium

Analysis of linkage disequilibrium (Linkage Disequilibrium, usually referred to as "LD") tests whether one or more genetic markers are linked and located in the vicinity of the locus involved in the disease, in family cohorts or within populations. Indeed, a given phenotype can be influenced by a specific allele variant, and polymorphisms in the neighborhood should also be linked to it, if they are in LD. In the absence of selection, population heterogeneity, or other confounding traits, LD should be theoretically correlated linearly with the distance between the disease locus and marker. According to the chromosomal regions and markers, the linkage disequilibrium varies. Markers segregate into blocks by specific combinations of alleles or haplotypes that have limited diversity and high linkage disequilibrium value. The genotyping of a few markers becomes sufficient to analyze a given locus. The markers studied in linkage disequilibrium studies, but also in association studies, are generally polymorphisms affecting only one nucleotide acid, called single nucleotide polymorphisms or SNPs, as they are very widespread (in the human genome every 600 nucleotides on average) and easy to genotype. The SNPs are selected based on their information – namely, their allelic frequency, frequency of heterozygosity, their location in the genome, their linkage disequilibrium in a population of a given ethnicity – which is available in the databanks of the international consortium HAPMAP (https://hapmap.ncbi.nlm.nih.gov/). The linkage disequilibrium study, as a genetic linkage, gives information on the physical position of the locus of dependence, more than it does information pertaining to a causal relationship between the marker and addiction, in contrast to association studies.

Association studies

The association studies compare the incidence in a population of patients with two or more variables, usually the phenotype and genotype, with the expected frequency by chance only (Figure 22.2). This expected frequency is given by the population of healthy individuals or controls in the event of a case-control association, or the frequency of transmission of alleles of parents towards the proband in the approaches of family associations. These two types of association studies may be applied to a genetic marker or a combination of several markers (haplotypes).

Case-control association studies compare the allele (or genotype) frequency of patients (with the disorder or trait under study) with that of matched controls (without the disorder or trait of interest). A significant excess of an allele (or a genotype) in patients means that this allele (or genotype) is associated with the disorder. A simple $\chi 2$ test determines the existence of a significant association with usually a risk value threshold of 5% error (called type I error). The measure of association may be discrete (i.e., presence or absence of a significant association) or continuous (i.e., relative risk and odds ratio). The relative risk (RR) is a ratio of the frequency of the "vulnerability" allele in the patient group to the frequency of this allele in the control group. The odds ratio (OR) is the ratio of the "vulnerability" allele in patients and controls to the ratio of the "non-vulnerability" allele in patients and controls. In conclusion, the study of case-control associations is a useful methodological tool that represents a powerful statistical test. However, the major limitation of this analysis comes through stratification of the population, for example when cases and controls are not of the same ethnic origin or sex. In this case, the difference of allele frequency in patients and healthy controls is not explained by the disorder

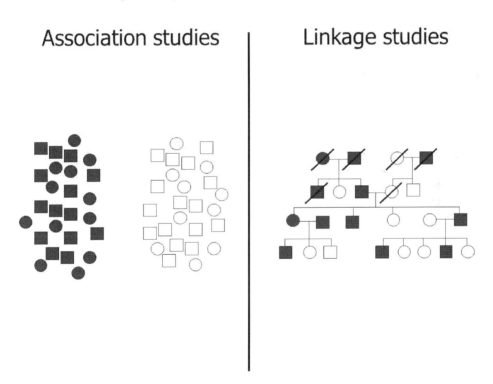

Figure 22.2 Case-control association studies (left) and family-based linkage studies (right). In genetics, the circle symbolizes women and the square represents men. Open circles and squares are healthy controls and filled in circles and squares are affected patients. Thus, families can be drawn such as in the example in the right part of the figure (including the deceased parents indicated with a diagonal bar). By counting the alleles of each subject and by comparing patients to controls using a simple chi-square analysis, the role of this allele (association study) can be demonstrated. Different approaches are used to solve the stratification bias, one of them being the family association study.

(present versus absent) but by an unequal distribution of subjects with a different genetic background in the two groups. For example, in France, subjects of Basque origin have many genetic specificities (it is a homogenous group of subjects with different allele frequencies) and cannot be phenotypically distinguished. Thus, if one detects a high frequency of a rare variant in a group that contains subjects of Basque origin, it will be difficult to ascertain whether the variant is attributable to the disorder under study rather than to the Basque origin.

The transmission disequilibrium test (TDT) is another statistical method used in association studies, which takes into account families whose parents are heterozygous for the genetic marker studied. The TDT measures an excess allele transmission from parents to affected offspring, compared with the distribution due to chance, when that chance would be one in two. The main limitation of the TDT is the mandatory use of complete families, or trios at minimum, because families missing a parent introduce bias. To solve this problem, a TDT of sibling pairs was designed (Sibs-TDT and S-TDT). The S-TDT compares allele frequencies among dependent children with siblings to those unaffected, to measure the difference (the discordant pairs to share the susceptibility allele less frequently than if this was due to chance alone). Since some complex pathologies, like substance use disorders, may be better described as a continuous variable, from normal (abstinence or recreative use) to the most severe form of addiction, an association test for quantitative traits (Quantitative-TDT) was established. Finally, the intrinsic information for families, such as their overall size, whether or not they contain a particularly large set of siblings, or whether or not they are extended so as to include several generations and/or cousins, was also taken into consideration in the siblings disequilibrium test (Pedigree Disequilibrium Test) or in software such as FBAT (Family-Based Association Test).

Pan-genomic screening: the Genome-Wide Association Study (GWAS)

Technological developments now allow an automatic genotyping at high throughput with 300,000 to 1.2 million SNPs simultaneously in a subject, thanks to DNA chips or microarrays. This method of pan-genome scan (GWAS) can be applied to large patient populations, of at least 1,000 affected subjects and controls, or family cohorts (patients and relatives) of the same size. The number of individuals included is critical to the statistical power of the associations that can be identified in the GWASs. Indeed, given the large number of SNPs genotyped, the significance level of the association must be corrected for the number of tests and preferably fixed at about 10^{-7} (without real consensus to date). Numerous GWASs have been made focusing on addiction, especially alcohol dependence and tobacco dependence, and among thousands of patients and controls that have been recruited through international consortia.

Sequencing and analysis of exome Genome-Wide Exome Sequencing (GWES)

The advent of the whole human genome sequencing in the early 2000s led to the development of new high-throughput sequencing technologies (Next-Generation Sequencing or NGS). This allows researchers to sequence all exons (about 190,000 in total) of the human genome. This strategy is called "exome sequencing". This technique allows researchers to identify genetic variations of genes, particularly mutations, specifically the de novo mutations that may appear in a patient while they are absent in the parents. This analysis of the exome or Genome-Wide Exome Sequencing (GWES) is proposed as a diagnostic tool in common pathologies. It has begun to be used in research in addiction. However, GWES can only sequence about 1.5%

of the entire human genome and other variations that it cannot sequence, for example those located in the regulatory regions of genes, may impact on the disorder.

First genes involved in an addiction: the saga of tobacco and nicotinic receptor genes

According to the World Health Organization (WHO), the total of tobacco consumers is 1.3 billion people leading to approximately 5 million deaths per year. In 2007, the first large-scale genetic analysis on nicotine dependence was carried out on 348 candidate genes, by genotyping 3,713 SNPs in 1,050 patients addicted to tobacco and 879 non-dependent smoker subjects (Saccone 2007). This study showed variants associated with nicotine dependence in the cluster of the genes CHRNA3, CHRNA5 and CHRNB4 coding respectively the subunits α3, α3 and β4, involved in the formation of nicotinic heteromeric receptors in acetylcholine (Figure 22.3).

The following year, a pan-genomic study by GWAS on several thousand patients with lung cancer, with and without tobacco dependence, also identified a genetic association with SNPs of these three genes. Exhaustive analysis of clinical data showed a much greater association between variants of these genes and the amount of smoking and tobacco addiction (Thorgeirsson 2008). In 2010, a GWAS of tens of thousands of smokers by three international consortiums confirmed the involvement of these genes in tobacco dependence (Liu 2010). The functional analysis of the SNP rs16969968, whose major G allele is changed in A that leads to the change of the aspartate amino acid at asparagine 398 in the α5 receptor, was associated with a significant effect. Indeed rodents carrying this allele showed a reduced activity of the nicotinic receptor (Bierut 2008). Finally, CHRNA3, CHRNA5 and CHRNB4 genes are expressed in the brain but also in the lungs, which would also explain their involvement in lung cancer. Thus, a common genetic pathway with different mutations on different receptors from the same family of nicotinic receptors, could lead to two phenotypes, addiction and lung cancer. It is probable that the same mechanism is implicated in both disorders as, in those carrying the vulnerability allele, tobacco both stimulates the cholinergic receptors in the nucleus accumbens, increasing the number of receptors (therefore requiring more smoking in order to relieve withdrawal symptoms), and also stimulates mitosis in the cholinergic cells of the lung, which might lead to cancer (which testifies to an absence of regulation of the mitosis phenomenon). A recent meta-analysis involving 94,000 smokers observed that the variant rs16969968 of the CHRNA5 gene is also associated with an earlier age of smoking, as well as higher consumption (Hartz 2012). Thus, individuals who carry one copy of this variant double their risk for tobacco addiction, and those with two copies have their risk multiplied by three.

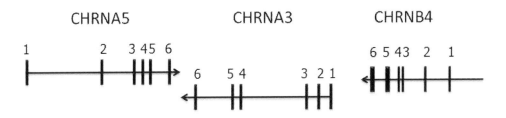

Figure 22.3 The genomic organization of the cluster of genes CHRNA3, CHRNA5 and CHRNB4 on chromosome 15q 24–25 region. The coding strand for CHRNA3 and CHRNB4 is the reverse chromosomal strand indicated by the head of the arrow. The vertical bars indicate the exons of each gene.

The cluster of associated genes (coding for α3, α5 and β4 subunits of cholinergic receptors) has an estimated attributable risk of 14% for tobacco dependence (Tobacco and Genetics Consortium 2010). This finding in relation to addiction to tobacco is remarkable for such a complex and multifactorial disease, with no equivalence in other addictions, or even in any psychiatric disorder. Challenging the involved phenotype (tobacco dependence) led to the demonstration that other potentially overlapping disorders, such as schizophrenia and Parkinson disease, could also be involved, and further modulated by parent monitoring or the existence of a smoking partner (Gorwood 2017).

Since these publications, these genes and other coding for the α and β subunits of nicotinic acetylcholine receptors, like *CHRNA4* and *CHRNB3* genes, have been sequenced in tobacco dependence, and some mutations and rare variants have been identified (Thorgeirsson 2010; Olfson 2016; Thorgeirsson 2016). Furthermore, genes have also been investigated in addiction to other substances, including alcohol, opioid and cocaine (Haller 2014). Several GWAS found an association between the initiation of smoking and the functional variant rs6265 of the *BDNF* gene that encodes the brain-derived neurotrophic factor. These studies also reported an association of the rs3733829 SNP of the *CYP2A6* gene (implicated in the metabolism of nicotine into cotinine) and tobacco consumption, but also the development of lung cancer (Thorgeirsson 2010, Tobacco and Genetics Consortium 2010).

Genes of the dopamine pathway in addiction

The addictive substances trigger pleasure, or may be used to relieve distress. Dopamine is the major neuromodulator for the balance between reward and aversion and it plays a key role in different components of drug addiction, e.g. the reinforcing effects of addictive drugs and their aversive effects. Several known genes are involved in the neurobiological dopaminergic pathway, leading different research teams to assess candidate gene variants associated with this pathway.

The most studied gene candidate for the dopaminergic pathway is the *DRD2* gene that encodes dopamine D2 receptor. Among the *DRD2* genes, one variant named TaqIA (SNP rs1800497) with A1 and A2 alleles, was found to be present in many addictions and especially in alcohol dependence. Thus, alcohol dependent patients carrying the A1 allele exhibit a hyposensitivity of dopamine receptors and an increased vulnerability. However, this variant has been considered as a vulnerability gene for alcoholism in more than 60 case-control studies, involving about 20,000 participants, with conflicting results. In fact, the variant of the A1 gene *DRD2* polymorphism TaqIA may be a risk factor for experience with alcohol and other psychotropic substances and impulsive or compulsive behavior (Gorwood 2012; Le Foll 2009). Finally, the TaqIA polymorphism was found to be located at the 3' region of the *DRD2* gene, in a new *ANKK1* gene, encoding an X-kinase protein that interacts with the D2 receptor to partly control its expression. The *ANKK1* gene contains variants also associated with alcohol dependence (Yang 2007). Other variants of the *DRD2* gene are also associated with alcohol, heroin and tobacco dependence (Gorwood 2012; Nelson 2013; Ma 2015).

Other dopamine receptor genes have been linked with drug addiction. Thus, several SNPs have been identified in the *DRD1* gene, and ten association studies have been conducted to date on different substance addictions (alcohol use disorder, heavy smoking, and methamphetamine use), most of them with negative results. For the *DRD3* gene, again most of the results are negative, except from an association with nicotine dependence that was detected in preliminary studies but requires further replication. The role of a variable number of tandem repeats (VNTR) polymorphism located in the third exon of the *DRD4* gene is most convincing in nicotine dependence and opioid dependence, while its association with alcohol use disorder is heterogeneous (Gorwood 2012; Le Foll 2009). The *DRD5* gene is an exon that encodes the receptor and only rs7655090 is associated with withdrawal severity score in alcohol dependence.

Another gene of the dopaminergic pathway is the *SLC6A3* (solute carrier family 6, member 3)/*DAT1* (dopamine active transport 1), which codes for the dopamine reuptake transporter. Alcohol-dependent patients, who were carriers of the allele of nine repeats of the VNTR located in the 15th exon in the 3' non-coding region of the *DAT1*, had significant rates of alcohol withdrawal complications (Le Strat 2008b).

The enzyme dopamine β-hydroxylase that converts the dopamine to noradrenaline is encoded by the *DBH* gene, which presents the SNP rs3025343 located in its upstream region that was found associated with smoking cessation (Thorgeirsson 2010).

Catechol-O-methyltransferase (COMT) metabolizes dopamine. The human *COMT* gene contains a functional polymorphism (Val158Met, SNP rs4680), with individuals carrying the Val allele having a 40% higher enzyme activity than Met homozygotes, at least in the frontal cortex. The COMT Val158Met variant represents a good candidate for association with addiction (Tunbridge 2012). Studies between COMT and drug-addiction-related phenotypes have showed inconsistent results, meta-analyses considering alcohol or illicit drugs being negative. However, these studies were conducted in relatively small samples, with a high heterogeneity.

Genome-Wide Association Studies on addiction

Genome-Wide Association Studies (GWASs) offer the possibility to analyze hundreds of thousands of single nucleotide polymorphisms in the absence of a priori assumptions in large populations of patients versus controls. The GWASs on alcohol dependence support the role of the gene *GABRA2*, which encodes the GABA receptor 2 subunit (Treutlein 2009; Gelernter 2014; Samochowiec 2014). These studies also confirm the association of genes involved in the metabolism of alcohol, especially alcohol deshydrogenase 1B & 1C encoded by the *ADH1B* and *ADH1C* genes and the aldehyde dehydrogenase 2, gene *ALDH2* (Wang 2012; Gelernter 2014).

The GWAS has been recently performed on three large independent cannabis dependence cohorts representing about 15,000 subjects (Sherva 2016). These studies found several SNPs significantly associated with cannabis dependence in novel candidate genes including a novel antisense transcript *RP11-206M11.7*, the solute carrier family 35 member G1, gene *SLC35G1*, and the CUB and Sushi multiple domains 1, gene *CSMD1*.

A GWAS on opioid dependence recently found an association, replicated in two independent cohorts, with several SNPs encompassing the *CNIH3* gene, which encodes the cornichon family AMPA receptor auxiliary protein 3 (Nelson 2016). Interestingly, the most highly associated SNP rs10799590 is predicted to be functional in fetal brain at the epigenetic level, modifying the binding of the transcription factor TAL1. In addition, it also influences brain response habituation, notably in the amygdala, as observed in vivo using functional IRM.

The functional non-synonymous SNP rs1799971 (A118G, Asn40Asp) located in the *OPRM1* that encodes the mu opioid receptor was investigated in a large meta-analysis combining 28,000 European-ancestry subjects, to search for an association with substance dependence including alcohol, opioid, cannabis, cocaine and nicotine (Schwantes-An 2016). This study reports a modest protective effect of the G allele on general addiction to substances, suggesting that this polymorphism contributes to biological mechanisms that are shared across the different dependence substances.

Finally, to date, no GWAS has shown a significant association of SNPs on genes involved in the dopamine pathway, especially in the *DRD2* gene. Thus, the meta-analysis of three GWASs on alcohol dependence did not indicate polymorphisms among dopamine receptor genes in the top 25 SNPs significantly associated with this addiction (Wang 2011).

Conclusion

The important development of the human genetic tools coupled with the creation of national and international consortia to recruit large samples of dependent patients and controls have been successfully used in the genetics on addiction (Psychiatric GWAS Consortium 2010). The genes involved in the brain dopamine pathway have been found to be associated with dependence to several substances. The GWASs have demonstrated the role of variants of nicotinic receptor genes in tobacco dependence and confirmed the involvement of the genes encoding the enzymes of degradation of ethanol in alcohol dependence. New biological pathways have also been identified by GWASs in cannabis and opioid dependences.

Furthermore, these high-throughput tools are now combined with new techniques to investigate the epigenetics in addiction (Robison 2011). Epigenetics is the study of gene expression modifications, in response to environmental factors or individual history but in absence of modifications of the DNA sequence. The genetic component in addictive disorders is about 60% to 70%. However, genetic studies are able to explain only half of it. The same explanatory gap has been observed for other complex genetic disorders, leading the scientific community to suggest that there is a "hidden" genetics. In fact, most recent studies have found that some of the genetic polymorphisms associated with an addictive disorder are strongly influenced by epigenetic mechanisms. Epigenetics (the change of the expression of the genome, beyond and over its genetic sequence) is therefore an important area for further question, with many studies currently underway (Nelson 2016; Zhang 2014). Indeed, this field of research may constitute a nice reconciliation (the missing link) between the important role of many environmental factors (for example, being a victim of a sexual trauma during childhood increases for decades the level of methylation of the receptor for glucocorticoids, which helps to regulate the stress axis in humans, meaning that its functioning is impaired) and genetic factors (as the substratum of such methylation sites is purely genetic, an equivalent trauma will not have the same impact on everyone, as it will be different according to each individual's genetic polymorphisms).

References

Bierut, L. J. (2011) "Genetic vulnerability and susceptibility to substance dependence", *Neuron* 69(4): 618–627.
Bierut, L. J., et al. (2008) "Variants in nicotinic receptors and risk for nicotine dependence", *American Journal of Psychiatry* 165(9): 1163–1171.
Ducci, F. and Goldman, D. (2012) "The genetic basis of addictive disorders", *Psychiatric Clinics of North America* 35(2): 495–519.
Fergusson, D. M., et al. (2003) "Early reactions to cannabis predict later dependence", *Archives of General Psychiatry* 60(10): 1033–1039.
Gelernter, J., et al. (2014) "Genome-wide association study of alcohol dependence: significant findings in African- and European-Americans including novel risk loci", *Molecular Psychiatry* 19(1): 41–49.
Gorwood, P., et al. (2012) "Genetics of dopamine receptors and drug addiction", *Human Genetics* 131(6): 803–822.
Gorwood, P., et al. (2017) "Genetics of addictive behavior: the example of nicotine dependence", *Dialogues in Clinical Neuroscience* 19(3): 237–245.
Haller, G., et al. (2014) "Rare missense variants in CHRNB3 and CHRNA3 are associated with risk of alcohol and cocaine dependence", *Human Molecular Genetics* 23(3): 810–819.
Hartz, S. M., et al. (2012) "Increased genetic vulnerability to smoking at CHRNA5 in early-onset smokers", *Archives of General Psychiatry* 69(8): 854–860.

Le Foll, B., *et al.* (2009) "Genetics of dopamine receptors and drug addiction: a comprehensive review", *Behavioural Pharmacology* 20(1): 1–17.

Le Strat, Y., *et al.* (2008a) "Molecular genetics of alcohol dependence and related endophenotypes", *Current Genomics* 9: 444–451.

Le Strat, Y., *et al.* (2008b) "The 3' part of the dopamine transporter gene DAT1/SLC6A3 is associated with withdrawal seizures in patients with alcohol dependence", *Alcoholism: Clinical and Experimental Research* 32(1): 27–35.

Liu, J. Z., *et al.* (2010) "Meta-analysis and imputation refines the association of 15q25 with smoking quantity", *Nature Genetics* 42(5): 436–440.

Ma, Y., *et al.* (2015) "The significant association of Taq1A genotypes in DRD2/ANKK1 with smoking cessation in a large-scale meta-analysis of Caucasian populations", *Translational Psychiatry* 5: e686.

Nelson, E. C., *et al.* (2013) "ANKK1, TTC12, and NCAM1 polymorphisms and heroin dependence: importance of considering drug exposure", *JAMA: Journal of the American Medical Association: Psychiatry* 70(3): 325–333.

Nelson, E. C., *et al.* (2016) "Evidence of CNIH3 involvement in opioid dependence", *Molecular Psychiatry* 21(5): 608–614.

Olfson, E., *et al.* (2016) "Rare, low frequency and common coding variants in CHRNA5 and their contribution to nicotine dependence in European and African Americans", *Molecular Psychiatry* 21(5): 601–607.

Psychiatric GWAS Consortium (2010) "A framework for interpreting genome-wide association studies of psychiatric disorders", *Molecular Psychiatry* 14(1): 10–17.

Ramoz, N. and Gorwood, P. (2015) "A genetic view of addiction", *Médecine Sciences (Paris)* 31(4): 432–438.

Robison, A. J. and Nestler, E. J. (2011) "Transcriptional and epigenetic mechanisms of addiction", *Nature Reviews Neuroscience* 12(11): 623–637.

Saccone, S. F., *et al.* (2007) "Cholinergic nicotinic receptor genes implicated in a nicotine dependence association study targeting 348 candidate genes with 3713 SNPs", *Human Molecular Genetics* 16(1): 36–49.

Samochowiec, J., *et al.* (2014) "Genetics of alcohol dependence: a review of clinical studies", *Neuropsychobiology* 70(2): 77–94.

Schwantes-An, T. H., *et al.* (2016) "Association of the OPRM1 variant rs1799971 (A118G) with non-specific liability to substance dependence in a collaborative de novo meta-analysis of European-ancestry cohorts", *Behavior Genetics* 46(2): 151–169.

Sherva, R., *et al.* (2016) "Genome-Wide Association Study of cannabis dependence severity, novel risk variants, and shared genetic risks", *JAMA: Journal of the American Medical Association: Psychiatry* 73(5): 472–480.

Thorgeirsson, T. E., *et al.* (2008) "A variant associated with nicotine dependence, lung cancer and peripheral arterial disease", *Nature* 452(7187): 638–642.

Thorgeirsson, T. E., *et al.* (2010) "Sequence variants at CHRNB3-CHRNA6 and CYP2A6 affect smoking behavior", *Nature Genetics* 42(5): 448–453.

Thorgeirsson, T. E., *et al.* (2016) "A rare missense mutation in CHRNA4 associates with smoking behavior and its consequences", *Molecular Psychiatry* 21(5): 594–600.

Tobacco and Genetics Consortium (2010) "Genome-wide meta-analyses identify multiple loci associated with smoking behavior", *Nature Genetics* 42(5): 441–447.

Treutlein, J., *et al.* (2009) "Genome-wide association study of alcohol dependence", *Archives of General Psychiatry* 66: 773–784.

Tunbridge, E. M., *et al.* (2012) "The role of catechol-O-methyltransferase in reward processing and addiction", *CNS & Neurological Disorders Drug Targets* 11(3): 306–323.

Wang, J. C., *et al.* (2012) "The genetics of substance dependence", *Annual Review of Genomics and Human Genetics* 13: 241–261.

Wang, K. S., *et al.* (2011) "A meta-analysis of two genome-wide association studies identifies 3 new loci for alcohol dependence", *Journal of Psychiatric Research* 45(11): 1419–1425.

Yang, B. Z., *et al.* (2007) "Association of haplotypic variants in DRD2, ANKK1, TTC12 and NCAM1 to alcohol dependence in independent case control and family samples", *Human Molecular Genetics* 16: 2844–2853.

Zhang, H., *et al.* (2014) "Identification of methylation quantitative trait loci (mQTLs) influencing promoter DNA methylation of alcohol dependence risk genes", *Human Genetics* 133(9): 1093–1104.

23
CHOICE IMPULSIVITY
A drug-modifiable personality trait

Annabelle M. Belcher, Carl W. Lejuez,
F. Gerard Moeller, Nora D. Volkow, and Sergi Ferré

Introduction

Impulsivity is defined as a predisposition toward rapid, unplanned reactions to internal or external stimuli with little regard for the negative consequences to the individual or others (Moeller et al. 2001). This facet of behavior is highly relevant for a host of psychiatric disorders where the ability to govern and refrain from committing rapid, unplanned reactions is objectively diminished; and substance use disorders (SUD) may represent the most prominent example. Considerable research has shown that individuals with SUD have poor impulse control in a variety of self-report measures and behavioral laboratory tasks (Kjome et al. 2010; MacKillop et al. 2011; Hamilton et al. 2015a, 2015b) and that the function of the neurobiological substrates of impulse control is altered in these individuals (Volkow et al. 2011; Ersche et al. 2012). Among a host of other mitigating factors that are broadly encompassed by environment and genetics, research on impulse control increases the understanding of the conditions that precipitate the maladaptive choice to use drugs. This research has provided evidence for arguments that addiction is not completely volitional, consistent with the definition of impulsivity as an uncontrolled reaction to salient stimuli.

Strong evidence suggests an important genetic contribution to SUD vulnerability, with epidemiological studies providing heritability estimates of ~50% (Kendler et al. 2012). Yet traditional molecular genetics approaches have been largely unsuccessful in identifying specific causal roles for genes, and many researchers have turned their search for genes to the identification of endophenotypes, understood as simpler clues to genetic underpinnings than the disease syndrome itself (Gottesman and Gould 2003). We recently reviewed evidence to support the notion that with their clear neurobiological and genetic underpinnings, three personality traits provide a tractable approach to studying vulnerability to SUD (Belcher et al. 2014). These traits, as described below, are positive emotionality/extraversion (*PEM/E*), negative emotionality/neuroticism (*NEM/N*), and constraint-disinhibition (*CON*).

An added benefit of using personality traits as end-points comes from the simplicity and cost-effectiveness of personality assessment tools. Furthermore, unlike the tools utilized in other human research paradigms (e.g., MRI or PET), a personality tool is less threatening or anxiety-inducing to subjects, a fact that facilitates data collection from heterogeneous populations. Additionally, in this age of big data, the ease and efficiency with which personality data can be collected allows for large-scale collection and analysis of personality in studies with a high number of subjects. Finally, with minimal required resources, personality assessment research offers a viable research opportunity for non-research hospitals and institutions—opportunities in which clinicians and staff could engage.

In our original analysis of the personality traits that confer vulnerability or resilience to the development of SUD, we included only canonical personality traits, captured by the most researched and accepted assessments: the Big-Three and the Big-Five scales of personality (Tellegen 1982; Costa and McCrae 1992). One element of impulsivity is captured by the personality trait *CON* (see below). However, as will be discussed, impulsivity is a multi-dimensional construct, having distinct identifiable neurobiological components. We argue that one of these dimensions of impulsivity, choice impulsivity (*CI*), is particularly relevant for SUD. We further argue that, as with the other three personality traits that figure heavily in determining vulnerability or resilience to SUD, *CI* offers a tractable approach to studying SUD. We submit that for various reasons, *CI* should be elevated to the status of a personality trait.

Impulsivity, a multi-dimensional construct with defined neural correlates

Plagued by significant differences of opinion regarding its definition and measurement, impulsivity is a complex construct. It is generally uncontested, however, that there are at least two different components to impulsivity: 'action' or 'rapid-response' impulsivity (herein referred to as *AI*), and 'choice' or 'cognitive' impulsivity (*CI*; Bari and Robbins 2013; Hamilton *et al.* 2015a, 2015b). Action impulsivity can be operationally defined as a diminished ability to inhibit prepotent responses, or a failure of volitional motor inhibition or disinhibition (Moeller *et al.* 2001). Action impulsivity can be further separated into two conceptually and neurobiologically distinct impairments in inhibitory processes: action initiation impulsivity, and ongoing or prepotent action impulsivity. Both types of *AI* can be operationally measured by specific tasks (Hamilton *et al.* 2015a).

Choice impulsivity implies a tendency to accept small immediate or likely rewards at the expense of large delayed or unlikely rewards (Moeller *et al.* 2001). Excessive *CI* overlaps conceptually with impairment in decision-making and particularly with delay or temporal discounting (DD) (McClure *et al.* 2004). A term derived from behavioral economics, DD offers one of the best constructs to study 'subjective valuation' or 'preference functions', the functions that relay objective values and subjective desirability (Kable and Glimcher 2007). Delay discounting is the phenomenon by which a delayed outcome of a choice reduces the subjective value of a reward and constitutes an operational measure of the degree of *CI* (Green and Myerson 2004; Peters and Büchel 2011).

In addition to disagreements regarding its terminology and definition, impulsivity research has been further clouded by generalizations made concerning the main circuits involved in *AI* and *CI* expression. The striatum, the main input structure of the basal ganglia, receives two major inputs: glutamatergic inputs from cortical, thalamic and limbic areas (hippocampus and amygdala), and dopaminergic inputs from the ascending dopamine system. In humans and non-human primates, three main functional striatal compartments can be distinguished according to their cortical inputs: the ventral, the rostral dorsal and the caudal dorsal striata. The ventral striatum receives glutamatergic inputs from the ventromedial prefrontal cortex (vmPFC), orbitofrontal cortex (OFC) and anterior cingulate cortex (ACC); the rostral dorsal striatum comprises the head and body of the caudate nucleus and anterior putamen and receives afferent projections from frontal and parietal association cortices; and the caudal dorsal striatum comprises the tail of the caudate nucleus and posterior part of the putamen and receives afferent projections from sensorimotor cortices (Lehéricy et al. 2004; Haber and Behrens 2014).

The ventral striatum is part of decision-making brain circuits involved in reward valuation. This system determines and stores the subjective value of rewards, maximizing utilities associated with different options to calculate the highest benefit/cost ratio (Kable and Glimcher 2009). Delay discounting (DD), 'effort discounting' (ED) and 'uncertainty discounting' (UD) refer to the empirical finding that both humans and animals value immediate, low-effort and high probability rewards more than delayed, high-effort and low probability rewards. A large number of behavioral and clinical studies indicate that DD, ED and UD are independent variables, possibly involving non-overlapping corticostriatal circuits with different ventral striatal compartments and prefrontal cortical areas that connect with other specific cortical areas (Peters and Büchel 2009; Prevost et al. 2010). For DD, several studies converge to suggest involvement of a circuit that includes the vmPFC, nucleus accumbens (NAc) and the posterior cingulate cortex (pCC) (Kable and Glimcher 2007; Peters and Büchel 2011), which is strongly interconnected with the vmPFC (Leech and Sharp 2014).

The main role of the ventral striatum, classically labeled as an interface between motivation and action (Mogenson et al. 1980), can be synthesized as determining "**whether to respond**," while that of the dorsal striatum determines "**how to respond**," to reward-associated stimuli. Based on elegant experiments on gaze orienting and learning of sequential motor responses in non-human primates, Kim and Hikosaka (2015) have provided an insightful model of basal ganglia function that highlights the simultaneous and differential processing of reward-oriented behaviors and reinforcement by the rostral and caudal dorsal striata. The model also implies that all dopamine-dependent functions, reward-oriented behavior and learning of stimulus–reward and reward–response associations (Wise 2004), are simultaneously processed by rostral dorsal and caudal dorsal striatal areas (Kim and Hikosaka 2015). With respect to learning (reinforcement), rostral dorsal areas are predominantly involved in an initial, more controlled, "volitional" (contingent on the outcome), accurate, and more labile learning; while caudal dorsal areas are involved in a slower and more "automatic" (non-contingent on the outcome) and long-lasting learning (Kim and Hikosaka 2015).

A distinct circuit included in the rostral dorsal striatum is specifically involved in volitional inhibition of response tendencies, or, more succinctly, in *CON*. This circuit involves two highly interconnected cortical areas: the right inferior frontal cortex (rIFC, and possibly the adjacent insular cortex) and the pre-supplementary motor area (preSMA), that project to the dorsal rostral striatum and the subthalamic nucleus (Swann et al. 2012; Aron et al. 2014). Thus, the dorsal rostral striatum also determines "**when to respond**" to reward-associated stimuli, and forms part of a brain circuit that constitutes a neural correlate of *CON* and, therefore, *AI*.

Choice impulsivity, a personality trait

A central issue in personality research concerns the latent structure of personality traits and their links to psychological constructs and specific neural systems. The trait approach assumes that rather stable and orthogonal variables determine behavior, and that these dimensions constitute the fundamental units of personality. *Stability* implies that although certain environmental factors can influence personality (e.g., stress), personality-trait consistency is more common, and when a personality-trait change occurs, it is seldom dramatic (Roberts and Caspi 2003). *Orthogonality* implies that traits are independent non-overlapping variables, which depend on the function of different brain circuits. Trait models of personality are also referred to as psychometric theories, emphasizing the notion that personality is measurable with the use of psychometric tests.

The most popular models of personality are the Big-Three and the Big-Five models, operationalized by the MPQ (Tellegen 1982) and the revised NEO-PI-R (Costa and McCrae 1992), respectively. The MPQ includes 11 primary dimension scales measuring facets that coalesce around three orthogonal traits, *PEM*, *NEM* and *CON* (Tellegen 1982, 1985). Positive emotionality and *NEM* are explicitly temperamental in nature. They incorporate dispositions toward positive and negative emotions, respectively, and are linked conceptually to the brain systems underlying appetitive-approach and defensive-withdrawal behaviors. Constraint-disinhibition (*CON*) encompasses dimensions related to behavioral restraint, the opposite end of which implies disinhibition—ostensibly the same concept as *AI*.

The NEO-PI-R (Costa and McCrae 1992) comprises 30 facets, six for each of five trait factors: neuroticism (*N*), extraversion (*E*), openness (*O*), agreeableness (*A*) and conscientiousness (*C*). Neuroticism and extraversion highly correlate with *NEM* and *PEM* respectively, generally constituting the same personality constructs (Church 1994; Clark and Watson 1999). Openness captures interest toward experience, and agreeableness implies an empathic personality. Finally, conscientiousness is a spectrum of constructs that describes individual differences in the propensity to be self-controlled, responsible toward others, hardworking, orderly and rule abiding. Roberts *et al.* (2014) have recently provided a compelling argument that many constructs that are not typically considered as personality traits have robust research paradigms that often run parallel to the work conducted in personality research. It was then hypothesized that many of these variables should be viewed as part of the family of *C* constructs, if not seen as measuring facets of the trait. Among those constructs, *C* encapsulates the two dimensions of impulsivity, *AI* and *CI* (with high impulse control, low impulsivity, scoring high in *C*; Roberts *et al.* 2014). In fact, a significant correlation has been reported between measures of *C* and *CON* (Church 1994; Clark and Watson 1999).

The validity of trait *C* as a psychological construct is arguably beyond question, based on its predictive value in terms of social achievement, mental and physical health and longevity (Roberts *et al.* 2014). However, the inclusion of *CI* and *AI*, two orthogonal constructs with different brain circuit correlates, implies that *C* provides a poor neurobiologically linked personality construct. This stands in stark contrast with the neurobiological validity of *PEM/E*, *NEM/N* and *CON* (Belcher *et al.* 2014). In fact, in the frame of Tellegen's model of personality, *AI* constitutes a personality trait (*CON*) in its own right. If, as mentioned above, we consider a personality trait as a stable and orthogonal variable that depends on the function of a specific brain circuit, it becomes obvious that *CI* should also be considered as a personality trait, and that efforts should be made to incorporate a validated instrument of DD into trait models of personality.

An additional support to the tenet of 'promoting' *CI* to the status of a personality trait comes from its *hereditability*. Behavior genetic studies have demonstrated substantial heritability from the various traits assessed by MPQ and NEO-PI-R. For all higher-order factors of the Big-Three and Big-Five models, the broad genetic influence ranges from 40% to 60% (Tellegen *et al.* 1988; Jang *et al.* 1996); an estimate that includes *CON* (Tellegen *et al.* 1988) and therefore *AI*. Studies of the relative contribution of genetic factors to inter-individual variability on *CI* (DD) indicate that, like *AI*, *CI* possesses substantial heritability (Anokhin *et al.* 2015). Complementary to these findings in familial genetics, animal studies demonstrate significant strain differences in DD (Wilhelm and Mitchell 2008). Additionally, a handful of studies provide strong evidence for the long-term stability of the individual differences in the degree of DD (Anokhin *et al.* 2015; for review, see Peters and Büchel 2011). Although further research is warranted, these data collectively argue that, like the more well-established traits of personality, *CI* has a strong genetic basis, and is a relatively stable trait over an individual's lifetime.

Several genes have been identified as possible moderators of *CI*. The two most prominent are the gene for catechol-O-methyltransferase (COMT) and the dopamine D_4 receptor (D4R) gene, inferred from the association of specific gene polymorphisms with differences in the degree of DD. COMT is a main dopamine-degrading enzyme in the prefrontal cortex. Different gene polymorphic variants lead to differential levels of COMT activity (Lachman *et al.* 1996; Chen *et al.* 2004). Low COMT activity (also demonstrated pharmacologically with the COMT inhibitor tolcapone) leads to an increased DD (Kayser *et al.* 2012). More specifically, there is compelling evidence for a U-shaped relationship between *CI* and COMT activity, as shown from experiments on genotype x drug (tolcapone) and genotype x age interaction experiments. Depending on the genotype and age, COMT inhibition can increase or decrease DD (Smith and Boettiger 2012; Farrell *et al.* 2012).

The gene encoding the human D4R contains a large number of polymorphisms in its coding sequence (LaHoste *et al.* 1996). The most extensive polymorphism is found in exon 3, a region that codes for the third intracellular loop of the receptor. This polymorphism consists of a variable number of tandem repeats in which a 48-bp sequence exists as several repeats, the most common being 4 or 7 repeats (D4.4R or D4.7R) (Wang *et al.* 2004). In a recent study, we demonstrated a very significant role of D4R in the control of ventral cortico-striatal glutamatergic transmission, with D4.7 conferring significantly stronger modulation (Bonaventura *et al.* 2017). The D4R polymorphic variants have been associated with numerous behavioral individual differences and neuropsychiatric disorders. The most reported association is the link between the D4.7R variant and attention deficit hyperactivity disorder (ADHD) (LaHoste *et al.* 1996; Faraone *et al.* 2005; Li *et al.* 2006; Gizer *et al.* 2009), a brain disorder marked by an ongoing pattern of inattention, hyperactivity and impulsivity. Several meta-analyses have clearly implicated *AI* and *CI*, as well as their neural correlates, in ADHD (Alderson *et al.* 2007; Lipszyc and Schachar 2010; Patros *et al.* 2016). In fact, D4.7R has been directly associated with both *AI* and *CI* (Congdon *et al.* 2008; Sweitzer *et al.* 2013).

In summary, we believe that *CI* possesses all the necessary characteristics sufficient to establish it as a personality trait: orthogonality, stability and heritability—while at the same time malleable with intervention—and with established genetic ties. As a personality trait, its potential use as an endophenotype of SUD is warranted.

Choice impulsivity, an endophenotype of SUD

Despite the clear demonstration that genetic factors contribute significantly to an individual's likelihood of manifesting a mental disorder at some point in his or her lifetime, traditional molecular genetics approaches have been largely unsuccessful in pinpointing specific gene associations. Traditional methods include the search for individual candidate genes and genome-wide association studies (GWAS) (Kendler *et al.* 2012). The candidate gene approach is hypothesis driven, studying genes thought to be involved in a central nervous system disorder. The association of the D4.7R polymorphic variant with ADHD is one of the few successful examples of this approach. Unfortunately, most attempts to find a clear genetic link with a neuropsychiatric disorder show associations that are difficult to replicate (Hosák 2007; Witte and Flöel 2012). An alternative, GWAS, is a hypothesis-free method that tests for the presence of genetic associations throughout the genome to investigate whether variation in any of the approximately 20,000 human genes might contribute to disease susceptibility or other individual differences. The GWAS approach analyzes anywhere from several hundred to more than one million SNPs in thousands of individuals, and represents a powerful tool for investigating the genetic architecture of complex diseases or traits (Manolio 2013). One main problem with GWAS is the high level of multiple statistical comparisons, with stringent analyses demanding individual SNPs with extremely low *p* values to achieve genome-wide significance (Hall *et al.* 2013).

It has been argued that molecular genetics will continue to be mostly unsuccessful in the search for genes associated with neuropsychiatric disorders for two reasons: (1) any observable direct genetic effects are subject to strong environmental influences; and (2) the causal link between the gene and the disease is likely to be too long and complex to be directly observable (Gottesman and Gould 2003). The introduction of the 'endophenotype' concept has provided an invaluable approach for the identification of genes that predispose or indemnify individuals from mental and psychiatric disorders. The endophenotype concept is understood as simpler clues to genetic underpinnings than the disease syndrome itself, and involves the genetic analysis of any of a variety of biological markers (cognitive, neurophysiological, anatomical, biochemical, etc.) of the disease. The concept promotes the view that psychiatric diagnoses can be decomposed or deconstructed into more tractable intermediate phenotypes by virtue of their assumed proximity to the genetic antecedents of the disease (Gottesman and Gould 2003; Bearden and Freimer 2006).

We have argued previously that personality traits can constitute endophenotypes of mental health disorders (Belcher *et al.* 2014). Several studies have linked the structure of psychopathology to the structure of personality, as defined by the MPQ or NEO-PI-R. Using the Big-Five traits plus *CON*, Kotov and colleagues (2010) conducted a large-scale meta-analysis of the relationship between personality traits and anxiety, depression, and SUD. The group identified high *N* and low *C* as common traits, with *CON* specifically low in SUD. Additionally, a large number of studies and meta-analyses have provided evidence for a significant positive correlation of *N* and significant negative correlation of *C* with depression, anxiety and SUD (Malouff *et al.* 2007; Terracciano *et al.* 2008; Kendler and Myers 2010). The negative correlation with *C*, which encapsulates *AI* and *CI*, may imply high *AI* and *CI*. Consistent with this notion, DD is significantly increased in depression (Imhoff *et al.* 2014; Pulcu *et al.* 2014) and is one of the main behavioral characteristics of patients with SUD, who typically discount delayed rewards much more steeply than control subjects (Peters and Büchel 2011; MacKillop *et al.* 2011).

We recently reviewed current evidence that personality traits can be used as endophenotypes of SUD, allowing for a better understanding of which individual differences in specific brain circuits provide vulnerability or resilience to SUD (Belcher *et al.* 2014). In our analysis, we identified Tellegen's three personality traits, *PEM*, *NEM* and *CON* (the inverse measure of *AI*), as tied to specific brain circuits and genes (Figure 23.1). Our heuristic model was based on a continuum of three independent variables (constituted by the three main orthogonal personality traits) that interact dynamically and with the environment to determine the degree of vulnerability to the development of SUD. Individuals with low *PEM*, high *NEM* and high *AI* would be most vulnerable (least resilient), and individuals with high *PEM*, low *NEM* and low *AI* would be least vulnerable (most resilient). To our previous analysis we now add *CI* as an additional personality trait in its own right, which is modulated by a brain circuit that includes the vmPFC, NAc and the posterior cingulate cortex (see above), the function of which is moderated by the COMT and the dopamine D4R genes (see above), as well as the α_{2A} adrenoceptor (α2AR) gene (see below; Figure 23.1).

Choice impulsivity, a drug modifiable endophenotype of SUD

The unquestionable association of *CI* with SUD and the fact that DD provides a well-characterized operational definition of *CI* that can also be applied to preclinical models provides a new frame for SUD preclinical and clinical research. With their ability to provide prospective

High Negative Emotionality/Neuroticism (*NEM/N*) increases, high Action Impulsivity (*AI*) and high Choice Impulsivity (*CI*) increases, but high Positive Emotionality/Extraversion (*PEM/E*) decreases vulnerability to SUD. We can link genes to these personality traits and to the brain circuits that modulate these traits, which can therefore be used as endophenotypes of SUD. *PEM/E* is modulated by the function of the central dopaminergic system and is moderated by the D_2 receptor gene. *NEM/N* is modulated by the glutamatergic outputs from the right anterior cingulate cortex (rACC) and ventromedial prefrontal cortex (vmPFC) to the amygdala and insula and is moderated by the serotonin transporter (5-HTT) gene. *AI* is modulated by a circuit including the pre-Supplementary Motor Area (preSMA) and right Inferior Frontal Gyrus (rIFG) to the striatum and the subthalamic nucleus (STN) and is moderated by the genes of the D_4 receptor, the α_{2A} receptor and the dopamine transporter (DAT). *CI* is modulated by a circuit including the ventromedial Prefrontal Cortex (vmPFC), the posterior Cingular Cortex (pCC) and the Nucleus Accumbens (NAc) and it is also moderated by the genes of the D_4 receptor, the α_{2A} receptor and by the enzyme cathecol-*O*-methyltransferase (COMT).

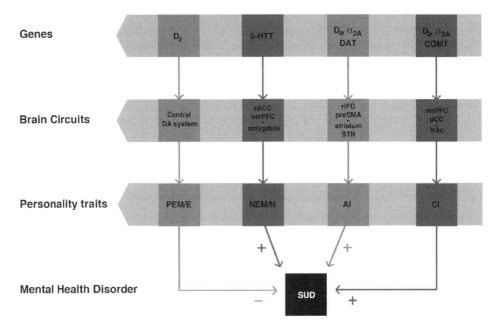

Figure 23.1 Personality traits as endophenotypes of SUD.

longitudinal studies, animal models in particular allow for interrogations regarding the premorbid brain (i.e., prior to drug exposure). The results indicate that the three elements—environment, genes and the drugs themselves—play significant and interacting roles in the manifestation of SUD. First, in agreement with the substantial heritability of *CI*, robust differences in DD across rodent strains have been reported (Isles *et al.* 2004; Wilhelm and Mitchell 2008). Second, the sensitivity of different strains of rodents to consume alcohol or self-administer psychostimulant drugs correlates with DD (Perry *et al.* 2005; Wilhelm and Mitchell 2008), effectively validating DD in the experimental animal model utilizing *CI* as a SUD endophenotype. These preclinical studies would then imply that high *CI* predates, and thus constitutes a vulnerability factor for, SUD. As part of the criteria for the endophenotype concept, proponents imply a "state independence" (presence of the endophenotype in the absence of the disorder), as well as the requirement of a higher rate of presence in non-affected family members than found in the general population (Gottesman and Gould 2003). In accord with this, two recent studies found that in healthy young adults, those with a family history of alcohol or another SUD had significantly higher DD (Acheson et al. 2011; Dougherty et al. 2014).

The next obvious question would be to address the additional factors that determine the development of SUD in the presence of the vulnerability endophenotype factor *CI*. We turn to gene-by-environment interactions, and to the concept of "plasticity genes" (Belsky et al. 2009). A long-standing literature established that the vulnerability of the genes that moderate a specific endophenotype are determined by the association with an adverse environment ("diathesis-stress" model), such as child abuse, negative life events, death of a parent and even maternal insensitivity (Rutter 2006). An alternative model proposes that these vulnerability genes actually promote more responsivity to both negative and positive environmental conditions, a "differential susceptibility" (Belsky et al. 2009). This concept provides a plausible explanation for the evolutionary advantage of those polymorphisms (by definition, common genetic variants), which potentiate positive outcomes in propitious environments, but may aggravate negative outcomes in impoverished/maladaptive environments. Thus, under specific conditions, these genes can constitute a significant vulnerability threat for the development of mental disorders. The D4.7R provides a particularly good example of a plasticity gene. Children carrying the D4.7R allele do worse in negative environments than control subjects without the "genetic risk," but they also profit most from positive environments. They were found more likely to exhibit pro-social behaviors and less likely to have non-clinical inattention as a function of maternal sensitivity and good quality of rearing environment than children without the allele (Bakermans-Kranenburg et al. 2011; Knafo 2011; Berry et al. 2013). Remarkably, a recent study by Sweitzer et al. (2013) reported that the presence of the D4.7R allele was associated with an opposite moderation of DD as a function of childhood socioeconomic status. Subjects with the allele that were raised in families with low parental education and occupational grade displayed significantly higher reward DD than like-reared counterparts without the D4.7R allele. In the absence of socioeconomic disadvantage, on the other hand, D4.7R-carrying subjects discounted future rewards less steeply (Sweitzer et al. 2013).

A corollary of the herein reviewed D4R gene-mediated moderation of the personality traits *AI* and *CI* is that it provides a clue for the apparent orthogonality of the *C* trait, as it encapsulates both traits (see above), which constitute endophenotypes of ADHD (Alderson et al. 2007; Lipszyc and Schachar 2010; Scheres et al. 2010; Patros et al. 2016) and SUD (present review and Belcher et al. 2014). Not surprisingly, as for SUD, low *C* is a consistent finding in ADHD (Gomez and Corr 2014)—a condition that, if left untreated, is a risk factor for SUD (Lee et al. 2011; Wilens et al. 2011; Van Emmerik-Van Oortmerssen et al. 2012). A situation similar to the D4R gene is emerging for the α2AR gene. Polymorphisms of this gene may confer vulnerability for developing ADHD as well as symptoms of impulse control disorders (Roman et al. 2003; Park et al. 2005). However, a large meta-analysis did not find a consistent significant association (Gizer et al. 2009). Yet when studied at the intermediate phenotype level (the endophenotype concept), a clear significant association was established between the α2AR gene polymorphisms and *AI* (Cummins et al. 2014). Additionally, the α2AR agonist guanfacine (currently used in the symptomatic treatment of ADHD) significantly decreases DD (Kim et al. 2012) in nonhuman primates, implying that *CI* can be pharmacologically manipulated.

A main conclusion from this review is that because of its personality trait and endophenotype characteristics, *CI* constitutes a main pre-existing vulnerability factor for SUD. But longitudinal animal studies have also demonstrated that chronic exposure to addictive drugs, and particularly to psychostimulants, increases DD, thereby modifying *CI* (Dandy and Gatch 2009; Setlow et al. 2009).

Although these types of prospective studies cannot be conducted in humans, a significant gene x drug interaction has been recently reported in which only cocaine users with risk α2AR gene polymorphisms demonstrated a significant increase in DD (Havranek *et al.* 2016). This represents a remarkable finding in humans that confirms the results of animal studies: that addictive drugs can significantly modify and contribute to the expression of the endophenotype *CI*.

Conclusion

We submit that as dissectible, orthogonal and measurable outputs of brain function, the study of personality offers a platform on which the study of mental health disorders can be based. As a behavioral construct with genetic conference (heritability), orthogonality and lifetime stability, *CI* presents with all of the major criteria that should qualify its elevation to the status of a personality trait. Further, with its identifiable neurobiological roots and tight genetic associations (particularly, polymorphisms of the D4R, COMT and α2AR genes), *CI* should be studied in its own right as an endophenotype of disorders of impulse control generally, and particularly as an endophenotype of SUD.

Acknowledgements

Support from the intramural funds of the National Institute on Drug Abuse (S.F.).

References

Acheson, A., Vincent, A. S., Sorocco, K. H., and Lovallo, W. R. (2011) "Greater discounting of delayed rewards in young adults with family histories of alcohol and drug use disorders: studies from the Oklahoma family health patterns project", *Alcoholism: Clinical and Experimental Research* 35(9): 1607–1613.

Alderson, R. M., Rapport, M. D., and Kofler, M. J. (2007) "Attention-deficit/hyperactivity disorder and behavioral inhibition: a meta-analytic review of the stop-signal paradigm", *Journal of Abnormal Child Psychology* 35(5): 745–758.

Anokhin, A. P., Grant, J. D., Mulligan, R. C., and Heath, A. C. (2015) "The genetics of impulsivity: evidence for the heritability of delay discounting", *Biological Psychiatry* 77(10): 887–894.

Aron, A. R., Robbins, T. W., and Poldrack, R. A. (2014) "Inhibition and the right inferior frontal cortex: one decade on", *Trends in Cognitive Sciences* 18(4): 177–185.

Bakermans-Kranenburg, M. J., van IJzendoorn, M. H., Caspers, K., and Philibert, R. (2011) "DRD4 genotype moderates the impact of parental problems on unresolved loss or trauma", *Attachment & Human Development* 13(3): 253–269.

Bari, A. and Robbins, T. W. (2013) "Inhibition and impulsivity: behavioral and neural basis of response control", *Progress in Neurobiology* 108: 44–79.

Bearden, C. E. and Freimer, N. B. (2006) "Endophenotypes for psychiatric disorders: ready for prime-time?" *Trends in Genetics* 22(6): 306–313.

Belcher, A. M., Volkow, N. D., Moeller, F. G., and Ferré, S. (2014) "Personality traits and vulnerability or resilience to substance use disorders", *Trends in Cognitive Sciences* 18(4): 211–217.

Belsky, J., Jonassaint, C., Pluess, M., Stanton, M., Brummett, B., and Williams, R. (2009) "Vulnerability genes or plasticity genes?" *Molecular Psychiatry* 14(8): 746–754.

Berry, D., Deater-Deckard, K., McCartney, K., Wang, Z., and Petrill, S. A. (2013) "Gene–environment interaction between dopamine receptor D4 7-repeat polymorphism and early maternal sensitivity predicts inattention trajectories across middle childhood", *Development and Psychopathology* 25(2): 291–306.

Bonaventura, J., Quiroz, C., Cai, N. S., Rubinstein, M., Tanda, G., and Ferré, S. (2017) "Key role of the dopamine D4 receptor in the modulation of coticostrial neurotransmission", *Science Advances* 3(1): e1601631.

Chen, J., Lipska, B. K., Halim, N., Ma, Q. D., Matsumoto, M., Melhem, S., Kolachana, B. S., Hyde, T. M., Herman, M. M., Apud, J., Egan, M. F., Kleinman, J. E., and Weinberger, D. R. (2004) "Functional analysis of genetic variation in catechol-O-methyltransferase (COMT): effects on mRNA, protein, and enzyme activity in postmortem human brain", *American Journal of Human Genetics* 75(5): 807–821.

Church, A. T. (1994) "Relating the Tellegen and five-factor models of personality structure", *Journal of Personality and Social Psychology* 67(5): 898–909.

Clark, L. A. and Watson, D. (1999) "Temperament: an organizing paradigm for trait psychology," in L. A. Pervin and O. P. John (eds), *Handbook of Personality: Theory and Research* (2nd Edition), Guilford, pp. 399–423.

Congdon, E., Lesch, K. P., and Canli, T. (2008) "Analysis of DRD4 and DAT polymorphisms and behavioral inhibition in healthy adults: implications for impulsivity", *American Journal of Medical Genetics Part B: Neuropsychiatric Genetics* 147B(1): 27–32.

Costa, P. T. and McCrae, R. R. (1992) *Revised NEO Personality Inventory (NEO-PI-R) and NEO Five-Factor Inventory Professional Manual*. Odessa, FL: Psychological Assessment Resources.

Cummins, T. D., Jacoby, O., Hawi, Z., Nandam, L. S., Byrne, M. A., Kim, B. N., Wagner, J., Chambers, C. D., and Bellgrove, M. A. (2014) "Alpha-2A adrenergic receptor gene variants are associated with increased intra-individual variability in response time", *Molecular Psychiatry* 19(9): 1031–1036.

Dandy, K. L. and Gatch, M. B. (2009) "The effects of chronic cocaine exposure on impulsivity in rats", *Behavioural Pharmacology* 20(5–6): 400–405.

Dougherty, D. M., Charles, N. E., Mathias, C. W., Ryan, S. R., Olvera, R. L., Liang, Y., and Acheson, A. (2014) "Delay discounting differentiates pre-adolescents at high and low risk for substance use disorders based on family history", *Drug and Alcohol Dependence* 143(1):105–111.

Ersche, K. D., Jones, P. S., Williams, G. B., Turton, A. J., Robbins, T. W., and Bullmore, E. T. (2012) "Abnormal brain structure implicated in stimulant drug addiction", *Science* 335(6068): 601–604.

Faraone, S. V., Perlis, R. H., Doyle, A. E., Smoller, J. W., Goralnick, J. J., Holmgren, M. A., and Sklar, P. (2005) "Molecular genetics of attention-deficit/hyperactivity disorder", *Biological Psychiatry* 57(11): 1313–1323.

Farrell, S. M., Tunbridge, E. M., Braeutigam, S., and Harrison, P. J. (2012) "COMT Val(158)Met genotype determines the direction of cognitive effects produced by catechol-O-methyltransferase inhibition", *Biological Psychiatry* 71(6): 538–544.

Gizer, I. R., Ficks, C., and Waldman, I. D. (2009) "Candidate gene studies of ADHD: a meta-analytic review", *Human Genetics* 126(1): 51–90.

Gomez, R. and Corr, P. J. (2014) "ADHD and personality: a meta-analytic review", *Clinical Psychology Review* 34(5): 376–388.

Gottesman, I. I. and Gould, T. D. (2003) "The endophenotype concept in psychiatry: etymology and strategic intentions", *American Journal of Psychiatry* 160(4): 636–645.

Green, L. and Myerson, J. (2004) "A discounting framework for choice with delayed and probabilistic rewards", *Psychological Bulletin* 130(5): 769–792.

Haber, S. N. and Behrens, T. E. (2014) "The neural network underlying incentive-based learning: implications for interpreting circuit disruptions in psychiatric disorders", *Neuron* 83(5): 1019–1039.

Hall, F. S., Drgonova, J., Jain, S., and Uhl, G. R. (2013) "Implications of genome wide association studies for addiction: are our a priori assumptions all wrong?" *Pharmacology & Therapeutics* 140(3): 267–279.

Hamilton, K. R., Littlefield, A. K., Anastasio, N. C., Cunningham, K. A., Fink, L. H., Wing, V. C., Mathias, C. W., Lane, S. D., Schütz, C. G., Swann, A. C., Lejuez, C. W., Clark, L., Moeller, F. G., and Potenza, M. N. (2015a) "Rapid-response impulsivity: definitions, measurement issues, and clinical implications", *Personality Disorders* 6(2): 168–181.

Hamilton, K. R., Mitchell, M. R., Wing, V. C., Balodis, I. M., Bickel, W. K., Fillmore, M., Lane, S. D., Lejuez, C. W., Littlefield, A. K., Luijten, M., Mathias, C. W., Mitchell, S. H., Napier, T. C., Reynolds, B., Schütz, C. G., Setlow, B., Sher, K. J., Swann, A. C., Tedford, S. E., White, M. J., Winstanley, C. A., Yi, R., Potenza, M. N., and Moeller, F. G. (2015b) "Choice impulsivity: definitions, measurement issues, and clinical implications", *Personality Disorders* 6(2): 182–198.

Havranek, M. M., Hulka, L. M., Tasiudi, E., Eisenegger, C., Vonmoos, M., Preller, K. H., Mössner, R., Baumgartner, M. R., Seifritz, E., Grünblatt, E., and Quednow, B. B. (2016) "α(2A)-Adrenergic receptor polymorphisms and mRNA expression levels are associated with delay discounting in cocaine users", *Addiction Biology* 22(2): 561–569, doi: 10.1111/adb.12324.

Hosák, L. (2007) "Role of the COMT gene Val158Met polymorphism in mental disorders: a review", *European Psychiatry* 22(5): 276–281.

Imhoff, S., Harris, M., Weiser, J., and Reynolds, B. (2014) "Delay discounting by depressed and non-depressed adolescent smokers and non-smokers", *Drug and Alcohol Dependence* 135: 152–155.

Isles, A. R., Humby, T., Walters, E., and Wilkinson, L. S. (2004) "Common genetic effects on variation in impulsivity and activity in mice", *The Journal of Neuroscience* 24(30): 6733–6740.

Jang, K. L., Livesley, W. J., and Vernon, P. A. (1996) "Heritability of the big five personality dimensions and their facets: a twin study", *Journal of Personality* 64(3): 577–591.

Kable, J. W. and Glimcher, P. W. (2007) "The neural correlates of subjective value during intertemporal choice", *Nature Neuroscience* 10(12): 1625–1633.

Kable, J. W. and Glimcher, P. W. (2009) "The neurobiology of decision: consensus and controversy", *Neuron* 63(6): 733–745.

Kayser, A. S., Allen, D. C., Navarro-Cebrian, A., Mitchell, J. M., and Fields, H. L. (2012) "Dopamine, corticostriatal connectivity, and intertemporal choice", *The Journal of Neuroscience* 32(27): 9402–9409.

Kendler, K. S. and Myers, J. (2010) "The genetic and environmental relationship between major depression and the five-factor model of personality", *Psychological Medicine* 40(5): 801–806.

Kendler, K. S., Chen, X., Dick, D., Maes, H., Gillespie, N., Neale, M. C., and Riley, B. (2012) "Recent advances in the genetic epidemiology and molecular genetics of substance use disorders", *Nature Neuroscience* 15(2): 181–189.

Kim, H. F. and Hikosaka, O. (2015) "Parallel basal ganglia circuits for voluntary and automatic behaviour to reach rewards", *Brain* 138(7): 1776–1800.

Kim, S., Bobeica, I., Gamo, N. J., Arnsten, A. F., and Lee, D. (2012) "Effects of α-2A adrenergic receptor agonist on time and risk preference in primates", *Psychopharmacology* 219(2): 363–375.

Kjome, K. L., Lane, S. D., Schmitz, J. M., Green, C., Ma, L., Prasla, I., Swann, A. C., and Moeller, F. G. (2010) "Relationship between impulsivity and decision making in cocaine dependence", *Psychiatry Research* 178(2): 299–304.

Knafo, A., Israel, S., and Ebstein, R. P. (2011) "Heritability of children's prosocial behavior and differential susceptibility to parenting by variation in the dopamine receptor D4 gene", *Development and Psychopathology* 23(1): 53–67.

Kotov, R., Gamez, W., Schmidt, F., and Watson, D. (2010) "Linking 'big' personality traits to anxiety, depressive, and substance use disorders: a meta-analysis", *Psychological Bulletin* 136(5): 768–821.

Lachman, H. M., Papolos, D. F., Saito, T., Yu, Y. M., Szumlanski, C. L., and Weinshilboum, R. M. (1996) "Human catechol-O-methyltransferase pharmacogenetics: description of a functional polymorphism and its potential application to neuropsychiatric disorders", *Pharmacogenetics* 6(3): 243–250.

LaHoste, G. J., Swanson, J. M., Wigal, S. B., Glabe, C., Wigal, T., King, N., and Kennedy, J. L. (1996) "Dopamine D4 receptor gene polymorphism is associated with attention deficit hyperactivity disorder", *Molecular Psychiatry* 1(2): 121–124.

Lee, S. S., Humphreys, K. L., Flory, K., Liu, R., and Glass, K. (2011) "Prospective association of childhood attention-deficit/hyperactivity disorder (ADHD) and substance use and abuse/dependence: a meta-analytic review", *Clinical Psychology Review* 31(3): 328–341.

Leech, R. and Sharp, D. J. (2014) "The role of the posterior cingulate cortex in cognition and disease", *Brain* 137: 12–32.

Lehéricy, S., Ducros, M., Van de Moortele, P. F., Francois, C., Thivard, L., Poupon, C., Swindale, N., Ugurbil, K., and Kim, D. S. (2004) "Diffusion tensor fiber tracking shows distinct corticostriatal circuits in humans", *Annals of Neurology* 55(4): 522–529.

Li, D., Sham, P. C., Owen, M. J., and He, L. (2006) "Meta-analysis shows significant association between dopamine system genes and attention deficit hyperactivity disorder (ADHD)", *Human Molecular Genetics* 15(14): 2276–2284.

Lipszyc, J. and Schachar, R. (2010) "Inhibitory control and psychopathology: a meta-analysis of studies using the stop signal task", *Journal of the International Neuropsychological Society* 16(6):1064–1076.

MacKillop, J., Amlung, M. T., Few, L. R., Ray, L. A., Sweet, L. H., and Munafò, M. R. (2011) "Delayed reward discounting and addictive behavior: a meta-analysis", *Psychopharmacology* 216(3): 305–321.

Malouff, J. M., Thorsteinsson, E. B., Rooke, S. E., and Schutte, N. S. (2007) "Alcohol involvement and the Five-Factor model of personality: a meta-analysis", *Journal of Drug Education* 37(3): 277–294.

Manolio, T. A. (2013) "Bringing genome-wide association findings into clinical use", *Nature Reviews Genetics* 14(8): 549–558.

McClure, S. M., Laibson, D. I., Loewenstein, G., and Cohen, J. D. (2004) "Separate neural systems value immediate and delayed monetary rewards", *Science* 306(5695): 503–507.

Moeller, F. G., Barratt, E. S., Dougherty, D. M., Schmitz, J. M., and Swann, A. C. (2001) "Psychiatric aspects of impulsivity", *American Journal of Psychiatry* 158(11): 1783–1793.

Mogenson, G. J., Jones, D. L. and Yim, C. Y. (1980) "From motivation to action: functional interface between the limbic system and the motor system", *Progress in Neurobiology* 14(2–3): 69–97.

Park, L., Nigg, J. T., Waldman, I. D., Nummy, K. A., Huang-Pollock, C., Rappley, M., and Friderici, K. H. (2005) "Association and linkage of alpha-2A adrenergic receptor gene polymorphisms with childhood ADHD", *Molecular Psychiatry* 10(6): 572–580.

Patros, C. H., Alderson, R. M., Kasper, L. J., Tarle, S. J., Lea, S. E., and Hudec, K. L. (2016) "Choice-impulsivity in children and adolescents with attention-deficit/hyperactivity disorder (ADHD): a meta-analytic review", *Clinical Psychology Review* 43: 162–174.

Perry, J. L., Larson, E. B., German, J. P., Madden, G. J., and Carroll, M. E. (2005) "Impulsivity (delay discounting) as a predictor of acquisition of IV cocaine self-administration in female rats", *Psychopharmacology* 178(2–3): 193–201.

Peters, J. and Büchel, C. (2009) "Overlapping and distinct neural systems code for subjective value during intertemporal and risky decision making", *The Journal of Neuroscience* 29(50): 15727–15734.

Peters, J. and Büchel, C. (2011) "The neural mechanisms of inter-temporal decision-making: understanding variability", *Trends in Cognitive Sciences* 15(5): 227–239.

Prévost, C., Pessiglione, M., Météreau, E., Cléry-Melin, M. L., and Dreher, J. C. (2010) "Separate valuation subsystems for delay and effort decision costs", *The Journal of Neuroscience* 30(42): 14080–14090.

Pulcu, E., Trotter, P. D., Thomas, E. J., McFarquhar, M., Juhasz, G., Sahakian, B. J., Deakin, J. F., Zahn, R., Anderson, I. M., and Elliott, R. (2014) "Temporal discounting in major depressive disorder", *Psychological Medicine* 44(9): 1825–1834.

Roberts, B. W. and Caspi, A. (2003) "The cumulative continuity model of personality development: striking a balance between continuity and change in personality traits across the life course" in R. M. Staudinger and U. Lindenberger (eds), *Understanding Human Development: Lifespan Psychology in Exchange with Other Disciplines*, Dordrecht, NL: Kluwer Academic Publishers, pp. 183–214.

Roberts, B. W., Lejuez, C., Krueger, R. F., Richards, J. M., and Hill, P. L. (2014) "What is conscientiousness and how can it be assessed?" *Developmental Psychology* 50(5): 1315–1330.

Roman, T., Schmitz, M., Polanczyk, G. V., Eizirik, M., Rohde, L. A., and Hutz, M. H. (2003) "Is the alpha-2A adrenergic receptor gene (ADRA2A) associated with attention-deficit/hyperactivity disorder?" *American Journal of Medical Genetics Part B: Neuropsychiatric Genetics* 120B(1): 116–120.

Rutter, M. (2006) *Genes and Behavior*, London, UK: Blackwell.

Scheres, A., Tontsch, C., Thoeny, A. L., and Kaczkurkin, A. (2010) "Temporal reward discounting in attention-deficit/hyperactivity disorder: the contribution of symptom domains, reward magnitude, and session length", *Biological Psychiatry* 67(7): 641–648.

Setlow, B., Mendez, I. A., Mitchell, M. R., and Simon, N. W. (2009) "Effects of chronic administration of drugs of abuse on impulsive choice (delay discounting) in animal models", *Behavioural Pharmacology* 20(5–6): 380–389.

Smith, C. T. and Boettiger, C. A. (2012) "Age modulates the effect of COMT genotype on delay discounting behavior", *Psychopharmacology (Berl)* 222(4): 609–617.

Swann, N. C., Cai, W., Conner, C. R., Pieters, T. A., Claffey, M. P., George, J. S., Aron, A. R., and Tandon, N. (2012) "Roles for the pre-supplementary motor area and the right inferior frontal gyrus in stopping action: electrophysiological responses and functional and structural connectivity", *Neuroimage* 59(3): 2860–2870.

Sweitzer, M. M., Halder, I., Flory, J. D., Craig, A. E., Gianaros, P. J., Ferrell, R. E., and Manuck, S. B. (2013) "Polymorphic variation in the dopamine D4 receptor predicts delay discounting as a function of childhood socioeconomic status: evidence for differential susceptibility", *Social Cognitive and Affective Neuroscience* 8(5): 499–508.

Tellegen, A. (1982) "Brief Manual for the Differential Personality Questionnaire", Unpublished manuscript, University of Minnesota, Minneapolis, MN.

Tellegen, A. (1985) "Structures of mood and personality and their relevance to assessing anxiety, with emphasis on self-report", in A. H. Tuma and J. D. Maser (eds), *Anxiety and the Anxiety Disorders*, Hillsdale, NJ: Erlbaum, pp. 681–706.

Tellegen, A., Lykken, D. T., Bouchard, T. J. Jr, Wilcox, K. J., Segal, N. L., and Rich, S. (1988) "Personality similarity in twins reared apart and together", *Journal of Personality and Social Psychology* 54(6): 1031–1039.

Terracciano, A., Löckenhoff, C. E., Crum, R. M., Bienvenu, O. J., and Costa, P. T., Jr (2008) "Five-Factor Model personality profiles of drug users", *BMC Psychiatry* 8(1): 22.

van Emmerik-van Oortmerssen, K., van de Glind, G., van den Brink, W., Smit, F., Crunelle, C. L., Swets, M., and Schoevers, R. A. (2012) "Prevalence of attention-deficit hyperactivity disorder in substance use disorder patients: a meta-analysis and meta-regression analysis", *Drug and Alcohol Dependence* 122(1–2): 11–19.

Volkow, N. D., Wang, G. J., Fowler, J. S., Tomasi, D., and Telang, F. (2011) "Addiction: beyond dopamine reward circuitry", *Proceedings of the National Academy of Sciences of the United States of America* 108(37): 15037–15042.

Wang, E., Ding, Y. C., Flodman, P., Kidd, J. R., Kidd, K. K., Grady, D. L., Ryder, O. A., Spence, M. A., Swanson, J. M., and Moyzis, R. K. (2004) "The genetic architecture of selection at the human dopamine receptor D4 (DRD4) gene locus", *American Journal of Human Genetics* 74(5): 931–944.

Wilens, T. E., Martelon, M., Joshi, G., Bateman, C., Fried, R., Petty, C., and Biederman, J. (2011) "Does ADHD predict substance-use disorders? A 10-year follow-up study of young adults with ADHD", *Journal of the American Academy of Child and Adolescent Psychiatry* 50(6): 543–553.

Wilhelm, C. J. and Mitchell, S. H. (2008) "Rats bred for high alcohol drinking are more sensitive to delayed and probabilistic outcomes", *Genes, Brain and Behavior* 7(7): 705–713.

Wise, R. A. (2004) "Dopamine, learning and motivation", *Nature Reviews Neuroscience* 5(6): 483–494.

Witte, A. V. and Flöel, A. (2012) "Effects of COMT polymorphisms on brain function and behavior in health and disease", *Brain Research Bulletin* 88(5): 418–428.

24
STRESS AND ADDICTION

Rajita Sinha

Introduction

Humans have long been known to use drugs and rewarding behaviors to modulate distress and emotions. Stress, trauma and negative emotional states challenge the individual to modulate and regulate these states. Perhaps you have even found yourself reaching for a cigarette, a joint, a beer or even that tub of chocolate chip ice-cream in the freezer. The central question of why some people may partake in these behaviors more than intended, and abuse and grow dependent on them, is at the heart of understanding the mechanisms that drive addiction vulnerability and resilience.

With the emergence of sophisticated brain imaging tools, along with biobehavioral controlled experimental studies and daily monitoring of stress, emotions, craving and rewarding behaviors in the real world with ecological momentary assessment (EMA) approaches, there has been a dramatic rise in research on understanding the underlying mechanisms for this association. Thus, this chapter focuses on these links between stress and addiction in humans, but also draws on supporting evidence from the broader animal literature. A definition of stress and its biological basis is presented with specific emphasis on brain and biobehavioral correlates of motivation and resilient and maladaptive coping. Epidemiological evidence linking early childhood and adult adversity with risk of addiction, and current research on the putative mechanisms underlying this association is presented. The prefrontal brain circuits play a critical role in adaptive learning and higher-order cognitive function, including controlling distress and desires/impulses, and in the association between stress and addiction risk. Binge and chronic drug use and abuse significantly affect both brain systems that support learning, motivation and reward, and also those involved in experiencing and regulating stress and emotions, and this chapter covers the effects of how premorbid and ongoing effects of repeated and chronic stress may promote addiction relapse risk. Finally, novel prevention and treatment approaches to addressing the deleterious effects of stress, and the combined effects of stress and addictive behaviors on relapse, is discussed.

Stress, emotions and adaptive behavior

Definition and conceptualization of stress

The term "stress" refers to our responses to harmful, threatening, overwhelming or challenging events. These responses start with, (1) perception of events or stimuli, along with (2) thoughts and cognitions, emotions, sensations and physiological arousal, and (3) adaptive or maladaptive behaviors and cognitions to resolve or reduce the stress and return to baseline or "homeostasis." Events may be external events like a person with a weapon, or internal events such as sleep deprivation or extreme hunger, or even chronic situations such as financial difficulties or illnesses. Such events that produce stress are referred to as stressors (Sinha 2008). For example, a conflict with a loved one may result in negative thoughts and emotions, a number of sensations of heart beating faster, muscle tightness, breathing faster, screaming, and also include motivation to leave the situation or throw something, and, finally, leaving to take a walk. Such a stressor creates stress involving the above three components along with making a choice of behavior to reduce or resolve the stress. Common types of emotional stressors include interpersonal conflict, loss of relationship, death of a close family member, or loss of a child. Highly threatening or stressful events also include emotional and physical traumas such as violent victimization and emotional, physical and sexual abuse. Common internal physiological states that are stressors include hunger or food deprivation, sleep deprivation or insomnia, psychoactive drug use, extreme hyper- or hypo-thermia and drug withdrawal states (see Table 24.1).

Table 24.1 Types of adverse life events, trauma, chronic stressors and individual-level variables predictive of addiction risk

Adverse life events	*Childhood and life trauma*	*Chronic stressors*	*Stressful internal states*
• Loss of parent; • Parental divorce and conflict; • Isolation and abandonment; • Single parent family structure; • Forced to live apart from parents; • Loss of child by death or removal; • Unfaithfulness of significant other; • Loss of home to natural disaster; • Death of significant other/close family member.	• Physical neglect; • Physical abuse by parent/caretaker/family member/spouse/significant other; • Emotional abuse and neglect; • Sexual abuse; • Rape; • Victim of gun shooting or other violent acts; • Observing violent victimization.	• Being overwhelmed; • Unable to manage life problems; • Difficulties with job, living situation; • Financial problems; • Interpersonal conflicts, loneliness; • Unfulfilled desires; • Problems with children; • Illness of loved ones; • Negative emotionality; • Poor behavioral control; • Poor emotional control.	• Hunger or food deprivation; • Food insecurity; • Extreme thirst; • Sleep deprivation, insomina; • Extreme hypo- or hyper-thermia; • Excessive drug use; • Drug withdrawal states; • Chronic illness.

Source: Updated and adapted from Sinha 2008.

Stress and emotions

Emotions refer to feelings of sadness, anger, fear, joy, disgust and surprise, which encompass the primary emotions, and also involve secondary emotions such as guilt and shame (Ekman and Davidson 1994). A key feature of emotions is that they are transient and may involve sensations and physiologic arousal (Lang and Bradley 2010) and resolve quickly with normal emotion regulation processes (Gross 2002). When they linger, they often involve multiple emotional states and overlap with stress as described above, thereby requiring adaptive or maladaptive responses or coping to resolve and reduce the distress state.

Stress, resilient coping and maladaptive behaviors

Stress can be mild/moderate in duration, or severe and extreme and these gradations in intensity contribute to the type of adaptation or coping behaviors that an individual may select to reduce the stress state. With mild/moderate stress that is limited in duration there are ample opportunities for utilizing social, biological and cognitive resources to appraise the stressors, and to engage in active coping processes to make decisions and select responses to decrease the stress in the service of goal-directed behaviors. Such experiences are often thought of as opportunities for learning and growth that reinforce resilient coping and promote mastery to support development and long-term needs. On the other hand, the more prolonged, repeated or chronic the stress, with greater uncontrollability or unpredictability of the stressors, the greater the risk of inflexible, ritualized and habitual behaviors in response to the stress. Examples of such behaviors may be avoidant behaviors like withdrawing from social networks, and binge or excessive drug use that, in turn, also have negative effects on brain and body systems involved in adaptive function. Thus, the dimensions of intensity, controllability, predictability, and flexibility are important in understanding the role of stress in increasing risk of maladaptive behaviors such as addiction.

Neural and physiological pathways involved in the stress response

On the basis of recent biobehavioral evidence, the acute stress response is manifested via three brain networks (Sinha *et al.* 2016). The first is the stress reaction network involving brain circuits that signal, monitor and process an acute distressing situation. This involves the brain's emotion (limbic) and motivational (striatal) regions such as the brainstem, hypothalamus, midbrain, amygdala, hippocampus, insula, thalamus, medial temporal cortical regions and the ventral and dorsal striatum. These limbic and striatal networks are activated to identify and learn about emotional stimuli and mobilize the cognitive and behavioral response based on their aversive and rewarding properties and contribute to the emotional experience of stress. Other cortical regions such as the anterior cingulate and dorsolateral prefrontal cortex (DLPFC) and precentral and inferior parietal regions further support the cognitive aspects of experiencing and understanding the stress state. The second network appears to be a dynamic adaptive process to reduce the experience of stress via avoidance, suppression and other neurobehavioral strategies, and involves brain areas such as the insula, hippocampus and ventrolateral prefrontal cortex. Finally, the third brain network is a flexible, dynamic set of regions involving the ventral striatum and ventromedial prefrontal cortex (VmPFC) and related circuits. This network underlies physiologic and active coping-related cognitive processes that supports goal-directed response selection and decision making, representative of resilient coping (Sinha *et al.* 2016). There are likely sex and developmental differences in the manifestation of these components and a disruption of any of these components and processes may well contribute to sustained and chronic stress states.

The above brain stress responses coordinate our well-known "fight" or "flight" response, which produce stress sensations such as the heart beating faster, breathing faster, mobilizing energy stores and inflammatory responses, via the hypothalamic-pituitary-adrenal (HPA) axis that signals to produce cortisol, and the autonomic nervous system that mobilizes heart rate and blood pressure arousal, and also immune responses. These physiological arousal pathways signal the body and the brain to coordinate and modulate the behavioral, cognitive and learning aspects of the stress response. The HPA axis is stimulated by the corticotropin releasing factor (CRF), released from the paraventricular nucleus (PVN) of the hypothalamus, to stimulate the adrenocorticotrophin hormone from the anterior pituitary that initiates the secretion of cortisol/corticosterone from the adrenal glands. The autonomic nervous system includes the sympathetic nervous system that mobilizes the cardiovascular and immune arousal responses, and the parasympathetic nervous system that is involved in regulating the sympathetic arousal by providing the 'brakes' to the sympathetic arousal and in regaining homeostasis via the sympathoadrenal medulary (SAM) pathways. Recent evidence also indicates that the sympathetic pathways provide further modulation of the adrenal glands for release of cortisol as well as release of norepinephrine and epinephrine. These are the core stress pathways involved in stress arousal and mobilization of the body and brain to respond to stress and in regulation of stress so as to regain homeostasis.

In addition, a number of neurochemical pathways have extensive influence in extrahypothalamic brain regions across the corticostriatal-limbic pathways and play a critical role in modulating the subjective and behavioral stress responses. For example, CRF, opioids gamma-aminobutyric acid (GABA) and glutamate are involved in excitatory and inhibitory functions to modulate the stress circuits identified above. Also, central catecholamines, particularly noradrenaline and dopamine, are involved in modulating brain motivational pathways (including the ventral tegmental area-VTA, ventral striatum/nucleus accumbens (NAc), and the ventromedial prefrontal (VmPFC) including orbitofrontal regions) that are important in regulating distress, exerting cognitive and behavioral control, and influencing choice, response selection and decision making. Furthermore, recent evidence also shows a role for feeding hormones in modulating brain circuits involved in energy regulation that may further contribute to motivation and behavioral responses to stress. Thus, hypothalamic and extrahypothalamic circuits involving multiple affective and energy regulating functions interact with brain motivational pathways to critically affect adaptive and homeostatic processes.

Neurobiological responses to high uncontrollable stress

Recent evidence from human brain imaging research shows that recent life stressors, trauma and chronic stress are associated with lower gray matter volume in medial prefrontal, hippocampus and insula regions of the brain. These are key regions contributing to each of the components of the stress response described above. Similarly, exposure to recent life stress and acute stress may decrease responses in the prefrontal regions such as the DLPFC and VmPFC associated with working memory, reward processing and resilient coping. Thus, with increasing levels of stress, there is a decrease in prefrontal functioning and increased limbic-striatal level responding, a brain pattern associated with low behavioral and cognitive control. Low behavioral and cognitive control linked to the prefrontal and insular cortex and high responding in limbic-emotional and striatal-motivation brain regions under stress provides one specific pattern for promoting addictive behavioral patterns where there is a decreased ability to control rewarding behaviors. Thus, motivational brain pathways are key targets of stress chemicals, which points to an important potential mechanism by which stress affects addiction vulnerability.

Stress and addiction risk

Many of the major theories of addiction also identify an important role of stress in addiction. These range from psychological models that view drug use and abuse as a coping strategy to deal with stress, reduce tension, self-medicate, or decrease withdrawal-related distress, to neurobiological models that propose incentive sensitization and stress allostasis concepts to explain how neuroadaptations in reward, learning and stress pathways may enhance the key features of addiction, namely, craving, loss of control and compulsion. This section reviews the converging lines of evidence that point to the critical role that stress plays in increasing addiction vulnerability.

Chronic stress and adversity and increased vulnerability to drug use

Considerable evidence from population-based and clinical studies indicate a positive association between psychosocial adversity, negative affect and chronic distress and addiction vulnerability. It is important to note that there are some negative studies that do not support the notion that stress increases drug use, and such studies highlight the need to consider a variety of measurement issues such as, measurement of stress (e.g., acute versus chronic; subjective or perceived stress versus repeated adverse life events, mood and stress induction versus a subjective rating of stress) and measurement of addictive behaviors (self-report, objective tests; assessment of one substance such as alcohol or multiple substances including nicotine), and measurement over a period of days (single assessment versus repeated real-life daily assessment, retrospective reporting versus prospective reports). Despite these caveats, the majority of studies show positive support for the association between stress and addictive behaviors.

The evidence in this area can be categorized into four broad types: (1) Prospective studies demonstrating that adolescents facing high recent negative life events show increased levels of drug use and abuse. (2) Studies linking trauma and maltreatment with addiction, and associated negative affect, chronic distress and risk of substance abuse. Furthermore, stress and trauma-related psychiatric illnesses such as mood and anxiety disorders, including post-traumatic stress disorder (PTSD), attention deficit hyperactivity disorder (ADHD) and behavioral conduct problems, are all associated with increased risk of substance use disorders. (3) Research on lifetime exposure to stressors and the impact of cumulative adversity on addiction vulnerability after accounting for a number of control factors such as race/ethnicity, gender, socioeconomic status, prior drug abuse, prevalence of psychiatric disorders, family history of substance use, behavioral and conduct problems. A high number of cumulative stressful life events is significantly predictive of alcohol and drug dependence in a dose-dependent manner, even after accounting for control factors. Furthermore, the dose-dependent effects of cumulative stressors on risk for addiction existed for both genders and for Caucasian, African-American and Hispanic race/ethnic groups (Sinha 2008). (4) Human laboratory experimental research and studies using ecological momentary assessment (EMA) techniques to assess ongoing drug use in daily life. Laboratory studies provoke emotional, cognitive or social stress in the context of an experiment followed by assessment of drug motivation and opportunities for drug self-administration. Such studies examining effects of stress exposure on drug use are limited to legal drugs such as alcohol and nicotine, due to ethical reasons of providing opportunities for illicit drug use to non-addicted or non-exposed samples. Findings from these studies generally support the notion that acute exposure to stress or negative mood or high ratings of negative mood and stress are associated with alcohol and nicotine use, but the effects of drinking history, history of adversity, social stress and expectancies are known to play a role in these studies. The types of adverse life events, chronic stressors, traumatic events and negative affective states that have been associated with risk of addiction are presented in Table 24.1.

Stress effects on motivation and adaptive circuits

There is growing evidence that high uncontrollable stress alters activity of the prefrontal and other cortical brain regions that coordinate higher cognitive or executive control functions (Sinha 2008; Arnsten *et al.* 2012). These functions include regulating distress and emotions, such as controlling and inhibiting impulses, refocusing and shifting attention, working memory, monitoring conflict and behavior, linking behaviors to possible future consequences, and flexible consideration of alternatives for response selection and decision making. Psychosocial and behavioral scientists have elegantly shown that under high levels of uncontrollable emotional and physiological stress or negative affect there is a decrease in working memory function, poor attention and flexibility, lower behavioral control and increases in impulsive and/or habitual responding, that, in turn, may compromise resilient coping. Chronic and repeated stress and early-life trauma are each associated with more maladaptive coping and poor executive control function.

The stress-related decrements in cognitive and behavioral functioning parallel brain-imaging studies showing a link between increasing numbers of stressful or traumatic life events and lower brain gray matter volume in orbitofrontal and ventromedial, dorsolateral prefrontal and rostral anterior cingulate cortices as well as insula and striatal regions. Early life stress and trauma and child maltreatment are associated with specific alterations in stress reactivity and signaling regions such as the amygdala, hypothalamus and hippocampus regions. This research is consistent with a number of basic science studies showing that early life stress and prolonged and repeated stress adversely affects development of the prefrontal cortex (PFC). Repeated and chronic stress-related changes in prefrontal cortical neuron spines, density, and retraction of synapses have been documented in animal studies (McEwen and Morrison 2013). Jointly such research indicates that the PFC and related circuits are highly dependent on environmental experiences for maturation and adaptive functioning and is a critical target of stress and adversity.

Other neurobiological evidence points to stress-impairing catecholamine modulation of prefrontal circuits, which, in turn, impairs executive functions like working memory and self-control. Such research is consistent with the growing evidence that adolescents with high exposure to life stress (as those listed in Table 24.1) and at-risk for substance abuse are more likely to show decreased emotional and behavioral control, and decreased self-control is associated with risk of substance abuse and other maladaptive behaviors. Adolescents at-risk for substance abuse are known to have decreased executive functioning, low behavioral and emotional control, poor decision making and greater levels of deviant behavior and impulsivity. The corticostriatal-limbic dopamine pathways have been associated with impulsivity and decision making and addiction risk, and, as discussed in previous sections, specific regions of this pathway, such as the VTA, NAc, PFC and the amygdala, are highly susceptible to stress-related signaling and plasticity associated with early life stress, repeated and chronic stress experiences.

Stress mechanisms affecting reward increase addiction risk

Both early life stress and chronic stress are also known to significantly affect the mesolimbic dopamine pathways and play a role in drug self-administration. Repeated and prolonged exposure to maternal separation in neonatal rats significantly alters the development of central CRF pathways. These animals as adults show exaggerated HPA and behavioral responses to stress. Such physiological and behavioral changes are associated with altered CRF mRNA expression in the PVN, increased CRF-like immunoreactivity in the locus coeruleus (LC) and increased CRF receptor levels in the LC and raphe nuclei. The adult animals also show decreased negative feedback sensitivity to glucocorticoids (GC) and these changes are accompanied by decreased GC receptor expression in the hippocampus and frontal cortex. Decreased GABA receptor levels in noradrenergic cell body regions in the LC and decreased central benzodiazepine (CBZ) receptor levels in the LC and the amygdala have also been reported. More importantly, rats exposed to maternal separation in childhood show significantly elevated DA responses to acute stress along with increased stress-induced behavioral sensitization and robust behavioral sensitization to psychostimulant administration, suggesting heightened behavioral responses to psychostimulants. Such cross-sensitization of stress and drugs of abuse is associated with enhanced release of DA in the NAc, lower NAc-core and striatal DA transporter sites, and reduced D3 receptor binding sites and mRNA levels in the NAc shell. In addition, changes in brain norepinephrine is also known to alter DA signaling and increase behavioral and psychostimulant sensitization. However, sex differences in sensitization patterns have been reported, suggesting that specific sex-specific alterations as they pertain to risk of addiction needs further attention.

The PFC, and particularly the right PFC, plays an important role both in activating the HPA axis and autonomic responses to stress and in regulating these responses. For example, lesions of the ventromedial PFC results in enhanced HPA and autonomic responses to stress. High levels of glucocorticoid receptors are also found in the PFC and chronic GC treatment results in a dramatic dendritic reorganization of PFC and hippocampal neurons. Human studies on the neurobiological effects of child maltreatment document neuroendocrine changes as well as alterations in size and volume of prefrontal, thalamic and cerebellar regions associated with maltreatment and with initiation of addiction. Similarly, adult human structural neuroimaging studies show stress-related decreases in gray matter volume in the medial PFC and reduced gray matter volume in prefrontal and anterior cingulate regions in samples of addicted populations. Together, the data presented in this section highlight the significance of stress effects on mesolimbic and prefrontal regions that are important in stress regulation, behavioral control and decision making.

Interestingly, biological stress responses with increases in the stress hormone cortisol (corticosterone in animals) can significantly affect dopamine signaling in the reward and motivation pathways. For example, stress exposure and increased corticosterone enhances dopamine release in the nucleus accumbens (NAc). Suppression of corticosterone by surgical removal of the adrenal glands (adrenalectomy) reduces extracellular levels of dopamine under basal conditions and in response to stress and psychostimulants. However, chronic corticosterone inhibits DA synthesis and turnover in the NAc. There is also evidence that, like drugs of abuse, stress and concomitant increases in CRF and corticosterone enhances the activity of the excitatory neurotransmitter glutamate in the VTA, which, in turn, enhances activity of dopaminergic neurons. Human brain imaging studies have further shown that stress-related increases in cortisol are associated with increased dopamine transmission in the ventral striatum, and some evidence also reveals that amphetamine-induced increases in cortisol are associated with both dopamine binding in the ventral striatum and with ratings of amphetamine-induced euphoria. Furthermore, both stress and drugs of abuse activate the mesolimbic pathways, and each result in synaptic adaptations in VTA dopamine neurons and in morphological changes in the medial prefrontal cortex.

In addition to a role in reward, a growing body of human imaging studies and preclinical data indicate that the ventral striatum/nucleus accumbens (VS/NAc) and the VTA are also involved in aversive conditioning, experience of aversive pain stimuli and in anticipation of aversive stimuli. Such evidence points to a role for the mesolimbic dopamine pathways beyond reward processing and includes its role in habit formation, motivation and attention to behavioral response during salient (aversive or appetitive) events. Furthermore, additional stress response regions, intricately connected to the mesolimbic DA pathways and involved in reward, learning and adaptive goal-directed behaviors, are the amygdala, hippocampus, insula and related corticolimbic regions. These regions along with the mesolimbic DA pathways play an important role in interoception, emotions and stress processing, impulse control and decision making, and in the addictive properties of drugs of abuse.

Drug use and abuse alters stress and reward pathways

Acute administration of the most commonly abused drugs such as alcohol, nicotine, cocaine, amphetamines and marijuana not only activate brain reward pathways (mesocorticolimbic dopaminergic systems) but also potently affect brain stress pathways (CRF-HPA axis and the autonomic nervous system pathways) with increases in plasma ACTH and corticosterone, changes in heart rate and blood pressure and skin conductance responses. Regular and chronic use of these drugs is also associated with adaptations in these stress systems that are specific by drug. For example, changes in heart rate and heart rate variability (HRV) are reported with regular and chronic alcohol use. Sustained increases in HPA axis function in the case of psychostimulants and tolerance to the inactivating effects of the drug in the case of morphine, nicotine and alcohol have also been demonstrated. These direct effects of drugs of abuse on major components of the neural and physiological stress responses raise the question of whether these stress-related adaptations may drive compulsive motivation and craving and addiction risk.

Drug abuse alters cortico-striatal reward/motivation pathways

Although acute administration of drugs increase mesolimbic dopamine, regular binge and chronic use of abusive drugs and acute withdrawal states decrease activity of the mesolimbic dopamine pathways with decreases in basal and stimulated dopamine reported in several preclinical and clinical studies. Basic science research shows chronic use of drugs dramatically alters central noradrenergic pathways in the ventral and dorsal striatum, other areas of the forebrain and the ventromedial prefrontal cortex. Human brain imaging studies corroborate these preclinical data, with reduced D2 receptors and dopamine transmission in the frontal and ventral striatum regions reported in alcohol dependent individuals and cocaine dependent individuals during acute withdrawal and protracted withdrawal (up to three–four months). Furthermore, blunted dopamine release in the ventral striatum and anterior caudate was also associated with the choice to self-administer drugs such as cocaine and alcohol over money in human addicts. These changes are similar to the effects of prolonged and repeated stressors on mesolimbic dopamine and norepinephrine deficiency noted in the previous section, and suggest that chronic drug effects on extrahypothalamic CRF, noradrenergic or glucocorticoid systems may directly impact the cortico-striatal limbic dopamine pathways.

On the other hand, regular and chronic exposure to drugs results in 'sensitization' or enhanced behavioral and neurochemical response to drugs and to stress. Synaptic alterations in the VTA, NAc and medial PFC modulated by glutamate effects on dopamine neurons, and CRF and noradrenergic effects on DA and non-DA pathways contribute to behavioral sensitization of stress and drugs of abuse. In addition, increased levels of brain-derived neurotrophic factor (BDNF) in the mesolimbic dopamine regions has been associated with increases in drug seeking during abstinence from chronic drug use. Furthermore, behavioral sensitization observed with drugs of abuse and with stress are associated with synaptic changes in mesolimbic dopamine regions, particularly the VTA, NAc and the amygdala, and such changes contribute to compulsive drug seeking. Thus, there are significant physiological, neurochemical and behavioral alterations in stress and dopaminergic pathways associated with chronic drug use, which, in turn, may affect craving and compulsive seeking, maintenance of drug use and relapse risk. It is not entirely clear how long these changes persist, and the extent to which there is recovery or normalization of these pathways and responses.

Chronic drug effects on stress responses, drug craving and relapse risk

Clinical symptoms of irritability, anxiety, emotional distress, sleep problems, dysphoria, aggressive behaviors and drug craving are common during withdrawal and early abstinence from alcohol, cocaine, opiates, nicotine and marijuana. Acute withdrawal states are associated with increases in CRF levels in CSF, plasma ACTH, cortisol, norepinephrine (NE) and epinephrine (EPI) levels. Early abstinence is associated with high basal ACTH and cortisol responses, and a blunted or suppressed ACTH and cortisol response to pharmacological and psychological challenges in addicted individuals. Active use states on the other hand may show greater reactivity to pharmacologic stress challenges in some types of addiction. These disruptions highlight the significant effects of drug use and abuse on physiological stress responses and suggest their possible effects on drug motivation and intake.

Drug craving or 'wanting' for drug is conceptually different from other anxiety and negative affect symptoms as it comes from 'desire' or a wish for rewarding or hedonic stimuli. However, with chronic drug use the terms craving and wanting often become associated with a physiological need, hunger and strong intent to seek out the desired object, thereby representative of the more compulsive aspects of craving and drug seeking identified by addicted patients, but also seen in non-dependent binge and heavy social drug users. In particular, craving and compulsive seeking is strongly manifested in the context of stress exposure, drug-related cues and drug itself and can become a potent trigger for relapse. Heightened craving or wanting of drug is associated with more severe drug use and is thought to represent the behavioral aspects of molecular and cellular changes in stress and dopamine pathways discussed in the previous section. Indeed, some support for this idea comes from laboratory and imaging studies summarized below.

The shift from normal healthy desire to drug craving with increased levels of drug use is also associated with changes in limbic, striatal and cortical brain systems. For example, changes in hypothalamic pituitary adrenal (HPA) axis responses, altered and blunted amygdala response to fear/threat-potentiated startle in heavy drinkers compared with light social drinkers, and autonomic imbalances in sympathetic/parasympathetic systems have been reported with increased drug use. A number of studies have assessed neural changes associated with the drug-craving state with correlations to subjective drug craving and to drug use/relapse. Brief exposure to cocaine cues, known to increase drug craving in cocaine-dependent (CD) individuals, increased activity in the amygdala and regions of the frontal cortex, and with gender differences reported in amygdala activity and the frontal cortex response in cocaine-dependent individuals. Cue-induced craving for nicotine, methamphetamine and opiates also activate regions of the prefrontal cortex, amygdala, hippocampus, insula and the VTA.

As stress also increases drug craving in addicted individuals relative to controls, brain activation during stress and neutral states using functional magnetic resonance imaging (fMRI) studies has been assessed in healthy controls and abstinent addicted individuals. For example, treatment-engaged abstinent cocaine-dependent (CD) individuals and controls both showed similar levels of distress and pulse changes during stress exposure, but brain response to emotional stress in paralimbic regions such as the anterior cingulate cortex (ACC), hippocampus and parahippocampal regions was greater in healthy controls during stress while CD patients showed a striking absence of such activation. In contrast, patients had increased activity in the caudate and dorsal striatum region during stress, activation that was significantly associated with stress-induced cocaine craving ratings. A larger follow-up study assessed brain response to stress, drug cues and neutral-relaxing cues using individualized guided imagery in CD versus healthy social drinking men and women. Findings indicated that CD versus healthy women showed greater activation in limbic-striatal regions such as the amygdala, caudate-putamen, insula and ACC during stress, while CD men relative to male controls showed significantly greater responsiveness in these regions to drug cues and in the neutral relaxing condition. These data are consistent with other evidence of increased reactivity to stress and drug cues during drug-craving states in patients relative to controls, and also highlight the importance of examining sex differences in drug-craving-related neuroadaptations.

As findings from multiple studies has shown that stress-induced and drug-cue-induced craving is significantly greater in addicted individuals relative to controls, we also assessed brain correlates of stress- and cue-induced alcohol craving in abstinent, treatment-engaged alcohol-dependent (AD) individuals. Results indicate a robust hyperactivity during the neutral relaxed state in the ventral striatum and the ventromedial PFC (VmPFC)/ACC, which correlated with provoked stress-induced and cue-induced drug craving. Stress- and drug-cue-induced craving was also associated with blunted responses in these regions in the stress and drug-cue conditions, and both hyperactivation of the VmpFC in the neutral relaxed state and hypoactivation of the VmPFC and the insula during stress was predictive of future time to alcohol relapse and severity of alcohol relapse during the subsequent recovery period. These findings identify neuroadaptations in the VmPFC, ventral striatum and insula networks that show disrupted functioning in the relaxed state, that, in turn, contributes to hypoactive responses during provoked/challenge conditions in addicted individuals.

Hypofrontal activation in the ACC and VmPFC during drug-cue and stress-induced craving states as reported above is consistent with a growing literature from neuropsychological and imaging studies examining prefrontal executive functions, including impulse control, decision making and set shifting, which has shown executive function deficits and hypo-frontal responses in addicted individuals compared with control volunteers. Interestingly, evidence from research on stress-related shifts in goal-direct motivation to habit-based responding also points to blunted responses in the medial PFC region. Together, the neuroadaptations associated with addiction as well as the neural responses under conditions of stress suggest that disruption of VmPFC and ACC function with blunted responses in these regions during stress, drug-cue and other challenge states may mediate increased drug craving and loss of behavioral control over drug craving. Whether it is blunted VmPFC/ACC responses due to stress/chronic drug exposure or stress-motivated habit responses or both that are required for increased risk of relapse are not fully understood. Certainly the human data from the alcohol-relapse findings described above point to disrupted top-down control of cravings that increase relapse risk.

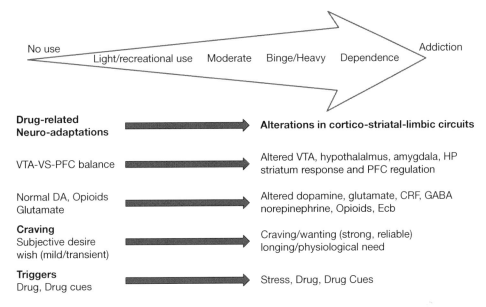

Figure 24.1 A schematic diagram representing drug craving and motivation on a dimensional continuum with increasing levels of drug exposure and history is presented. Drug-related neuroadaptations in cortico-striatal limbic networks and neurochemical adaptations that may be examined for their contribution to the drug-craving state are highlighted. VTA = ventral tegmental area; HP = hippocampus; CRF = corticotrophin releasing factor; VS = ventral striatum; DA = dopamine; PFC = prefrontal cortex; GABA = Gamma-aminobutyric acid; Ecb = endocannabinoids.

Source: Reprinted from Sinha 2013, *Current Opinion in Neurobiology*.

A schematic representing drug motivation and craving on a dimensional continuum that is based on level of exposure and drug history and associated neuroadaptations in stress and reward circuitry and in neurochemistry is presented in Figure 24.1 (adapted from Sinha 2013).

Conclusions and future directions

This chapter focuses on the accumulating evidence from preclinical, clinical and population studies that highly stressful situations and chronic stress increases addiction vulnerability, facilitating continued drug use and abuse and ultimately chronicity of addiction severity and relapse risk. The types of stressors that increase addiction risk are identified in Table 24.1. These tend to be highly emotionally distressing events that are uncontrollable and unpredictable. The themes range from loss, violence and aggression, poor support, interpersonal conflict, and isolation and trauma. There is also evidence for a dose-dependent relationship between accumulated adversity and addiction risk, with the greater the number of stressors an individual is exposed to, the higher the risk of developing substance use disorders. Work-related stressors have weaker support but individual level variables such as trait negative emotionality and poor self-control (possibly similar to poor executive function) appear to also contribute uniquely to addiction risk. Exposure to such stressors early in life and accumulation of stress (chronicity) results in neuroendocrine, physiological, behavioral and subjective changes that tend to be long lasting and adversely affect development of brain systems involved in learning and motivation and stress-related adaptive behaviors. Research that directly addresses stress-related neurobiological changes and its association with behavioral outcomes is sorely needed. Evidence to clarify the contribution of stress on alterations in mesolimbic dopamine activity and its association to drug use is also needed.

A review of evidence indicating the effects of drug use and abuse on stress responses and dopamine transmission is presented. Altered stress/emotional and motivational responses that are associated with craving and relapse to drug use is reviewed. While substance abuse results in changes in stress and dopaminergic pathways involved in motivation, self-control and adaptive processes necessary for survival, whether such changes enhance drug seeking or craving and drug-use behaviors is lacking. For example, studies on whether prior exposure to licit and illicit drugs modifies the association between stress and drug self-administration are rare. While there are specific neuroadaptations in reward and associated regions, it is also important to examine which of these changes are involved in increasing drug intake and supportive of addictive processes such as progressive loss of control, persistence of craving and escalating drug self-administration. As stress also increases risk of mood and anxiety disorders that are highly co-morbid with addiction, it is important to examine whether there are specific stress-related factors that contribute to risk for co-morbid mood and anxiety disorders and addiction risk. Exploration of gene–environment interactions could be particularly helpful in answering such questions.

Drug motivation and craving is also presented as a dimensional construct that grows with increasing levels of drug use. With increased salience of drug, there is greater motivation or 'wanting' of drug as posited by the incentive sensitization model of addiction. Evidence from human experimental and neuroimaging studies also shows increases in stress-induced wanting along with higher drug-cue-induced craving in the addicted state, which is accompanied by increased levels of stress-related arousal, anxiety and an expanded network of corticolimbic-striatal activation under stress, drug, drug cue and relaxed exposure conditions. Interestingly, several studies show disruption of medial prefrontal activity during both craving or cognitive challenge that is associated with drug use/relapse in addicted individuals. As the medial prefrontal region has been identified as an important region in self-control, it appears that compromised self-control may be a key aspect of the compulsive motivation for drug intake and addiction risk and relapse.

Despite the growing evidence for stress and reward adaptations that strengthen drug motivation, there is a need to identify the behavioral, neurobiological and physiological adaptations that occur with increased levels of drug use even in the absence of addictive disorders, and, specifically, the key components that predict development of craving and enhanced motivation for drug. Furthermore, identification of the neuro-biological components specific to the drug-craving state may result in development of valid craving-related neural markers that may be targeted for future prevention and treatment efforts. For example, as data are accruing on disrupted prefrontal control over urges as a key component of the drug-craving state, targeting such compromised prefrontal function as a biomarker for normalization with agents that rescue the prefrontal cortical neurons may represent a useful strategy in treatment development. Alternatively, strategies that reduce activity in the limbic-striatal network in response to drug, cues or stress may also be useful in reducing drug craving, drug use and relapse. Studies on increasing understanding of the neurobiology of aversive states that promote sensitization and approach behaviors as well as those on the neurobiology of inhibitory control processes and compulsive habits may contribute to improving strategies for reducing drug craving and relapse prevention in addiction. Finally, it is not known if the human brain recovers from addiction-related neuroadaptations and can return to healthy levels of desire. It may well be that a realistic goal is not to decrease limbic-striatal reactivity and related alterations associated with craving, but to improve self-control and regulation of drug craving, which, with long-term recovery, may diminish craving circuits that drive habit, automaticity and relapse risk.

Abbreviations

HPA Axis: hypothalamic-pituitary-adrenal; SAM: Sympathoadrenal medulary; VTA: Ventral Tegmental Area; NAc: Nucleus Accumbens; mPFC: Medial Prefrontal Cortex; CRF: Corticotrophin Releasing Factor.

Acknowledgements

Preparation of this review was supported by grants from the National Institutes of Health, R01-AA13892, R01-DA18219, P50-DA016556, R01-020504, U01-DE019589, PL1-DA024859.

References

Adinoff, B. and Stein, E. A. (2011) *Neuroimaging in Addiction*, 1st Edition, West Sussex, UK: Wiley-Blackwell.

Arnsten, A. F. T. (2009) "Stress signaling pathways that impair prefrontal cortex structure and function", *Nature Reviews Neuroscience* 10(6): 410–422.

Arnsten, A., Mazure, C. M. and Sinha, R. (2012) "This is your brain in meltdown", *Scientific American* 306(4): 48–53.

Ekman, P. and Davidson, R. J. (1994) *The Nature of Emotion*, New York, NY: Oxford University Press.

Epstein, D. H., Willner-Reid. J., Vahabzadeh, M., Mezghanni, M., Lin, J. L. and Preston, K. L. (2009) "Real-time electronic diary reports of cue exposure and mood in the hours before cocaine and heroin craving and use", *Archives of General Psychiatry* 66(1): 88–94.

Goldstein, R. Z. and Volkow, N. D. (2011) "Dysfunction of the prefrontal cortex in addiction: neuroimaging findings and clinical implications", *Nature Reviews Neuroscience* 12(11): 652–669.

Gross, J. J. (2002) "Emotion regulation: affective, cognitive and social consequences", *Psychophysiology* 39(3): 281–291.

Koob, G. F. and Volkow, N. D. (2016) "Neurobiology of addiction: a neurocircuitry analysis", *The Lancet Psychiatry* 3(8): 760–773.

Lang, P. J. and Bradley, M. M. (2010) "Emotion and the motivational brain", *Biological Psychology* 84(3): 437–450.

McEwen, B. S. (2007) "Physiology and neurobiology of stress and adaptation: central role of the brain", *Physiological Reviews* 87(3): 873–904.

McEwen, B. S. and Morrison, J. H. (2013) "The brain on stress: vulnerability and plasticity of the prefrontal cortex over the life course", *Neuron* 79(1): 16–29.

Schwabe, L. and Wolf, O. T. (2009) "Stress prompts habit behavior in humans", *The Journal of Neuroscience* 29(22): 7191–7198.

Sinha, R. (2008) "Chronic stress, drug use and vulnerability to addiction", *Annals of the New York Academy of Sciences: Addiction Reviews* 1141(1):105–130.

Sinha, R. (2013) "The clinical neurobiology of drug craving", *Current Opinion in Neurobiology* 23(4): 649–654.

Sinha, R. and Jastreboff, A. M. (2013) "Stress is a common risk factor in obesity and in addiction", *Biological Psychiatry* 73(9): 827–835.

Sinha, R., Lacadie, C., Constable, T. C. and Seo, D. (2016) "Dynamic neural activity during stress signals resilient coping", *Proceedings of the National Academy of Sciences* 113(31): 8837–8842.

SECTION C

Psychological and neural mechanisms

25
MECHANISTIC MODELS FOR UNDERSTANDING ADDICTION AS A BEHAVIOURAL DISORDER

Dominic Murphy and Gemma Lucy Smart

The very understanding of addiction as a disease is controversial, as is the best way to explain it. Psychiatry is dominated by a version of the medical model that sees psychopathologies as diseases whose observable symptoms are causally explained by abnormalities in underlying neurobiological systems. Although addictions seem similar in terms of symptoms, they can vary widely in their consequences and pathology. Many conceptions of addiction see the symptoms as primarily behavioural or psychological, defining it in terms of phenomena like craving (Elster 1999) or in economic terms (Ross et al. 2008). However, even if these psychological or economic theories of the phenomena are correct, there is still room for exploring the logic of the causal explanations that comport with the medical model. This is because the medical model can be seen as an application to psychiatry of the commitments of cognitive neuroscience (Murphy 2006) and the dominant approach in the cognitive neurosciences is that human behaviour consists of capacities that can be analysed into other personal level capacities (Cummins 2000). In turn, these can then be situated within a hierarchy of biological processes. Personal-level phenomena can be broken down into their component processes and these processes can be understood, typically in representational terms, as the outputs of sub-personal systems that do things like assign a meaning to a phonological representation or compute visual edges.

This decomposition of the task and the allocation of the subtasks to interacting physical entities gives the general form of a mechanistic explanation. In this chapter we will discuss the application of this mechanistic perspective to addiction. It is important to note at the outset that mechanistic explanations are a specific type of causal explanation. Not all phenomena admit of mechanistic explanations, and not all causal explanations can be expressed in terms of mechanisms. What distinguishes a mechanistic explanation is that it depends on the spatial arrangement of component parts and the nature of their interactions.

In the neurosciences we expect to identify processes that can be assigned to parts of the brain, where they will be revealed as the outcome of mechanisms – interacting systems of biological components. If it is the right approach, it will work throughout cognitive neuroscience. So, even if addiction is best understood as neither a disorder nor a fundamentally neurological phenomenon, proponents of the mechanistic approach would still expect to explain it in the terms we will present, provided that we can explain the features of addiction as the result of interacting neurological systems. Addiction does not have to be a pathological phenomenon in order to receive a mechanistic explanation.

We will try to set out the mechanistic approach in general terms in the first section, distinguishing it from functional explanation, which makes fewer biological commitments. We then discuss a mechanistic approach to addiction in more detail in the second section.

Mechanistic explanation

What is a mechanism?

In recent years philosophers have stressed the way in which explanation in the life sciences depends on finding mechanisms (Bechtel and Richardson 1993; Craver 2007; Tabery 2009). Rather than seeing explanation as a search for laws, mechanists seek the components within a system whose entities and activities are organised to produce the phenomena we want to explain. Philosophers disagree over exactly how to characterise mechanisms, but it is agreed that mechanisms comprise component parts and these parts have characteristic activities. There is a debate over whether the activities, as well as the physical structures within the mechanism, should be counted as primitive parts of the mechanism, but we can put that debate aside.

Let's look at an illustrative example. Perhaps the first and most enduring mechanistic explanation in the life sciences was Harvey's (1628) discussion of the pumping of the blood by the heart. The heart has anatomically distinguishable components: these include the left and right atria, the left and right ventricle and the valves that lie between them. We might also include the veins and arteries, as they reach the heart, as part of the system, or we might not. (We will discuss the drawing of boundaries below.) If we were to consider the circulatory system as a whole we would certainly include the pulmonary vein and artery, the vena cava, the aorta and the capillaries. Bechtel (2006, p. 30) argues that the blood is also part of the mechanism, though a passive one.

We explain the circulation of the blood as the result of the contraction of the ventricles, which pushes blood into the large arteries. The valves close to prevent the blood flowing the wrong way. The left ventricle pumps blood through the aorta into smaller arteries that conduct it through the organs via the capillaries. It returns to the right atrium through the veins and passes to the right ventricle. When the right ventricle contracts, the blood is pushed through the pulmonary artery to the lungs where it takes in fresh oxygen and returns via the pulmonary vein.

This explanation relies on 1) identifying physical structures, such as chambers, valves and blood vessels and 2) showing how they interact to explain the phenomena involved in circulation, including the passage of the blood around the body, its oxygen levels and other chemical properties. Unless the physical components were in those spatial relations, the explanation would not work: if the pulmonary artery did not physically interact with the respiratory system, for example, blood would not be re-oxygenated.

For a mechanistic explanation, then, we need a causal story that depends on interaction between spatially related components. Spatial and causal relations alone are not enough. It may be correct, for instance, to explain the founding and prosperity of Singapore as due to its being situated so as to dominate trade through the straits of Malacca. That's a causal explanation and it relies on spatial relations (between the Malay Peninsula and the island of Sumatra) but it is not a mechanistic explanation, because the entities it mentions are arrayed in space but are not doing anything.

The first part of the process of mechanistic explanation is the identification of the phenomena at issue. The psychological hallmarks of addiction, like craving, tolerance or relapse, count as phenomena. We discover the phenomena in the laboratory or through epidemiological study, and the clinical picture defines the phenomena we need to explain. The clinical picture may mix up many different sorts of sign and symptom, including things we think of as paradigmatically psychological and others that seem more physical. We will focus on the sorts of explanations of these diverse phenomena that can be offered by the cognitive neurosciences. The point of such explanations is that they employ information processing systems in the brain, not that they are limited to explaining psychological phenomena. Mechanistic explanations of addiction, then, aim at understanding phenomena by showing how they arise as the result not of the operation of natural laws, but via the interaction of components within brain systems. These mechanisms can themselves be further decomposed into their own subcomponents. They are composed of cells, and the functions of those cells would receive in their turn a mechanistic explanation in cell biological theory.

So the basic picture is that a mechanism has parts within it, and they in turn will have their own components. As Bechtel (2015) stresses, identifying the boundaries of the mechanism involves a strong element of idealisation. Biological mechanisms are as a rule causally connected with surrounding systems. The boundaries we draw reflect hypotheses about the best way to delineate natural phenomena for some explanatory purpose. What count as the boundaries of a mechanism relative to one explanation might be ignored in another explanation, in which the overall system gets partitioned very differently.

As an example, suppose a delusion is a psychotic episode that depends on relations between different psychological processes in different brain systems. Those brain systems might be involved in more than one mechanism. Suppose we think that a delusion involves a perceptual error, propagated through systems that are involved in decision making. The visual systems would also be a source of input into many other psychological processes, and lapses in decision making might be implicated in addiction or OCD as well as in a delusion. So, although a mechanistic explanation in the cognitive neurosciences will often refer to systems that we can discriminate anatomically, like the hippocampal complex, it is not those parts that count as the mechanisms. A mechanism is a team of parts chosen for their explanatory prowess with respect to one set of phenomena. A given pair of physiologically connected systems might be part of a mechanism in one context but not in another. A mechanism is not necessarily (though in practice it might often turn out to be) a physical system that is recognised by anatomical or physiological criteria prior to the start of our enquiry.

Mechanisms and levels

So on this account, explanation in neuroscience, as in biology more generally, involves describing mechanism(s) at levels of explanation in ways that make apparent the relationships between causally relevant variables at different levels (Woodward 2010). So we have a progressive decomposition that is always relative to the behaviour of the overall system that needs to be explained.

How do levels fit in? There are many different ways in which levels talk is used (for a field guide, see Craver 2007, Ch. 5) Central to Craver's influential account of mechanistic explanation in the neurosciences is the relation of causal relevance between phenomena at different levels of explanation (Craver 2002, 2007). Causal relevance is defined in terms of manipulability and intervention. Events at one level are causally relevant in so far as they make a difference at another level.

One picture of levels descends from Oppenheim and Putnam (1958) and is ontological – the biological world is made up of entities (say organisms or cells) that have other entities within them (say, organs within the organism or organelles within the cell) all the way down to the subatomic level. In so far as mechanistic explanations have a metaphysics, this is it. However, the details of a mechanistic explanation often involve a different conception of levels, in which an identical process is described or explained in more than one way. These levels are not ontological but epistemic or descriptive. They mention not different phenomena but the same phenomena described in different scientific terms.

The conception of levels that is of interest to us presently belongs to what Cummins (2000, p. 129) calls 'top-down computationalism' – as he puts it, the idea that the brain is a computer and the mind is what it does. The account of levels usually associated with top-down computationalism is that, as Sterelny (1990, p. 43) puts it, there are three domains in psychology, and a level for each. The top level specifies what the system does, one level down specifies the information-processing means by which it does it, and the base level specifies the physical realisation. Sterelny calls the top level ecological; it specifies the cognitive capacities we are interested in. This picture of levels is not metaphysical in the sense outlined above. It does not describe different structures in a natural hierarchy: it describes different ways of understanding the same process. Identifying a potential mate, say, is the same as performing a certain computation, which is also a particular biochemical response.

Sterelny's presentation accords with the consensus among philosophers of psychology that the topmost level should be understood intentionally. Marr (1982), who originated the picture, understood the top level as abstractly mathematical, not ecological (Egan 1995 summarises the debate). As Sterelny notes, the ecological interpretation (which describes what the system is doing) does not have to be limited to the personal level, since it can describe subcomponents within a system. Nor need it be expressed in folk-psychological terms. However, the top-down computational picture, applied to addiction, identifies clinical variables at the topmost level, and then looks for their physical implementation.

Marr was interested in understanding vision in the abstract, as a process that could be multiply realised in diverse physical systems. But vision, like anything else biological, may be comprehensible in the abstract whilst also having a particular realisation in a species. It is this particular biological form taken by the abstract task that we can break down into subtasks that we assign to structural components of the organism. The computational description shows how the task is done, and the biological description shows how the computation is physically realised.

The domains of such a theory may be explanatory rather than ontological, but the level of physical realisation in the brain hooks up the explanatory picture with the metaphysical one we mentioned earlier. The mechanist assumes that the phenomena, identified in intentional or computational terms, require and can receive a physical explanation. This physical explanation will cite a biological mechanism at some appropriate level within the ontological hierarchy. This is what tethers the explanatory strategy to the biological world and makes the picture a mechanistic one. It does not just explain psychological processes but situates them within a broader picture of the world by showing how cognitive processes can be understood, via levels of explanation, as situated in another hierarchy of levels, that of ontological composition.

The mechanistic picture also fits with an idea of explanation as involving our showing how things work – taking the bigger system apart to reveal the workings within it. Breaking a problem down and showing how the bits are solved, and then breaking down those bits and showing how they are solved, is fundamental to how explanation works. This vision of explanation as decomposition or analysis is not tied to any particular metaphysics. But the multi-level picture of the mind is a way to take the analytic picture of explanation and tie it to a physicalist metaphysics that also trades in levels. The very generality of this picture of explanation as breaking problems down in to sub-problems can seem to make mechanistic approaches very attractive. But there are other types of explanation that work similarly, notably functional explanation, and it is important to see how they differ.

Functional explanation and mechanisms

Mechanistic explanation is not the only type of explanation with the property of epistemic decomposition. Functional explanations work similarly, and also fit in with an abstract picture of the medical model as the investigation of abnormal relations among cognitive-biological processes. A functional explanation breaks down some capacity of a system into a set of simpler capacities and shows how the organisation of these simpler capacities enables the overall capacity to obtain. But it does not specify any physical realisation. Therefore, functional analyses are not mechanistic. They identify a system or phenomenon in terms of the processes that make it happen, while ignoring its physical makeup.

Roth and Cummins (forthcoming) note that natural and artificial hearts can be given a common functional analysis even if they have nothing physical in common. They also point to circuit diagrams as representing processes that can be understood even if one knows nothing about what the items represented by the diagram are made of. They note how a functional analysis of the capacity for multiplication can be given in terms of the partial products algorithm even though specifying the algorithm provides no information at all about the physical set up or processes that implement it. Roth and Cummins argue that the functional analysis provided by the algorithm gives us a complete explanation for the capacity to multiply numbers because it shows us that any system able to compute the algorithm is thereby able to multiply. And what more could you ask of an explanation?

Plenty, if you're a mechanist. We can put Cummins and Roth's conclusion as a challenge to mechanists in the philosophy of psychiatry. If we were able to understand the behaviour of addicts as the product of pathological sub-dispositions, and provide a successful functional analysis of addiction, then why would we need a lot of information about physical mechanisms? Put another way, if the logic of the medical model is just that we can explain pathologies by showing how they arise via the interaction of abnormal psychological dispositions, then functional analysis is enough. If we go further and insist on connecting the functional analysis to the physiology, what have we gained?

Piccinini and Craver (2011) argue that functional analyses are insufficient as explanations because they only give us sketches of mechanisms. The sketches need to be filled out with the details of the implementation by appeal to the physical constitution and operations of the system. Roth and Cummins, in contrast, contend that the functional analysis is not a sketch of anything, but a complete autonomous explanation. They do acknowledge that we might gain a deeper understanding of how systems work by grasping their physical details. But they insist that such understanding does not mean that the original functional explanation was incomplete. What they think a mechanistic explanation can do is provide what they call a "vertical explanation" which supplies details of how the "horizontal explanation" provided by a functional explanation is implemented in a given physical set up.

To insist on a mechanistic explanation of addiction, then, would be to argue that a functional explanation, were it possible, would not be enough. A mechanistic explanation would show how the functional analysis of addiction into psychological capacities and dispositions actually takes place in human beings. There are at least two reasons why one might push for a mechanistic account. First, in showing us the physical specifications of the addicted mind, it would open up the possibility of physical intervention and treatment. Second, it would enable us to integrate our understanding of addiction into a wider picture of how the human brain works. This might help us to reconceive our understanding of addictive phenomena by showing their connections to other kinds of pathology, and also their relations to one another. If some addictions are physically similar to each other in unexpected ways, and others seem closer to other phenomena, we can enrich our understanding of addiction. Mechanistic explanations of addiction thus offer considerable payoffs. However, they might also revise our picture. If addiction is to be explained mechanistically, it needs to be explicable in certain ways, and some approaches to addiction may not fit. In the next section we explore these ramifications and look at some examples.

Addiction

Defining the phenomena

The first step in a mechanistic explanation, we have noted, is to identify the phenomena that we need to explain, and there are divergences among theorists of addiction over what these are. Advocates of mechanistic explanations urge a naturalistic approach. They recommend that if we wish to identify a kind or set of kinds, it is to science that we must turn. There are different sciences, though, that claim to understand addiction. Perhaps the dominant clinical and research approach at present is that of the DSM, which primarily discusses substance abuse disorders but also reclassified pathological gambling as an addiction in its fifth edition. It says (American Psychiatric Association 2013, p. 483) that the "essential feature of a substance use disorder is a cluster of cognitive, behavioral, and physiological symptoms indicating that the individual continues using the substance despite significant substance-related problems." The DSM-5 sees each class of substance as essentially connected to a syndrome (Criterion A) featuring "impaired control, social impairment, risky use, and pharmacological criteria" (p. 483).

In very broad terms, addiction can be described as a set of behaviours relating to consumption patterns, as Becker and Murphy (1988) asserted. They claimed people could become addicted to substances but also to a wide range of behaviours like eating, working, practising religion and watching television. There are two problems with such a broad view. Conceptually, many theorists have reservations about asserting all there is to an addiction is a pattern of irrational preferences, as the economic approach suggests. And as a charter for mechanistic explanation, it seems implausible: it is unlikely that we could provide a unified mechanistic explanation of addiction across such heterogeneous activities. This does not mean that there will never be a unified account of addiction at all. But it is unlikely that everything that seems to fit a purely behaviourally based definition of addiction would share the same neurological/mechanistic explanation. A mechanistic explanation, if it is to be a general explanation of addiction, needs to apply to a phenomenon that unifies all the syndromes of interest.

The dominant neuroscientific enterprise concentrates on addiction to substances, and so seems a good place to start when looking for a phenomenon suitable for a mechanistic explanation of addiction. Primarily this is because the action of potentially addictive drugs can produce specific and generalisable physical effects on the brain, including on brain systems that are involved in the processes of motivation and reward. The DSM-5 (American Psychiatric Association 2013, p. 481) mentions ten classes of drugs as relevant to substance-related disorders. Some such as Goldstein (2001) argue that humans can only be addicted to a set of seven substances: nicotine, caffeine, cannabis, opiates, psycho stimulants, alcohol and hallucinogens. It is Goldstein's view that addiction is a brain disease and that no matter how much one craves a habitual behaviour, it is not an addiction (Goldstein 2001; Leshner 1997).

Elster (1999) and Orford (1985) adopt an intermediate position in which addiction is characterised by a group of experiences shared by addicts. This general view allows for some though not all behavioural regularities to be candidates for addictions. Elster (1999) argues that the most important element of addiction is craving, although he doesn't provide a strong empirical account of why. Impaired control has stronger empirical support, which we will expand on. At this point though it's worth noting that just craving or impaired control alone will not work. There also has to be actual engagement in an addictive activity or imbibing of a substance, i.e., a psychological/behavioural component.

Experimental models often focus on an addict's impaired control, and such models suit a mechanistic account well. They seek to explain impaired control in terms of the neurological mechanisms that regulate behaviours related to control, and this is combined with a behavioural account that explains the desire to use drugs or engage in a repetitive behaviour at a psychological level. Each experimental model cashes out what has gone awry in addicts in a different way, with the focus being on impaired control (Pober 2013). At a fine grained level, impaired control is a heterogeneous property. It is unlikely to provide the kind of unification a mechanistic explanation requires for an overall account of addiction. But when considered at a coarse grain it can be argued that it could be a causally basic property. So – 'having desire for' and 'having impaired control' are both functionally relevant properties of addiction, which could receive a mechanistic explanation. They are connected to each other causally, but one cannot be reduced to the other. Furthermore 'impaired control' or 'loss of control' is clearly a broader category itself than addiction, so we're only talking here about the kind of impaired control present in addiction: behaviour that cannot be controlled (or only partly controlled) by an agent, and that contributes to an explanation of what we call addiction.

Sinnott-Armstrong and Pickard (2013) argue that harm is another necessary component. They argue that it's the combination of impaired control and harm that provides a complete account of addiction. This is correct if you are tied to the idea that addiction is a pathology. But, as we argued earlier, it is possible to provide a mechanistic account of addiction without making that commitment.

The reward system as mechanism

Addiction appears to involve a direct manipulation of the 'reward system' of the brain – a collection of neural structures that regulate behaviour through reinforcement (American Psychiatric Association 2013, p. 481). The circuit is controlled by the action of dopamine-containing neurons in the ventral tegmental (TA) area, the nucleus accumbens (NA), and the prefrontal cortex (PFC). To understand how appeals to the midbrain system can explain addictive behaviour it is important to note that, neurologically, there is a difference in the brain between pleasure and reward that does not necessarily fit with our common use of the terms. Pleasure signals evolved in primates with much less complicated reward systems than our own and are really limited to a set of basic cues: air temperature, sexual contact, dampness, etc. Everything else we normally associate with pleasure is actually associations between pleasure signals and reward signals.

Our folk-psychological concept of pleasure and reward does not correlate with our neuroscientific understanding of them. Pleasure, it turns out, is not a reliable proxy for reward. This is because the main evolutionary adaptation of the human brain has been adaptive intelligence. We have reward systems that were selected to prefer interest and curiosity about changes in our environment, rather than pleasure. The reward system requires supervision from frontal circuits to avoid excessive impulses, such as those seen in addictive behaviour. However, because the limbic system evolved before the cognitive system, it did not evolve special adaptations for the cognitive system. Ross *et al.* (2008) argue that addiction is a result of imperfect communication between the two systems that permits impaired control. That is, disorders such as addictive behaviour involve neurological vulnerabilities as a result of the imperfect adaption of the midbrain (limbic) system in relation to the cognitive system.

We have evolved to pursue what stimulates our reward systems more conscientiously than what is pleasurable. This fits well with the distinction that Robinson and Berridge make extensively in their work between 'wanting' versus 'liking' (Robinson and Berridge 2008; Robinson and Berridge 2001; Robinson and Berridge 1993). Their theory rests on the idea of incentive-sensitisation, that is, that neural systems are sensitised by addictive drugs. Such sensitisation directly affects motivation, and therefore produces compulsive patterns of drug-seeking behaviour. Important for their account is that the brain systems sensitised – the mesotelecephalic dopamine (DA) system and DA projections in the NE and accumbens-related circuitry – do not mediate pleasure-producing effects (i.e., drug 'liking'). Rather they mediate a subcomponent of the reward system they term incentive salience (i.e., drug 'wanting'). The theory holds that compulsive drug-related behaviour may occur when incentive salience due to drug-taking becomes pathologically amplified (Robinson and Berridge 2001, p. 109). It is not simply that addiction involves needing more of the same stuff: as a result of the amplification the agent doesn't have complete control over their behaviour, because neurologically they are compelled to engage in drug use or addictive behaviour. Predictions from this theory have been confirmed in several drug addiction categories, though not in cannabis addiction, which seems to show the opposite effect (Pober 2013). Despite this anomaly, such a distinction goes some way in explaining why addicts can both persist in addictive behaviour and not find it enjoyable.

Two other mechanistic accounts of addiction are worth mentioning. The first centres on aberrant learning and the dorsal striatum. Everitt, Robbins and colleagues (2005, 2008) agree with Robinson and Berridge that the mesolimbic DA system is affected by regular drug use, but differ in their explanation of the function of that system and the neuroadaptions involved in addiction. They suggest that behavioural control shifts to habit-based learning functions in the dorsal striatum, away from normal reward-based learning functions in the mesolimbic DA system. In addiction drug-seeking and drug-taking behaviour that is normally regulated becomes habitual and unmediated, leading to impaired control. This lack of control explains involuntary responses to drug-related stimuli and in particular relapse: drug-seeking behaviour in response to stimuli after a period of abstinence. Experimental evidence supports this for most substances, except once again for cannabis.

The second is a frontostriatal dysfunction model posited by Volkow and colleagues (Volkow et al. 2013; Goldstein and Volkow 2011; Goldstein and Volkow 2002; Volkow and Fowler 2000) that does better at explaining cannabis addiction, but not opiates. They focus on the PFC areas, proposing that compromises to the PFC in terms of overall functional capacity directly relate to the loss of control in addiction. That is, the normal self-control systems that regulate drug-taking and drug-seeking behaviour are no longer operating optimally. This compromised inhibition is the biological substrate of impaired self-control in the model. It involves several distinct subregions of the PFC and the ventral striatum. No single component or neurocircuit is responsible for the addictive phenotype – a combination of subregions is involved.

These three models exemplify mechanistic explanations for pathological drug-taking that have some empirical support. They aim to show how the behavioural phenomena and experimental effects of interest arise from interactions among components of the brain. None of these models provide an adequate explanation for all addictive substances. They also do not aim to explain behavioural addictions. To argue for the possibility of genuine behavioural addictions, Ross et al. (2008) make a strong case for gambling. Indeed, they see it as the fundamental case of addiction because it best illuminates what it is about the neural systems in the brain that are affected in addiction. Their account relies on economic explanations of behaviour both at a neurological and psychological level, combined with empirical evidence from the psychosciences. It can explain most substance addictions well, though it cannot adequately explain nicotine addiction. Ross *et al.* see their account as identifying and explaining the paradigm case of addiction; if we accept that claim, then we would have a way of identifying addiction based on shared physical structure. This would obviously comport with the requirements of mechanistic explanation very neatly; we provide an explanation of the paradigm case and other addictions will resemble it mechanistically.

However, even on a narrow, neurochemical perspective, addictions may not be generally explained by one mechanistic account. Pober (2013) argues against the idea that the category of addiction can be adequately explained by any singular mechanistic account currently available (and hence is not a natural kind). Perhaps this is correct, but there is as yet no compelling evidence to show that we cannot find a mechanism whose physical interactions explain a loss of control across at least a wide range of addictions.

So, whether we can find a mechanistic explanation of the mental and physical compulsion that covers all the poorly controlled drug-taking and drug-seeking behaviour we wish to call addiction is an open question. We can distinguish two claims. First, a unificationist claim to the effect that all addictions are explicable via a common underlying mechanism. This is a very strong claim that would rely on two successes: first, generalising a phenomenon like loss of control to cover all instances of addiction; second, explaining that phenomenon in terms of the physical interactions of the components of the reward system (or some other neurological mechanism).

A second, less general claim would be that kinds of addiction are to be understood in terms of the mechanisms that separately explain them; in that case we might appeal to a different conception of addiction that unifies the overall category quite loosely on the basis of diverse phenomena, or we might just conclude that we cannot find a general mechanism for all forms of addiction. That would mean that if addiction is a natural kind, it is not a natural kind in virtue of a common causal-mechanical structure.

References

American Psychiatric Association (2013) *Diagnostic and Statistical Manual of Mental Disorders, 5th Edition*, Washington, DC: American Psychiatric Association.
Bechtel, W. (2006) *Discovering Cell Mechanisms: The Creation of Modern Cell Biology*. Cambridge, UK: Cambridge University Press.
Bechtel, W. (2015) "Can mechanistic explanation be reconciled with scale-free constitution and dynamics?" *Studies in History and Philosophy of Biological and Biomedical Science* 53: 84–93.
Bechtel, W. and Richardson, R. C. (1993) *Discovering Complexity: Decomposition and Localization as Strategies in Scientific Research*. Princeton, NJ: Princeton University Press.
Becker, G. and Murphy, K. (1988) "A theory of rational addiction", *Journal of Political Economy* 96(4): 675–700.
Craver, C. (2002) "Interlevel experiments and multilevel mechanisms in the neuroscience of memory", *Philosophy of Science* 69(S3): 83–97.
Craver, C. (2007) *Explaining the Brain*. New York, NY: Oxford University Press.
Cummins, R. (2000) "'How does it work?' versus 'what are the laws?': two conceptions of psychological explanation", in F. Keil and R. A. Wilson (eds), *Explanation and Cognition*, Cambridge, MA: MIT Press, pp. 117–145.
Egan, F. (1995) "Computation and content", *Philosophical Review* 104(2): 181–203.
Elster, J. (1999) *Strong Feelings: Emotion, Addiction and Human Behaviour*. Cambridge, MA: The MIT Press.
Everitt, B. J., Belin, D., Economidou, D., Pelloux, Y., Dalley, J. W. and Robbins, T. W. (2008) "Neural mechanisms underlying the vulnerability to develop compulsive drug-seeking habits and addiction", *Philosophical Transactions of the Royal Society London B: Biological Sciences* 363: 3125–3135.

Everitt, B. J. and Robbins, T. W. (2005) "Neural systems of reinforcement for drug addiction: from actions to habits to compulsion", *Nature Neuroscience* 8(11): 1481–1488.
Goldstein, A. (2001) *Addiction, 2nd Edition*, New York, NY: Oxford University Press.
Goldstein, R. Z. and Volkow, N. D. (2002) "Drug addiction and its underlying neurobiological basis: neuroimaging evidence for the involvement of the frontal cortex", *American Journal of Psychiatry* 159(10): 1642–1652.
Goldstein, R. Z. and Volkow, N. D. (2011) "Dysfunction of the prefrontal cortex in addiction: neuroimaging findings and clinical implications", *Nature Reviews Neuroscience* 12(11): 652–669.
Harvey, W. (1628) *An Anatomical Disputation Concerning the Movement of the Heart and Blood in Living Creatures*, trans. G. Whitteridge, Oxford, UK: Blackwell.
Leshner, A. I. (1997) "Addiction is a brain disease, and it matters", *Science* 278(5335): 45–47.
Marr, D. (1982) *Vision: A Computational Investigation into the Human Representation and Processing of Visual Information*. San Francisco, CA: W. H. Freeman and Co.
Murphy, D. (2006) *Psychiatry in the Scientific Image*. Boston, MA: MIT Press.
Oppenheim, P. and Putnam, H. (1958) "The unity of science as a working hypothesis", in G. Maxwell, H. Feigl and M. Scriven (eds), *Concepts, Theories, and the Mind-Body Problem*, Minneapolis, MN: Minnesota University Press, pp. 3–36.
Orford, J. (1985) *Excessive Appetites*, Chichester, UK: Wiley.
Piccinini, G. and Craver, C. (2011) "Integrating psychology and neuroscience: functional analyses as mechanism sketches", *Synthese* 183(3): 283–311.
Pober, J. M. (2013) "Addiction is not a natural kind", *Frontiers in Psychiatry* 4: 1–11.
Robinson, T. and Berridge, K. (1993) "Incentive-sensitization as the basis of drug craving", *Behavioural Pharmacology* 4(4): 443.
Robinson, T. and Berridge, K. (2000). "The psychology and neurobiology of addiction: an incentive-sensitization view", *Addiction* 95(8s2): 91–117.
Robinson, T. and Berridge, K. (2001) "Incentive-sensitization and addiction", *Addiction* 96(1): 103–114.
Robinson, T. and Berridge, K. (2008) "The incentive sensitization theory of addiction: some current issues", *Philosophical Transactions of the Royal Society B* 363: 3137–3146.
Ross, D., Sharp, C., Spurrett, D. and Vuchinich, R. (2008) *Midbrain Mutiny: The Picoeconomics and Neuroeconomics of Disordered Gambling*, Cambridge, MA: MIT Press.
Roth, M. and Cummins, R. (2017) "Neuroscience, psychology, reduction and functional analysis", in D. Kaplan (ed.), *Explanation and Integration in Mind and Brain Science*, Oxford, UK: Oxford University Press.
Sinnott-Armstrong, W. and Pickard, H. (2013) "What is addiction?" in K. W. M. Fulford, M. Davies, R. Gipps, G. Graham, J. Sadler, and G. Stanghellini (eds), *Oxford Handbook of Philosophy of Psychiatry*, New York, NY: Oxford University Press, pp. 851–864.
Sterelny, K. (1990) *The Representational Theory of Mind*. Oxford, UK: Blackwell.
Tabery, J. (2009) "Difference mechanisms: explaining variation with mechanisms", *Biology and Philosophy* 24(5): 645–664.
Volkow, N. D. and Fowler, J. S. (2000) "Addiction is a disease of compulsion and drive: involvement of the orbitofrontal cortex", *Cerebral Cortex* 10(3): 318–325.
Volkow, N. D., Wang, G. J., Tomasi, D. and Baler, R. D. (2013) "Unbalanced neuronal circuits in addiction", *Current Opinion in Neurobiology* 23(4): 639–648.
Woodward, J. (2010) "Causation in biology: stability, specificity, and the choice of levels of explanation", *Biology and Philosophy* 25(3): 287–318.

26
CONTROLLED AND AUTOMATIC LEARNING PROCESSES IN ADDICTION

Lee Hogarth

The nature of controlled and automatic processes

Human and animal behaviour is purposeful in meeting biological needs. But the learning mechanisms underpinning behaviour cannot be deduced from simple observation. Behaviour may be governed by intentional decision making available to consciousness or by automatic learning processes below consciousness. Comparative psychology has long wrestled with the question of whether animals have a mental life or are mere automata that react to stimuli. The same question has been applied to addicts (Wise and Koob 2014). The dominant theory of addiction claims that drug-seeking is automatic, which explains why addiction persists despite heavy costs and the desire to quit. Yet craving is a core construct in addiction theory, suggesting that conscious desires play an important role. The controlled account of addiction is gaining ground, arguing that drug-seeking persists because addicts assign abnormally high value to the drug outweighing the costs, and they relapse despite reporting a desire to quit simply because they change their mind when priorities later change.

Addictive behaviour may not be exclusively controlled or automatic, but a combination of the two. The common dual-process account argues that early recreational drug use is controlled and becomes automatic with practice. A variant of this claim is that controlled and automatic processes operate concurrently and interact. They may be competitive, with each dominating under different conditions, or they may be summative, complementing each other, conferring a double hazard. Finally, the two processes may be hierarchical, with controlled choices initiating automatic response chains that are completed with little executive supervision. The purpose of this chapter is to scrutinise the state of empirical evidence supporting the controlled and automatic accounts of addiction.

Distinguishing controlled and automatic processes with cognitive paradigms

Cognitive psychology has characterised automatic processes underpinning addictive behaviour as fast, stimulus-bound, difficult to control, effortless and unconscious. Controlled processes have the opposite characteristics (Tiffany 1990). Cognitive procedures seek to test whether drug-related behaviour better matches one set of characteristics. This chapter provides only a flavour of the vast corpus of research in this area.

Interference procedures identify drug consumption as automatic if it does not interfere with another task. In these procedures, drug users consume the drug while performing a target detection task, and if target reaction times are slowed, drug consumption is identified as being controlled insofar as it occupies (competes for) cognitive resources required for target detection. However, if consumption does not slow target detection it is identified as automatic because it does not occupy cognitive resources. Two studies have found that smoking slowed target detection in less dependent smokers but not in more dependent smokers, suggesting that smoking is initially controlled but becomes automatic with dependence (Motschman and Tiffany 2016; Baxter and Hinson 2001).

This conclusion seems convincing until one considers that another interference paradigm (the drug Stroop task) claims the opposite – that the cognitive bias for drug-related cues is automatic if it does interfere with another task. In the Stroop task, drug users have to rapidly name the ink colour of drug and neutral words. At least two studies have shown that drug words slowed ink-colour naming to a greater extent in more dependent users, suggesting that drug words automatically command cognitive resources in dependent users, reducing their availability for colour naming (Field 2005; Cousijn et al. 2013). This conclusion directly contradicts the two earlier studies where smoking produced less interference with target detection in more dependent smokers. How could interference effects in opposite directions both provide evidence that addiction is automatic? In conclusion, cognitive paradigms cannot provide definitive evidence that addictive behaviour is controlled or automatic until empirical effects have been extensively replicated and all alternative explanations for the effects have been properly explored.

Distinguishing controlled and automatic processes with learning paradigms

Methods devised by comparative psychologists to study the mental life of animals have been translated to help determine whether addiction is controlled or automatic. Support for each position will be evaluated in four categories of learning: Pavlovian conditioning, Pavlovian to instrumental transfer, instrumental conditioning, and incentive learning/self-medication. The majority of this work has been conducted with human smokers.

Pavlovian conditioning

Reactivity to external drug cues (such as cigarettes, alcohol, cocaine packaging or places, etc.) is usually explained by Pavlovian conditioning (the section below on p. 332 discusses internal states). On this view, the drug-cue conditioned stimulus (CS) predicts drug reinforcement (the unconditioned stimulus US), which elicits unconditioned responses (URs), so the CS comes to elicit conditioned responses (CRs) that are similar to the URs. One view of Pavlovian conditioning is that it is underpinned by an automatic stimulus–response (S–R) structure, in which the CS is directly linked to the UR, enabling the CS to elicit the UR (CR) automatically without any intervening cognitive processes. Early theories of drug-cue reactivity favoured automatic Pavlovian S–R type accounts. However, learning research has established that CS retrieve an expectation or representation of the US that drives the CR, revealed, for example, by the CR being altered if the value of the expected food US is increased or decreased by hunger or satiety. This suggests that Pavlovian conditioning is underpinned by a controlled stimulus–stimulus (S–S) structure (Rescorla 1988). Cue reactivity theories have followed suit and now suggest that cognitive expectancy of the drug generates the CR, and the historical development of this view is outlined below.

Wikler's early account of cue reactivity argued that drug cues elicit conditioned withdrawal, which automatically primes drug-seeking to ameliorate this state. To quote: 'abstinence distress . . . may be reactivated long after cure . . . providing an unconscious motivation to relapse' (Wikler 1984, p. 280). Key support for this claim is that drug cues elicit conditioned withdrawal (O'Brien et al. 1977), and compensatory responses (Siegel et al. 1982), but the assumption that these CRs exert unconscious motivational influences has not been supported (see the section below on p. 332 on incentive learning and self-medication). Another complication was added by positive reinforcement theorists who found that drug cues could elicit dopamine and locomotor activity akin to the drug effect. This positive state arguably energises drug-related cognitions driving behaviour, i.e., an S–S account (Stewart et al. 1984). To quote: 'Conditioned stimuli associated with these drugs arouse neural states that mimic features of those produced by the drugs themselves and thereby serve to increase . . . the probability of drug-related thoughts and actions' (p. 251). Later studies found that physiological CRs are not consistently drug-like or drug-opposite and so do not distinguish withdrawal and positive reinforcement accounts (Carter and Tiffany 1999). By contrast, studies that have taken subjective measures reliably suggest that drug cues elicit positive emotional states and craving (Niaura et al. 1988) and that cue-elicited craving is the best predictor of cue-elicited drug consumption (Hogarth et al. 2010), again supporting a controlled S–S positive reinforcement account of cue reactivity.

The S–S account of cue reactivity is substantiated by the finding that drug-paired CS only elicit craving and drug consumption if participants are aware of the predictive relationship between the CS and drug availability (for a review see Hogarth and Duka 2006). The causal role of drug expectancies in driving behaviour is supported by studies that have manipulated expectancies through verbal instructions. Field and Duka (2001) found that instructing participants that the CS–drug contingency was no longer in force immediately abolished conditioned craving to the CS. Dols et al. (2000) found that telling subjects that a CS predicted drug availability immediately generated conditioned craving to the CS. General human conditioning studies also support these claims, showing that CS only elicit CRs in subjects who report explicit awareness of CS–US contingencies, and CRs can be generated or abolished immediately by instructions that modify CS–US expectancies (Mitchell et al. 2009). Together, these data favour a controlled Pavlovian S–S positive-reinforcement-based account of drug conditioning, where drug cues elicit an expectation of the drug that generates explicit craving, drug-seeking and physiological changes. The findings contradict automatic Pavlovian S–R accounts of drug conditioning.

Finally, it is crucial to note that the capacity of drug cues to increase craving above baseline is not reliably associated with level of dependence (Perkins 2009; Perkins 2012), and is not modified by reducing the value of the drug through satiety or pharmacotherapy (Drobes and Tiffany 1997; Hitsman et al. 2013). These data suggest that drug cues elicit an expectation that the drug is likely to be available, which increases craving in proportion with expected drug availability. Knowledge about drug availability arguably reaches asymptote quickly (i.e., it does not take long to learn that drugs are normally available when drug cues are present) and so does not vary with dependence. Consequently, cue-induced craving does not vary with dependence. Finally, the expected availability and the expected reward value of the drug arguably exert additive effects on craving, and so cue-elicited craving above baseline is not modulated by satiety or pharmacotherapy (Hogarth 2012; Sayette and Tiffany 2013). This analysis supports a controlled account of cue-reactivity in which a cue-elicited expectation of drug availability plays a crucial role in driving behaviour.

Pavlovian to instrumental transfer

The role of expectancies underpinning cue-elicited drug-seeking has been studied further using the Pavlovian to instrumental transfer (PIT) procedure. The PIT procedure was first established in rats (Colwill and Rescorla 1988), and later applied to human addiction (Hogarth *et al.* 2007). In a typical drug PIT design (see Table 26.1), subjects first learn that one instrumental response earns the drug outcome, whereas another response earns a food outcome (R1-O1, R2-O2). In a separate phase, subjects learn that two Pavlovian stimuli differentially predict those same outcomes (S1-O1, S2-O2). In the transfer test, the Pavlovian stimuli are presented while subjects are free to choose between the two responses in extinction, where the outcomes are no longer delivered (S1:R1/R2, S2:R1/R2). It is typically found that each cue augments choice of the response that earns the same outcome (S1:R1>R2, S2:R1<R2). This specific transfer effect indicates that the drug cues do not modify instrumental drug-seeking by eliciting a general excitatory CR, because if this were the case, each S would excite both Rs equally. Rather, each S must retrieve an expectation of its associated O in order to selectively enhance performance of the R that was also associated with that O.

There are two accounts of PIT, one automatic and one controlled, which will be discussed in turn. The automatic S-O-R account of PIT argues that each S retrieves a representation of its associated O, which triggers the associated R automatically (de Wit and Dickinson 2009). Pavlovian conditioning establishes the S-O link. In instrumental training, agents learn the relationship between the R and the O in a forward manner (R-O), and this association is bidirectional (O-R), so when the S retrieves its associated O this elicits the associated R automatically. This O-R claim is based on William James' ideomotor theory that the mental representation of an outcome, e.g. water, is sufficient to generate the muscle sequence to obtain that drink. The O-R connection is automatic in that, although one may be aware of the idea of the outcome, there is little conscious insight into the generation of the muscle sequence required to obtain the outcome.

The automaticity of the S-O-R links arguably explain why cue-elicited drug-seeking persists despite heavy costs and desire to quit (Watson *et al.* 2014). The original evidence for the S-O-R position came from the insensitivity of the PIT effect to outcome devaluation (see Table 26.2). If one outcome is devalued prior to the transfer test in the PIT design, it is typically found that both Ss selectively activate the R with the shared O, to an equal extent irrespective of whether the O has been devalued or not, in both animals (Corbit *et al.* 2007) and humans (Watson *et al.* 2014; Hogarth 2012; Hogarth and Chase 2011; Hogarth *et al.* 2010; but see Eder and Dignath 2016). The S-O-R theory explains this finding by suggesting that the S activates a perceptual representation of its associated O that does not encode the reward value of the O, such that the level of activation of the R is equivalent irrespective of the O's value. Thus, cue-elicited drug-seeking occurs even when the drug is not desired.

Table 26.1 The Pavlovian to instrumental transfer procedure used to test if stimulus control of choice is mediated by an expectation of the signalled outcome. S = stimulus, R = response, O = outcome

Pavlovian training	Instrumental training	Extinction test
S1 – O1	R1 – O1	S1:R1/R2
S2 – O2	R2 – O2	S2:R1/R2

Table 26.2 A combination of Pavlovian to instrumental transfer and outcome devaluation procedures, used to test if a reduction in the expectation value of the outcome modifies the capacity of stimuli control to control choice

Pavlovian training	Instrumental training	Devaluation treatment	Extinction test
S1 – O1	R1 – O1	O1 or O2	S1:R1/R2
S2 – O2	R2 – O2		S2:R1/R2

Hierarchical instrumental learning theory offers the alternative, controlled, account of PIT (Rescorla 1991). The hierarchical account argues that participants acquire explicit knowledge of the R-O relations in instrumental learning (R1-O1, R2-O2), and S-O relations in the Pavlovian phase (S1-O1, S2-O2). In the transfer test, the presentation of each S evokes an explicit inference (belief) that the R-O relationship with the shared O is more likely to be reinforced (S:R-O), and so R is performed (Hogarth *et al.* 2014; Seabrooke *et al.* 2015). The insensitivity of PIT to devaluation arguably occurs because the S-elicited expectation that a particular R-O relation is likely to effectively produce the O primes the R in proportion to the R-O likelihood, independently of the current value of the O (Hogarth 2012). In other words, the expectation that a response has a high probability of being effective drives up choice of that response irrespective of how valuable the outcome is.

The hierarchical account predicts that participants' cue-elicited beliefs about R-O effectiveness drive the PIT effect. Support for this claim comes from the finding that instructing participants during the transfer test that 'Pictures do not indicate which key is more likely to be rewarded!' abolished the PIT effect (Hogarth *et al.* 2014). Furthermore, Seabrooke *et al.* (2015) showed that the PIT effect could be completely reversed by instructing participants that 'Pictures indicate which response will *not* be rewarded'. These studies also found that participants in the control group who strongly believed that cues indicated which response was more likely to be reinforced showed a larger PIT effect. Finally, the cue-availability paradigm has shown that telling participants the probability that a drug-seeking response will produce the drug at the start of each trial reliably modulates the latency of the drug-seeking response (Carter and Tiffany 2001). These studies suggest that cue-elicited drug-seeking is mediated by hierarchical beliefs concerning the S:R-O relationships, rather than automatic S-O-R links.

The hierarchical account is corroborated by the finding that the PIT effect is larger when the R-O contingencies are partially reliable (33%) compared with fully reliable (100%) (Cartoni *et al.* 2015). The S-O-R theory anticipates the opposite finding because the O-R link should be weaker in the unreliable condition. By contrast, the hierarchical account anticipates this finding, because PIT effects should be greater when cues resolve uncertainty about the effectiveness of these R-O contingencies.

Finally, it is crucial to note that the drug PIT effect has no reliable association with drug dependence (Martinovic *et al.* 2014; Hogarth and Chase 2011; Hogarth 2012; Hogarth and Chase 2012; Hogarth *et al.* 2015; Garbusow *et al.* 2014). This corresponds with the finding that cue-elicited craving has no association with dependence (Perkins 2009; Perkins 2012). These findings contradict the S-O-R account, which argues that the automatic process underpinning cue effects is important for dependence (Watson *et al.* 2014). The hierarchical account anticipates these findings because knowledge that drug cues signal when drug-seeking will be effective should reach asymptote quickly, and thus show no association with dependence. To conclude, the past two sections have charted a transition in thinking about external drug cue reactivity from an automatic Pavlovian S–R model to a controlled S–S model in which the cue-elicited expected efficacy of drug-seeking drives responding.

Instrumental learning

Addictive behaviour may be a deliberate choice driven by the expected value of the drug (Heyman 2009; MacKillop 2016; Ahmed 2010; Bickel et al. 2014; Hogarth 2012; Hogarth and Troisi 2015). On this view, addicts experience drugs as having a high reward value for a variety of biopsychosocial reasons, and choose the drug intentionally because they value it so highly, and are willing to bear the heavy costs. Although addicts may express a wish to quit, they simply change their mind when priorities later change, for example, when drugs become available, when encouraged by friends, in a state of withdrawal, depression, loneliness, at a celebration, or at favourable times of day or week, etc. Although this view sees drug-seeking as an intentional choice, the constitutional factors that cause the drug to be assigned such high value are outside intentional control.

The choice account is supported by the finding that willingness to work or pay for drugs decreases with the effort or price, for tobacco (Chase et al. 2013; MacKillop et al. 2012), alcohol (MacKillop 2016), heroin (Greenwald and Steinmiller 2009; Greenwald 2008), cocaine (Grossman and Chaloupka 1998) and amphetamine (Kirkpatrick et al. 2012). Greater drug demand is also reliably associated with greater dependence and poorer treatment outcomes (Bickel et al. 2014; MacKillop 2016). Similarly, preferential choice of drugs over food reinforcers is reliably associated with drug demand (Chase et al. 2013), dependence (Hogarth 2012; Hogarth and Chase 2011; Hogarth and Chase 2012; Moeller et al. 2009) and treatment outcomes (Moeller et al. 2013; Perkins et al. 2002; Murphy et al. 2015), and can identify animals who are vulnerable to dependence (Ahmed 2010; Panlilio et al. 2015). Another important finding is that paying addicts to remain abstinent (contingency management) is one of the most effective behavioural interventions (Zajac et al. in this volume). Finally, most drug users eventually quit unaided, and commonly cite changing priorities such as jobs and family as their reason for quitting, suggesting drug use is a choice (Heyman 2013).

Although preferential drug choice looks like controlled decision-making, it could be automatic. The goal-directed account claims that addicts choose to seek drugs because they know what responses produce drugs and how valuable the drugs are, and if the relative value placed on the drug outcome exceeds alternative reinforcers, the drug-seeking response is selected for action (Olmstead et al. 2001). The habitual account claims that drug reward strengthens the link between drug cues present at that time and drug-seeking responses (an S–R habit association), enabling cues to elicit drug-seeking automatically. Thus, dependence may give rise to greater drug choice because the constitutionally determined experience of greater drug reinforcement either establishes higher rates of goal-directed drug-seeking, or stronger S–R habitual drug-seeking.

The outcome devaluation procedure can distinguish whether drug-seeking is goal-directed or habitual (Dickinson 1985) (see Table 26.3). Subjects learn that one response produces the drug (R1-O1) and another response produces a food reward (R2-O2). One outcome is devalued by pairing it with sickness, specific satiety or health warnings, depending on species and procedure. Finally, subjects choose between the two responses in extinction. If choice is reduced for the devalued outcome, it must be controlled by knowledge of the R-O contingencies and the current value of the outcomes, i.e., is goal-directed. The S–R habit theory could not explain this devaluation effect because the outcomes are omitted from the test, so experience of the now devalued outcome could not modify the strength of the S–R association controlling the response. If there is no reduction in choice for the devalued outcome, choice is deemed habitual in being elicited automatically by contextual cues without knowledge of the value of the outcomes.

Table 26.3 The outcome-devaluation procedure used to test if choice is governed by the expected value of the outcome (goal-directed) or not (habitual). R = response, O = outcome

Instrumental training	Devaluation treatment	Extinction test
R1 – O1	O1 or O2	R1/R2
R2 – O2		

The outcome devaluation procedure with animals has shown that drug-seeking behaviour can be both goal-directed (Olmstead et al. 2001; Hutcheson et al. 2001) and habitual (Dickinson et al. 2002; Miles et al. 2003; Corbit et al. 2012), suggesting that either process could be responsible for drug dependence. There are also many demonstrations that although food-seeking is initially goal-directed, it can become habitual with extensive training (Dickinson and Balleine 2010), casting doubt that habit demonstrated in the outcome-devaluation assay is diagnostic of the pathological process unique to addiction. However, two studies have shown that combined drugs+food reinforcers give rise to habitual responding more rapidly than food alone, suggesting that habit formation may be disproportionate in addiction (Dickinson et al. 2002; Miles et al. 2003). Countering this claim, it has been found that if foods are increased in value, they give rise to habitual responding more rapidly (Nordquist et al. 2007), suggesting that the drug+food reinforcers may generate quicker habits simply because they have higher value than food alone. Two other general criticisms further degrade the habit account of addiction. First, it is possible that animals continue to respond for the drug following devaluation (appear habitual) because they expect the drug delivered in the test context to have the same value it has always had, and is different from the drug delivered in the separate devaluation context (Jonkman et al. 2010). Discrimination between drugs presented in the test and devaluation contexts could occur more readily for drugs than food because drugs have higher value or are more novel reinforcers (Rosas et al. 2013). Thus, the insensitivity of drug-seeking to devaluation may not reflect habit formation. Second, habitual responding only develops in simple procedures with minimal response options and outcomes, and is not found in more complex environments (Kosaki and Dickinson 2010). One can question whether the habitual behaviour seen in simple animal designs can generalise to complex human drug-seeking environments (Sjoerds et al. 2014). In conclusion, these animal data provide little evidence that addictive behaviour is underpinned by habit learning, and are equally or more compatible with a goal-directed explanation of addiction.

Crucially, human drug dependence is not associated with greater habit formation as assessed by the outcome devaluation procedure (Hogarth 2012; Hogarth and Chase 2011; Hogarth et al. 2012b). In this procedure, smokers learned that two responses earned tobacco and food points respectively, before tobacco was devalued by satiety, health warnings or pharmacotherapy (in different studies). In the extinction test that followed, tobacco choice was decreased indicating that this behaviour was goal-directed (mediated by an expectancy of the devalued tobacco outcome). If more dependent smokers were more habitual, the devaluation effect should have been smaller in the more dependent sample. However, there was no association between dependence and the magnitude of the devaluation effect in any of the studies, suggesting that human dependence is not associated with a predilection for habitual drug-seeking. Rather, dependence in these studies was strongly associated with higher rates of preferential choice of the drug, and this choice was demonstrably goal-directed as indicated by its sensitivity to devaluation. Moreover, dependence is associated with self-reported willingness to pay for drugs (Bickel et al. 2014; MacKillop 2016), which must reflect the expected value of the drug. These data provide compelling support for the claim that dependence is mediated by hyper-valuation of the drug driving higher rates of goal-directed drug-seeking, and provide no support for the habit account.

Several other human outcome devaluation studies provide indirect support for a habit account of addiction. These studies have shown that habitual responding for a drug reinforcer is more pronounced during exposure to drug cues (Hogarth *et al.* 2013), and that habitual responding for nondrug reinforcers is more pronounced following alcohol administration (Hogarth *et al.* 2012a), exposure to acute stress (Schwabe and Wolf 2010), and in more impulsive individuals (Hogarth *et al.* 2012b). Because more dependent individuals commonly experience these conditions, one might expect them to be more prone to habit. However, dependence itself is not associated with habit in comparable outcome devaluation procedures as noted above. The implication is that the demonstrations of habit listed in this paragraph reflect a general interference effect on memory resources, rather than the fundamental process underpinning addiction.

The loudest claim that human addiction is underpinned by habit learning has come from studies that have explored the habitual status of responding for nondrug reinforcers in various computer tasks (Ersche *et al.* 2016; Sjoerds *et al.* 2013; McKim *et al.* 2016; Reiter *et al.* 2016). These studies have shown that addicts compared with controls have performance deficits in tasks that require correct responses to be selected dependent on stimuli presented and following instructions about the changed value of nondrug outcomes. However, these tasks do not provide definitive metrics of habit. Impaired performance in addicts could be attributed to trivial group characteristics such as reduced motivation to engage in the task or general cognitive impairments. Addicts have shown a wide variety of task deficits that are not interpreted as evidence for habit (e.g., Bechara and Damasio 2002; Garavan and Stout 2005; Hart *et al.* 2012; Griffiths *et al.* 2012; Verdejo-Garcia *et al.* 2006). So there is no compelling reason to cite habit as a key cause of addiction in preference to any other class of cognitive deficit. Finally, those studies that claim to have demonstrated habit learning in addicts report no correlations between dependence level and the behavioural index of habit (Ersche *et al.* 2016; Sjoerds *et al.* 2013; McKim *et al.* 2016; Reiter *et al.* 2016). Thus, the performance deficit is demonstrably not involved in the aetiology of dependence. To conclude, this section showed that dependence was reliably associated with goal-directed drug choice, whereas studies supporting habit accounts of addiction were either insignificant, indirect or ambiguous.

Incentive learning/self-medication

The self-medication hypothesis claims that drugs are taken to alleviate adverse states (Pomerleau and Pomerleau 1984) and depression and anxiety do indeed prime drug-seeking (Leventhal and Zvolensky 2015). However, adverse emotional states could prime drug-seeking via controlled or automatic processes. Several theories claim that adverse states prime drug-seeking via automatic S–R associations (Baker *et al.* 2004; Belin *et al.* 2013; Wikler 1984). They argue that greater drug reinforcement in the adverse state strengthens the direct link between the adverse state and drug-seeking, allowing automatic control. However, adverse emotional states might instead undergo incentive learning (Hutcheson *et al.* 2001). That is, adverse states predict that drugs have higher value so may evoke this expectation promoting goal-directed drug-seeking. Indeed, negative mood induction reliably primes corresponding effects on craving and drug consumption (Heckman *et al.* 2015), favouring the goal-directed account, but S–R theory could dismiss craving as an epiphenomenon rather than causal.

A recent outcome revaluation (see Table 26.3) study tested whether negative mood modulates goal-directed drug-seeking. Smokers learned that two responses produced different outcomes (R1-tobacco, R2-food reward). A negative mood induction was then used to raise the expected value of tobacco before choice between the two responses was assessed in extinction. It was found that participants who showed an increase in negative mood increased their tobacco choice in the extinction test. This effect must be mediated by the negative mood state raising the expected value of the drug, priming goal-directed drug-seeking, and cannot be explained by automatic S–R theories because no outcomes were presented at test and so they could not modify the strength of any S–R association controlling the responses. Furthermore, negative mood could not generalise from other drug-seeking responses to the novel drug-seeking responses tested in the procedure because there was no difference between the drug and food-seeking responses apart from the outcome these responses earn, so any selective generalisation would have to be based on a representation of the outcome, which is outside the scope of S–R theory (Smyth *et al.* 2008). Supporting the goal-directed interpretation, Owens *et al.* (2014) found that negative mood induction raised self-reported willingness to pay for the drug, which is arguably due to an increase in the expected value of the drug. These data suggest that adverse states prime goal-directed drug-seeking rather than exert automatic control via S–R associations. Indeed, there is now good evidence that depression confers increased sensitivity to negative affective states priming goal-directed drug-seeking, suggesting that vulnerability to dependence may be conferred by excessive incentive learning effects on goal-directed drug-seeking (Hogarth *et al.* 2018; Hogarth and Hardy 2018; Hogarth *et al.* 2017).

Conclusion

The bulk of evidence presented in this chapter has supported the controlled account of drug-seeking behaviour. The first section concluded that cognitive paradigms have failed to adequately discriminate between controlled and automatic processes and so are inconclusive. The section on Pavlovian conditioning concluded that drug cue reactivity relies on an explicit expectation of drug availability, consistent with the controlled account. Similarly, the section on Pavlovian to instrumental transfer concluded that external drug cues prime drug-seeking behaviour through an expectation that the drug-seeking response is more likely to be effective, also consistent with the controlled account. The section on instrumental learning concluded that individual differences in dependence are reliably associated with greater expected value of the drug driving higher rates of goal-directed drug choice and economic demand, whereas evidence for excessive habitual drug-seeking was insignificant, indirect or ambiguous. Finally, the section on self-medication concluded that internal adverse states augment goal-directed drug-seeking by raising the expected value of the drug, again supporting the controlled account. Overall, then, addicts may differ from recreational users in ascribing higher value to the drug and experiencing more adverse states (withdrawal, comorbid psychiatric illness) that further raises the expected drug value, but the two groups are apparently equally reactive to external cues because knowledge of easy drug availability is equally effecting. What then for the role of automatic processes? Although drug-seeking itself appears to be goal-directed, the biopsychosocial factors that shape an individual's constitution such that they experience the drug as having high value are largely outside their deliberative control. The individual cannot deny the pleasure, relief or escape the drug provides, they cannot change their neurochemistry through best intentions, and they cannot easily change their life circumstances (poverty, urban tedium, etc.), all of which elevate the drug above other options. It may be through changing direct experience of the relatively high value of drugs that psychological science has its best hope of achieving the long promised, and long overdue, impact on addictive behaviour (Hall *et al.* 2015).

References

Ahmed, S. H. (2010) "Validation crisis in animal models of drug addiction: beyond non-disordered drug use toward drug addiction", *Neuroscience and Biobehavioral Reviews* 35(2): 172–184.

Baker, T. B., Piper, M. E., McCarthy, D. E., Majeskie, M. R. and Fiore, M. C. (2004) "Addiction motivation reformulated: an affective processing model of negative reinforcement", *Psychological Review* 111(1): 33–51.

Baxter, B. W. and Hinson, R. E. (2001) "Is smoking automatic? Demands of smoking behavior on attentional resources", *Journal of Abnormal Psychology* 110(1): 59–66.

Bechara, A. and Damasio, H. (2002) "Decision-making and addiction (part I): impaired activation of somatic states in substance dependent individuals when pondering decisions with negative future consequences", *Neuropsychologia* 40: 1675–1689.

Belin, D., Belin-Rauscent, A., Murray, J. E. and Everitt, B. J. (2013) "Addiction: failure of control over maladaptive incentive habits", *Current Opinion in Neurobiology* 23(4): 564–572.

Bickel, W. K., Johnson, M. W., Koffarnus, M. N., Mackillop, J. and Murphy, J. G. (2014) "The behavioral economics of substance use disorders: reinforcement pathologies and their repair", *Annual Review of Clinical Psychology* 10: 641–677.

Carter, B. L. and Tiffany, S. T. (1999) "Meta-analysis of cue-reactivity in addiction research", *Addiction* 94(3): 327–340.

Carter, B. L. and Tiffany, S. T. (2001) "The cue-availability paradigm: the effects of cigarette availability on cue reactivity in smokers", *Experimental and Clinical Psychopharmacology* 9(2): 183–190.

Cartoni, E., Moretta, T., Puglisi-Allegra, S., Cabib, S. and Baldassarre, G. (2015) "The relationship between specific Pavlovian instrumental transfer and instrumental reward probability", *Frontiers in Psychology* 6: 1697.

Chase, H. W., Mackillop, J. and Hogarth, L. (2013) "Isolating behavioural economic indices of demand in relation to nicotine dependence", *Psychopharmacology* 226(2): 371–380.

Colwill, R. M. and Rescorla, R. A. (1988) "Associations between the discriminative stimulus and the reinforcer in instrumental learning", *Journal of Experimental Psychology: Animal Behavior Processes* 14(2): 155–164.

Corbit, L. H., Janak, P. H. and Balleine, B. W. (2007) "General and outcome-specific forms of Pavlovian-instrumental transfer: the effect of shifts in motivational state and inactivation of the ventral tegmental area", *European Journal of Neuroscience* 26(11): 3141–3149.

Corbit, L. H., Nie, H. and Janak, P. H. (2012) "Habitual alcohol seeking: time course and the contribution of subregions of the dorsal striatum", *Biological Psychiatry* 72(5): 389–395.

Cousijn, J., Watson, P., Koenders, L., Vingerhoets, W. A. M., Goudriaan, A. E. and Wiers, R. W. (2013) "Cannabis dependence, cognitive control and attentional bias for cannabis words", *Addictive Behaviors* 38(12): 2825–2832.

De Wit, S. and Dickinson, A. (2009) "Associative theories of goal-directed behaviour: a case for animal–human translational models", *Psychological Research PRPF* 73(4): 463–476.

Dickinson, A. (1985) "Actions and habits – the development of behavioral autonomy", *Philosophical Transactions of the Royal Society of London Series B: Biological Sciences* 308: 67–78.

Dickinson, A. and Balleine, B. W. (2010) "The cognitive/motivational interface", in M. L. Kringelbach and K. C. Berridge (eds), *Pleasures of the Brain. The Neural Basis of Taste, Smell and Other Rewards*, Oxford, UK: Oxford University Press.

Dickinson, A., Wood, N. and Smith, J. W. (2002) "Alcohol seeking by rats: action or habit?" *Quarterly Journal of Experimental Psychology Section B: Comparative and Physiological Psychology* 55(4): 331–348.

Dols, M., Willems, B., van den Hout, M. and Bittoun, R. (2000) "Smokers can learn to influence their urge to smoke", *Addictive Behaviors* 25(1): 103–108.

Drobes, D. J. and Tiffany, S. T. (1997) "Induction of smoking urge through imaginal and in vivo procedures: physiological and self-report manifestations", *Journal of Abnormal Psychology* 106(1): 15–25.

Eder, A. B. and Dignath, D. (2016) "Asymmetrical effects of post-training outcome revaluation on outcome-selective Pavlovian-to-instrumental transfer of control in human adults", *Learning and Motivation* 54: 12–21.

Ersche, K. D., Gillan, C. M., Jones, P. S., Williams, G. B., Ward, L. H. E., Luijten, M., De Wit, S., Sahakian, B. J., Bullmore, E. T. and Robbins, T. W. (2016) "Carrots and sticks fail to change behavior in cocaine addiction", *Science* 352(6292): 1468–1471.

Field, M. (2005) "Cannabis 'dependence' and attentional bias for cannabis-related words", *Behavioural Pharmacology* 16(5–6): 473–476.
Field, M. and Duka, T. (2001) "Smoking expectancy mediates the conditioned responses to arbitrary smoking cues", *Behavioural Pharmacology* 12: 183–194.
Garavan, H. and Stout, J. C. (2005) "Neurocognitive insights into substance abuse", *Trends in Cognitive Sciences* 9(4): 195–201.
Garbusow, M., Schad, D. J., Sommer, C., Jünger, E., Sebold, M., Friedel, E., Wendt, J., Kathmann, N., Schlagenhauf, F., Zimmermann, U. S., Heinz, A., Huys, Q. J. M. and Rapp, M. A. (2014) "Pavlovian-to-instrumental transfer in alcohol dependence: a pilot study", *Neuropsychobiology* 70: 111–121.
Greenwald, M. K. (2008) "Behavioral economic analysis of drug preference using multiple choice procedure data", *Drug and Alcohol Dependence* 93(1–2): 103–110.
Greenwald, M. K. and Steinmiller, C. L. (2009) "Behavioral economic analysis of opioid consumption in heroin-dependent individuals: effects of alternative reinforcer magnitude and post-session drug supply", *Drug and Alcohol Dependence* 104(1–2): 84–93.
Griffiths, A., Hill, R., Morgan, C., Rendell, P. G., Karimi, K., Wanagaratne, S. and Curran, H. V. (2012) "Prospective memory and future event simulation in individuals with alcohol dependence", *Addiction* 107(10): 1809–1816.
Grossman, M. and Chaloupka, F. J. (1998) "The demand for cocaine by young adults: a rational addiction approach", *Journal of Health Economics* 17(4): 427–474.
Hall, W., Carter, A. and Forlini, C. (2015) "The brain disease model of addiction: is it supported by the evidence and has it delivered on its promises?" *The Lancet Psychiatry* 2: 105–110.
Hart, C. L., Marvin, C. B., Silver, R. and Smith, E. E. (2012) "Is cognitive functioning impaired in methamphetamine users? A critical review", *Neuropsychopharmacology* 37(3): 586–608.
Heckman, B. W., Carpenter, M. J., Correa, J. B., Wray, J. M., Saladin, M. E., Froeliger, B., Drobes, D. J. and Brandon, T. H. (2015) "Effects of experimental negative affect manipulations on ad libitum smoking: a meta-analysis", *Addiction* 110(5): 751–760.
Heyman, G. M. (2009) *Addiction: A Disorder of Choice*, Cambridge, MA: Harvard University Press.
Heyman, G. M. (2013) "Addiction and choice: theory and new data", *Frontiers in Psychiatry* 4.
Hitsman, B., Hogarth, L., Tseng, L.-J., Teige, J. C., Shadel, W. G., Dibenedetti, D. B., Danto, S., Lee, T. C., Price, L. H. and Niaura, R. (2013) "Dissociable effect of acute varenicline on tonic versus cue-provoked craving in non-treatment-motivated heavy smokers", *Drug and Alcohol Dependence* 130: 135–141.
Hogarth, L. (2012) "Goal-directed and transfer-cue-elicited drug-seeking are dissociated by pharmacotherapy: evidence for independent additive controllers", *Journal of Experimental Psychology: Animal Behavior Processes* 38(3): 266–278.
Hogarth, L., Attwood, A. S., Bate, H. A. and Munafò, M. R. (2012a) "Acute alcohol impairs human goal-directed action", *Biological Psychology* 90: 154–160.
Hogarth, L. and Chase, H. W. (2011) "Parallel goal-directed and habitual control of human drug-seeking: implications for dependence vulnerability", *Journal of Experimental Psychology: Animal Behavior Processes* 37(3): 261–276.
Hogarth, L. and Chase, H. W. (2012) "Evaluating psychological markers for human nicotine dependence: tobacco choice, extinction, and Pavlovian-to-instrumental transfer", *Experimental and Clinical Psychopharmacology* 20(3): 213–224.
Hogarth, L., Chase, H. W. and Baess, K. (2012b) "Impaired goal-directed behavioural control in human impulsivity", *Quarterly Journal of Experimental Psychology* 65(2): 305–316.
Hogarth, L., Dickinson, A. and Duka, T. (2010) "The associative basis of cue elicited drug taking in humans", *Psychopharmacology* 208(3): 337–351.
Hogarth, L., Dickinson, A., Wright, A., Kouvaraki, M. and Duka, T. (2007) "The role of drug expectancy in the control of human drug seeking", *Journal of Experimental Psychology: Animal Behavior Processes* 33(4): 484–496.
Hogarth, L. and Duka, T. (2006) "Human nicotine conditioning requires explicit contingency knowledge: is addictive behaviour cognitively mediated?" *Psychopharmacology* 184(3–4): 553–566.
Hogarth, L., Field, M. and Rose, A. K. (2013) "Phasic transition from goal-directed to habitual control over drug-seeking produced by conflicting reinforcer expectancy", *Addiction Biology* 18(1): 88–97.
Hogarth, L. and Hardy, L. (2018) "Depressive statements prime goal-directed alcohol-seeking in individuals who report drinking to cope with negative affect", *Psychopharmacology* 235: 269–279.

Hogarth, L., Hardy, L., Mathew, A. R. and Hitsman, B. (2018) "Negative mood-induced alcohol-seeking is greater in young adults who report depression symptoms, drinking to cope, and subjective reactivity", *Experimental and Clinical Psychopharmacology* 26(2): 138–146.

Hogarth, L., Mathew, A. R. and Hitsman, B. (2017) "Current major depression is associated with greater sensitivity to the motivational effect of both negative mood induction and abstinence on tobacco-seeking behavior", *Drug and Alcohol Dependence* 176: 1–6.

Hogarth, L., Maynard, O. M. and Munafò, M. R. (2015) "Plain cigarette packs do not exert Pavlovian to instrumental transfer of control over tobacco-seeking", *Addiction* 110: 174–182.

Hogarth, L., Retzler, C., Munafò, M. R., Tran, D. M. D., Troisi II, J. R., Rose, A. K., Jones, A. and Field, M. (2014) "Extinction of cue-evoked drug-seeking relies on degrading hierarchical instrumental expectancies", *Behaviour Research and Therapy* 59: 61–70.

Hogarth, L. and Troisi, J. R. I. (2015) "A hierarchical instrumental decision theory of nicotine dependence", in D. J. K. Balfour and M. R. Munafò (eds), *The Neurobiology and Genetics of Nicotine and Tobacco*, Berlin: Springer International Publishing.

Hutcheson, D. M., Everitt, B. J., Robbins, T. W. and Dickinson, A. (2001) "The role of withdrawal in heroin addiction: enhances reward or promotes avoidance?" *Nature Neuroscience* 4(9): 943–947.

Jonkman, S., Kosaki, Y., Everitt, B. J. and Dickinson, A. (2010) "The role of contextual conditioning in the effect of reinforcer devaluation on instrumental performance by rats", *Behavioural Processes* 83(3): 276–281.

Kirkpatrick, M. G., Gunderson, E. W., Johanson, C.-E., Levin, F. R., Foltin, R. W. and Hart, C. L. (2012) "Comparison of intranasal methamphetamine and d-amphetamine self-administration by humans", *Addiction* 107(4): 783–791.

Kosaki, Y. and Dickinson, A. (2010) "Choice and contingency in the development of behavioral autonomy during instrumental conditioning", *Journal of Experimental Psychology: Animal Behavior Processes* 36(3): 334–342.

Leventhal, A. M. and Zvolensky, M. J. (2015) "Anxiety, depression, and cigarette smoking: a transdiagnostic vulnerability framework to understanding emotion–smoking comorbidity", *Psychological Bulletin* 141(1): 176–212.

Mackillop, J. (2016) "The behavioral economics and neuroeconomics of alcohol use disorders", *Alcoholism: Clinical and Experimental Research* 40(4): 672–685.

Mackillop, J., Few, L. R., Murphy, J. G., Wier, L. M., Acker, J., Murphy, C., Stojek, M., Carrigan, M. and Chaloupka, F. (2012) "High-resolution behavioral economic analysis of cigarette demand to inform tax policy", *Addiction* 107(12): 2191–2200.

Martinovic, J., Jones, A., Christiansen, P., Rose, A. K., Hogarth, L. and Field, M. (2014) "Electrophysiological responses to alcohol cues are not associated with Pavlovian-to-instrumental transfer in social drinkers", *PLOS ONE* 9(4): e94605.

McKim, T. H., Bauer, D. J. and Boettiger, C. A. (2016) "Addiction history associates with the propensity to form habits", *Journal of Cognitive Neuroscience* 28(7): 1024–1038.

Miles, F. J., Everitt, B. J. and Dickinson, A. (2003) "Oral cocaine seeking by rats: action or habit?" *Behavioral Neuroscience* 117(5): 927–938.

Mitchell, C. J., De Houwer, J. and Lovibond, P. F. (2009) "The propositional nature of human associative learning", *Behavioral and Brain Sciences* 32(2): 183–198.

Moeller, S. J., Beebe-Wang, N., Woicik, P. A., Konova, A. B., Maloney, T. and Goldstein, R. Z. (2013) "Choice to view cocaine images predicts concurrent and prospective drug use in cocaine addiction", *Drug and Alcohol Dependence* 130(1–3): 178–185.

Moeller, S. J., Maloney, T., Parvaz, M. A., Dunning, J. P., Alia-Klein, N., Woicik, P. A., Hajcak, G., Telang, F., Wang, G. J., Volkow, N. D. and Goldstein, R. Z. (2009) "Enhanced choice for viewing cocaine pictures in cocaine addiction", *Biological Psychiatry* 66(2): 169–176.

Motschman, C. A. and Tiffany, S. T. (2016) "Cognitive regulation of smoking behavior within a cigarette: automatic and nonautomatic processes", *Psychology of Addictive Behaviors* 30(4): 494–499.

Murphy, J. G., Dennhardt, A. A., Yurasek, A. M., Skidmore, J. R., Martens, M. P., MacKillop, J. and McDevitt-Murphy, M. E. (2015) "Behavioral economic predictors of brief alcohol intervention outcomes", *Journal of Consulting and Clinical Psychology* 83(6): 1033–1043.

Niaura, R. S., Rosenhow, D. J., Binkoff, J. A., Monti, P. M., Pedraza, M. and Abrams, D. B. (1988) "Relevance of cue reactivity to understanding alcohol and smoking relapse", *Journal of Abnormal Psychology* 97(2): 133–152.

Nordquist, R. E., Voorn, P., Malsen, J., Joosten, R., Pennartz, C. M. A. and Vanderschuren, L. (2007) "Augmented reinforcer value and accelerated habit formation after repeated amphetamine treatment", *European Neuropsychopharmacology* 17(8): 532–540.

O'Brien, C. P., Testa, T., O'Brien, T. J., Brady, J. P. and Wells, B. (1977) "Conditioned narcotic withdrawal in humans", *Science* 195(4282): 1000–1002.

Olmstead, M. C., Lafond, M. V., Everitt, B. J. and Dickinson, A. (2001) "Cocaine seeking by rats is a goal-directed action", *Behavioral Neuroscience* 115(2): 394–402.

Owens, M. M., Ray, L. A. and Mackillop, J. (2014) "Behavioral economic analysis of stress effects on acute motivation for alcohol", *Journal of the Experimental Analysis of Behavior* 103(1): 77–86.

Panlilio, L. V., Hogarth, L. and Shoaib, M. (2015) "Concurrent access to nicotine and sucrose in rats", *Psychopharmacology* 232(8): 1451–1460.

Perkins, K. A. (2009) "Does smoking cue-induced craving tell us anything important about nicotine dependence?" *Addiction* 104(10): 1610–1616.

Perkins, K. A. (2012) "Subjective reactivity to smoking cues as a predictor of quitting success", *Nicotine and Tobacco Research* 14(4): 383–387.

Perkins, K. A., Broge, M., Gerlach, D., Sanders, M., Grobe, J. E., Cherry, C. and Wilson, A. S. (2002) "Acute nicotine reinforcement, but not chronic tolerance, predicts withdrawal and relapse after quitting smoking", *Health Psychology* 21(4): 332–339.

Pomerleau, O. F. and Pomerleau, C. S. (1984) "Neuroregulators and the reinforcement of smoking: towards a biobehavioral explanation", *Neuroscience and Biobehavioral Reviews* 8(4): 503–513.

Reiter, A. M. F., Deserno, L., Wilbertz, T., Heinze, H.-J. and Schlagenhauf, F. (2016) "Risk factors for addiction and their association with model-based behavioral control", *Frontiers in Behavioral Neuroscience* 10: 26.

Rescorla, R. A. (1988) "Pavlovian conditioning: it's not what you think it is", *American Psychologist* 43(3): 151–160.

Rescorla, R. A. (1991) "Associative relations in instrumental learning – The 18 Bartlett memorial lecture", *Quarterly Journal of Experimental Psychology Section B: Comparative and Physiological Psychology* 43: 1–23.

Rosas, J. M., Todd, T. P. and Bouton, M. E. (2013) "Context change and associative learning", *Wiley Interdisciplinary Reviews: Cognitive Science* 4(3): 237–244.

Sayette, M. A. and Tiffany, S. T. (2013) "Peak provoked craving: an alternative to smoking cue-reactivity", *Addiction* 108(6): 1019–1025.

Schwabe, L. and Wolf, O. T. (2010) "Socially evaluated cold pressor stress after instrumental learning favors habits over goal-directed action", *Psychoneuroendocrinology* 35(7): 977–986.

Seabrooke, T., Hogarth, L. and Mitchell, C. (2015) "The propositional basis of cue-controlled reward seeking", *The Quarterly Journal of Experimental Psychology* 69(12): 2452–2470.

Siegel, S., Hinson, R. E., Krank, M. D. and McCully, J. (1982) "Heroin overdose death – contribution of drug-associated environmental cues", *Science* 216(4544): 436–437.

Sjoerds, Z., De Wit, S., van den Brink, W., Robbins, T. W., Beekman, A. T., Penninx, B. W. and Veltman, D. J. (2013) "Behavioral and neuroimaging evidence for overreliance on habit learning in alcohol-dependent patients", *Translational Psychiatry* 3: e337.

Sjoerds, Z., Luigjes, J., van den Brink, W., Denys, D. and Yücel, M. (2014) "The role of habits and motivation in human drug addiction: a reflection", *Frontiers in Psychiatry* 5.

Smyth, S., Barnes-Holmes, D. and Barnes-Holmes, Y. (2008) "Acquired equivalence in human discrimination learning: the role of propositional knowledge", *Journal of Experimental Psychology: Animal Behavior Processes* 34(1): 167–177.

Stewart, J., De Wit, H. and Eikelboom, R. (1984) "Role of conditioned and unconditioned drug effects in self-administration of opiates and stimulants", *Psychological Review* 63: 251–268.

Tiffany, S. T. (1990) "A cognitive model of drug urges and drug-use behaviour: role of automatic and nonautomatic processes", *Psychological Review* 97: 147–168.

Verdejo-Garcia, A., Bechara, A., Recknor, E. C. and Perez-Garcia, M. (2006) "Executive dysfunction in substance dependent individuals during drug use and abstinence: an examination of the behavioral, cognitive and emotional correlates of addiction", *Journal of the International Neuropsychological Society* 12(3): 405–415.

Watson, P., Wiers, R. W., Hommel, B. and De Wit, S. (2014) "Working for food you don't desire. Cues interfere with goal-directed food-seeking", *Appetite* 79: 139–148.

Wikler, A. (1984) "Conditioning factors in opiate addiction and relapse", *Journal of Substance Abuse Treatment* 1(4): 279–285.

Wise, R. A. and Koob, G. F. (2014) "The development and maintenance of drug addiction", *Neuropsychopharmacology* 39(2): 254–262.

Zajac, K., Alessi, S. M. and Petry, N. M. (this volume, pp. 455–463) "Contingency management approaches".

27
DECISION-MAKING DYSFUNCTIONS IN ADDICTION

Antonio Verdejo-Garcia

Introduction

The purpose of this chapter is to provide an overview of the decision-making dysfunctions identified in people with substance addictions, as compared with non-drug-using groups, via cognitive tasks that challenge different aspects of choice. I will specifically review empirical evidence regarding five types of cognitive tasks, related to five different aspects of decision-making: reflection impulsivity, delay discounting, decision-making under risk, decision-making under ambiguity and social decision-making. Each of these aspects has specific implications for the development and/or maintenance of addiction. Reflection impulsivity refers to the amount of information that is collected before making a decision, either in open situations or within a risk–reward trade-off scenario (i.e., gathering more information is linked to less reward). Therefore, reflection impulsivity would be relevant to making decisions that are adequately informed (e.g., pursuing knowledge about the positive/negative effects of a certain drug, or drug dosage) versus decisions made on the spur of the moment (e.g., taking whatever drug/dosage is available). Delay discounting relates to choices between lower immediate rewards and larger more delayed rewards (e.g., between $5 now and $50 in 90 days). It is specifically relevant to weighing the pros and cons of potential rewards taking into account both the present and the future. People showing high levels of delay discounting would focus on the immediate lure of drug-related reward, and would be less mindful about the long-term effects of drug use. Decision-making under risk, or risk-taking, consists of choosing between options that contain explicit information about their outcomes. Conversely, decision-making under ambiguity involves choices between options with unknown (or hardly accessible) outcomes. Decision-making under risk and ambiguity are both relevant to evaluating decision-making options in relation to their outcomes; hence, they would be implicated in conflict choices, such as drugs versus work/study the next day, or continuing drug use versus seeking treatment. Sometimes these choices will be framed in risky scenarios (e.g., the risk of choosing drugs versus school/work is explicit), but some other times they will be framed in ambiguous scenarios (e.g., treatment success is not guaranteed; taking drugs during treatment will only have negative consequences if you are caught). Finally, social decision-making is broadly defined as choices made within a social context, by virtue of the presence of others, or the need to interact with others. This aspect of decision-making is particularly relevant for decisions that involve other people, either during evaluation of options (e.g., peer pressure to take drugs) or during evaluation of consequences (e.g., the negative impact of drug use on loved ones). I will focus on data from validated behavioral choice measures of the above-described mechanisms (see Table 27.1 for a summary of the paradigms employed), along with neuroimaging findings that speak of the neural correlates of these mechanisms, in populations with substance addictions that are drug-abstinent at the time of testing. Therefore, the findings that I will review speak to the (non-acute) residual and/or persistent decision-making dysfunctions associated with substance use. I will start with the most highly used legal substance (i.e., alcohol), and then progress with the most frequent drugs of concern among addiction treatment settings: cannabis, stimulants and opiates. Since this sequence (i.e., alcohol, cannabis, and "hard drugs") echoes the most typical individual trajectory of drug use from a developmental standpoint, findings concerning early onset drugs such as alcohol and cannabis will also be used to illustrate premorbid and early risk factors for decision-making dysfunctions among individuals with addiction. Although findings are mainly interpreted in relation to consumption of the above-described substances, I will also emphasize the relevance of sociodemographic, personality, and clinical factors whenever they seem relevant to interpret decision-making profiles. Ultimately, the decision-making profile of individuals with substance use disorders reflects a combination of their genetic and environmental background, their personality, the neuro-psycho-pharmacological effects of the drugs taken and the current clinical and contextual situation that they are experiencing. While case-control comparisons between people with substance addictions and non-using groups typically match the two groups in terms of socio-demographic characteristics including ancestry, age, education,

Table 27.1 Decision-making paradigms used in neurocognitive research

Cognitive mechanism	Task	Task description
Reflection impulsivity	Beads Task	Participants are shown two jars holding 100 beads, with different ratios of red and blue beads found within them (e.g., 85:15 red to blue and 20:80 blue to red). Participants are told that beads are going to be drawn one at a time with replacement from one of these two jars, and that each jar has the same chance to be chosen. Participants are required to determine from which jar the beads have been drawn from. They are told that they can request as many beads as necessary to decide, and are asked after each bead was drawn whether they require more draws, or if they have made a decision.
	Information Sampling Task	Participants are presented with a 5x5 array of gray boxes on the screen, and two larger colored panels below these boxes. They are instructed that they are playing a game for points, which consists of two parts (*free* and *costly* conditions). Participants are informed that they can win points by making a correct decision about which of the two colors is in the majority under the gray boxes. They are told that they can open as many boxes as they wish before making a decision. In the *free* condition, they can open as many boxes as they like. In the *costly* condition, each new box opened discounts points.

Decision-making under risk	Cups Task	Comprises two domains: *Gain* and *Loss*. In each trial, participants are shown two arrays of cups, and asked to choose between a safe and a risky option. The safe option is represented by a single cup with a 100% probability of either winning (*Gain* domain) or losing (*Loss* domain) a small amount of money. The risky option is represented by two, three or five cups with different probabilities (0.20, 0.33 and 0.50) of winning or losing a larger amount of money. Outcomes are displayed after choices are made.
	Coin Flipping Task	In each trial, participants are instructed to decide whether to accept or reject gambles that offer a 0.50 probability of either gaining an amount of money or losing another amount. Participants are asked to indicate one of four responses to each gamble (strongly accept, weakly accept, weakly reject, or strongly reject). The magnitudes of the potential gain and loss range from $10 to $40 and $5 to $20, respectively.
	Balloon Analogue Risk Task	Participants are presented with balloons on a screen, and instructed to earn monetary rewards by pumping the balloon up with mouse clicks (5 cents per pump). Each click causes the balloon to progressively inflate and monetary reward to be added to a counter. However, if the balloon is pumped past its individual explosion point, a "pop" sound effect is generated from the computer, and participants lose the money earned on that balloon. Participants are given a series of balloon trials in which to earn money, and the only money earned is from trials in which they stopped pumping the balloon before it popped.
	Cambridge Gambling Task	Participants are presented with a row of ten boxes on the computer screen appeared in varying ratios of red and blue. At the bottom of the screen are rectangles containing the words 'Red' and 'Blue'. Participants are instructed to guess whether a yellow token is hidden in a red box or a blue box. They are informed that the goal is to accumulate as many points as possible. In the gambling stages of the task, participants start with a number of points, and can select a proportion of these points to gamble on their confidence in this judgement.
	Randomised Lottery Task	Participants are presented with a series of binary lotteries and they are instructed to earn as much money as possible by making a decision between two options in each trial. Each lottery consists of: a lottery characterized by a randomized winning probability of 50 points (risky option), and a guaranteed payoff that is randomly distributed between zero and the 50 points (safe option). For example, participants have to decide between 50 points with a probability of 25% or a guaranteed payoff of 25 points.
Decision-making under ambiguity	Iowa Gambling Task	Participants are presented with four decks of cards (A, B, C and D). They are told that the task consists of a series of card selections, and that the goal is to earn as much money as possible. Each time a participant selects a card, an immediate monetary reward is awarded ($100 in decks A and B and $50 in decks C and D). However, the selection of some cards carries probabilistic monetary losses (which is large in decks A and B and small in decks C and D). Selecting cards mostly from decks A and B leads to an overall loss and selecting cards from decks C and D leads to an overall gain.

(continued)

Table 27.1 (continued)

Cognitive mechanism	Task	Task description
Delay Discounting	Delay Discounting	Participants are instructed to make a series of choices between a smaller amount of money immediately (smaller-sooner reward) and a larger amount after a specific delay (larger-delayed reward). For example, participants are asked whether they would prefer to receive $5 now or $80 in 30 days.
Social Decision-making	Facial Emotional Decoding	Participants are presented with a series of faces, for five seconds each, on the screen. Participants are instructed to identify which emotion (anger, disgust, fear, happiness, sadness and surprise) best describes the facial expression displayed. There is no time limit for responding.
	Moral Decision-making Task	This task consists of four categories of dilemmas: non-moral dilemmas (involve non-moral decisions); impersonal moral dilemmas (involve non-emotional salient moral decisions); and personal moral dilemmas (involve emotional salient moral decisions), which are further classified into low-conflict (high inter-subject response agreement) and high-conflict dilemmas (characterized by lower levels of response agreement). For each dilemma, participants are instructed to answer whether or not they would execute the proposed action and indicate how difficult the question is.
	Ultimatum Game	This task involves one proposer and one responder, and participants always play the responder's role. In each trial, the proposer receives a sum of money and proposes how to divide the sum between the proposer and the responder. The amount of money that the proposer offers to share is indicated on the computer screen, and participants have three seconds to accept or reject the offer. They are informed that if they accept the proposer's offer, both players are supposed to be paid in the specified way. However, if they reject the offer, none of them would get the money.
	Distribution and Dictator Game	Participants are informed that each of the games has only one trial. The Distribution game involves two players and participants play the proposer's role (player A). In each trial, player A proposes how to divide the points with player B, who is a passive recipient and is informed about which distribution player A chooses. Player A can distribute the points in different ways: ranging from a fair distribution where both players would receive the same amount of points each, to the most opportunistic distribution where player A would receive double the amount of points of player B. In the Dictator game, player A receives an endowment of 50 points and can give any amount from 0 to 50 points to player B.

Alcohol

Heavy alcohol users, compared with non-heavy users, have been shown to display dysfunctions in reflection impulsivity (they collect less information before making decisions in the Beads task) (Banca et al. 2016), decision-making under risk in both the Cups and the Coin Flipping tasks (Brevers et al. 2014), decision-making under ambiguity indicated by the Iowa Gambling Task (IGT) (Loeber et al. 2009), and social decision-making indicated by Facial Emotional Decoding and Moral Dilemmas tasks (Carmona-Perera et al. 2013; Carmona-Perera et al. 2014). Poorer performance in these measures has been generally associated with heavier use and other severity indices, such as the number of alcohol detoxifications experienced (Loeber et al. 2009; Brevers et al. 2014; Carmona-Perera et al. 2014), but also with stable personality traits such as non-planning impulsivity (i.e., a general tendency to plan less), general disinhibition linked to Cluster B personality disorders (i.e., more frequent impulsive behaviors), and antisocial characteristics (i.e., less empathy and remorse) (Dom et al. 2006; Cantrell et al. 2008; Miranda et al. 2009; Tomassini et al. 2012). Interestingly, in contrast to the general decision-making findings, heavy alcohol users seem to have normal delay discounting rates (Kirby and Petry 2004; Banca et al. 2016).

The relationship between decision-making dysfunctions and trait characteristics among individuals with alcohol addiction suggests that some of these dysfunctions may be pre-morbid and hence act as potential vulnerability factors for the development of alcoholism. Specifically, dysfunctions in decision-making under uncertainty (IGT) have been reported in early adolescents with relatively low exposure to alcohol, and these deficits longitudinally predict future involvement in binge-drinking behavior (Goudriaan et al. 2007; Johnson et al. 2008). Conversely, a twin study that has examined monozygotic twin pairs discordant for alcohol use has demonstrated that disadvantageous choices in the IGT are directly linked to patterns of alcohol consumption (Malone et al. 2014). However, the same study found a genetically shared non-alcohol related dysfunction in one of the brain regions most importantly involved in decision-making, namely the orbitofrontal cortex (Malone et al. 2014). This finding is consistent with another imaging study (using electroencephalography (EEG)), which has shown that high-risk offspring of alcoholics have premorbidly weaker activity in prefrontal cortical regions (note: it is difficult to ascertain the precise location of the prefrontal region with EEG) during both loss and gain conditions in a monetary gambling task similar to the IGT (Kamarajan et al. 2015). Altogether, the findings suggest a dynamic interaction whereby subtle neural abnormalities may predispose to decision-making alterations, which are further significantly exacerbated by alcohol use.

In sum, individuals with alcohol addiction have been shown to have deficits in four of the five nominated mechanisms of decision-making (reflection impulsivity, risk-taking, decision-making under ambiguity and social decision-making) while spared in delay discounting. Deficits in decision-making under ambiguity, measured with the IGT, have been linked to a dysfunctional orbitofrontal cortex. Since at least part of this orbitofrontal cortex dysfunction is genetically-driven, it is likely that premorbid biological alterations contribute to the decision-making dysfunctions found in alcohol users. Overall, decision-making deficits have been linked to both trait characteristics (impulsivity, disinhibition, and antisociality) and estimates of severity of alcohol use, suggesting a dynamic interaction.

Cannabis

Heavy cannabis/marijuana users have been shown to display dysfunctions in reflection impulsivity (they gather less information in both open situations and risk–reward trade-off situations in the Information Sampling Task (IST)) (Clark et al. 2009), decision-making under risk in the Balloon Analogue Risk Task (BART) (Hanson et al. 2014), and decision-making under ambiguity indicated by the Iowa Gambling Task (IGT) (Whitlow et al. 2004; Verdejo-Garcia et al. 2007). Computational modelling of the IGT has been used to disentangle the different mechanisms leading to disadvantageous choices in the task (Busemeyer and Stout 2002). This approach has revealed that chronic cannabis users show greater attention to gains and lesser attention to losses, greater reliance on recent (versus past) outcomes, and poorer consistency with expected payoffs (Fridberg et al. 2010). These results indicate that cannabis users have deficits in some of the micro-processes that contribute to performance in complex ambiguous decision-making tasks, such as valuing the gains more than the losses (motivation), focusing on recent outcomes rather than in the overall tally (memory-recency), and being erratic rather than consistent across series of choices (consistency). This account is consistent with imaging findings, which have linked cannabis-related decision-making deficits with reduced orbitofrontal cortex activity and increased cerebellar activity (Bolla et al. 2005). While the orbitofrontal cortex is importantly involved in higher-order coding of reward value, the cerebellum is implicated in more simple forms of reward-based motivation and learning (Moreno-Lopez et al. 2015). In contrast to research in other domains of decision-making, current research has not shown consistent evidence of cannabis-related dysfunctions in delay discounting (Mejia-Cruz et al. 2016) or social decision-making. However, it has been shown that adolescent cannabis users display greater activation of striatal regions during choice tasks performed under peer pressure (Gilman et al. 2016a; Gilman et al. 2016b). This discrepancy between behavioral and brain imaging findings may suggest that adolescent cannabis users apply "compensatory" brain activation to function normally, or that existing behavioral paradigms are yet not sensitive enough to capture social decision-making deficits in the population of cannabis users.

Although heavy marijuana use (~20 joints per day) and related dependence are clearly linked to decision-making dysfunctions, these deficits are not noticeable in less severe, more functional cohorts, suggesting that factors other than cannabis can play an important role in explaining alterations in choice behavior. In a study with moderate cannabis users (~3 joints per day), we found no evidence of decision-making deficits in the IGT when examining the whole sample. However, the subsample of cannabis users carrying the *short/short* genotype of the SLC6A4 (5HTTLPR) gene, linked to reduced availability of serotonin (i.e., less serotonin is available for brain synapses, and hence there is poorer baseline function of serotonin-related processes such as affective processing), displayed significantly poorer performance in the IGT task (Verdejo-Garcia et al. 2013). Interestingly, the *short/short* genotype of the SLC6A4 gene has only been associated with deleterious effects in brain function when it interacts with life adversity (e.g., developmental trauma and/or personality disorders) (Belsky and Pluess 2009), a notion that warrants further investigation among cannabis-using populations.

In sum, individuals with cannabis addiction are shown to be impaired in three of the five identified mechanisms of decision-making (reflection impulsivity, risk-taking, and decision-making under ambiguity) and spared in delay discounting and social decision-making. Behavioral deficits have been linked to lesser activity in the orbitofrontal cortex and greater activity in the cerebellum during ambiguous choices involving reward and punishment. Collectively, these deficits have been linked to both genetic characteristics (a common genetic variation in the serotonin transporter gene) and indicators of severity of cannabis use, namely, heavy daily use of the drug.

Stimulants

Chronic cocaine users have been shown to display dysfunctions in reflection impulsivity (in both open situations and information vs reward scenarios in the IST) (Stevens *et al.* 2015), delay discounting (Kirby and Petry 2004), decision-making under risk indicated by the Randomised Lottery Task (RLT) (Wittwer *et al.* 2016), and decision-making under ambiguity indicated by the Iowa Gambling Task (IGT) – the latter accounted by greater attention to gains and lower choice consistency (Stout *et al.* 2004; Verdejo-Garcia *et al.* 2007; Hulka *et al.* 2014). Poorer performance in the RLT and the IGT have been specifically associated with heavier cocaine use measured with self-report and concentration of cocaine metabolites in hair (Verdejo-Garcia *et al.* 2014; Wittwer *et al.* 2016). Similar deficits have been described in chronic users of methamphetamine, who have a higher tendency to win/stay in probabilistic reward tasks, and dose-related effects of stimulants on attention to gains in the IGT (the motivational mechanism) (Ahn *et al.* 2014; Harle *et al.* 2015). Cocaine users have also been shown to display social decision-making deficits indicated with social interaction tasks. Specifically, they show more self-serving behavior in the Distribution Game and the Dictator Game, although these social choice patterns are noticeable in both recreational users and cocaine dependents, suggesting that they could play an important role in the vulnerability for stimulant use disorders (Hulka *et al.* 2014).

Imaging findings have shown that a dysfunctional orbitofrontal cortex is importantly involved in both ambiguous and social decision-making deficits among stimulant users (Bolla *et al.* 2003; Tanabe *et al.* 2009; Preller *et al.* 2014). In cocaine dependent users, this region appears to be specifically associated with an alteration of the affective tagging of social and non-social decision outcomes (Verdejo-Garcia 2014). Reduced dorsolateral prefrontal cortex (DLPFC) and medial prefrontal cortex activity have also been associated with cocaine-related effects on decision-making under ambiguity, and social decision-making in Moral Dilemmas (Verdejo-Garcia *et al.* 2014; Verdejo-Garcia *et al.* 2015). Decreased DLPFC and anterior insula/rostral anterior cingulate cortex activity (the latter linked to lesser sensitivity to losses) have been linked to increased risk-taking in methamphetamine users (Gowin *et al.* 2014; Kohno *et al.* 2014).

A relevant aspect of decision-making findings in stimulant dependent users is their robust association with meaningful clinical outcomes. We have shown that a decision-making profile characterized by insensitivity to future consequences, defined as impaired ability to anticipate both reward and punishment outcomes in the IGT, is associated with future relapse in 90% of affected individuals (Verdejo-Garcia *et al.* 2014). Poorer performance in the IGT has also been associated with greater degree of social dysfunction among cocaine users (Cunha *et al.* 2011). These findings suggest that decision-making under ambiguity is a reliable predictor of negative treatment and psychosocial outcomes in this population.

In sum, individuals with stimulant addiction have shown to have deficits in all the nominated mechanisms of decision-making (reflection impulsivity, delay discounting, decision-making under risk, decision-making under ambiguity and social decision-making). Non-social and social decision-making deficits have been linked to overlapping alterations in the orbitofrontal cortex and the medial prefrontal cortex. Overall, deficits are linked to estimates of severity of cocaine and methamphetamine use and seem to be linked to neuroadaptive effects of stimulants in medial frontostriatal and frontolimbic pathways.

Heroin and other opioids

Chronic heroin users have been shown to display dysfunctions in reflection impulsivity (in open situations and information vs reward trade-off situations in the IST) (Clark et al. 2006), delay discounting (Kirby and Petry 2004), decision-making under risk indicated by the Cambridge Gamble Task (CGT) (Fishbein et al. 2007; Baldacchino et al. 2015), and decision-making under ambiguity indicated by the Iowa Gambling Task (IGT) (Verdejo-Garcia and Perez-Garcia 2007). The CGT deficits have been specifically associated with patterns of heroin use in a sample of relatively "pure" opioid-only users (Fishbein et al. 2007), and linked to orbitofrontal cortex deficits in current and former users (Ersche et al. 2005). The IGT deficits seem to be associated with some trait characteristics, such as general impulsivity and reduced loss aversion (Ahn et al. 2014). Reduced reflection impulsivity has been demonstrated in both current and former opioid users (Clark et al. 2006), and steeper delay discounting and poorer decision-making under ambiguity are noticeable even after protracted periods of abstinence (Li et al. 2013), suggesting that they are stable characteristics, somehow disconnected from the residual neuropharmacological effects of the opioid drugs. Although there is not much literature on social decision-making in this population, a recent study using a social interaction task (the Ultimatum Game) has shown heroin-related increases in rejection of unfair offers of low reward magnitude, coupled with *decreased* rejection of unfair offers of high reward magnitude (Hou et al. 2016). It is as yet unclear to what extent these deficits are clinically significant, and if they relate to heroin consumption or, more generally, to socioeconomic disadvantage.

A specific area of interest within the opioid use literature is that of decision-making deficits associated with different opioid pharmacotherapies, including heroin substitution pharmacotherapies (methadone, buprenorphine) and analgesic opioids used in the treatment of patients with chronic pain (oxycodone). An important consideration in this instance is that, unlike the bulk of the literature reviewed so far – focused on abstinent users and residual/protracted correlates of drug use – participants on opioid substitution therapies can be under acute pharmacological effects of the opioid prescription drugs. Current research indicates that methadone use is associated with a tendency to place higher-magnitude bets in the CGT, although it does not have an impact on risk-taking choice behavior (Baldacchino et al. 2015). Conversely, methadone use has been significantly associated with poorer decision-making under ambiguity indicated by the IGT whereas, comparatively, buprenorphine seems to be linked to preserved IGT performance (Pirastu et al. 2006). These deficits have been linked to altered midbrain, striatal and hippocampal coding of reward and loss signals during decision-making (Gradin et al. 2014). Finally, the one study available concerning decision-making in chronic pain patients taking analgesic opioids has not shown evidence of a significant impact of these drugs on decision-making performance (Baldacchino et al. 2015).

In sum, individuals with heroin addiction have been shown to have deficits in reflection impulsivity, delay discounting, decision-making under risk, and decision-making under ambiguity. There is a need for more studies on social decision-making. The association between patterns of drug use and decision-making decrements is not as clear as in stimulant users. Overall, decision-making deficits are quite stable even in long-term abstinent ex-users, suggesting that sociodemographic variables and trait characteristics could play an important role in these deficits, on top of heroin consumption. In support of this notion, treatment with pharmacologically similar opioid analgesics has not been linked to decision-making deficits in available studies. Methadone, however, has been shown to have a significant impact on reward processing during risk-taking and in decision-making under ambiguity. Deficits have been linked to altered risk-taking evoked orbitofrontal activation in current and former heroin users, and to altered midbrain, striatal and hippocampal activation in current methadone users.

Discussion and concluding remarks

The above-described studies illustrate that substance addictions are associated with significant decision-making deficits in comparison with non-drug-using cohorts. Decision-making deficits are generally characterized by greater motivation for reward and increased tolerance for uncertainty, risk and potential losses. Alcohol and cocaine users are comparatively more impaired (in breadth and depth), and deficits have been clearly associated with patterns of drug use. Heroin and methadone users have also shown significant deficits in most decision-making mechanisms, although the acute and residual neuro-pharmacological effects of opioid drugs on these mechanisms is less clear, and trait and socioeconomic variables, such as high impulsivity and low loss aversion, which can be considered "normal" in the context of users' lifestyles, are likely to play an important role. Only heavy cannabis users show significant dysfunctions of decision-making; for most recreational and moderate users, the degree of cognitive deficits is likely to be determined by the interaction between common genetic variations related to serotonin and catecholamine systems and life adversity including neurodevelopmental stressors and/or personality disorders.

In regard to neuroanatomical substrates, there is remarkable consistency in the link between decision-making deficits and alterations in the orbitofrontal cortex. Medial orbitofrontal cortex alterations have been associated with blunted affective tagging of non-drug related rewards (e.g., social rewards) and lateral orbitofrontal cortex alterations with differential coding of gains and losses in risk-taking tasks. A broader circuitry involving the medial prefrontal cortex and the anterior cingulate cortex, the insula, and thalamic and brain stem regions implicated in core emotional processes have also shown to impact social-related decisions. Interestingly, at least part of the link between orbitofrontal dysfunction and alcohol/drug use is accounted for by genetic factors. Therefore, premorbid factors (e.g., genes, temperament/personality, adversity) and patterns of alcohol and/or drug use should both be taken into account when explaining the neurobiological correlates of decision-making deficits in substance addictions.

A longstanding debate, discussed across the different sections above, relates to whether these deficits are premorbid or substance induced. Evidence differs across substances: alcohol and stimulants (cocaine and methamphetamine) have demonstrated clear neurotoxic and neuroadaptive effects, respectively. Conversely, background and trait characteristics and socioeconomic adversity seem to play a more important role in the decision-making deficits of cannabis- and opioid-dependent users. Regardless of the intensity of neuro-pharmacological effects of drug usage, the developmental course of addiction seems to imply a dynamic, synergistic process, whereby premorbid risk factors (genetic influences, poor parenting, early trauma and stress) have demonstrated capacity to alter the neural circuitry of decision-making, and drugs of abuse have demonstrated capacity to exacerbate such damage. Behavioral manifestations of decision-making dysfunctions are likely to reflect different forms of interaction between these multiple antecedent factors.

References

Ahn, W. Y., Vasilev, G., et al. (2014) "Decision-making in stimulant and opiate addicts in protracted abstinence: evidence from computational modeling with pure users", *Frontiers in Psychology* 5: 849.

Baldacchino, A., Balfour, D. J., et al. (2015) "Impulsivity and opioid drugs: differential effects of heroin, methadone and prescribed analgesic medication", *Psychological Medicine* 45(6): 1167–1179.

Banca, P., Lange, I., et al. (2016) "Reflection impulsivity in binge drinking: behavioural and volumetric correlates." *Addiction Biology* 21(2): 504–515.

Belsky, J. and Pluess, M. (2009) "Beyond diathesis stress: differential susceptibility to environmental influences", *Psychological Bulletin* 135(6): 885–908.

Bolla, K. I., Eldreth, D. A., et al. (2003) "Orbitofrontal cortex dysfunction in abstinent cocaine abusers performing a decision-making task", *Neuroimage* 19(3): 1085–1094.

Bolla, K. I., Eldreth, D. A., et al. (2005) "Neural substrates of faulty decision-making in abstinent marijuana users", *Neuroimage* 26(2): 480–492.

Brevers, D., Bechara, A., et al. (2014) "Impaired decision-making under risk in individuals with alcohol dependence", *Alcoholism: Clinical and Experimental Research* 38(7): 1924–1931.

Busemeyer, J. R. and Stout, J. C. (2002) "A contribution of cognitive decision models to clinical assessment: decomposing performance on the Bechara gambling task", *Psychological Assessment* 14(3): 253–262.

Cantrell, H., Finn, P. R., et al. (2008) "Decision making in alcohol dependence: insensitivity to future consequences and comorbid disinhibitory psychopathology", *Alcoholism: Clinical and Experimental Research* 32(8): 1398–1407.

Carmona-Perera, M., Clark, L., et al. (2014) "Impaired decoding of fear and disgust predicts utilitarian moral judgment in alcohol-dependent individuals", *Alcoholism: Clinical and Experimental Research* 38(1): 179–185.

Carmona-Perera, M., Reyes Del Paso, G. A., et al. (2013) "Heart rate correlates of utilitarian moral decision-making in alcoholism", *Drug and Alcohol Dependence* 133(2): 413–419.

Clark, L., Robbins, T. W., et al. (2006) "Reflection impulsivity in current and former substance users", *Biological Psychiatry* 60(5): 515–522.

Clark, L., Roiser, J. P. et al. (2009) "Disrupted 'reflection' impulsivity in cannabis users but not current or former ecstasy users", *Journal of Psychopharmacology* 23: 14–22.

Cunha, P. J., Bechara, A., et al. (2011) "Decision-making deficits linked to real-life social dysfunction in crack cocaine-dependent individuals", *The American Journal on Addictions* 20(1): 78–86.

Dom, G., De Wilde, B., et al. (2006) "Decision-making deficits in alcohol-dependent patients with and without comorbid personality disorder", *Alcoholism: Clinical and Experimental Research* 30(10): 1670–1677.

Ersche, K. D., Fletcher, P. C., et al. (2005) "Abnormal frontal activations related to decision-making in current and former amphetamine and opiate dependent individuals", *Psychopharmacology (Berl)* 180(4): 612–623.

Fishbein, D. H., Krupitsky, E., et al. (2007) "Neurocognitive characterizations of Russian heroin addicts without a significant history of other drug use", *Drug and Alcohol Dependence* 90(1): 25–38.

Fridberg, D. J., Queller, S., et al. (2010) "Cognitive mechanisms underlying risky decision-making in chronic cannabis users", *Journal of Mathematical Psychology* 54(1): 28–38.

Gilman, J. M., Lee, S., et al. (2016a) "Variable activation in striatal subregions across components of a social influence task in young adult cannabis users", *Brain and Behavior* 6(5): e00459.

Gilman, J. M., Schuster, R. M., et al. (2016b) "Neural mechanisms of sensitivity to peer information in young adult cannabis users", *Cognitive, Affective, & Behavioral Neuroscience* 16(4): 646–661.

Goudriaan, A. E., Grekin, E. R., et al. (2007) "Decision making and binge drinking: a longitudinal study", *Alcoholism: Clinical and Experimental Research* 31(6): 928–938.

Gowin, J. L., Harle, K. M., et al. (2014) "Attenuated insular processing during risk predicts relapse in early abstinent methamphetamine-dependent individuals", *Neuropsychopharmacology* 39(6): 1379–1387.

Gradin, V. B., Baldacchino, A., et al. (2014) "Abnormal brain activity during a reward and loss task in opiate-dependent patients receiving methadone maintenance therapy", *Neuropsychopharmacology* 39(4): 885–894.

Hanson, K. L., Thayer, R. E., et al. (2014) "Adolescent marijuana users have elevated risk-taking on the balloon analog risk task", *Journal of Psychopharmacology* 28(11): 1080–1087.

Harle, K. M., Stewart, J. L., et al. (2015) "Bayesian neural adjustment of inhibitory control predicts emergence of problem stimulant use", *Brain* 138(11): 3413–3426.

Hou, Y., Zhao, L., et al. (2016) "Altered economic decision-making in abstinent heroin addicts: evidence from the ultimatum game', *Neuroscience Letters* 627: 148–154.

Hulka, L. M., Eisenegger, C., et al. (2014) "Altered social and non-social decision-making in recreational and dependent cocaine users", *Psychological Medicine* 44(5): 1015–1028.

Johnson, C. A., Xiao, L., et al. (2008) "Affective decision-making deficits, linked to a dysfunctional ventromedial prefrontal cortex, revealed in 10th grade Chinese adolescent binge drinkers", *Neuropsychologia* 46(2): 714–726.

Kamarajan, C., Pandey, A. K., et al. (2015) "Reward processing deficits and impulsivity in high-risk offspring of alcoholics: a study of event-related potentials during a monetary gambling task", *International Journal of Psychophysiology* 98(2 Pt 1): 182–200.

Kirby, K. N. and Petry, N. M. (2004) "Heroin and cocaine abusers have higher discount rates for delayed rewards than alcoholics or non-drug-using controls", *Addiction* 99(4): 461–471.

Kohno, M., Morales, A. M., et al. (2014) "Risky decision making, prefrontal cortex, and mesocorticolimbic functional connectivity in methamphetamine dependence", *JAMA: The Journal of the American Medical Association: Psychiatry* 71(7): 812–820.

Li, X., Zhang, F., et al. (2013) "Decision-making deficits are still present in heroin abusers after short- to long-term abstinence", *Drug and Alcohol Dependence* 130(1–3): 61–67.

Loeber, S., Duka, T., et al. (2009) "Impairment of cognitive abilities and decision making after chronic use of alcohol: the impact of multiple detoxifications", *Alcohol and Alcoholism* 44(4): 372–381.

Malone, S. M., Luciana, M., et al. (2014) "Adolescent drinking and motivated decision-making: a cotwin-control investigation with monozygotic twins", *Behavior Genetics* 44(4): 407–418.

Mejia-Cruz, D., Green, L., et al. (2016) "Delay and probability discounting by drug-dependent cocaine and marijuana users", *Psychopharmacology (Berl)* 233(14): 2705–2714.

Miranda, R. Jr, MacKillop, J., et al. (2009) "Influence of antisocial and psychopathic traits on decision-making biases in alcoholics", *Alcoholism: Clinical and Experimental Research* 33(5): 817–825.

Moreno-Lopez, L., Perales, J. C., et al. (2015) "Cocaine use severity and cerebellar gray matter are associated with reversal learning deficits in cocaine-dependent individuals", *Addiction Biology* 20(3): 546–556.

Pirastu, R., Fais, R., et al. (2006) "Impaired decision-making in opiate-dependent subjects: effect of pharmacological therapies", *Drug and Alcohol Dependence* 83(2): 163–168.

Preller, K. H., Herdener, M., et al. (2014) "Functional changes of the reward system underlie blunted response to social gaze in cocaine users", *Proceedings of the National Academy of Sciences of the United States of America* 111(7): 2842–2847.

Stevens, L., Roeyers, H., et al. (2015) "Impulsivity in cocaine-dependent individuals with and without attention-deficit/hyperactivity disorder", *European Addiction Research* 21(3): 131–143.

Stout, J. C., Busemeyer, J. R., et al. (2004) "Cognitive modeling analysis of decision-making processes in cocaine abusers", *Psychonomic Bulletin & Review* 11(4): 742–747.

Tanabe, J., Tregellas, J. R., et al. (2009) "Medial orbitofrontal cortex gray matter is reduced in abstinent substance-dependent individuals", *Biological Psychiatry* 65(2): 160–164.

Tomassini, A., Struglia, F., et al. (2012) "Decision making, impulsivity, and personality traits in alcohol-dependent subjects", *The American Journal on Addictions* 21(3): 263–267.

Verdejo-Garcia, A. (2014) "Social cognition in cocaine addiction", *Proceedings of the National Academy of Sciences of the United States of America* 111(7): 2406–2407.

Verdejo-Garcia, A., Albein-Urios, N., et al. (2014) "Decision-making impairment predicts 3-month hair-indexed cocaine relapse", *Psychopharmacology (Berl)* 231(21): 4179–4187.

Verdejo-Garcia, A., Benbrook, A., et al. (2007) "The differential relationship between cocaine use and marijuana use on decision-making performance over repeat testing with the Iowa Gambling Task", *Drug and Alcohol Dependence* 90(1): 2–11.

Verdejo-Garcia, A., Clark, L., et al. (2015) "Neural substrates of cognitive flexibility in cocaine and gambling addictions", *British Journal of Psychiatry* 207(2): 158–164.

Verdejo-Garcia, A., Contreras-Rodriguez, O., *et al.* (2014) "Functional alteration in frontolimbic systems relevant to moral judgment in cocaine-dependent subjects", *Addiction Biology* 19(2): 272–281.

Verdejo-Garcia, A., Fagundo, A. B., *et al.* (2013) "COMT val158met and 5-HTTLPR genetic polymorphisms moderate executive control in cannabis users", *Neuropsychopharmacology* 38(8): 1598–1606.

Verdejo-Garcia, A. and Perez-Garcia, M. (2007) "Profile of executive deficits in cocaine and heroin polysubstance users: common and differential effects on separate executive components", *Psychopharmacology (Berl)* 190(4): 517–530.

Verdejo-Garcia, A. J., Perales, J. C., *et al.* (2007) "Cognitive impulsivity in cocaine and heroin polysubstance abusers", *Addictive Behaviors* 32(5): 950–966.

Whitlow, C. T., Liguori, A., *et al.* (2004) "Long-term heavy marijuana users make costly decisions on a gambling task", *Drug and Alcohol Dependence* 76(1): 107–111.

Wittwer, A., Hulka, L. M., *et al.* (2016) "Risky decisions in a lottery task are associated with an increase of cocaine use", *Frontiers in Psychology* 7(1): 640.

28

THE CURRENT STATUS OF THE INCENTIVE SENSITIZATION THEORY OF ADDICTION

Mike J. F. Robinson, Terry E. Robinson, and Kent C. Berridge

What is the nature of addiction?

Most adults have at one point in their life used a potentially addictive substance, if legal, illicit or prescription drugs are included. Yet relatively few develop sufficient problems to meet the formal criteria for drug addiction, even for potent illegal drugs such as cocaine or heroin. The same is true for substances and activities that carry a similar potential for abuse and addiction, such as gambling and highly processed sugary foods. For all of these, addiction is characterized as a compulsive behavior consisting of overwhelming involvement with procurement and consumption of the reward, persistence in the face of increasing costs, and a loss of control and narrowing of interests away from other forms of reward besides the reward of choice. The essence of the incentive-sensitization hypothesis is that the core features of addiction result from mesocorticolimbic hypersensitivity to drugs and to drug-related cues, resulting from neuroadaptations, which produces excessive and narrowly focused 'wanting' or incentive salience to obtain and consume the addictive target. Individuals differ in susceptibility to drug-induced sensitization, and differ in patterns of drug-taking that can determine whether and how much incentive-sensitization they eventually incur (Allain *et al.* 2015). For those who do become sensitized, mesolimbic interactions with corticolimbic circuitry focus excessive desire specifically on that target of addiction, resulting in surges of enhanced 'wanting' of the addictive target upon encountering related cues or vivid imagery. Incentive-sensitization persists even after drug consumption stops and after withdrawal symptoms end, contributing to the chief problem in treating drug addiction, namely relapse. Even after prolonged periods of withdrawal and abstinence, a high percentage of addicted individuals in treatment programs eventually relapse to drug taking. For example, in a study of heroin users, relapse rates to re-use after cessation were approximately 60% within 3 months and at least 75% within 12 months. For this reason, drug addiction is characterized as a chronic relapsing disorder; relapse is the rule rather than the exception, and often occurs repeatedly. The persistence of incentive-sensitization may be a reason why successful abstinence from addiction can require years or even decades of effort. Many who are addicted in their teens or 20s eventually do escape in their 30s (Heyman 2009). Those successes may also involve major changes to life contexts that previously acted as triggering cues for sensitized 'wanting', so that remaining sensitization is no longer so detrimentally expressed. Conceivably, a decline in the intensity of sensitized 'wanting' peaks may also be aided by the natural age-related decay of brain dopamine systems that occurs after the 20s (Mukherjee *et al.* 2002). And still, some addicts never do fully escape, not even in their 40s, 50s, 60s or beyond – a potential testimony to the enduring persistence of incentive sensitization.

Traditionally there have been three alternative explanations to incentive-sensitization that are suggested to explain relapse: 1) *Drug euphoria* – that addicts resume drug taking to experience the intense pleasure (euphoria) they remember the drug producing; 2) *Overlearning habits or predictions* – drug taking becomes such a well-entrenched habit that relapse is almost inevitable, or that learning becomes distorted in other ways to create false predictions about drug rewards; and finally 3) *Withdrawal escape* – that the withdrawal syndrome that accompanies the cessation of drug intake is so unpleasant an addict would do anything to stop it, and so relapse occurs as an escape from withdrawal. All three of these explanations certainly play a role in relapse, yet several considerations suggest that they leave out many situations where relapse occurs.

First, drug pleasure or euphoria certainly accounts for the initial pattern of drug use and abuse, but may have more difficulty accounting for relapse. In fact, drug self-administration can be maintained in the absence of any subjective pleasure (Fischman and Foltin 1992), which supports the view that subjective pleasure does not play a necessary causal role in drug-taking behavior. In addition, some individuals actually have been described as experiencing a decrease in drug pleasure after prolonged use, due to development of tolerance, yet experience a simultaneous increase in drug craving. Thus relapse may occur at a moment of reduced drug pleasure but elevated drug craving. Also, relapse happens even in situations when addicts know their drug will fail to lead to intense pleasure, but rather to more misery.

It also has been suggested that the repeated use of drugs creates a learning disorder or makes drug taking an overly ritualized habitual act. This may be true of drug consumption, but perhaps less so regarding the preceding motivation to obtain drug (Singer *et al.* 2018). Learned habits alone cannot account for the excessive motivational attraction that drugs and their cues develop through the course of addiction. The idea that addiction is merely a rigid stimulus-response habit does not account for how motivation imbues the act of drug taking with characteristic flexibility and innovation of new means of obtaining drugs when required, nor with compulsive overtones that cannot readily be overridden by the resolution to abstain. Other extremely well-learned habits such as tying one's shoes and brushing one's teeth are not compulsive in a motivational sense – those habits can easily be left undone or stopped midway if one wishes, without experiencing a compulsive urge to continue, despite the cost. Furthermore, recent evidence suggests that the learning of strong associations does not mean that well-learned predictors carry the most incentive motivation. Instead, as can be seen with gambling, highly uncertain and therefore weak predictors can invigorate motivation and at times sensitize reward pathways (Robinson, Fischer, *et al.* 2015).

Finally, many addictive drugs surely do induce homeostatic responses that oppose the primary drug effects, and produce the phenomena of tolerance (when drug is present) and withdrawal (when drug is absent). Withdrawal in particular is typically described as an intense negative emotional state accompanied by dysphoria, anxiety and irritability. Withdrawal may indeed be a potent reason why many addicts relapse and take drugs, at least while the withdrawal lasts (Koob 1996). Yet withdrawal is a relatively short-lived phenomenon that decays substantially within days to weeks. It is increasingly well managed by pharmaceutical treatments (e.g., methadone, buprenorphine, nicotine patch, etc.), which may relieve negative feelings of withdrawal distress and are useful in helping the addict manage daily functions, but they are rarely a magic cure for the underlying addiction. By contrast, while relapse frequently occurs during withdrawal it also often occurs much later, and even in fully 'detoxified' addicts who are no longer experiencing any strong negative symptoms of withdrawal. In fact, the motivational impact of drug cues may even increase as withdrawal symptoms abate (Pickens *et al.* 2011). Even if one adds conditioned withdrawal that sometimes occurs later – symptoms resurrected by drug cues – to the category of withdrawal, long-term feelings of withdrawal remain relatively infrequent and weak as a cause of relapse. For example, McAuliffe reported that only 11 out of 40 (27.5%) heroin addicts reported experiencing conditioned withdrawal feelings at all, and only 2 (5%) said it led them to ever resume drug use (McAuliffe 1982). In addition, addicts will often voluntarily undergo the unpleasant process of withdrawal in detoxification clinics, sometimes again and again, to reduce tolerance and the monetary cost of their addiction, and to possibly regain some of the euphoria that comes with taking the drug. In fact, the traditional belief that the avoidance of withdrawal was a critical factor on the road to addiction was in large part due to reliance on observations of opiate abusers, and it has since then been noted that drug addiction and even relapse can occur with little to no physical withdrawal symptoms. In short, withdrawal does not always occur, is not always highly avoided or so bad as to be the chief cause of addictive drug-taking, and the end of withdrawal does not signal the end of addiction.

In contrast to these suggestions, the Incentive-Sensitization Theory (T. E. Robinson and Berridge 1993) proposes that relapse in addicts typically occurs as a result of drug-induced brain changes that lead to intense incentive motivation for drugs, or pulses of 'wanting' often triggered by drug cues or vivid imagery about drug-taking, which may control behavior implicitly, or sometimes may be experienced as feelings of drug craving. Craving is defined as pathologically intense feelings of wanting, which can be produced when incentive salience (or core 'wanting') is translated into conscious awareness. Sensitized 'wanting' is phasic and targeted, rather than expressed as a constant craving for all life's rewards. Triggers of sensitized 'wanting' are typically particularly specific cues and contexts (Robinson *et al.* 1998; Wyvell and Berridge 2001; Tindell *et al.* 2005; Robinson, Fischer, *et al.* 2015). The narrow focus of sensitized 'wanting' on an addicted target results from mesolimbic dopamine interactions with corticolimbic circuitry including the amygdala, which normally assign incentive salience to particular learned reward targets, but which can be stimulated into generating intense, narrowly focused 'wanting' of one thing at the expense of all others as occurs in addiction (Nocjar and Panksepp 2002; Mahler and Berridge 2009; Robinson *et al.* 2014). This targeted focus may come from associative learning, whereby intense drug effects become focused on the stimuli present while experiencing such effects. At its core, the motivation to take drugs is due to the over-attribution of incentive salience to drug-related stimuli. It is important to note that incentive salience is a distinct psychological process from withdrawal and drug pleasure.

In particular, the Incentive Sensitization Theory suggests that craving and relapse are governed by a sensitized neural system (mesocorticolimbic dopamine and related systems) that normally functions to attribute incentive salience to reward cues. This system transforms ordinary stimuli, such as cues associated with rewards, into incentive stimuli, making them attractive and able to trigger an urge to pursue and consume their reward. Repeated drug use produces sensitization of this brain system, which leads to increased 'wanting', which in turn leads to excessive control of behavior by drug-related incentive stimuli. The mechanism that assigns incentive salience to particular cues rather than others still remains to be determined. However, evidence shows that sensitization specifically magnifies 'wanting' for the addictive target, and does so in an all-or-none manner that is dependent on the presence of these cues, rather than as a generalized increase in overall motivation. Addiction can therefore be described as a problem of excessive focused 'wanting'. Importantly, excessive 'wanting' can occur even in the absence of excessive liking for drugs. In fact, the increasing dissociation that addicts exhibit between how much they 'want' drugs and the pleasure drugs produce, explains many of the irrational features of behavior behind their drug-taking and drug-seeking habits. Irrational 'wanting' for addictive drugs can lead to persistent taking of drugs again and again despite the adverse consequences and even if the euphoria of drug consumption declines (Reed et al. 2009; Small et al. 2009). Incentive-sensitization sees unconscious and irrational 'wanting' to be the essence of addiction, while viewing most other diagnostic criteria to be sensitization consequences or cognitive counter-reactions to excessive craving, such as conscious denial of the problem, inability to cut down, disruption of life obligations, etc. The basic mechanisms of the excessive attribution of incentive salience to drugs and drug-related stimuli can even occur as a mostly unconscious process, creating urges to take drugs whether or not a strong subjective feeling of craving is simultaneously present. Such dissociation between acted-on motivation and confusing subjective feelings is what often renders the compulsive quality of an addict's own behavior astonishing even to him or her.

Incentive salience and the Incentive Sensitization Theory

A stunning anecdote often arising from clinical settings is that as drugs become wanted more and more, they may become liked less and less. For example, several reports has even shown tolerance to the euphoric effects of psychostimulant drugs in cocaine-dependent abusers despite enhanced drug-seeking (Bartlett et al. 1997; Reed et al. 2009; Small et al. 2009). This is compatible with the idea that the basic brain mechanisms of reward 'liking' and reward 'wanting' are dissociable. In fact, the incentive sensitization theory put forward by Robinson and Berridge partitions drug reward into three components, 'liking', 'wanting' and learning. Incentive-sensitization suggests that each of these components plays a role in the development of drug use, but that it is primarily distorted amplification of the 'wanting' component alone that makes drug addiction so compulsive and resistant to recovery. The incentive sensitization theory does not deny that drugs produce pleasure, or are learned about, or that drug cessation produces a period of unpleasant withdrawal that prompts some individuals to relapse while it lasts. It simply suggests that the attribution of incentive salience is the critical process that gives rise to 'wanting', and that addictive drug use in susceptible people creates very long-lasting brain changes, such as neural sensitization of dopamine-related systems that connect the ventral tegmental area (VTA) to targets in the nucleus accumbens, neostriatum, amygdala, ventral pallidum and prefrontal cortex. The attribution of incentive salience is a psychological process mediated by brain mesocorticolimbic systems that helps direct behavior towards naturally sought after rewards, such as food, water and sex. It heightens perception and focuses attention towards the particular sights, sounds and smells associated with these rewards in a way that normally promotes well-being and survival (Hickey and Peelen 2015). Incentive salience may have evolved to guide behavior in the right direction even prior to having experienced the rewards, but also came to be recruited by pleasurable experiences and combined with learning to add additional guidance. Incentive salience obeys what are sometimes called Bolles-Bindra-Toates principles of incentive motivation (Berridge 2004), which posit that Pavlovian cues that predict pleasant rewards trigger motivation, interact with current physiological states, and become wanted and liked in much the same way as the reward itself. However, incentive salience theory further dissociates 'wanting' (incentive salience itself) from 'liking' (hedonic impact), as different psychological processes with different neural mechanisms even when those two processes occur simultaneously. 'Wanting' in quotation marks is distinct from more cognitive forms of desire, which are meant by the word wanting in the ordinary sense, which involve explicit goals and declarative expectations of the desired outcome. At times, awareness only trickles down from the activation of interpretive cognitive processes that usually translate implicit activation ('wanting') into explicit subjective feelings (wanting). In addicts, this would explain why they often have so little insight into their apparent hunger for drugs and their cues, and why they may persist in drug taking despite an array of adverse consequences. Yet core 'wanting' and cognitive wanting usually go together. Cue-triggered peaks in 'wanting' are sudden, intense, temporary, reversible yet also repeatable.

Addictive cravings often wax and wane, a feature that contradicts notions of addiction as a constant unchanging habit. Some modern researchers may have mistaken intense 'wanting' peaks to be simply habitual, because a primary criterion for habit in modern studies has been simple perseveration of seeking in the face of punishment or degraded outcome – which is equally well produced by compulsive 'wanting'. Further, the intensity of 'wanting' peaks is variable, being modulated by state-related changes in the reactivity of mesocorticolimbic systems. For example, 'wanting' peaks can be magnified by states of stress or emotional excitement, and by priming doses of drugs themselves. Motivation priming by drugs is one reason why many alcoholic treatment programs recommend their members to avoid all subsequent use of alcohol: an attempt to take even a single drink in a social context entails the danger of precipitating an intense binge of further consumption. Although pleasure may normally help assign incentive salience, brain stimulation and certain drugs of abuse that directly or indirectly activate primarily dopamine systems can skip this step by directly activating and sensitizing the neural substrate of incentive salience, causing in essence a form of sham reward ('wanting' without 'liking'). Amplified further by neural sensitization and focused onto drug rewards as the target, these 'wanting' peaks transform ordinary levels of cue-triggered 'wanting' into excessive addicted-levels of urges to take drugs. The incentive sensitization theory also highlights the fact that the neuroadaptations responsible for the sensitization of incentive salience are a long-lasting if not permanent phenomenon, potentially persisting for years after the individual stops taking drugs. This we suggest is why relapse is so prevalent and persistent despite recovery and regardless of withdrawal, and even when strong pleasure is not to be expected from taking a drug.

Sensitization of mesolimbic dopamine, the common currency for incentive salience

Sensitization has also been demonstrated in the brains of ordinary people, as direct elevation of the amount of mesolimbic dopamine released in response to an addictive drug. For example, significantly greater amphetamine-induced ventral striatal dopamine release was observed two weeks and again one year after the administration of three drug doses over a one-week period (Boileau et al. 2006). Compelling evidence for neural sensitization of dopamine release has also been shown in Parkinson patients with dopamine dysregulation syndrome (DDS) (Evans et al. 2006). This leads to compulsive use of dopaminergic drug medications, with increased reports of drug wanting but not drug liking, and increased dopamine release in the ventral striatum especially in the combined presence of cues and the dopamine-stimulating drug. Consequent incentive-sensitization may also manifest itself by pathological gambling, hypersexuality, food bingeing and punding (a form of complex behavioral stereotypy). Conversely, there was no apparent sensitization to how much subjects liked amphetamine or other dopamine-stimulating drugs in the aforementioned studies.

Neurochemically, sensitization leads to an enhanced dopamine elevation produced by an addictive drug in the synapses of the nucleus accumbens in the face of a drug challenge.

Anatomically, there are also persistent changes in the brain cells and circuits of the mesolimbic system that respond to drugs and control incentive salience. These include structural changes in the morphology of neurons in brain structures of the nucleus accumbens and prefrontal cortex. The length of dendrites on medium spiny neurons in the nucleus accumbens and on pyramidal neurons in the prefrontal cortex is increased, accompanied by an increase in spine density. Changes seem to occur both pre- and post-synaptically in connectivity in brain reward systems. Increased release of dopamine is a pre-synaptic consequence of drug sensitization, seen both in vitro and in vivo, whereas post-synaptically, dopamine D1 receptors show increased sensitivity. In spite of this, there have been reports showing a reduction in the availability of dopamine D2 receptors in cocaine addicts, often suggested to mean either that they have fewer dopamine receptors or that more of their receptors are already occupied by dopamine. Yet, recent findings in animals indicate that cocaine causes an increase in the subpopulation of D2 receptors that are in a state of high-affinity, which can occur even in spite of a reduction in overall D2 receptors (Briand *et al.* 2008). An increase in the proportion of dopamine D2 receptors in the high-affinity state would not be evident in human studies because the ligands used in human imaging studies do not discriminate between low and high affinity states, but, if it does occur in addicts, it would result in dopamine supersensitivity. Psychologically, these neuroadaptations may combine to create pathological levels of 'wanting' for drugs and their associated stimuli, although specific cause–effect relations have not yet been elucidated. In animal experiments, sensitization with amphetamine facilitates the later acquisition of a self-administration behavior of taking drugs such as cocaine and amphetamine, and in a number of ways increases the incentive motivational power of drug rewards and their cues.

Factors that influence sensitization

Sensitization is a complex phenomenon that is influenced by the dose, timing, and spacing of the drug, along with the context in which it is taken, and individual features of the person who is taking it (including genes, sex, hormonal status, prior stress, etc.). Sensitization is strongest when drugs are taken at high doses and intermittently in spaced pulses (rather than as a continuous stream) (Allain *et al.* 2015; Zimmer *et al.* 2012). Once induced in a brain, sensitization may last for years even if no more drug is taken during that period. The phenomenon of sensitization also displays a tremendous amount of individual variation, with some individuals developing rapid and robust sensitization in contrast to others who sensitize very little if at all. So far it has been shown in animals that there are genetic differences in the propensity of individual brains to undergo sensitization, and in the functioning of the mesolimbic dopamine system. There is also evidence suggesting that the genetic variation in acute responsiveness to drugs is different to that responsible for differences in sensitization. Nonetheless, most animals and likely humans do show some degree of psychomotor sensitization, although few may still reach the levels sufficient to trigger compulsive drug seeking and taking. It is important to recognize that a major distinction exists between psychomotor sensitization and incentive sensitization. Although they share several features, and can overlap together, they can also separate. They are therefore essentially distinct processes, whereby psychomotor sensitization may be far more pervasive, easy to induce, and often less intense and less complex in mechanism than focused incentive-sensitization – which is therefore rarer to develop. Thus, on the whole, only some people who take drugs might actually develop sufficient incentive sensitization of brain mesolimbic systems to produce enduring pathological 'wanting', and only those who develop this degree of sensitization might become true addicts with compulsive levels of 'wanting' to take drugs that persist long after withdrawal is over. That relative rareness may help account for why although 55% of 18–34 year olds have at one point sampled a potentially addictive drug, only a relative few become addicts.

Cross-sensitization

Mechanisms of induction and expression of sensitization may differ across drugs, but sensitization to one drug often will produce a sensitized response to other drugs, an effect otherwise known as cross-sensitization (such as between cocaine and heroin). The implication is that as an individual's drug history increases, the incentive value attributed to the act of drug taking and to drug-related stimuli will progressively be enhanced, which will increase the probability of repeating drug-seeking and drug-taking behavior in the future.

Non-drug addictions? Cross-sensitization has also been reported between drugs and other addictive behaviors, such as sugar consumption and gambling. Pathological gamblers, for example, show greater amphetamine-induced striatal dopamine release than healthy controls (Boileau *et al.* 2013), and in animals, repeated exposure to uncertainty produces hypersensitivity to the locomotor effects of amphetamine (Singer *et al.* 2012). Similarly for food, prior sensitization to amphetamine results in sugar overconsumption, whereas a sugar-bingeing regimen produces hypersensitivity to amphetamine (Avena *et al.* 2008). Furthermore, cross-sensitization has even been reported between drugs and stress. Notably, pretreatment with either amphetamine, cocaine or morphine will cause hyper-responsiveness to stress, whereas initial sensitization to stress or administration of corticosterone will produce a heightened response to a later drug challenge. Evidence suggests that addictive drugs, sugar, gambling and stress all activate and sensitize dopamine systems. This highlights a critical role for stress in the occurrence of relapse, whereby stressful life events may act as powerful triggers of drug and reward craving, and, if sufficient to produce sensitization, could possibly predispose individuals to subsequent drug addiction, food overconsumption or gambling disorder. It also explains the high rates of poly-drug use (abuse and addiction to more than one drug) and the high rates of comorbidity between drug addiction, gambling, food, and stress/anxiety disorders.

Context and craving

Environmental cues associated with the drug-taking experience become powerful triggers of sensitized 'wanting' and relapse into addictive drug taking. The hypersensitive response that addicts may display towards the incentive properties of drug-related cues may be strongest in contexts previously related to drug taking in the past. The context may therefore act as an occasion-setter to modulate the incentive power of drug cues, dramatically raising the probability that a sensitized elevated neural response will be expressed in the future in that same place or situation. Different occasion-setting stimuli can either facilitate or inhibit the expression of sensitization in an environment-specific manner. Facilitators do not necessarily elicit conditioned responses but instead control the ability of other stimuli to do so. Conditioned inhibitor properties may be developed by contexts not associated with drug use, thereby preventing the expression of the sensitized response in that context. For example, the presence of relevant stimuli to drug-abusing subjects during testing gives rise to a positive correlation between past psychostimulant drug use and the subjects' striatal dopamine response. In contrast, the absence of these cues when the drug was administered meant that drug history predicted a smaller striatal dopamine response. This means that the presence of drug-related cues help unmask the hypersensitivity of their dopaminergic brain systems, induced by prior drug use, which otherwise can be hidden in their absence.

The presentation of drug-associated cues alone produces an increase in dopamine release in addicted humans, and the magnitude of the dopamine release is closely related to that individual's degree of addiction severity. Consistent with the incentive sensitization theory, there is also evidence that drug addicts display a bias of attention towards visual drug-associated cues, as if the cues were more attractive and attention-grabbing. Similarly, abstinence from smoking for only 24 h can dramatically potentiate neural responses to smoking-related cues. This highlights the importance of drug cues, and the exacerbated salience attributed to them. It also emphasizes how the completeness of the contextual picture, by the presence of the myriad of components that make up the drug-associated context (mood, location, drug paraphernalia, etc.), may culminate in uncontrollable levels of craving and further increase the risk of relapse. However, the strong reliance of sensitization on context might be argued to contribute to the decrease in addiction rates with age. Since age, life situations and the passage of time can function as strong contexts, it is possible that the expression of incentive-sensitization is modified by these context shifts and helps reduce addictions rates in older generations. Similarly, the small yet progressive decays in the dopaminergic system with age might also act as a contributing factor.

Beyond purely drug addiction, incentive-sensitization may also manifest itself by food bingeing, pathological gambling, hypersexuality, and other compulsive motivations (Robinson, Fischer, et al. 2015). For example, the idea of food addiction has been recently postulated (Ahmed et al. 2013), in part due to reports of overeating as a chief cause of growing obesity trends. However, unlike drugs, food fulfills a bodily need. This may make food addiction particularly hard to treat, since it would be impossible to achieve total abstinence. Instead, food addiction might be best reframed as an addiction to particular types of foods and substances, such as refined sugars and flour, from which an individual could selectively refrain. Changes in 'wanting' alone could be responsible for such cases of overeating. Sensitization of mesolimbic dopamine systems by exposure to cycles of binging alternating with dieting has been suggested to occur (Avena et al. 2008). Enhanced sensitivity of the mesolimbic reward system could attribute high levels of incentive motivation to the sights and smells related to food and drive excessive consumption, without necessarily producing comparable levels of 'liking' when the food is itself consumed. In animal studies, only some individuals develop obesity when exposed to a junk-food diet, and those who do express an enhanced attraction to food cues despite a blunting of their 'liking' response to sweetness (Robinson, Burghardt, et al. 2015). It has also been suggested that individuals that carry genes promoting elevated dopamine function might experience stronger cue-triggered urges in response to food cues, making them more liable to develop obesity.

In some other individuals, food addiction could result from exaggerated 'wanting' combined with exaggerated 'liking'. Davis and colleagues have found that certain individuals who are both obese and binge eaters are far more likely to carry both alleles for a gene that codes a gain of function for mu-opioids, and another allele for a gene that may be associated with higher binding for the dopamine D2 receptor (Davis and Carter 2009). Together these genetic traits have been suggested to combine to simultaneously increase 'liking' and 'wanting' for foods in a manner that strongly promotes binge eating and gives rise to addictive-like features, including loss of control and relapse. This may be a rare example where 'wanting' and 'liking' increase in tandem, making it exceptionally hard to adhere to levels of food intake that conform to physiological needs.

Gambling

Gambling might also involve special recruitment of incentive salience brain systems. Uncertainty may especially promote incentive salience under some conditions, which mirror many of the hallmarks of gambling (Anselme et al. 2013). It has even been shown that not knowing how hard one must work to be rewarded may even sensitize the brain in ways similar to addictive drugs (Singer et al. 2012). Sensitization of dopaminergic systems may originate from the fact that uncertain reward cues produce a greater dopamine signal during reward anticipation (Fiorillo et al. 2003), which in turn may promote risk-seeking behavior, as evidenced in gambling. In fact, the degree of striatal dopamine release during a gambling task correlates with the severity of problem gambling, and problem gamblers also exhibit attentional bias towards gambling-related cues as compared with healthy controls, suggesting that these stimuli also take on increased incentive salience in human gamblers. During losses, the brains of problem gamblers also recruit areas related to wins and show increased dopamine release in the ventral striatum as compared with healthy controls, implying that loss still generates motivation in problem gamblers, despite not being enjoyed. In a recent PET imaging study, problem gamblers also reported lower feelings of euphoria in response to an amphetamine challenge as compared with healthy controls. These studies provide further evidence for a dissociation of 'liking' and 'wanting' in individuals addicted to gambling. Individuals with gambling disorder sometimes seem driven by cues to gamble, in all cases at a global monetary loss, for only a moderate experienced euphoria. It seems that despite a conscious awareness that, for example, slot-machine gambling will repeatedly end with a loss, showing a cognitive awareness that playing will fail to fulfill any monetary need, players want to gamble, sometimes expressly to satisfy an urge to play.

Summary

The persistence of the neuroadaptations that underlie incentive sensitization suggests that recovering from addiction may be a long and slow process. Cognitive, behavioral and mindfulness therapies may gradually reduce some layers of responsiveness to drug and reward cues, but other layers may persist. Drug medications that would selectively reverse the expression of mesolimbic sensitization without undesirable side effects have yet to be developed. As a result, research is increasingly turning its focus towards the factors that determine an individual's susceptibility to pathological levels of sensitization and therefore put them at risk for addiction, as well as to ways of coping with a sensitized 'wanting' system.

The incentive sensitization theory helps explain why the development of addiction is a gradual and incremental process, but also such a persistent problem once established. The sensitization of 'wanting' independent of pleasure can produce an addict who wants drugs more and more – without necessarily liking them proportionately as much, or even liking them less and less. The incentive sensitization theory does not deny a role for pleasure-seeking in initial experimenting with drugs or casual use, nor deny that ritualized habits contribute to daily addictive drug consumption in many addicts, or that withdrawal avoidance may motivate a great deal of addictive drug-taking. These explanations are not mutually exclusive. However, the incentive sensitization theory suggests that the critical change underlying addictive relapse is the sensitization or hypersensitivity of brain mesolimbic systems to the incentive motivational effects of rewards and their associated stimuli. The result is a bias in attentional processing and incentive motivation value attributed to the cues and the addictive behavior, which, when combined with impaired executive control, gives rise to the symptoms of addiction and the bad choices that addicts make regarding drugs, food and even gambling.

References

Ahmed, S. H., et al. (2013) "Food addiction", in Donald W. Pfaff (ed.), *Neuroscience in the 21st Century*, New York, NY: Springer, pp. 2833–2857.

Allain, F., et al. (2015) "How fast and how often: the pharmacokinetics of drug use are decisive in addiction", *Neuroscience and Biobehavioral Reviews* 56: 166–179.

Anselme, P., Robinson, M. J. F. and Berridge, K. C. (2013) "Reward uncertainty enhances incentive salience attribution as sign-tracking", *Behavioural Brain Research* 238: 53–61.

Avena, N. M., Rada, P. and Hoebel, B. G. (2008) "Evidence for sugar addiction: behavioral and neurochemical effects of intermittent, excessive sugar intake", *Neuroscience and Biobehavioral Reviews* 32(1): 20–39.

Bartlett, E., et al. (1997) "Selective sensitization to the psychosis-inducing effects of cocaine: a possible marker for addiction relapse vulnerability?" *Neuropsychopharmacology* 16: 77–82.

Berridge, K. C. (2004) "Motivation concepts in behavioral neuroscience", *Physiology and Behavior* 81(2): 179–209.

Boileau, I. et al. (2006). Modeling sensitization to stimulants in humans: an [11C]raclopride/positron emission tomography study in healthy men. *Archives of General Psychiatry*, 63(12): 1386–1395. http://doi.org/10.1001/archpsyc.63.12.1386

Boileau, I., et al. (2013) "In vivo evidence for greater amphetamine-induced dopamine release in pathological gambling: a positron emission tomography study with [^{11}C]-(+)-PHNO", *Molecular Psychiatry* 19(12): 1305–1313.

Briand, L. A., et al. (2008) "Cocaine self-administration produces a persistent increase in dopamine D2 High receptors", *European Neuropsychopharmacology* 18(8): 551–556.

Davis, C. and Carter, J. C. (2009) "Compulsive overeating as an addiction disorder. A review of theory and evidence", *Appetite* 53(1): 1–8.

Evans, A. H. et al. (2006). Compulsive drug use linked to sensitized ventral striatal dopamine transmission. *Annals of Neurology*, 59(5): 852–858. http://doi.org/10.1002/ana.20822

Fiorillo, C. D., Tobler, P. N. and Schultz, W. (2003) "Discrete coding of reward probability and uncertainty by dopamine neurons", *Science* 299(5614): 1898–1902.

Fischman, M. W. and Foltin, R. W. (1992) "Self-administration of cocaine by humans: a laboratory perspective", *Ciba Foundation Symposium* 166: 165–180.

Heyman, G. M. (2009) *Addiction: A Disorder of Choice*, Cambridge, MA: Harvard University Press.

Hickey, C. and Peelen, M. V. (2015) "Neural mechanisms of incentive salience in naturalistic human vision", *Neuron* 85(3): 512–518.

Koob, G. F. (1996) "Drug addiction: the yin and yang of hedonic homeostasis", *Neuron* 16(5): 893–896.

Mahler, S. V. and Berridge, K. C. (2009) "Which cue to 'want?' Central amygdala opioid activation enhances and focuses incentive salience on a prepotent reward cue", *The Journal of Neuroscience* 29(20): 6500–6513.

McAuliffe, W. E. (1982) "A test of Wikler's theory of relapse: the frequency of relapse due to conditioned withdrawal sickness", *The International Journal of the Addictions* 17(1): 19–33.

Mukherjee, J., et al. (2002) "Brain imaging of 18F-fallypride in normal volunteers: blood analysis, distribution, test-retest studies, and preliminary assessment of sensitivity to aging effects on dopamine D-2/D-3 receptors", *Synapse* 46(3): 170–188.

Nocjar, C. and Panksepp, J. (2002) "Chronic intermittent amphetamine pretreatment enhances future appetitive behavior for drug- and natural-reward: interaction with environmental variables", *Behavioural Brain Research* 128(2): 189–203.

Pickens, C. L., et al. (2011) "Neurobiology of the incubation of drug craving", *Trends in Neurosciences* 34(8): 411–420.

Reed, S. C., et al. (2009) "Cardiovascular and subjective effects of repeated smoked cocaine administration in experienced cocaine users", *Drug and Alcohol Dependence* 102(1–3): 102–107.

Robinson, M. J. F., Anselme, P., et al. (2015) "Amphetamine-induced sensitization and reward uncertainty similarly enhance incentive salience for conditioned cues", *Behavioral Neuroscience* 129(4): 502–511.

Robinson, M. J. F., Burghardt, P. R., et al. (2015) "Individual differences in cue-induced motivation and striatal systems in rats susceptible to diet-induced obesity", *Neuropsychopharmacology* 40(9): 2113–2123.

Robinson, M. J. F., Fischer, A. M., et al. (2015) "Roles of 'wanting' and 'liking' in motivating behavior: gambling, food, and drug addictions", in P. D. Balsam and E. H. Simpson (eds), *Current Topics in Behavioral Neurosciences*, Vol. 27, New York, NY: Springer, pp. 105–136.

Robinson, M. J. F., Warlow, S. M. and Berridge, K. C. (2014) "Optogenetic excitation of central amygdala amplifies and narrows incentive motivation to pursue one reward above another", *Journal of Neuroscience* 34(50): 16567–16580.

Robinson, T. E. and Berridge, K. C. (1993) "The neural basis of drug craving: an incentive-sensitization theory of addiction", *Brain Research: Brain Research Reviews* 18(3): 247–291.

Robinson, T. E., et al. (1998) "Modulation of the induction or expression of psychostimulant sensitization by the circumstances surrounding drug administration", *Neuroscience and Biobehavioral Reviews* 22(2): 347–354.

Singer, B.F., et al. (2018) "Are cocaine-seeking 'habits' necessary for the development of addiction-like behavior in rats?" *The Journal of Neuroscience* 38(1): 60–73.

Singer, B. F., Scott-Railton, J. and Vezina, P. (2012) "Unpredictable saccharin reinforcement enhances locomotor responding to amphetamine", *Behavioural Brain Research* 226(1): 340–344.

Small, A. C., et al. (2009) "Tolerance and sensitization to the effects of cocaine use in humans: a retrospective study of long-term cocaine users in Philadelphia", *Substance Use and Misuse* 44(13): 1888–1898.

Tindell, A. J., et al. (2005) "Ventral pallidal neurons code incentive motivation: amplification by mesolimbic sensitization and amphetamine", *The European Journal of Neuroscience* 22(10): 2617–2634.

Wyvell, C. L. and Berridge, K. C. (2001) "Incentive sensitization by previous amphetamine exposure: increased cue-triggered 'wanting' for sucrose reward", *The Journal of Neuroscience* 21(19): 7831–7840.

Zimmer, B. A., Oleson, E. B. and Roberts, D. C. (2012) "The motivation to self-administer is increased after a history of spiking brain levels of cocaine", *Neuropsychopharmacology* 37(8): 1901–1910.

29
RESTING-STATE AND STRUCTURAL BRAIN CONNECTIVITY IN INDIVIDUALS WITH STIMULANT ADDICTION

A systematic review

Anna Zilverstand, Rafael O'Halloran, and Rita Z. Goldstein

Introduction

Addiction is a complex disease process, encompassing a relapsing cycle of intoxication, bingeing, withdrawal and craving. While neuroimaging studies describing brain functioning at the level of single brain regions have decisively shaped our understanding of each of these stages, the importance of moving beyond this approach and investigating how brain regions assemble into brain networks, which support complex psychological processes such as reward seeking or inhibitory self-control, has long been recognized (Goldstein and Volkow 2002; Koob and Volkow 2010; Goldstein and Volkow 2011). In recent years, numerous studies have therefore started to explore the role of disrupted brain networks in addiction by investigating structural and functional connectivity between and within these large-scale brain networks. Neuroimaging can be used to evaluate structural brain connectivity through 'diffusion-weighted imaging', which can assess the integrity of the white matter tracts (comprised of nerve cell projections) connecting brain regions to one another. Second, functional neuroimaging can be employed to compute the strength of 'functional connectivity', which uses co-activation of brain regions as an indicator of shared neuronal activity between them (Shmuel and Leopold 2008). The latter is usually acquired during resting-state, a baseline state in which the brain is not actively involved in a task and drugs have not been administered. Resting-state functional connectivity therefore captures a default brain state independent of external input/challenges, providing insight into the intrinsic 'functional architecture' of the brain that underlies functioning during task demands (Fox and Raichle 2007). Both structural and resting-state neuroimaging are therefore useful tools for understanding the fundamental makeup of the brain and assessing how this is changed in addiction.

This systematic review discusses a growing body of studies that have used these approaches in populations of individuals with stimulant addiction, expanding the scope of a previous review that included four studies of the 23 resting-state studies reviewed here (Sutherland *et al.* 2012). Further, this is the first comprehensive review of 17 studies that have employed diffusion-weighted imaging for structural connectivity. In particular, we focus on summarizing aberrant brain connectivity of individuals with stimulant addiction as compared with a control group, as well as brain–behavior correlations. Four major networks/brain systems implicated in stimulant addiction are discussed: the reward system, the salience network, the executive control network and learning networks.

Study selection

We performed a two-stage systematic literature search to identify studies investigating structural and resting-state functional connectivity in individuals with stimulant addiction. First, we searched Medline/Pubmed using a search term comprised of the method ((("fMRI" OR "magnetic resonance imaging") AND (resting-state OR resting OR rest)) OR ("diffusion" AND ("MRI" OR "DTI")) combined with a term related to stimulant use (cocaine OR methamphetamine OR amphetamine OR stimulant). Additionally, we performed a second manual search for relevant papers based on the reference lists of the included papers. Studies adhering to the following criteria were included:

Studies published in English, in a peer-reviewed journal.

Studies comparing adults with a (DSM-III/IIIR/IV/V) diagnosis to a matched control group (>10 participants per group).

Studies investigating resting-state functional connectivity (excluding data derived from task-fMRI) without additional manipulations (e.g., pharmacological interventions) OR studies investigating structural brain connectivity using diffusion-weighted imaging.

We identified 23 resting-state studies published from 2010 to 2016, as well as 17 studies investigating structural connectivity by diffusion MRI published from 2002 to 2016, conducted in individuals with cocaine, amphetamine, methamphetamine or poly-substance addiction (in which one of the substances was cocaine).

Data extraction

Table 29.1 (resting-state functional connectivity) and Table 29.2 (structural connectivity) summarize the results and methods employed by the reviewed studies, including the specific analysis used, the population, state of addiction (e.g., urine positive or negative) and the number of participants included. The tables present group differences in functional and structural connectivity ("increased connectivity", "decreased connectivity"), as well as brain–behavior correlations ("brain*behavior correlations") reported in these studies. As findings from resting-state studies may not necessarily be comparable because of different labelling systems used, all originally reported results were relabeled using a single labeling system. Specifically, the originally reported peak coordinates were transformed into a common space, the Montreal Neurological Institute (MNI) standard brain, using Brett's algorithm (www.sdmproject.com/utilities) and then relabeled using the MRIcron Brodmann template (www.nitrc.org/projects/mricron). All relabeling was performed by the same person using the following delineation: ventrolateral prefrontal cortex (vlPFC=BA 44/45/47), dorsolateral prefrontal cortex (dlPFC=BA 9/46), premotor cortex (lateral BA 6), supplementary motor area (SMA=medial BA 6), parietal cortex (BA 7/39/40), orbitofrontal cortex (OFC=BA 11), rostral/dorsal anterior cingulate cortex (rACC/dACC=BA 24/25/32) and posterior cingulate cortex (PCC=BA 23). See Box 29.1 for "Measuring resting-state connectivity" and Box 29.2 for "Measuring structural connectivity" for details on the methods employed by the reviewed studies. When summarizing and discussing results, and to enhance this review's clarity, we focused on the brain regions belonging to the four major networks depicted in Figure 29.1.

The brain regions belonging to the four discussed networks, the salience network (OFC, rACC, dACC, anterior insula), reward system (VTA, NAcc), learning (caudate, putamen, amygdala, hippocampus), and executive control network (vlPFC, dlPFC, premotor, SMA, parietal), are depicted on the top. On the bottom, a visual summary of aberrant connectivity in individuals with stimulant addiction is represented. Overall, the four discussed networks show enhanced connectivity with each other (solid lines), while connectivity with other brain systems such as the visual system is decreased (non-solid lines). Specifically, there was consistent evidence for enhanced connectivity between a) the reward system and salience network, b) both the reward system/salience network with the executive control network, and c) between the reward system and the learning network.

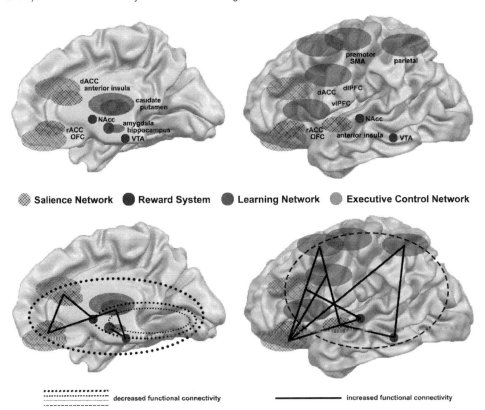

Figure 29.1 Resting-state connectivity in stimulant addiction.

Box 29.1 Measuring resting-state connectivity

The first resting-state functional connectivity studies employed a seed-based functional connectivity analysis, in which the activation level time course of one (or several) seed brain region(s) is correlated with the activation level time course of every other brain voxel, generating a functional connectivity map for each region of interest. To expand on this and analyze whole-brain functional connectivity, independent component analysis was introduced as a data-driven method for defining co-activated networks by maximizing statistical independence between a given number of networks. Often only select networks of interest are analyzed with this method. Other, more novel whole-brain connectivity methods include voxel-wise methods (correlating the time course of each voxel with all other voxels; interhemispheric connectivity = correlating each voxel with the corresponding voxel in the other hemisphere) and Graph Theory/Functional Connectivity Density (FCD) mapping. Graph Theory/FCD derive a whole-brain functional connectivity matrix (correlating the activation level time courses of all voxels/brain regions with all others) to perform mathematical complex network analysis on this matrix. Prominent measures include degree, participation coefficient, global FCD and nodal strength, which all estimate "global connectivity" to measure the functional integration of a voxel/brain region with the rest of the brain. Other Graph Theory methods include modularity analysis, which is a data-driven approach for defining modules to estimate within and between module connectivity.

The reward system and salience network

A hallmark of drug addiction is the dysregulation of the brain's dopaminergic reward system, manifested by an increase of the incentive salience attributed to the consumed drug at the expense of decreased salience attributed to non-drug-related reinforcers and associated with a concomitant compromise in self-control especially in a drug related context (Koob and Volkow 2016; Goldstein and Volkow 2011). Chronic drug use is associated with this modified incentive function supported by a reduction of tonic (yet enhanced cue-induced phasic) dopamine in the core areas of the reward system: the midbrain ventral tegmental area (VTA) and nucleus accumbens (NAcc = ventral striatum) (Koob and Volkow 2016; Everitt and Robbins 2005). In contrast to the baseline/tonic state, and consistent with phasic dopamine increases, neuroimaging has demonstrated that the NAcc is hyperactivated during drug cue exposure, with NAcc activation levels correlating with self-reported craving (Kühn and Gallinat 2011). During resting-state, the NAcc is part of an extended 'reward system', which includes the rACC and the OFC (Amft et al. 2016). Beyond its role in processing incentive salience/craving (NAcc), this extended reward system is also involved in the representation of personal relevance (rACC) (Moeller and Goldstein 2014) and the representation of expected value (OFC) (Chase et al. 2015). While the VTA, NAcc, rACC and OFC thus form the extended reward system, the OFC is also an anchor of a 'salience network' of regions co-activated during resting-state, which encompasses the OFC, the dACC and the anterior insula (Corbetta and Shulman 2002; Corbetta et al. 2008; Seeley et al. 2007). This salience network has been ascribed a reorienting function, directing attentional resources towards a salient stimulus (Corbetta et al. 2008). While the anterior insula may have a crucial role in representing drug urges (Naqvi and Bechara 2009), the dACC has been ascribed a role in integrating the expected value of allocating control (Shenhav et al. 2013).

Table 29.1 Resting-state connectivity in individuals with stimulant addiction

First author	Year	Analysis	Drug	State	Group	Increased connectivity (group difference)	Decreased connectivity (group difference)	Brain*behavior correlations
WHOLE BRAIN ANALYSES								
Kelly	2011	interhemispheric RSFC/ROI	COC	uri−	EG=25 HC=24	n.s.	interhemispheric: Frontoparietal (vlPFC/dlPFC/premotor/dACC/SMA/parietal)	ROI: Frontoparietal*attentional lapses
Ray	2015	ICA/ROI (13 components/61 ROI)	COC	uri−	EG=20 HC=17	within network: Frontoparietal (vlPFC/dlPFC/parietal), Sensorimotor; ROI: dACC-parietal, dlPFC-parietal	ROI: dACC-visual, precuneus-visual, visual-visual	within network: Sensorimotor*years use; ROI: dACC-visual*years use; visual-visual*frequency/money use
Wang	2015	graph theory (degree)	COC	uri−	EG=20 HC=19	global: insula, parahippocampus, SMA, visual, auditory, somatosensory	n.s.	n.s.
Liang	2015	graph theory: intermodule connectivity, participation coefficient; ROI	COC	uri−/uri+	EG=47 HC=47	n.s.	between network: SN (dACC/insula)-DMN; global (participation): rACC, PCC, insula; ROI-Network: rACC-SN, insula-DMN, PCC-ECN	within network: SN (dACC/insula)*self-referential emotional processing
Konova	2015	FCD (functional connectivity density mapping)	COC	uri−/uri+	EG=19 HC=15	local/global: DMN (OFC, PCC/Precuneus), auditory/visual; global: Putamen, Parahippocampus, Premotor, Superior Parietal; local: Amygdala, vlPFC/dlPFC, somatosensory	n.s.	local/global in DMN (OFC/PCC)*years use
Meunier	2012	nodal strength, seed (OFC)	STI	uri+	EG=18 HC=18	n.s.	nodal strength: OFC; seed: OFC-premotor, OFC-dACC, OFC-visual, OFC-somatosensory	nodal strength: OFC*compulsive symptoms

NETWORK ANALYSES

Author	Year	Method	Substance	uri-/uri+	Sample				
McHugh	2017	ICA (DMN, ECN, SN)	COC	uri-	EG=45 HC=21	n.s.		n.s.	within network: ECN*perseverative errors
Ding	2013	ICA (DMN)	COC	uri-	EG=24 HC=24	n.s.		between network: DMN-vlPFC/dlPFC, DMN-hippocampus	n.a.
Wisner	2013	ICA (insula)	COC	uri-/ uri+	EG=33 HC=32	n.s.		between network: Insula/ dACC-putamen/NAcc/ Globus Pallidus	between network: Insula/ dACC-putamen/NAcc/ Globus Pallidus*impulsivity
Cisler	2013	ICA (frontal)/ seed (insula)	COC	uri-/ uri+	EG=41 HC=19	seed: insula-vlPFC/dlPFC, insula-dACC/SMA		Seed: mid insula-posterior insula	n.a.

REGIONAL ANALYSES

Author	Year	Method	Substance	uri-/uri+	Sample				
Viswan-ath	2015	interhemispheric RSFC (vlPFC, dlPFC, premotor, insula, striatum, ACC)	any drug/ AMP in PSY	uri-	EG=100 PC=51	interhemispheric: insula (COC users), vlPFC (AMP users)		n.s.	interhemispheric: insula*substance dependence (COC users), vlPFC*substance dependence (AMP users)
Cam-chong	2013	seed (NAcc, rACC)	STI	uri-	EG=36 PC=23	seed: Nacc-insula; NAcc-dACC-dlPFC; rACC-putamen		n.s.	seed: Nacc-insula*abstinence duration (COC users)
Wilcox	2011	seed (NAcc, OFC)	COC	uri-	EG=14 HC=16	seed: NAcc-OFC/rACC		seed: NAcc-PCC precuneus/visual, OFC-PCC/precuneus/ parietal	n.a.
McHugh	2013	seed (NAcc, caudate, putamen)	COC	uri-	EG=45 HC=22	n.s.		seed: putamen-insula	seed: putamen-insula* impulsiveness
Hu	2015	seed (NAcc, caudate, putamen)	COC	uri-/ uri+	EG=56 HC=56	seed: caudate-dlPFC, putamen-visual		seed: NAcc/putamen-dACC	seed: caudate-dlPFC*current use/ impulsivity, NAcc-vlPFC/OFC/dACC > compulsions

(continued)

Table 29.1 (continued)

First author	Year	Analysis	Drug	State	Group	Increased connectivity (group difference)	Decreased connectivity (group difference)	Brain*behavior correlations
Gu	2010	seed (VTA, NAcc, amygdala, hippocampus, thalamus, rACC)	COC	uri-/ uri+	EG=39 HC=39	n.s.	seed: VTA-NAcc, VTA-thalamus, amygdala-rACC/dlPFC, hippocampus-rACC/dlPFC/SMA, rACC-insula, thalamus-putamen	seed: VTA-NAcc*years use, VTA-thalamus*years use
Konova	2013	seed (VTA, NAcc, amygdala, hippocampus, thalamus, rACC)	COC	uri-/ uri+	EG=18 HC=16	seed: NAcc-putamen/pallidus, rACC-parietal	seed: hippocampus-SMA/motor, NAcc-operculum/visual/auditory	seed: NAcc-putamen/pallidus*addiction severity
Kohno	2016	seed (VTA)	MET	uri-/ uri+	EG=19 HC=26	seed: VTA-putamen, VTA-caudate, VTA-insula, VTA-vlPFC	n.s.	seed: VTA-NAcc*impulsivity
Cam-chong	2011	seed (rACC)	COC	uri+	EG=27 HC=24	seed: rACC-vlPFC/dlPFC/SMA	n.s.	seed: rACC-dlPFC*reversal learning
Kohno	2014	seed (VTA, NAcc, dlPFC)	MET	uri-	EG=25 HC=27	seed: VTA-putamen, VTA-amygdala, VTA-hippocampus, VTA-insula, VTA-vlPFC, VTA-parietal	n.s.	n.a.
Verdejo-Garcia	2014	seed (dACC, PAG, insula)	COC	uri-	EG=10 HC=14	n.s.	seed: dACC-thalamus, PAG-insula/putamen	n.a.
McHugh	2014	seed (amygdala)	COC	uri-	EG=45 HC=22	n.s.	seed: amygdala-OFC/rACC, amygdala-visual	n.a.
Adinoff	2015	seed (hippocampus)	COC	uri-	EG=40 HC=20	n.s.	n.s.	seed: hippocampus-PCC/precuneus > time to relapse

Results are represented such that, for example, increased connectivity "ROI 1-ROI 2" indicates increased connectivity between these two regions.

RSFC = resting-state functional connectivity, ROI = region of interest, ICA = independent component analysis; EG = experimental group, HC = healthy controls, PC = controls with psychiatric disease; COC = cocaine, STI = stimulants, AMP = amphetamine, MET = methamphetamine; PSY = psychiatric inpatients; uri+ = urine positive, uri- = urine negative; Brain regions: OFC = orbitofrontal cortex, vlPFC = ventrolateral prefrontal cortex, dlPFC = dorsolateral prefrontal cortex, ACC = anterior cingulate cortex, NAcc = nucleus accumbens, rACC = rostral anterior cingulate cortex, VTA = ventral tegmental area, dACC = dorsal anterior cingulate, PAG = periaqueductal gray, SMA = supplementary motor area, PCC = posterior cingulate cortex.

Brain networks: DMN = default mode network, ECN = executive control network, SN = salience network; n.a. = not applicable, n.s. = not significant.

Table 29.2 Structural connectivity in individuals with stimulant addiction.

First Author	Year	Analysis	Drug	State	Group	Increased connectivity (group difference)	Decreased connectivity (group difference)	Brain*behavior correlations
Lim	2002	ROI	COC	uri+	EG=12 HC=13	n.s.	FA: Frontal WM	n.a.
Moeller	2005	ROI	COC	uri+/ uri-	EG=18 HC=18	n.s.	FA: Genu and Rostral CC	Immediate and Delayed Memory *1/FA
Moeller	2007	ROI	COC	uri+/ uri-	EG=13 HC=18	n.s.	FA, λT: CC	n.a.
Lim	2008	TBSS	COC	uri+	EG=21 HC=21	n.s.	FA: IC, CC, Frontal WM	Impulsivity * 1/FA
Ma	2009	ROI	COC	uri-	EG=19 HC=18	n.s.	FA, λT, D: Isthmus, Splenium CC	n.a.
Lane	2010	TBSS	COC	n.r.	EG=15 HC=18	n.s.	FA, λT: SLF, CST, SCR	Decision making * 1/λT
Romero	2010	ROI	COC	uri-	EG=32 HC=33	FA: dACC	FA: frontal WM, CC	Impulsivity * 1/FA, frontal WM
Kaag	2017	TBSS	MULTI	n.r.	EG=67 HC=67	n.s.	FA, λT: ATR, SCR, dACC, IFOF, SLF, UF, CC	Number of substances * 1/FA * λT
Tang	2015	ROI	COC, MET	uri+/ uri-	EG=21 (HIV) HC=22	n.s.	FA: Genu CC, ALIC, AC	Cognitive performance * FA
Bell	2011	TBSS	COC	uri-	EG=43 HC=43	FA: Right SLF, Splenium CC	FA: CC, SLF, SCR	Abstinence duration both * FA and *1/ FA for various regions
Viswanath	2015	TBSS	MULTI	n.r.	EG=90 PC=61	n.s.	FA: CC, IC, SCR, frontal WM, CST, EC	n.a.
Kim	2009	ROI	MET	uri-	EG=11 HC=13	n.s.	FA: Genu CC	Cognitive flexibility *1/FA in Genu CC
Salo	2009	ROI	MET	uri-	EG=37 HC=17	n.s.	FA: Genu, Splenium CC	Inhibitory control *1/FA
Tobias	2010	ROI	MET	uri+	EG=23 HC=18	n.s.	FA: Right frontal WM, Genu CC, midcaudal corona radiata, right perforant fibers	Brief Symptom Inventory Positive Symptom Test, Depression * FA, left mid-caudal corona radiata

(continued)

Table 29.2 (continued)

First Author	Year	Analysis	Drug	State	Group	Increased connectivity (group difference)	Decreased connectivity (group difference)	Brain*behavior correlations
Chung	2007	ROI	MET	uri-	EG=32 HC=30	n.s.	FA: frontal WM (in males, not females)	n.a.
Lin	2015	ROI	MET	uri+	EG=18 HC=22	n.s.	No differences in caudate or putamen	n.a.
Alicata	2009	ROI	MET	uri+/ uri-	EG=30 HC=30	n.s.	D: caudate, putamen. FA: frontal WM	Cumulative lifetime dose ⋆ D in putamen. Age onset ⋆ 1/D in putamen.

Abbreviations: EG = experimental group, HC = healthy controls, PC = controls with psychiatric disease, ROI = region of interest, TBSS = tract-based spatial statistics, HIV = human immunodeficiency virus, COC = cocaine, STI = stimulants, AMP = amphetamine, MET = methamphetamine, MULTI = multiple substances studied including cocaine; uri+ = urine positive, uri- = urine negative, SLF = superior longitudinal fasciculus, CST = corticospinal tract, SCR = superior corona radiata, CC = corpus callosum, IC = internal capsule, n.r. = not reported, n.a. = not applicable, n.s. = not significant, WM = white matter, Cing = cingulum, ATR = anterior thalamic radiation, IFOF = inferior fronto-occipital fasciculus, UF = uncinate fasciculus, ALIC = anterior limb of the internal capsule, AC = anterior commisure, EC = external capsule, FA = fractional anisotropy, λT, Radial diffusivity, D = mean diffusivity.

The literature reviewed in this chapter demonstrates that both the extended reward system (VTA/NAcc/rostralACC/OFC) and the salience network (OFC/dorsalACC/anterior insula) show aberrant connectivity in stimulant addiction.

Resting-state connectivity of the reward system and salience network

Within the reward system (VTA-NAcc), resting-state functional connectivity was found to be decreased in cocaine-addicted individuals (Gu et al. 2010), with a stronger decrease being associated with longer cocaine use (Gu et al. 2010) and higher impulsivity in healthy controls (Kohno et al. 2016). Further, global connectivity between the reward system/salience network (OFC/rACC/dACC/insula) and the rest of the brain was reduced in individuals with stimulant addiction (Meunier et al. 2012; Liang et al. 2015), with a stronger decrease linked to both more compulsive drug taking (Meunier et al. 2012) and more impairment of self-referential emotional processing (Liang et al. 2015). Moreover, a number of studies found that functional connectivity between the reward system/salience network and several other important brain systems, such as the Default Mode Network (DMN) that is involved in self-referential processing, but also visual and auditory networks, was reduced in individuals with stimulant addiction (Wilcox et al. 2011; Konova et al. 2013; Liang et al. 2015; Ray et al. 2015; Meunier et al. 2012), while connectivity specifically with the executive control and learning networks was enhanced (discussed below). Overall, chronic drug use and the associated decrease in dopamine in the reward system thus seem to be associated with a partial decoupling of the reward system/salience network from the rest of the brain (decreased global connectivity), particularly from perceptual and self-referential processing systems (visual/auditory/DMN). Importantly, this result was independent of the state of the participants (urine positive or negative as indicative of recent use vs abstinence, respectively) (Table 29.1).

Within the salience network, functional connectivity was increased, as indicated by increased insular interhemispheric connectivity in cocaine users, correlated with addiction severity (Viswanath et al. 2015). Further, functional connectivity between the reward system (NAcc) and the salience network (NAcc-OFC/rACC/dACC/insula) was enhanced (Camchong et al. 2013; Wilcox et al. 2011). Importantly, this was only observed when the analysis was conducted by placing the seed directly in the NAcc, excluding the putamen (Camchong et al. 2013; Wilcox et al. 2011; Wisner et al. 2013; Hu et al. 2015). Enhanced connectivity between the reward system and the salience network was correlated with lower ability to remain abstinent (Camchong et al. 2013).

In summary, both group comparisons as well as brain–behavior correlations support a general decrease in functional connectivity between the reward system/salience network and the rest of the brain (excluding the executive control/learning systems), while co-activation between the two networks was enhanced (Figure 29.1). This partial decoupling of the reward system/salience network from other important brain systems with concomitant enhanced connectivity between the reward and salience systems were correlated with impulsivity, drug use, compulsive drug taking, addiction severity and ability to remain abstinent.

Structural connectivity of the reward and salience networks

Structural connectivity specifically of the NAcc/VTA has not been investigated so far. Within the salience network, increases in fractional anisotropy (FA) (see Box 29.2 for definition) in the dACC white matter were reported in 32 cocaine-addicted individuals (Romero et al. 2010).

The authors speculated that these changes may reflect hyperactivity of the dACC due to its role in assessing error in reward prediction. In contrast, however, a more recent study utilizing a whole brain analysis approach in 67 poly-substance abusers found decreases in FA in the dACC that were correlated with the number of substances abused (Kaag et al. 2017).

Box 29.2 Measuring structural connectivity

Metrics: A simple approach to analyzing diffusion-weighted MRI data is to fit it with a diffusion tensor model, which characterizes diffusion in the tissue along a preferred axis, known as axial diffusivity (AD), as well as diffusion perpendicular to the preferred axis, known as radial diffusivity (RD). The mean diffusivity (D), an average of diffusivity in all directions, is also computed. The extent to which AD and RD differ can be quantified with a metric known as fractional anisotropy (FA). Fractional anisotropy is the most commonly used metric derived from diffusion-weighted MRI and is a measure of the directionality of diffusion in the tissue. The FA values are dimensionless and range from 0 to 1, with values close to zero indicative of tissue with homogeneous diffusion, such as gray matter, and values approaching 1 indicative of tissue with highly directional diffusion, such as white matter.

Analysis: Group differences in D, FA, AD, and RD can be assessed by calculating the mean values in a region of interest (ROI). An alternative that eliminates the need to select a ROI a priori is voxel-wise statistical analysis carried out using Tract-Based Spatial Statistics (TBSS) (Smith et al. 2006). In TBSS data from all subjects is projected onto a mean FA tract skeleton before applying voxel-wise cross-subject statistics.

Interpretation: RD is interpreted to be sensitive to myelination, while AD is interpreted to be sensitive to axonal damage. Fractional anisotropy is a combination of both and is generally interpreted as a measure of white matter integrity. Diffusivity is sensitive to edema and general loss of tissue structure.

The executive control network

The dysregulation of the reward system in chronic drug abusers is associated with compromised inhibitory self-control pronounced especially during stages of craving and drug consumption (Goldstein and Volkow 2011). Neuroimaging studies have long implicated the prefrontal cortex in impaired self-control in drug addiction (Goldstein and Volkow 2011), particularly highlighting the vlPFC and dlPFC as core regions of a cognitive control network (Zilverstand et al. 2017; Buhle et al. 2014). In addition to the vlPFC and dlPFC, the dACC has also been implicated in impaired behavioral inhibitory control in addiction (Luijten et al. 2014). During resting-state these prefrontal regions (vlPFC/dlPFC/dACC) co-activate with regions involved in motor planning (premotor/SMA), as well as with parietal regions involved in the allocation of attention (Cole and Schneider 2007), to form the cognitive control network (Cole and Schneider 2007) also named the dorsal frontoparietal executive control network (Seeley et al. 2007; Corbetta and Shulman 2002; Corbetta et al. 2008). The studies presented in this review indicate fundamentally changed connectivity of this network (vlPFC/dlPFC/dACC/premotor/SMA/parietal) in stimulant addiction.

Resting-state connectivity of the executive control network

Interhemispheric connectivity within the frontoparietal executive control network was decreased in individuals with cocaine addiction (Kelly *et al.* 2011), indicating stronger lateralization within the network, with lower interhemispheric connectivity/higher lateralization being correlated with more attentional lapses (Kelly *et al.* 2011). Further, lower within-network connectivity was associated with worse performance on adaptive learning tasks (McHugh *et al.* 2017). Third, increased local connectivity within single brain regions of the executive control network, indicating greater segregation within the network, was reported in the prefrontal cortex (vlPFC/dlPFC), the motor (premotor/SMA) and parietal cortex (Konova *et al.* 2015; Ray *et al.* 2015; Wang *et al.* 2015). However, in contrast to expectations, this increase in local connectivity in the motor cortex was negatively correlated with duration of cocaine use, with lower – more normalized – local connectivity being associated with longer duration of cocaine use (Ray *et al.* 2015). To summarize, the reviewed studies did not provide clear evidence for a direct link between increased drug use and abnormal within-network functional connectivity of the executive control network, independently of impaired neurocognitive functioning (attention/adaptive learning).

Further, analog to the reward system/salience network, the frontoparietal executive control network was decoupled from other important systems of the brain, such as the DMN (Ding and Lee 2013), while showing enhanced connectivity with both the reward system and the salience network. This enhanced connectivity between these three major brain systems/networks, the reward system, the salience network and the executive control network, was supported by a large number of studies (Kohno *et al.* 2016; Camchong *et al.* 2011; Kohno *et al.* 2014; Hu *et al.* 2015; Konova *et al.* 2013; Camchong *et al.* 2013; Cisler *et al.* 2013). The reward system (VTA/NAcc) showed increased coupling with frontal (vlPFC/dACC) and parietal executive control regions (Kohno *et al.* 2014; Kohno *et al.* 2016; Camchong *et al.* 2013), with the increased coupling between the NAcc and frontal executive regions (vlPFC/dACC) being associated with more compulsive drug use (Hu *et al.* 2015). Similarly, the salience network (rACC/insula) showed heightened functional connectivity with frontal (vlPFC/dlPFC/dACC/SMA) and parietal executive control regions (Camchong *et al.* 2011; Camchong *et al.* 2013; Konova *et al.* 2013; Cisler *et al.* 2013), with this increased coupling (rACC-dlPFC) being related to worse performance in adaptive learning tasks (Camchong *et al.* 2011). Importantly, this result was independent both of the state of participants (urine positive or negative) and drug of choice (methamphetamine or cocaine) (Table 29.1).

In summary, while changes in local/within-network connectivity of the executive control network were not correlated with drug use (but with neurocognitive performance), worse adaptive learning and more compulsive drug use were associated with enhanced connectivity of the executive control network with the reward system/salience network and a partial decoupling from other important brain networks, such as the DMN (Figure 29.1).

Structural connectivity of the executive control network

Overall, the reviewed diffusion-weighted MRI studies suggest that stimulant addiction is associated with lower FA in many brain regions, driven by increases in radial diffusivity (increased diffusion of water perpendicular to axon direction), which indicates white matter degradation driven by damage to or loss of the axon's myelin sheath (Moeller *et al.* 2007). The first diffusion-weighted MRI studies in cocaine users found lower FA in the inferior frontal white matter (Lim *et al.* 2002) and corpus callosum (Moeller *et al.* 2005). Changes in the FA in the corpus callosum were later shown to be anti-correlated with impulsivity (Lim *et al.* 2008) and seem to be more severely reduced in subjects that smoked crack cocaine compared with those who used cocaine intranasally (Ma *et al.* 2009). Findings in the corpus callosum of methamphetamine abusers appear to be similar to those in cocaine abuse with reduced FA in the genu and splenium of the corpus callosum correlated with worse performance on tasks of cognitive flexibility (Kim *et al.* 2009) and inhibitory control (Salo *et al.* 2009).

A later voxel-wise comparison of FA between cocaine-addicted individuals and controls found additional areas of deficit that included many white matter tracts (Lane *et al.* 2010). In this study increases in radial diffusivity in the corticospinal tract and posterior corona radiata (both originating in the motor system) were correlated with poor performance on a gambling task suggesting myelin damage in impaired decision making.

Learning networks

The increase of habitual and compulsive behavior, which characterizes drug addiction, depends on pavlovian and instrumental learning (Everitt and Robbins 2005). Brain structures in the dorsal striatum, the caudate and putamen, support this automatization of behavior, with the caudate being more involved in initiation of behavior and the putamen specifically subserving habit learning (Grahn *et al.* 2008). During habit learning, the amygdala/hippocampal cluster provides the contextual information necessary for associative learning (Everitt and Robbins 2005). During resting-state this amygdala/hippocampal cluster has also been shown to be part of the extended social-affective DMN (Amft *et al.* 2016). In stimulant addiction, resting-state connectivity of these learning networks is altered.

Resting-state connectivity of learning networks

In cocaine users, functional connectivity between the putamen and the dorsal parts of the salience network (dACC/insula) was decreased (Hu *et al.* 2015; McHugh *et al.* 2013; Wisner *et al.* 2013), with two studies reporting that a larger decrease in connectivity was linked to higher impulsivity (McHugh *et al.* 2013; Wisner *et al.* 2013). Decreases in functional connectivity were also reported between the amygdala and the salience network (McHugh *et al.* 2014), between the putamen and periaqueductal gray, which as the insula is involved in interoceptive processing (Verdejo-Garcia *et al.* 2014) and between the amygdala/hippocampal cluster and the executive control network (Gu *et al.* 2010; Konova *et al.* 2013).

In contrast, the coupling with the reward system (VTA/NAcc/rACC) was increased for both the caudate and putamen (Konova *et al.* 2013; Kohno *et al.* 2014; Kohno *et al.* 2016; Camchong *et al.* 2013), with a larger increase in connectivity associated with higher addiction severity (Konova *et al.* 2013). Further, functional connectivity between the caudate and the executive control network was found to be increased, with higher connectivity being correlated with higher impulsivity and current use (Hu *et al.* 2015). Also, increased connectivity between the hippocampus and the reward system was reported (Kohno *et al.* 2014). Finally, stronger coupling between the hippocampus and DMN predicted time to relapse in cocaine addiction (Adinoff *et al.* 2015) and increased local/global connectivity in the DMN was associated with longer duration of cocaine use (Konova *et al.* 2015).

In summary, decreases in functional connectivity were reported between the learning and salience network, with this decrease being associated with impulsivity. Further, evidence across studies supported that the caudate and putamen, involved in initiation of behavior and habit learning respectively, showed increased coupling with the reward system, with this increase being associated with addiction severity (Figure 29.1).

Structural connectivity of learning networks

The uncinate fasciculus, a white matter tract that connects the hippocampus and amygdala with the prefrontal cortex including the OFC, has been shown to have lower FA in poly-substance abusers (Kaag et al. 2017). A study in methamphetamine addiction found lower FA in the perforant fibers implicating the hippocampus, and perhaps the learning network (Tobias et al. 2010). Little work in structural connectivity of the caudate and putamen has been done to date. One study found higher diffusivity within the caudate and putamen in methamphetamine users (Alicata et al. 2009). This higher diffusivity was correlated with lifetime dose and anti-correlated with age of onset, implying that lower connectivity within the caudate and putamen is associated with chronicity of methamphetamine use. A later study, yet with a smaller sample, failed to replicate these results (Lin et al. 2015).

Discussion

Resting-state functional and structural imaging are novel tools that can deepen our understanding of the structural and functional architecture of the brain at baseline, independent of task performance. The results reviewed here strongly indicate that this baseline state is fundamentally reconfigured in stimulant addiction, suggesting that individuals with stimulant addiction interact with the world coming from a somewhat different state.

The reviewed findings on resting-state functional connectivity demonstrate that stimulant addiction is associated with an increase in co-activation (as an indicator of shared neuronal activity) specifically between the four discussed networks: the reward system, the salience, executive control, and learning network (Figure 29.1). There was consistent evidence for enhanced connectivity between a) the reward and salience system, b) both the reward system/salience network with the executive control network and c) between the learning and the reward system. Importantly, this pattern of enhanced functional connectivity was not driven by a general increase of functional connectivity across all brain networks, as there was also consistent evidence for a decrease of functional connectivity between the discussed networks (e.g., reward system/salience network/executive control network) and other important brain systems (e.g., DMN, visual/auditory networks). Finally, the observed pattern in aberrant resting-state connectivity seemed to depend neither on the drug of choice (cocaine, amphetamine or methamphetamine), nor on the state of the participants (urine positive or negative), suggesting that the observed reconfiguration arises through a general change in the baseline functional architecture, independent of abstinence/current use states/type of stimulant drug.

A possible interpretation of these results is that in stimulant addiction the baseline state of the brain is reconfigured such that the reward system/salience network show increased 'neuronal interaction' with both the learning and executive functioning networks because they are 'driving' these two networks involved in cognitive/inhibitory control, learning, and behavior selection. Increased interaction between these systems during a baseline state may thus prompt a stronger involvement of the brain's reward/salience system in learning and behavior selection and may bias the individual's response when encountering a challenging situation. This could predispose an individual to a premature breakdown of self-control during a challenge, or to a heightened drug-related responding, as indeed suggested by the direction of associations with behavioral measures (including impulsivity, compulsive drug use and severity of use).

The results presented here extend a previous review on nicotine and drug addiction, which provided evidence for aberrant connectivity in all the major networks discussed here, but was inconclusive regarding the direction of the observed changes (Sutherland et al. 2012).

This previous review focused on nicotine addiction and included only the four oldest of the 23 resting-state studies on stimulant addiction discussed here. It also used much broader delineations for summarizing results, mostly discussing the functional connectivity between a region and broad 'cortical areas', rather than between two regions/networks. Finally, it collapsed findings across different classes of drugs, including stimulants and opioids. Importantly, however, this previous review proposed a model predicting that acute administration of nicotine would lead to an increased coupling between salience and executive control networks as well as a decreased coupling between the salience network and the DMN (Sutherland *et al.* 2012). The results presented here provide support to this model, suggesting that chronic drug use is associated with a shift in balance towards enhanced coupling between reward system/salience network, learning, and executive control networks and concurrent decoupling from other large brain systems during resting-state.

A limitation of the presented resting-state results is that while we were able to identify 23 resting-state studies in stimulant addiction, most studies only investigated a small part of the brain by using either a region of interest (ROI) approach and performing seed-based functional connectivity analyses, or by performing a network analysis but selecting only one component of interest. Overall, there is a lack of unbiased whole-brain studies, so the connectivity patterns for less frequently studied brain regions (e.g., prefrontal cortex regions) are less clear. Results therefore remain somewhat fragmented, which necessitated that we summarized results through virtually assembling regions into larger resting-state networks. Another limitation pertains to the nature of the cross-sectional, correlational research performed by all of the reviewed studies; longitudinal within-subject results could allow the study of the impact of individual variables including response to treatment on the described shift in brain state.

The majority of diffusion-weighted imaging studies in individuals with stimulant (both single substance and poly substance) addiction found lower FA in large white matter tracks throughout the brain. It has been postulated that reductions in FA are the result of myelin reduction caused by hypo-perfusion related to stimulant exposure. Some studies point to neuroinflammation as a potential mechanism for changes in structural connectivity metrics in drug abuse (Alicata *et al.* 2009; Lin *et al.* 2015; Kaag *et al.* 2017). Both vascular and neuroinflammation effects may contribute to the preponderance of changes in large fiber pathways such as the corpus callosum. However, this result could possibly also be attributed to low sensitivity in imaging parameters.

A potential confounder is the poly-substance use that characterizes most participants with drug addiction. A recent study of structural connectivity in multiple substance users found that structural changes (decreases in FA in many tracts) were related to the number of substances abused and not to the amount of use (Kaag *et al.* 2017). Sex differences may also play a role in structural connectivity, with some studies finding decreases in FA in male methamphetamine-addicted individuals but not in females (Chung *et al.* 2007). Future studies will no doubt take these confounds into careful consideration.

The reduction in structural connectivity indicated by white matter reductions could point to a potential mechanism by which the changes in functional connectivity associated with stimulant addiction could come about. Specifically, the reductions in FA and increases in radial diffusivity may indicate that demyelination provides a structural basis for the observed changes in the functional architecture. However, given that linking structural and functional imaging is still an area of developing research, this link remains speculative. A further limitation is that causality cannot be ascertained with human imaging, hence a causal link between structural connectivity, functional connectivity and substance abuse cannot be established.

Despite these limitations, these results may inform future development of treatment approaches by pointing towards potential neural targets. Specifically, the results presented in this review converge with the results from a previous review on imaging the impact of therapeutic cognitive interventions in addiction, including cognitive behavioral therapy, motivational interventions and mindfulness training (amongst others), on brain functioning. Both reviews identified the reward system/salience network and the executive control network as primary potential treatment targets in stimulant addiction (Zilverstand et al. 2016). This convergent result underscores the contribution of the presented research on resting-state brain functioning in stimulant addiction by suggesting that the aberrant connectivity patterns observed during resting-state may influence brain functioning beyond the mere state of non-involvement in a task, to, for instance, affecting an individual's responsiveness to therapeutic cognitive interventions.

Conclusion

In conclusion, results support a fundamentally altered baseline state of brain functioning in individuals with stimulant addiction. Specifically, a reconfiguration of the functional architecture of the addicted brain seems to be associated with enhanced prominence of the reward system/salience network that may become a major driving factor, possibly influencing important brain systems involved in action selection, habit learning and executive control. Underlying structural impairments encompassing large white matter tracts, especially frontal tracts, could contribute to some of the functional results as further associated with reductions in cognitive control and more compulsive drug use patterns. Irrespective of the underlying biological mechanism or causality, this review highlights the merits of performing imaging studies targeting brain connectivity in clinical populations, which could help shape our understanding of the fundamental changes occurring in the addicted brain.

References

Adinoff, B., et al. (2015) "Basal hippocampal activity and its functional connectivity predicts cocaine relapse", *Biological Psychiatry* 78(7): 496–504.
Alicata, D., et al. (2009) "Higher diffusion in striatum and lower fractional anisotropy in white matter of methamphetamine users", *Psychiatry Research: Neuroimaging* 174(1): 1–8.
Amft, M., et al. (2016) "Definition and characterization of an extended social-affective default network", *Brain Structure and Function* 220(2): 1031–1049.
Bell, R. P., et al. (2011) "Assessing white matter integrity as a function of abstinence duration in former cocaine-dependent individuals", *Drug and Alcohol Dependence* 114(2): 159–168.
Buhle, J. T., et al. (2014) "Cognitive reappraisal of emotion: a meta-analysis of human neuroimaging studies", *Cerebral Cortex* 24(11): 2981–2990.
Camchong, J., et al. (2011) "Frontal hyperconnectivity related to discounting and reversal learning in cocaine subjects", *Biological Psychiatry* 69(11): 1117–1123.
Camchong, J., Stenger, V. A. and Fein, G. (2013) "Resting state synchrony in long-term abstinent alcoholics with versus without comorbid drug dependence", *Drug and Alcohol Dependence* 131(1–2): 56–65.
Chase, H. W., et al. (2015) "Reinforcement learning models and their neural correlates: an activation likelihood estimation meta-analysis", *Cognitive, Affective, and Behavioral Neuroscience* 15(2): 435–459.
Chung, A., et al. (2007) "Decreased frontal white-matter integrity in abstinent methamphetamine abusers", *International Journal of Neuropsychopharmacology* 10(6): 765–775.
Cisler, J. M., et al. (2013) "Altered functional connectivity of the insular cortex across prefrontal networks in cocaine addiction", *Psychiatry Research: Neuroimaging* 213(1): 39–46.
Cole, M. W. and Schneider, W. (2007) "The cognitive control network: integrated cortical regions with dissociable functions", *NeuroImage* 37(1): 343–360.
Corbetta, M., Patel, G. and Shulman, G. L. (2008) "The reorienting system of the human brain: from environment to theory of mind", *Neuron* 58(3): 306–324.
Corbetta, M. and Shulman, G. L. (2002) "Control of goal-directed and stimulus-driven attention in the brain", *Nature Reviews Neuroscience* 3(3): 215–229.

Ding, X. and Lee, S. W. (2013) "Cocaine addiction related reproducible brain regions of abnormal default-mode network functional connectivity: a group ICA study with different model orders", *Neuroscience Letters* 548: 110–114.

Everitt, B. J. and Robbins, T. W. (2005) "Neural systems of reinforcement for drug addiction: from actions to habits to compulsion", *Nature Neuroscience* 8(11): 1481–1489.

Fox, M. D. and Raichle, M. E. (2007) "Spontaneous fluctuations in brain activity observed with functional magnetic resonance imaging", *Nature Reviews Neuroscience* 8(9): 700–711.

Goldstein, R. Z. and Volkow, N. D. (2002) "Drug addiction and its underlying neurobiological basis: neuroimaging evidence for the involvement of the frontal cortex", *American Journal of Psychiatry* 159(10): 1642–1652.

Goldstein, R. Z. and Volkow, N. D. (2011) "Dysfunction of the prefrontal cortex in addiction: neuroimaging findings and clinical implications", *Nature Reviews Neuroscience* 12(11): 652–669.

Grahn, J. A., Parkinson, J. A. and Owen, A. M. (2008) "The cognitive functions of the caudate nucleus", *Progress in Neurobiology* 86(3): 141–155.

Gu, H., et al. (2010) "Mesocorticolimbic circuits are impaired in chronic cocaine users as demonstrated by resting-state functional connectivity", *NeuroImage* 53(2): 593–601.

Hu, Y., et al. (2015) "Impaired functional connectivity within and between frontostriatal circuits and its association with compulsive drug use and trait impulsivity in cocaine addiction", *JAMA Psychiatry* 72(6): 584–592.

Kaag, A. M., et al. (2017) "White matter alterations in cocaine users are negatively related to the number of additionally (ab) used substances", *Addiction Biology* 22(4): 1048–1056.

Kelly, C., et al. (2011) "Reduced interhemispheric resting state functional connectivity in cocaine addiction", *Biological Psychiatry* 69(7): 684–692.

Kim, I. S., et al. (2009) "Reduced corpus callosum white matter microstructural integrity revealed by diffusion tensor eigenvalues in abstinent methamphetamine addicts", *Neurotoxicology* 30(2): 209–213.

Kohno, M., et al. (2014) "Risky decision making, prefrontal cortex, and mesocorticolimbic functional connectivity in methamphetamine dependence", *JAMA Psychiatry* 71(7): 812–820.

Kohno, M., et al. (2016) "Midbrain functional connectivity and ventral striatal dopamine D2-type receptors: link to impulsivity in methamphetamine users", *Molecular Psychiatry* 21(11): 1554–1560.

Konova, A. B., et al. (2013) "Effects of methylphenidate on resting-state functional connectivity of the mesocorticolimbic dopamine pathways in cocaine addiction", *JAMA Psychiatry* 70(8): 857–868.

Konova, A. B., et al. (2015) "Effects of chronic and acute stimulants on brain functional connectivity hubs", *Brain Research* 1628: 147–156.

Koob, G. F. and Volkow, N. D. (2010) "Neurocircuitry of addiction", *Neuropsychopharmacology* 35: 217–238.

Koob, G. F. and Volkow, N. D. (2016) "Neurobiology of addiction: a neurocircuitry analysis", *The Lancet Psychiatry* 3(8): 760–773.

Kühn, S. and Gallinat, J. (2011) "Common biology of craving across legal and illegal drugs – a quantitative meta-analysis of cue-reactivity brain response", *European Journal of Neuroscience* 33(7): 1318–1326.

Lane, S. D., et al. (2010) "Diffusion tensor imaging and decision making in cocaine dependence", *PLOS One* 5(7): e11591.

Liang, X., et al. (2015) "Interactions between the salience and default-mode networks are disrupted in cocaine addiction", *The Journal of Neuroscience* 35(21): 8081–8090.

Lim, K. O., et al. (2002) "Reduced frontal white matter integrity in cocaine dependence: a controlled diffusion tensor imaging study", *Biological Psychiatry* 51(11): 890–895.

Lim, K. O., et al. (2008) "Brain macrostructural and microstructural abnormalities in cocaine dependence", *Drug and Alcohol Dependence* 92(1–3): 164–172.

Lin, J. C., et al. (2015) "Investigating the microstructural and neurochemical environment within the basal ganglia of current methamphetamine abusers", *Drug and Alcohol Dependence* 149: 122–127.

Luijten, M., et al. (2014) "Systematic review of ERP and fMRI studies investigating inhibitory control and error processing in people with substance dependence and behavioural addictions", *Journal of Psychiatry and Neuroscience* 39(3): 149–169.

Ma, L., et al. (2009) "Diffusion tensor imaging in cocaine dependence: regional effects of cocaine on corpus callosum and effect of cocaine administration route", *Drug and Alcohol Dependence* 104(3): 262–267.

McHugh, M. J., et al. (2013) "Striatal-insula circuits in cocaine addiction: implications for impulsivity and relapse risk", *The American Journal of Drug and Alcohol Abuse* 39(6): 424–432.

McHugh, M. J., et al. (2014) "Cortico-amygdala coupling as a marker of early relapse risk in cocaine-addicted individuals", *Frontiers in Psychiatry* 5(16): 1–13.

McHugh, M. J., et al. (2017) "Executive control network connectivity strength protects against relapse to cocaine use", *Addiction Biology* 22(6): 1790–1801.

Meunier, D., et al. (2012) "Brain functional connectivity in stimulant drug dependence and obsessive–compulsive disorder", *NeuroImage* 59(2): 1461–1468.

Moeller, F. G., et al. (2005) "Reduced anterior corpus callosum white matter integrity is related to increased impulsivity and reduced discriminability in cocaine-dependent subjects: diffusion tensor imaging", *Neuropsychopharmacology* 30(3): 610–617.

Moeller, F. G., et al. (2007) "Diffusion tensor imaging eigenvalues: preliminary evidence for altered myelin in cocaine dependence", *Psychiatry Research: Neuroimaging* 154(3): 253–258.

Moeller, S. J. and Goldstein, R. Z. (2014) "Impaired self-awareness in human addiction: deficient attribution of personal relevance", *Trends in Cognitive Sciences* 18(12): 635–641.

Naqvi, N. H. and Bechara, A. (2009) "The hidden island of addiction: the insula", *Trends in Neurosciences* 32(1): 56–67.

Narayana, P. A., et al. (2009) "Diffusion tensor imaging of cocaine-treated rodents", *Psychiatry Research: Neuroimaging* 171(3): 242–251.

Ray, S., Gohel, S. R. and Biswal, B. B. (2015) "Altered functional connectivity strength in abstinent chronic cocaine smokers compared to healthy controls", *Brain Connectivity* 5(8): 476–486.

Romero, M. J., et al. (2010) "Cocaine addiction: diffusion tensor imaging study of the inferior frontal and anterior cingulate white matter", *Psychiatry Research: Neuroimaging* 181: 57–63.

Salo, R., et al. (2009) "Cognitive control and white matter callosal microstructure in methamphetamine-dependent subjects: a diffusion tensor imaging study", *Biological Psychiatry* 65(2): 122–128.

Seeley, W. W., et al. (2007) "Dissociable intrinsic connectivity networks for salience processing and executive control", *The Journal of Neuroscience* 27(9): 2349–2356.

Shenhav, A., Botvinick, M. M. and Cohen, J. D. (2013) "The expected value of control: an integrative theory of anterior cingulate cortex function", *Neuron* 79(2): 217–240.

Shmuel, A. and Leopold, D. A. (2008) "Neuronal correlates of spontaneous fluctuations in fMRI signals in monkey visual cortex: implications for functional connectivity at rest", *Human Brain Mapping* 29(7): 751–761.

Smith, S. M., et al. (2006) "Tract-based spatial statistics: voxelwise analysis of multi-subject diffusion data", *Neuroimage* 31(4): 1487–1505.

Sutherland, M. T., et al. (2012) "Resting state functional connectivity in addiction: lessons learned and a road ahead", *NeuroImage* 62(4): 2281–2295.

Tang, V. M., et al. (2015) "White matter deficits assessed by diffusion tensor imaging and cognitive dysfunction in psychostimulant users with comorbid human immunodeficiency virus infection", *BMC Research Notes* 8: 1.

Tobias, M. C., et al. (2010) "White-matter abnormalities in brain during early abstinence from methamphetamine abuse", *Psychopharmacology* 209(1): 13–24.

Verdejo-Garcia, A., et al. (2014) "Functional alteration in frontolimbic systems relevant to moral judgment in cocaine-dependent subjects", *Addiction Biology* 19(2): 272–281.

Viswanath, H., et al. (2015) "Interhemispheric insular and inferior frontal connectivity are associated with substance abuse in a psychiatric population", *Neuropharmacology* 92: 63–68.

Wang, Z., et al. (2015) "A hyper-connected but less efficient small-world network in the substance-dependent brain", *Drug and Alcohol Dependence* 152: 102–108.

Wilcox, C. E., et al. (2011) "Enhanced cue reactivity and fronto-striatal functional connectivity in cocaine use disorders", *Drug and Alcohol Dependence* 115(1–2): 137–144.

Wisner, K. M., et al. (2013) "An intrinsic connectivity network approach to insula-derived dysfunctions among cocaine users dysfunctions among cocaine users", *The American Journal of Drug and Alcohol Abuse* 39(6): 403–413.

Zilverstand, A., et al. (2016) "Cognitive interventions for addiction medicine: understanding the underlying neurobiological mechanisms", *Progress in Brain Research* 224: 285–304.

Zilverstand, A., Parvaz, M. A. and Goldstein, R. Z. (2017) "Neuroimaging cognitive reappraisal in clinical populations to define neural targets for enhancing emotion regulation. A systematic review", *NeuroImage* 151: 105–116.

30
IMAGING DOPAMINE SIGNALING IN ADDICTION

Diana Martinez and Felipe Castillo

Positron Emission Tomography (PET) radioligand imaging

Positron Emission Tomography (PET) is an imaging modality that uses radiotracers to measure receptor levels in the brain. Radiotracers are molecules that bind to a specific receptor and are labeled with a positron-emitting radionuclide, usually carbon-11 (^{11}C) or fluorine-18 (^{18}F). This allows investigators to quantify neurotransmitter and other related receptors in the human brain. There are radiotracers available to image a number of different brain receptors, including dopamine receptors and transporters, serotonin receptors and transporters, GABA and glutamate receptors, opioid receptors, and others. In addition to PET, Single Photon Emission Computed Tomography (SPECT) can also be used to image receptors.

The main outcome measure used in PET and SPECT imaging studies of clinical populations is called "binding potential" (BP). Binding potential is obtained from the ratio of specific binding to nonspecific binding. This is illustrated in Figure 30.1, where specific binding refers to the radiotracer bound to the receptor or cellular target of interest, while nonspecific binding is the radiotracer bound to other non-target proteins in the brain. Nonspecific binding is generally

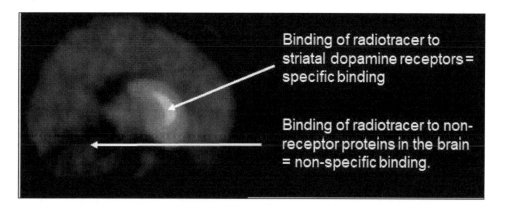

Figure 30.1 Using PET and [^{11}C]raclopride to image dopamine receptor density in the brain. Increased PET signal activity correlates with specific binding of the radiotracer to the receptor.

low, and using this ratio as an outcome measure provides a method for normalizing the signal across subjects and across conditions.

Most studies in addiction will compare two groups, one that is defined by meeting criteria for a specific substance use disorder and a group of matched controls. This serves to identify the neurotransmitter systems that are altered in addiction. However, imaging studies can also make correlations between the neurobiology of addiction and clinically relevant behaviors in subjects with a substance use disorder.

Using PET to image dopamine and dopamine receptors

In addition to measuring receptors, PET can also be used to image changes in neurotransmitter levels in the brain. The PET radiotracer most frequently used to image dopamine is [^{11}C]raclopride, which binds to the D2 family of receptors (referred to as D2 for simplicity). In addition to measuring D2 receptor levels, [^{11}C]raclopride can be used to measure changes in extracellular dopamine in specific regions in the brain. This is shown in the top panel of Figure 30.2, where

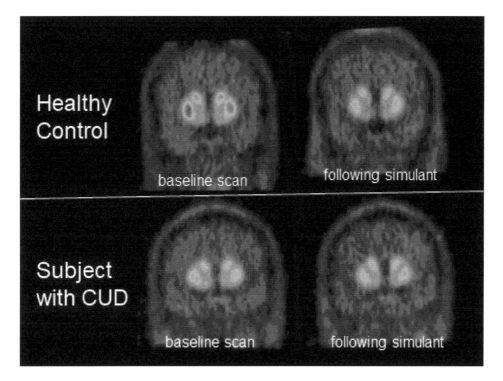

Figure 30.2   Using PET and [^{11}C]raclopride to measure changes in endogenous dopamine in the striatum. The top left panel shows baseline D2 binding to the radiotracer in a healthy control, and the top right panel shows D2 binding following the administration of methylphenidate (60 mg PO). Methylphenidate blocks the dopamine transporter on the dopamine nerve terminals in the striatum, resulting in a large increase in extracellular dopamine levels. As a result, fewer D2 receptors are available to bind to [^{11}C]raclopride. Thus, the decrease in [^{11}C]raclopride binding provides an indirect measure in stimulant-induced increases in endogenous dopamine. The bottom two panels show the same paradigm in a patient with cocaine use disorder. The large increase in extracellular dopamine levels seen in healthy controls are not appreciated in cases of cocaine use disorder.

the highest uptake of [^{11}C]raclopride is seen in the striatum, which contains the caudate, putamen, and nucleus accumbens. The striatum has the highest density of D2 receptors in the brain, and dopamine is released in high levels in this brain region.

When a subject is administered a psychostimulant, such as amphetamine or methylphenidate, large amounts of dopamine are released in the brain. The release of pre-synaptic dopamine results in fewer D2 receptors being available to bind to [^{11}C]raclopride. Therefore, in the same individual, a comparison of BP prior to and following stimulant administration provides an indirect measure of dopamine release. This is depicted in Figure 30.2, where an individual subject's scan is shown at baseline (top left) and following the administration of methylphenidate (top right). Following methylphenidate, the [^{11}C]raclopride signal is reduced, due to the reduction in the D2 receptors available to bind to the radiotracer. The decrease in [^{11}C]raclopride binding is related to the increase in extracellular dopamine elicited by the stimulant.

Figure 30.2 also shows the same imaging outcomes (D2 receptor BP and dopamine release) in cocaine use disorder (bottom panel). The clinical relevance of this finding in cocaine dependence is determined by the role of dopamine in the brain, and the effect of chronic cocaine exposure on dopamine signaling. This is discussed further below in this chapter.

Role of dopamine in the brain

The dopamine system is closely associated with modulating motivated, reward-driven behavior and it is also directly implicated in addiction. The brain's dopamine system originates in the midbrain, with the dopamine neuron cell bodies located in the substantia nigra and ventral tegmental area. The dopamine projections from the midbrain consist of four pathways: 1) mesolimbic, which projects from the ventral tegmental area to the nucleus accumbens of the striatum, and plays a crucial role in reward-driven behavior and addiction; 2) mesostriatal, consisting of projections from the substantia nigra to the dorsal striatum (caudate and putamen), which largely plays a role in modulating cognitive processes and motor output; 3) mesocortical, which projects from the ventral tegmental area to the prefrontal cortex, and directs behavior toward abstract goals; and 4) tubero-infundibular, which regulates the hypothalamus–pituitary system (Cools 2008). While the function of these projection systems can be segregated based on the brain regions they project to, it is important to recognize that there is overlap across these pathways. For example, although the limbic pathway is crucial for reward-driven behavior, both animal and human studies show that the dopamine projections involved in cognitive processing are involved in addiction as well.

Essentially, dopamine signaling makes the behavior required to obtain a reward more likely to be repeated. Salamone (2009) showed that increased dopamine signaling in the nucleus accumbens increases effort-related behavior, and that dopamine depletion results in an animal being less likely to exert effort to receive a food reward. In this context, dopamine can be understood as mediating the behavioral economics of motivated behavior and regulating the reinforcement value of a reward (i.e., the extent to which a given reward is worth the effort required). Dopamine can also be viewed as mediating the "incentive salience" of a reward, meaning the extent to which the reward is wanted by the animal (Berridge and Robinson 1998).

In a series of electrophysiological studies in non-human primates, Schultz (2010) showed that dopamine serves as a reward prediction error, and codes for reward as it differs from prediction, and thus plays a crucial role in reward-based learning. While there is debate over the mechanism by which dopamine mediates the reinforcing effects of a reward, whether it is by signaling the willingness to work for a reward, the degree to which a reward is wanted, or reward-related learning, *the fact that drugs of abuse, such as cocaine, are associated with dysregulation of dopamine transmission that affects decision making is a crucial starting point for understanding the neurobiology of this disorder.*

Cocaine at the synapse

We will use cocaine as an example of how drugs of abuse work on the dopamine system, as cocaine has a very direct effect on dopamine signaling. At the dopamine nerve terminal of the striatum, cocaine binds the dopamine transporter (DAT) and blocks the re-uptake of dopamine by the pre-synaptic dopaminergic neurons. Since DAT re-uptake is the major mechanism for terminating the dopamine signal, cocaine administration results in a significant increase in extracellular dopamine, several orders of magnitude over baseline levels, based on microdialysis studies (Di Chiara and Imperato 1988; Bradberry 2000). The excess in extracellular dopamine is dependent on the firing of the neuron, since dopamine is released from the pre-synaptic dopamine neuron as it fires, but then is not re-absorbed by the dopamine transporter.

Cocaine blocks the re-uptake of dopamine, and increases dopamine levels, but it is not selective for dopamine. Cocaine blocks the uptake of serotonin and norepinephrine as well, since it also blocks these transporters (Carroll et al. 1992; Lewin et al. 1992). In fact, cocaine has a similar affinity for each of these monoamine transporters. However, it is the effect on extracellular dopamine that has been shown to produce most of the behavioral effects that are associated with addiction. Thus, it is the increase in dopamine, rather than cocaine's effect on serotonin or norepinephrine, that is most directly associated with the reinforcing effects of cocaine. A crucial study of cocaine analogues in monkeys demonstrated that the potency of DAT inhibition (rather than inhibition of the norepinephrine or serotonin transporters) correlated with their ability to support cocaine self-administration (Ritz et al. 1987). Similarly, mice that express a DAT that is insensitive to cocaine, but still functional, do not self-administer cocaine (Thomsen et al. 2009). Inhibition of the norepinephrine and serotonin transporters has less of an effect on the addictive aspect of cocaine in animal models, while drugs that block the DAT inhibit cocaine's effects (Spealman et al. 1989; Tella 1995). In addition, cocaine self-administration is dependent on the mesolimbic dopamine pathway being intact, which is not the case for noradrenergic or serotonergic neurons (Kuhar et al. 1991).

PET radioligand imaging in cocaine use disorder

The most studied addiction using PET imaging is cocaine use disorder, and many of these studies have focused on imaging the D2 receptor and dopamine release. Overall, this research shows that cocaine use disorder is associated with a decrease in D2 receptor binding and dopamine release. The first study, published in 1990, showed that cocaine dependence was associated with a 35% decrease in D2 receptor BP in the striatum compared with healthy control subjects (Volkow et al. 1990). A number of subsequent studies have replicated this finding, and imaging studies performed with [^{11}C]raclopride show a decrease in D2 receptor binding of about 15% in individuals with cocaine use disorder compared with control subjects (Volkow et al. 1993; Volkow et al. 1997; Martinez et al. 2004).

Questions that arise from this finding include: 1) is this decrease is reversible with abstinence; 2) does it serve as a biomarker for addiction; 3) does low D2 receptor represent a risk factor for the development of addiction? With respect to abstinence, one study in cocaine dependent subjects has addressed this issue by imaging subjects at baseline and after three months of inpatient treatment (Volkow et al. 1993). The results showed that the decrease in D2 receptors did not change even after three months of abstinence, indicating that low D2 receptor binding persists for at least three months (Volkow *et al.* 1993). This finding is in agreement with a study in rhesus monkeys, which showed that D2 receptor availability was decreased by 15–20% within one week of cocaine exposure, and that this decrease persisted for up to one year of abstinence (Nader *et al.* 2006).

Although the original work imaging D2 receptors was performed in cocaine dependence, the decrease in D2 receptor binding in the striatum has since been demonstrated in addiction to different substances and alcohol. This has been shown in alcohol, heroin, methamphetamine, and nicotine dependence as shown in Table 30.1 (Trifilieff and Martinez 2014). Additionally, as shown in Table 30.1, the decrease in D2 receptor binding has been associated with alterations in reward-related behaviors. This data indicates that low D2 receptor binding serves as a biomarker for addiction in research studies. In other words, while any one individual person's PET scan does not predict addictive behaviors, the analysis of PET scans performed at a group level consistently shows that D2 receptors are decreased in substance use disorders.

A question that arises is whether low D2 receptor binding precedes the development of addiction, and serves as a risk factor. Studies in both rhesus monkeys and human subjects have sought to address this question. In rhesus monkeys exposed to a social hierarchy, social dominance is associated with a higher striatal D2 receptor binding compared with subordinate animals (Grant *et al.* 1998; Morgan *et al.* 2002). In addition, low D2 receptor binding, combined with low social status, was predictive of subsequent cocaine self-administration in monkeys (Nader *et al.* 2006), although no competing reinforcer was available. Nonetheless, these results suggest that low D2 receptor levels in the striatum preceded the animals' willingness to take cocaine, and could serve as a risk factor for the development of addiction.

In human volunteers, imaging studies have investigated this question by imaging non-addicted subjects who have a high risk for developing an addiction, such as those with a family history of addiction. In a study of social drinkers, subjects with a strong family history of alcohol dependence had higher D2 receptor BP in the striatum compared with social drinkers with no family history of alcoholism (Volkow *et al.* 2006). Since the family-history-positive subjects would be expected to have a high risk for alcohol use disorder, but are not dependent themselves, these findings suggest that increased D2 receptor BP may be protective (Volkow *et al.* 2006). Similar results have been reported in non-addicted siblings of subjects with cocaine addiction, who had higher D2 receptor binding compared with controls (Volkow *et al.* 2006).

Table 30.1 Imaging studies showing that D2 receptor binding potential is reduced across addictions and correlates with clinically relevant behavior

Drug abused	Number of studies showing decreased striatal D2 receptor binding	Behavior
Cocaine	7 out of 8	Heavy cocaine use, years of use, tolerance to the effects of cocaine, cocaine-seeking behavior, treatment response
Alcohol	8 out of 10	Craving for alcohol, attention to alcohol cues, future alcohol use, severity of addiction
Opiates	3 out of 4	Severity and heaviness of opiate use
Methamphetamine	4 out of 4	Impulsive behavior, severity of disease, response to treatment
Nicotine	5 out of 5	Severity and heaviness of smoking, depressive symptoms

Another group of subjects who are expected to have a high risk of developing an addiction are those who report a positive experience in response to drug exposure. In a study of non-addicted individuals, Volkow et al. (1999; 2002) demonstrated that high striatal D2 receptor BP was predictive of an unpleasant reaction to the psychostimulant methylphenidate, whereas low D2 binding was associated with a pleasurable experience. This finding suggests that high D2 receptor binding may confer a resilience to the development of addictive behaviors, while low D2 BP may reflect a vulnerability (Volkow et al. 1999; Volkow et al. 2002). However, not all human PET studies show results that are in agreement with this theory, and have shown no difference in D2 receptor BP in family-history-positive and negative social drinkers or in the reaction to psychostimulant administration (Munro et al. 2006). In addition, while some studies have shown that low D2 receptor binding is associated with a risk for addiction, and suggest that this neurobiologic marker might occur prior to the onset of addiction, other studies in non-human primates have also shown that chronic exposure to cocaine itself also reduce D2 receptor binding (Farfel et al. 1992; Moore et al. 1998; Nader et al. 2002; Nader et al. 2006).

Imaging dopamine release in cocaine use disorder

As described above, PET imaging with [^{11}C]raclopride and induction of a pre-synaptic dopamine release can be used to image changes in the level of endogenous dopamine. This entails using a stimulant challenge, where a drug that increases dopamine release in the brain is administered and PET scans are obtained before (at baseline) and after the administration of the medication (usually amphetamine or methylphenidate). Using these methods, Volkow et al. (1997) first showed that cocaine dependence is associated with a decrease in [^{11}C]raclopride displacement in the striatum following methylphenidate administration (Volkow et al. 1997). Thus, this study shows that cocaine dependence is associated with a loss of dopamine release and blunted dopamine transmission. The subjects with cocaine use disorder also reported a decrease in the positive effects of the stimulant compared with the controls. Using SPECT and an amphetamine challenge (0.3 mg/kg IV), Malison et al. (1999) performed a similar study and also showed blunted dopamine release in the striatum. More recent PET studies using higher resolution PET scanners show similar results: dopamine signaling is blunted across the different subdivisions that represent the different dopamine projection systems described above controls (Martinez et al. 2004; 2007).

These studies show that cocaine dependence is associated with a decrease in pre-synaptic dopamine release, and this hypothesis is supported by a PET study that imaged pre-synaptic dopamine stores in the striatum. Using the levodopa analogue 6-[^{18}F]-fluoro-L-DOPA (FDOPA), which provides a measure of presynaptic dopamine activity, Wu et al. (1997) showed that subjects with cocaine use disorder who had been abstinent 11 to 30 days had lower uptake compared with controls (Wu et al. 1997). Notably, the time frame of the decrease in pre-synaptic dopamine corresponds with the reported peak time of cocaine craving and dysphoria during abstinence, and a higher risk of relapse (Gawin and Kleber 1986; Satel et al. 1991).

Functional significance of imaging findings in cocaine use disorder

The imaging studies described here consistently show that cocaine use disorder is associated with both a reduction in D2 receptor binding potential as well as a decrease in dopamine transmission. The next question is what behavioral significance do these findings have?

Dopamine and drug seeking behavior

Using a laboratory model of cocaine self-administration in human volunteers, a previous PET imaging study showed that blunted dopamine release in the ventral striatum correlated with cocaine seeking behavior (Martinez et al. 2007). In this study subjects with cocaine use disorder were scanned with [^{11}C]raclopride and an amphetamine challenge, in order to obtain both D2 receptor BP and dopamine release. The imaging results showed that both of these measures of dopamine transmission were blunted compared with healthy controls. Following the PET scans, the cocaine dependent subjects underwent self-administration sessions, where they chose between a dose of smoked cocaine and money, which serves as a competing reinforcer. The results, summarized in Table 30.2, showed that the subjects with cocaine use disorder with the greatest deficit in dopamine release were more likely to choose the cocaine over money, while those with higher dopamine release chose money.

Thus, these results show that, in subjects with cocaine use disorder, those with the greatest blunting of dopamine release were more likely to choose cocaine over the monetary reinforcer. The doses of cocaine used were low, and produced little in the way of pleasurable effects. The self-administration sessions were developed as a laboratory model of relapse and are based on animal studies showing that a low, priming dose of cocaine reinstates cocaine self-administration (Self et al. 1996; Khroyan et al. 2000; Shaham et al. 2003).

The failure of cocaine use disorder subjects with low dopamine release to alter their behavior can be viewed as a reward system that is unable to shift between alternative sources of reward. In other words, subjects with low dopamine release tend to make more impulsive choices, and chose a smaller, immediate reward over a larger, delayed reward (Trifilieff and Martinez 2014). To an extent, motivation can be thought of as the inverse of impulsivity, as the ability to exert greater effort in order to obtain a more valuable reward. These findings in human studies are consistent with rodent studies of motivation, which demonstrate that motivated behavior can be modulated by dopamine signaling in the ventral striatum, where increasing D2 receptor signaling enhances motivation, whereas impaired D2 signaling lessens it (Salamone et al. 2007).

Cognition and loss of control

Addiction is characterized as a relapsing cycle of intoxication, withdrawal, and relapse that results in excessive drug use despite adverse consequences (Goldstein and Volkow 2011). While dysfunction of the dopamine system modulates much of the disruption in reward-driven behavior, there are also a number of cognitive processes subsumed by the prefrontal cortex that play a role in addiction. Thus, imbalance with the prefrontal cortical regions also negatively affect a wide range of behaviors that modulate the severity of addiction.

Imaging data reviewed by Goldstein and Volkow (2011) shows that addicted individuals exhibit an impaired response inhibition and attribute excessive attention to drugs and drug-related cues. These behaviors are modulated by the prefrontal cortex, which has extensive

Table 30.2 Behavioral self-administration data from subjects with cocaine use disorder, differences seen across the cohort that exhibits raclopride displacement in the striatum and the cohort that does not

	Average cocaine self-administrations (6 mg dose)
Cases with raclopride displacement in striatum (n=14)	0.86 ± 1.1
Cases without raclopride displacement in striatum (n=9)	2.89 ± 1.7

connections to the striatum. Studies imaging D2 receptor binding and brain activation in the prefrontal cortex show that, in cocaine dependence, low D2 receptor binding correlates with a decrease in glucose metabolism (measured with [^{18}F]FDG and PET) in the prefrontal cortex of cocaine abusers (Volkow et al. 1992). Subsequent studies have shown that cocaine users have lower rates of glucose usage in regions of the prefrontal cortex, such as the orbitofrontal and cingulate cortex, which persists even after months of abstinence (Volkow et al. 1993; Goldstein et al. 2004; Volkow et al. 2005; Beveridge et al. 2008). The orbitofrontal cortex plays a pivotal role in expectation and self-evaluation and the cingulate is crucial for mediating higher-order, cognitive control over behavior (Goldstein and Volkow 2011).

Therefore, given the crucial role that these pre-frontal cortical brain regions play in maintaining inhibitory control, this research shows that dysregulation of striatal dopamine transmission affects the prefrontal cortical circuits, which act together to modulate drive and impulsivity (Dalley et al. 2011; Groman and Jentsch 2012).

Imaging dopamine transmission in other addictions

As mentioned above, many PET studies have investigated cocaine use disorder, and while fewer imaging studies have been published in other addictions, these studies also show alterations in striatal dopamine. Imaging studies investigating striatal transmission have shown that blunted signaling has effects on clinically relevant behaviors. This includes methamphetamine dependence, where low D2 receptor binding and blunted dopamine release has been associated with impulsivity, emotional dysregulation, and relapse (Ballard et al. 2015; Volkow et al. 2015; Kohno et al. 2016; Okita et al. 2016). Additionally, studies in methamphetamine dependence show that striatal dopamine is associated with treatment response. Methamphetamine abusers who have low D2 receptor binding and blunted dopamine release were more likely to relapse in the setting of treatment compared with those who responded (Wang et al. 2012). Recently, a study in methamphetamine-dependent subjects showed that eight weeks of exercise training increased striatal D2 receptor binding (Robertson et al. 2016). The mechanism for this in humans is not known, but rodent studies have shown that exercise increases levels of striatal, dopamine receptor binding, and tyrosine hydroxylase mRNA (Greenwood et al. 2011; MacRae et al. 1987). Other studies in rodents show that exercise attenuates methamphetamine-induced damage to dopaminergic terminals.

In alcohol dependence, PET imaging studies have shown that low D2 receptor BP is associated with greater alcohol craving and greater cue-induced activation of the prefrontal cortex using functional magnetic resonance imaging (fMRI) (Heinz et al. 2004). These findings led the authors to hypothesize that dopaminergic dysfunction in the ventral striatum may attribute incentive salience to alcohol-associated stimuli via two proposed mechanisms. Alcohol cues may elicit craving by either excessive activation of the networks associated with attention and behavior control (Heinz et al. 2004), or by too little activation of prefrontal networks when alternative rewards (non-alcohol) are presented to subjects (Reiter et al. 2016).

Imaging dopamine and response to treatment

Two PET imaging studies have investigated whether dopamine signaling in the striatum, measured as D2 receptor binding and pre-synaptic dopamine release, can predict response to treatment. In the first study, treatment-seeking cocaine-dependent subjects were scanned with [^{11}C] raclopride and a stimulant challenge, to obtain BPND and delta BPND, prior to entry in a treatment program. The treatment used contingency management, which uses positive reinforcement (monetary vouchers) as incentive for abstinence from cocaine (Higgins et al. 2003).

The results showed that the cocaine-dependent subjects with higher D2 receptor levels and higher dopamine release responded to treatment, while those with lower levels did not. Similar results were seen in a study of methamphetamine abusers, where subjects were scanned with [^{11}C]raclopride to obtain BPND and delta BPND, and then enrolled in treatment. The methamphetamine-abusing subjects who successfully maintained abstinence in treatment had higher values for both D2 receptor binding and dopamine release, compared with methamphetamine abusers who did not respond to treatment (Wang et al. 2012).

In both studies, the subjects with the higher D2 binding in the striatum and higher dopamine release showed less impulsive behavior and chose the larger, delayed reward (treatment) over a smaller, immediate reward (drug use). In fact, both studies showed that the subjects who responded to treatment did not differ from matched controls. Thus, substance-use-disordered subjects with intact dopamine transmission were able to expand effort to obtain the larger goal, which was their pursuit of recovery from addiction.

Conclusion

Imaging studies in humans as well as animal models have helped in understanding the neurobiological effects of cocaine and identify substrates of cocaine, and by proxy, other addictions. Cocaine use induces a massive increase in extracellular dopamine in the mesoaccumbens and nigrostriatal pathways, an effect that is likely to enhance the rewarding properties of cocaine. This leads to the abnormal consolidation of distinct learning processes that depend on interconnected striatal subdivisions, and facilitates the transition from a recreational use to a substance use disorder. Imaging data in humans clearly demonstrate that addiction to cocaine is associated with long-lasting modifications of the dopaminergic transmission and more particularly with a decrease in dopamine release and D2 receptor availability in the striatum. Surprisingly, despite the extensive characterization of the molecular and cellular mechanisms responsible for the induction of cocaine-induced neuronal alterations and addictive-like behaviors in animal models, little is known about their precise role for long-lasting behavioral adaptations induced by cocaine. More importantly, there is a lack of evidence for a possible reversibility or prevention of these changes, which would be candidate targets for treatment therapies.

Abbreviations

Positron Emission Tomography (PET); Single Photon Emission Computed Tomography (SPECT); Dopamine type 2 receptor (D2); Binding Potential (BP); Gamma-Aminobutyric acid (GABA); Per Os (PO); Intravenous (IV).

References

Ballard, M. E., Mandelkern, M. A., Monterosso, J. R., Hsu, E., Robertson, C. L., Ishibashi, K., Dean, A. C. and London, E. D. (2015) "Low dopamine D2/D3 receptor availability is associated with steep discounting of delayed rewards in methamphetamine dependence", *The International Journal of Neuropsychopharmacology* 18(7): pyu119.

Berridge, K. C. and Robinson, T. E. (1998). "What is the role of dopamine in reward: hedonic impact, reward learning, or incentive salience?" *Brain Research. Brain Research Reviews* 28(3): 309–369.

Beveridge, T. J., Gill, K. E., Hanlon, C. A. and Porrino, L. J. (2008) "Review. Parallel studies of cocaine-related neural and cognitive impairment in humans and monkeys", *Philosophical Transactions of the Royal Society of London. Series B, Biological Sciences* 363(1507): 3257–3266.

Bradberry, C. W. (2000) "Acute and chronic dopamine dynamics in a nonhuman primate model of recreational cocaine use", *Journal of Neuroscience* 20(18): 7109–7115.

Carroll, F. I., Lewin, A. H., Boja, J. W. and Kuhar, M. J. (1992) "Cocaine receptor: biochemical characterization and structure-activity relationships of cocaine analogues at the dopamine transporter", *Journal of Medicinal Chemistry* 35(6): 969–981.

Cools, R. (2008) "Role of dopamine in the motivational and cognitive control of behavior", *The Neuroscientist* 14(4): 381–395.

Dalley, J. W., Everitt, B. J. and Robbins, T. W. (2011). "Impulsivity, compulsivity, and top-down cognitive control", *Neuron* 69(4): 680–694.

Di Chiara, G. and Imperato, A. (1988) "Drugs abused by humans preferentially increase synaptic dopamine concentrations in the mesolimbic system of freely moving rats", *Proceedings of the National Academy of Sciences of the United States of America* 85(14): 5274–5278.

Farfel, G. M., Kleven, M. S., Woolverton, W. L., Seiden, L. S. and Perry, B. D. (1992) "Effects of repeated injections of cocaine on catecholamine receptor binding sites, dopamine transporter binding sites and behavior in rhesus monkey", *Brain Research* 578(1–2): 235–243.

Gawin, F. H. and Kleber, H. D. (1986) "Abstinence symptomatology and psychiatric diagnosis in cocaine abusers. Clinical observations", *Archives of General Psychiatry* 43(2): 107–113.

Goldstein, R. Z., Leskovjan, A. C., Hoff, A. L., Hitzemann, R., Bashan, F., Khalsa, S. S., Wang, G. J., Fowler, J. S. and Volkow, N. D. (2004) "Severity of neuropsychological impairment in cocaine and alcohol addiction: association with metabolism in the prefrontal cortex", *Neuropsychologia* 42(11): 1447–1458.

Goldstein, R. Z. and Volkow, N. D. (2011) "Dysfunction of the prefrontal cortex in addiction: neuroimaging findings and clinical implications", *Nature Reviews Neuroscience* 12(11): 652–669.

Grant, K. A., Shively, C. A., Nader, M. A., Ehrenkaufer, R. L., Line, S. W., Morton, T. E., Gage, H. D. and Mach, R. H. (1998) "Effect of social status on striatal dopamine D2 receptor binding characteristics in cynomolgus monkeys assessed with positron emission tomography", *Synapse* 29(1): 80–83.

Greenwood, B. N., Foley, T. E., Le, T. V., Strong, P. V., Loughridge, A. B. and Day, H. E. (2011) "Long-term voluntary wheel running is rewarding and produces plasticity in the mesolimbic reward pathway", *Behavioural Brain Research* 217(2): 354–362.

Groman, S. M. and Jentsch, J. D. (2012) "Cognitive control and the dopamine D(2)-like receptor: a dimensional understanding of addiction", *Depression and Anxiety* 29(4): 295–306.

Heinz, A., Siessmeier, T., Wrase, J., Hermann, D., Klein, S., Grusser, S. M., Flor, H., Braus, D. F., Buchholz, H. G., Grunder, G., Schreckenberger, M., Smolka, M. N., Rosch, F., Mann, K. and Bartenstein, P. (2004) "Correlation between dopamine D(2) receptors in the ventral striatum and central processing of alcohol cues and craving", *American Journal of Psychiatry* 161(10): 1783–1789.

Higgins, S. T., Sigmon, S. C., Wong, C. J., Heil, S. H., Badger, G. J., Donham, R., Dantona, R. L. and Anthony, S. (2003) "Community reinforcement therapy for cocaine-dependent outpatients", *Archives of General Psychiatry* 60(10): 1043–1052.

Khroyan, T. V., Barrett-Larimore, R. L., Rowlett, J. K. and Spealman, R. D. (2000) "Dopamine D1- and D2-like receptor mechanisms in relapse to cocaine-seeking behavior: effects of selective antagonists and agonists", *Journal of Pharmacology and Experimental Therapeutics* 294(2): 680–687.

Kohno, M., Okita, K., Morales, A. M., Robertson, C. L., Dean, A. C., Ghahremani, D. G., Sabb, F. W., Rawson, R. A., Mandelkern, M. A., Bilder, R. M. and London, E. D. (2016) "Midbrain functional connectivity and ventral striatal dopamine D2-type receptors: link to impulsivity in methamphetamine users", *Molecular Psychiatry* 21(11): 1554–1560.

Kuhar, M. J., Ritz, M. C. and Boja, J. W. (1991) "The dopamine hypothesis of the reinforcing properties of cocaine", *Trends in Neurosciences* 14(7): 299–302.

Lewin, A. H., Gao, Y. G., Abraham, P., Boja, J. W., Kuhar, M. J. and Carroll, F. I. (1992) "2ß-substituted analogues of cocaine – synthesis and inhibition of binding to the cocaine receptor", *Journal of Medicinal Chemistry* 35: 135–140.

MacRae, P. G., Spirduso, W. W., Cartee, G. D., Farrar, R. P. and Wilcox, R. E. (1987) "Endurance training effects on striatal D2 dopamine receptor binding and striatal dopamine metabolite levels", *Neuroscience Letters* 79(1–2): 138–144.

Malison, R. T., Mechanic, K. Y., Klumpp, H., Baldwin, R. M., Kosten, T. R., Seibyl, J. P. and Innis, R. B. (1999) "Reduced amphetamine-stimulated dopamine release in cocaine addicts as measured by [123I]IBZM SPECT", *Journal of Nuclear Medicine* 40(5): 110P.

Martinez, D., Broft, A., Foltin, R. W., Slifstein, M., Hwang, D. R., Huang, Y., Perez, A., Frankle, W. G., Cooper, T., Kleber, H. D., Fischman, M. W. and Laruelle, M. (2004) "Cocaine dependence and d2 receptor availability in the functional subdivisions of the striatum: relationship with cocaine-seeking behavior", *Neuropsychopharmacology* 29(6): 1190–1202.

Martinez, D., Narendran, R., Foltin, R. W., Slifstein, M., Hwang, D. R., Broft, A., Huang, Y., Cooper, T. B., Fischman, M. W., Kleber, H. D. and Laruelle, M. (2007) "Amphetamine-induced dopamine release: markedly blunted in cocaine dependence and predictive of the choice to self-administer cocaine", *American Journal of Psychiatry* 164(4): 622–629.

Moore, R. J., Vinsant, S. L., Nader, M. A., Porrino, L. J. and Friedman, D. P. (1998) "Effect of cocaine self-administration on dopamine D2 receptors in rhesus monkeys", *Synapse* 30(1): 88–96.

Morgan, D., Grant, K. A., Gage, H. D., Mach, R. H., Kaplan, J. R., Prioleau, O., Nader, S. H., Buchheimer, N., Ehrenkaufer, R. L. and Nader, M. A. (2002) "Social dominance in monkeys: dopamine D2 receptors and cocaine self-administration", *Nature Neuroscience* 5(2): 169–174.

Munro, C. A., McCaul, M. E., Oswald, L. M., Wong, D. F., Zhou, Y., Brasic, J., Kuwabara, H., Kumar, A., Alexander, M., Ye, W. and Wand, G. S. (2006) "Striatal dopamine release and family history of alcoholism", *Alcoholism: Clinical and Experimental Research* 30(7): 1143–1151.

Nader, M. A., Daunais, J. B., Moore, T., Nader, S. H., Moore, R. J., Smith, H. R., Friedman, D. P. and Porrino, L. J. (2002) "Effects of cocaine self-administration on striatal dopamine systems in rhesus monkeys. Initial and chronic exposure", *Neuropsychopharmacology* 27(1): 35–46.

Nader, M. A., Morgan, D., Gage, H. D., Nader, S. H., Calhoun, T. L., Buchheimer, N., Ehrenkaufer, R. and Mach, R. H. (2006) "PET imaging of dopamine D2 receptors during chronic cocaine self-administration in monkeys", *Nature Neuroscience* 9(8): 1050–1056.

Okita, K., Ghahremani, D. G., Payer, D. E., Robertson, C. L., Dean, A. C., Mandelkern, M. A. and London, E. D. (2016) "Emotion dysregulation and amygdala dopamine D2-type receptor availability in methamphetamine users", *Drug and Alcohol Dependence* 161: 163–170.

Reiter, A. M., Deserno, L., Kallert, T., Heinze, H. J., Heinz, A., and Schlagenhauf, F. (2016) "Behavioral and neural signatures of reduced updating of alternative options in alcohol-dependent patients during flexible decision-making", *Journal of Neuroscience* 36(43): 10935–10948.

Ritz, M. C., Lamb, R. J., Goldberg, S. R. and Kuhar, M. J. (1987) "Cocaine receptors on dopamine transporters are related to self-administration of cocaine", *Science* 237(4819): 1219–1223.

Robertson, C. L., Ishibashi, K., Chudzynski, J., Mooney, L. J., Rawson, R. A., Dolezal, B. A., Cooper, C. B, Brown, A. K., Mandelkern, M. A. and London, E. D. (2016) "Effect of exercise training on striatal dopamine D2/D3 receptors in methamphetamine users during behavioral treatment", *Neuropsychopharmacology* 41(6): 1629–1636.

Salamone, J. D. (2009) "Dopamine, effort, and decision making: theoretical comment on Bardgett et al. (2009)", *Behavioral Neuroscience* 123(2): 463–467.

Salamone, J. D., Correa, M., Farrar, A. and Mingote, S. M. (2007) "Effort-related functions of nucleus accumbens dopamine and associated forebrain circuits", *Psychopharmacology (Berl)* 191(3): 461–482.

Satel, S. L., Price, L. H., Palumbo, J. M., McDougle, C. J., Krystal, J. H., Gawin, F., Charney, D. S., Heninger, G. R. and Kleber, H. D. (1991) "Clinical phenomenology and neurobiology of cocaine abstinence: a prospective inpatient study", *American Journal of Psychiatry* 148(12): 1712–1716.

Schultz, W. (2010) "Dopamine signals for reward value and risk: basic and recent data", *Behavioral and Brain Functions* 6: 24.

Self, D. W., Barnhart, W. J., Lehman, D. A. and Nestler, E. J. (1996) "Opposite modulation of cocaine-seeking behavior by D1- and D2-like dopamine receptor agonists", *Science* 271(5255): 1586–1589.

Shaham, Y., Shalev, U., Lu, L., De Wit, H. and Stewart, J. (2003) "The reinstatement model of drug relapse: history, methodology and major findings", *Psychopharmacology (Berl)* 168(1–2): 3–20.

Spealman, R. D., Madras, B. K. and Bergman, J. (1989) "Effects of cocaine and related drugs in nonhuman primates. II. Stimulant effects on schedule-controlled behavior", *The Journal of Pharmacology and Experimental Therapeutics* 251(1): 142–149.

Tella, S. R. (1995) "Effects of monoamine reuptake inhibitors on cocaine self-administration in rats", *Pharmacology, Biochemistry, and Behavior* 51(4): 687–692.

Thomsen, M., Han, D. D., Gu, H. H. and Caine, S. B. (2009) "Lack of cocaine self-administration in mice expressing a cocaine-insensitive dopamine transporter", *Journal of Pharmacology and Experimental Therapeutics* 331(1): 204–211.

Trifilieff, P. and Martinez, D. (2014) "Imaging addiction: D2 receptors and dopamine signaling in the striatum as biomarkers for impulsivity", *Neuropharmacology* 76(Pt-B): 498–509.

Volkow, N. D., Fowler, J. S., Wang, G. J., Hitzemann, R., Logan, J., Schlyer, D. J., Dewey, S. L. and Wolf, A. P. (1993) "Decreased dopamine D2 receptor availability is associated with reduced frontal metabolism in cocaine abusers", *Synapse* 14(2): 169–177.

Volkow, N. D., Fowler, J. S., Wolf, A. P., Schlyer, D., Shiue, C. Y., Alpert, R., Dewey, S. L., Logan, J., Bendriem, B., Christman, D., et al. (1990) "Effects of chronic cocaine abuse on postsynaptic dopamine receptors", *American Journal of Psychiatry* 147(6): 719–724.

Volkow, N. D., Hitzemann, R., Wang, G. J., Fowler, J. S., Wolf, A. P., Dewey, S. L. and Handlesman, L. (1992) "Long-term frontal brain metabolic changes in cocaine abusers", *Synapse* 11(3): 184–190.

Volkow, N. D., Wang, G. J., Begleiter, H., Porjesz, B., Fowler, J. S., Telang, F., Wong, C., Ma, Y., Logan, J., Goldstein, R., Alexoff, D. and Thanos, P. K. (2006) "High levels of dopamine D2 receptors in unaffected members of alcoholic families: possible protective factors", *Archives of General Psychiatry* 63(9): 999–1008.

Volkow, N. D., Wang, G. J., Fowler, J. S., Logan, J., Gatley, S. J., Gifford, A., Hitzemann, R., Ding, Y. S., and Pappas, N. (1999) "Prediction of reinforcing responses to psychostimulants in humans by brain dopamine D2 receptor levels", *American Journal of Psychiatry* 156(9): 1440–1443.

Volkow, N. D., Wang, G. J., Fowler, J. S., Logan, J., Gatley, S. J., Hitzemann, R., Chen, A. D., Dewey, S. L. and Pappas, N. (1997) "Decreased striatal dopaminergic responsiveness in detoxified cocaine-dependent subjects", *Nature* 386: 830–833.

Volkow, N. D., Wang, G. J., Fowler, J. S., Thanos, P., Logan, J., Gatley, S. J., Gifford, A., Ding, C., Wong, Y. S. and Pappas, N. (2002) "Brain DA D2 receptors predict reinforcing effects of stimulants in humans: replication study", *Synapse* 46(2): 79–82.

Volkow, N. D., Wang, G. J., Ma, Y., Fowler, J. S., Wong, C., Ding, Y. S., Hitzemann, R., Swanson, J. M. and Kalivas, P. (2005) "Activation of orbital and medial prefrontal cortex by methylphenidate in cocaine-addicted subjects but not in controls: relevance to addiction", *The Journal of Neuroscience* 25(15): 3932–3939.

Volkow, N. D., Wang, G. J., Smith, L., Fowler, J. S., Telang, F., Logan, J. and Tomasi, D. (2015) "Recovery of dopamine transporters with methamphetamine detoxification is not linked to changes in dopamine release", *Neuroimage* 121: 20–28.

Wang, G. J., Smith, L., Volkow, N. D., Telang, F., Logan, J., Tomasi, D., Wong, C. T., Hoffman, W., Jayne, M., Alia-Klein, N., Thanos, P. and Fowler, J. S. (2012) "Decreased dopamine activity predicts relapse in methamphetamine abusers", *Molecular Psychiatry* 17(9): 918–925.

Wu, J. C., Bell, K., Najafi, A., Widmark, C., Keator, D., Tang, C., Klein, E., Bunney, B. G., Fallon, J. and Bunney, W. E. (1997) "Decreasing striatal 6-FDOPA uptake with increasing duration of cocaine withdrawal", *Neuropsychopharmacology* 17(6): 402–409.

31

THE NEUROBIOLOGY OF PLACEBO EFFECTS

Elisa Frisaldi, Diletta Barbiani, and Fabrizio Benedetti

Introduction

The placebo effect is a psychobiological phenomenon whereby treatment cues trigger improvement. While traditionally viewed as a nuisance variable to be controlled for, the past three decades have seen a surge in interest in the placebo effect in light of some remarkable clinical and laboratory discoveries that have demonstrated its potential power to improve patient outcomes.

Many attempts have been made to conceptualize the placebo effect and most of the research on placebos has focused on expectations and learning as the two main factors involved in placebo responsiveness. Expectations are representations of future outcomes, and are held by each individual about his or her own emotional and physiological responses. Expectations are unlikely to operate alone and several factors have been identified and described, such as memory, motivation and meaning of the illness experience. There could be at least two ways by which positive expectation can lead to clinical improvement: modulation of anxiety and activation of reward mechanisms (Benedetti 2014). On the other hand, learning mechanisms, such as behavioral conditioning and social learning, are also crucial. For example, the previous exposure to a biologically effective treatment leads to substantial placebo responses that are capable of mimicking the pharmacological action of the previously administered active drug (Colloca and Benedetti 2006).

It is important to point out that expectation and learning are not mutually exclusive, since learning can lead to the reinforcement of expectations or can even create *de novo* expectations. Specifically, in light of a reinterpretation of conditioning in cognitive terms, a conditioning procedure would lead to the expectation that a certain event will follow another event, and this would occur on the basis of the information that the conditioned stimulus provides about the unconditioned stimulus (Kirsch et al. 2004). Thus, expectation and learning may overlap and blend together in a number of different conditions, and may represent two sides of the same coin. In line with this concept, Colloca and Miller (2011) point out that the placebo effect can be conceived as a learned response: various types of cues (verbal, conditioned, and social) trigger expectancies, which, in turn, generate placebo effects.

Expectation, reward and conditioning have also been shown to modulate the reinforcing effects of drugs of abuse, with reward being of crucial importance. For example, in drug abusers, responses to a drug are more pleasurable when subjects expect to receive the drug than when they do not (Kirk *et al.* 1998). Also, the ability of drugs to increase the release of dopamine in the nucleus accumbens, an effect associated with their reinforcing value, is larger when animals are given cocaine in an environment where they had previously received it – and therefore expected it – rather than in an environment where they had not (Duvauchelle *et al.* 2000). Similarly, cocaine-induced changes in regional brain metabolism differ when animals self-administer cocaine from when administration is involuntary (Graham and Porrino 1995).

By taking all these things into account, it is not surprising to find placebo effects related to reward mechanisms in different types of addiction. In this review, we first outline what is currently known about the neurobiology of placebo effects, with particular emphasis on reward mechanisms and motivated behaviors, and then focus our attention on some of the most significant studies that uncovered both psychological and biological mechanisms underlying placebo and placebo-like effects in addiction.

The neurobiology of placebo effects

Most of our knowledge on the neurobiological mechanisms of the placebo effect comes from the field of pain and analgesia. It is well established that placebos modulate pain perception through different endogenous neuronal networks. In particular, the activation of the descending pain-modulating network from the cerebral cortex to the brainstem and spinal cord has been described, with the involvement of opioid, cannabinoid, cholecystokinin (CCK) and dopamine systems (Benedetti 2014). What has emerged over the past few years is that while the endogenous opioids and endocannabinoids are involved in placebo analgesia in different circumstances (Benedetti *et al.* 2011c), CCK plays a key role in the opposite effect, namely, in nocebo hyperalgesia (Benedetti *et al.* 2006). In this latter case, negative verbal suggestions induce anticipatory anxiety about the impending pain increase, and this verbally-induced anxiety triggers the activation of CCK which, in turn, facilitates pain transmission.

The opioid system activated by placebos is the most studied and widely understood. The mu opioid antagonist, naloxone, prevents some types of placebo analgesia, thus indicating that this system plays a crucial role (Amanzio and Benedetti 1999). By contrast, the CCK-antagonist, proglumide, enhances placebo analgesia on the basis of the CCK anti-opioid action (Benedetti *et al.* 1995), and the activation of the CCK type-2 receptors by means of the agonist pentagastrin disrupts placebo analgesia (Benedetti *et al.* 2011a). Thus, the balance between CCKergic and opioidergic systems is a key element in placebo analgesia. Moreover, the cerebral cortex and the brainstem are affected by the administration of both placebo and the opioid agonist remifentanil, thus suggesting a related mechanism in placebo-induced and opioid-induced analgesia (Petrovic *et al.* 2002).

Following the exposure to nonopioid drugs the placebo response is not mediated by mu opioid receptors but by the CB1 cannabinoid receptors. Indeed, when the nonopioid drug ketorolac is administered for two consecutive days and then replaced with a placebo on the third day, the placebo analgesic response is not reversed by naloxone. However, the CB1 cannabinoid receptor antagonist, rimonabant, blocks this placebo analgesic response completely (Benedetti *et al.* 2011b). Moreover, the whole lipidic pathway, involving arachidonic acid, endogenous cannabinoid ligands (e.g., anandamide), and the synthesis of prostaglandins and thromboxane, seems to be important in the modulation of the placebo response in pain. In particular, cyclooxygenase, which is involved in prostaglandins and thromboxane synthesis, has been found to be modulated by both placebo and nocebo in hypobaric hypoxia headache (Benedetti *et al.* 2014).

Many neuroimaging studies have been performed to understand the functional neuroanatomy of placebo analgesia (Eippert *et al.* 2009; Petrovic *et al.* 2002; Wager *et al.* 2004; Zubieta *et al.* 2005). A meta-analysis of brain imaging data using the activation likelihood estimation method (ALE) identified two distinct phases: the expectation phase of analgesia and the pain inhibition phase (Amanzio *et al.* 2013). During expectation, areas of activation were found in the anterior cingulate, precentral and lateral prefrontal cortex, and in the periaqueductal gray. During pain inhibition, deactivations were present in the mid- and posterior cingulate cortex, superior temporal and precentral gyri, anterior and posterior insula, claustrum and putamen, thalamus and caudate body. Overall, many of the regions that are activated during expectation are likely to belong to a descending pain inhibitory system that inhibits pain processing.

Dopamine has also been found to be involved in placebo responses in both pain and Parkinson's disease. Indeed, there is extensive evidence for a prominent placebo effect in Parkinson's disease, which makes this motor disorder an excellent model to better investigate and understand placebo responses (Benedetti 2014). Differently from placebo analgesia, where an increase in dopamine binding to D2/D3 receptors and in opioid binding to μ-receptors occurs in the nucleus accumbens, in nocebo hyperalgesia a decreased binding to the same receptors is observed (Scott *et al.* 2008). Likewise, dopamine receptors are activated in both ventral (nucleus accumbens) and dorsal striatum when a placebo is administered to patients with Parkinson's disease (de la Fuente-Fernández *et al.* 2001, 2002). Dopaminergic activation in the nucleus accumbens in both pain and Parkinson's disease suggests that reward mechanisms could play an important role across different conditions.

Expectation and reward in placebo effects

From a neuroscientific point of view, expecting a future event may involve several brain mechanisms aimed at preparing the body to anticipate that event. For example, expectations may induce changes through reward mechanisms, which ensure future reward acquisition. These mechanisms are mediated by the mesolimbic dopaminergic system that has been found to be active in a number of reward-seeking behaviors. The general organization of this system involves the ventral tegmental area that sends dopaminergic projections to the nucleus accumbens (ventral striatum) as well as to other regions, such as the amygdala.

For almost a half-century, it has been shown that striatal dopamine is a crucial component of reward and reward-based learning and, in turn, the nucleus accumbens serves as a hub of the brain's reward pathways and plays a central role in selecting adaptive, motivated behavior (Sesack and Grace 2010). The reward system is activated by hunger, sex, electrical self-stimulation, and drug self-administration. Even though all these behaviors are present in both animals and humans, it is worth noting that there are reward-seeking behaviors that are typical of human beings, such as monetary reward, that typically activate the mesolimbic dopaminergic system (Scott *et al.* 2007). Motivated behavior that aims to suppress discomfort from sickness and to seek clinical improvement can be approached within the context of motivation/reward mechanisms as well (Benedetti 2010). There is compelling experimental evidence that the mesolimbic dopaminergic system is activated when a subject expects therapeutic benefit. Most, if not all, of this evidence comes from the placebo literature, particularly from pain and Parkinson's disease.

For example, in 2001 de la Fuente-Fernández *et al.* (2001) conducted the first brain imaging study of the placebo effect by means of positron emission tomography. These researchers assessed the release of endogenous dopamine by using raclopride, a radiotracer that binds to dopamine D2 and D3 receptors and competes with endogenous dopamine. In this study, Parkinsonian patients were aware that they would be receiving an injection of either active drug (apomorphine, a dopamine receptor agonist) or placebo (a subcutaneous injection of saline solution that the patient believed to be apomorphine), according to classical clinical trial methodology. After placebo administration, it was found that dopamine was released in the striatum, corresponding to a change of 200% or more in extracellular dopamine concentration and comparable to the response to amphetamine in subjects with an intact dopamine system. Although in the studies by de la Fuente–Fernández *et al.* (2001, 2002) all patients showed dopamine placebo responses, only half of them reported concomitant motor improvement. These patients released larger amounts of dopamine in the dorsal motor striatum (putamen and caudate), suggesting a relationship between the amount of dorsal striatal dopamine release and clinical benefit. Instead, this relationship was not present in the ventral striatum, namely in the nucleus accumbens, in which all patients showed increased dopamine release, irrespective of whether they perceived any improvement. Accordingly, the investigators proposed that the dopamine released in the nucleus accumbens was associated with the patients' expectation of improvement in their symptoms, which could in turn be considered a form of reward.

In order to determine to what extent the strength of expectation of clinical improvement would influence the amount of striatal dopamine release after placebo administration in patients with Parkinson's disease, Lidstone *et al.* (2010) manipulated patients' expectations by telling them that they had a probability of 25%, 50%, 75%, or 100% of receiving active medication when they in fact received a placebo. Significant dopamine release occurred when the declared probability of receiving active medication was 75%, but not for other probabilities. Whereas response to prior medication was the major determinant of placebo-induced dopamine release in the dorsal motor striatum, expectation of clinical improvement was additionally required to drive dopamine release in the ventral striatum. Therefore, the strength of belief of improvement can directly modulate dopamine release in Parkinsonian patients, and this emphasizes the importance of uncertainty and/or salience both in the design of clinical trials and in clinical practice.

In line with these previous brain imaging studies, in which an involvement of the ventral striatum was found after placebo administration, in 2002 another brain imaging study was carried out (Mayberg *et al.* 2002) to measure changes in brain glucose metabolism in male patients with unipolar depression who were treated with either placebo or fluoxetine for six weeks. Common and unique responses were described. In fact, both placebo and fluoxetine treatment induced regional metabolic increases in the prefrontal, anterior cingulate, premotor, parietal, posterior insula, and posterior cingulate, while metabolic decreases were observed in the subgenual cingulate, parahippocampus and thalamus. Fluoxetine response, however, was associated with additional subcortical and limbic changes in the brainstem, striatum, anterior insula, and hippocampus, sources of efferent input to the response-specific regions identified with both agents. Interestingly, there were unique ventral striatal and orbital frontal changes in both placebo and drug responders at one week of treatment, that is, well before the clinical benefit could be seen. Therefore, these changes seem to be associated with expectation and anticipation of the clinical benefit. It is also worth noting that such changes were seen neither in the eventual drug nonresponders nor at six weeks when the antidepressant response was well established (Mayberg *et al.* 2002; Benedetti *et al.* 2005).

Placebo and placebo-like effects in addiction

Many drugs of abuse act at the level of the mesolimbic dopaminergic system and, as already pointed out, this system represents the neurochemical grounding of addiction. For example, nicotine acts at the level of the ventral tegmental area, whose dopaminergic neurons also express cholinergic nicotinic receptors, whereas cocaine acts on the nucleus accumbens, prolonging dopaminergic activity (Wise 1996; Benedetti 2010).

In humans, stimulants (Drevets et al. 2001), nicotine (Brody et al. 2009), alcohol (Boileau et al. 2003), and marijuana (Bossong et al. 2009) increase dopamine release in the dorsal and ventral striatum, which is consistent with animal studies showing that most drugs of abuse increase dopamine release from the presynaptic neurons in the striatum, regardless of their primary mechanism of action (Di Chiara and Imperato 1988). In addition, some of these studies have reported that participants who display the greatest dopamine increases with the drug also report the most intense "high" or "euphoria" experiences (Volkow et al. 1996; Drevets et al. 2001; Volkow et al. 2003a).

Today we know some mechanisms, both psychological and biological, underlying placebo and placebo-like effects in addiction. Indeed, in light of what we have described in the previous sections, it is not surprising to find placebo effects that are related to reward mechanisms in different types of addiction.

In a study, the effects of methylphenidate on brain glucose metabolism have been analyzed in cocaine abusers in different conditions by adopting a "balanced placebo design:" in one condition, cocaine abusers expected to receive the drug and indeed received the drug; in a second condition, they expected to receive a placebo but actually received the drug; in a third condition, they expected to receive a placebo and indeed received placebo; in a fourth condition, they expected to receive the drug but actually received placebo (Volkow et al. 2003b). The increases in metabolism were about 50% larger, particularly in the cerebellum and the thalamus, when methylphenidate was expected than when it was not. By contrast, methylphenidate induced larger increases in the left lateral orbitofrontal cortex when it was unexpected compared with when it was expected. In addition, the self-reports of "high" were also 50% greater when methylphenidate was expected than when it was not. Volkow et al. (2003b) also found a correlation between the subjectively reported "high" and the metabolic activity in the thalamus, but not in the cerebellum. This study strongly suggests that expectations enhance the drug effects and that the thalamus may mediate this kind of drug enhancement, whereas the orbitofrontal cortex mediates the unexpected response to the drug.

Using again the balanced placebo design, the same authors repeated the experiment in non-drug-abusing subjects (Volkow et al. 2006). It was found that methylphenidate-induced metabolism decreases in the striatum were larger when the subjects expected the stimulant than when they did not. The effect of expectation in these non-drug-abusing subjects differs from that of cocaine abusers, in whom expectation enhanced both methylphenidate behavioral effects and methylphenidate-induced increases in thalamic and cerebellar metabolism (Volkow et al. 2003b). Failure to see such an effect of expectation in non-drug-abusing subjects suggests that the enhanced thalamic and cerebellar responses in cocaine abusers reflected conditioned responses. In addition, it was found that the effect of expectation in non-drug-abusing subjects affected their responses to placebo: when they expected to receive methylphenidate, but actually received a placebo, increases in the ventral cingulate gyrus and in the nucleus accumbens were found. Because subjects were told that methylphenidate could be experienced as pleasant, unpleasant or devoid of subjective effects, these results suggest that activation of the ventral cingulate gyrus and the nucleus accumbens after placebo administration reflects both the expectation and the uncertainty around the effects of a novel stimulus, regardless of it being rewarding or not. Thus, expectation needs to be considered as a variable modulating the reinforcing and therapeutic effects of drugs even in subjects who have no prior experience with the drug itself.

In a more recent study, Looby and Earleywine (2011) showed that expectation to receive methylphenidate enhances subjective arousal rather than cognitive performance. On one visit subjects orally ingested what they believed to be methylphenidate, though actually a placebo, and on the other visit they received no medication. The control group received no medication on either visit. During the administration visit, experimental participants, compared with the control subjects, reported feeling significantly more high and stimulated than in the non-administration visit. However, there were no differences in cognitive enhancement between visits or groups, which indicates that in this condition placebos affect mood but not cognition.

Smoking is another good example of how both pharmacological and nonpharmacological factors are at work in tobacco dependence. Although nicotine is necessary for maintaining tobacco dependence, it is generally not sufficient. In fact, despite the fact that nicotine replacement therapy is capable of delivering as much nicotine as a cigarette (and almost as rapidly), many smokers relapse to smoking (Perkins et al. 2003). In one study using the balanced placebo design to investigate smoking placebo effects, briefly abstinent smokers were given a stressor (embarrassing speech) while receiving a nicotine or denicotinized cigarette, along with verbal instructions about its content ("told nicotine" or "told no nicotine") (Juliano and Brandon 2002). The researchers found an interaction of instructions and nicotine on craving: smokers who received denicotinized cigarettes and were told there was nicotine inside reported a decrease in smoking urge after smoking the cigarette. Conversely, smokers who received nicotine cigarettes reported a decrease in urge, regardless of what they were told. Therefore, according to this study, expectations of getting nicotine-induced placebo responses when placebo cigarettes were smoked, but expectations of receiving denicotinized cigarettes did not reduce the pharmacological effect of nicotine.

In a subsequent study with the balanced placebo design, in which low/high dose nicotine instructions were used, Perkins et al. (2004) found that expectations affected both denicotinized and nicotine cigarettes. In other words, expecting high-dose nicotine but getting a denicotinized cigarette induced placebo effects, whereas expecting a denicotinized cigarette but receiving a high dose reduced the effects of nicotine. Interestingly, according to Perkins et al. (2004), verbal instructions did not affect craving and withdrawal but only the number of puffs earned for those smokers who were given low nicotine, thus indicating that verbally induced expectations may affect only some responses to smoking.

Sensory cues including the sight and smell of smoke and handling of a cigarette have also been found to relate to placebo and placebo-like effects. For example, reduction in liking, satisfaction, and self-administration has been described after local anesthesia of the respiratory pathways (Rose et al. 1985) and blockade of olfactory and visual cues (Baldinger et al. 1995). This evidence suggests that salient sensory cues alone, during the consumption of a denicotinized cigarette, can produce substantial craving and withdrawal relief.

It is also worth mentioning the importance of placebo and expectation effects in the context of classical clinical trials. In a trial of nicotine replacement treatment for smoking reduction, smokers were randomly assigned to receive nicotine, matching placebo products, or no intervention. After six months, participants were asked to guess which group they were believed to belong to (either nicotine or placebo). Regardless of the actual treatment received, smokers who believed they had received nicotine had significantly better outcomes than those who believed they had received the placebo (Dar et al. 2005).

Contrary to what should be expected, there is no clear-cut evidence of placebo and placebo-related effects in alcohol abuse. Placebo effects are most consistently found in studies on alcohol-induced sexual arousal (George and Stoner 2000), but there is only modest evidence to support powerful placebo effects across different domains of social behavior. There may be several explanations for this phenomenon. For example, subjects who received placebo usually report that they consumed alcohol but do not typically report feeling as intoxicated as those who received alcohol (Maisto et al. 2002). Therefore, the alcohol and the placebo conditions may not be as equivalent as researchers would like to believe. In other words, it is often difficult to manipulate verbal instructions and to deceive subjects in the setting of alcohol consumption. Within the context of the balanced placebo design, it has been proposed to replace the drug/no drug instructions with the high/low dose instructions, in order to make the instructions more credible (Ross and Pihl 1989; Martin and Sayette 1993). Nonetheless, some studies indicate that expectations do play a role in the response to alcohol. Even when subjects receive the same dose of alcohol and attain the same blood alcohol concentration, some people show a high degree of impairment whereas others show little or none. This interindividual variability may in part be due to different drinkers' expectations about the alcohol effect. In fact, studies of alcohol effects on motor and cognitive performance have shown that those drinkers who expect the least impairment are actually least impaired while those who expect the most impairment are mostly impaired after alcohol consumption (Fillmore and Vogel-Sprott 1995). The more recent study of the US National Institute on Alcohol Abuse and Alcoholism, called COMBINE (Combining Medications and Behavioral Interventions), examined 1,383 alcohol-dependent patients. They found a significant placebo effect related to pill intake and meeting a healthcare professional. Contributing factors to the placebo response were the repeated advice to attend Alcoholics Anonymous and the optimistic view of medication effects (Weiss et al. 2008).

Conclusions

Placebo responses and expectations appear to be of crucial importance when associated with the effects of different drugs of abuse. Advances in neuroimaging and the use of the "balanced placebo design" have significantly increased our understanding of the neurobiology of placebo and placebo-like effects in addiction. A future challenge of placebo research will be to better characterize the neuronal circuits in charge of top-down control. In fact, restoring the proper function of these circuits could represent a crucial factor in the management of addiction and/or its symptoms, like craving. Furthermore, disentangling the effects of drugs of abuse from the associated placebo effects may allow clinicians to maximize therapeutic outcomes in the context of medical practice and minimize placebo responses in the setting of clinical trials.

References

Amanzio, M. and Benedetti, F. (1999) "Neuropharmacological dissection of placebo analgesia: expectation-activated opioid systems versus conditioning-activated specific subsystems", *The Journal of Neuroscience* 19: 484–494.

Amanzio, M., Benedetti, F., Porro, C. A., Palermo, S. and Cauda, F. (2013) "Activation likelihood estimation meta-analysis of brain correlates of placebo analgesia in human experimental pain", *Human Brain Mapping* 34(3): 738–752.

Baldinger, B., Hasenfratz, M. and Battig, K. (1995) "Switching to ultralow nicotine cigarettes: effects of different tar yields and blocking of olfactory cues", *Pharmacology Biochemistry and Behavior* 50(2): 233–239.

Benedetti, F. (2010) *The Patient's Brain: The Neuroscience Behind the Doctor–Patient Relationship*, Oxford, UK: Oxford University Press.

Benedetti, F. (2014) *Placebo Effects, 2nd Edition*, Oxford, UK: Oxford University Press.

Benedetti, F., Amanzio, M. and Maggi, G. (1995) "Potentiation of placebo analgesia by proglumide", *The Lancet* 346(8984): 1231.

Benedetti, F., Mayberg, H. S., Wager, T. D., Stohler, C. S. and Zubieta, J. K. (2005) "Neurobiological mechanisms of the placebo effect", *Journal of Neuroscience* 25(45): 10390–10402.

Benedetti, F., Amanzio, M., Vighetti, S. and Asteggiano, G. (2006) "The biochemical and neuroendocrine bases of the hyperalgesic nocebo effect", *The Journal of Neuroscience* 26(46): 12014–12022.

Benedetti, F., Amanzio, M. and Thoen, W. (2011a) "Disruption of opioid-induced placebo responses by activation of cholecystokinin type-2 receptors", *Psychopharmacology* 213(4): 791–797.

Benedetti, F., Amanzio, M., Rosato, R. and Blanchard, C. (2011b) "Non-opioid placebo analgesia is mediated by CB1 cannabinoid receptors", *Nature Medicine* 17(10): 1228–1230.

Benedetti, F., Carlino, E. and Pollo, A. (2011c) "How placebos change the patient's brain", *Neuropsychopharmacology* 36(1): 339–354.

Benedetti, F., Durando, J. and Vighetti, S. (2014) "Nocebo and placebo modulation of hypobaric hypoxia headache involves the cyclooxygenase-prostaglandins pathway", *Pain* 155(5): 921–928.

Boileau, I., Assaad, J. M., Pihl, R. O., Benkelfat, C., Leyton, M., Diksic, M., Tremblay, R. E. and Dagher, A. (2003) "Alcohol promotes dopamine release in the human nucleus accumbens", *Synapse* 49(4): 226–231.

Bossong, M. G., van Berckel, B. N., Boellaard, R., Zuurman, L., Schuit, R. C., Windhorst, A. D., van Gerven, J. M., Ramsey, N. F., Lammertsma, A. A. and Kahn, R. S. (2009) "Delta 9-tetrahydrocannabinol induces dopamine release in the human striatum", *Neuropsychopharmacology* 34(3): 759–766.

Brody, A. L., Mandekern, M. A., Olmstead, R. E., Allen-Martinez, Z., Scheibal, D., Abrams, A. L., Costello, M. R., Farahi, J., Saxena, S., Monterosso J. and London, E. D. (2009) "Ventral striatal dopamine release in response to smoking a regular vs a denicotinized cigarette", *Neuropsychopharmacology* 34(2): 282–289.

Colloca, L. and Benedetti, F. (2006) "How prior experience shapes placebo analgesia", *Pain* 124(1–2): 126–133.

Colloca, L. and Miller, F. G. (2011) "How placebo responses are formed: a learning perspective", *Philosophical Transactions of The Royal Society B: Biological Sciences* 366(1572): 1859–1869.

Dar, R., Stronguin, F. and Etter, J. F. (2005) "Assigned versus perceived placebo effects in nicotine replacement therapy for smoking reduction in Swiss smokers", *Journal of Consulting and Clinical Psychology* 73(2): 350–353.

De la Fuente-Fernández, R., Ruth, T. J., Sossi, V., Schulzer, M., Calne, D. B. and Stoessl, A. J. (2001) "Expectation and dopamine release: mechanism of the placebo effect in Parkinson's disease", *Science* 293(5532): 1164–1166.

De la Fuente-Fernández, R., Phillips, A. G., Zamburlini, M., Sossi, V., Calne, D. B., Ruth, T. J. and Stoessl, A. J. (2002) "Dopamine release in human ventral striatum and expectation of reward", *Behavioural Brain Research* 136(2): 359–363.

Di Chiara, G. and Imperato, A. (1988) "Drugs abused by humans preferentially increase synaptic dopamine concentrations in the mesolimbic system of freely moving rats", *Proceedings of the National Academy of Sciences of the United States of America* 85(14): 5274–5278.

Drevets, W. C., Gautier, C., Price, J. C., Kupfer, D. J., Kinahan, P. E., Grace, A. A., Price, J. L. and Mathis, C. A. (2001) "Amphetamine-induced dopamine release in human ventral striatum correlates with euphoria", *Biological Psychiatry* 49(2): 81–96.

Duvauchelle, C. L., Ikegami, A., Asami, S., Robens, J., Kressin, K. and Castaneda, E. (2000) "Effects of cocaine context on NAcc dopamine and behavioral activity after repeated intravenous cocaine administration", *Brain Research* 862(1–2): 49–58.

Eippert, F., Bingel, U., Schoell, E. D., Yacubian, J., Klinger, R., Lorenz, J. and Büchel, C. (2009) "Activation of the opioidergic descending pain control system underlies placebo analgesia", *Neuron*, 63(4): 533–543.

Fillmore, M. T. and Vogel-Sprott, M. (1995) "Expectancies about alcohol-induced motor impairment predict individual differences in responses to alcohol and placebo", *Journal of Studies on Alcohol* 56(1): 90–98.

George, W. H. and Stoner, S. A. (2000) "Understanding acute alcohol effects on sexual behavior", *Annual Review of Sexual Research* 11(1): 92–124.

Graham, J. and Porrino, L. J. (1995) "Neuroanatomical substrates of cocaine self-administration", in R. Hammer (ed.), *Neurobiology of Cocaine*, Boca Raton, FL: CRC Press, pp. 3–14.

Juliano, L. M. and Brandon, T. H. (2002) "Effects of nicotine dose, instructional set, and outcome expectancies on the subjective effects of smoking in the presence of a stressor", *Journal of Abnormal Psychology* 111(1): 88–97.

Kirk, J. M., Doty, P. and De Wit, H. (1998) "Effects of expectancies on subjective responses to oral delta9-tetrahydrocannabinol", *Pharmacology Biochemistry and Behaviour* 59(2): 287–293.

Kirsch, I., Lynn, S. J., Vigorito, M. and Miller, R. R. (2004) "The role of cognition in classical and operant conditioning", *Journal of Clinical Psychology* 60(4): 369–392.

Lidstone, S. C., Schulzer, M., Dinelle, K., Mak, E., Sossi, V., Ruth, T. J., de la Fuente-Fernández, R., Phillips, A. G. and Stoessl, A. J. (2010) "Effects of expectation on placebo-induced dopamine release in Parkinson disease", *Archives of General Psychiatry* 67(8): 857–865.

Looby, A. and Earleywine, M. (2011) "Expectation to receive methylphenidate enhances subjective arousal but not cognitive performance", *Experimental and Clinical Psychopharmacology* 19(6): 433–444.

Maisto, S. A., Carey, M. P., Carey, K. B. and Gordon, C. M. (2002) "The effects of alcohol and expectancies on risk perception and behavior skills relevant to safer sex among heterosexual young adult women", *Journal of Studies on Alcohol* 63(4): 476–485.

Martin, C. S. and Sayette, M. A. (1993) "Experimental design in alcohol administration research: limitations and alternatives in the manipulation of dosage-set", *Journal of Studies on Alcohol* 54(6): 750–761.

Mayberg, H. S., Silva, J. A., Brannan, S. K., Tekell, J. L., Mahurin, R. K., McGinnis, S. and Jerabek, P. A. (2002) "The functional neuroanatomy of the placebo effect", *American Journal of Psychiatry* 159(5): 728–737.

Perkins, K. A., Sayette, M., Conklin, C. and Caggiula, A. (2003) "Placebo effects of tobacco smoking and other nicotine intake", *Nicotine & Tobacco Research* 5(5): 695–709.

Perkins, K. A., Jacobs, L., Ciccocioppo, M., Conklin, C., Sayette, M. and Caggiula, A. (2004) "The influence of instructions and nicotine dose on the subjective and reinforcing effects of smoking", *Experimental and Clinical Psychopharmacology* 12(2): 91–101.

Petrovic, P., Kalso, E., Petersson, K. M. and Ingvar, M. (2002) "Placebo and opioid analgesia – imaging a shared neuronal network", *Science* 295(5560): 1737–1740.

Rose, J. E., Tashkin, D. P., Ertle, A., Zinser, M. C. and Lafer, R. (1985) "Sensory blockade of smoking satisfaction", *Pharmacology Biochemistry and Behavior* 23(2): 289–293.

Ross, D. F. and Pihl, R. O. (1989) "Modification of the balanced-placebo design for use at high blood alcohol levels", *Addictive Behaviors* 14(1): 91–97.

Scott, D. J., Stohler, C. S., Egnatuk, C. M., Wang, H., Koeppe, R. A. and Zubieta, J. K. (2007) "Individual differences in reward responding explain placebo-induced expectations and effects", *Neuron* 55(2): 325–336.

Scott, D. J., Stohler, C. S., Egnatuk, C. M., Wang, H., Koeppe, R. A. and Zubieta, J. K. (2008) "Placebo and nocebo effects are defined by opposite opioid and dopaminergic responses", *Archives of General Psychiatry* 65(2): 220–231.

Sesack, S. R. and Grace, A. A. (2010) "Cortico-basal ganglia reward network: microcircuitry", *Neuropsychopharmacology* 35(1): 27–47.

Volkow, N. D., Wang, G. J., Fowler, J. S., Gatley, S. J., Ding, Y. S., Logan, J., Dewey, S. L. and Liberman, R. (1996) "Relationship between psychostimulant-induced 'high' and dopamine transporter occupancy", *Proceedings of the National Academy of Sciences of the United States of America* 93(19): 10388–10392.

Volkow, N. D., Fowler, J. S. and Wang, G. J. (2003a) "The addicted human brain: insights from imaging studies", *The Journal of Clinical Investigation* 111(10): 1444–1451.

Volkow, N. D., Wang, G. J., Ma, Y., Fowler, J. S., Zhu, W., Maynard, L., Telang, F., Vaska, P., Ding, Y. S., Wong, C. and Swanson, J. M. (2003b) "Expectation enhances the regional brain metabolic and the reinforcing effects of stimulants in cocaine abusers", *Journal of Neuroscience* 23(36): 11461–11468.

Volkow, N. D., Wang, G. J., Ma, Y., Fowler, J. S., Wong, C., Jayne, M., Telang, F. and Swanson, J. M. (2006) "Effects of expectation on the brain metabolic responses to methylphenidate and to its placebo in non-drug abusing subjects", *NeuroImage* 32(4): 1782–1792.

Wager, T. D., Rilling, J. K., Smith, E. E., Sokolik, A., Casey, K. L., Davidson, R. J., Kosslyn, S. M., Rose, R. M. and Cohen, J. D. (2004) "Placebo-induced changes in FMRI in the anticipation and experience of pain", *Science* 303(5661): 1162–1167.

Weiss, R. D., O'Malley, S. S., Hosking, J. D., Locastro, J. S., Swift, R. and COMBINE Study Research Group (2008) "Do patients with alcohol dependence respond to placebo? Results from the COMBINE Study", *Journal of Studies on Alcohol and Drugs* 69(6): 878–884.

Wise, R. A. (1984) "Neural mechanisms of the reinforcing action of cocaine", in J. Grabowski (ed.), *Cocaine: Pharmacology, Effects, and Treatment of Abuse*, Rockville, MD: NIDA Research Monography, pp. 15–28.

Zubieta, J. K., Bueller, J. A., Jackson, L. R., Scott, D. J., Xu, Y., Koeppe, R. A., Nichols, T. E. and Stohler, C. S. (2005) "Placebo effects mediated by endogenous opioid activity on mu-opioid receptors", *The Journal of Neuroscience: The Official Journal of the Society for Neuroscience*, 25(34): 7754–7762.

32

BRAIN MECHANISMS AND THE DISEASE MODEL OF ADDICTION

Is it the whole story of the addicted self? A philosophical-skeptical perspective

Şerife Tekin

Introduction

Contemporary scientific and philosophical debates on addiction center on two models. In the brain disease model, drug addiction is a "chronic and relapsing brain disease that results from the prolonged effects of drugs on the brain" (Leshner 1997: 45). In contrast, the self or person model individuates addiction in folk psychological terms, as a kind of behavior marked by the repeated use of a drug of choice (DoC) and the difficulty involved in quitting (Satel and Lilienfeld 2013; Flanagan 2013a; Tekin et al. 2017). On the one hand, the brain disease model promotes cellular-level understanding, painting the addict as a biological organism. It takes seriously recent advances in brain imaging methods and the increased sophistication in the neuroscientific modeling of the brain's reward system—the mesolimbic dopamine system—and promotes pharmaceuticals to target and rebalance the chemical anomalies in it. On the other hand, the self model sees the addict as a person or intentional system (Dennett 1971) who acts according to beliefs and desires and responds to complex historical, environmental, and interpersonal aspects of life. It points to the personal-socio-cultural markers of addiction and promotes social, psychological, and behavioral intervention strategies, e.g., encouraging the addict to change the interpersonal environment that triggers and perpetuates addictive behavior, to receive intensive psychotherapy, to cultivate meditation, and so on.[1]

Although each model accepts the validity of the other, to a certain point, there is a *prima facie* tension between them. The brain disease model acknowledges the environmental, social, and cultural factors in the development of addiction, but its proponents often favor neuroscientific work in *lieu* of other research paradigms, e.g., social sciences and cognitive sciences. The self model acknowledges the importance of the cellular-level understanding of addiction, but its adherents warn of the dangers of neuro-enthusiasm (Satel and Lilienfeld 2013), i.e., the overpromotion of scientific and clinical approaches to target the person. In this chapter, I propose using Ken Schaffner's framework for thinking about reductionism in science (Schaffner 2013) to resolve the issue. In this framework, the brain disease model moves in the direction of what

Schaffner calls "sweeping reductionism," with the self model moving towards "creeping reductionism." Because of the virtues of creeping reductionism, I argue the self model has more promise for addiction research.

I start the chapter with a brief overview of Schaffner's framework for reduction in science, describing his notions of "sweeping" and "creeping" reductionism. I apply this framework to the brain disease and the self models, noting the epistemic advantages of committing to the latter. In my analysis, I draw on scientific research and clinical work on addiction; to this I add first-person accounts of those influenced by addiction, a seldom used but rich resource (Tekin 2016).

Reductionism in science and psychiatry

A common debate among scientists and philosophers is whether human sciences, such as psychology and psychiatry, involve phenomena distinct from those targeted in the physical sciences. According to reductionism, target phenomena in human sciences are only *prima facie* distinct from those in the physical sciences, lending themselves to explanation or even replacement by phenomena in the physical and chemical sciences. In this respect, mental disorders can be reduced to phenomena in physics and chemistry.

For Schaffner, reductionism exists on a spectrum. On the one extreme, human phenomena "are nothing but aggregates of physicochemical entities," a view he labels "sweeping reductionism" (Schaffner 2013: 1003). For "sweeping" reductionists, "there is a theory of everything" and "there is nothing but those basic elements—for example, a very powerful biological theory that explains all of psychology and psychiatry" (Schaffner 2013: 1003). At the other extreme is "creeping reductionism" (Schaffner 2013). A "creeping" reductionist will argue that the different models of the target phenomena—in this case addiction—provide fragmentary explanations that must be combined for a fuller understanding. In Schaffner's view, a full commitment to sweeping reductionism is impossible, except perhaps as a metaphysical claim. Furthermore, sweeping reductionism has little value in the biological and psychological sciences.

The brain disease model leans towards a sweeping reductionist commitment, if not fully embodying it, and this is cause for concern. Creeping reductionists, for Schaffner, do not typically commit to a nothing-but approach in scientific explanations. Rather, they embrace a pragmatic and pluralistic parallelism, working at several levels of aggregation and discourse at once. Insofar as the self model I discuss here proposes to work on addiction at different levels (ranging from cellular and molecular, to individual and social) simultaneously, it embodies creeping reductionism and is, thus, a scientifically and clinically resourceful approach, enabling cross-fertilization across different levels.

The brain disease model of addiction

In the brain disease model, all drugs of abuse have common direct or indirect effects on a single pathway in the brain—the brain's reward system, i.e., the mesolimbic dopamine system. (Leshner 1997).[2] The mesolimbic dopamine pathway is individuated at the molecular, cellular, structural, and functional levels; it extends from the ventral tegmentum to the nucleus accumbens, with connections to such areas in the brain as the limbic system and the orbitofrontal cortex (Hyman 1996; Ortiz *et al.* 1995). It is associated with motivation, decision-making, and inhibitory control and is thought to be a key detector of rewarding stimuli, as it controls responses to natural rewards such as food, sex, and social interactions. According to the disease model, drugs interact with the mesolimbic pathway to produce addictive effects and "reset" the brain's reward system by causing sharp increases in the release of dopamine (Volkow *et al.* 2016; Di Chiara 2002). Such increases elicit a reward signal that triggers associative learning, leading individuals to associate certain stimuli with the drug use, e.g., the environment of drug taking, persons with whom it has been taken, etc. These environmental cues may all trigger craving for and use of the drug. Such conditioned responses become deeply ingrained, often lasting long after use has stopped.

Though there is no scientific consensus on the matter, for some scientists the activation of the brain's reward system is significantly different from that associated with natural rewards, such as food, sex, and social relationships (Wise 2002).[3] Dopamine cells stop firing after repeated consumption of a "natural reward," satiating the drive to continue, but addictive drugs circumvent the natural satiation and continue to directly increase dopamine levels (Wise 2002). Accordingly, compulsive behaviors are more likely to emerge when people use drugs than when they pursue a natural reward (Volkow *et al.* 2016).

A well-known metaphor for the disease model is the broken switch (Leshner 1997). Drug use begins as a voluntary behavior, "but when that switch is thrown, the individual moves into the state of addiction, characterized by compulsive drug seeking and use" (Leshner 1997: 46). Put otherwise, addiction is a consequence of fundamental changes in brain function, and the goal of treatment must be either to reverse or to compensate for those brain changes, through pharmaceuticals or behavioral treatments (Leshner 1997). Elucidation of the biology underlying the metaphorical switch is seen as the key to the development of more effective treatments, particularly anti-addiction medications.

It would be unfair to characterize the brain disease model as fully embodying a sweeping reductionist framework according to which addiction is "nothing but" anomalies in the brain's dopamine system, as its proponents acknowledge the importance of social and environmental factors in its development. That said, however, the brain disease model leans towards a "sweeping reductionist" framework insofar as it promotes research into the neurobiology of addiction and does not offer an explicit framework for investigating the social and environmental factors contributing to addiction. A case in point is the National Institute of Mental Health (NIMH)'s Research and Domain Criteria (RDoC) initiative, i.e., its development of an alternative to the DSM-5 in guiding research for mental disorders. It does not fund research projects unless they explicitly analyze the neurobiology of mental disorder (Vaidyanathan 2016, personal communication). It encourages research programs that advance initiatives to find pharmaceuticals to target and rebalance the neurobiological breakdown in the brain. An exclusive focus on the neurobiology of addiction discourages alternative initiatives, including, for instance, research into the relationship between homelessness and addiction or intervention strategies involving housing programs for the homeless.

Critics have voiced a number of concerns with the brain disease model. One is that brain chemistry is plastic and responds to more than simply repeated drug use; a wide range of factors come into play, including changes in the physical and social environment. Arguably, therefore, anomalies in the brain's reward system that contribute to the development of addictive behavior can be corrected not only by anti-addiction medications but also by psychological and behavioral interventions (Satel and Lilienfeld 2013). Another, perhaps more important, limitation is the model's apparent lack of focus on the complexity and multi-aspectuality of the lives of individuals with addiction, despite the need to develop clinically effective strategies for recovery (Tekin *et al.* in press). This implies that the brain is the most important and useful level of analysis for understanding and treating addiction, even though drug use and abuse do not emerge in a vacuum independent of an individual's history, interpersonal relationships, socio-economic status, etc., and its successful treatment requires targeting and analyzing these dimensions. For example, addictive behavior only affects people who have sufficient access to drugs to abuse, and this might be relevant to understand the demographics of the population with a drug abuse problem. Furthermore, the individual reasons people start drug use, e.g., to remedy shyness in social contexts, to cope with a dysfunctional relationship, to adopt the social expectations of a particular group, etc., are as important in understanding addiction as are the neurochemical underpinnings.

In short, the personal, social, and cultural context of addiction is as equally important as the neurobiology of addiction, and the brain disease model does not easily lend itself to the study of the former. Despite the general recognition of the importance of these factors, the conceptual and empirical frameworks for studying addiction continue to target the brain mechanisms, at the expense of a conceptual and empirical framework that benefits from person-level understanding, using resources, say, from cognitive psychology, sociology, anthropology, etc. A plethora of treatment strategies that go beyond drug-related interventions thus remain under-explored. Scientific psychiatry may be missing valuable opportunities to study the complex factors that contribute to the development of addiction and other mental disorders.

The self model of addiction

As noted above, the self model considers addiction a person-level phenomenon; it represents addiction in reference to an individual's personal history, socio-economic status, race, gender, and other identity-constituting factors, her interpersonal relationships, and her conception of herself, in addition to her biological make-up. It differs from the brain disease model in a fundamental way: the target of inquiry is the self or the person or the intentional system (Dennett 1971), not just the brain. In this respect, the self model is "holistic" and aims to explain addiction in reference to the full complexity of an individual. In contrast, the brain disease model embodies "smallism" (Wilson 1999), over-emphasizing the building blocks of human bodies, such as neurons, and underestimating more integrated and complex phenomena, such as addiction.

The self model takes the self, or the person, to be a dynamic, complex, relational configuration with a more or less integrated cluster of person-level properties (Tekin *et al.* in press; Neisser 1988; Jopling 2000; Thagard 2014; Bechtel 2008; Tekin 2014, 2015). Under what I previously called the multitudinous self view (Tekin 2014), the self is individuated and empirically tractable—using the resources provided by cognitive, developmental, social, and biological psychology—through five different but complementary dimensions: (1) the ecological aspect of the self, or the embodied self in the physical world, which perceives, acts, and interacts with the physical environment; (2) the interpersonal aspect of the self, or the self embedded in the social world, which constitutes and is constituted by inter-subjective relationships; (3) the temporally extended aspect of self, or the self in time, grounded in memories of the past and anticipation of the future; (4) the private aspect of the self, which is exposed to experiences available only to the first person and not to others; (5) the conceptual aspect of the self, which represents the self to that individual by drawing on her properties or characteristics and the social and cultural context to which she belongs.[4] The different aspects of the self connect the individual to herself and to the physical, social, and cultural environment in which she is situated. A condition such as addiction affects each dimension of the self, and the more-or-less integrated unity of all dimensions.

Because each aspect is experienced from the first-person point of view and can also be tracked from third-person points of view, the model can represent addiction phenomenologically, empirically, and conceptually. To clarify, let me start with the ecological dimension of the multitudinous self. It represents the individual's embodiment in the physical world, including her biological features, genetic make-up, and the constraints of her body. In this sense, the ecological dimension is specified by the body, the physical conditions of a particular environment, and the active perceptual exploration of and response to these conditions. This particular aspect of the multitudinous self tracks addictive behavior in a number of ways. First, there is something going on in the body—in the central nervous system, brain cells, brain's reward system, hormones, genes, etc.—of an individual when she becomes addicted to a certain drug. As discussed in the previous section, recent work on the brain's mesolimbic dopamine system, on which the brain disease model is grounded, indicates correlations between addictive behavior and dopamine activity (Nestler 2013). We cannot dismiss such findings; nor should we wish to, and future research should shed more light on the connections. Second, the symptoms of withdrawal can be traced through the ecological dimension. The addict's hands shake. She gets anxious, restless, and irritable. She wakes up in the middle of the night with nausea or wanting the DoC. In a worst-case scenario, she has delirium and hallucinations. This continues until she takes the DoC.

The manifestation of addiction in the ecological dimension of the self is not only accessible through a first-person perspective (the addict herself) but also through the third-person perspective (e.g., scientists who study her behavior, brain, hormones, genes, interpersonal relationship, childhood, socio-economic class, etc., and clinical practitioners, including doctors, nurses, social workers, etc.). The addict experiences the craving; her partner observes her restlessness, anger and frustration in the absence of the DoC; her psychotherapist notices she is more tense and quieter than usual; scientists gather evidence about the level of the DoC in her blood; social workers try to help her get connected to communities where she can find fulfillment and flourish. Knowledge of the ecological aspect of the self can and does facilitate a number of effective (albeit limited) interventions. For example, through cognitive behavior therapy, an addict may learn self-monitoring to recognize cravings early, identify situations putting her at risk, develop strategies for coping with cravings—sometimes with the support of medications—and avoid high-risk situations (Carroll and Onken 2005).

Now consider the intersubjective aspect of the self, the part handling the "species-specific signals of emotional rapport and communication between the self and other people" (Neisser 1988: 387; Bechtel 2008). Through intersubjectivity, an individual begins the interpersonal relationships of care and concern through which her identity is formed, then enriched or impoverished, depending on the level and the kind of care she receives. This aspect of the self tracks addictive behavior patterns in multiple ways. First, forms of addictive behavior in the consumption of the DoC progress in a particular kind of social environment. For example, in a lifestyle Flanagan calls "the male life of public and gregarious heavy drinking," "social drinking" is widely encouraged and becomes the context through which individuals socialize in their professional lives to talk "business" (Flanagan 2013b: 870). Over time, drinking becomes the sole reason to take part in these events, not business. Second, the kind and the quality of an individual's interpersonal relationships are major factors in the development of addictive behavior. Individuals with addiction often have a complex history of family relationships. Being subjected to physical or sexual abuse as a child is strongly linked to addiction (Marcenko *et al.* 2000; Langeland *et al.* 2002). Sometimes people are in relationships they do not want to be in, and intoxication becomes an easy escape. It ultimately becomes a problem on its own, however, harming not only the self, but others (Graham 2013).

Tracking the intersubjective dimension of the self may provide explanations of why a person becomes and stays addicted. At the same time, it may facilitate the development of effective interventions. If a particular lifestyle is enabling addiction, as in the "male life of public and gregarious heavy drinking," interventions could include helping the individual change his social environment. Other forms of interventions include helping him get rid of a relationship causing distress or develop more effective coping strategies.

The temporally extended aspect of the self also tracks addiction. This aspect individuates the person in time; she is shaped by her experiences and memories of the past and her anticipation of the future. The development of addiction over time may take many different trajectories, but there are some commonalities. A drug is sampled first, often with no intention of making it a regular activity. There is no fear of the harm that may ensue. It is found to be enjoyable, invites further consumption, and becomes increasingly frequent. At some point, the individual may notice the behavior is harmful and contradicts her vision of her future self. She may recognize the necessity of not using the DoC to realize her goals, and this may help her stop the addictive behavior.

Understanding the temporal dimension of the self may facilitate an understanding of how and why people get addicted, e.g., what kind of developmental environment contributes to the development of addictive behavior, and why they have difficulty keeping their promises to themselves and others, e.g., growing up in families where people have not kept promises. The study of this aspect of the self may facilitate the development of effective interventions. The ability to recognize temporal extendedness is already used as a resource in cognitive behavior therapy when patients are taught stimulus control strategies, for example. In this type of therapy, addicts learn to avoid situations associated with drug use and to spend more time in activities incompatible with drug use. They learn to practice "urge control" by recognizing and changing the thoughts, feelings, and plans that lead to drug use. Patients' past attempts and failures are used as a benchmark to customize therapy (Azrin et al. 1994).

The private aspect of the self traces the individual's conscious awareness of felt experiences—what William James takes to be uniquely ours (James 1890). The private aspect of the self is not phenomenologically available to anyone else (e.g., feelings of pain or disappointment) but, with the help of language, it can be communicated (Bechtel 2008: 260). Addiction is traceable in the private aspect of the self, including the felt experience of being addicted, cravings for the DoC, the distress of not using it when consumption is delayed, and regrets about consuming it despite resolves to the contrary. It is extremely difficult for others, e.g., caregivers, clinicians, etc., to appreciate the complexity of the various phenomena experienced by the addict. However, linguistic representations of these experiences, say, when the individual is describing what it is like verbally or in writing, provide substantive information, and these data can be used for explanations and interventions. Some say the memoirs of addiction are helpful, not only to the experts wanting to fathom addiction, but also to the addicts themselves (Flanagan 2013a, 2013b). They argue the private dimension of the self offers excellent resources to scientifically investigate addiction. As we better understand this aspect of the self, we will better understand what addiction is and what successful interventions look like.

Last but not least, we have the conceptual aspect of the self. Self-concepts selectively represent the self to the self. They are the products of the dynamic interaction of the other four aspects of the self with the external social and cultural environment. Self-concepts include ideas about our physical bodies (ecological aspect), interpersonal experiences (intersubjective aspect), the kinds of things we have done in the past and are likely to do in the future (temporally extended aspect), and the quality and meaning of our thoughts and feelings (private aspect) (Jopling 1997, 2000; Neisser 1988; Bechtel 2008). For instance, an individual's self-concept as a "responsible person" is the product of the intersubjective aspect of her selfhood and of the norms of responsibility in the culture of which she is a part. Self-regarding feelings and attitudes, such as self-confidence, security, self-esteem, self-respect, and social trust, emerge as she develops self-concepts and as the different dimensions of the self interact with her social and cultural world. Self-concepts are informed by the other four aspects of the multitudinous self and by the individual's unique embodied experiences in the world (Neisser 1988; Jopling 1997; Bechtel 2008; Tekin 2011). In turn, self-concepts inform and shape the other aspects of the self.

Self-concepts are also informed by pathologies to which the individual is subjected. For example, an addict may develop "self-regarding reactive attitudes of bewilderment, disappointment, and shame" about her addiction (Flanagan 2013a: 6). In what Flanagan calls the twin normative failure model, a fundamental aspect of addiction is a conflict in the addict's self-concepts (Tekin *et al.* in press). She suffers from twin normative failures, in that she recognizes (1) she can't successfully moderate or quit the DoC on her own because she fails to "execute normal powers of effective rational agency," and her excess consumption of the drug of (DoC) leads to (2) a failure to "live up to the hopes, expectations, standards, and ideals she has for a good life for herself" (Flanagan 2013a: 1). In the first kind of failure, she conceptualizes herself, correctly, as unable to control her drug abuse. She wants to quit, resolves to take steps that will help her quit, yet fails to do so. In the second kind of failure, she conceptualizes herself as a person with multiple roles: e.g., professor, friend, daughter, activist. She understands herself as a person through these multiple roles, and she has standards for how to play these roles. She has a vision, for instance, of herself as a good scholar, or a dedicated human rights activist, or a responsible colleague, etc. However, due to her addiction, she is unable to live up to her standards, ideals or aspirations, or to the norms of the multiple roles she occupies. The successful negotiation of these roles is part of what it will mean for her to be the person she aims to be. In her self-narrative, the addict realizes the meaning, worth, and success of her life depend on her not using, but she continues to use. Given the two failures, addiction undermines both the rationality of the addict's life and its goodness. The addict is not the person she could or should be; nor is she the person she wants to be.

Self-concepts are also action-guiding (Tekin 2014, 2015). They inform how individuals behave and can motivate them to change. In the context of addiction, the formation or alteration of self-concepts will influence future actions. Hopelessness in the face of repeated relapses and self-concepts such as being weak-willed may diminish an addict's ability to quit the addictive behavior. Alternatively, she may express conflict and heightened distress because of a strong resolution to quit drinking, especially if she is unable to do so. From this conflict, she may be able to redefine her self and enumerate new behavioral goals. She may take a step towards change by altering a self-concept and stabilizing a new behavior pattern. Or perceiving herself as someone who needs help, she may reach out to the communities of other individuals with addiction. The success of Alcoholics Anonymous programs partially owes to this.

Because of their plastic nature, self-concepts offer great opportunities for successful clinical interventions. Clinicians may work towards helping the addict develop more positive and resourceful self-concepts, strengthening her self-esteem, as well as her self-control capacities, and help her flourish by stopping her use of the DoC. By working with her self-concepts, clinicians can motivate her to think, act, and behave in certain ways, expanding her possibilities for action (Tekin 2010, 2011, 2014, 2015; Jopling 1997).

The model of the self, as described above, fits easily into a creeping reductionist framework, with myriad advantages for psychiatry. Such a model could enhance, for instance, etiological psychiatric research using an interventionist account of causality (Woodward 2003), as recently applied by Kenneth Kendler (2014). In the interventionist account, factors that are "difference makers" are causal in the development of a phenomenon, for example, addiction. As one goal of psychiatric research is to unpack the risk factors, i.e., the difference makers (Kendler 2014), researchers into mental disorders may be well advised to take a closer look at the different aspects of the self. With the interventionist model of causality, we can identify a wide range of etiological factors spawning the ecological, interpersonal, temporal, private and conceptual dimensions of a phenomenon. If, for instance, empirical data show variations in the socio-economic status of the community are directly associated with variations in the risk of addiction, we can develop interventions, e.g., increasing social capital in neighborhoods through community interventions. Another example may be studying addictive behavior through the ecological aspect of the self using neuroscientific methods (e.g., the anomalies in the mesolimbic reward system) to see if, say, pharmaceutical interventions targeting the dopamine system are the difference makers. As the above suggests, we can identify risk factors and, ultimately, reduce rates of addiction by studying the multiple dimensions of the self, each of which may specify multiple risk factors.

The self model's creeping reductionist framework also embraces pragmatic goals of psychiatry. Researchers and clinicians have diverse interests: sometimes the goal of research is to reveal the disease pathways at a molecular level; at other times, the goal is to motivate behavioral and habitual changes in a short period in a clinical setting. But different aspects of the multitudinous self model can be used for these various purposes by identifying the different research targets.

Another virtue of the self model is that it transcends the deep debate about the relative values of explanations in psychiatry grounded in the biological sciences versus those in the psychological and social sciences. The multitudinous self model, because it encourages research from cellular and molecular levels to cognitive, personal, and social levels, promises substantial scientific progress in clarifying the etiology of psychiatric illness by working at diverse levels and dimensions. It promotes multiple complementary research programs in psychiatry. Finally, with the help of the multitudinous self model, we can reach scientific insights and develop a better capacity to make specific interventions, without having to make larger assumptions about the relative superiority of biological, psychological, or epidemiological variables.

Conclusion

The self model offers a rich way to understand and address addiction, largely because of its creeping reductionist framework. Creeping reductionism is open to pluralistic explanations, including the effect on addiction of the various aspects of the self, and encourages the development of a diverse portfolio of intervention strategies. It tames unfettered (and unwise) optimism about the use of pharmaceuticals to treat addiction, embraces pluralistic approaches, and refuses to privilege one explanation over another. It does not deny the efficacy of pharmacological treatment methods, but it accepts that, while one individual may benefit from a strictly pharmacological treatment, another may flourish with person-centered psychotherapy. In brief, the notion of creeping reductionism is a conceptually productive way of making sense of addiction, with great practical import.

Notes

1 Note that there are different subspecies of the brain disease models and the self models, each offering a version of the models discussed in this chapter; for brevity, I discuss only these two.
2 Nora Volkow, current head of the National Institute of Drug Abuse (NIDA), and George Koob, head of the National Institute of Alcohol Abuse and Alcoholism (NIAAA), are more cautious than Alan Leshner, who was the head of NIDA in 1997. Their arguments are more nuanced; they now write about the "brain disease model of addiction." They refrain from stating directly that addiction is a brain disease and note the biological and social factors involved in addiction (Volkow et al. 2016).
3 See Foddy (2011) to see some objections to the argument that the activation of the brain's reward system is significantly different from that associated with natural rewards, such as food, sex, and social relationships.
4 For a detailed development of this model, see Tekin (2014); Tekin, Flanagan, Graham (in press); Tekin (forthcoming).

References

Azrin, N. H., McMahon, P. T., Donahue, B., Besalel, V., Lapinski, K. J., Kogan, E., Acierno, R. and Galloway, E. (1994) "Behavioral therapy for drug abuse: a controlled treatment outcome study", *Behavioral Research and Therapy* 32(8): 857–866.

Baumann, M. H., Milchanowski, A. B. and Rothman, R. B. (2004) "Evidence for alterations in α2-adrenergic receptor sensitivity in rats exposed to repeated cocaine administration", *Neuroscience* 125(3): 683–690.

Bechtel, W. (2008) *Mental Mechanisms: Philosophical Perspectives on Cognitive Neuroscience*, London, UK: Routledge.

Belluzzi, J. D., Wang, R. and Leslie, F. M. (2005) "Acetaldehyde enhances acquisition of nicotine self-administration in adolescent rats", *Neuropsychopharmacology* 30(4): 705–712.

Carroll, K. M. and Onken, L. S. (2005) "Behavioral therapies for drug abuse", *American Journal of Psychiatry* 162(8): 1452–1460.

Dennett, D. C. (1971) "Intentional systems", *Journal of Philosophy* 68(4): 87–106.

Di Chiara, G. (2002) "Nucleus accumbens shell and core dopamine: differential role in behavior and addiction", *Behavioural Brain Research* 137(1–2): 75–114.

Flanagan, O. (2013a) "The shame of addiction", *Frontiers in Psychiatry* 120(4): 1–11.

Flanagan, O. (2013b) "Identity and addiction: what alcoholic memoirs teach", in K. W. M. Fulford, M. Davies, R. Gipps, G. Graham, J. Sadler, G. Stanghellini and T. Thornton (eds), *The Oxford Handbook of Philosophy and Psychiatry*, Oxford, UK: Oxford University Press, pp. 865–888.

Foddy, B. (2011) "Addicted to food, hungry for drugs", *Neuroethics* 4(2): 79–89.

Graham, G. (2013) *The Disordered Mind: Philosophy of Mind and Mental Illness*, 2nd Edition, London, UK: Routledge.

Hyman, S. E. (1996) "Addiction to cocaine and amphetamine", *Neuron* 16(5): 901–904.

James, W. (1890) *The Principles of Psychology*, New York, NY: H. Holt.

Jopling, D. (1997) "A self of selves?" in U. Neisser and D. Jopling (eds), *The Conceptual Self in Context*, Cambridge, UK: Cambridge University Press, pp. 249–269.

Jopling, D. (2000) *Self-Knowledge and The Self*, New York, NY: Routledge.

Kendler, K. (2014) "The structure of psychiatric science", *American Journal of Psychiatry* 171: 931–938.

Koob, G. F. (1992) "Drugs of abuse: anatomy, pharmacology and function of reward pathways", *Trends in Pharmacological Sciences* 13(5): 177–184.

Koob, G. F. (1996) "Drug addiction: the yin and yang of hedonic homeostasis", *Neuron* 16(5): 893–896.

Langeland, W., Draijer, N. and van den Brink, W. (2002) "Trauma and dissociation in treatment-seeking alcoholics: towards a resolution of inconsistent findings", *Comprehensive Psychiatry* 43(3): 195–203.

Lee, B., Tiefenbacher, S., Platt, D. M. and Spealman, R. D. (2004) "Pharmacological blockade of α2-adrenoreceptors induces reinstatement of cocaine-seeking behavior in squirrel monkeys", *Neuropsychopharmacology* 29(4): 686–693.

Leshner, A. I. (1997) "Addiction is a brain disease, and it matters", *Science* 278(5335): 45–47.

Marcenko, M. O., Kemp, S. P. and Larson, N. C. (2000) "Childhood experiences of abuse, later substance use, and parenting outcomes among low-income mothers", *American Journal of Orthopsychiatry* 70(3): 316–326.

Neisser, U. (1988) "Five kinds of self-knowledge", *Philosophical Psychology* 1(1): 35–59.

Nestler, E. J. (2013) "Cellular basis of memory for addiction", *Dialogues in Clinical Neuroscience* 15(4): 431–443.

Ortiz, J., Fitzgerald, L. W., Charlton, M., Lane, S., Trevisan, L., Guitart, X., et al. (1995) "Biochemical actions of chronic ethanol exposure in the mosalimbic dopamine system", *Synapse* 21: 289–298.

Satel, S. and Lilienfeld, S. O. (2013) *Brainwashed*, New York, NY: Basic Books.

Schaffner, K. (2013) "Reduction and reductionism in psychiatry", in K. W. M. Fulford, M. Davies, R. Gipps, G. Graham, J. Sadler, G. Stanghellini and T. Thornton (eds), *The Oxford Handbook of Philosophy and Psychiatry*, Oxford, UK: Oxford University Press.

Szumlinski, K. K., et al. (2004) "Homer proteins regulate sensitivity to cocaine", *Neuron* 43(3): 401–413.

Tekin, Ş. (2010) "Mad narratives: self-constitutions through the diagnostic looking glass", PhD Dissertation, York University, Toronto, Canada.

Tekin, Ş. (2011) "Self-concept through a diagnostic looking glass: narrative and mental disorder", *Philosophical Psychology* 24(3): 357–380.

Tekin, Ş. (2014) "The missing self in Hacking's looping effects", in H. Kincaid and J. A. Sullivan (eds), *Mental Kinds and Natural Kinds*, Cambridge, MA: MIT Press, pp. 227–256.

Tekin, Ş. (2015) "Against hyponarrating grief: incompatible research and treatment interests in the DSM-5", in S. Singy and P. Demazeux (eds), *The DSM-5 in Perspective: History, Philosophy and Theory of the Life Sciences*, Dordrecht, Netherlands: Springer Press.

Tekin, Ş. (2016) "Are mental disorders natural kinds? A plea for a new approach to intervention in psychiatry", *Philosophy, Psychiatry, and Psychology* 23(2): 147–163.

Tekin, Ş. (forthcoming) "The missing self in scientific psychiatry", *Synthese*.

Tekin, Ş., Flanagan, O. J. and Graham, G. (2017) "Against the drug cure model: addiction, identity, pharmaceuticals", in D. Ho (ed.), *Philosophical Issues in Pharmaceutics: Development, Dispensing, and Use*, New York, NY: Springer Press.

Thagard, P. (2014) "The self as a system of multilevel interacting mechanisms", *Philosophical Psychology* 27(2): 145–163.

Vitaro, F., Wanner, B., Ladouceur, R., Brendgen, M. and Tremblay, R. E. (2004) "Trajectories of gambling during adolescence", *Journal of Gambling Studies* 20(1): 47–69.

Volkow, N. D., Wang, G. J., Fowler, J. S., *et al.* (1997) "Decreased striatal dopaminergic responsiveness in detoxified cocaine-dependent subjects", *Nature* 386(6627): 830–833.

Volkow, N. D., Koob, G. and McLellan, T. (2016) "Neurobiologic advances from the brain disease model of addiction", *New England Journal of Medicine* 374(4): 363–371.

West, R. and Brown, J. (2013) *Theory of Addiction*, 2nd Edition, Chichester, UK: John Wiley and Sons.

Wilson, R. (1999) "The individual in biology and psychology", in V. Hardcastle (ed.), *Biology meets Psychology: Philosophical Essays*, Cambridge, MA: MIT Press.

Wise R. A. (2002) "Brain reward circuitry: insights from unsensed incentives", *Neuron* 36(2): 229–240.

Woodward, J. (2003) *Making Things Happen*, New York, NY: Oxford University Press.

Zhang, Y., Schlussman, S. D., Rabkin, J., Butelman, E. R., Ho, A. and Kreek, M. J. (2013) "Chronic escalating cocaine exposure, abstinence/withdrawal, and chronic re-exposure: effects on striatal dopamine and opioid systems in C57BL/6J mice", *Neuropharmacology* 67: 259–266.

PART III

Consequences, responses, and the meaning of addiction

SECTION A

Listening and relating to addicts

33
THE OUTCASTS PROJECT

Humanizing heroin users through documentary photography and photo-elicitation

Aaron Goodman

For decades, many documentary photographers and photojournalists have consistently produced stigmatizing images of injection drug users. These images have helped shape the perceptions of decision makers and the public about drug users, harm reduction and healthcare for some of the most vulnerable people in our society. This chapter focuses on a research study that I conducted from 2014 to 2015 involving documentary photography and photo-elicitation aimed at disrupting hegemonic representations of chronic heroin users in Vancouver's Downtown Eastside (DTES).

The impetus for the study was the growing heroin "epidemic" that is sweeping across North America (Seelye 2015). Since 2000, the number of deaths related to opioid pain relievers and heroin in the US has soared by 200% (www.cdc.gov). This is a shocking rate of increase, which is in fact only worsening, largely due to the introduction of fentanyl and carfentanil, both extremely powerful synthetic opioids, to the market.

In Canada, 60,000 to 90,000 people are affected by opioid addiction (University of British Columbia 2011). Between 2000 and 2010, approximately 200 people died every year from drug overdoses in British Columbia (BC) (Bailey 2017), the province that has seen the greatest number of opioid overdoses in Canada. In 2017, 1,422 people experienced fatal overdoses in BC. Across Canada, it is estimated that more than 4,000 people died from opioid-related overdoses in the same year (Lupick 2018).

Frontline workers, clinicians, first responders and governments are struggling to respond to the surging number of opioid overdoses. Photographers who are documenting the situation are helping to raise awareness of the crisis. However, there is a need for photographers to focus on people who use drugs in ways that do not further marginalize them. As a visual researcher and documentary photographer, I attempted to do this by collaborating with three long-term heroin users taking part in North America's first heroin-assisted treatment program in Vancouver's DTES. It was known as the Study to Assess Long-term Opioid Maintenance Effectiveness (SALOME) and was in effect from 2011 to 2015. The clinical trial was led by Providence Health Care and the University of British Columbia. It tested whether hydromorphone, a licensed orally administered pain medication, also known as Dilaudid, is as effective as injectable diacetylmorphine (the active ingredient of heroin) in treating vulnerable and long-term heroin users (Providence Health Care n.d.).

People who are addicted to heroin often receive oral methadone as an opiate replacement. However, many individuals with chronic heroin addiction do not sufficiently respond to it. Studies have shown that injectable diacetylmorphine has helped heroin-dependent people decrease their consumption of illicit drugs, improve their housing situations and reduce homelessness. Heroin-assisted treatment has also helped drug users lower their risk of HIV, Hepatitis B and Hepatitis C infection, and reduce their engagement in delinquent behavior, leading to reduced costs associated with judicial processes (Uchtenhagen 2002).

My goal in this study was to document the daily lives of three SALOME participants and help communicate their personal stories to the public and decision makers. I hoped to give viewers an opportunity to learn some of the reasons why the participants had started using drugs and remain drug dependent. I discovered that on their own, photos are unable to communicate the full story of drug users' lives and that addictions are often connected to hurtful experiences and traumas. In order to help amplify the participants' voices and experiences, I conducted photo-elicitation interviews and asked them to express their thoughts about the images I had taken. I then paired photos of the participants with their own words in a web-based interactive project. I invite you to view The Outcasts Project to see photos of Cheryl, Johnny, and Marie and to hear their own voices: www.outcastsproject.com.

I was also motivated to conduct this study because I was concerned when, in October 2013, Canada's former Health Minister Rona Ambrose ruled that diacetylmorphine was a restricted substance through the Food and Drug Act. The decision ended her department's authorization of doctors to prescribe diacetylmorphine, which meant SALOME would have to shut down. In response, five former SALOME patients launched a constitutional challenge in BC's Supreme Court. Their goal was to overturn the federal government's regulation of diacetylmorphine. In May 2014, the court sided with the plaintiffs. The decision allowed SALOME study participants to access ongoing diacetylmorphine treatment after they exited the study. The only requirement was that they had to have not sufficiently responded to other treatments and were being treated by a physician who had or could make an application to Health Canada to continue receiving the drug (Woo 2014).

I was also concerned about the Conservative Party's rhetoric during the legislative and judicial processes outlined above. The government was encouraging its supporters to back its efforts to prevent physicians from "giving heroin to addicts." It encouraged the public to sign an online petition against programs such as SALOME. "If the NDP or Liberals are elected in 2015, you can bet they would make this heroin-for-addicts program permanent" (Werry 2015), the Conservative government warned.

I was also troubled by many examples of inflammatory and derogatory language that people were using to describe heroin users in online comment sections connected to news stories about SALOME. Some people were supportive of SALOME. Amidst the heated debate taking place online, my impression was that the voices of SALOME participants were not sufficiently being heard. I approached Providence Health Care with the idea of producing a photo-based project about SALOME participants. The organization was supportive of the concept and granted me access to the Crosstown Clinic in Vancouver's DTES where SALOME was located. Over the following days I had informal discussions with several SALOME participants at the clinic and I listened to them closely. Many told me that taking part in SALOME had allowed them to rebuild parts of their lives after years and often decades of struggle as chronic drug users. Some said becoming SALOME participants had allowed them to gain a level of health and stability that had permitted them to reconnect with family members whom they had not seen in many years. Others said SALOME had enabled them to find jobs and acquire secure housing, often after years of being homeless. Many told me SALOME had effectively saved their lives.

In my conversations with SALOME participants, I explained my intention of producing a photo-based project about people taking part in the clinical trial over the course of a year. I indicated that my goal was to challenge and shape dominant representations of chronic heroin users. Approximately 20 SALOME participants expressed interest in taking part in the project. I ultimately invited three participants—Cheryl, Marie and Johnny—to participate in the study and they enthusiastically agreed. They told me they have all used heroin for decades and have been unable to stop, in spite of repeatedly trying methadone and other treatments. Some SALOME participants whom I met with told me that, aside from receiving their treatment at the Crosstown Clinic three times a day, as was prescribed, they did not do many other things. Some told me they remained focused on acquiring and using illicit drugs. Meanwhile, Cheryl, Marie and Johnny said they were engaged in a relatively wide range of activities such as becoming reacquainted with family members, doing part-time jobs, spending time with friends and more. This was important information for me and contributed to my decision to invite them to take part in the study. I knew if I were to succeed in creating humanizing images of long-term heroin users, I needed opportunities to photograph the study's participants engaging in activities other than exclusively using drugs.

Over the course of the study, Cheryl, Marie and Johnny shared intimate details of their lives with me. Marie, who is in her mid-thirties, told me that when she was a girl she dreamed of being a professional dancer and auditioned for the National Ballet School of Canada. However, her early life was very challenging because her mother was an alcoholic and had an abusive partner. As a teenager, Marie said she started using drugs and was in an extremely abusive relationship in which she was practically held hostage for nearly a decade. Cheryl, also in her mid-thirties, is a sex worker and told me that since she became a SALOME participant she had significantly reduced her illicit drug use. Cheryl told me about physical and sexual abuse she had experienced and that her father, whom she loved, had recently died. Johnny, who is in his forties, said that SALOME had enabled him to stop using illicit drugs and stealing from local grocery stores and tourist shops in order to support his heroin addiction. He told me that SALOME had enabled him to find stable housing, start eating healthy food, establish close friendships, and reconnect with his mother who lived in another city. At the same time, he was experiencing serious physical challenges, losing weight, and would later learn he needed to have hip replacement surgery.

Limits and risks of documentary photography

Scholars from a range of disciplines are increasingly using visual methods to produce knowledge about communities across the globe (Prosser and Loxley 2008: 4). Visual methods, including documentary photography, can allow researchers to access and produce information that cannot otherwise be communicated (Pink 2004: 361).

This study was designed to determine if I could create humanizing images of long-term heroin users. I spent a year attempting to do so. Throughout the study, as I developed a professional relationship with the participants, they told me more details about their lives and personal histories. I learned about challenging experiences and traumas they had endured, which, in their views, have contributed to their drug use. The difficulty for me, I discovered, was finding ways of communicating these details to the viewer through the medium of photography. I recognized that, on their own, photographs could not perform this crucial task.

Visual researchers and photographers cannot represent subjects 100% accurately either. Neil Postman (1985: 72) writes: "By itself, a photograph cannot deal with the unseen, the remote, the internal, the abstract." Photography is incapable of accessing or representing a subject's internal processes (i.e., their thoughts, feelings and physical sensations).

In order to provide an extra layer of narration that could allow the photographs to communicate more fully (Sontag 1977: 23), I conducted photo-elicitation interviews with the participants. In photo-elicitation interviews, photos are used in order to invite participants to speak about the photos and engage in dialogue about their personal responses to them (Prosser and Loxley 2008: 19). These responses can allow researchers to learn nuanced information about participants' thinking and experiences (ibid). Photo-elicitation is also used to foster greater connections between researchers and participants who often inhabit very different worlds (Harper 2012: 157). It allows participants to communicate what Michel Foucault (1980: 82) calls "subjugated knowledges," which are frequently dismissed by researchers and officials. I then paired photos of the participants with excerpts from their audio-recorded photo-elicitation interviews and published them online in an interactive format.

In many photo-elicitation projects, participants respond to images they have taken themselves. This can allow participants to develop a sense of agency that comes from playing active roles in the research or production process and having feelings of ownership of the photos. It can also help participants feel they are on equal footing with researchers (Prosser and Loxley 2008). For this study, I took the photos because I was interested in exploring how, as a documentary photographer, I could create counter-narratives. Sarah Pink (2013: 28) writes that it is important to recognize there are different types of photographers, such as domestic, amateur, professional and ethnographic. She adds that each kind of photographer has her or his own intentions and produces unique types of information. I discovered that taking photos provided opportunities for me to closely observe the participants' lives and learn about them by spending time with them (Schwartz 1992).

Drug genre photography: "dark, seedy, secret worlds"

The first documentary photographers in North America began working in the early 20th century. Two of the most influential and early photographers of this period, Jacob Riis and Lewis Hine, photographed disadvantaged people. By creating awareness about disenfranchised communities, they hoped to encourage social reform. Riis focused on living conditions for slum dwellers in New York's Lower East Side and Hine documented exploitative child labour practices (Fitzgerald 2002: 373). The 1955 "Family of Man" exhibition at New York's Museum of Modern Art, which later toured the world, featured the work of 273 documentary photographers, including Diane and Allan Arbus, Robert Capa and Henri Cartier-Bresson. The exhibition's theme, the "paradoxes of the human experience," continues to inspire documentary photographers to this day.

Although documentary photographers have often been praised for their work, many have been accused of exploiting and sensationalizing their subjects, benefiting financially from others' suffering, and using controversial work to bolster their careers.

A number of influential photographers such as Larry Clark, Nan Goldin, Eugene Richards and Lincoln Clarkes, have taken images of heroin users for decades. Their work has been widely celebrated. While preparing to conduct this study and over the course of the year in which I produced it, I looked critically at many of these photographers' images and recognized they have represented drug users in problematic ways. John Fitzgerald writes that photographers often portray heroin users as residents of "dark, seedy, secret worlds." This can "Other the subject, or make them appear different through eroticizing or exoticising them (Fitzgerald 2002: 374). Stuart Hall writes that by representing subjects in stereotypical ways, photographers exert a form of "symbolic violence" and have the power to "mark, assign and classify" that can amount to *"ritualized expulsion"* (Hall 1997: 259).

Through their work, documentary photographers have collectively created a genre of photography in which heroin users are predominantly represented in stigmatizing ways. These images often mask subjects' individual personalities and life circumstances. People who use drugs are frequently represented solely as drug users. Although drug addiction can be all-consuming, people who use drugs, like all of us, have unique personal histories and personalities. By exploring the work of several photographers who have focused on drug users, I felt compelled to attempt to represent the participants of this study in more balanced and humanizing ways. Although photographers are inevitably influenced by recognizable images that others have taken, Pink (2013: 27) writes that photographers have the capacity to create counter-narratives that challenge dominant modes of construction.

In his 1971 book of realist photos titled *Tulsa*, Larry Clark focused his lens on young people experimenting with sex, drugs and guns. His images have played a significant role in shaping negative public perceptions about people who use drugs. Kerr Houston (2014) writes that one of Clark's photos from the *Tulsa* series, in which a shirtless young man appears to be preparing to self-inject drugs into his arm (Figure 33.1), reflects the photographer's overall style. Houston describes it as "raw or coarse . . . grainy . . . violent and brash" (ibid). In this image, it is impossible to see the subject's eyes and the details of his silhouetted face. The trope of the faceless drug user has often been replicated, each time depriving subjects of agency and failing to communicate their unique physical traits and identities. At times, concealing a drug user's facial features could be a sign of respect or indication of a photographer's ethical approach. Some photographers may not want to increase the risks of social isolation, prejudice or legal challenges for the people they photograph. When it comes to drug users who have died from overdoses, it would be irresponsible to reveal their faces in photographs. However, photographers' decisions to hide drug users' faces in their images often have little to do with protecting subjects' privacy and are commonly rooted in a desire to create sensational images.

The cover of Eugene Richard's 1994 book, *Cocaine True, Cocaine Blue*, features a close-up image of a woman with a syringe in her mouth. Charles Hagen (1994) writes that Richard's "sensationalistic" photos lead the viewer to become passive and not advocate for change. The photo of the woman is disturbing in the way that it exoticizes her and makes her look primitive. Photojournalists working for news organizations such as AP (AP Photo/Linsley 2009), Getty Images (Martínez Kempin 2012), the Denver Post (Amon 2012), and many others, have often taken photos of drug users with syringes in their mouths. These images have similar effects as Richard's cover photo. In most of these photos, the subjects' eyes are partially or completely out of frame, or people are photographed in soft focus and it is difficult to see their faces in detail.

In 2012, former fashion photographer Lincoln Clarkes released a series of over 300 portraits that he had taken of female heroin users in Vancouver's DTES between 1997 and 2001. The photos were published in a book titled *Heroines*, published by Anvil Press. The images helped create awareness about the need for harm reduction services and the risks the women faced. At least five of the women in Clarkes' photos went missing or were discovered on serial killer Robert Pickton's farm at Port Coquitlam, BC (Canadian Press 2014). In 2007, Pickton was convicted of six counts of second-degree murder and is believed to be responsible for the deaths of dozens of others (Canadian Press 2013). However, Paul Ugor (2007) writes that Clarkes' approach was exploitative, noting the photographer paid his subjects with cigarettes, biscuits or five dollars to pose. Ugor adds that the women who were photographed likely did not know how widely the images would be distributed and who would ultimately own the photos.

Many of Clarkes' photos are compelling and gritty. However, he replicates many of the tropes of earlier drug genre photography. In one of his images, a woman, whose face is hidden

Figure 33.1 Larry Clark, *Untitled*, 1971.

by her hat, pushes a needle into her arm inside a bus shelter in front of an enormous Calvin Klein poster featuring model Kate Moss. The viewer is unable to see the subject's face. As a result, she becomes anonymous and powerless in the photo. In another image from Clarkes' series, another woman injects drugs into her neck in an alley in Vancouver's DTES. Clarkes captures the intensity of the moment—the woman's pursed lips and intense gaze. However, he fails to provide any information about the woman other than the fact that she is a drug user. What do we gain by seeing photos of drug users injecting into their necks? Unless more information is presented about the subjects and the wider contexts in which they are situated in the form of captions or narration, images like Clarkes' risk stigmatizing subjects.

From the outset of this study, I was not interested in solely highlighting the participants' drug use. My intention was to take images that reflected different elements of their lives and who they are as individuals. When I look at Clarkes' images, I get the sense that he missed an opportunity to tell deeper and meaningful stories about the women he photographed. Drug use was his prominent interest, not the women's individual backstories and the social and economic conditions that affected them.

Photographic methodology

The central intention behind The Outcasts Project was to produce a body of practice-based visual work that could counter dominant modes in documentary photography of representing heroin users. I had no illusions that the study would lead to widespread changes in the way photographers approach their work. I hope, however, the project can contribute to discussions about how photographers and visual researchers could possibly engage with practices that allow them to humanize drug users.

I adopted what Pink (2013: 24) calls a "realist" approach to photography that is often practiced by ethnographers. By photographing the participants for over a year, I had many opportunities to watch what occurred in their lives, to engage in dialogue and listen to them. I was able to ask them many questions during the photographic process and photo-elicitation interviews that allowed me to understand them. This helped me reflect on their experiences in study through a "first-hand empirical investigation" (Hammersley and Atkinson 1995: 1). The photos are not an objective representation that communicates absolute truths about the participants' experiences. As Pink (2013: 18) writes, ethnographic knowledge is based on the researcher's experiences. By collaborating with participants, using photo-elicitation as a method, and engaging in self-reflexivity, my aim was to present a version of reality that communicates knowledge about the situations, environments and experiences of the participants as authentically as possible (ibid).

Martyn Hammersley and Paul Atkinson outline four theoretical social roles for photo-fieldwork. These include "complete participant, participant as observer, observer as participant and complete observer" (2007: 82). I adopted the role of participant as observer, which is characterized by "comparative involvement: subjectivity and sympathy" (ibid: 104). Irving Goffman and Lyn H. Lofland (1989: 130) suggest that scholars who conduct this type of research should spend no less than a year in the field in order to produce in-depth projects and become deeply familiar with subjects. It was only by spending a year engaging with the participants that I was able to understand them well enough to make informed choices about how to shape the project in ways that can contribute to new knowledge about them and photography (Becker 1981: 11).

It was important for me to maintain my role as researcher and documentary photographer throughout the study. However, over the course of the year, a degree of closeness developed between the participants and me. This helped me gain greater insights into their experiences and develop empathy for them. This also shaped my interactions and the photos I took of them (Fairclough 2013: 16). However, building a rapport with the participants was not straightforward or easy. For the first weeks of the study they were relatively guarded and only allowed me to photograph them self-injecting at the Crosstown Clinic. It was not until several weeks later, after they saw I was fully committed to the study, that they allowed me to take photos of them outside the clinic. I considered this initial period an important phase in which the participants developed a level of trust in me. Their initial reluctance to grant me access to document their lives was understandable. They were aware of the long history of stigmatizing journalism coverage of Vancouver's DTES and its residents (Goodman 2001) and the legacy of stigmatizing photography focused on heroin users in the community and beyond. In any life history research project, Adra Cole and Gary Knowles (2001: vii) write, it is natural for those who are "researched" to experience anxiety since their lives will be closely examined and, in this case, photographed. In the DTES, this sense of apprehension is often understandably heightened.

Collecting data: curating the photos

Over the course of a year I took dozens of photos of each of the three participants. The study could have go on indefinitely. However, I felt it had reached its natural conclusion when I found myself no longer being able to photograph the participants taking part in new or different activities. At that point I began to examine all of the images. My goal was to choose ten photos of each participant that I would then make the focus of photo-elicitation interviews and publish as part of the study. Future research could involve drug users in curating images, which would give subjects greater control over the creative process. Since the study focused on work that I conduct as a researcher and photographer, I felt it was appropriate for me to select the photos and rely on my own judgment to do so. My decision to curate the images was also a pragmatic one. I considered involving participants in the process, however I needed the process to be as efficient as possible. I write below that, over the course of the year, it was challenging finding the participants to photograph them. Also, they had short attention spans and curating the images required considerable time and focus. It was a process that ultimately took several hours over many days and one I repeated many times in order to ensure that I had made choices that aligned with the overall goals of the study. I also felt it was appropriate for me to curate the photos because I anticipated that the participants would be deeply engaged in the study during their photo-elicitation interviews. At the same time, I recognized that, by curating the photos, I exerted a "guiding influence" (ibid: 10) on this process and all others of the study.

While I curated the images, I prioritized photos that I felt effectively illustrate interesting or diverse types of composition. I considered where participants are situated in the photos, how much or how little of the images they fill up with their bodies, and their relationship to other objects in the frame (Caple 2013: 80). I also chose images that convey participants' positive or negative emotions. I selected images that help communicate information about challenging situations they were facing and highlight their particular and unique physical attributes. I deliberately left out images that I thought could viewed as overly stigmatizing. I also chose photos that I hoped the participants would find interesting and inspire them to offer detailed feedback during their photo-elicitation interviews. I recognized that the more details the participants could possibly provide in these interviews, the more nuanced information the project would be able to communicate to the viewer.

Reflections and challenges

In what follows, I outline the two primary difficulties that I faced while conducting the study.

Access

One of the most significant obstacles was scheduling regular photo sessions with the participants. They did not own cell phones or use email or social media, so the only way I was able to find them was by waiting at the Crosstown Clinic at their scheduled treatment times. Initially I hoped this would be a productive method. However, like many other SALOME participants, Cheryl, Marie and Johnny frequently missed their treatments at the clinic. A few times during the year, for example, Cheryl did not come to the clinic for days at a time. During these periods she was likely exclusively using street drugs. There were many times when progress on the study was stalled because I was not able to find participants and plan photo sessions with them.

Once participants began to trust me, they allowed me to photograph different parts of their lives. For example, they allowed me to take photos of them inside their apartments in Vancouver's DTES. However, throughout the year, there were some things they did not let me document. For instance, Marie denied my requests to photograph her with her mother and other family members with whom she had recently reconnected. On one occasion, I travelled with Marie across the city for nearly two hours by bus. It was Thanksgiving and she was going to visit her mother for one of the first times in two years without any contact with her. Marie allowed me to photograph her on the bus, as she walked to her mother's apartment building and pressed the intercom buzzer, but she did not let me come through the front door with her. I understood and respected her wishes for privacy that were likely shared by her mother and family. At the same time, this prevented me from highlighting the fact that Marie has a family that cares about her. This is potentially one of the most humanizing elements of her life and I could not include it in the study.

Difficulty capturing a variety of photographs

The greatest challenge was photographing the participants engaging in a wide range of activities in a variety of settings. The principal reason was that Cheryl and Marie continued to use illicit drugs outside of the Crosstown Clinic while they were SALOME participants. Acquiring and using street drugs, self-injecting drugs in their own apartments, and self-injecting their treatments at the clinic occupied most of their time. I did not want the majority of photos in the study to be of participants using drugs. In spite of these limitations, by the end of the year, I was moderately satisfied with the range of photos I had been able to take of the participants. Some of the images that I believe communicate unique and compelling things about the participants, beyond the fact that they use drugs, include a photo of Cheryl crying in the courtyard of a church where her father's funeral was held, Johnny in a grocery store where he used to steal to support his heroin addiction before joining SALOME, and Marie at a Vancouver park stretching in a dancer's pose (Figure 33.2).

Benefits of photo-elicitation

In this section I outline my rationale for adopting photo-elicitation as a method. I then highlight how it allowed participants to communicate nuanced information about their lives that could not be conveyed through images alone.

I believe some of the photos that I took of the participants are humanizing. On their own, images of Cheryl and Marie using illicit drugs in their apartments could potentially be viewed as stigmatizing. However, showing the images to participants and recording their feedback in photo-elicitation interviews allowed them to communicate important information about their lives. Pairing their words with the images online helped me to present photos with critical context. Philippe Bourgois and Jeffrey Schonberg (2009: 14) write that photographs can lead viewers to develop negative views about subjects. They argue photographers risk further marginalizing subjects and perpetuating their suffering. They write: "Letting a picture speak its thousand words can result in a thousand deceptions" (ibid). Benjamin Ball (2016: 7–8) writes that pairing images with audio can "expand our vision beyond the photographic frame, provide us with the necessary context and narrative to see property."

Each photo-elicitation interview lasted approximately an hour. I asked each participant the same set of eight questions along with other spontaneous questions. These are the principal questions that I asked:

1. What do you think people will see in this photo?
2. Do you think the photo accurately represents you?
3. How would you have liked me to take the photo differently and why?
4. Is there anything missing?
5. Is there any other information you would like people to know?
6. What do you hope people get out of this photograph?
7. What do you want to show?
8. Could you imagine a different photograph that help people understand your life?

One of my photos of Marie stretching in a dance posture in a park near the Crosstown Clinic (Figure 33.2) may seem unexceptional to viewers. However, I hope the photo reflects Marie's determination to follow her passion in life, namely to dance, in spite of the significant hurdles she has faced. As noted above, when she was a girl, Marie auditioned for the National Ballet School of Canada. However, simply getting to the audition was difficult because her mother, an alcoholic, had been drinking and got lost on the way. As a result, Marie experienced anxiety during her audition, did not perform the way she wanted to, and was not admitted into the program. In her photo-elicitation interview, Marie refers to her inner child and the joy and freedom she can still feel. These emotions, she says, make her the unique person she is, and that she is more than a drug user.

The following is an excerpt from Marie's photo-elicitation interview about this image and I hope it helps the viewer learn about her:

AARON: What do you think people on the outside see in this photo?
MARIE: Just a girl who is playful, carefree, and fun-loving. Maybe had some sort of dance background or some gymnast background or something.
AARON: Does the photo accurately represent you?

Figure 33.2 Aaron Goodman, *Marie stretches in a dance posture in a park in Vancouver's Downtown Eastside. When she was a girl, she auditioned for the National Ballet of Canada,* 2014.

MARIE: Sure, because you know, a needle in my arm is only 10% of who I am. The other parts are going to a park and playing, having fun outside and watching children play. Yea, being as much a part of as I can be in the community.

AARON: How would you have liked me to take photo differently and why? What's missing?

MARIE: No, actually, that's a very good picture. I actually want a copy.

AARON: Of course. I will give you the photo. Is there any other information you would like people to know?

MARIE: That my addiction isn't the whole of me. There's a lot more to me than that and this picture shows it.

AARON: Can you be more specific? What does it show?

MARIE: It shows me goofing off, and playing, and having fun. Maybe the child inside coming out. Maybe, who knows?

AARON: And what do you hope people get out of the photograph? What do you want to show?

MARIE: That I'm not just some dirty, mistrusting, drug addict from the skid row.

I think my photo of Johnny leaning on his walker and making his way down the aisle of a grocery store in Vancouver's DTES (Figure 33.3), along with the details he shared in his photo-elicitation interview, help to humanize him. The interactive and multimedia nature of The Outcasts Project allows the viewer to listen to Johnny speak of his thoughts and emotions about the image. He explains that before he joined SALOME he often stole from this particular grocery store. He says he is proud because he has stopped shoplifting and buys all his own food and supplies.

Figure 33.3 Aaron Goodman, *Johnny walks in a store where he used to shoplift before he joined SALOME*, 2014.

The following is an excerpt from Johnny's photo-elicitation interview about this image:

JOHNNY: What this photo does accurately represent is the fact that I used to go into the shops and shoplift. I shoplifted quite a bit of food from this shop to feed my drug addictions, and that's one thing you don't see in this picture right now.
AARON: Is there anything missing?
JOHNNY: I don't think there's anything missing here. I came and stole from this place and yet, a year later, I'm welcome to come in that store because I made an immense change and do not steal in there now. I come in and buy food like any other individual and it makes me so proud to be able to do that.
AARON: What do you hope people get from this photograph?
JOHNNY: What I hope people get out of this photograph is that even though, once the road of drug addiction takes you down, you can always turn your path to the positive. And this just proves to me that I can walk into the store and the same manager who took me in back and aggressively told me he never wanted to see me in the store, knows me by my first name today and asks me how I'm doing as I do with him. And that means the world to me.

Without having an opportunity to listen to Cheryl speak about my photo of her preparing to self-inject drugs in her apartment (Figure 33.4), the image would not convey a lot of nuanced information. In her photo-elicitation interview, Cheryl makes an emotional and compelling case for heroin-assisted treatment. By listening to her, I believe the viewer has a chance to develop greater understanding and compassion for long-term and vulnerable drug users like Cheryl.

Figure 33.4 Aaron Goodman, *Cheryl prepares to self-inject drugs in her apartment in Vancouver's Downtown Eastside*, 2014.

The following is an excerpt from Cheryl's photo-elicitation interview:

CHERYL: I just want to show people that, you know, I'm a drug addict, I'm a recovering addict. This [heroin-assisted treatment] program I'm on is a great program. It's helped me in many, many angles. I just hope you see . . . there are addicts out there in the world that need a little bit more help. That others out there . . . they need help. We need for you people to see that we're not just stereotyped monsters. We're people just like you, just with an addiction. Something that we do a little bit more than others. I just want you to know that, you know. When you look at this, take it with a grain of salt, because it could be your daughter, it could be your son out there doing exactly what I'm doing, but they had the door closed because you guys look down on it, so they don't want share that with you . . . I just hope that you have a better perspective on what a heroin addict's life or what the person does in order to get their drugs or whatever they're doing in their drug addict lives. That [if] we can get better help for addicts, our world would be a safer, more serene place.

Some people may consider the photos of SALOME participants using illicit drugs in their apartments as problematic and believe that they send conflicting messages about heroin-assisted treatment. This was not my intention and the science is conclusive about SALOME's benefits for long-term and vulnerable drug users (Woo 2016).

Illicit drug use was a significant part of Cheryl and Marie's lives during the period in which the study was conducted and I could not ignore it. By pairing images of participants using illicit drugs with excerpts from their photo-elicitation interviews, I was able to create opportunities for the viewer to learn about the participants' backstories and reasons for using and becoming dependent on drugs. Schonberg (cited in Ryška 2015: 92) points to Paulo Freire, who refers to this process as "conscientization," whereby "one becomes conscious of the conditions surrounding their predicament, their lives."

The three participants explained that trauma, including physical injury and emotional or physical abuse, had contributed to their drug use. Gabor Maté is a physician who has worked with drug users in Vancouver's DTES for decades. He writes: "Not all addictions are rooted in abuse or trauma, but I do believe they can all be traced to painful experience. A hurt is at the center of all addictive behaviors" (Maté 2008: 73). Other factors including family history of addiction, having a mental disorder, peer pressure, lack of family involvement, anxiety, depression and loneliness, and being male can contribute to addiction (Mayo Clinic n.d.). I attempted to communicate these factors as well through the study.

Ethical issues and approaches

It is beyond the scope of this chapter to outline all of the ethical issues connected to the study. However, since the participants were highly vulnerable, it was imperative to me to conduct the project in ways that did not cause additional emotional distress or psychological harm for them. The ethical foundation of the project and the approaches I adopted emerged out of a desire to treat and represent chronic heroin users with respect. In what follows, I briefly highlight some of the primary ethical concerns and the steps I took in order to mitigate the risks of harming participants.

Voluntary and informed consent, anonymity, and confidentiality were important elements that I considered and discussed with participants. I often explained that photos of them would be published on the internet and that police, future employers and others could potentially learn they are heroin users. I informed them that this could affect their reputation and create economic, social and legal risks for them. I made sure they knew they could withdraw from the study at any time and get support if they experienced elevated levels of distress from social workers at the Crosstown Clinic. In spite of the risks, the three participants repeatedly told me they were interested in taking part in the study because they wanted to raise awareness about heroin-assisted treatment. They said they had benefited from it in significant ways and hoped other drug users would have opportunities to access the program and others like it. Marie and Johnny asked that I refer to them by pseudonyms in the study.

Conclusion

In October 2015 the Conservative party of Canada was defeated in a federal election. Liberal leader Justin Trudeau formed a government as prime minister. In September 2016 the Liberals reversed laws enacted by the previous Conservative administration in order to allow physicians to treat people with chronic heroin addiction with diacetylmorphine. The decision made it possible for all 110 former SALOME participants, including Cheryl, Marie and Johnny, to continue receiving heroin-assisted treatment.

As the opioid overdose crisis escalates, the Liberal government is facing increasing calls to address the situation in comprehensive ways. Officials have stated they will implement several actions in order to prevent further overdose deaths. One possibility, the government has said, is to grant permits for additional heroin-assisted treatment programs. As of November 2017, the only heroin-assisted treatment in Canada remains Vancouver's Crosstown Clinic (Lunn 2017).

Debate among politicians, clinicians, scientists, harm reduction workers, drug users, their families and the general public will inevitably continue about the prospect of making heroin-assisted treatment more widely available. Visual researchers, documentary photographers and photojournalists could help inform discussions in creative and productive ways by attempting to humanize drug users.

In this study, my intention was to create counter-narratives about long-term and vulnerable heroin users. I engaged in documentary photography and conducted photo-elicitation interviews with participants in order to obtain and share nuanced information about their lives in an online and interactive format. I view the project as a preliminary exploration of this methodology and its possibilities. Future work could employ PhotoVoice and focus on drug users who take their own photos and curate images. Further studies could also involve conducting photo-elicitation interviews with drug users about photos featuring other drug users as subjects.

Acknowledgements

I am grateful to the study's participants, Cheryl, Marie and Johnny, who shared their time to take part in this study and trusted me to document a period of their lives. I would like to thank Dr Hanna Pickard, Dr Serge Ahmed, Dr Alice Baroni and Dr Scott McMaster who read drafts of this chapter and provided feedback. I am also grateful to Luiz Lopes who designed the project's interactive website, Sheetal Reddy who transcribed photo-elicitation interviews, and Providence Health Care for their support of the project. This work was supported by a Katalyst Grant at Kwantlen Polytechnic University under grant number SPF 661414.

References

Amon, J. (2012) "Alice cooks her shot of heroin in the bathroom of a Taco Bell", Image, *Denver Post*, 8 October, www.denverpost.com/heroinindenver/ci_21721849/heroin-addicts-have-hard-time-turning-lives-around, accessed 14 November 2017.

AP Photo/Linsley, B. (2009) "A heroin addict holds a used syringe in his mouth after shooting up in an abandoned lot in San Juan", Image, *Boston.com*, http://archive.boston.com/bigpicture/2009/10/2009_un_world_drug_report.html, accessed 13 April 2016.

Bailey, I. (2017) "More than 400 overdose deaths anticipated in Vancouver by the end of 2017, city says", *Globe and Mail*, 21 August, https://beta.theglobeandmail.com/news/british-columbia/more-than-400-overdose-deaths-anticipated-in-vancouver-by-end-of-2017-city-says/article36046087/?ref=http://www.theglobeandmail.com Accessed 9 October 2017.

Ball, B. (2016) "Multimedia, slow journalism as process, and the possibility of proper time", *Digital Journalism* 4(4): 1–13.

Becker, H. (1981) *Exploring Society Photographically*, Evanston, IL: Mary and Leigh Block Gallery.

Bourgois, P. I. and Schonberg, J. (2009) *Righteous Dopefiend*, Berkeley, CA: University of California Press.

Canadian Press (2013) "Robert Pickton denies killing women in civil suit defence", *CBC News*, 21 December, www.cbc.ca/news/canada/british-columbia/robert-pickton-denies-killing-women-in-civil-suit-defence-1.2473053, accessed 12 February 2016.

Canadian Press (2014) "Robert Pickton's victims' families to get $50k each", *CBC News*, 17 March, www.cbc.ca/news/canada/british-columbia/robert-pickton-s-victims-families-to-get-50k-each-1.2575932, accessed 6 June 2016.

Caple, H. (2013) *Photojournalism: A Social Semiotic Approach*, London, UK: Palgrave Macmillan.

Centers for Disease Control and Prevention (CDC) (2016) *Morbidity and mortality weekly report. Increases in drug and opioid overdose deaths – United States, 2000–2014*, Centers for Disease Control and Prevention, 1 January, www.cdc.gov/mmwr/preview/mmwrhtml/mm6450a3.htm, accessed 15 January 2016.

Cole, A. L. and Knowles, G. J. (2001) *Lives in Context: The Art of Life History Research*, Lanham, MD: AltaMira Press.

Fairclough, C. M. (2013) "Visualising the migrant's experience: one woman's story", Doctoral dissertation, University of Salford, UK.

Fernandez, J. (1995) "Amazing grace: meaning deficit, displacement and new consciousness in expressive interaction", in A. Cohen and N. Rapport (eds), *Questions of Consciousness*, London, UK: Routledge.

Fitzgerald, J. L. (2002) "Drug photography and harm reduction: reading John Ranard", *International Journal of Drug Policy* 13(5): 369–385.

Foucault, M. (1980) *Power/Knowledge: Selected Interviews and Other Writings, 1972–1977*, New York, NY: Pantheon.

Goffman, E. and Lofland, L. H. (1989) "On fieldwork", *Journal of Contemporary Ethnography* 18(2): 123.

Hagen, C. (1994) "Review/photography; 'cocaine true': art or sensationalism", *New York Times*, 11 March, www.nytimes.com/1994/03/11/arts/review-photography-cocaine-true-art-or-sensationalism.html, accessed 11 January 2016.

Hall, S. (1997) *Representation: Cultural Representations and Signifying Practices, Volume 2*, London, UK: Sage.

Hammersley, M. and Atkinson, P. (2007) *Ethnography: Principles in Practice*, London, UK: Routledge.

Harper, D. A. (2012) *Visual Sociology*, Abingdon, UK: Routledge.

Houston, K. (2014) "Larry Clark in Amsterdam by Kerr Houston", *Bmore Art*, last modified 17 August, www.bmoreart.com/2014/08/tulsa-teenage-lust-and-the-domestication-of-the-outrageous.html, accessed 17 February 2016.

Karstens-Smith, G. (2017) "British Columbia boosts drug-checking service in fight against overdose deaths", *Canadian Press/Victoria Times Colonist*, 10 November, www.theguardian.pe.ca/news/british-columbia-boosts-drug-checking-service-in-fight-against-overdoses-160876, accessed 14 November 2017.

Lincoln Clarkes Photography (2007) *Worldwide Greeneyes*, http://worldwidegreeneyes.com/heriones-essay, accessed 12 March 2016.

Lunn, S. (2017) "Liberals say they'll back prescription heroin, drug checking services to fight opioid crisis", *CBC News*, 15 November, www.cbc.ca/news/politics/opioid-prescription-heroin-1.4403709, accessed 18 November 2017.

Lupick, T. (2018) "Photos: 2018 National Day of Action on the Overdose Crisis sees protests in Vancouver and cities across Canada", *Georgia Straight*, 20 February, www.straight.com/news/1035176/2018-national-day-action-overdose-crisis-sees-protests-vancouver-and-cities-across, accessed 2 March 2018.

Martínez, K. (2012) *Man with Hypodermic Needle Preparing to Give Himself an Injection*, Image, *Picture Engine*, Christian/Getty Images, www.gettyimages.ca/detail/photo/drug-abuse-royalty-free-image/155441155?esource=SEO_GIS_CDN_Redirect, accessed 23 March 2015.

Maté, G. (2008) *In the Realm of Hungry Ghosts*, Toronto, Canada: Alfred A. Knopf.

Mayo Clinic (n.d.) "Diseases and conditions: drug addiction", *Mayo Clinic*, www.mayoclinic.org/diseases-conditions/drug-addiction/symptoms-causes/syc-20365112, accessed 12 June 2016.

Pink, S. (2004) "Visual methods" in C. Seale, *et al.* (eds), *Qualitative Research Practice*, London, UK: Sage, pp. 361–378.

Pink, S. (2013) *Doing Visual Ethnography: Images, Media and Representation in Research*, London, UK: Sage.

Postman, N. (1985) *Amusing Ourselves to Death: Public Discourse in the Age of Show Business*, London, UK: Penguin.

Prosser, J. and Loxley, A. (2008) *ESRC National Centre for Research Methods Review Paper. Introducing Visual Methods*, National Centre for Research Methods, October.

Providence Health Care (n.d.) "SALOME Clinical Trial Questions and Answers", *Providence Health Care*, www.providencehealthcare.org/salome/faqs.html, accessed 6 April 2016.

Ryška, T. (2015) "Interview with Jeffrey Schonberg", *Cargo Journal* 10: 1–2.

Schwartz, D. (1992) *Waucoma Twilight*, Washington, DC: Smithsonian Institution Press.

Seelye, K. Q. (2015) "Obituaries shed euphemisms to chronicle toll of heroin", *New York Times*, 11 July, www.nytimes.com/2015/07/12/us/obituaries-shed-euphemisms-to-confront-heroins-toll.html?ref=topics&_r=2, accessed 14 April 2016.

Sontag, S. (1977) *On Photography*, London, UK: Macmillan.

Uchtenhagen, A. (2002) "Heroin assisted treatment for opiate addicts: the Swiss experience", *Parliament of Canada*, www.parl.gc.ca/Content/SEN/Committee/371/ille/presentation/ucht1-e.htm, accessed 15 April 2016.

Ugor, P. (2007) "Heroines essay: imaging the invisible, naming suffering", *West Coast Line* (53), Vancouver, BC: Simon Fraser University.

University of British Columbia (2011) "UBC-providence health research to examine new treatments for heroin addiction", University of British Columbia Public Affairs, 12 October, https://news.ubc.ca/2011/10/12/ubc-providence-health-research-to-examine-new-treatments-for-heroin-addiction, accessed 3 February 2016.

Werry, A. (2015) "A real test of the Conservative party's new tone", *Maclean's*, 19 November, www.macleans.ca/politics/ottawa/a-real-test-of-the-conservative-partys-new-tone, accessed 13 May 2016.

Woo, A. (2014) "Court rules to allow patients to continue supervised heroin use", *Globe and Mail*, 29 May, www.theglobeandmail.com/news/british-columbia/heroin-prescriptions-for-addicts-okayed-in-court-ordered-injunction/article18913042, accessed 11 May 2016.

Woo, A. (2016) "Vancouver study finds new treatment path for chronic heroin addiction", *The Globe and Mail*, 6 April, www.theglobeandmail.com/news/british-columbia/vancouver-study-finds-new-treatment-path-for-chronic-heroin-addiction/article29537912, accessed 17 June 2017.

34
OUR STORIES, OUR KNOWLEDGE

The importance of addicts' epistemic authority in treatment

Peg O'Connor

Introduction

William James and his great work, *The Varieties of Religious Experience* (1902), had an enormous influence in addiction treatment. In 1934, Bill Wilson was at the Charles B. Towns hospital trying to dry out again. Full of despair and hopelessness, he cried out, "If there is a God, let Him show Himself! I am ready to do anything, anything!" Suddenly the room was filled with a white light and he felt a wind of spirit blowing. At that point he realized he was a free man (AA World Services 1957: 63). Very soon after this experience, he worried he was losing his mind. A friend, however, thought differently and gave Wilson a copy of *Varieties* as a way to understand this remarkable experience. James was most concerned in *Varieties* with people who experienced what he called a conversion or rebirth. These are people for whom religion exists "not as a dull habit but as an acute fever" (James 2012: 14). Bill W seemed to have caught that fever; he later went on to found Alcoholics Anonymous (AA).

In *Varieties*, Bill Wilson found a treasure trove of first person stories that resonated with his own experiences as a hopeless and incurable alcoholic. Several of the conversion experiences that James highlights and returns to throughout *Varieties* involve reformed drunkards or inebriates, to use the language of his time. These experiences were all respectfully and compassionately treated by James, which is not surprising for several reasons. Perhaps most relevant was that James had intimate knowledge of the ravages of drunkenness. His own brother, Robertson (Bob), had several stints in asylums for the inebriate and spent his last days living with William and his wife.

As a physician, philosopher, and psychologist before psychology was its own discipline, James was and remains a valuable and insightful companion to addicts and those who work professionally with us. James's work is especially relevant today with increasing rates of addiction to increasingly dangerous substances often accompanied by mental disorders requiring more effective and nuanced treatment plans. James was an astute student of human nature and human suffering in large part because he observed and listened to people; he sought out their stories. First person narratives are the primary source of evidence for spiritual matters and conversions that James marshals in *Varieties*. This chapter explores three dimensions of these first person stories that are important for the recovery of individuals (just as Bill W discovered) and for the field of addiction and treatment research especially as it becomes more medicalized.

The first intriguing strand in James's treatment of first person narratives is that he lets people speak for themselves and, by so doing, accords them a great deal of epistemic authority. In other words, James does not position himself as a third party playing the role of expert on these individuals' experiences and the understanding and meaning they were able to make of them. The inebriates—people who are often judged to be among the most unreliable or morally suspect—are taken by James to be reliable witnesses to and interpreters of their own experiences. These experiences are side by side with experiences of saints, literary giants and intellectuals such as Lev Tolstoy, Henry David Thoreau, and John Stuart Mill, and religious leaders such as George Fox and Jonathan Edwards. James later revealed he included his own experience of a deep pathological melancholy in *Varieties*.

It is important to see this approach as democratizing epistemic authority. To use an expression from James in *The Will to Believe*, when we respect each other's mental freedoms, which includes the ability to tell one's own story, we can create an "intellectual republic" (James 1956: 30). This is what James creates in *Varieties* and it is something that needs to be carried over to addiction and treatment research programs now.

The second intriguing dimension of first person narratives is the way James underscores the social nature of self-knowledge. Reading these accounts in *Varieties*, Bill Wilson saw himself clearly. He saw his own experiences—all the pain, suffering, self-loathing, attempts and failures to stop, and despair—reflected back to him in a way that enabled him to make sense of what had happened to him, what he had done to himself, and how he might live differently. James's concepts such as conversion, misery threshold, higher and friendly power, in conjunction with the first person stories provided Bill W with a framework for understanding himself. Bill W adopted and modified these concepts as he developed the Twelve Steps of Alcoholics Anonymous.

Many of us share our stories of addiction both as a way to help ourselves maintain our sobriety and to help others achieve sobriety. Many feel as if we have a responsibility to share our stories and listen to others. The way one tells her story and how others hear it matter enormously. This leads to the third interesting dimension of the first person narratives. Almost without fail, especially with the "reformed drunkards," an attitude of humility runs throughout the narratives. There is a very interesting relationship between humility as an epistemic virtue and the possibility for self-knowledge and the achievement of sobriety that these narratives reveal. Humility in telling must be met with humility in listening. James himself embodies that humility in *Varieties* and so provides a good model for physicians, researchers, and treatment professionals working with people struggling with addictions.

These three features of *The Varieties of Religious Experience*—epistemic authority democratized through first person narratives, self-knowledge as social, and humility as an epistemic virtue—are all crucial for understanding addiction and the possibilities for recovery. I will focus primarily on humility, though the first two strands will begin to emerge. I will end the chapter with a concern about the decentering of addicts' stories and their self-knowledge, and the concentration of epistemic authority in the hands of physicians and insurance companies that together pose a very serious challenge to the "intellectual republic" of addiction and treatment programs. A lack of humility on that level runs the risk of creating conditions where it might be more difficult for individuals to recover.

Humility as an epistemic virtue

There is a gap in the humility literature on the relationship between humility and self-knowledge. Treatments of humility most often conceive it as a relational virtue that is primarily concerned with attitudes towards one's own and others' achievements or accomplishments. The achievements and accomplishments generally include expertise, awards, career success, and material wealth. Neither humility itself nor self-knowledge nor sobriety is seen as an achievement. I claim humility is itself an achievement and that it is a necessary condition for self-knowledge in relation to one's addiction. A lack of humility sharply limits if not precludes self-knowledge, which may make recovery far more difficult to achieve and maintain.

Humility as a virtue is a mean between two extremes of vanity and self-abasement. Much of the humility literature focuses on vanity. Roberts and Wood primarily define humility in the negative; it is the absence of vanity and arrogance. Vanity, on their view, is the concern of a person to make a good appearance only (though I would say primarily) for the positive social status it confers. Arrogance is more an illicit entitlement claim based on one's alleged superiority. Entitlement is crucial; an arrogant person may feel justified not to see others as knowledgeable, and hence disregard anything they say. Arrogant people tend to think that others cannot know as much or what they know is inconsequential. Arrogance and vanity are inseparable; they tend to feed off one another (Roberts and Wood 2003: 259).

More difficult to identify than overt arrogance are instances where one person's dismissal of another's knowledge or interpretation of his own experiences springs from good intentions and/or a belief that one knows what's in the best interest of the other. This is covert arrogance or arrogance in drag. Whether this belief comes from academic or other professional training, life experiences or something else, the result is the same. The one does not regard the other as an equal who may himself have knowledge, insight, or valuable experience.

Humility, on Roberts and Wood's view, involves avoiding overestimation of yourself and your accomplishments. Additionally, humility is a lack of concern or inattentiveness to how others regard you and your accomplishments and how you regard yourself (Roberts and Wood 2003: 261). There is something right about the avoidance of overestimation, but what about the other end of the spectrum?

Humility must also avoid the underestimation of oneself. At the far end of the spectrum is self-abasement, which is well beyond low self-esteem. Self-abasement is a set of beliefs and attitudes that one has no worth and is in some senses repulsive. Having no worth, a person deserves nothing good and has nothing positive to contribute. A person who suffers from self-abasement may believe that she knows everything about herself and all of it is bad.

J. L. A. Garcia offers a more robust definition of humility that is a mean between the two extremes and that complements features of good recovery. Someone is humble about her having talents, skills, achievements, etc. if and only if she

1. downplays her being talented, skillful, accomplished,
2. doesn't stress in her self-identity her liberties, privileges, etc.,
3. doesn't dismiss the talents and achievements of others,
4. recognizes how others have helped her to have those talents, skills, and accomplishments,
5. commits to her moral improvement,
6. recognizes her own shortcomings, and
7. recognizes her duties and obligations to others.

Garcia 2006: 418

Garcia concludes that we can say of the person who meets all seven of these conditions (they constitute one big conjunction) that "she is unimpressed *with herself* (emphasis in original).[1]

Conditions 1 through 3 align well with the overestimation characterization offered by Roberts and Wood. Vain and arrogant people tend to overly stress their achievements and tend not to be able to recognize the talent, experience, and knowledge of others. People who do not meet any one of the first three conditions will have a harder time meeting the next four, which guard against a person progressing too far down the spectrum toward self-abasement. To recognize that others have helped you is to implicitly recognize that someone sees something worthwhile in you even as you yourself may struggle to do so. Moving away from self-abasement is a form of moral improvement in the most basic sense of recognizing your humanity and agency. To recognize something as a shortcoming as opposed to abject failure is a form of moral progress. Shortcomings can be transformed into strengths.

The two extremes get in the way of overcoming addiction in different but complementary ways. Self-abasement breeds a kind of self-deception that in its worst form becomes a form of fatalism: why even bother to try because someone like me can never get better. Overt and covert vanity may prompt a professional to minimize or disregard what a person struggling with addiction has to say. This brings me back to James's *Varieties*; James's treatment of first person stories of addicts' struggles is instructive for avoiding both the extremes and cultivating a genuine humility.

Meet the drunkards of *Varieties*

The first person stories in *Varieties*, especially the ones related to drunkenness, embody Garcia's dimensions of humility. These stories are powerfully evocative because of the despair, loneliness, and hopelessness they convey. James offers two compelling cases. The first is a Christian Evangelist named Henry Alline, who struggled with drinking and "carnal mirth." Alline kept making promises to himself not to engage in these activities. But as is familiar to many addicts, he broke those promises when tempted by wine, music, and the company of friends. Full of self-loathing, he became "wild and rude." After the debauchery, he describes himself as "guilty as ever . . . I was one of the most unhappy creatures on earth" (James 2012: 138).

The second story James offers is S. H. Hadley, who described himself as "a homeless, friendless, dying, drunkard" (James 2012: 159). Hadley describes the night that he is sitting in a Harlem saloon, shaking with the delirium tremens and feeling utterly broken and penniless. He writes

> As I sat there thinking, I seemed to feel some great and mighty presence. I did not know then what it was. I did learn afterwards that it was Jesus, the sinner's friend. I walked up to the bar and pounded it with my fist till I made the glass rattle. Those who stood by drinking looked on with scornful curiosity. I said I would never take another drink, if I died on the street, and really I felt as though that would happen before morning. Something said, "if you want to keep that promise, go and have yourself locked up." I went to the nearest station house and had myself locked up.

Once he is released from his cell, he returns to his brother's home and then visits a mission where "the apostle to the drunkard and outcast—that man of God—Jerry M'Auley" was praying. Hadley listened to the testimony of 25 other drunkards who had given up the rum. As he was listening, he recalls, "I felt I was a free man . . . From that moment till now, I have never wanted a drink of whiskey" (James 2012: 160). Hadley's soul was no longer riven by the inner conflict over his drinking. For this reason, James describes Hadley as an example of a Divided Self becoming unified.

Hadley's story is one of deep despair and desperation as is Alline's. There is no bragging about his past drinking exploits. Rather, there is a frankness and honesty about his past failures and transgressions. Any self-deception about his drinking and its effects on him and others had long been stripped away; he knew himself to be completely miserable. So long as he lied to himself or otherwise deceived himself, nothing was going to change. Both Hadley and Alline are brutally honest; there are no attempts to gloss over or excuse their past actions. There's no braggadocio; their behavior was the source of great shame and regret.[2]

Both Alline and Hadley find what James calls their "misery thresholds" (James 2012: 109–110). This concept is an excellent way to understand the depth and degree of emotional suffering that a person can tolerate. Each person's misery threshold is subjective and unique to him.[3] Each person must find this threshold for himself, but that does not entail that he must find it alone. Knowing how much suffering a person can take and tolerate is a hard-won form of self-knowledge that is critical for people struggling with addiction and living in recovery. It is also critical knowledge for professionals to recognize in the creation of treatment plans.

Returning to the last four of Garcia's criteria for humility, Hadley recognizes the help he had from the jailer who locked him up, a brother who took him in, the "apostle of the drunkard," and the 25 men whose stories he heard. By seeing himself in their stories, he started to identify his own shortcomings and, perhaps most importantly, saw a different way to live. By not drinking and helping others in the same ways he had been helped, Hadley made a commitment to his moral improvement by recognizing and acting on an obligation to help others.

By the time he read the stories in *Varieties*, Bill W had already been stripped of his self-deception. Arrogance and vanity were long since gone. He saw himself in these stories; sure, the particulars were different but the trajectory was largely the same. This provided him with great insight into what he needed to do differently if he truly wanted to stay sober. One was that he could not do it alone. The self-knowledge that Alline, Hadley, and Wilson came to possess is deeply social and acquired in the company of others. Just as important, their self-knowledge is not just a cognitive state involving propositions. Rather, it is a matter of actions and how one shows up in the world. Self-knowledge is ultimately practical. It is seen not just in what we say and believe, but rather in what we do.

Assault on an intellectual republic of recovery

The field of addiction treatment most assuredly does not function as an epistemic democracy or intellectual republic. This brings me to my final Jamesian consideration about first person narratives and addiction. It is fair to say that as a medical doctor, James would have respected the potential contributions that sciences could make to the treatment of addiction. It is just as accurate to say that James had an equal or greater concern about the expansion of positivist science, or what he called "medical materialism," into all realms of human living (James 2012: 19). James worried that materialist/physicalist explanations along with their reductionist tendencies bring along a brand of determinism. As he writes in the "Dilemma of Determinism,"

> What does determinism profess? It professes that those parts of the universe already laid down absolutely appoint and decree what the other parts shall be. The future has no ambiguous possibilities hidden in its womb . . . the whole is in each and every part, and welds it with the rest into an absolute unity, an iron block, in which there can be no equivocation or shadow of turning.
>
> *James 1956: 150*

Determinism would call free will into doubt and fly in the face of human freedom, which is the gravest concern for James. Determinism may breed a sense of fatalism; why bother to resist something that far outstrips an individual's ability to control it and that is inevitable? This is why it is appropriate to revisit these warnings now in the context of addiction and recovery.

Addiction studies is presently undergoing a seismic shift, with neuroscience beginning to overtake the field and trump all other forms of knowledge. The view is that our brains—our pleasure circuitry—is the organ working in overdrive. The language of dependency and addiction is changing to reflect the new neuroscientific advanced claim that "addiction is a chronic brain disease" (Volkow, Koob and McLellan 2016). The strongest version of this view may imply the only acceptable treatment for patients will be of the pharmacological sort. Even more worrisome is the fact that in the United States, the National Institute of Mental Health now requires that clinical researchers focus on neural circuits or biomarkers. As a consequence, only certain treatment modalities based on a physicalist account will have the opportunity to be "evidenced based," and eligible for insurance coverage. There's legitimate concern that pharmacology is triumphing. As psychiatrist Bessel van der Kolk notes about the category of mental disorders, "the brain-disease model takes control over people's fates out of their own hands and puts doctors and insurance companies in charge of fixing their problems" (van der Kolk 2014: 37).

The shifting of the terrain in addiction studies poses a significant risk to a balance of epistemic authority. Medical authority will always both claim and be accorded greater epistemic standing and thereby increasing the likelihood of epistemic injustice. Epistemic injustice, as Miranda Fricker defines it, is of two sorts. The first is testimonial injustice, which she describes as prejudice in the economy of credibility (Fricker 2007: 21). Some people, because of their race, class, sex, or some other group membership such as drug users, are not taken as reliable, credible, or truthful. The lack of reliability may extend to nearly everything, especially to their own experiences. Those whose identities involve the intersection of multiple suspect groups—addict, poor, nonwhite, female, for example—will be subject to far greater testimonial injustice than others. In other words, the stories of addicts will be distrusted and discounted.

The second sort of epistemic injustice is hermeneutical injustice, which exists at a structural level when a community lacks the appropriate conceptual framework to interpret certain experiences. One obvious example of hermeneutical injustice is the lack of understanding many people had about women's experiences in the workforce that we now (thanks to feminist frameworks) identify as sexual harassment (Fricker 2007). With respect to addiction, in a very short amount of time, the brain disease model has become dominant and has displaced other models. With its language of brains being hijacked or commandeered, it offers a picture of how addictions progress physiologically in all addicts. Sameness of physiological progression becomes the expectation and norm. There is little room left to discuss difference and specificity. One of the consequences of the brain disease model is that addicted brains rather than persons living with addictions are the focus of research. Addicted persons become objectified and become a type—an addict.

Hermeneutical injustice creates stereotypes and warped images of groups of people that are then more easily seen, recognized, and taken as true than the actual patients sitting before them in an office. Along with stereotypes come all sorts of judgments and questions. If a person is not willing to use medication assisted treatment, will he be judged not to be trying hard enough or not taking his sobriety seriously enough? If a person seems to want medication assisted treatment too much, will he be accused of drug seeking?

Those who have privileged positions within the dominant framework will have/be accorded the most epistemic authority; they will have a credibility excess. They will have the institutional power to make diagnoses and then decide what counts as an effective or appropriate course of action.

To be very clear, I am not dismissing the importance of recent developments in neuroscience and advances in medication assisted treatment. Nor am I imputing bad motives to scientists, physicians, and treatment professionals working in the field. To the contrary, the vast majority of people working in the field have the best intentions. However, those intentions can provide cover for what I called covert arrogance that can adversely affect treatment. I am concerned about the epistemic injustices (often connected to social, political, and economic injustices) that create conditions where some people will have a much harder time getting sober in the very contexts that should give them the best chances.

To see the importance of first person narratives, self-knowledge, and humility of both people struggling with addiction and treatment professionals, consider someone addicted to heroin who has the good fortune to have access to a physician who is authorized to treat opioid addiction with medications. Imagine this person has been free from heroin and other opioid drugs for at least seven days. A physician may offer Vivitrol (injectable naltrexone given monthly) to help reduce cravings. Cravings are both physical (the body wants what it has become accustomed to) and psychological (the mind wants relief or pleasure). Vivitrol blocks the euphoria of heroin but does not block the other effects of heroin on the body's respiratory and cardiac systems. The assumption is that, without the high, cravings will subside and so too will use. The fact that Vivitrol is effective for a month relieves a person of having to make a daily decision about taking this prescription medication and gives him time to work on treatment, counseling, and other activities related to sobriety.

Vivitrol offers great promise and many people have used it to very good effect. It is not intended for long-term use as a maintenance drug such as Suboxone or methadone. However, for it to be effective and not life-threatening, Vivitrol must be administered only under certain conditions and these warrant serious consideration. The medication guide explicitly states that the risk of opioid overdose is high if a person is taking Vivitrol and then uses opioid drugs. Because Vivitrol blocks the euphoria, a person may try to shoot through its blocking effects. Additionally, Vivitrol's blocking effects decrease to the point of ceasing over the course of a month. Using opioid drugs in the same amounts of past usage may cause an overdose because a person's tolerance has decreased.

This patient faces a profound existential question: should he try to save his life by risking his life? How is he to decide what to do? How is a course of treatment offered and then chosen? Imagine that person is much like Henry Alline, whom James introduces in *Varieties*, who described himself, "In a storm, yet I continued to be the chief contriver and ringleader of the frolics for many months . . . though it was a total toil and torment to attend them" (James 2012: 138). Alline clearly recognized the pull his friends had on him; not only was he vulnerable to the allures of the parties, he became the main instigator, all the while being tormented by his own actions. This is crucial self-knowledge that should also count as important information for a treatment provider.

It is more than sharing a timeline of using history; it is hearing the reasons he used and the understanding and the meaning he made of his use and progression to addiction. It is hearing about how it made him feel. It is hearing about the physical, social, and economic environments in which he used and the environments in which he will be trying to be sober. Knowing how and why a person came to treatment is one way to come to understand an addict's misery threshold. Knowing a person's treatment goals matters enormously as well. One needs to know the *person* and not just the medical condition.

Someone like S. H. Hadley might realize that effective treatment for him requires being "locked up" in some way or in a very rigid structure such as in-patient treatment or intense out-patient treatment. He might realize that he needs the supervision and high degree of accountability that a drug court provides. In a drug court setting, participants often form bonds with other participants and also with the drug court team comprising judges, probation officers, therapists, police officers, and physicians. Making connections, fostering relationships, helping others to get and stay sober, making different (and better) life choices are all ways that people embody the final four criteria of Garcia—recognizing how others have helped, committing to moral improvement, recognizing shortcomings, and meeting obligations to self and others.

Effective treatment requires a great deal of time and dedication on the parts of both patients and professionals. In the United States it is not possible for the majority of people who need treatment (especially those with opioid dependence) to get it. The number of physicians who are certified to provide medication assisted treatment (MAT) is stunningly low compared with the number of people seeking help. The number of healthcare claims for people with opioid addiction increased 3000% between 2007 and 2014. Recent changes to the Controlled Substances Act has increased the number of MAT patients a physician can see from 100 to 275 but this falls far short of present need.

In theory, best practices for MAT resist the pull of "medical materialism" and reductionism by recommending a combination of medication, counseling, and behavior therapy. The reality is that sky rocketing demand, time constraints, and insurance coverage complexities make it far more difficult to implement best practices for effective treatment. While it is true there are commonalities to addictions' trajectories, each person is unique as is the development and progression of his addiction. This is precisely why first person stories must infuse any treatment program, but especially medically assisted treatment.

Conclusion

William James provides anyone interested in addiction with powerful stories and conceptual resources. He embodies epistemic virtues that lead to better knowing and, he would acknowledge, better ways of living. There is an important role for James in contemporary discussions of addiction. Adopting a much more thorough going Jamesian approach provides a powerful counter balance to the scientization of addiction research and treatment. The more these questions are scientized and brought under an umbrella of scientific expertise, the more epistemic injustice may thrive. James invites and authorizes his readers to embrace their own epistemic authority with a deep and abiding humility. It is an invitation we ought all to accept.

Notes

1 Contra Julia Driver, ignorance of one's achievement of sobriety seems misguided.
2 See Flanagan 2013.
3 This is preferable to the more contemporary notion of "rock bottom," which many people use in a way that seems to mean losing it all. There's a belief that one will be able to quit only when one loses it all and a companion belief that unless one has lost it all, one isn't an addict. That's bad logic.

References

AA World Services (1957) *Alcoholics Anonymous Comes of Age: A Brief History*, New York, NY: AA World Services.
Driver, J. (1989) "The virtues of ignorance", *The Journal of Philosophy* 86(7): 373–384.
Flanagan, O. (2013) "The shame of addiction", *Frontiers in Psychiatry* 4: 120.
Fricker, M. (2007) *Epistemic Injustice: Power and the Ethics of Knowing*, Oxford, UK: Oxford University Press.
Garcia, J. L. A. (2006) "Being unimpressed with ourselves: reconceiving humility", *Philosophia* 34(4): 417–435.

James, W. (1956) *The Will to Believe and Other Essays in Popular Philosophy and Human Immortality*, New York, NY: Dover.
James, W. (2012) *The Varieties of Religious Experience*, Oxford, UK: Oxford University Press.
Roberts, R. and Wood, W. (2003) "Humility and epistemic goods", in M. DePaul and L. Zagzebski (eds), *Intellectual Virtue*, Oxford, UK: Clarendon.
van der Kolk, B. (2014) *The Body Keeps Score: Brain, Mind, and Body in the Healing of Trauma*, New York, NY: Penguin.
Volkow, N., Koob, B. and McLellan, T. (2016) "Neurobiologic advances from the brain disease model of addiction", published online in *The New England Journal of Medicine* 374: 363–371.

35

REACTIVE ATTITUDES, RELATIONSHIPS, AND ADDICTION

Jeanette Kennett, Doug McConnell, and Anke Snoek

Introduction

What happens to close relationships of love and friendship in addiction? How does the impact of addiction on those relationships affect the addicted person's view of themselves and of their capacities for change and recovery? In this chapter we focus on the structure of close personal relations and diagnose how these relationships are disrupted by addiction. We draw upon Peter Strawson's landmark paper 'Freedom and Resentment' (2008, first published 1962) to argue that loved ones of those with addiction veer between, (1) reactive attitudes of blame and resentment generated by disappointed expectations of goodwill and reciprocity, and (2) the detached objective stance from which the addicted person is seen as less blameworthy but also as less fit for ordinary interpersonal relationships. We examine how these responses, in turn, shape the addicted person's view of themselves, their character and their capacities, and provide a negative narrative trajectory that impedes recovery. We close with a consideration of how these effects might be mitigated by adopting less demanding variations of the participant stance.

Psychological visibility, self-knowledge, and interpretation

As social, embodied creatures we build our view of ourselves through physical and social interactions and through the interpretive practices that characterise different social relationships. Our interactions with inanimate physical objects, for example, can reveal and develop in us a variety of qualities, e.g. the strength to lift an object or the agility to climb over it. Our interactions with other animals, such as pets, who can respond to our intentions and attitudes go further than this by rendering us *psychologically visible* to ourselves, e.g. as threatening, playful, or affectionate – as possessing certain personal qualities, capacities and temperament (Branden 1993). Psychological visibility, however, is largely a function of our interactions with other human agents and the feedback that those interactions provide. Consider first the multitude of trivial and complex interactions that fill our day. We buy bus tickets, go through the supermarket checkout, greet colleagues, run meetings, ask for or provide directions, send texts, update our status on social media, order coffee. Through these interactions and the responses of others to us within them, we gain information both about what we are like and what is expected of us.

Much of this information is revealed via the other's range of emotional reactions to us: a yawn suggests I am boring, while a hurt expression might reveal that I was rude. Such responses may challenge or confirm the view we have of ourselves. I thought my joke was funny but others see it as lame. I thought I was being admirably blunt but I am coming across as rude and arrogant. Where there is a mismatch between the view we have of ourselves and the view we get through others' eyes we might try to reconcile them in a number of ways. I might modify my behaviour and responses so that you will see me the way I wish to be seen. Your raised eyebrow response to my miserly tip might trigger shame and prompt me to be more generous next time. In doing so, I bring my behaviour into line with the person I want to be, or at least with prevailing social norms. Alternatively, I might challenge or dismiss your interpretation of me. In one-off interactions perhaps this is easily done (she just can't take a joke) but, in interactions over time, a consistently negative response from others will be hard to ignore. Then I might internalise the other's view of me. If I come to see myself as clumsy in word or deed I will likely not attempt tasks that require finesse or dexterity. If I see myself as boring I may be less inclined to engage socially for fear of rejection. If I have a stigmatised social identity – e.g., that of an addict – the range of interactions available to me narrows and the negative feedback, even in casual interactions, becomes consistent. This also can have profound effects on my agency – my view of what I can do and how I can do it.

The participant stance, the reactive attitudes, and effective agency

Peter Strawson (2008) has pointed out the very great importance that we place on the attitudes of others in our interactions with them. In particular, Strawson focuses on a subset of our emotional reactions to each other that specifically communicate our *moral* expectations – the reactive attitudes. We expect people to act towards us with goodwill; when they do, we respond with gratitude and with forgiveness for unintentional harms. When they violate those expectations, we respond with resentment and hurt feelings. The reactive attitudes are, on this account, central to praise and blame. When we are prepared to expose others to the reactive attitudes we take what Strawson calls the 'participant stance' towards them.

Adults, he suggests, take the participant stance by default because they assume each other to be morally responsive and responsible agents capable of adjusting their behaviour and attitudes in the light of the demand for goodwill and the feedback on their behaviour provided within this stance.[1] They must assume, that is, that the other has the set of agential capacities associated with autonomy, notably for moral understanding, self-reflection and self-control. The participant stance thus helps render the agent *morally visible* to herself as an equal moral agent and a member of the moral community and subject to mutual obligations within it. Specific instances of reactive attitudes help her to learn about her particular qualities that have a moral valence, such as degrees of generosity, trustworthiness, and consideration of others. Unexpected reactive attitudes alert her to the fact that she has lost control of a moral aspect of her appearance. When the reactions of others are as the agent expects, they allow her to proceed with confidence because they indicate that she appears as she hoped to.

By contrast, we adopt what Strawson calls the 'objective stance' when the other's actions are consistently unmodified by our responses to them. We come to believe that they are incapable of ordinary adult relationships either because they lack the relevant agential capacities or refuse to exercise them. Strawson gives the example of the 'hopeless schizophrenic' (2008, 8), others include those with late-stage dementia. When faced with such individuals, Strawson thinks we give up the disposition to resent or blame them for their harmful actions; we move, *and should move*, to the objective stance (2008, 9). From the objective stance, the other is seen as a problem "to be managed or handled or cured or trained; perhaps simply to be avoided" (2008, 10). The objective attitude may involve some emotions, such as repulsion, fear and pity, but not "the range of reactive feelings and attitudes which belong to involvement or participation with others in inter-personal human relationships" (2008, 10). However, Strawson also notes that we may also use the objective attitude as an escape from the strain of involvement with others (2008, 18). Certainly it is a stance that can characterise, in particular, professional interactions with groups such as delinquent adolescents, or, in the present case, people who are addicted. In these cases, the objective attitude is damaging insofar as it limits the kind of self-visibility available to those targeted by it. Those targeted may come to see themselves as things to be managed, handled, cured, trained, feared or pitied, rather than as capable and morally responsible agents.[3]

Close relationships – families and friends

In normal circumstances, the effects of brief negative interactions in the public sphere on identity and sense of agency are outweighed by positive experiences in our more intimate relationships. Friends and family typically provide more charitable interpretations of the agent protecting her against rude shopkeepers, thoughtless co-workers, distant doctors or patronising social workers. My clumsiness might be interpreted to me as endearing rather than irritating; my labouring a point is seen by those who love me as indicative of an admirably deep concern about an issue rather than as pedantic and boring. Furthermore, friends and family know the agent's history and interact with her in various intimate situations so they can provide richer, more insightful, diachronic interpretations of the agent than others.[2] It is therefore not surprising that friends and family tend to shape the agent's self-concept more thoroughly than others do. Familial relationships and close friendships, however, contribute to our identities in different ways, and so these relationships are likely to be affected in somewhat different ways when one party develops an addiction.

We claim, following Cocking and Kennett (1998), that close friendships are marked by a heightened mutual receptivity to our friends' ways of seeing us and to being directed by their interests. So, for example, I will typically be more open to activities that I have no antecedent interest in, such as going to the opera, if asked by my close friend than by my parents or a casual acquaintance. Similarly, we are more receptive to close friends' interpretations of us than those of mere acquaintances and our friend's interpretations and enthusiasms may encourage us to develop in new ways and strike out in new directions (e.g., "You have a great voice. You should join a choir"). Indeed, it is arguably an essential characteristic of close relationships that each remains responsive to a degree of shaping by the other. It is this characteristic of close friendship that explains the particularity and intimacy of friendship (Cocking and Kennett 1998). Conversely, friendships become more distant and prone to the display of negative reactive attitudes to the extent that one or both participants feels that the other is insufficiently responsive to her interests and fixed in her interpretations. In such cases, the agent may feel that her friend no longer sees *her*.

By contrast, our identities within the family – who we are within the 'us' of the family – is less open to mutual shaping by the participants. These identities are in significant part a function of shared history and propinquity and, as David Velleman (2005) points out, of family resemblance. We inherit our family history. We grow up with our parents and siblings. There is of course interpretation and direction in families, lots of it, but it tends to be guided and constrained by that same family history, family resemblances, and role relations within the family.[3]

Identity within the family, therefore, tends to be more fixed than in friendships. I am the clever one, or the sporty one, the one who can be relied upon. This is usually – where the identity is positive – a good thing. Family relationships provide a necessary anchor for my agency – and a port to which I can return – whereas friendships (especially newer ones) can provide a space for exploration, for the trying on of new identities and new characteristics.

Given the importance of these close relationships to our self-understanding, self-concept, and sense of efficacy, loss or damage to a close relationship can affect our agency and identity in multiple ways. When we lose a close relationship, we may lose a sense of who we are, of our place in the world and our capacities to realise our plans and achieve our goals. There is no guarantee, for example, that others will see me as my family member or friend does. So, if our relationship founders, my view of myself as feisty rather than aggressive, or as having a subversive sense of humour rather than being simply obscure, may be lost and my identity will be less robust in the face of less charitable interpretations or the social invisibility that a stigmatised group membership may bestow.

We argue that in severe cases of addiction the addicted person's ability to play their part in the relationships just described may be quite limited; they may lose not one relationship but many, and so they are particularly vulnerable to the threats to identity, visibility, self-esteem and self-efficacy that the loss of relationships carry and to the imposition of a new addict identity.

How does addiction do this? Severe addiction is characterised by a narrowing of interests and attention to the pursuit and consumption of the drug of choice. Marc Lewis talks about the hourglass shape of addiction:

> They start out unique: each person begins with his or her own specific culture, family environment, level of education, personality, social network, personal secrets, and all the rest. But then, when addiction takes hold, these lives start to look exactly the same. Regardless of whether it's cocaine, opiates, alcohol, or even food, that wide range of individual differences shrinks to a narrow tube – the middle of the hourglass.
>
> *Lewis 2012*

In the middle of the hourglass, the addicted person is not (for the most part) directly responsive to others' interests or to their interpretations and has little to offer in return. They are impaired for forms of intimacy that require an interest in and capacity to focus upon the other person for their own sake, to undertake joint activities and develop joint interests. And they are unfit for family relationships that require fulfilling a certain role and abiding by the norms that govern the role – playing one's part as a son or daughter, sibling or parent. They are no longer the sporty one, or the reliable one, as previous interests and pursuits are lost. Moreover, this narrowing and rigidifying of interests is exacerbated when they treat their friends and families instrumentally, as a source of money or free shelter and food, and abuse their trust. When this happens repeatedly, loved ones typically respond to the addicted person with anger, blame and resentment. Ultimately they may move to the objective stance in order to protect themselves, and the individualised interpretations of personality, character and actions previously offered in their intimate relationships may be replaced by the stereotype of the addict. We now explore in more detail this process and its effects.

The participant stance and the shaping of self-concept in addiction

As noted above, in many (but obviously not all) cases of addiction, the addicted person treats their friends and family poorly, making unreasonable demands for money, stealing from them, breaking promises, and generally acting with disregard for their interests. They thus violate the most basic interpersonal expectation of goodwill as well as the relational expectations of particularised care and concern. Friends and family initially respond from within the participant stance with resentment and blame – holding the addicted person responsible for their actions. The addicted person also often exposes his family to negative reactive attitudes when they refuse his demands.

> He would go to my mum 'shut up' and then it would get to the stage he would actually lose it, shout and bawl. If my mum went like that, 'right, just get out of the house' he would shout the whole way down the close [tenement stairwell] . . . and the whole way down the street he would be shouting 'ya fucking bitch' (Sibling, Martina).
>
> *Barnard 2007, 32*

Typically, a feedback loop of resentment develops because family members resent being the target of the addicted person's unreasonable resentment. It is also common for the addicted person to seed resentment between other family members and then subsequently be resented for having had that effect on the family.[4]

> I was very angry with them [his brothers and sisters], very resentful towards them. It works both ways because they become very resentful towards you. . . . In harming himself he was effecting the whole family . . . everyone's emotions are all shot up (Parent: Ms Garvey).
>
> *Barnard 2007, 29*

Eventually the addicted person's actions are suspected of being manipulative by default. The lack of trust severely limits the kinds of interactions the addicted person can have with their friends and family.

> It's always money, money, money. And I mean it makes me cringe when he puts his arm around me and he'll say 'I love you ma' and I push him away because I know it's not genuine . . . And it's terrible to do that with . . . But I can't help it (Parent: Mrs Blain).
>
> *Barnard 2007, 33*

When an embrace with one's mother is returned in kind, one usually becomes visible to oneself as someone who expresses love and is a worthy recipient of love. However, if the embrace is rejected, that makes one visible to oneself as someone who is not loved, someone who is manipulative and doesn't express love.[5]

As the content of the addict's self-concept becomes entrenched, the addicted person begins to draw on it to frame her experience and her plans.

> I was worried about what other people were thinking about me and you know being an addict and . . . I come from a good family and you know (. . .) I can fit in that group if I want to, but I know that I'm an addict (. . .) I think that they can see it, you know? (Nicole).[6]

Once one sees oneself as an addict, that self-concept more easily makes sense of experiences that fit with being an addict, e.g. others seeing one as an addict. Furthermore, plans that fit with being an addict make more sense so that recovery seems less achievable for someone like her.[7]

Consistent negative reactive attitudes from others also erodes belief in the effectiveness of one's agency more generally and undermines self-efficacy.

> People's perception of me is so important now because I've ruined half of it. And I don't deny that that's my doing but to have them believe in me again is so important. And I can't even look at you in the eye 'cause I wonder if you think I'm even telling you the truth. This is how I feel . . . I'm so scared that people are going to judge me that I don't know whether . . . what I'm saying is worth it or not (Diana).[8]

Diana exhibits low self-esteem (I don't know whether what I'm saying is worth it); a suspicion that she is seen as untrustworthy (I wonder if you think I'm telling the truth), which might show a lack of self-trust; and a lack of self-respect (I can't look you in the eye). Such negative self-evaluations are reinforced by the negative attitudes directed at her by others and they are particularly detrimental because they undermine the self-authority needed for recovery and agency in general (Mackenzie 2014). Diana admits a role in ruining people's perceptions of her but that doesn't mean she can repair those perceptions on her own. In fact, she recognises that she needs others to "believe in her again", to believe that she is worthy of some trust, respect, and esteem. If they do, they will help her to re-establish the self-authority she needs to direct her own recovery.

Exposure to negative reactive attitudes within the participant stance accelerates relationship difficulties as the addicted person is motivated to distance himself from negative reactive attitudes and the associated disvalued self-image.

> You start to use and you turn your back on everybody for fear of being caught. And then you just become so selfish about wanting to feel better rather than . . . and so you turn on all your morals, all your principles and values, yeah (Tom).[9]

The addicted agent often feels intense shame and guilt because, between periods of intoxication, they are well aware that their behaviour conflicts with their values and those of the people in their close relationships (Flanagan 2013). Because our close relationships make us particularly visible to ourselves, participation in these relationships will tend to elicit shame and so the addicted person is motivated to withdraw from them.

Close friends and family can often see that guilt and shame are compounding the problems the addicted person faces. This recognition can motivate a different reactive attitude – forgiveness. Forgiving involves

> ceasing to have towards the wrongdoer the retributive reactive attitudes that her wrongdoing supports, without changing our judgments about the wrongness and culpability of her action. . . . Forgiving makes sense when our concern is with giving them a chance to be different in the future.
>
> *Allais 2008, 20*

Therefore, forgiveness should help alleviate guilt and support self-forgiveness, which should help alleviate shame. Indeed, forgiveness and self-forgiveness do seem to be important for an ultimately successful recovery from addiction (Weaver, Turner and O'Dell 2000). However, it seems that, typically, forgiveness is unhelpful when the other is still in the neck of the hourglass of addiction. By forgiving before the addicted person is ready to begin to "be different in the future", friends and family set the addicted person up for failure and fresh resentment. Families may thus cycle between forgiveness and resentment, with it becoming progressively harder to forgive each time. Friends who lack the ties of obligation to the addicted person may simply end the relationship as the addicted person fails to engage in the mutual interpretation and shaping of interests and concerns that are constitutive of close friendship. In summary, close relationships become more distant as the non-addicted parties protect themselves from the destructive behaviour of the addicted person and the addicted person avoids the painful self-visibility provided by friends and family.

The addicted person's participation in their close relationships is often largely replaced by participation in relationships with drug-using acquaintances. This is partly pragmatic; these relationships help the agent secure and use drugs. However, many addicted people also report feeling more comfortable with these acquaintances.

> The only time I feel comfortable is around other drug addicts and other people who are on substance because they accept [me] no matter what and I'm a pretty hard person to tolerate at times (Julia).[10]

We would expect these interactions to be more comfortable because they generate a less confronting self-visibility. Addicted people are less likely to expose each other to negative reactive attitudes for using drugs and they might be more sympathetic, understanding, resigned, or apathetic to behaviours that draw resentment in non-addicted circles. Despite being more comfortable, however, these relationships often lack an important defining feature of close friendship: participants in these relationships are not usually open to shaping each other in ways outside of their shared interest in securing and using drugs. Indeed, people trying to cease using drugs often report negative feedback from their drug-using acquaintances.

> The other addicts aren't really . . . they don't want to see someone get on with their life 'cause then . . . oh this is what I think, then . . . it's saying to them, maybe you can do this but they don't want to . . . they're comfortable. . . .'cause I notice when I'm going well, no-one's that happy and it's like no-one wants to give you a shot when you're hanging out but when you've been clean for six months everyone wants to give you a shot (John).[11]

A shift to the objective stance

When an addicted person displays consistent disregard for herself and others despite resentment or forgiveness this appears to make her an appropriate target of the objective stance, i.e., as someone to be managed, handled, cured, trained or simply avoided. We certainly see examples of people taking the objective stance to addicted people:

> Oh I know the person that's using the drugs is not the son I had, I understand that, that's a shell of the person I knew . . . I mean all I see is an addict, I don't see my son anymore because I know that's not my son, definitely isn't my son (Parent: Mr Bell).
>
> Barnard 2007, 50

The effect of such invisibility and exclusion from social interaction on the agent's self-authority is particularly severe.

> You're a drug addict person. People look to you . . . a different way . . . they judge you, they're scared of you, lots of things. Different from . . . normal people. . . . You're really low, you're just like nobody (Hien).[12]

Indeed, as Strawson recognises, taking the objective stance for too long is incompatible with human interactions. "Being human, we cannot, in the normal case, do this for long, or altogether. If the strains of involvement, say, continue to be too great, then we have to do something else – like severing a relationship" (Strawson 2008, 10). So the objective stance too, leads to a breakdown of the relationship and certainly to a loss of the intimacy and security that close relationships provide.

Moreover, the objective stance may be based upon, and nurture, the belief that the outcomes of addiction will be determined by something that is independent of the efforts of the addicted person.[13] If addiction is seen, for example, as an illness or a disease, the progress of which is outside the voluntary control of the addicted person, then friends, family and the addicted person themselves should defer to the experts on treatment and management of the condition. Such a view of addiction can provide relief to all parties from painful blame and shame, but it comes at a cost both to the relationships and to recovery. In taking the objective attitude, the addicted person's behaviour may be viewed entirely mechanistically – as caused by forces external to their agency to which they are largely helpless onlookers. But if recovery actually requires at least some effort from the addicted person (Pickard 2012), the objective stance might compound addiction by discouraging significant others from both the belief that such efforts can be effective and/or that the addicted person is capable of making them. This leads to a sense of fatalism in their attitude towards the addicted person as they give up on the prospects of re-engagement and recovery. While they may continue to love their friend or family member they step back from more active forms of engagement.

> I'd be very surprised if any of my boys get through this, now, the two that's using. I think the best I can hope for is burying them . . . I don't think they'll get through this, you know, it's been too long now (Parent: Mr Merrick).
>
> *Barnard 2007, 50*

Similarly, addicted persons frequently end up adopting a resigned and fatalistic stance towards their lives and their addiction (Kennett 2013). Repeated experiences of failure to live up to their obligations, of other's disappointment and anger with them, and their own shame at their behaviour, results in a loss of confidence in their ability to exert control over their circumstances, to play their part, and to shape the lives they would value having and the people they would value being. They become onlookers rather than participants in their own lives, sharing others' bleak assessments of the likely outcome of their addiction. We can see evidence of addicted people internalising the objective stance in the fatalistic stories they tell about their pasts and futures. When asked where he saw himself in one year's time, one study participant said, "Probably exactly where I am now. Exactly where I am now" (Howard). Another respondent said:

> When I'm in the throes of addiction and I'm trying to stop and I can't stop, my head's going this is who you are. I can accept that you know what I mean. That's . . . as weird as that sounds I can accept that I'm a junkie. I'm . . . my life is over and this is what I'll be until I die. It's the only way I can stop is to die" (Dan).[14]

A modified stance: rebuilding relationships and resolving the dichotomy

For the addicted person, the participant and objective stances present a detrimental dichotomy that makes maintaining relationships difficult and recovery less likely. The participant stance as laid out by Strawson demands a level of responsiveness to others and responsibility in discharging obligations that is beyond that which the addicted person can consistently achieve. Conversely, the objective stance assumes the other lacks the capacities needed to play their part in inter-personal relationships *tout court*, when, for the most part, addicted people retain this capacity to a significant extent. As Jennifer Radden notes, it is cruel and unjust to treat someone as having more autonomy than they do but, equally, to treat them as having less than they do (2002, 399). The above discussion has explained how both injustices have compounding, detrimental effects.

The mistake has been to think that engagement from within the participant stance must presume and require equal responsibility and capacity of the parties or else it is somehow defective (Kennett 2007). This is not how human relationships (or human–animal relationships) work. It is possible to navigate a middle path where one participates with the other in a way that is sensitive to their present capacities for engagement and responsiveness and so keeps the door of the relationship open to both parties. There are a multitude of valued ways in which we can, and ordinarily do, engage with and shape each other that don't depend on particularly high levels of autonomy or moral responsibility. I can value your appreciation of slap-stick, your melodramatic involvement in watching your sports team, your ironic commitment to a brand of beer, your clumsiness, your warmth, your laugh, your particular facial expressions. None of these things depend strongly on the kinds of developed agential capacities that underwrite and justify the reactive attitudes of indignation and blame with which we began our discussion. We commonly adjust our responses and expectations in the light of facts about the other, their sensitivities, capacities, and so on, without abandoning the participant stance. A father, for example, chooses a slap-stick comedy rather than a political satire to watch because he is tailoring the activity to his son. He chooses something they can enjoy together and there is nothing defective about such a choice or such an interaction. Even among friends at the bar a similar thing can happen: if the conversation happens to drift into one person's area of expertise she doesn't hold her friends to the same standards as her colleagues (and her friends might only have a partial grasp of the way she is holding back).

In addiction-affected relationships, the potential for mutual shaping and responsiveness can often be improved by choosing (and limiting) times for interaction to when the addicted person is neither intoxicated or craving, and by selecting activities that will highlight the positive aspects of their character and be mutually enjoyable. Perhaps mention of drug policy will set my friend off on a rant about his mistreatment by the state and, I suspect, this shifting of responsibility and associated self-pity is unhelpful in his recovery. Better to watch a movie that appeals to our shared sense of humour or shoot a few hoops in the backyard. Families and friends can try to scaffold experiences that maintain narrative threads with the pre-addicted self, draw upon retained skills and capacities, and connect to different possible futures, even while recognising the addicted person's currently limited capacity for reciprocity. Where they do succeed in sharing a pleasant meal, a cup of coffee or half an hour shooting hoops that can be enough for that time. Each positive interaction provides positive visibility for the addicted person. In that moment they can see themselves as holding up their end of the relationship, as valued, and as loved.

Despite efforts to tailor the scope of the interaction appropriately, the addicted person might still sometimes act in blameworthy ways, particularly when they are in the neck of the hourglass of addiction. Furthermore, even when they have emerged from this phase and are seeking paths to recovery there is likely to be a history of blameworthy events that haven't yet been fully resolved. In response, the non-addicted person can be tempted back towards either pole of the dichotomy. The objective stance would relieve them from the effort of resenting and may justify (or rationalise) withdrawal from the relationship. The participant stance appeals because it defends one's own value as someone who deserves goodwill. It also communicates the hope that the other could be better and, outside of addiction, this pressure often works. Unfortunately, for those suffering from addiction it often asks too much of them too soon.

In the face of blameworthy behaviour, perhaps we can modify the participant stance so that we communicate the breach of our expectations but without the negative effects of the usual reactive attitudes surrounding blame. This seems to be the approach that Hanna Pickard (2017) suggests when she distinguishes affective blame from our practices of judging and holding others responsible for their actions. Affective blame is the normal response in the participant stance and involves resentment. In practices of responsibility without blame, however, resentment is suspended, but the addicted person is still held to be accountable for their actions and explicitly seen as capable of doing better. This, Pickard argues, is essential to recovery since the addict must also see themselves as the agent of their own recovery.

Conclusion

The dynamic between an addicted person and their friends and family is often an important contributor to the chronicity of addiction. Friends and family take the participant stance by default and expose the addicted person to consistent resentment. As a result, the addicted person suffers from guilt and shame and becomes visible to himself as a drug-user who deliberately treats others with ill will. As he internalises this self-concept, although he maintains his status as a person capable of moral responsibility, he sees himself as a bad person who is undeserving of a better life and for whom continued drug-use makes more sense than recovery.

The addicted person's consistent irresponsiveness (if not insensitivity) to reactive blame and resentment may then lead friends and family to take the objective stance, treating the addicted person as a problem to be managed. This undermines the addicted person's status as a moral equal and encourages him to develop a fatalistic self-concept at odds with any exercise of agency including that required for recovery. Both participant and objective stances encourage a mutual withdrawal from close relationships, which tends to leave the addicted person with just their drug-using relationships. This unbalanced social environment provides few opportunities to develop and maintain the variety of non-drug-using identities and capacities necessary for recovery.

However, we are neither forced nor normatively required to choose between the full participant stance or the objective stance in our dealings with addicted people. Not all valuable intimate relations need to presuppose full autonomous agency and we have argued that it is possible to establish or maintain areas of genuine reciprocity with persons whose agency and social responsiveness is underdeveloped, impaired, or intermittent. This is a worthwhile endeavour for families and friends of addicted persons that supports the addicted person's agency and self-concept while providing some protection from the worst impacts of a loved one's addiction.

Notes

1 There is a considerable debate about the order of explanation in Strawson (see Todd 2016). Is responsibility constituted by our actual practices of praise and blame or are these practices appropriately constrained by features of the agent themselves? We assume the latter for the purposes of this chapter.
2 Of course, the interpretations provided by friends and family are not always more accurate than that of others. Friends and family can develop set views of a person rooted in their shared history so that their interpretations are less responsive to changes in the agent. Similarly, friends and family may be more prone to collude with a person's false self-image because they want to support her. Those less involved with the agent will be less affected by these sources of bias; however, the relatively narrow, less intimate nature of their interactions with the agent prevents their interpretations from having a particularly pervasive influence on the agent's self-concept.
3 Long-running friendships also exhibit these characteristics to some extent because, here too, the shared history constrains how the other is prepared to see us. There may not be, therefore, such a sharp distinction between family relationships and longer-running friendships. However, familial relationships involve the longest shared histories (often from birth) and are influenced by family traditions and expectations that predate one's birth. So, identities within the family are typically less malleable than in even our longest friendships.
4 The addicted person may also be resented for harming the family's reputation. The family becomes the target of stigma. "It is [shameful] … because some of the times she's went in and she's stole other people's kids toys and they come up and say 'She stole out the house.' And you go out and you feel it, you see the heads going and maybe you'll walk up and they're going on 'Aw see these f'ing junkies. The bane of our life.' (Parent: Ms Nugent)" (Barnard 2007, 30).
5 It might even suggest a self-concept of being unlovable and incapable of expressing love. That, more serious, psychological move is indicative of the objective stance as we explain below.
6 From interviews conducted by Anke Snoek for the Australian Research Council funded Project: Addiction and Moral Identity.
7 For a detailed discussion of this phenomenon see McConnell (2016).
8 From interviews conducted by Anke Snoek for the Australian Research Council funded Project: Addiction and Moral Identity.
9 From interviews conducted by Anke Snoek for the Australian Research Council funded Project: Addiction and Moral Identity.
10 From interviews conducted by Anke Snoek for the Australian Research Council funded Project: Addiction and Moral Identity.
11 From interviews conducted by Anke Snoek for the Australian Research Council funded Project: Addiction and Moral Identity.
12 From interviews conducted by Anke Snoek for the Australian Research Council funded Project: Addiction and Moral Identity.
13 In most cases, it is clear that the addicted person desires to recover and so his failure to consistently respond to the reactive attitudes may be seen by him and others as evidence that he lacks the agential capacity, not that he refuses to use that capacity.
14 From interviews conducted by Anke Snoek for the Australian Research Council funded Project: Addiction and Moral Identity.

References

Allais, L. (2008) "Dissolving reactive attitudes: forgiving and understanding", *South African Journal of Philosophy* 27(3): 179–200.
Barnard, M. (2007) *Drug Addiction and Families*, London: Jessica Kingsley Publishers.
Branden, N. (1993) "Love and psychological visibility", in Badhwar, N. K. (ed.), *Friendship, A Psychological Visibility*, Ithaca, NY: Cornell University Press, pp. 65–72.
Cocking, D. and Kennett, J. (1998) "Friendship and the self", *Ethics* 108(3): 502–527.
Flanagan, O. (2013) "The shame of addiction", *Frontiers in Psychiatry* 4: 120–131.
Kennett, J. (2007) "Mental disorder, moral agency and the self", in B. Steinbok (ed.), *The Oxford Handbook of Bioethics*, Oxford, UK: Oxford University Press, pp. 90–113.
Kennett, J. (2013) "'Just say no?' Addiction and the elements of self-control", in N. Levy (ed.), *Addiction and Self-Control*, Oxford, UK: Oxford University Press.

Lewis, M. (2012) "The hourglass shape of addiction and recovery", in *Understanding Addiction: A New Perspective Linking Brain, Behaviour, and Biography*, available at: www.memoirsofanaddictedbrain.com/connect/the-hourglass-shape-of-addiction-and-recovery-2/ (accessed 26 February 2018).

Mackenzie, C. (2014) "Three dimensions of autonomy: a relational analysis", in *Autonomy, Oppression, and Gender*, New York, NY: Oxford University Press, pp. 15–41.

McConnell, D. (2016) "Narrative self-constitution and recovery from addiction", *American Philosophical Quarterly* 53(3): 307–322.

Pickard, H. (2012) "The purpose in chronic addiction", *AJOB Neuroscience* 3(2): 40–49.

Pickard, H. (2017) "Responsibility without blame for addiction", *Neuroethics* 10: 169–180.

Radden, J. (2002) "Psychiatric ethics", *Bioethics* 16(5): 397–411.

Strawson, P. (2008) *Freedom and Resentment and Other Essays*, London: Routledge.

Todd, P. (2016) "Strawson, moral responsibility, and the 'order of explanation': an intervention", *Ethics* 127(1): 208–240.

Velleman, D. J. (2005) "Family history", *Philosophical Papers* 34(3): 357–378.

Weaver, G. D., Turner, N. H. and O'Dell, K. J. (2000) "Depressive symptoms, stress, and coping among women recovering from addiction", *Journal of Substance Abuse Treatment* 18(2): 161–167.

SECTION B

Prevention, treatment, and spontaneous recovery

36
CONTINGENCY MANAGEMENT APPROACHES

Kristyn Zajac, Sheila M. Alessi, and Nancy M. Petry

Overview

Contingency management (CM) is an evidence-based addictions treatment that provides tangible reinforcers for engagement in positive behaviors (e.g., abstinence from substance use, attending treatment). The primary goal of CM is to reward behaviors that are inconsistent with continued substance use. This technique is based on the principles of operant conditioning or the idea that an individual is more likely to repeat a behavior that is rewarded. Contingency management has been successfully applied to a range of substance use disorders in a variety of settings, as described in detail below.

This chapter provides a brief overview of the principles of effective CM and the types of reinforcers that have been shown to be effective in treating addictions. It also describes the current state of the science, summarizing results of key studies evaluating CM. Because CM has not been widely studied with other addictive behaviors (e.g., gambling, food addiction), the primary focus is drug and alcohol addiction. Finally, issues related to implementation of CM in real-world settings are discussed.

Principles of effective CM

Contingency management for substance use disorders typically involves providing a reinforcer for abstinence from substance use or for attending treatment sessions. Reinforcers are typically either vouchers that can be exchanged for goods or services or draws from a prize bowl that represent chances to win prizes. The pros and cons of these two options are discussed below. If the patient attends treatment (in CM for attendance) or provides a negative drug screen (in CM for abstinence), he or she earns the reinforcer. If the patient misses treatment or provides a drug screen indicating substance use, he or she does not earn the reinforcer and may receive a mild punisher, such as a lower value reinforcer the next time attendance or abstinence occurs.

Each of the principles described below are empirically related to better patient outcomes. Thus, adherence to these principles in designing a CM program for a given treatment center or patient is key to realizing positive effects on substance use. For a detailed step-by-step guide to CM implementation, see Petry (2012).

First, *the target behavior must be objectively defined*. The behavior must be feasible to assess in a clinical setting and verifiable by the clinic staff. If the CM plan is designed to reinforce abstinence, biological measures, such as urine drug screens or breathalyzers for alcohol, should be used. It is also important to take into account the testing windows of biological measures. For example, urine tests for cocaine or opioids generally need to be administered two or three times per week at least 48 hours apart, to ensure no use occurred during the week. Breath tests for smoking or alcohol typically require multiple tests daily due to shorter windows of detection. Another easily measured behavior is attendance. Other behaviors can be targets for CM; however, a common pitfall in designing CM plans is choosing behaviors that can be difficult to track and verify (e.g., use of cognitive behavioral skills between sessions) and, therefore, are not ideal for reinforcement.

Second, the target behavior must be measured and reinforced with *high frequency*. An ideal CM plan would withhold reward following every instance of substance use and, conversely, provide frequent access to reward when abstinence is achieved and maintained. Tests must be strategically scheduled so that carryover effects of substance use from one test to the next are minimized and reinforcement opportunities are maximized. Thus, most substance use must be assessed at least twice per week and preferably three times per week in the early stages of treatment (Cone and Dickerson 1992; Saxon *et al.* 1998). In addition to ensuring higher accuracy in reinforcing actual abstinence throughout the week, high frequency measurement and reinforcement, especially early in treatment, allows patients to learn the relationship between abstinence, the drug screen result, and the reinforcement.

Third, *immediacy*, or the reinforcement of a behavior as soon as possible after it occurs, is important to effective CM. In an ideal scenario, a decision to refrain from substance use would be reinforced right after it occurred in the real world, as immediate reinforcement is the most powerful motivator in learning new behaviors (Zeiler 1977) and is more effective than delayed reinforcement (Roll *et al.* 2000). Although this degree of immediacy is typically not possible, steps can be taken to reduce the time between abstinence and reward. These include frequent testing, as noted above, and use of instant office-based urine testing kits rather than sending urine samples out to the lab for testing, which can take several days.

In general, *higher magnitude* rewards increase treatment response rates (Lussier *et al.* 2006), and rewards must be of sufficient magnitude to compete with substance use. At the same time, the costs of providing a CM program must be reasonable and sustainable for treatment facilities. When using voucher-based CM (i.e., patients earn vouchers that can be exchanged for desired goods for each negative urine screen), patients can typically earn up to $1,200 in rewards, with average earnings around $600 over 12 weeks (Higgins *et al.* 1994; Silverman *et al.* 1996). Reducing the value of vouchers is not recommended, as this practice decreases CM's effectiveness (e.g., Higgins *et al.* 2007). Prize-based CM is an alternative to voucher-based CM, in which patients earn chances to win prizes for submitting negative drug screens. Because patients do not earn prizes for every negative test, costs can be lower than in voucher-based CM, and studies have found that these two approaches have similar efficacy (e.g., Petry *et al.* 2005c).

Another key feature of effective CM is *escalating rewards* for sustained abstinence. For example, a patient may win one chance for a prize the first time they provide a negative screen and then two for the second consecutive screen, three for the third consecutive screen, et cetera. When implementing escalating rewards it is recommended to have a "reset" back to the lowest value when patients provide a positive screen or fail to submit a scheduled sample (Holtyn and Silverman 2016). This serves as a mild punisher and increases the incentive to maintain abstinence for extended periods. Although escalating rewards add cost to CM programs, they also improve abstinence duration (Roll *et al.* 1996), which predicts long-term abstinence (e.g., Higgins *et al.* 2000a).

Types of reinforcers

A number of different types of reinforcers have been found to be effective with CM. As a general rule, reinforcers should be motivating to individual patients, be able to compete with the rewards associated with substance use, and be practical to provide in clinical settings. Here, we briefly review the pros and cons of a variety of potential reinforcers.

Vouchers have been widely used in CM studies (e.g., Higgins *et al.* 1994, 2000b; Silverman *et al.* 1996). As noted above, in this type of CM, patients earn vouchers for every negative drug screen, and these vouchers can be exchanged for useful or desirable items, such as gift cards to restaurants, bus tokens, or electronic equipment. The value of the vouchers escalates with consecutive negative screens. The advantages of vouchers are: 1) the items can be individualized to be appealing to each patient; 2) cash is not provided, which reduces the possibility of exchanging rewards for drugs; and 3) programs can refuse to provide requested rewards that are not consistent with a drug-free lifestyle (e.g., gift cards to restaurants/stores that sell alcohol, drug paraphernalia, cigarettes, or weapons). However, a disadvantage of typical voucher programs may be the high per-patient cost for incentives, although high-cost programs may be needed and justified in some cases (Silverman *et al.* 2002). Additional costs include staff time in purchasing rewards requested by patients, which adds considerably to the overall program cost. Thus, many clinics may find voucher-based CM difficult to finance.

Cash incentives can be used in the same manner as vouchers by providing the cash equivalent of vouchers and escalating the value with consecutive negative screens. In fact, cash incentives can reduce the overall cost of CM programs, as it eliminates the need for staff time to shop for individualized rewards. In addition, patients generally prefer cash over vouchers of the same value, which may make it a stronger motivator. However, there are some arguments against using cash incentives, including the concern that patients will use their rewards to purchase drugs or alcohol. These fears have been unfounded in the research on cash-based CM. Festinger *et al.* (2005) randomly assigned patients receiving outpatient drug treatment to receive either cash or vouchers in the amount of $10, $40, or $70 for completing a follow-up assessment. When patients were assessed again three days later, there were no effects of payment type or amount on new substance use. In a subsequent study, participants were randomized to either 12 weeks of voucher or cash CM for cocaine-negative urine screens or a non-CM control condition (Festinger *et al.* 2014). Cash and voucher CM both increased longest duration of abstinence compared with the control group, and cash CM was not associated with risk of harm, including increased craving, gambling, or drinking. In addition, the frequent monitoring and reinforcement of abstinence that is part of CM diminishes the risk of negative outcomes related to providing cash.

Prize CM can be less costly than voucher-based CM. Rather than earning a reward for each negative screen, patients earn the chance to draw cards from a prize bowl. A standard bowl contains 500 cards; 250 cards are labelled with a supportive statement (e.g., "good job!"), 210 are labelled 'small prize' and can be exchanged for a prize worth about $1 (e.g., toiletries, snacks), 39 are labelled 'large prize' and can be exchanged for a $20 prize (e.g., gift cards), and 1 card is labelled 'jumbo prize' and can be exchanged for a prize worth $100 (e.g., DVD or mp3 player). The number of draws escalates with each consecutive negative screen. The value of rewards for prize-based CM is generally much lower (maximum total expected value of about $400 for a 12-week CM program) compared with voucher-based CM (maximum values of $1,000–$1,200). Prize cabinets are kept onsite, which allows for the patient to receive incentives immediately and reduces staff time for shopping. Studies have demonstrated that prize and voucher CM have similar efficacy when the amount earned is comparable (e.g., Petry *et al.* 2005c), and prize CM can be more cost-effective than voucher CM (Olmstead and Petry 2009).

Research evidence

There have been decades of research supporting the efficacy of CM for both abstinence and treatment attendance across a range of substance use disorders. Randomized controlled trials have found positive results of CM for treating individuals with addictions to alcohol, opioids, benzodiazepines, marijuana, cocaine, nicotine, methamphetamine, as well as individuals with polysubstance abuse (see Prendergast *et al.* 2006 and Lussier *et al.* 2006 for reviews). Contingency management has also been shown to have positive effects on other important patient outcomes, including quality of life, HIV risk behaviors, and psychiatric symptoms (Hanson *et al.* 2008; Petry *et al.* 2007, 2013). Although a comprehensive review of the CM literature is beyond the scope of this chapter, a few key studies are summarized.

The largest clinical trial of CM to date randomized over 400 patients with stimulant use disorders to usual care or usual care plus CM (Petry *et al.* 2005b). The trial recruited from six outpatient community-based treatment centers across the US. The CM sessions occurred twice weekly for 12 weeks. Patients assigned to the CM group earned draws from the prize bowl for stimulant-negative screens, whereas those in the usual services group simply dropped off urine samples twice weekly. The CM patients achieved longer durations of abstinence (4.4 vs 2.6 weeks, respectively) and stayed in treatment longer than those in usual care. Another study with the same design recruited 338 methadone-maintained patients with stimulant use disorders (Peirce *et al.* 2006). Patients in the CM condition were twice as likely as those in usual care to submit stimulant-negative samples and achieved longer durations of sustained abstinence on average.

Contingency management has also demonstrated consistent efficacy in improving treatment attendance, an important outcome due to high rates of missed sessions and dropout among patients with addictions. Contingency management for attendance allows patients to earn reinforcement for simply coming to treatment sessions, and reinforcement escalates for sustained periods of attendance. This approach can be readily implemented in either individual or group settings. In one study, methadone-maintained patients who abused cocaine were randomly assigned to standard treatment (which included group counseling) or standard treatment plus CM (Petry *et al.* 2005a). Patients randomized to CM could earn draws for attending group sessions and submitting negative screens for cocaine. Compared with controls, patients in the CM condition submitted a greater proportion of negative cocaine samples (34.6% versus 16.8%) and attended more treatment sessions (6.6 weeks versus 3.0 weeks). Another study randomly assigned patients with substance use disorders to standard care plus frequent urine screenings or the same care plus CM delivered in a group setting (Petry *et al.* 2011). The CM patients earned chances to put their names in a hat by submitting negative screens and attending group. Therapists then drew names from the hat and awarded those patients chances to win prizes ranging from $1 to $100. Compared with controls, patients in the CM group attended more days of treatment, stayed in treatment for more weeks, and had longer durations of abstinence.

Contingency management in real-world settings

As discussed above, there is strong empirical evidence for the efficacy of CM from well-designed randomized controlled trials where treatment is delivered with a high level of fidelity. There is also growing interest in the translation of CM from research to practice through large-scale dissemination efforts. One of the largest of these efforts has been accomplished by the Veteran's Administration, which initiated a nationwide rollout of CM in their substance abuse treatment programs in 2011, resulting in successful implementation of CM in nearly 100 clinics by 2014 (Petry *et al.* 2014).

A recent study evaluated the efficacy of CM when delivered entirely by community-based clinicians, rather than research staff (Petry *et al.* 2012). Twenty three clinicians from three methadone maintenance clinics were trained on CM using a didactics seminar and a period of supervision during which clinicians delivered CM to pilot participants and received feedback from CM experts. Therapists were readily trained to high levels of CM treatment fidelity. There were some observed mistakes in the treatment delivery early on, but these improved quickly with feedback and weekly supervision. Following the training stage, 130 patients were randomized to CM with one of the community therapists or to treatment as usual. Patients in the CM condition earned the opportunity to win prizes by submitting negative screens for cocaine and alcohol. Compared with the control group, CM patients stayed in treatment longer, achieved longer durations of abstinence, and submitted a higher percentage of negative urine drug screens. Thus, results indicate that community-based therapists can deliver CM with positive effects on patient outcomes.

Although this research indicates that CM can be delivered in the community with a high level of effectiveness, there are important considerations and potential barriers that need attention when implementing CM programs. The first is related to *funding* of the CM programs. As noted earlier, there are several costs associated with CM in addition to the costs of prizes. These include urine drug screens, administrative time for shopping for prizes, and costs related to training and supervision of clinicians. It is not recommended that agencies implement a version of CM that significantly diminishes the value of the prizes, as this will also reduce the effectiveness of the program. However, choosing a prize-based CM program may help to reduce costs. It may also be possible to solicit community donations for gift cards and goods that can serve as prizes. Available data on the cost-effectiveness of CM is favorable, particularly in light of the costs associated with continued substance abuse (Olmstead and Petry 2009). Ultimately, treatment communities, insurers, political forces, treatment consumers, and communities will determine the acceptability of the cost–benefit ratio of CM.

A second consideration for CM implementation is the provision of sufficient *training and supervision of clinicians*. As highlighted in the CM implementation studies as well as the larger literature on evidence-based treatments, adherence to the treatment model is key to producing favorable patient outcomes. Further, research suggests that counselors rarely implement CM as intended (42% of the time) when only minimal feedback was provided (Andrzejewski *et al.* 2000). In that study, CM adherence was greatly improved by either simply providing feedback to therapists about whether or not they met adherence criteria in delivering CM (71% adherence to CM protocol) or allowing them to earn drawings for cash prizes if they met performance criteria (81% adherence). These results underscore the importance of ongoing quality assurance. Thus, as with any evidence-based approach, treatment settings must be committed to investing the time and resources necessary to ensure that clinicians have the training and oversight to deliver CM with high fidelity.

Another potential barrier to implementation is a *mismatch between the way CM is most commonly delivered in research studies* (individual sessions, focus on single substances) *and the way in which many substance abuse treatment clinics operate* (a preponderance of group-based sessions, targets polysubstance abuse). As discussed earlier in this chapter, CM can be delivered effectively in group settings with some minor adaptations (e.g., Petry *et al.* 2005a). Contingency management provides some flexibility when it comes to treating polysubstance abuse as well. In treatment settings where complete abstinence is the focus, vouchers or chances to win prizes can be made contingent upon provision of negative urine drugs screens for multiple drugs and can be expanded to include negative breathalyzers. In settings that offer more flexibility, CM is most effective when it targets abstinence for a single drug at a time, because complete abstinence is too difficult to achieve for some and would result in no opportunities for reinforcement. In this case, the CM plan can be modified to make rewards contingent upon abstinence from the primary drug of choice, with bonuses awarded for negative screens for other substances.

Finally, *political and ideological barriers* to CM implementation may be the most challenging to overcome, and these appear to persist despite overwhelming evidence of CM's efficacy. One study surveyed 383 counselors, supervisors, and other clinic staff members in five states on their opinions and perceptions of CM (Kirby *et al.* 2006). Although 77% of those surveyed were open to including "social incentives" (i.e., praise) in their treatment approach, far fewer (54%), although still the majority, responded favorably towards tangible incentives. The most common barrier identified was funding, but over half of participants felt that tangible incentives did not address the underlying causes of addiction, over a quarter viewed incentives as "bribes," and a little less than a quarter believed that incentives would prevent the development of internal motivation for abstinence. However, there is no evidence that CM has a negative effect on motivation compared with standard care (e.g., Ledgerwood and Petry 2006). Another common objection to CM is that addictive behaviors will return more quickly or to a greater extent when the incentive period ends. This, too, has not been supported by research, and increased abstinence following CM compared with standard care has been observed (e.g., Alessi *et al.* 2007; Higgins *et al.* 2000b; Petry and Martin 2002; Petry *et al.* 2005a). Interestingly, concerns about enduring treatment effects are typically not as strongly expressed in discussions about other interventions for addictions, including pharmacotherapies, or for behavioral treatments for other chronic illnesses (e.g., diabetes).

An overarching attitude expressed not only by clinicians but by communities and society-at-large is the idea that we should not be paying patients for "what they should be doing already" (i.e., abstaining from drugs and alcohol). This common assertion shows an underlying bias against those suffering from addictions, suggesting that they do not deserve to receive a treatment that has repeatedly been shown to be effective by rigorous scientific research. This point is further highlighted by the lack of protest against similar treatments based on CM principles that are used to treat a wide range of other behavioral issues, such as intellectual disabilities, autism, and disruptive behaviors (Matson and Boisjoli 2009; McCart and Sheidow 2016).

One potential approach to negative reactions to CM is to provide more in-depth education about the purpose and proposed underlying mechanisms of CM. Several researchers have proposed that the effectiveness of CM relies on humans' decision-making skills (e.g., Regier and Redish 2015). In other words, CM provides alternative rewards to drug use that help individuals struggling with addictions to engage in more deliberate decision-making processes. Individuals with addictions, particularly those seeking treatment, can typically recognize the long-term benefits of abstinence. However, these benefits are abstract and far in the future, making it difficult for them to compete with the immediate benefits of drug use. Contingency management provides immediate and concrete rewards, which can increase an individual's ability to use deliberative decision making to choose non-drug rewards.

This perspective on underlying mechanisms provides a potential way to present CM to appeal to community practitioners by, for example, addressing concerns that CM provides "bribes" to patients, is coercive, or does not help the patient to develop internal motivation. Instead, CM can be presented as: 1) a way to help the patient to practice and strengthen the skill of deliberate decision making about substance use with the aid of a CM plan; and 2) a way to help the patient to set short-term goals (e.g., providing a negative drug screen at their next session) that can be monitored and rewarded with the help of the clinician. In other words, instead of a system for external reward, CM could be reframed as a skills-based intervention. Both practitioners and society may find this reframing to be more palatable and in line with their own thinking about addictive behaviors.

Despite an abundance of evidence for the efficacy of CM and the effectiveness of its application in real-world settings, it has yet to be fully disseminated into community-based practice (McGoven et al. 2004). Some of this can be accounted for by true barriers to implementation, and suggestions provided in this chapter can serve as a starting point for overcoming such barriers. However, larger societal and political biases against CM principles will need to be addressed in order for widespread adoption of this effective treatment to occur.

Summary

Contingency management is an evidence-based approach to addiction treatment that has been tested in numerous rigorous studies with a variety of populations. Effective implementation of CM requires careful attention to treatment fidelity and adherence to its core behavioral principles. At the same time, various adaptations of CM have been tested to create a range of options for clinicians who want to adapt CM for their real-world setting. Despite these efforts, CM is not widely offered to the vast majority of patients who could benefit from it, likely due to a range of logistical and ideological barriers. Further, many clinicians who do offer CM are not providing optimal CM procedures, due to lack of full understanding of the treatment and financial barriers. This disconnect is unfortunate given CM's potential to help individuals with substance use disorders to achieve and maintain abstinence and engage in substance use treatment. Future efforts will be needed to overcome barriers to dissemination of CM, including educational efforts aimed at clinicians and policy makers.

Acknowledgments

This publication was supported by the National Institutes of Health through Grant Numbers K23DA034879, R01DA013444, R01AA021446, R01AA023502, R01DA027615, DP3DK097705, R01HD075630, P50DA009241, P60AA003510.

References

Alessi, S. M., Hanson, T., Wieners, M. and Petry, N. M. (2007) "Low-cost contingency management in community clinics: delivering incentives partially in group therapy", *Experimental and Clinical Psychopharmacology* 15(3): 293–300.

Andrzejewski, M. E., Kirby, K. C., Morral, A. R. and Iguchi, M. Y. (2000) "Technology transfer through performance management: the effects of graphical feedback and positive reinforcement on drug treatment counselors' behavior", *Drug and Alcohol Dependence* 63: 179–186.

Cone, E. J. and Dickerson, S. L. (1992) "Efficacy of urinalysis in monitoring heroin and cocaine abuse patterns: implications in clinical trials for treatment of drug dependence", in R. B. Jain (ed.), *Statistical Issues in Clinical Trials for Treatment of Opioid Dependence*, Washington, DC: US Government Printing Office, pp. 46–58.

Festinger, D. S., Marlow, D. B., Croft, J. R., Dugosh, K. L., Mastro, N. K., Lee, P. A. and Patapis, N. S. (2005) "Do research payments precipitate drug use or coerce participation?" *Drug and Alcohol Dependence* 78(3): 275–281.

Festinger, D. S., Dugosh, K. L., Kirby, K. C. and Seymour, B. L. (2014) "Contingency management for cocaine treatment: cash vs. vouchers", *Journal of Substance Abuse Treatment* 47(2): 168–174.

Hanson, T., Alessi, S. M. and Petry, N. M. (2008) "Contingency management reduces drug-related human immunodeficiency virus risk behaviors in cocaine-abusing methadone patients", *Addiction* 103(7): 1187–1197.

Higgins, S. T., Budney, A. J., Bickel, W. K., Foerg, F. E., Donham, R. and Badger, G. J. (1994) "Incentives improve outcome in outpatient behavioral treatment of cocaine dependence", *Archives of General Psychiatry* 51(7): 568–576.

Higgins, S. T., Badger, G. J. and Budney, A. J. (2000a) "Initial abstinence and success in achieving longer term cocaine abstinence", *Experimental and Clinical Psychopharmacology* 8(3): 377–386.

Higgins, S. T., Wong, C. J., Badger, G. J., Ogden, D. E. H. and Dantona, R. L. (2000b) "Contingent reinforcement increases cocaine abstinence during outpatient treatment and 1 year of follow-up", *Journal of Consulting and Clinical Psychology* 68(1): 64–72.

Higgins, S. T., Heil, S. H., Dantona, R., Donham, R., Matthews, M. and Badger, G. J. (2007) "Effects of varying the monetary value of voucher-based incentives on abstinence achieved during and following treatment among cocaine-depended outpatients", *Addiction* 102(2): 271–281.

Holtyn, A. F. and Silverman, K. (2016) "Effects of pay resets following drug use on attendance and hours worked in a therapeutic workplace", *Journal of Applied Behavioral Analysis* 49(2): 377–382.

Kirby, K. C., Benishek, L. A., Dugosh, K. L. and Kerwin, M. E. (2006) "Substance abuse providers' beliefs and objections regarding contingency management: implications for dissemination", *Drug and Alcohol Dependence* 85(1): 19–27.

Ledgerwood, D. M. and Petry, N. M. (2006) "Does contingency management affect motivation to change substance use?" *Drug and Alcohol Dependence* 83(1): 65–72.

Lussier, J. P., Heil, S. H., Mongeon, J. A. and Badger, G. J. (2006) "A meta-analysis of voucher-based reinforcement therapy for substance use disorders", *Addiction* 101(2): 192–203.

Matson, J. L. and Boisjoli, J. A. (2009) "The token economy for children with intellectual disability and/or autism: a review", *Research in Developmental Disabilities* 30(2): 240–248.

McCart, M. R. and Sheidow, A. J. (2016) "Evidence-based psychosocial treatments for adolescents with disruptive behavior", *Journal of Clinical Child and Adolescent Psychology* 45(5), 529–563.

McGovern, M. P., Fox, T. S., Xie, H. and Drake, R. E. (2004) "A survey of clinical practices and readiness to adopt evidence-based practices: dissemination research in an addiction treatment system", *Journal of Substance Abuse Treatment* 26(4): 305–312.

Olmstead, T. A. and Petry, N. M. (2009) "The cost-effectiveness of prize-based and voucher-based contingency management in a population of cocaine- or opioid-dependent patients", *Drug and Alcohol Dependence* 102(1–3): 108–115.

Peirce, J. M., Petry, N. M., Stitzer, M. L., Blaine, J., Kellogg, S., Satterfield, F. and Li, R. (2006) "Effects of lower-cost incentives on stimulant abstinence in methadone maintenance treatment: a National Drug Abuse Treatment Clinical Trials Network study", *Archives of General Psychiatry* 63(2): 201–208.

Petry, N. M. (2012) *Contingency Management for Substance Abuse Treatment: A Guide to Implementing Evidence-Based Practice*, New York, NY: Routledge/Taylor and Francis Group.

Petry, N. M. and Martin, B. (2002) "Lower-cost contingency management for treating cocaine-abusing methadone patients", *Journal of Consulting and Clinical Psychology* 70: 398–405.

Petry, N. M., Martin, B. and Simcic, F. (2005a) "Prize reinforcement contingency management for cocaine dependence: integration with group therapy in a methadone clinic", *Journal of Consulting and Clinical Psychology* 73(2): 354–359.

Petry, N. M., Peirce, J. M., Stitzer, M. L., Blaine, J., Roll, J. M., Cohen, A. and Kirby, K. C. (2005b) "Effect of prize-based incentives on outcomes in stimulant abusers in outpatient psychosocial treatment programs: a National Drug Abuse Treatment Clinical Trials Network study", *Archives of General Psychiatry* 62(10): 1148–1156.

Petry, N. M., Alessi, S. M., Marx, J., Austin, M. and Tardif, M. (2005c) "Vouchers versus prizes: contingency management treatment of substance abusers in community settings", *Journal of Consulting and Clinical Psychology* 73: 1005–1014.

Petry, N. M., Alessi, S. M. and Hanson, T. (2007). "Contingency management improves abstinence and quality of life in substance abusers", *Journal of Consulting and Clinical Psychology* 75: 307–315.

Petry, N. M., Weinstock, J. and Alessi, S. M. (2011) "A randomized trial of contingency management delivered in the context of group counseling", *Journal of Consulting and Clinical Psychology* 79(5): 686–696.

Petry, N. M., Alessi, S. M. and Ledgerwood, D. M. (2012) "A randomized trial of contingency management delivered by community therapists", *Journal of Consulting and Clinical Psychology* 80(2): 286–298.

Petry, N. M., Alessi, S. M. and Rash, C. J. (2013) "Contingency management treatments decrease psychiatric symptoms", *Journal of Consulting and Clinical Psychology* 81(5): 926–931.

Petry, N. M., DePhilippis, D., Rash, C. J., Drapkin, M. and McKay, J. R. (2014) "Nationwide dissemination of contingency management: the Veterans Administration initiative", *American Journal on Addictions* 23(3): 205–210.

Prendergast, M., Podus, D., Finney, J., Greenwell, L. and Roll, J. (2006) "Contingency management for treatment of substance use disorders: a meta-analysis", *Addiction* 101(11): 1546–1560.

Regier, P. S. and Redish, A. D. (2015) "Contingency management and deliberative decision-making processes", *Frontiers in Psychiatry* 6: 1–13.

Roll, J. M., Higgins, S. T. and Badger, G. J. (1996) "An experimental comparison of three different schedules of reinforcement of drug abstinence using cigarette smoking as an exemplar", *Journal of Applied Behavior Analysis* 29(4): 495–505.

Roll, J. M., Reilly, M. P. and Johanson, C. E. (2000) "The influence of exchange delays on cigarette versus money choice: a laboratory analog of voucher-based reinforcement therapy", *Experimental and Clinical Psychopharmacology* 8(3): 366–370.

Saxon, A. J., Calsyn, D. A., Wells, E. A. and Stanton, V. V. (1998) "The use of urine toxicology to enhance patient control of take-home doses in methadone maintenance: effects on reducing illicit drug use", *Addiction Research* 6(3): 203–214.

Silverman, K., Wong, C. J., Higgins, S. T., Brooner, R. K., Montoya, I. D., Contoreggi, C. and Preston, K. L. (1996) "Increasing opiate abstinence through voucher-based reinforcement therapy", *Drug and Alcohol Dependence* 41(2): 157–165.

Silverman, K., Svikis, D., Wong, C. J., Hampton, J., Stitzer, M. L. and Bigelow, G. E. (2002) "A reinforcement-based therapeutic workplace for the treatment of drug abuse: three-year abstinence outcomes", *Experimental and Clinical Psychopharmacology* 10(3): 228–240.

Zeiler, M. D. (1977) "Elimination of reinforced behavior: intermittent schedules of not-responding", *Journal of the Experimental Analysis of Behavior* 25: 23–32.

37
TWELVE-STEP FELLOWSHIP AND RECOVERY FROM ADDICTION

John F. Kelly and Julie V. Cristello

Introduction

Alcohol and other substance use disorders (SUD) are highly prevalent psychiatric conditions in most developed and many developing nations, conferring a prodigious burden of disease, disability, and premature mortality (World Health Organization 2014). During the past 50 years, increased substance-related harms have been accompanied by developments in clinical science and effective psychosocial and pharmacological treatments. While the advancements in professional care have been immensely valuable, the feasibility of managing these chronic disorders on a purely professional basis has been economically challenging. Professional services are typically only available during weekday business hours and delivered in an acute care format ending after a few weeks or months. The cost, rigidity, and time-limited nature of professional services, together with growing awareness of the long-term vulnerability to relapse for individuals suffering from SUD, has seen subsequent expansion of freely available community-based twelve-step mutual-help organizations (TSMHOs), such as Alcoholics Anonymous (AA) and Narcotics Anonymous (NA) and many others, during the past fifty years (Kelly and White 2012). These organizations can serve as flexible long-term addiction recovery management resources and, from an intervention dissemination and impact perspective (e.g., Glasgow *et al.* 2003), have a wide reach available in most communities, appear to be readily adopted and implemented, and have been shown to have evident staying power existing and growing for more than 80 years (Kelly and Yeterian 2014).

The TSMHOs have proven to be popular ubiquitous resources in many countries, and while these facts represent one kind of evidence that TSMHOs are helpful in addiction recovery, a lingering clinical and public health question is whether they are shown to help when subjected to more systematic and scientific scrutiny. From a scientific standpoint, it is only recently that these peer-led, recovery-focused groups and professionally-delivered treatments designed to facilitate their use (i.e., Twelve-Step Facilitation [TSF]), have undergone rigorous scientific inquiry to determine whether, who, and how, they may help. The principal goal of this chapter is to provide an overview of TSMHOs, the new science pertaining to them, and their relationship to addiction recovery.

Given the often confusing array of terms pertaining to "12-step" interventions, the chapter begins by briefly clarifying terminology. This is followed by a brief overview of the history and origin of 12-step MHOs, and the evidence for the clinical and public health utility of MHOs and related clinical interventions (i.e., TSF) designed to facilitate involvement in them, in preventing relapse, and enhancing SUD remission rates and reducing health care costs. This is followed by a review of the theoretical and empirical research that has focused on how exactly MHOs, like AA, confer benefits that support long-term recovery.

Twelve-step terminology

There is an array of terminology surrounding "12-step" interventions for SUD. Alcoholics Anonymous was the original 12-step entity, which began in 1935 in the United States (see more below) and is a free standing, community-based fellowship that provides help through a network of informal gatherings, convened at rented venues such as community centers, churches, and hospitals. The "TSF" is the name given to the professional clinical intervention designed to facilitate active engagement in this (e.g., AA) community organization. On the other hand, when the phrase "12-step treatment" is used, it typically (but not always) refers to a professional residential treatment program in which patients are educated in depth about the AA fellowship and the 12-steps, exposed to 12-step meetings, and work through the first few of the 12-steps during treatment (McElrath 1997). Because of the potential to conflate these three 12-step-related resources, it is important to be mindful of terminology when reviewing research studies, as it is not always immediately apparent to which of these different entities the research pertains.

History, origin, and growth

The TSMHOs trace their beginnings to AA and a meeting between William G. Wilson ("Bill W.") and Dr Robert H. Smith ("Dr Bob") in Akron, Ohio, in the United States. Bill W. had a long history of severe alcohol addiction, but with the help of an alcohol treatment hospital in New York City and a Christian religious group designed to help people with alcohol addiction (the Oxford Group), he had been able to achieve a limited amount of sobriety. During this period of sobriety in 1935, Bill W. was on a business trip in Akron and was about to begin drinking again after a failed business negotiation. He had learned, however, that if he could find another "alcoholic" to try to help, it could help him take his mind off drinking and re-orient himself towards recovery. In search of someone actively using alcohol, he was directed to a local physician, Dr Bob, who was known for having a severe case of alcohol addiction. Through this meeting and sharing of common experiences, Bill W. was able to win this physician's trust. Bill W.'s sobriety, and example, coaching, and support, helped the physician break his own addiction and begin recovery. Of particular significance was that his redirected, recovery-focused activity enabled him to forget about his earlier disappointment and remain sober. From this, Bill W. confirmed an essential ingredient of AA's model of mutual support that was later formalized in AA's 12-step program. This key ingredient (embodied in most MHOs) was that helping others helps oneself.

Since its inception in 1935, AA's 12-step program and expanding fellowship has remained popular and been adopted widely throughout North America and in more than 150 countries. AA's original alcohol-specific model has been adapted successfully also to address other drug problems as they have emerged (e.g., NA, Cocaine Anonymous (CA), Marijuana Anonymous (MA)). The TSMHOs promote a particular philosophy of recovery characterized by an emphasis on total abstinence, service to others, personal and spiritual growth, and reciprocal ("mutual") helping (Humphreys 2004; Kelly and Yeterian 2014). They are comprised of millions of individuals who possess the lived experience of successful recovery from addiction and who typically provide around-the-clock support "on demand" as needed. The TSMHOs also strongly encourage and provide daily personal recovery monitoring and mentoring via a TSMHO mentor known as a "sponsor", who provides more intensive support and accountability. They also recommend members stay in frequent contact between formal meetings by phone or text, and most TSMHOs offer online meetings to increase opportunities for recovery-specific support at any time.

Meetings are organized and run by volunteer members and they are available free of charge, but ask for voluntary contributions to cover the costs of rented space and refreshments (Humphreys 2004). The TSMHOs have some unique advantages over professional treatment, such as being readily available in the community (i.e., accessible immediately without a waiting list), holding meetings on holidays, evenings, and weekends when professional care is often unavailable, and being free and accessible without health insurance. The limited availability, inflexibility, and cost of professional services that are covered by insurance or government-funded agencies means that these free, flexible, and readily available community resources can serve a valuable role in addressing major public health problems like SUD.

TSMHO and TSF clinical research

Much has been learned about the clinical and public health utility of TSMHOs over the past 40 years. During the past 25 years, TSMHOs and clinical interventions designed to facilitate their use (i.e., TSF interventions) have been subjected to rigorous scientific scrutiny (Kelly and Yeterian 2014; McCrady and Miller 1993; Kelly 2017; Laudet et al. 2014). Outcomes of these initiatives have been the accumulation of a body of scientific research that has supported the clinical and public health utility of TSMHOs and clinical interventions to facilitate their use, as well as elucidation of the mechanisms through which TSF treatments and TSMHOs confer benefits. As noted in detail below, many of the mechanisms through which MHOs have been found to work include changing the nature and structure of individuals' social networks (Groh et al. 2008) as well as other social-cognitive and spiritual constructs (Buckingham et al. 2013; Kelly 2017; Kelly et al. 2009).

Are TSMHOs effective recovery management resources?

Most of the research conducted on TSMHOs has examined their effectiveness by studying attendance rates and substance use outcomes as they naturally occur. While providing one kind of evidence regarding effectiveness, it has been somewhat difficult to estimate true causal effects of TSMHO participation because people "self-select" into these groups, and these individuals may be somehow different and, potentially, better off/have a better prognosis than those not choosing to participate. To help overcome this self-selection bias, studies have used increasingly rigorous methods (i.e., randomized controlled trials [RCTs], instrumental variable analyses, and propensity score matching adjustments to control for major confounds) to estimate the unique and independent influence of TSMHO participation on subsequent substance use outcomes. As described in detail below, efficacy studies of TSMHO participation have been approximated by comparing professionally-delivered TSF treatments that attempt to enhance TSMHO participation and thereby improve substance-related outcomes. These have been tested against other more established treatments, such as cognitive-behavioral therapy (CBT).

Emrick et al. (1993) conducted the first systematic meta-analytic study of 107 individual studies of AA participation and found that greater frequency of AA attendance and better drinking outcomes were significantly and modestly correlated ($r \sim 0.2$), but the overall methodological quality of the studies conducted up until that point (circa 1990) was generally poor. This finding, together with a call from the United States Institute of Medicine (IOM) of the National Academy of Sciences for more rigorous research on AA and its mechanisms of action in 1990 (Institute of Medicine 1990), subsequently produced a flurry of federally funded research in the United States (Kelly and Yeterian 2014) and in other countries as well (Manning et al. 2012; Vederhus et al. 2014).

Effectiveness studies

Longitudinal studies show that among individuals with SUD, both the intensity and duration of AA attendance is associated with better remission rates, even when initial abstinence and other confounds (e.g., prior treatment) are statistically controlled (Timko et al. 2000; Moos and Moos 2006; Ouimette et al. 1998). It is important to note that when considering both the use of TSMHOs and formal treatment services (including SUD and mental health services), TSMHO attendance is shown to confer an additional benefit on substance use and related outcomes (Moos and Moos 2006; Bergman et al. 2014; Moos et al. 2001; Fiorentine 1999; Tonigan et al. 1998).

Even though effectiveness studies are more representative of "real world" TSMHO participation, they too have some important limitations. For instance, in the existing literature, the samples for these studies are often drawn from professionally treated populations, rather than from community TSMHO populations, meaning that the samples consist of people who have received some sort of professional treatment for SUD in addition to attending TSMHO. Thus, it becomes difficult to know if the findings of these studies would generalize to individuals who only attend TSMHOs and to estimate the independent contribution of TSMHOs to outcomes (Tonigan et al. 1996).

One long-term prospective study took this potential confound into account by comparing outcomes in treated and untreated problem drinkers over a 16-year follow-up period (Timko et al. 2000; Moos and Moos 2006). The treated subsamples were those in formal (i.e., professional) and/or informal (i.e., TSMHO) treatments. This study found that individuals who self-selected into AA-only were more likely to be abstinent at one- and three-year follow-ups than those who self-selected into formal treatment-only (48% and 50% vs 21% and 26%, respectively), although by the eight-year follow-up, these groups were similar (49% vs 46%). Those who self-selected into both AA and formal treatment were also more likely than those with formal treatment-only to be abstinent at years one and three (42% and 51% vs 21% and 26%) and again were not significantly different by year eight (58% vs 46%). Those who received formal treatment plus AA did not differ significantly from those in AA-only across the follow-up in terms of abstinence rates (Timko et al. 2000). Additionally, a longer duration of AA attendance in years one–three independently predicted abstinence, as well as a lower likelihood of drinking problems, at year 16 (Moos and Moos 2006). These findings indicate that for some individuals, TSMHO participation alone can serve as an effective intervention for problem drinking.

Another naturalistic study conducted with a large (n = 3,018), male, inpatient sample drawn from 15 United States' Department of Veterans Affairs (VA) treatment programs found that, in contrast to patients who received only outpatient treatment during a one-year follow-up, patients who attended only TSMHOs were more likely to be abstinent and free of alcohol dependence symptoms and were less likely to be depressed, after controlling for major confounds (Ouimette et al. 1998). However, patients attending both TSMHOs and outpatient treatment had the best substance use outcomes, a finding that has been demonstrated elsewhere (Fiorentine and Hillhouse 2000). A two-year follow-up of the VA sample (n = 2,319) used structural equation modeling to examine the causal links between AA involvement and substance use (McKellar et al. 2003). Findings showed that AA involvement led to decreased alcohol consumption and fewer alcohol-related problems after controlling for the level of patient motivation, comorbid psychopathology, and demographic variables. Similar findings were demonstrated in a community outpatient study with both male and female participants, which found that participation in addiction TSMHOs led to subsequent improvement in alcohol-related outcomes after controlling for relevant pre- and post-treatment variables (e.g., addiction severity, motivation, use of professional treatment during the follow-up period; Kelly et al. 2006). Another prospective longitudinal study with a drug-dependent inpatient sample in the UK found that, across the five-year follow-up period, those who attended AA/NA after discharge were three–four times more likely to be abstinent from heroin and four–five times more likely to be abstinent from alcohol than those who did not attend AA/NA, although there were no differences between these groups in the likelihood of being abstinent from stimulants after the one-year follow-up (Gossop et al. 2008). In addition, more frequent AA/NA attendance (>1x/week) was associated with greater odds of abstinence from alcohol and heroin than less frequent attendance (<1x/week). Together, these naturalistic studies suggest that 12-step groups can serve as an important adjunct to professional care, especially after discharge, and can provide a buffer against SUD relapse.

Other naturalistic effectiveness studies have used even stronger analytic approaches to establish a true causal impact of TSMHO on SUD recovery. For example, prospective time-lagged analyses (e.g., examining the association between TSMHO participation at one-time point and outcomes at a subsequent time point) have shown that AA attendance and active involvement (e.g., obtaining a "sponsor"/recovering mentor, verbally participating in meetings, socializing with other TSMHO members outside of meetings) leads to decreased alcohol consumption and fewer alcohol-related problems. Importantly, the reverse association (decreased alcohol consumption and alcohol-related problems leading to increased participation) has not been supported, and analyses have controlled for confounds such as level of initial patient motivation, addiction severity, engagement with treatment during follow-up, and comorbid psychopathology (McKellar et al. 2003; Kelly et al. 2006).

Although less is known about TSMHO benefits among patients with primary drug use disorders, available evidence using longitudinal studies and controlling for a host of possible confounds suggests that TSMHO participation, and active involvement in particular, promotes increased abstinence rates among primary drug patients as well (Weiss et al. 2005; Kelly et al. 2014). Taken together, longitudinal methodologically rigorous studies indicate TSMHO participation has salutary effects on SUD outcomes and can be added to formal SUD and mental health treatment to positively affect abstinence and remission. Despite these advances in our understanding of the impact of community-based TSMHO participation on SUD outcomes, noteworthy limitations in the TSMHO literature are the dearth of studies on TSMHOs other than AA (e.g., NA) and the reporting of positive quality-of-life outcomes beyond abstinence/remission.

Efficacy studies: twelve-step facilitation

In contrast to studies of effectiveness, which examine the clinical utility and impact of interventions under real-world conditions, "efficacy" studies examine interventions under tightly controlled, optimal, conditions. As such, efficacy studies often have a stronger ability to infer cause and effect (that the benefit to patients was truly caused by the treatment) as randomization to treatment conditions (assuming the sample size is above 20 per treatment) evens out any pre-existing differences between comparison groups that could cloud true effects. As noted above, TSF is a professionally-delivered intervention designed to systematically encourage and support engagement with AA. The term was first used in a large study called Project MATCH in the 1990s (Project MATCH Research Group 1993) and the approach was manualized to provide a replicable intervention procedure that addiction professionals could deliver (Project MATCH Research Group 1993; Nowinski *et al.* 1992). The TSF can be delivered individually or in groups and aims to facilitate TSMHO attendance and involvement. In addition to its use as a stand-alone treatment (Brown *et al.* 2002; Litt *et al.* 2009; Project MATCH Research Group 1997; Litt *et al.* 2016), TSF has been studied in several other formats, including as an integrated part of an existing treatment such as CBT (Walitzer *et al.* 2009), a distinct component of a multi-session treatment package (Kaskutas *et al.* 2009), and a modular appendage add-on to standard treatment (Kahler *et al.* 2004; Timko and DeBenedetti 2007). As well as being utilized among addiction specialty providers, TSF has been used also in non-specialty primary care services (Pettinati *et al.* 2005; Kelly and McCrady 2008). Thus, clinicians wishing to utilize TSF have a variety of empirically-supported methods at their disposal, which provide several options for using TSF with varying degrees of intensity.

The TSF tends to produce as good, or better, alcohol and other drug-use outcomes (Walitzer *et al.* 2009; Litt *et al.* 2009). This is especially striking when one looks at sustained abstinence and remission (Litt *et al.* 2009; Project MATCH Research Group 1998; Humphreys and Moos 2001; Humphreys and Moos 2007). In the large multi-site Project MATCH clinical trial, for example (Project MATCH Research Group 1997), all three treatments studied tried to encourage patients to remain sober, yet, relative to cognitive behavior therapy (CBT) and motivational enhancement therapy (MET), TSF had 60% and 71% more cases, respectively, in full sustained remission during the first year following treatment; at a three-year follow-up the proportion completely abstinent during the past 90 days was 50% higher in TSF relative to CBT (Longabaugh *et al.* 1998). Other RCTs focusing on drugs other than alcohol have found similar results (e.g., Crits-Christoph *et al.* 1999; Donovan *et al.* 2013). As mentioned, these TSF clinical interventions are not tests of 'AA' or TSMHOs per se, but their sole aim is to stimulate AA participation, and where mediational tests have been conducted to determine why TSF produces these better outcomes (e.g. compared with CBT), in keeping with TSF theory, it is found that it is because of the greater AA meeting attendance and active AA involvement among patients in the TSF conditions (Walitzer *et al.* 2009; Litt *et al.* 2009; Longabaugh *et al.* 1998; Subbaraman and Kaskutas 2012).

Further sophisticated causal analyses using propensity score matching (Ye and Kaskutas 2009) and instrumental variable analyses (Humphreys *et al.* 2014) have added additional evidence (even beyond randomized clinical trials, which are assumed to be the ultimate tests of whether a treatment is superior or inferior to another treatment) that TSMHO participation confers a clear causal benefit.

Studies have also found that TSF treatments that engage patients with TSMHOs not only produce significantly higher rates of abstinence post-treatment compared with comparison treatments, but result in lower health-care costs (Humphreys and Moos 2001; Humphreys and Moos 2007; Mundt *et al.* 2012; Kelly 2017). In a large two-year follow-up study of patients treated in ten inpatient treatment facilities in the US VA system (Humphreys and Moos 2001; Humphreys and Moos 2007) for example, compared with patients receiving SUD treatment in programs that did not focus extensively on TSMHO participation during and following treatment, patients receiving SUD treatment in programs that did had significantly higher continuous abstinence rates, higher TSMHO participation, and substantially lower formal treatment service utilization post-inpatient discharge, resulting in a lowered cost of approximately $8,000 per patient for those treated in those TSMHO-facilitating programs. Extrapolating these cost savings across the ten facilities, if all study participants were to have obtained the TSMHO treatment, it would have resulted in approximately $10–15 million in health-care cost savings while simultaneously producing better SUD outcomes.

For whom are TSMHOs particularly helpful or not helpful?

A question often asked is whether TSMHOs are helpful for individuals with co-occurring psychiatric disorders (Bogenschutz et al. 2006), those taking psychotropic medications including anti-relapse medication (Rychtarik et al. 2000), atheists or agnostics (Winzelberg and Humphreys 1999), women (Del Boca and Mattson 2001), and young people (Kelly et al. 2012b). However, the available empirical evidence suggests that, in general, individuals with co-occurring psychiatric illnesses in addition to their SUD can benefit from participation in traditional AA or NA meetings as much or more (e.g., Timko et al. 2013) when compared with their SUD-only counterparts. An exception to these findings, however, are those with more socially impairing mental illness such as schizophrenia or severe unipolar depression (Kelly et al. 2003; Aase et al. 2008), who may be better suited for dual-diagnosis-focused MHOs, such as Double Trouble in Recovery (Magura 2008; Laudet et al. 2004). Research also suggests that less religiously inclined individuals are less likely to become engaged, but those that do participate have outcomes as good as those who are more religiously inclined (Ye and Kaskutas 2009; Winzelberg and Humphreys 1999). Despite the fact that AA's recovery program and early successes were based on nearly all-male samples, later research has confirmed that females attend as frequently as men, actually become more involved than men, and derive as much recovery benefit as men (Del Boca and Mattson 2001; Kelly and Hoeppner 2013). Also, research has revealed that young people may need a greater degree of facilitation and systematic encouragement to engage with TSMHOs, but those that do connect with such groups derive as much benefit as their older adult counterparts (Hoeppner et al. 2014). Further research has found that initial TSMHO engagement and outcomes for young people might be enhanced by counseling them to attend meetings with a greater proportion of same-aged individuals (e.g., young people AA meetings; (Kelly et al. 2005; Labbe et al. 2013)). Over the long-term, however, research suggests they may benefit less from these young-person meetings, indicating exposure to 12-step meetings with a greater proportion of older adults with more life and recovery experience may be needed to continue to enhance recovery (Kelly et al. 2005; Labbe et al. 2013).

How does TSMHO participation promote better outcomes?

A relatively novel and intriguing area of investigation in the realm of treatment and TSMHO participation has revealed noteworthy results regarding how TSMHOs confer benefits. These have not always been consistent with these organizations' own theory regarding how salutary recovery-related change is purported to occur (Kelly 2017). In TSMHOs' own literature it is stated that members achieve recovery through a combination of factors including a "spiritual awakening," service to others, sponsorship, and working the 12 steps (Alcoholics Anonymous 2001; Narcotics Anonymous 2008). The empirical literature overlaps to some degree with these 12-step-specific mechanisms of change, but is accounted for mostly by mechanisms similar to those mobilized by formal interventions.

Research on the mechanisms of behavior change in TSMHOs has focused on two broad areas: common change processes, or those processes that are likely to underlie or be common to treatment in general (e.g., recovery motivation, abstinence self-efficacy, social network changes), and 12-step-specific practices, which are unique to 12-step groups and not likely to function in other types of treatment (e.g., working the 12 steps, spiritual awakening). While these categories of mechanisms have often been separated in research for the sake of clarity, it is likely that, in reality, they are highly intertwined and reciprocally influential (e.g., working the 12 steps may increase self-efficacy and vice versa) or may simply represent different terms that describe the same underlying process (e.g., AA's focus on avoiding "people, places, and things" is almost identical to the CBT notion of avoiding triggers/cues; McCrady 1994). Although direct evidence does not exist for neurobiological change, extrapolating from addiction relapse studies conducted with primates, it may be possible to say that AA involvement accelerates adaptive brain changes (e.g., increases in dopamine D2 receptor levels; Morgan et al. 2002) that may in turn protect against relapse.

Cognitively, increased abstinence self-efficacy – the confidence to handle risky or difficult situations without alcohol/drug use – has been shown to be a partial mediator of the beneficial effects of AA participation on reduced alcohol and other drug use (Forcehimes and Tonigan 2008). Participation also appears to work by increasing one's commitment (i.e., motivation) to abstain from alcohol and other drugs (Kelly et al. 2002; Morgenstern et al. 1997). Enhanced abstinence-focused coping efforts (e.g., cognitive reappraisal and stimulus control) also explain, in part, how AA works (Morgenstern et al. 1997). Despite support for these common psychological processes as mediators of TSMHO's effectiveness, affective mechanisms (i.e., those pertaining to mood/emotion) are less certain. For example, in two studies (Kelly et al. 2010b; Kelly et al. 2010a), the mediating role of reduced depression became non-significant when concurrent alcohol use was accounted for (suggesting reduced alcohol use, not depression, explains that effect) and reduced anger was not associated with AA attendance during the first year. Other studies have found that AA works to promote recovery via reducing cravings (Kelly and Greene 2013) and by reducing impulsivity (Blonigen et al. 2011; Blonigen et al. 2013).

Socially, changes in support networks – particularly building an abstinence supportive network – are among the strongest and most consistent mediators for the effects of TSMHO attendance and active involvement on better substance use outcomes. Explanatory models where social networks and substance use are measured concurrently (Humphreys et al. 1999; Kaskutas et al. 2002) have been enhanced by more recent studies where social network changes mediate AA's benefit when the variables are time lagged (i.e., social network changes are measured after AA participation and before drinking outcomes; Bond et al. 2003; Kelly et al. 2011a). These changes include both reducing pro-drinkers and increasing pro-abstainers in the network, and AA-specific network changes appear to be particularly key. For example, among outpatients in a large clinical trial known as Project MATCH, attending three meetings per week versus none was associated with a 20–30% decrease in pro-drinking ties and 15% increase in pro-abstinent ties and, in turn, each pro-drinking tie was associated with a 7% decrease in abstinence days and each pro-abstinence tie was associated with a 4% increase in abstinent days (Kelly et al. 2011a).

While there is substantial evidence for the role of these common psychological factors and processes as mediators of AA's effectiveness, support for 12-step-specific mediators has been more mixed and studied less extensively. Although increases in subjective levels of spirituality and related constructs are often associated with AA participation, in three studies neither degree of spiritual/religious beliefs (among members of 12-step and non-12-step MHOs; Atkins and Hawdon 2007), purpose in life (Oakes 2008), or having had a "spiritual awakening" (Owen et al. 2003) were significant predictors of subsequent abstinence rates. On the other hand, when using a more comprehensive measure (the Religious Background and Behavior Scale; Connors et al. 1996), Kelly et al. (2011b) found that increased spirituality/religiosity did mediate AA's effects. Tonigan et al. (2013) reported a similar finding in a separate sample using a modified version of a Religious Background and Behavior subscale called God consciousness (meditation, thinking about God, direct experiences of God). Also, Krentzman and colleagues (2013) found that spirituality is also a mechanism of behavior change through which AA confers addiction recovery-related benefit.

The vast majority of mechanisms studies noted above investigated single mediators (e.g., spirituality or abstinence self-efficacy or social network changes). Consequently, if recovery motivation, social network changes, spirituality, and so on, are all found separately to be significant mechanisms of behavior change through which TSMHOs confer benefit, one is left wondering which are most important. To try to determine the relative importance of these different mediators, a series of "multi-mediator" investigations have been conducted.

In a multi-mechanism analysis that tested the effects of six purported mediators simultaneously, it was found that AA's beneficial effects on future abstinence were explained in both less severely addicted and more severely addicted outpatient samples through its ability to stimulate increases in pro-abstainers and decreases in pro-drinkers in individuals' social networks, as well as in increases in self-efficacy to cope with high risk social situations without drinking; increased spirituality/religiosity was also a significant mediator, but only among the more clinically severe patients. Reduced depression and self-efficacy to cope with negative affect were not significant mediators in either group of patients (Kelly et al. 2012a). Second, how AA and other TSMHOs work may differ across the life course. Among adolescents, for example, lagged multiple mediation model testing found motivation for abstinence, but not self-efficacy or coping, mediated the effect of AA and NA participation in increasing abstinence, with active involvement a stronger predictor of increased motivation than attendance (Kelly et al. 2002; Kelly et al. 2000). Similarly, in a study comparing adults 18–29 with those 30 and older in Project MATCH, it has been found that young adults benefitted as much, but in different ways, from TSMHO participation. Specifically, only reductions in pro-drinkers in one's social network and increased abstinence self-efficacy to handle risky social situations mediated AA's effects in the young adult group (Hoeppner et al. 2014). Also, a smaller amount of the overall effect of AA participation on subsequent alcohol use outcomes was captured by the mediating variables in this analysis, suggesting that young adults benefit from AA in ways not yet specified and that require further investigation. Another analysis in the same sample focusing on whether men and women benefitted in different ways from TSMHO participation found that men and women both benefitted from AA-facilitated recovery adaptive changes in their social networks, but women benefitted much more from AA by its ability to boost their capacity to cope with negative affect without drinking intensively; for men this was not the case. Rather, men were helped by AA's ability to boost their confidence in coping with high-risk social drinking situations without consuming alcohol (Kelly and Hoeppner 2013).

In sum, TSMHOs appear to help people attain and sustain remission and reduce alcohol and other drug use by facilitating recovery-focused changes in their social networks, boosting and maintaining recovery motivation and cognitive and behavioral coping skills and reducing negative affect, craving and impulsivity.

Conclusion

Organizations such as AA and NA are widely available in many countries and have strong empirical support as effective recovery support resources for those suffering from SUD (United States Surgeon General 2016). Rigorous scientific studies have shown also that professional 12-step treatments including clinical linkage to these freely available community resources during treatment can enhance clinical outcomes, particularly continuous abstinence and sustained

SUD remission, while simultaneously reducing the burden on health care systems, thereby reducing health care costs. The way TSMHOs have been found to confer benefit is through mechanisms similar to those found in professional interventions; namely, by boosting and maintaining motivation for recovery, enhancing recovery coping skills and abstinence self-efficacy, facilitating recovery adaptive changes in individuals' social networks, and reducing craving and impulsivity. The TSMHOs have been found to enhance abstinence also by boosting spirituality, particularly among those with more severe addiction problems (Kelly 2017). The TSMHOs' free availability in most communities mean that these ubiquitous recovery support services can play a valuable role in ameliorating the public health harms attributable to SUD.

References

Aase, D. M., Jason, L. A. and Robinson, W. L. (2008) "12-step participation among dually-diagnosed individuals: a review of individual and contextual factors", *Clinical Psychology Review* 28(7), 1235–1248.

Alcoholics Anonymous (2001) *Alcoholics Anonymous: The Story of How Many Thousands of Men and Women have Recovered from Alcoholism*, 4th Edition, New York, NY: Alcoholics Anonymous World Services.

Atkins, R. G. Jr and Hawdon, J. E. (2007) "Religiosity and participation in mutual-aid support groups for addiction", *Journal of Substance Abuse Treatment* 33(3): 321–331.

Bergman, B. G., *et al.* (2014) "Psychiatric comorbidity and 12-step participation: a longitudinal investigation of treated young adults", *Alcoholism: Clinical and Experimental Research* 38(2): 501–510.

Blonigen, D. M., *et al.* (2011) "Impulsivity is an independent predictor of 15-year mortality risk among individuals seeking help for alcohol-related problems", *Alcoholism: Clinical and Experimental Research* 35(11): 2082–2092.

Blonigen, D. M., Timko, C. and Moos, R. H. (2013) "Alcoholics Anonymous and reduced impulsivity: a novel mechanism of change", *Substance Abuse* 34(1): 4–12.

Bogenschutz, M. P., Geppert, C. M. A. and George, J. (2006) "The role of 12-step approaches in dual-diagnosis treatment and recovery", *The American Journal on Addictions* 15: 50–60.

Bond, J., Kaskutas, L. A. and Weisner, C. (2003) "The persistent influence of social networks and alcoholics anonymous on abstinence", *Journal of Studies on Alcohol* 64(4): 579–588.

Brown, T. G., *et al.* (2002) "Process and outcome changes with relapse prevention versus 12-step aftercare programs for substance abusers", *Addiction* 97(6): 677–689.

Buckingham, S. A., Frings, D. and Albery, I. P. (2013) "Group membership and social identity in addiction recovery", *Psychology of Addictive Behaviors* 27(4): 1132–1140.

Connors, G. J., Tonigan, J. S. and Miller, W. R. (1996) "A measure of religious background and behavior for use in behavior change research", *Psychology of Addictive Behaviors* 10(2): 90–96.

Crits-Christoph, P., *et al.* (1999) "Psychosocial treatments for cocaine dependence: National Institute on Drug Abuse collaborative cocaine treatment study", *Archives of General Psychiatry* 56(6): 493–502.

Del Boca, F. K. and Mattson, M. E. (2001) "The gender matching hypothesis", in R. Longabaugh and P. Wirtz (eds), *Project MATCH Hypotheses: Results and Causal Chain Analysis*, Project MATCH Monograph Series, Bethesda, MD: National Institute on Alcohol Abuse and Alcoholism.

Donovan, D. M., *et al.* (2013) "Stimulant abuser groups to engage in 12-step: a multisite trial in the National Institute on Drug Abuse clinical trials network", *Journal of Substance Abuse Treatment* 44(1): 103–114.

Emrick, C. D., *et al.* (1993) "Alcoholics Anonymous: what is currently known?" in B. S. McCrady and W. R. Miller (eds), *Research on Alcoholics Anonymous: Opportunities and Alternatives*, Piscataway, NJ: Rutgers Center of Alcohol Studies, pp. 41–76.

Fiorentine, R. (1999) "After drug treatment: are 12-step programs effective in maintaining abstinence?" *American Journal of Drug and Alcohol Abuse* 25(1): 93–116.

Fiorentine, R. and Hillhouse, M. P. (2000) "Drug treatment and 12-step program participation: the additive effects of integrated recovery activities", *Journal of Substance Abuse Treatment* 18(1): 65–74.

Forcehimes, A. A. and Tonigan, J. S. (2008) "Self-efficacy as a factor in abstinence from alcohol/other drug abuse: a meta-analysis", *Alcoholism Treatment Quarterly* 26(4): 480–489.

Glasgow, R. E., Lichtenstein, E. and Marcus, A. C. (2003) "Why don't we see more translation of health promotion research to practice? Rethinking the efficacy-to-effectiveness transition", *American Journal of Public Health* 93(8): 1261–1267.

Gossop, M., Stewart, D. and Marsden, J. (2008) "Attendance at Narcotics Anonymous and Alcoholics Anonymous meetings, frequency of attendance and substance use outcomes after residential treatment for drug dependence: a 5-year follow-up study", *Addiction* 103(1): 119–125.

Groh, D. R., Jason, L. A. and Keys, C. B. (2008) "Social network variables in alcoholics anonymous: a literature review", *Clinical Psychology Review* 28(3): 430–450.

Hoeppner, B., Hoeppner, S. S. and Kelly, J. (2014) "Do young people benefit from AA as much, and in the same ways, as older people? A moderated multiple mediation analysis", *Drug and Alcohol Dependence* 143: 181–188.

Humphreys, K. (2004) *Circles of Recovery: Self-help Organizations for Addictions*, Cambridge, UK: Cambridge University Press.

Humphreys, K., et al. (1999) "Do enhanced friendship networks and active coping mediate the effect of self-help groups on substance abuse?" *Annals of Behavioral Medicine* 21(1): 54–60.

Humphreys, K. and Moos, R. H. (2001) "Can encouraging substance abuse patients to participate in self-help groups reduce demand for health care? A quasi-experimental study", *Alcoholism: Clinical and Experimental Research* 25(5): 711–716.

Humphreys, K. and Moos, R. H. (2007) "Encouraging post-treatment self-help group involvement to reduce demand for continuing care services: two-year clinical and utilization outcomes", *Alcoholism: Clinical and Experimental Research* 31(1): 64–68.

Humphreys, K., Blodgett, J. C. and Wagner, T. H. (2014) "Estimating the efficacy of Alcoholics Anonymous without self-selection bias: an instrumental variables re-analysis of randomized clinical trials", *Alcoholism: Clinical and Experimental Research* 38(11): 2688–2694.

Institute of Medicine (1990) *Broadening the Base of Treatment for Alcohol Problems*, Washington, DC: The National Academies Press.

Kahler, C. W., et al. (2004) "Motivational enhancement for 12-step involvement among patients undergoing alcohol detoxification", *Journal of Consulting and Clinical Psychology* 72(4): 736–741.

Kaskutas, L. A., Bond, J. and Humphreys, K. (2002) "Social networks as mediators of the effect of Alcoholics Anonymous", *Addiction* 97(7): 891–900.

Kaskutas, L. A., et al. (2009) "Effectiveness of making Alcoholics Anonymous easier: a group format 12-step facilitation approach", *Journal of Substance Abuse Treatment* 37(3): 228–239.

Kelly, J. F. (2017) "Tens of millions successfully in long-term recovery – let us find out how they did it", *Addiction* 112(5): 762–763.

Kelly, J. F., Myers, M. G. and Brown, S. A. (2000) "A multivariate process model of adolescent 12-step attendance and substance use outcome following inpatient treatment", *Psychology of Addictive Behaviors* 14(4): 376–389.

Kelly, J. F., Myers, M. G. and Brown, S. A. (2002) "Do adolescents affiliate with 12-step groups? A multivariate process model of effects", *Journal of Studies on Alcohol* 63(3): 293–304.

Kelly, J. F., McKellar, J. D. and Moos, R. (2003) "Major depression in patients with substance use disorders: relationship to 12-step self-help involvement and substance use outcomes", *Addiction* 98(4): 499–508.

Kelly, J. F., Myers, M. G. and Brown, S. A. (2005) "The effects of age composition of 12-step groups on adolescent 12-step participation and substance use outcome", *Journal of Child & Adolescent Substance Abuse* 15(1): 63–72.

Kelly, J. F., et al. (2006) "A 3-year study of addiction mutual-help group participation following intensive outpatient treatment", *Alcoholism: Clinical and Experimental Research* 30(8): 1381–1392.

Kelly, J. F. and McCrady, B. S. (2008) "Twelve-step facilitation in non-specialty settings", in M. Galanter and L. A. Kaskutas (eds), *Recent Developments in Alcoholism: Research on Alcoholics Anonymous and Spirituality in Addiction Recovery*, Totowa, NJ: Springer, pp. 321–346.

Kelly, J. F., Magill, M. and Stout, R. L. (2009) "How do people recover from alcohol dependence? A systematic review of the research on mechanisms of behavior change in Alcoholics Anonymous", *Addiction Research & Theory* 17(3): 236–259.

Kelly, J. F., et al. (2010a) "Negative affect, relapse, and Alcoholics Anonymous: does AA work by reducing anger?" *Journal of Studies on Alcohol and Drugs* 71(3): 434–444.

Kelly, J. F., et al. (2010b) "Mechanisms of behavior change in Alcoholics Anonymous: does Alcoholics Anonymous lead to better alcohol use outcomes by reducing depression symptoms?" *Addiction* 105(4): 626–636.

Kelly, J. F., et al. (2011a) "The role of Alcoholics Anonymous in mobilizing adaptive social network changes: a prospective lagged mediational analysis", *Drug and Alcohol Dependence* 114(23): 119–126.

Kelly, J. F., et al. (2011b) "Spirituality in recovery: a lagged mediational analysis of Alcoholics Anonymous' principal theoretical mechanism of behavior change", *Alcoholism: Clinical and Experimental Research* 35(3): 454–463.

Kelly, J. F., et al. (2012a) "Determining the relative importance of the mechanisms of behavior change within Alcoholics Anonymous: a multiple mediator analysis", *Addiction* 107(2): 289–299.

Kelly, J. F., Stout, R. L. and Slaymaker, V. (2012b) "Emerging adults' treatment outcomes in relation to 12-step mutual-help attendance and active involvement", *Drug and Alcohol Dependence* 129(1–2): 151–157.

Kelly, J. F. and White, W. (2012) "Broadening the base of addiction recovery mutual aid", *Journal of Groups in Addiction & Recovery* 7(2–4): 82–101.

Kelly, J. F. and Greene, M. C. (2013) "The twelve promises of Alcoholics Anonymous: psychometric validation and mediational testing as a 12-step specific mechanism of behavior change", *Drug and Alcohol Dependence* 133(2): 633–640.

Kelly, J. F. and Hoeppner, B. B. (2013) "Does Alcoholics Anonymous work differently for men and women? A moderated multiple-mediation analysis in a large clinical sample", *Drug and Alcohol Dependence* 130(1–3): 186–193.

Kelly, J. F., Greene, M. C. and Bergman, B. (2014) "Do drug dependent patients attending Alcoholics Anonymous rather than Narcotics Anonymous do as well? A prospective, lagged, matching analysis", *Alcohol and Alcoholism* 49(5): 1–9.

Kelly, J. F. and Yeterian, J. D. (2014) "Mutual-help groups for alcohol and other substance use disorders" in B. S. McCrady and E. E. Epstein (eds), *Addictions: A Comprehensive Guidebook*, New York, NY: Oxford University Press, pp. 500–525.

Krentzman, A. R., Cranford, J. A. and Robinson, E. A. (2013) "Multiple dimensions of spirituality in recovery: a lagged mediational analysis of Alcoholics Anonymous' principal theoretical mechanism of behavior change", *Substance Abuse* 34(1): 20–32.

Labbe, A. K., et al. (2013) "The importance of age composition of 12-step meetings as a moderating factor in the relation between young adults' 12-step participation and abstinence", *Drug and Alcohol Dependence* 133(2): 541–547.

Laudet, A., Timko, C. and Hill, T. (2014) "Comparing life experiences in active addiction and recovery between veterans and non-veterans: a national study", *Journal of Addictive Diseases* 33(2): 148–162.

Laudet, A. B., et al. (2004) "The effect of 12-step based fellowship participation on abstinence among dually diagnosed persons: a 2-year longitudinal study", *Journal of Psychoactive Drugs* 36(2): 207–216.

Litt, M. D., et al. (2009) "Changing network support for drinking: Network Support project 2-year follow-up", *Journal of Consulting and Clinical Psychology* 77(2): 229–242.

Litt, M. D., et al. (2016) "Network Support II: randomized controlled trial of Network Support treatment and cognitive behavioral therapy for alcohol use disorder", *Drug and Alcohol Dependence* 165: 203–212.

Longabaugh, R., et al. (1998) "Network Support for drinking, Alcoholics Anonymous and long-term matching effects", *Addiction* 93(9): 1313–1333.

Magura, S. (2008) "Effectiveness of dual focus mutual aid for co-occurring substance use and mental health disorders: a review and synthesis of the 'double trouble' in recovery evaluation", *Substance Use and Misuse* 43(12–13): 1904–1926.

Manning, V., et al. (2012) "Does active referral by a doctor or 12-step peer improve 12-step meeting attendance? Results from a pilot randomised control trial", *Drug and Alcohol Dependence* 126(1–2): 131–137.

McCrady, B. S. (1994) "Alcoholics Anonymous and behavior therapy: can habits be treated as diseases? Can diseases be treated as habits?" *Journal of Consulting and Clinical Psychology* 62(6): 1159–1166.

McCrady, B. S. and Miller, W. R. (1993) *Research on Alcoholics Anonymous: Opportunities and Alternatives*, Piscataway, NJ: Rutgers Center of Alcohol Studies.

McElrath, D. (1997) "The Minnesota Model", *Journal of Psychoactive Drugs* 29(2): 141–144.

McKellar, J., Stewart, E. and Humphreys, K. (2003) "Alcoholics Anonymous involvement and positive alcohol-related outcomes: cause, consequence, or just a correlate? A prospective 2-year study of 2,319 alcohol-dependent men", *Journal of Consulting and Clinical Psychology* 71(2): 302–308.

Moos, R., et al. (2001) "Outpatient mental health care, self-help groups, and patients' one-year treatment outcomes", *Journal of Clinical Psychology* 57(3): 273–287.

Moos, R. H. and Moos, B. S. (2006) "Participation in treatment and Alcoholics Anonymous: a 16-year follow-up of initially untreated individuals", *Journal of Clinical Psychology* 62(6): 735–750.

Morgan, D., et al. (2002) "Social dominance in monkeys: dopamine D2 receptors and cocaine self-administration", *Nature Neuroscience* 5(2): 169–174.

Morgenstern, J., et al. (1997) "Affiliation with Alcoholics Anonymous after treatment: a study of its therapeutic effects and mechanisms of action", *Journal of Consulting and Clinical Psychology* 65(5): 768–779.

Mundt, M. P., et al. (2012) "12-step participation reduces medical use costs among adolescents with a history of alcohol and other drug treatment", *Drug and Alcohol Dependence* 126(1–2): 124–130.

Narcotics Anonymous (2008) *Narcotics Anonymous, 6th Edition*, Chatsworth, CA: Narcotics Anonymous World Services.

Nowinski, J., Baker, S. and Carroll, K. (1992) *Twelve Step Facilitation Therapy Manual: A Clinical Research Guide for Therapists Treating Individuals with Alcohol Abuse and Dependence*, Washington, DC: Government Printing Office: NIAAA Project MATCH Monograph, Vol. 1, DHHS Publication No. (ADM), pp. 92–1893.

Oakes, K. (2008) "Purpose in life: a mediating variable between involvement in Alcoholics Anonymous and long-term recovery", *Alcoholism Treatment Quarterly* 26(4): 450–463.

Ouimette, P. C., Moos, R. H. and Finney, J. W. (1998) "Influence of outpatient treatment and 12-step group involvement on one-year substance abuse treatment outcomes", *Journal of Studies on Alcohol* 59(5): 513–522.

Owen, P. L., et al. (2003) "Participation in Alcoholics Anonymous: intended and unintended change mechanisms", *Alcoholism: Clinical and Experimental Research* 27(3): 524–532.

Pettinati, H. M., et al. (2005) "A structured approach to medical management: a psychosocial intervention to support pharmacotherapy in the treatment of alcohol dependence", *Journal of Studies on Alcohol. Supplement* 15: 170–178.

Project MATCH Research Group (1993) "Project MATCH: rationale and methods for a multisite clinical trial matching patients to alcoholism treatment", *Alcoholism: Clinical and Experimental Research* 17(6): 1130–1145.

Project MATCH Research Group (1997) "Matching alcoholism treatments to client heterogeneity: Project MATCH posttreatment drinking outcomes", *Journal of Studies on Alcohol and Drugs* 58(1): 7–29.

Project MATCH Research Group (1998) "Matching alcoholism treatments to client heterogeneity: Project MATCH three-year drinking outcomes", *Alcoholism: Clinical and Experimental Research* 22(6): 1300–1311.

Rychtarik, R. G., et al. (2000) "Alcoholics Anonymous and the use of medications to prevent relapse: an anonymous survey of member attitudes", *Journal of Studies on Alcohol* 61(1): 134–138.

Subbaraman, M. S. and Kaskutas, L. A. (2012) "Social support and comfort in AA as mediators of 'Making AA easier' (MAAEZ), a 12-step facilitation intervention", *Psychology of Addictive Behaviors* 26(4): 759–765.

Timko, C., et al. (2000) "Long-term outcomes of alcohol use disorders: comparing untreated individuals with those in alcoholics anonymous and formal treatment", *Journal of Studies on Alcohol and Drugs* 61(4): 529–540.

Timko, C. and DeBenedetti, A. (2007) "A randomized controlled trial of intensive referral to 12-step self-help groups: one-year outcomes", *Drug and Alcohol Dependence* 90(2–3): 270–279.

Timko, C., et al. (2013) "Dually diagnosed patients' benefits of mutual-help groups and the role of social anxiety", *Journal of Substance Abuse Treatment* 44(2): 216–223.

Tonigan, J. S., Toscova, R. and Miller, W. R. (1996) "Meta-analysis of the literature on Alcoholics Anonymous: sample and study characteristics moderate findings", *Journal of Studies on Alcohol and Drugs* 57(1): 65–72.

Tonigan, J. S., Connors, G. J. and Miller, W. R. (1998) "Participation and involvement in Alcoholics Anonymous", in T. Babor and F. DelBoca (eds), *Treatment Matching in Alcoholism*, New York, NY: Cambridge University Press, pp. 184–204.

Tonigan, J. S., Rynes, K. N. and McCrady, B. S. (2013) "Spirituality as a change mechanism in 12-step programs: a replication, extension, and refinement", *Substance Use and Misuse* 48(12): 1161–1173.

United States Department of Health and Human Services (HHS), Office of the Surgeon General (2016) *Facing Addiction in America: The Surgeon General's Report on Alcohol, Drugs, and Health*, Washington, DC: HHS, November.

Vederhus, J. K., et al. (2014) "Motivational intervention to enhance post-detoxification 12-step group affiliation: a randomized controlled trial", *Addiction* 109(5): 766–773.

Walitzer, K. S., Dermen, K. H. and Barrick, C. (2009) "Facilitating involvement in Alcoholics Anonymous during out-patient treatment: a randomized clinical trial", *Addiction* 104(3): 391–401.

Weiss, R. D., et al. (2005) "The effect of 12-step self-help group attendance and participation on drug use outcomes among cocaine-dependent patients", *Drug and Alcohol Dependence* 77(2): 177–184.

Winzelberg, A. and Humphreys, K. (1999) "Should patients' religiosity influence clinicians' referral to 12-step self-help groups? Evidence from a study of 3,018 male substance abuse patients", *Journal of Consulting and Clinical Psychology* 67(5): 790–794.

World Health Organization (2014) *Global Status Report on Alcohol and Health, 2014 Edition*, Geneva: WHO.

Ye, Y. and Kaskutas, L. A. (2009) "Using propensity scores to adjust for selection bias when assessing the effectiveness of Alcoholics Anonymous in observational studies", *Drug and Alcohol Dependence* 104(1–2): 56–64.

38
OPIOID SUBSTITUTION TREATMENT AND HARM MINIMIZATION APPROACHES

Mark K. Greenwald

Introduction

Harm minimization (HM) policies and interventions aim to reduce substance abuse-related problem behaviors (Ritter and Cameron 2006). Addressing harms of substance use with evidence-based methods is increasingly accepted as *pragmatically necessary and philosophically preferable* to the moral hazards of condoning substance use as 'social evil' (Hammett et al. 2008; UNODC 2016).

Goals of HM include improving access to care, treatment adherence, reductions in substance use, preventing diversion and misuse of pharmacotherapeutic and illegal drugs, reducing consequences of substance use, and integration of care. Effectiveness of HM interventions can be defined in terms of *public health* (reducing aggregate harm), *clinical practice* (delivering treatments that improve patient outcomes), and *patient-centeredness* (quality of life).

Opioid substitution treatment (OST) is a well-validated therapeutic approach founded on *pragmatic* and *pharmacological* principles that replacing illegal, non-medical or injection opioid use with safer modes of opioid administration can engage and retain patients in treatment, leading to less drug use and greater health benefits.

Direct pharmacological outcomes of OST include suppressing opioid withdrawal symptoms, and blocking the reinforcing effects of abused opioids if the patient self-administers them; this is termed pharmacological substitution. Achieving opioid blockade requires higher medication doses – leading to higher plasma levels and occupancy of brain *mu*-opioid receptors – relative to opioid withdrawal suppression (Greenwald et al. 2014). *Indirect* outcomes of OST include reduction of drug use-related unemployment, criminal activity, unsafe drug injection practices and risky sexual behavior, and less morbidity and mortality (Gibson et al. 2008; Lawrinson et al. 2008; Sun et al. 2015).

Table 38.1 summarizes features of OST options. Treatment with methadone and buprenorphine, integrated with psychosocial care to improve efficacy (Amato et al. 2012; CSAT 2005; McLellan et al. 1993), are the primary approaches used worldwide.

Novel longer-acting formulations of buprenorphine are emerging: an injectable subcutaneous depot that can suppress opioid withdrawal signs/symptoms and block opioid abuse-related effects for one month (RBP-6000, Sublocade®) that received FDA approval November 30, 2017; an injectable subcutaneous depot that can suppress opioid withdrawal and need for

Table 38.1 Opioid substitution therapy (OST) modalities: current and anticipated future options

OST medication	Route of administration	Dosing frequency	Legal status/availability
Methadone	Oral	Once daily	Approved in many nations
Buprenorphine/ naloxone	Sublingual tablets and buccal film	Typically once daily but also alternate days	Approved (Suboxone™ and generics)
Buprenorphine	Subcutaneous depot	Once monthly	Approved in USA (Sublocade®)
		Once weekly or once monthly	Phase III (CAM2038)
	Subdermal implant	Once per 6 months (with surgical removal of prior implant)	Approved in USA (Probuphine®)
Heroin	Clients' preferred route of use	Typically twice daily	Approved in several European nations

rescue sublingual doses for one–four weeks (CAM2038), undergoing phase III study; and surgically implanted rods that can suppress withdrawal and attenuate opioid use for six months (Probuphine®), recently approved in the USA.

Medically supervised heroin use, or heroin-assisted treatment (HAT), has been observed to improve patient-level outcomes (less criminal activity, risky behaviors and drug use, better employment and physical health, retention and transition to stable lifestyle) and community-level outcomes (reduction in drug markets, lowering costs to health care, social welfare and criminal justice systems) (Blanken et al. 2010; Haasen et al. 2007; Verthein et al. 2008). Heroin-assisted treatment is official policy in several European nations (EMCDDA 2012).

Harm-minimization in the context of opioid substitution treatment

Harms of opioid use are avoidable and HM/OST interventions are cost-effective (Alistar et al. 2011; Clark et al. 2015). Once established, HM interventions should be sustainable because dissolution of such programs has led to worsening of outcomes (Bach et al. 2015; Hartzler et al. 2010). Key HM interventions are summarized in Table 38.2.

Increase access: promote treatment entry and engagement by reducing barriers

Most contemporary addiction treatment systems suffer from structural barriers including social stigma, inadequate public education about treatment benefits, too few healthcare personnel with specialty training, gaps in treatment access and affordability, and limited reimbursement. Enrolling more individuals in OST requires lowering the threshold for entry (Islam et al. 2013; Strike et al. 2013).

In a recent review, Kourounis et al. (2016) concluded that low-threshold program features promoting treatment *access* include: reducing time to entry; making treatment available in primary care; decreasing cost of treatment, especially for economically disadvantaged persons, which typically implies greater government subsidy; addressing needs of pregnant women and adolescents; and flexibility of admission criteria to accommodate comorbidities, assuming behavioral healthcare for such conditions is integrated with treatment for OUD.

Table 38.2 Summary of harm minimization approaches

Objective	Specific approaches
Increase treatment access	1) Reduce time to entry 2) Expand availability of sites (e.g., primary care) 3) Decrease financial cost of treatment 4) Expand availability to subpopulations (e.g., pregnant, adolescents, psychiatric comorbidities) 5) Increase training of addiction specialty physicians
Increase treatment retention/adherence	1) Adequate and flexible medication doses 2) Continuum of care (multidisciplinary) 3) Incentives to improve attendance
Decrease substance use	1) Opioids (e.g., substitution treatment, counseling) 2) Other illegal drug use, alcohol, smoking (e.g., contingency management, brief motivational interventions, other medications) 3) Prescription drug monitoring program utilization (e.g., reduce misuse and diversion) 4) Reduce or improve safety of injection drug use (e.g., needle/syringe exchange, monitored self-administration) 5) Reduce overdoses and fatalities (e.g., naloxone provision, education)
Integration of care (holistic)	1) Reduce structural barriers (e.g., program philosophy, education to offset negative staff attitudes) 2) Personalized care (e.g., demographically sensitive, case management, technology-based treatment) 3) Mental health care (e.g., psychotropic medications, incentives for on-site counseling attendance) 4) Physical health care (e.g., directly observed therapy for HIV/HCV, pain management, and other comorbidities coordinated with internal medicine) including pregnant women and neonates (coordinated with obstetrics/gynecology), and safer-sex interventions (e.g., condoms, education) 5) Coordination/integration with other settings (e.g. emergency department, community pharmacies, prisons) to provide continuum of care

Interim OST without counseling can also promote treatment access and retention (Friedman et al. 1994), and reduce opioid use (Krook et al. 2002; Schwartz et al. 2006), arrests (Schwartz et al. 2009) and HIV risk behaviors (Wilson et al. 2010).

Increase retention and adherence

Evidence-based guidance for improving retention in OST includes: improving staff knowledge and attitudes about OST; clarifying program goals and treatment plans; addressing patients' financial needs; providing best-practice services as early as possible; reducing attendance burden; enhancing staff–patient interactions; individualizing medication doses; and not limiting length of treatment (CSAT 2005; Kourounis et al. 2016).

Although dosing should be flexible, it is well established that higher doses related to improved OST retention (Faggiano et al. 2003; Fareed et al. 2012), which is associated with less opioid use (Bao et al. 2009; Cao et al. 2014).

Incentive-based (contingency management) interventions have been used to improve attendance at counseling sessions (Brooner *et al.* 2004; Rhodes *et al.* 2003), which may improve overall treatment "dose" and thereby outcomes.

Decrease drug use

The principal goal of HM within OST is to decrease opioid use. A secondary goal is to reduce non-opioid substance use, because OST outcomes are worse among patients who abuse cocaine (Levine *et al.* 2015; Sofuoglu *et al.* 2003), alcohol (Chatham *et al.* 1997; Roux *et al.* 2014) and non-prescribed benzodiazepines (Brands *et al.* 2008; White *et al.* 2014), whereas there is no reliable deleterious effect of marijuana use on OST outcomes.

Unfortunately, few effective medications exist for treating abuse of non-opioid drugs. Clinical study findings indicate sustained-release *d*-amphetamine can reduce cocaine use among OUD individuals maintained on methadone (Grabowski *et al.* 2004), buprenorphine (Greenwald *et al.* 2010) or HAT (Nuitjen *et al.* 2016); however, safety/diversion concerns regarding agonist pharmacotherapy for cocaine may limit its use. Alcohol use disorder medications have limited efficacy/acceptability (acamprosate, disulfiram) or are incompatible with concurrent OST (naltrexone). Smoking cessation medications can be safely integrated with OST (Reid *et al.* 2011); however, quit rates are lower among OST patients than the general population (Miller and Sigmon 2015). During OST, adjunctive medications may be useful in attenuating stress-related drug use (Kowalczyk *et al.* 2015).

Fortunately, behavioral interventions have been found to decrease non-opioid use during OST. These include contingency management approaches for reducing cocaine use (Blanken *et al.* 2016; Petry *et al.* 2015). Brief interventions can reduce hazardous alcohol use (Darker *et al.* 2012), which in OUD patients increases risks of opioid relapse (Clark *et al.* 2015) and hepatitis C-related mortality (Hser *et al.* 2001). Some OST clinics prohibit benzodiazepine use to avoid toxicity in combination with opioids (Lee *et al.* 2014).

Although all USA states now have prescription drug monitoring program (PDMP) databases, most do not mandate that clinicians use these data. Lack of PDMP utilization is problematic in the context of OST, given the high rates of polysubstance use. Using state-level data from 1999–2013, Patrick *et al.* (2016) found that PDMP implementation reduced annual opioid overdose death rates, and programs that monitored at least four drug schedules and updated data at least weekly observed greater overdose reductions than programs without these characteristics.

There is minimal literature on PDMP utilization in the setting of OST; thus, we pilot-tested a protocol in our methadone program involving convergent use of PDMP, pill count, and urine drug screen data (Christensen *et al.* 2010). Among all patients (n=158), 33% had a positive PDMP report; 8% were using prescribed medications appropriately, and 25% had a positive PDMP report unaccounted for by pill count, suggesting misuse or diversion. Within the latter subgroup, 61% had several prescriptions, especially opioids and benzodiazepines, and 35% had prescriptions for both drug classes. This protocol detected high rates of misuse/diversion and should be considered for use in OST programs.

Integration of holistically oriented care

Reduce structural barriers

Treatment programs vary along a continuum from HM vs rehabilitation. Relative to rehabilitation-oriented programs, HM facilities have higher patient/staff ratios, tolerate ongoing drug use, conduct urine drug testing less frequently, use less case management, and serve patients with more socioeconomic problems (Gjersing *et al.* 2010). Opioid substitution treatment staff attitudes are important for supporting HM services: negative staff attitudes include concerns about excess workload (Hobden and Cunningham 2006), whereas favorable attitudes are associated with higher education levels and knowledge about HIV/AIDS (Deren *et al.* 2011).

Personalized care

Some data suggest that OST retention might be improved by making care sensitive to gender (Levine *et al.* 2015; Liebrenz *et al.* 2014), race (Mancino *et al.* 2010; Saxon *et al.* 1996), ethnicity (Grella *et al.* 1995; Liebrenz *et al.* 2014) and age (Degenhardt *et al.* 2008; Strike *et al.* 2005).

Use of case management and peer counselors can facilitate OST service delivery (Treloar *et al.* 2015). In one study, MMT clients receiving enhanced case management (links to services and drop-in centers with access to staff) had fewer no-shows and reduced social, family and psychiatric problems – but not drug use – than patients receiving standard psychosocial support (Hesse and Pedersen 2008). These services are particularly important for patients in lower socioeconomic strata.

Technology-based treatment (e.g. texting, internet, and interactive voice response) can also be used to provide tailored, real-time communication in high-risk situations (Mitchell *et al.* 2015; Moore *et al.* 2013).

Integrated psychiatric care

Psychiatric conditions are common among OST patients and complicate clinical management of OUD (Brooner *et al.* 2013; Marienfeld and Rosenheck 2015). Most patients do not receive psychiatric care during OST or they receive community treatment referral that often results in poor attendance (King *et al.* 2014; McGovern *et al.* 2006). Evidence indicates that co-located and integrated psychiatric services within OST can improve attendance at psychiatrist appointments and reduce psychiatric distress, but may not reduce substance use (Brooner *et al.* 2013; Kidorf *et al.* 2015).

Integrated physical healthcare

Opioid substitution treatment is de facto HIV prevention (Bruce 2010; Metzger and Zhang 2010). Directly observed therapy during OST can improve adherence to medications for treating HIV (Bach *et al.* 2015; Nosyk *et al.* 2015), hepatitis (Grebely *et al.* 2016; Masson *et al.* 2013), and tuberculosis (Gourevitch *et al.* 1996; Snyder *et al.* 1999).

Coordination of care should ideally include on-site screening and treatment of other comorbidities (Bachireddy *et al.* 2014; Fareed *et al.* 2010). This is important because heroin users consume expensive emergency services at higher rates than the general public (Chen *et al.* 2015). Safer-sex education and brief interventions should be folded into on-site services (Calsyn *et al.* 2009).

Many individuals entering OST suffer from chronic pain (Becker *et al.* 2015), which complicates management of opioid addiction. In a recent meta-analysis, chronic non-cancer pain (which was related to the presence of psychiatric disorders) did not affect illegal *opioid* use, but had a significant protective effect for reducing illegal *non-opioid* substance use (Dennis *et al.* 2015).

Reduce substance use during pregnancy

Pregnant women in OST can benefit from interventions to reduce non-opioid substance abuse, which can yield future health benefits for the child. Although evidence quality in this field of investigation is limited (Terplan et al. 2015), some studies suggest brief interventions or contingency management can reduce cigarette smoking in OST patients (Akerman et al. 2015; Ram et al. 2016). Addiction care should continue postpartum and always be coordinated with obstetric care (Jones et al. 2014).

Decrease injection drug use

Heroin-assisted treatment embraces the practice that injection should occur in a safe setting with sanitary needles/syringes, leading to greater contact with medical personnel toward reducing morbidities of injection drug use (EMCDDA 2012). Overlap in clientele between OST and needle/syringe exchange programs for injection drug users (IDUs) presents the opportunity to integrate services (Kidorf et al. 2009; Strathdee et al. 2006). Methadone and buprenorphine also help to reduce injection drug use, HIV transmission, and hepatitis C incidence among IDUs (MacArthur et al. 2014; Tsui et al. 2014).

Reduce overdose

Given the alarming rise in opioid overdoses and fatalities (CDC 2015), overdose prevention has become a critical goal. Aside from education, a major initiative has been to increase distribution of the opioid antagonist naloxone to first responders (Mueller et al. 2015), and enacting Good Samaritan state laws that encourage overdose witnesses to respond. Injectable naloxone has been available for decades, but has limited reach. Recently, medical use of intranasal naloxone was approved in the USA.

Community pharmacies can contribute to opioid overdose risk reduction. Naloxone dispensing from pharmacies has increased recently (Jones et al. 2016). Pharmacists can access PDMP data to assess concurrent medications, and can offer training and counseling, but there are also medico-legal barriers (Nielsen and Van Hout 2016).

Emergency departments are an important venue for addressing overdose prevention and connection to services. This setting offers a "reachable moment" for overdose prevention, brief intervention and transition to treatment that can potentially improve longer-term outcomes (Bohnert et al. 2016; D'Onofrio et al. 2015). Although emergency department physicians mostly support HM interventions, barriers include lack of knowledge, time, training, and institutional support (Samuels et al. 2016).

Reducing relapse, overdose, and harms among prisoners

When opioid-dependent individuals enrolled in OST are incarcerated in USA jails or prisons, most are denied OST (Fiscella et al. 2004); in contrast, OST is typically made available to prisoners in Europe, Canada and Australia (Larney and Dolan 2009). Discontinuation of OST decreases opioid tolerance, which can increase overdose risk after correctional facility release (Møller et al. 2010). Offering OST during incarceration is associated with improved HIV treatment adherence (Wolfe et al. 2010), engagement in OST after community re-entry, less post-release relapse, but not all studies indicate successful long-term outcomes (Hedrich et al. 2012).

Conclusions and reflections

Opioid substitution treatment using methadone or buprenorphine maintenance is a proven modality for improving retention, reducing drug use and associated harms, and several nations have instituted HAT as an alternative approach for treatment-resistant individuals. Regardless of the opioid agonist, this population requires comprehensive and integrated care to reduce the personal and public health burden of OUD. Viewed from ethical or economic perspectives, societies cannot afford silo-based approaches that disaggregate care because it is also well established that service fragmentation leads to worse treatment adherence and outcomes.

Harm minimization approaches are focused on holistic care that is pragmatic in its dual emphasis on safety and effectiveness. The HM interventions are compatible with OST and available evidence indicates they are cost-effective. Scaling up and synthesizing these treatment frameworks may produce synergistic effects on health-economic outcomes (Murphy and Polsky 2016; Wilson *et al.* 2015). On the other hand, it is also clear that HM initiatives, despite supporting empirical data, face widespread resistance to implementation. Societies and cultures that historically adopt moralistic, rather than medical, views of addiction are usually refractory to accepting HM approaches. The USA – whose drug policies have lagged behind those of many European countries – offers an instructive example, as it stands at a pragmatic (if not philosophical) crossroads to these issues. The ongoing opioid epidemic, which has led to unprecedented numbers of overdoses and deaths, has forced some softening of drug policy, e.g. many states have increased implementation of certain HM approaches such as naloxone distribution and use. It remains to be seen what further steps might be adopted.

Another challenge is the *context* in which HM approaches are implemented. This author oversees an opioid (primarily methadone) treatment research clinic, and there is a useful example to be offered about these challenges. We recently tested a pilot program in this setting; in collaboration with our treatment funding authority, patients who would otherwise have been discharged from other local MMT clinics for failure to comply with their policies (e.g., non-abstinent, diversion, non-adherence to program rules) were instead administratively transferred to our clinic. Our "Second Chance" program was designed with HM principles, and enabled patients to be retained despite providing positive drug urinalysis results. These patients were assessed at intake and presented with high rates of psychiatric comorbidities, many had prescription opioids or benzodiazepines, half were injection drug users, and about half received disability benefits. These factors are viewed as barriers to patients' prior inability to become abstinent. Nonetheless, in our program, higher methadone doses predicted longer retention and less opioid use, suggesting this complex population is responsive to OST and other services that were made available to them (Tavakoli *et al.* in preparation). Notably, the "Second Chance" program was implemented within our standard clinic, directly alongside patients who were not subject to similar contingencies (e.g. there was higher expectation of abstinence among standard patients). Existing patients and several staff members had difficulty accepting this new subgroup of patients, particularly the less-restrictive approach of allowing transferred patients to test drug-positive repeatedly without being discharged. In short, even seasoned addiction treatment professionals working in an academic treatment research clinic who are accustomed to the typical challenges presented by these patients, found it difficult to bridge this philosophical gap. We had to re-educate our staff regarding goals of this program and, even then, some remained resistant. This experience suggests that necessary components of a successful OST + HM program may include a sensitive leadership structure, staff members that embrace HM principles, clearly articulated program rules, surveillance of outcomes, ongoing discussion, and possible re-calibration of attitudes and approaches based on evidence.

In summary, the landscape of opioid addiction is changing worldwide and is complex to manage. Opioid substitution treatment with HM approaches are part of an evolution in care of this population, but implementation of these programs are neither inevitable nor without significant challenges. Success of such programs should be driven primarily by empirical data, but it is also evident that bridging philosophical and political divides will be a rate-limiting step in the adoption of these important interventions. Future initiatives should focus on methods that can improve: cost-effectiveness and sustainability of specific HM approaches, integration of services, and access/reach of these interventions, toward achieving personal and public health gains in outcomes for individuals with OUD.[1]

Acknowledgements

Financial support: NIH 2 R01 DA015462, Helene Lycaki/Joe Young, Jr Funds (State of Michigan), and the Detroit Wayne Mental Health Authority supported preparation of this manuscript.

Note

1 Conflict of interest statement: The author has received compensation from Indivior PLC for research consulting related to Phase II and III development of RBP-6000 (Sublocade®), and from Titan Pharmaceuticals as site investigator for Phase II development of Probuphine®.

References

Akerman, S. C., et al. (2015) "Treating tobacco use disorder in pregnant women in medication-assisted treatment for an opioid use disorder: a systematic review", *Journal of Substance Abuse Treatment* 52: 40–47.

Alistar, S. S., et al. (2011) "Effectiveness and cost effectiveness of expanding harm reduction and antiretroviral therapy in a mixed HIV epidemic: a modeling analysis for Ukraine", *PLOS Medicine* 8(3): e1000423.

Amato, L., et al. (2012) "Should psychosocial intervention be added to pharmacological treatment for opiate abuse/dependence? An overview of systematic reviews of the literature", *Italian Journal of Public Health* 3: 15–20.

Bach, P., et al. (2015) "Association of patterns of methadone use with antiretroviral therapy discontinuation: a prospective cohort study", *BMC Infectious Diseases* 15: 537.

Bachireddy, C., et al. (2014) "Integration of health services improves multiple healthcare outcomes among HIV-infected people who inject drugs in Ukraine", *Drug and Alcohol Dependence* 134(1): 106–114.

Bao, Y.-P., et al. (2009) "A meta-analysis of retention in methadone maintenance by dose and dosing strategy", *American Journal of Drug and Alcohol Abuse* 35(1): 28–33.

Becker, W. C., et al. (2015) "Buprenorphine/naloxone dose and pain intensity among individuals initiating treatment for opioid use disorder", *Journal of Substance Abuse Treatment* 48(1): 128–131.

Blanken, P., et al. (2010) "Outcome of long-term heroin-assisted treatment offered to chronic, treatment-resistant heroin addicts in the Netherlands", *Addiction* 105(2): 300–308.

Blanken, P., et al. (2016) "Efficacy of cocaine contingency management in heroin-assisted treatment: results of a randomized controlled trial", *Drug and Alcohol Dependence* 164: 55–63.

Bohnert, A. S. B., et al. (2016) "A pilot randomized clinical trial of an intervention to reduce overdose risk behaviors among emergency department patients at risk for prescription opioid overdose", *Drug and Alcohol Dependence* 163: 40–47.

Brands, B., et al. (2008) "The impact of benzodiazepine use on methadone maintenance treatment outcomes", *Journal of Addictive Diseases* 27(3): 37–48.

Brooner, R. K., et al. (2004) "Behavioral contingencies improve counseling attendance in an adaptive treatment model", *Journal of Substance Abuse Treatment* 27(3): 223–232.

Brooner, R. K., et al. (2013) "Managing psychiatric comorbidity within versus outside of methadone treatment settings: a randomized and controlled evaluation", *Addiction* 108(11): 1942–1951.

Bruce, R. D. (2010) "Methadone as HIV prevention: high volume methadone sites to decrease HIV incidence rates in resource limited settings", *International Journal of Drug Policy* 21(2): 122–124.

Calsyn, D. A., et al. (2009) "Motivational and skills training HIV/sexually transmitted infection sexual risk reduction groups for men", *Journal of Substance Abuse Treatment* 37(2): 138–150.

Cao, X., et al. (2014) "Retention and its predictors among methadone maintenance treatment clients in China: a six-year cohort study", *Drug and Alcohol Dependence* 145: 87–93.

Center for Substance Abuse Treatment (2005) "Medication-assisted treatment for opioid addiction in opioid treatment programs", in *Treatment Improvement Protocol (TIP) Series 43, HHS publication No. (SMA) 12-4214*, Rockville, MD: Substance Abuse and Mental Health Services Administration.

Chatham, L. R., et al. (1997) "Heavy drinking, alcohol-dependent vs. nondependent methadone-maintenance clients: a follow-up study", *Addictive Behaviors* 22(1): 69–80.

Chen, I. M., et al. (2015) "Health service utilization of heroin abusers: a retrospective cohort study", *Addictive Behaviors* 45: 281–286.

Christensen, C., et al. (2010) "Use of the Michigan Automated Prescription System (MAPS) to monitor and manage diversion in an urban methadone clinic", in *Problems of Drug Dependence 2010: Proceedings of the 72nd Annual Scientific Meeting*, The College on Problems of Drug Dependence, Inc., www.cpdd.org/Pages/Meetings/Meetings_PDFs/2010AbstractBook.pdf, p. 28.

Clark, R. E., et al. (2015) "Risk factors for relapse and higher costs among Medicaid members with opioid dependence or abuse: opioid agonists, comorbidities, and treatment history", *Journal of Substance Abuse Treatment* 57: 75–80.

Darker, C. D., et al. (2012) "Brief interventions are effective in reducing alcohol consumption in opiate-dependent patients: results from an implementation study", *Drug and Alcohol Review* 31(3): 348–356.

Degenhardt, L., et al. (2008) "Drug use and risk among regular injecting drug users in Australia: does age make a difference?" *Drug and Alcohol Review* 27(4): 357–360.

Dennis, B., et al. (2015) "Impact of chronic pain on treatment prognosis for patients with opioid use disorder: a systematic review and meta-analysis", *Substance Abuse: Research and Treatment* 9(9): 59–80.

Deren, S., et al. (2011) "Attitudes of methadone program staff toward provision of harm-reduction and other services", *Journal of Addiction Medicine* 5(4): 289–292.

D'Onofrio, G., et al. (2015) "Emergency department-initiated buprenorphine/naloxone treatment for opioid dependence: a randomized clinical trial", *JAMA: The Journal of the American Medical Association* 313(16): 1636–1644.

EMCDDA (2012) *New Heroin-Assisted Treatment. Recent Evidence and Current Practices of Supervised Injectable Heroin Treatment in Europe and Beyond*. European Monitoring Centre for Drugs and Drug Addiction, Luxembourg: Publication Office of the European Union.

Faggiano, F., et al. (2003) "Methadone maintenance at different dosages for opioid dependence", *Cochrane Database of Systematic Reviews* 3: CD002208.

Fareed, A., et al. (2010) "On-site basic health screening and brief health counseling of chronic medical conditions for veterans in methadone maintenance treatment", *Journal of Addiction Medicine* 4: 160–166.

Fareed, A., et al. (2012) "Effect of buprenorphine dose on treatment outcome", *Journal of Addictive Diseases* 31(1): 8–18.

Fiscella, K., et al. (2004) "Jail management of arrestees/inmates enrolled in community methadone maintenance programs", *Journal of Urban Health* 81(4): 645–654.

Friedman, P., et al. (1994) "Retention of patients who entered methadone maintenance via an interim methadone clinic", *Journal of Psychoactive Drugs* 26(2): 217–221.

Gibson, A., et al. (2008) "Exposure to opioid maintenance treatment reduces long-term mortality", *Addiction* 103(3): 462–468.

Gjersing, L., et al. (2010) "Staff attitudes and the associations with treatment organisation, clinical practices and outcomes in opioid maintenance treatment", *BMC Health Services Research* 10: 194.

Gourevitch, M. N., et al. (1996) "Successful adherence to observed prophylaxis and treatment of tuberculosis among drug users in a methadone program", *Journal of Addictive Diseases* 15(1): 93–104.

Grabowski, J., et al. (2004) "Agonist-like or antagonist-like treatment for cocaine dependence with methadone for heroin dependence: two double-blind randomized clinical trials", *Neuropsychopharmacology* 29(5): 969–981.

Grebely, J., et al. (2016) "Treatment for hepatitis C virus infection among people who inject drugs attending opioid substitution treatment and community health clinics: the ETHOS study", *Addiction* 111(2): 311–319.

Greenwald, M. K., et al. (2010) "Sustained release d-amphetamine reduces cocaine- but not 'speedball'-seeking behavior in buprenorphine-maintained volunteers: a test of dual agonist pharmacotherapy for heroin/cocaine polydrug abusers", *Neuropsychopharmacology* 35(13): 2624–2637.

Greenwald, M. K., et al. (2014) "Buprenorphine maintenance and *mu*-opioid receptor availability in the treatment of opioid use disorder: implications for clinical use and policy", *Drug and Alcohol Dependence* 144: 1–11.

Grella, C. E., et al. (1995) "Ethnic differences in HIV risk behaviors, self-perceptions, and treatment outcomes among women in methadone maintenance treatment", *Journal of Psychoactive Drugs* 27: 421–433.

Haasen, C., et al. (2007) "Heroin-assisted treatment for opioid dependence", *British Journal of Psychiatry* 191(1): 55–62.

Hammett, T. M., et al. (2008) "'Social evils' and harm reduction: the evolving policy environment for human immunodeficiency virus prevention among injection drug users in China and Vietnam", *Addiction* 103(1): 137–145.

Hartzler, B., et al. (2010) "Dissolution of a harm reduction track for opiate agonist treatment: longitudinal impact on treatment retention, substance use and service utilization", *International Journal of Drug Policy* 21(1): 82–85.

Hedrich, D., et al. (2012) "The effectiveness of opioid maintenance treatment in prison settings: a systematic review", *Addiction* 107(3): 501–517.

Hesse, M. and Pedersen, M. U. (2008) "Easy-access services in low-threshold opiate agonist maintenance", *International Journal of Mental Health and Addiction* 6(3): 316–324.

Hobden, K. L. and Cunningham, J. A. (2006) "Barriers to the dissemination of four harm reduction strategies: a survey of addiction treatment providers in Ontario", *Harm Reduction Journal* 3: 35.

Hser, Y.-I., et al. (2001) "A 33-year follow-up of narcotics addicts", *Archives of General Psychiatry* 58(5): 503–508.

Islam, M. M., et al. (2013) "Defining a service for people who use drugs as low threshold: what should be the criteria?" *International Journal of Drug Policy* 24: 220–222.

Jones, C. M., et al. (2016) "Increase in naloxone prescriptions dispensed in US retail pharmacies since 2013", *American Journal of Public Health* 106(4): 689–690.

Jones, H. E., et al. (2014) "Clinical care for opioid-using pregnant and post-partum women: the role of obstetric providers", *American Journal of Obstetrics and Gynecology* 210(4): 302–310.

Kidorf, M., et al. (2009) "Improving substance abuse treatment enrollment in community syringe exchangers", *Addiction* 104(5): 786–795.

Kidorf, M., et al. (2015) "Substance use and response to psychiatric treatment in methadone-treated outpatients with comorbid psychiatric disorder", *Journal of Substance Abuse Treatment* 51: 64–69.

King, V. L., et al. (2014) "Challenges and outcomes of parallel care for patients with co-occurring psychiatric disorder in methadone maintenance treatment", *Journal of Dual Diagnosis* 10(2): 60–67.

Kourounis, G., et al. (2016) "Opioid substitution therapy: lower the treatment thresholds", *Drug and Alcohol Dependence* 161: 1–8.

Kowalczyk, W. J., et al. (2015) "Clonidine maintenance prolongs opioid abstinence and decouples stress from craving in daily life: a randomized controlled trial with ecological momentary assessment", *American Journal of Psychiatry* 172(8): 760–767.

Krook, A. L., et al. (2002) "A placebo-controlled study of high dose buprenorphine in opiate dependents waiting for medication-assisted rehabilitation in Oslo, Norway", *Addiction* 97(5): 533–542.

Larney, S. and Dolan, K. (2009) "A literature review of international implementation of opioid substitution treatment in prisons: equivalence of care?" *European Addiction Research* 15(2): 107–112.

Lawrinson, P., et al. (2008) "Key findings from the WHO collaborative study on substitution therapy for opioid dependence and HIV/AIDS", *Addiction* 103(9): 1484–1492.

Lee, S. C., et al. (2014) "Comparison of toxicity associated with nonmedical use of benzodiazepines with buprenorphine or methadone", *Drug and Alcohol Dependence* 138: 118–123.

Levine, A. R., et al. (2015) "Gender-specific predictors of retention and opioid abstinence during methadone maintenance treatment" *Journal of Substance Abuse Treatment* 54: 37–43.

Liebrenz, M., et al. (2014) "Ethnic- and gender-specific differences in the prevalence of HIV among patient in opioid treatment – a case register analysis", *Harm Reduction Journal* 11: 23.

MacArthur, G. J., et al. (2014) "Interventions to prevent HIV and hepatitis C in people who inject drugs: a review of reviews to assess evidence of effectiveness", *International Journal of Drug Policy* 25(1): 34–52.

Mancino, M., et al. (2010) "Predictors of attrition from a national sample of methadone maintenance patients", *American Journal of Drug and Alcohol Abuse* 36(3): 155–160.

Marienfeld, C. and Rosenheck, R. A. (2015) "Psychiatric services and prescription fills among veterans with serious mental illness in methadone maintenance treatment", *Journal of Dual Diagnosis* 11(2): 128–135.

Marsch, L. A., et al. (2005) "Comparison of pharmacological treatments for opioid-dependent adolescents: a randomized controlled trial", *Archives of General Psychiatry* 62(10): 1157–1164.

Masson, C. L., et al. (2013) "A randomized trial of a hepatitis care coordination model in methadone maintenance treatment", *American Journal of Public Health* 103(10): e81–88.

McGovern, M. P., et al. (2006) "Addiction treatment services and co-occurring disorders: prevalence estimates, treatment practices, and barriers", *Journal of Substance Abuse Treatment* 31(3): 267–275.

McLellan, A. T., et al. (1993) "The effects of psychosocial services in substance abuse treatment", *JAMA: The Journal of the American Medical Association* 269(15): 1953–1959.

Metzger, D. S. and Zhang, Y. (2010) "Drug treatment as HIV prevention: expanding treatment options", *Current HIV/AIDS Reports* 7(4): 220–225.

Miller, M. E. and Sigmon, S. C. (2015) "Are pharmacotherapies ineffective in opioid-dependent smokers? Reflections on the scientific literature and future directions", *Nicotine & Tobacco Research* 17(8): 955–959.

Mitchell, S. G., et al. (2015) "The use of technology in participant tracking and study retention: lessons learned from a Clinical Trials Network study", *Substance Abuse* 36(4): 420–426.

Møller, L. F., et al. (2010) "Acute drug-related mortality of people recently released from prisons", *Public Health* 124(11): 637–639.

Moore, B. A., et al. (2013) "The Recovery Line: a pilot trial of automated, telephone-based treatment for continued drug use in methadone maintenance", *Journal of Substance Abuse Treatment* 45(1): 63–69.

Mueller, S. R., et al. (2015) "A review of opioid overdose prevention and naloxone prescribing: implications for translating community programming into clinical practice", *Substance Abuse* 36(2): 240–253.

Murphy, S. M. and Polsky, D. (2016) "Economic evaluations of opioid use disorder interventions", *PharmacoEconomics* 34: 863–887.

Nielsen, S. and Van Hout, M. C. (2016) "What is known about community pharmacy supply of naloxone? A scoping review", *International Journal of Drug Policy* 32: 24–44.

Nosyk, B., et al. (2015) "The causal effect of opioid substitution treatment on HAART medication refill adherence", *AIDS* 29(8): 965–973.

Nuitjen, M., et al. (2016) "Sustained-release dexamfetamine in the treatment of chronic cocaine-dependent patients on heroin-assisted treatment: a randomised, double-blind, placebo-controlled trial", *The Lancet* 387(10034): 2226–2234.

Patrick, S. W., et al. (2016) "Implementation of prescription drug monitoring programs associated with reductions in opioid-related death rates", *Health Affairs* 35(7): 1324–32.

Petry, N. M., et al. (2015) "Standard magnitude prize reinforcers can be as efficacious as larger magnitude reinforcers in cocaine-dependent methadone patients", *Journal of Consulting and Clinical Psychology* 83(3): 464–472.

Ram, A., et al. (2016) "Cigarette smoking reduction in pregnant women with opioid use disorder", *Journal of Addiction Medicine* 10(1): 53–59.

Reid, M. S., et al. (2011) "Smoking cessation treatment among patients in community-based substance abuse rehabilitation programs: exploring predictors of outcome as clues toward treatment improvement", *American Journal of Drug and Alcohol Abuse* 37: 472–478.

Rhodes, G. L., et al. (2003) "Improving on-time counseling attendance in a methadone treatment program: a contingency management approach", *American Journal of Drug and Alcohol Abuse* 29(4): 759–773.

Ritter, A. and Cameron, J. (2006) "A review of the efficacy and effectiveness of harm reduction strategies for alcohol, tobacco and illicit drugs", *Drug and Alcohol Review* 25(6): 611–624.

Roux, P., et al. (2014) "Predictors of non-adherence to methadone maintenance treatment in opioid-dependent individuals: implications for clinicians", *Current Pharmaceutical Design* 20(25): 4097–4105.

Samuels, E. A., et al. (2016) "Emergency department-based opioid harm reduction: moving physicians from willing to doing", *Academic Emergency Medicine* 23(4): 455–465.

Saxon, A. J., et al. (1996) "Pre-treatment characteristics, program philosophy and level of ancillary services as predictors of methadone maintenance treatment outcome", *Addiction* 91(8): 1197–1209.

Schwartz, R. P., et al. (2006) "A randomized controlled trial of interim methadone maintenance", *Archives of General Psychiatry* 63(1): 102–109.

Schwartz, R. P., et al. (2009) "Interim methadone treatment: impact on arrests", *Drug and Alcohol Dependence* 103(3): 148–154.

Snyder, D. C., et al. (1999) "Tuberculosis prevention in methadone maintenance clinics. Effectiveness and cost-effectiveness", *American Journal of Respiratory and Critical Care Medicine* 160: 178–185.

Sofuoglu, M., et al. (2003) "Prediction of treatment outcome by baseline urine cocaine results and self-reported cocaine use for cocaine and opioid dependence", *American Journal of Drug and Alcohol Abuse* 29(4): 713–727.

Strathdee, S. A., et al. (2006) "Facilitating entry into drug treatment among injection drug users referred from a needle exchange program: results from a community-based behavioral intervention trial", *Drug and Alcohol Dependence* 83(3): 225–232.

Strike, C. J., et al. (2005) "Factors predicting 2-year retention in methadone maintenance treatment for opioid dependence", *Addictive Behaviors* 30(5): 1025–1028.

Strike, C. J., et al. (2013) "What is low threshold methadone maintenance treatment?" *International Journal of Drug Policy* 24: e51–56.

Sun, H.-M., et al. (2015) "Methadone maintenance treatment programme reduces criminal activity and improves social well-being of drug users in China: a systematic review and meta-analysis", *BMJ Open* 5(1): e005997.

Tavakoli, E., Deal, E., Rhodes, G. L., Mischel, E., Christensen, C., Amirsadri, A., and Greenwald, M. K. (in preparation) "Predictors of retention and drug use among previously treatment-resistant patients transferred to a specialty 'Second Chance' methadone maintenance program".

Terplan, M., et al. (2015) "Psychosocial interventions for pregnant women in outpatient illicit drug treatment programs compared to other interventions", *Cochrane Database of Systematic Reviews* 4: CD006037.

Treloar, C., et al. (2015) "Evaluation of two community-controlled peer support services for assessment and treatment of hepatitis C virus infection in opioid substitution treatment clinics: the ETHOS study, Australia", *International Journal of Drug Policy* 26(10): 992–998.

Tsui, J. I., et al. (2014) "Association of opioid agonist therapy with lower incidence of hepatitis C virus infection in young adult injection drug users", *JAMA: Internal Medicine* 174(12): 1974–1981.

UNODC (2016) "Reducing the adverse health consequences of drug abuse: a comprehensive approach", www.unodc.org/documents/prevention/Reducing-adverse-consequences-drug-abuse.pdf (accessed June 28 2016).

Verthein, U., et al. (2008) "Long-term effects of heroin-assisted treatment in Germany", *Addiction* 103(6): 960–966.

White, W. L., et al. (2014) "Patterns of abstinence or continued drug use among methadone maintenance patients and their relation to treatment retention", *Journal of Psychoactive Drugs* 46(2): 114–122.

Wilson, D. P., et al. (2015) "The cost-effectiveness of harm reduction", *International Journal of Drug Policy* 26(s1): S5–S11.

Wilson, M. E., et al. (2010) "Impact of interim methadone maintenance on HIV risk behaviors", *Journal of Urban Health* 87(4): 586–591.

Wolfe, D., et al. (2010) "Treatment and care for injecting drug users with HIV infection: a review of barriers and ways forward", *The Lancet* 376(9738): 355–366.

39
SELF-CHANGE
Genesis and functions of a concept

Harald Klingemann and Justyna I. Klingemann

Traditionally, addiction has been understood from the clinical perspective as an inexorably progressive 'disease' with addiction-related problems getting more severe and damaging in the absence of therapeutic intervention. The clinical focus on severely dependent populations, representing only the tip of the iceberg when considering epidemiological studies, supported such a deterministic disease concept (Room 1977), while in fact self-change is the major path out of addiction (Dawson *et al.* 2005, 2006; Smart 2007). The concept of self-change (called also "natural recovery", "spontaneous remission" or "self-healing"), that is overcoming addiction without professional help or exiting an addiction career without formal intervention (Klingemann 2004; Stall and Biernacki 1986), has challenged this paradigm and received broad empirical support as numerous reviews of the literature have shown (Klingemann and Sobell 2007; Klingemann *et al.* 2010; Smart 1976; Sobell *et al.* 2000; Waldorf and Biernacki 1979). This contribution positions the self-change concept within a broader theoretical context and highlights its central assumption in comparison. Moreover, the diffusion of the self-change concept as an innovation in the addiction field is described from a socio-historical perspective. Finally the ideological, normative implications of this concept and research are highlighted.

Anatomy of the concept

The explanation and dynamics of types of social behavior considered as "deviant" or problematic have been discussed within general frameworks of social sciences including constructivist and structural/functional theoretical traditions (Clinard and Meier 2011; Hammersley this volume). Models and theories of addiction represent specific applications of these general approaches – even though this is often not recognized – and share their basic assumptions and limitations. From a structural point of view, research on social problems traditionally has focused on societal risk factors and individual vulnerabilities to explain previously defined deviant behavior. Phenomenological constructivist approaches, e.g. in criminology, have explored the genesis of the public perception of social problems (Gusfield 1986) and forwarded the concept of deviant careers (Becker 1963; Faupel 2011). However, both perspectives assume a passive role of the individual being subject of various risk factors and societal opportunity structures (Merton 1938). Both frameworks imply that, without societal response or intervention, deviant behavior patterns would persist or get worse. From that perspective promoting self-change at best would imply structural changes in society, redefining societal norms underpinning the definition of deviant behaviors and giving up control policies, which intensify deviant careers and reinforce deviant self-images (Becker 1963). The perception of addiction as an individual failure and moral weakness potentially was countered by the adoption of the disease concept, which attributes a passive patient role to the individual (Parsons 1966) and tries to reduce stigma. "Loss of control" is serving as a central definitional element of addiction implying a deterministic course of the disease. From that perspective, the possibility of self-change has been questioned by the tautological claim that "an ability to cease addictive behaviors on one's own, suggests that the individual was not addicted in the first place" (Chiauzzi and Liljegren 1993: 306). Furthermore, the definition of recovery from addiction focuses mainly on lifelong abstinence and becomes a dominating treatment objective, persisting independently of scientific evidence with the controlled drinking debate serving as a prominent example (Sobell and Sobell 1995). The trajectory of deviant drinking careers exclusively leading to "rock bottom" excludes a positive holistic biographical approach. Instead, it frames the therapeutic process as destroying the "old addicted identity" and building a new one as "the recovering addict." This perspective undermines the autonomy of the addict, and ignores the whole spectrum of skills necessary to pursue addiction careers (Klingemann 1999), which was a classic topic of the sociological Chicago school when deconstructing the concept of "anti-social" behavior and exploring "deviant" sub- and counter-cultures (Anderson 1923; Sutherland 1924). Moreover, the process of bio-medicalization fuelled by behavioral neuroscience has further reduced the necessity to include socio-cultural factors into models of recovery from addiction, claiming that addiction is not a choice for severely affected individuals (Leyton 2013).

The self-change perspective offers a paradigm change, claiming that the addicted individual is able to make informed choices and has the potential to regain control over his/her life by developing individual recovery strategies and proactively seeking support. In fact, most individuals facing addiction problems try to cope with it on their own, and many of them do succeed. There are many reasons for not seeking professional support, but some self-changers simply believed that they can overcome addiction on their own, wanted to do it on their own terms and felt proud of succeeding (Klingemann 1991, 1992). Self-changers often differ from clinical populations by having less severe problems, a more stable social situation, and more resources (Bischof et al. 2004; Granfield and Cloud 2001; Russel et al. 2001; Sobell et al. 1993). Exploration of the self-change pathways showed that severe negative consequences of addiction and the "hitting the rock bottom" phase is not a necessary element of recovery: some self-changers are influenced by meaningful changes in their social life or significant others; there are also some who overcame addiction by drifting out harmoniously without specific turning points (Cunningham et al. 2005; Granfield and Cloud 2001; Klingemann 1991; Klingemann 2011). Moreover, it seems that not the specific components, but the multidimensionality of change emerges as the most important feature of maintaining self-change (Klingemann 2012).

Addicted individuals are able to pursue proactively their personal change agenda and choose between abstinence and reduced risk consumption as their goals (Klingemann et al. 2010). King and Tucker (2000) found three major ways in which self-changers were coping with alcohol addiction: (1) stable abstinence; (2) a short period of stable abstinence followed by low risk drinking; and (3) gradually lowered consumption to the level of low risk drinking. Humphreys suggests that self-changers with higher socio-economic status and less severe addiction problems tend to choose low risk drinking, while the others choose abstinence (Humphreys et al. 1995).

Cloud and Granfield (2008) specified resources (social, physical, human and culture) in individuals' social environments that are relevant for change of addictive behavior, conceptualizing it as the "recovery capital." Those resources are unequally distributed across society, therefore the treatment providers could learn from self-changers and provide 'support on request' facilitating self-change. The "assisted self-change" concept enhances individual stock taking and decisional balancing (e.g. by motivational interviewing) and provides tools for self-monitoring (Cohn et al. 2011). Public health models stop short of this vision, promoting interventions to reduce health illiteracy in order to empower people (Kickbusch and Maag 2008) and discarding the idea of "a world of smart people" (Klingemann and Bergmark 2006). Consequently, this strength-based approach implies a dramatic shift in the relationship between expert and lay referral systems (Freidson 1961), currently mirrored in the general trend towards patient-driven health care models with the empowered individual at the center of action-taking and the physician's role as care consultant and health collaborator (Swan 2009).

Diffusion and rejection of the concept

While the first pioneer studies on natural recovery from drug use (Winick 1962), alcoholism (Drew 1968), and smoking and obesity (Schachter 1982) had little impact in the field (Blomqvist 2007), evidence from large-scale consumption surveys in the 1970s, demonstrating the variability of consumption patterns and recovery pathways (Smart 2007), questioned the assumption of the homogeneity of addiction. Yet, resistance and skepticism among the major actors in the field, including the addiction research community, persisted. Barber's classic contribution, "Resistance by scientists to scientific discovery," to the sociology of science (Barber 1961) highlights, among others, the role of "substantive concepts," "methodological conception," and "religious ideas" when it comes to the dynamics of the diffusion of innovations and discoveries in science. Dominating substantive concepts and theories are not simply given up even when contradicted by new empirical evidence, as Barber illustrates with the resistance to the heliocentric theory from the astronomer-scientists at the time. Furthermore, problems to integrate new ideas into established models and methodological conceptions represent barriers to change. Barber exemplifies this with the case of radioactive methods of measurement: "scientists . . . have resisted discovery because of their preference for the evidence of the senses" (Barber 1961: 598). Similarly, most self-changers constitute a hidden population compared with patients of addiction treatment who have been the major focus of research in this field. Studying the process of self-change requires qualitative unconventional approaches such as media recruitment (Klingemann 2010) and mixed methods including holistic life-history approaches, which do not easily fit quantitative statistical survey methods.

Treatment providers also are hardly receptive to the idea of self-change. Would treatment no longer be needed? Is the professional turf threatened? From their daily practice they are focused on the narrow selection of help-seeking and severely addicted patients and consequently view the world as they experience it. Furthermore, treatment providers tend to be influenced by their own convictions and beliefs. Particularly in the United States, we observe the immersion of faith-based treatments. More specifically, addiction counselors, often with little formal training (Sobell and Sobell 2011), share beliefs in a higher power as part of 12-step programs, which promote individual powerlessness as a central feature and consequently discard self-change (Klingemann et al. 2013; Orford this volume). Along the same lines Barber refers to Copernicus' colleagues resisting his ideas because of their own religious beliefs (Barber 1961: 599).

Results of the self-change studies have not translated during the last decade into specific policy making and strength-based empowerment campaigns. Consequently, images of addiction as a chronic and severe disease, as conveyed by the media, continue to dominate in the general population as well (Klingemann and Klingemann 2007; Klingemann et al. 2017).

Theoretical range and blind spots

Limitations of the self-change concept as it has been described here include most prominently a potential individualistic perspective. Even most frequently used terms, such as "self-change," "natural recovery," or "spontaneous remission" imply cognitive coping of the addicted individual and decisional balancing processes weighing costs and benefits. This perspective is compatible with the idea of "assisted self-change" and clinical behavioral approaches, but narrows the initially much wider perspective of the natural trajectory of changes in addiction including "maturing out processes" in which recovery from addiction has been perceived as a consequence of changes of social roles in the life course (Snow 1973; Winick 1962). Survey research showing the non-deterministic changeability of consumption patterns from an epidemiological perspective (Dawson et al. 2005, 2006; Smart 2007) did not make any assumption concerning the nature and causality of the underlying change processes. In this context Cunningham points to the limitations of media-recruited samples regarding reasons for change: "while some people provide reasons that they have really thought about, others may provide reasons that they think they should give, and yet many others provide reasons out of the blue, because they have never even thought about them" (Cunningham et al. 2005: 82). Moreover, the focus on cognitive change processes and a strength-based concept potentially might carry a risk of the revival of moral and guilt concepts – "they can do it on their own but they don't" – if used by liberal policies reducing professional help and implementing financial cuts in social work and counseling. As a more recent antidote to this, and as a complementary element to the approach of assisted self-change from clinical psychology ("self-change clinics"), research from a sociological perspective has stressed physical and socio-environmental structures, cultural context and related life circumstances, which affect one's capacity to reach a stable recovery from addiction. The latest stage of self-change research focuses increasingly on the dimensions of a "self-change friendly society" (Blomqvist et al. 2016; Klingemann and Klingemann 2007; Klingemann et al. 2017). From this macro-societal perspective, images of the nature of addiction in the general population have an impact on the course of change processes. So far, research shows that society seems to be far more pessimistic regarding the chance of self-change than is suggested by epidemiological research on the prevalence of that phenomenon. Although self-change is a major path out of addiction, the governing perception of addiction as a disease that must be treated professionally prevails and tends to be very persistent (Blomqvist 2009; Klingemann et al. 2017). Moreover, studies show that addiction-related stigma – a central part of which are negative, misinformed stereotypes – is extremely persistent and proves to be damaging for the risk groups. Addicts are perceived as lacking willpower, being dangerous and unpredictable and experience less understanding and social support than individuals suffering from other disorders (Schomerus et al. 2011). Self-change studies identified stigma as one of the most important reasons for not seeking treatment – people fear being labeled and discriminated against as "addicts" (Klingemann 1991; Tucker and Vuchinich 1994). Consequently, addiction-related stigma impedes the process of seeking help, aggravating the course of addiction (Moskalewicz and Klingemann 2015; Room 2005).

Another challenge for the self-change concept is the assumption of the universalistic nature of non-assisted change processes, claiming that change processes are driven by the same elements across various types of addictions and, furthermore, that change can be characterized as self-change within a comprehensive model valid within and without treatment contexts (DiClemente 2007). Self-change or desistance from licit and illicit drugs, non-substance-related addictive behaviors, crime and eating disorders, all seem to follow the same pattern. However, most studies have been conducted in Western European and Anglophone countries and models of self-change suffer from an ethnocentric bias. Addiction is a heterogonous phenomenon and mirrors the fact that no social or cultural group is homogeneous. To illustrate this, Barker and Hunt (2007) discuss cultural issues in addiction recovery by contrasting specialist cultures (Western, highly individualist, technologically innovative, egalitarian and secular) with generalist cultures (non-Western, collectivist, hierarchical, sacred/religious, traditional with a strong focus on family and group interconnections) as Weberian' ideal types (Hall 1956, 1959; Oyserman *et al.* 2002; Weber 1904). Specialist cultures prize health because it is seen as linked with individual success. They tend to distinguish physical health problems ("correctable" through appropriate diagnosis and technological treatment) from mental health problems (fundamentally disruptive to the social fabric). Therefore, an addiction diagnosis leads to the marginalization of the addict (Barker and Hunt 2007). Health is valued also in generalist cultures but illness is often accepted with a degree of stoicism (or fatalism), and mental health is not separated from physical health. Because the ability to control life is not a major concern in generalist societies, an addicted individual is likely to be less disruptive of personal, familial, or community life and might remain a valued member still able to contribute at least something to the family (Barker and Hunt 2007; Lemert 1958; Partanen 1991). In many non-Western societies, recovery may be seen primarily as a social event, an integrative process, restoring both group cohesion and individual integration – a process Kleinman (1980) calls "cultural healing." Some self-change studies employing a non-Western paradigm show that seeking cultural understanding and increasing sense of identity, and also through learning cultural practices (songs, legends, prayers, ceremonies, art), was fundamental in the process of recovery and regaining wellbeing (Mohatt *et al.* 2008; Morjaria-Keval 2006; Tempier *et al.* 2011). While the family or community might condemn the behavior, the addicted individual remains within the family circle receiving public support for the efforts to deal with addiction (Barker and Hunt 2007). One should note, however, that while some studies stress the role of the social world of kinship responsibilities among Alaska Natives (Mohatt *et al.* 2008; Tempier *et al.* 2011), others show that closeness of association can work against Aboriginal people who try to disassociate themselves from a previously heavy drinking lifestyle (Brady 1993). Moreover, in non-Western cultures, spirituality cannot be separated from other aspects of self and cultural being and therefore spirituality and the process of meaning-making play a crucial role in recovery. Those research findings support the holistic view of recovery that transcends the individual, happening within the contextual framework of perceiving, understanding and experiencing the world (Morjaria-Keval 2006; Tempier *et al.* 2011). A closer look at the few studies from non-Western cultures shows that while full abstinence is a required characteristic of recovery in many Western countries, significant moderation of use of the addictive substance is an acceptable endpoint in many other cultures (Everett *et al.* 1976). This is rarely evident from Western-culture self-change studies, which often excluded individuals who were not abstainers. Finally, the processes of maturing out might have different dimensions in non-Western cultures: studies among Australian Aborigines (Brady 1993), urban American Indians (Barker and Kramer 1996) and the male population of Fiji (Walter 1982) showed that drunkenness is considered to be an unacceptable state at old age, especially as a person achieves increased social status (e.g. becoming a grandparent).

Future perspectives

The concept of change without treatment has matured out and proceeded from a provocative and political discussion at the peak of epidemic illicit drug use in the 1980s and 1990s, e.g. in Switzerland and Germany (Happel 1986; Klingemann and Klingemann 2007), to the development of a more general model of individual change processes, and finally incorporating societal and cultural conditions. "Most addicts quit and drug-induced neural plasticity does not prevent quitting" (Heymann 2013: 9): The self-change framework demonstrates that alternative models of addiction, beyond the polarization between moral failure and the medical model of brain disease, provide a new and useful conceptualization embracing research findings that were usually neglected in the past. Future research perspectives should attempt a better validation of findings, comparing problems and cultures, the inclusion of non-cognitive change mechanisms, and a longitudinal analysis of individual change of the addictive behavior over the life course (Klingemann et al. 2010). A model embracing the respect of the user perspective and the idea of addicts as choosing beings has met in practice with resistance from various policy and treatment stakeholders, impeding its diffusion and acceptance. According to Kuhn "paradigms gain their status because they are more successful than their competitors in solving a few problems that the group of practitioners has come to recognize as acute" (Kuhn 1970: 23). It remains to be seen how long it will take until epidemiological evidence and fundamental problems in societies' treatment response (Klingemann and Bergmark 2006) lead to a crisis of deterministic medical models of addiction discarding the idea of addiction as a choice.

References

Anderson, N. (1923) *The Hobo. The Sociology of the Homeless Man*, Chicago, IL: University of Chicago Press.
Barber, B. (1961) "Resistance by scientists to scientific discovery", *Science* 134(3479): 596–602.
Barker, J. C. and Hunt, G. (2007) "Natural recovery: a cross-cultural perspective", in H. Klingemann and L. C. Sobell (eds), *Promoting Self-Change from Addictive Behaviors. Practical Implications for Policy, Prevention and Treatment*, New York, NY: Springer, pp. 213–238.
Barker, J. C. and Kramer, B. J. (1996) "Alcohol consumption among older urban American Indians", *Journal of Studies on Alcohol* 57(2): 119–124.
Becker, H. S. (1963) *Outsiders*, Toronto, ON: The Free Press of Glencoe.
Bischof, G., Rumpf, H. J., Meyer, C., Hapke, U. and John, U. (2004) "What triggers remission without formal help from alcohol dependence? Findings from the TACOS study", in P. Rosenqvist, J. Blomqvist, A. Koski-Jännes and L. Öjesjö (eds), *Addiction and Life Course*, Helsinki, Finland: Nordic Council for Alcohol and Drug Research, pp. 85–101.
Blomqvist, J. (2007) "Self-change from alcohol and drug abuse: often-cited classics", in H. Klingemann and L. C. Sobell (eds), *Promoting Self-Change from Addictive Behaviors. Practical Implications for Policy, Prevention and Treatment*, New York, NY: Springer, pp. 31–57.
Blomqvist, J. (2009) "What is the worst thing you could get hooked on? Popular images of addiction problems in contemporary Sweden", *Nordic Studies on Alcohol and Drugs* 26(4): 373–398.
Blomqvist, J., Raitasalo, K., Melberg, H. O., Schreckenberg, D., Peschel, C., Klingemann, J., Koski-Jännes A. (2016) "Popular images of addiction in five European countries", in M. Hellman, V. Berridge, K. Duke and A. Mold (eds), *Concepts of Addictive Substances and Behaviours across Time and Place*, Oxford, UK: Oxford University Press, pp. 193–212.
Brady, M. (1993) "Giving away the grog: an ethnography of Aboriginal drinkers who quit without help", *Drug and Alcohol Review* 12(4): 401–411.
Chiauzzi, E. J. and Liljegren, S. (1993) "Taboo topics in addiction treatment – an empirical review of clinical folklore", *Journal of Substance Abuse Treatment* 10(3): 303–316.
Clinard, M. B. and Meier, R. F. (2011) *Sociology of Deviant Behavior*, Belmont, CA: Wadsworth.
Cloud, W. and Granfield, R. (2008) "Conceptualising recovery capital: expansion of a theoretical construct", *Substance Use and Misuse* 43(12/13): 1971–1986.
Cohn, A. M., Hunter-Reel, D., Hagman, B. T. and Mitchell, J. (2011) "promoting behavior change from alcohol use through mobile technology: the future of ecological momentary assessment", *Alcoholism: Clinical and Experimental Research* 35(12): 2209–2215.
Cunningham, J. A., Blomqvist, J., Koski-Jannes, A. and Cordingley, J. (2005) "Maturing out of drinking problems: perceptions of natural history as a function of severity", *Addiction Research & Theory* 13(1): 79–84.

Dawson, D. A., Grant, B. F., Stinson, F. S. and Chou, P. S. (2006) "Estimating the effect of help-seeking on achieving recovery from alcohol dependence", *Addiction* 101(6): 824–834.

Dawson, D. A., Grant, B. F., Stinson, F. S., Chou, P. S., Huang, B. and Ruan, W. J. (2005) "Recovery from DSM-IV alcohol dependence: United States, 2001–2002", *Addiction* 100(3): 281–292.

DiClemente, C. C. (2007) "Mechanisms, determinants and processes of change in the modification of drinking behavior", *Alcoholism: Clinical and Experimental Research* 31(10 Suppl): 13S–20S.

Drew, L. R. H. (1968) "Alcoholism as a self-limiting disease", *Quarterly Journal of Studies on Alcohol* 29(4): 956–967.

Everett, M. W., Waddel, J. O. and Heath, D. B. (1976) *Cross-Cultural Approaches to the Study of Alcohol: An Interdisciplinary Perspective*, The Hague, The Netherlands: Mouton.

Faupel, C. (2011) "The deviant career", in C. D. Bryant (ed.), *Routledge Handbook of Deviant Behavior*, New York, NY: Routledge, pp. 145–202.

Freidson, E. (1961) "Organization of medical-practice and patient behavior", *American Journal of Public Health and the Nation's Health* 51: 43–52.

Granfield, R. and Cloud, W. (2001) "Social context and 'natural recovery': the role of social capital in the resolution of drug-associated problems", *Substance Use & Misuse* 36(11): 1543–1570.

Gusfield, J. R. (1986) *Symbolic Crusade: Status Politics and the American Temperance Movement*, Chicago, IL: University of Illinois Press.

Hall, E. T. (1959) *The Silent Language*, Garden City, NY: Doubleday.

Hall, E. T. (1956) *The Hidden Dimension*, Garden City, NY: Doubleday.

Hammersley, R. (this volume, pp. 220–228) "Sociology of addiction".

Happel, H.-V. (1986) "Ausstieg aus der Drogenabhängigkeit am Beispiel der Selbstheiler. Projektbeschreibung", *Suchtgefahren* 32: 367–368.

Heyman, G. M. (2013) "Addiction and choice: theory and new data", *Frontiers in Psychiatry/Addictive Disorders and Behavioral Dyscontrol* 4: art.31.

Humphreys, K., Moos, R. H. and Finney, J. W. (1995) "Two pathways out of drinking problems without professional treatment", *Addictive Behaviors* 20(4): 427–441.

Kickbusch, I. and Maag, D. (2008) "Health literacy", in K. Heggenhougen and S. Quah (eds), *International Encyclopedia of Public Health*, San Diego, CA: Academic Press, pp. 204–211.

King, M. P. and Tucker, J. A. (1998) "Natural resolution of alcohol problems without treatment: environmental contexts surrounding the initiation and maintenance of stable abstinence or moderation drinking", *Addictive Behaviors* 23(4): 537–541.

King, M. P. and Tucker, J. A. (2000) "Behavior change patterns and strategies distinguishing moderation drinking and abstinence during the natural resolution of alcohol problems without treatment", *Psychology of Addictive Behaviors* 14(1): 48–55.

Kleinman, A. (1980) *Patients and Healers in the Context of Culture: An Exploration of the Borderland between Anthropology, Medicine, and Psychiatry*, Berkeley, CA: University of California Press.

Klingemann, H. (1991) "The motivation for change from problem alcohol and heroin use", *British Journal of Addiction* 86(6): 727–744.

Klingemann, H. (1992) "Coping and maintenance strategies of spontaneous remitters from problem use of alcohol and heroin in Switzerland", *The International Journal of the Addictions*, 27(12): 1359–1388.

Klingemann, H. (1999) "Addiction careers and careers in addiction", *Substance Use & Misuse* 34(11): 1505–1526.

Klingemann, H. (2004) "Natural recovery from alcohol problems", in N. Heather and T. Stockwell (eds), *The Essential Handbook of Treatment and Prevention of Alcohol Problems*, Chichester, UK: John Wiley & Sons Ltd, pp. 161–175.

Klingemann, H. and Bergmark, A. (2006) "The legitimacy of addiction treatment in a world of smart people", *Addiction* 101(9): 1230–1237.

Klingemann, H. and Klingemann, J. (2007) "Hostile and favorable societal climates for self-change: some lessons for policy makers", in H. Klingemann and L. C. Sobell (eds), *Promoting Self-Change from Addictive Behaviors. Practical Implications for Policy, Prevention and Treatment*, New York, NY: Springer, pp. 187–212.

Klingemann, H., Schlaefli, K. and Steiner, M. (2013) "What do you mean by spirituality? Please draw me a picture! Complementary faith-based addiction treatment in Switzerland from the client's perspective", *Substance Use & Misuse* 48(12): 1187–1202.

Klingemann, H. and Sobell, L. C. (2007) *Promoting Self-Change from Addictive Behaviors. Practical Implications for Policy, Prevention and Treatment*, New York, NY: Springer.

Klingemann, H., Sobell, M. B. and Sobell, L. C. (2010) "Continuities and changes in self-change research", *Addiction* 105(9): 1510–1518.

Klingemann, J. I. (2010) "Ja muszę Pani powiedzieć szczerze, że o tych problemach to NIKT nie wiedział. . . Metody badania populacji ukrytych – przykład zjawiska samowyleczenia z uzależnienia", *Profilaktyka Społeczna i Resocjalizacja* 16: 227–249.

Klingemann, J. I. (2011) "Lay and professional concepts of alcohol dependence in the process of recovery from addiction among treated and non-treated individuals in Poland – a qualitative study", *Addiction Research & Theory* 19(3): 266–275.

Klingemann, J. I. (2012) "Mapping the maintenance stage of recovery. A qualitative study among treated and non-treated former alcohol dependents in Poland", *Alcohol and Alcoholism* 47(3): 296–303.

Klingemann, J., Klingemann, H. and Moskalewicz, J. (2017) "Popular views on addictions and on prospects for recovery in Poland", *Substance Use & Misuse* 52(13): 1765–1771.

Kuhn, T. S. (1970) *The Structure of Scientific Revolutions*, Chicago, IL: University of Chicago Press.

Lemert, E. M. (1958) "The use of alcohol in three Salish tribes", *Quarterly Journal of Studies on Alcohol* 19: 90–107.

Leyton, M. (2013) "Are addictions diseases or choices?", *Journal of Psychiatry & Neuroscience* 38(4): 219–221.

Merton, R. K. (1938) "Social structure and anomie", *American Sociological Review* 3(5): 672–682.

Mohatt, G. V., Rasmus, S. M., Thomas, L., Allen, J., Hazel, K. and Marlatt, G. A. (2008) "Risk, resilience, and natural recovery: a model of recovery from alcohol abuse for Alaska Natives", *Addiction* 103(2): 205–215.

Morjaria-Keval, A. (2006) "Religious and spiritual elements of change in Sikh men with alcohol problems: a qualitative exploration", *Journal of Ethnicity in Substance Abuse* 5(2): 91–118.

Moskalewicz, J. and Klingemann, J. (2015) "Addictive substances and behaviours and social justice", in P. Anderson, J. Rehm and R. Room (eds), *Impact of Addictive Substances and Behaviours on Individual and Societal Well-being*, Oxford, UK: Oxford University Press, pp. 143–160.

Orford, J. (this volume, pp. 209–219) "Power and addiction".

Oyserman, D., Coon, H. M. and Kemmelmeier, M. (2002) "Rethinking individualism and collectivism: evaluation of theoretical assumptions and meta-analyses", *Psychological Bulletin* 128(1): 3–72.

Parsons, T. (1966) *The Social System (3rd Edition)*, New York, NY, and London, UK: The Free Press, Collier-Macmillan limited.

Partanen, P. (1991) *Sociability and Intoxication: Alcohol and Drinking in Kenya, Africa, and the Modern World*, Helsinki, Finland: The Finnish Foundation for Alcohol Studies.

Room, R. (1977) "Measurement and distribution of drinking patterns and problems in general populations", in G. Edwards, M. M. Gross, M. Keller, J. Moser and R. Room (eds), *Alcohol-Related Disabilities*, Geneva: World Health Organization, pp. 61–87.

Room, R. (2005) "Stigma, social inequality and alcohol and drug use", *Drug and Alcohol Review* 24(2): 143–155.

Russel, M., Peirce, R. S., Chan, A. W., Wieczorek, W. F., Moscato, B. S. and Nochajski, T. H. (2001) "Natural recovery in a community-based sample of alcoholics: study design and descriptive data", *Substance Use and Misuse* 36(11): 1417–1441.

Schachter, S. (1982) "Recidivism and self-cure of smoking and obesity", *American Psychologist* 37(4): 436–444.

Schomerus, G., Lucht, M., Holzinger, A., Matschinger, H., Carta, M. G. and Angermeyer, M. C. (2011) "The stigma of alcohol dependence compared with other mental disorders: a review of population studies", *Alcohol and Alcoholism* 46(2): 105–112.

Smart, R. G. (1976) "Spontaneous recovery in alcoholics – review and analysis of available research", *Drug and Alcohol Dependence* 1(4): 277–285.

Smart, R. G. (2007) "Natural recovery or recovery without treatment from alcohol and drug problems as seen from survey data", in H. Klingemann and L. C. Sobell (eds), *Promoting Self-Change from Addictive Behaviors. Practical Implications for Policy, Prevention and Treatment*, New York, NY: Springer, pp. 59–71.

Snow, M. (1973) "Maturing out of narcotic addiction in New York City", *International Journal of the Addictions* 8(6): 921–938.

Sobell, L. C., Ellingstad, T. P. and Sobell, M. B. (2000) "Natural recovery from alcohol and drug problems: methodological review of the research with suggestions for future directions", *Addiction* 95(5): 749–764.

Sobell, L. C., Sobell, M. B., Toneatto, T. and Leo, G. I. (1993) "What triggers the resolution of alcohol problems without treatment", *Alcoholism: Clinical and Experimental Research* 17(2): 217–224.

Sobell, M. B. and Sobell, L. C. (1995) "Controlled drinking after 25 years – how important was the great debate?" *Addiction* 90: 1149–1153.

Sobell, M. B. and Sobell, L. C. (2011) "It is time for low-risk drinking goals to come out of the closet", *Addiction* 106(10): 1715–1717.

Stall, R. and Biernacki, P. (1986) "Spontaneous remission from the problematic use of substances – an inductive model derived from a comparative analysis of the alcohol, opiate, tobacco, and food obesity literatures", *International Journal of the Addictions* 21(1): 1–23.

Sutherland, E. H. (1924) *Principles of Criminology*, Chicago, IL: University of Chicago Press.

Swan, M. (2009) "Emerging patient-driven health care models: an examination of health social networks, consumer personalized medicine and quantified self-tracking", *International Journal of Environmental Research and Public Health* 6(2): 492–525.

Tempier, A., Dell, C. A., Papequash, E. C., Duncan, R. and Tempier, R. (2011) "Awakening: 'spontaneous recovery' from substance abuse among aboriginal peoples in Canada", *International Indigenous Policy Journal* 2(1): 1–18.

Tucker, J. A. and Vuchinich, R. E. (1994) "Environmental events surrounding natural recovery from alcohol-related problems", *Journal of Studies on Alcohol* 55(4): 401–411.

Waldorf, D. and Biernacki, P. (1979) "Natural recovery from heroin-addiction – review of the incidence literature", *Journal of Drug Issues* 9(2): 281–289.

Walter, M. A. H. B. (1982) "Drink and be merry for tomorrow we preach: alcohol and the male menopause in Fiji", in M. Marshall (ed.), *Through a Glass Darkly: Beer and Modernization in Papua New Guinea*, Boroko: Institute of Applied Social and Economic Research Monograph, p. 434.

Weber, M. (1904) *Die Objektivität sozialwissenschaftlicher und sozialpolitischer Erkenntnis*, Tübingen: J.C.B. Mohr.

Winick, C. (1962) "Maturing out of narcotic addiction", *Bulletin on Narcotics* 14: 1–7.

SECTION C

Ethics, law, and policy

40

ADDICTION

A structural problem of modern global society

Bruce K. Alexander

The alarming spread of alcohol and drug addiction in the late modern era has provoked a vast amount of moralistic and medical analysis (e.g., White 1998). However, neither the moralistic nor the medical paradigm has generated interventions that have curtailed the continuing spread of addiction. Moreover, despite loud proclamations to the contrary, neither the moralistic nor the medical paradigm has led to a compelling and robust understanding of addiction (Szasz 1973/1985; Peele, Brodsky, and Arnold 1992: 19–46; Kalant 2009; Heyman 2009; Pickard 2012; Hart 2013; Ahmed, Lenoir, and Guillem 2013; Levy 2013; Satel and Lilienfeld 2013; Helm *et al.* 2014; Hall, Carter, and Forlini 2015; Lewis 2015; Alexander 2016). Unfortunately, no widely accepted alternative has yet taken the place of these two played-out paradigms.

One flourishing line of thought in this theoretically unsettled interval stresses the importance of community support in the recovery of addicted people. A vigorous, widespread "recovery movement" has emerged to organize and celebrate community support (White 2000; 2011; Livingston 2012; Best and Savic 2014). This burgeoning movement is identified with phrases like "recovery community," "recovery model," "recovery capital," "building community," "restoring community," "building relationships," and "support groups."

I believe that this current enthusiasm for community thinking foreshadows a long overdue paradigm shift in the field of addiction that will entail turning away from *both* the moralistic and medical paradigms. The new paradigm will not seek the cause of addiction problems in individual vulnerabilities, weaknesses, disorders, reinforcement histories, or pathologies, but in the intrinsic vicissitudes of life in the globalizing world society of the modern age.

The "dislocation theory of addiction" (Alexander 2008/2010) summarized here builds on this societal paradigm. Dislocation theory leads to a radical proposition for controlling addiction through large-scale social change. This proposition would have been unthinkable when I entered the addiction field a half-century ago, but the world has changed. I believe it must be taken very seriously now.

The dislocation theory of addiction

The dislocation theory stands outside the box of most current addiction thinking in at least three ways: First, dislocation theory does not focus initially on addicted *individuals*, as do *both* the moralistic and the medical paradigms. Instead, it focuses on an emerging global *society* within

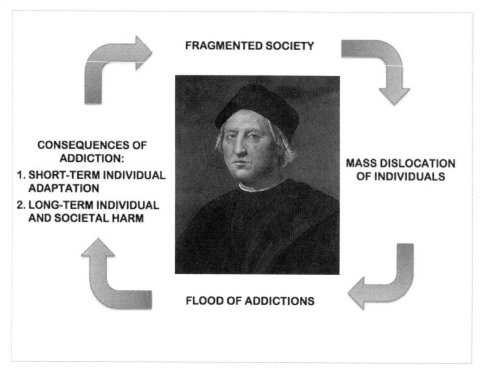

Figure 40.1 An outline of a societal view of severe addiction as a positive feedback loop or "vicious cycle" with an old portrait of Christopher Columbus as the central image. Painting by Ridolfo del Ghirlandaio (1483–1561) circa 1520, Wikimedia commons, public domain.

which addiction has spread irrepressibly in the five centuries since Columbus' voyages, and which many social scientists refer to as "modernity" or "the modern era" (Dussel 1995; Berman 2010; Mann 2011). The modern era is marked by rejection of tradition, priority of individualism, globalization of commerce, faith in scientific and technological progress, and steady growth of capitalism, industrialization, urbanization – and addiction.

Second, dislocation theory does not privilege drug and alcohol addiction. The traditional English language meaning of "addiction" (definition 1a in the current *Oxford English Dictionary*) refers to a state of "being dedicated or devoted *to* a thing, esp. an activity or occupation; adherence or attachment, esp. of an immoderate or compulsive kind." Dislocation theory encompasses severe addictions to wealth, power, sex, work, drugs, alcohol, gambling, love, eating, shopping, hoarding, dieting, Internet games, fantasy literature, social media, exercise, sports, narcissistic status seeking, or whatever. Severe instances of these addictions can be as destructive and intractable as severe drug and alcohol addiction (e.g., Sussman, Lisha, and Griffiths 2011).

Third, dislocation theory is drawn as much from historical, anthropological, sociological, economic, biographical, and clinical analysis as from quantitative medical research (see Alexander 2008/2010). A theory of addiction must align with quantitative findings about known risk factors of course, but it must also fit careful observations of the relationship between risk factors and social and historical reality.

Fragmentation

Figure 40.1 starts with the fact of fragmentation. From the time of Christopher Columbus onward, colonization of the globe by European powers fragmented local societies by conquest, disease, enslavement, economic exploitation, enticement, religious domination, and ecological devastation (Hobsbawm 1989: chap. 3; Dussel 1995; Wright 2004; Mann 2011).

As the colonizing European powers fragmented distant societies, they also fragmented subcultures within their own homelands, although with somewhat more restraint. The agricultural and industrial revolutions that accompanied world colonization devastated stable peasant farms and commons throughout Europe (Polanyi 1944; El Saffar 1994: 62–68; Bollier 2014).

The fragmentation of world society that began in the early modern era continues in the 21st century, amidst further globalization of free-market capitalism, neoliberalism, consumerism, corporate culture, "enterprise culture," "metacolonialism," high-tech surveillance, rapid technical change, ecological devastation, real estate bubbles, international "free-trade" agreements, austerity, relentlessly efficient manufacturing and agribusiness, unending financial crises, and continuing plunder of aboriginal peoples (Chossudovsky 2003; Dufour 2003; Harvey 2011: 66, 176; Hickinbottom-Brawn 2013; Snowden 2014; Klein 2014; McWilliams 2015; Nikiforuk 2015; Levitin 2015; Bulhan 2015; Rodrik 2016; Daley 2016). Today's global fragmentation is not just propagated by European nations, but by global powers on every inhabited continent.

Beneath the steamroller of modernity, families and communities are crushed; children are traumatized and neglected; local economies are pulverized; traditional values and religions are trivialized; corrupt governments oppress their own people; and cultural arts are reduced to mass-produced trinkets sold in tourist towns. Some authors speak of today's fragmentation in terms of "postmodernity" rather than "modernity" (e.g., Berardi 2009), but the analysis remains essentially the same.

Societal fragmentation has come to seem an inescapable consequence of the increasing industrial productivity and technical advancement that have enabled the earth to support seven billion people. However, this brave new world is in deep – possibly terminal – trouble, in large part because of unforeseen side effects of fragmentation, which include dislocation and addiction.

Dislocation

I use the word "dislocation" to designate the *individual, psychological* effects that follow from societal fragmentation. Dislocation has been described on many levels. For example, sociologists speak of the absence of sustaining interactions between individuals and their local reference groups. Psychologists speak of a meager sense of belonging, identity, meaning, and purpose, resulting in anxiety and depression. Existentialists speak of alienation, "nothingness," or "non-being." Christians speak of waning faith in the "God of our fathers" and of "the poor in spirit." Evolutionary biologists speak of the unsatisfied, innate social needs of the human species. Dislocation has not yet been adequately described in the language of neuroscience – but it needs to be (Hellig, Epstein, Nader, and Shaham 2016).

Mass dislocation has come to seem normal. The modern market system for production and distribution of goods requires that people – both rich and poor – must perform competitively and efficiently, unimpeded by sentimental ties to families, friends, traditional values, or religious laws. Stringent economic rationality is said to make the law of supply and demand function efficiently, and thus to "clear the markets" each day. Nations that have embraced the global market system in recent decades, including China and other BRIC nations, have been able to join the geopolitical superpowers.

Seen in a positive light, dislocation can provide a space for enjoyable personal initiative and individual creativity. However, prolonged, severe dislocation eventually leads to an agonizingly empty life (Fromm 1941; Polanyi 1944; Barrett 1962; Frankl 1963; Erikson 1968; Berry 2009: 35–48; Tolman 2013; Klein 2014: 158–160; Hickinbottom-Brawn 2013; Verhaeghe 2014; Rosin 2015). Severe, prolonged dislocation causes anguish, suicide, depression, disorientation, ill health, and domestic violence (Durkheim 1897/1951; Polanyi 1944; Barrett 1962; Chandler, Lalonde, Sokol and Hallet 2003; Berardi 2009). Because it is unbearable, prolonged dislocation has been imposed as a dreaded punishment (e.g., solitary confinement, exile, ostracism, banishment, shunning, and excommunication) from ancient times to the present (White 2014). Dislocation in the form of extreme social isolation is indispensable in today's terrifyingly scientific technology of torture (Klein 2007: chap. 1).

Describing the growth of dislocation from societal fragmentation in modernity is more than a lament of existentialists, theologians, and social workers. Specific linkages between societal fragmentation and individual dislocation have been experimentally and neurologically demonstrated for every stage of life, beginning before birth. For example:

> Severe stress endured by pregnant women in a fragmented society can render their children socially fearful and dislocated years later. Some brain mechanisms underlying this causal relationship have been discovered. Traumatic postnatal stress that is commonplace in a fragmented society has also been linked directly with dislocation and addiction (see Maté 2008).

> Inadequate attachment to a mother, other significant adults, and age-mates has been shown experimentally and clinically to drastically undermine a developing child's social and emotional capabilities later in life (Erikson 1963; 1968; Bowlby 1969; Blum 2002: chaps 6, 7, 10).

> Real estate markets dominated by speculators and foreign investors can make settled family and neighborhood life excruciatingly difficult for parents raising children. I witness this firsthand among young friends and relatives dislocated by today's insanely inflated and volatile housing market in Vancouver (see also Surowiecki 2014; Tencer 2015; O'Neil 2015).

> Dehumanizing factory systems, such as Foxconn, where many of the world's cell phones are made, can leave worker's lives so empty that suicide becomes epidemic (Tharoor 2014; Kantor and Streitfeld 2015).

> Corrupted political systems generate profound apathy and vulnerability to political fanaticism in powerless citizens (Wolin 2008; Risen 2014; Rodrik 2016; Porter 2016).

> Lack of family and neighborhood support in a mobile society can push elderly people into incapacitating despair (McLaren 2014).

> The Internetification of life for people of all ages produces confusion, impossible demands for continuous multitasking and measurable stress responses in the brain (Levitin 2015).

Of course, individuals can also be severely dislocated by natural disasters that cannot be attributed to societal fragmentation (e.g., Deraniyagala 2013). Nonetheless, fragmentation is a massive source of dislocation in the modern era, and many thinkers, in both capitalist and socialist countries, believe that dislocation has become essentially universal because of it (see Taylor 1991; Dufour 2003; Berardi 2009; Albrecht 2012; Donskis 2014; Welch 2015).

The word "dislocation" defies precise definition. For example, dislocation is experienced on a psychological level as a meager sense of belonging, identity, meaning, and purpose. But can a strong identity make up for a weak sense of purpose? Can a strong sense of belonging and love make up for other lacks? It is impossible to say. Nonetheless, the modern spread of dislocation – under various names – has been recognized by acclaimed thinkers for centuries. It is easy to overlook its existence in an age dominated by positivist epistemology. Unfortunately, positivist epistemology cannot eliminate dislocation; it can only obscure it.

Addiction: adaptation to dislocation

Just as widespread dislocation historically follows social fragmentation, severe addiction historically follows dislocation. Extensive historical, anthropological, and clinical evidence documents this predictable sequence in Europe, Asia, and North America (Alexander 2008/2010: chap. 6).

Abundant clinical and biographical evidence shows *why* addictions track dislocation so closely. Addictions can provide dislocated people with some real relief and compensation for bleak, empty lives, when nothing else works (Alexander 2008/2010: chaps 6–8; Hart 2013: 74–95; Fetting 2016).

Addictions can be adaptive in a fragmented world because – in psychological terms – they can provide severely dislocated people with a sense of belonging, identity, meaning, and purpose, at least in the short term. Without their addictions, many people would have terrifyingly little reason to live and risk falling into incapacitating anxiety, depression, or suicide.

For example, rather than being overwhelmed by emptiness, drug-addicted people can keep frantically busy chasing drugs. At the same time, they can identify with the fascinating mythology of William S. Burroughs, Curt Cobain, Phillip Seymour Hoffman, Amy Winehouse, Robin Williams, or Prince and the imagery of the "tragically hip" or "the coolest" (Burroughs 1967; Pryor 2003).

Similarly, many people who are severely addicted to horserace gambling incessantly exchange information and hunches within a colorful subculture of characters at the track, and share a mythology of famous gamblers and legendary horses (Ryan 2014a; 2014b).

People who use drugs or gamble recreationally, rather than addictively, have found more fulfilling and reliable sources of psychosocial integration. However, countless millions of people have such desperate needs for belonging, identity, meaning, and purpose that they seize onto one or more addictive pursuits, striving to build lives around them. When addictions prove to be the only visible hope for severely dislocated people to endure their existence, they hold onto them with the same iron grip that they would use to seize a piece of floating junk in a stormy sea.

The personal utility of severe addiction in adapting to dislocation explains its dangerously high prevalence in a fragmented, dislocated world. Moreover, some addictions also serve economic functions in a modern world that requires overwork and overconsumption to keep the wheels turning and the profits growing (Slater 1980; Vance 2015; Kantor and Streitfeld 2016). Therefore, addictions to work and consumer goods are subtly encouraged by mass advertising. Addiction is as intrinsic to the modern era as competitiveness, loneliness, apathy, anxiety, and depression. However, severe addiction is terribly damaging, *even though* it serves an adaptive function for individuals and an economic function for society.

The adaptiveness of addiction is often obscured. Many severely addicted people deny their chronic dislocation, because they feel embarrassed that they do not "have a life." They may not know that dislocation bedevils most people in the modern world. They may not fully realize the adaptive functions of their own addiction. In moments of insight, however, they can explain the functions of their addiction with surprising candor (Alexander 2008/2010: 158–160; Pond and Palmer 2016: 21–22).

Parents of addicted people often insist that their offspring's addictions were caused by addictive drugs, genetic predispositions, or brain pathology. Recognizing addiction's adaptive functions would require facing the inadequacies of the environment they were able to provide their children, generating unbearable feelings of parental failure.

Mass media endlessly and authoritatively proclaim that addiction is a chronic disease caused by drug use in people with susceptible brains and stressful personal histories, rather than an adaptation to dislocation in a fragmented world. This Official View (as I call it) is backed by endless authoritative repetition and media dramatization (e.g., *Nature* editorial 2014; Seelye 2015; Pond and Palmer 2016; Volkow, Koob, and McLellan 2016). It is lavishly funded and sponsored by the National Institute of Drug Abuse in the United States, despite its glaring deficiencies (Alexander 2016).

Another complication is that whereas dislocation theory explains the high prevalence of severe, damaging kinds of addiction, addiction has milder forms as well. People can fit the *Oxford English Dictionary* definition of addiction when they fall head-over-heels in love, undertake a binge of work to finish an important project, or devote themselves to a life of compassionate community service.

Fragmenting consequences of addiction: the cycle continues

Addiction helps many people adapt to the stress of chronic dislocation, and fuels the economic engines of the global economy, but there is still another reason for its high prevalence in modern society. The long-term harmful consequences of many addictions exacerbate the fragmentation of modern society, thereby increasing the dislocation that flows from fragmentation and the prevalence of addiction. The vicious cycle takes another turn.

Fragmenting consequences of addiction include: Environmental and social destruction administered by wealth and power addicts pursuing profits in their executive suites; environmental and social destruction caused by wasteful addictive consumption by millions of more-and-less severely addicted customers; all the talented people who are lost from productive citizenship and supportive family function because of severe addictions and protracted recoveries; social insecurity produced by the depredations of overtly criminal street addicts supporting their habits; and elders who cannot contribute stabilizing wisdom to succeeding generations because they are consumed by addictive involvements with television or psychoactive prescription drugs (Alexander 2015).

Therefore, because of its long-term harmful consequences to society, severe addiction is not only a downstream *adaptation* to societal fragmentation but also an upstream *cause* of fragmentation. Addiction perpetuates itself by exacerbating the fragmentation of the modern world.

A radical proposition

Because modernity mass produces fragmentation and dislocation, it is, by itself, a *sufficient cause* of a widespread and intractable flood of addiction in today's world. If modern global society is to bring addiction under control, it will have to be re-structured, just as it will have to be re-structured to control environmental destruction and economic injustice (Klein 2014).

This radical proposition goes far beyond the comfortable conclusion that "environmental and social factors" along with "genetic factors" can explain why "addictive drugs" cause addiction in some people but not in others (e.g., Volkow, Koob, and McLellan 2016). It asserts that modernity *by itself* is sufficient cause of a flood of addiction, notwithstanding individual differences in susceptibility.

Please note that this radical proposition does not deny the role that individual events play in the lives of individual addicted people. However, no theory of individual pathology, and no synthesis of such theories can explain the flood of addiction that is produced by modernity. Here is a thought experiment: Envision a modern world in which all human traumas, genetic anomalies, character defects, cognitive errors, destructive reinforcement histories, illegal drug selling, and drug overprescribing that have been identified as causes of addiction were magically eliminated. If the dislocation theory is correct, the flood of addictions would scarcely recede, because 1) severe addiction is a natural, non-pathological way of adapting to the chronic dislocation of modernity and 2) most severe addiction does not involve drugs or alcohol.

Nobody can draw the blueprint for the less fragmented global society that must somehow be fabricated from an unprecedented blending of humanistic values and ultra-modern technology. However, we can work towards this future nonetheless, in conjunction with countless environmentalist, spiritual, social justice, and social recovery groups that are all seeking to interrupt the vicious cycle in their own ways, in their own localities (Hawken 2007; Klein 2014). The eventual melding of this enormous pool of localized human energy can lead into a less fragmented world civilization whose psychosocially integrated citizens will not encounter dislocation and addiction as major problems. The macro-level changes that will be needed go beyond the obviously important ecological issues of keeping the biosphere of planet earth habitable (because planetary destruction is not the only source of dislocation and addiction) and beyond issues of eliminating social injustice and gross income inequality (because dislocation and addiction are not only problems of the poor and marginalized). The changes must somehow provide opportunities for people, rich and poor, to fulfill the basic social and spiritual needs that make us whole as human beings while maintaining a viable global society.

This radical restructuring would have seemed preposterous and even unthinkable in the past and is still unthinkable within the institutions of wealth and power that govern the world. But it is now evident that *the emerging global society of the twenty-first century is going to be restructured radically no matter what happens in the field of addiction. The restructuring is already visible* in the irrepressible reshaping of modern life by technology, changing weather patterns, the radical political/economic changes envisioned in the Paris Climate Agreement of 2015, the gender revolution, and the "free trade" agreements that are reshaping relationships between citizens, nations, and corporations around the globe. If nightmare outcomes are to be avoided, the inevitable reshaping must be informed at every stage by the need to control dislocation, addiction, and other psychological components of ever-increasing societal fragmentation.

I believe that the global addiction disaster cannot be solved solely by people acting in the role of treatment professionals, researchers, policy experts, public-spirited billionaires, or superheroes. If it can be solved, the solution will emerge from very large numbers of people acting together as neighbors, family members, and members of local action groups over sustained periods and the new mentality that eventually emerges from this process. The overarching emergent qualities of this piecemeal revolution are yet to be determined, but they are the hope for the future.

None of us will have to look very far to find a group near us that is addressing some aspect of the current disastrous situation. The best hope for solving the modern addiction problem – and a host of interrelated problems – begins with them, and they need our help now.

References

Ahmed, S. H., Lenoir, M. and Guillem, K. (2013) "Neurobiology of addiction versus drug use driven by lack of choice", *Current Opinion in Neurobiology* 23(4): 581–587.

Albrecht, G. (2012) "The age of Solastalgia", *The Conversation*, August 7, retrieved January 1 2015 from: http://theconversation.com/the-age-of-solastalgia-8337.

Alexander, B. K. (2008/2010) *The Globalization of Addiction: A Study in Poverty of the Spirit*, Oxford, UK: Oxford University Press.
Alexander, B. K. (2015) "Addiction, environmental crisis, and global capitalism", retrieved August 14 2015 from www.brucekalexander.com/articles-speeches/283-addiction,-environmentalcrisis,-and-global-capitalism.
Alexander, B. K. (2016) "Replacing the official view of addiction", in Davis, J. E. and Gonzales, A. M. (eds), *To Fix or to Heal: Patient Care, Public Health, and the Limits of Biomedicine*, New York, NY: NYU Press, pp. 208–237.
Barrett, W. (1962) *Irrational Man: A Study in Existential Philosophy*, Garden City, NY: Doubleday Anchor Books.
Berardi, F. (2009) *Precarious Rhapsody: Semiocapitalism and the Pathologies of the Post-Alpha Generation*, Brooklyn, NY: Autonomedia.
Berman, M. (2010) *All That is Solid Melts into Air: The Experience of Modernity (9th Edition)*, London, UK: Verso.
Berry, T. (2009) *The Sacred Universe: Earth, Spirituality, and Religion in the Twenty-First Century*, New York, NY: Columbia University Press.
Best, D. and Savic, M. (2014) "Substance use and offending: risk factors and addiction recovery", in Sheehan, R. and Ogloff, J. (eds), *Working Within the Forensic Paradigm: Cross-Discipline Approaches for Policy and Practice*, Abingdon, UK: Taylor and Francis.
Blum, D. (2002) *Love at Goon Park: Harry Harlow and the Science of Affection*, Cambridge, MA: Perseus.
Bollier, D. (2014) *Think Like a Commoner: A Short Introduction to the Life of the Commons*, Gabriola Island, BC: New Society Publishers.
Bowlby, J. (1969) *Attachment: Attachment and Loss, Vol. 1*, New York, NY: Basic Books.
Bulhan, H. A. (2015) "Stages of colonialism in Africa: from occupation of land to occupation of being", *Journal of Social and Political Psychology* 3(1): 239–256.
Burroughs, W. S. (1967) "Kicking drugs: a very personal story", *Harper's* July, 235: 39–42.
Chandler, M. J., Lalonde, C. E., Sokol, B. W. and Hallet, D. (2003) "Personal persistence, identity development, and suicide: a study of native and non-native North American adolescents", *Monographs of the Society for the Study of Child Development* 68(2): Serial No. 273.
Chossudovsky, M. (2003) *The Globalization of Poverty: Impact of IMF and World Bank Reforms*, London, UK: Zed.
Daley, S. (2016) "Outcry echoes up to Canada", *The New York Times*, April 3, A1.
Deraniyagala, S. (2013) *Wave: A Memoir*, Toronto, ON: McClelland and Stewart: pp. 55–59.
Donskis, L. (2014) "Clashing sensibilities in politics and literature: the cases of Rex Warner and Czeslaw Milosz", *Homo Oeconomicus* 31(3): 369–396.
Dufour, D.-R. (2003) *L'Art de Réduire les Têtes; Sur la Nouvelle Servitude de L'Homme Libéré à L'Ere du Capitalisme Total*, Paris, France: Édition Noël.
Durkheim, E. (1951) *Suicide: A Study in Sociology*, trans. J. A. Spaulding and G. Simpson, Glencoe, IL: Free Press (original work published 1897).
Dussel, E. (1995) *The Invention of the Americas: Eclipse of the "Other" and the Myth of Modernity*, trans. M. D. Barber, New York, NY: Continuum.
El Saffar, R. A. (1994) *Rapture Encaged – The Suppression of the Feminine in Western Culture*, New York, NY: Routledge.
Erikson, E. H. (1963) *Childhood and Society (2nd Edition)*, New York, NY: Norton.
Erikson, E. H. (1968) *Identity, Youth and Crisis*, New York, NY: Norton.
Fetting, M. (2016) *Perspectives on Substance Use, Disorders, and Addiction: with Clinical Cases (2nd Edition)*, London, UK: Sage.
Frankl, V. E. (1963) *Man's Search for Meaning: An Introduction to Logotherapy*, New York, NY: Pocket Books.
Fromm, E. (1941) *Escape from Freedom*, New York, NY: Farrar and Rinehart.
Hall, W., Carter, A. and Forlini, C. (2015) "The brain disease model of addiction: is it supported by the evidence and has it delivered on its promises?", *The Lancet: Psychiatry* 2(1): 105–110.
Hart, C. (2013) *High Price: A Neuroscientist's Voyage of Self-Discovery that Challenges Everything you Know about Drugs and Society*, New York, NY: HarperCollins.
Harvey, D. (2011) *The Enigma of Capital and the Crises of Capitalism*, London, UK: Profile Books.
Hawken, P. (2007) *Blessed Unrest: How the Largest Social Movement in the World came into Being*, New York, NY: Viking.

Hellig, M., Epstein, D. H., Nader, M. A. and Shaham, Y. (2016) "Time to connect: bringing social context into addiction neuroscience", *Nature Reviews Neuroscience* 17(9): 592–599.

Helm, D. et al. (2014) "Addiction: not just brain malfunction", *Nature* 507: 40.

Heyman, G. M. (2009) *Addiction: A Disorder of Choice*, Cambridge, MA: Harvard University Press.

Hickinbottom-Brawn, S. (2013) "Brand 'you': the emergence of social anxiety disorder in the age of enterprise", *Theory and Psychology* 23(6): 732–751, retrieved October 31 2015 from http://tap.sagepub.com/content/23/6/732.

Hobsbawm, E. J. (1989) *The Age of Empire: 1875–1914*, New York, NY: Random House (Vintage).

Kalant, H. (2009) "What neurobiology cannot tell us about addiction", *Addiction* 105(5): 780–789.

Kantor, J. and Streitfeld, D. (2015) "Inside Amazon: wrestling big ideas in a bruising workplace", *The New York Times*, August 15, retrieved June 14 2016 from: www.nytimes.com/2015/08/16/technology/inside-amazon-wrestling-big-ideas-in-a-bruisingworkplace.html?_r=0.

Klein, N. (2007) *The Shock Doctrine: The Rise of Disaster Capitalism*, Toronto, ON: Knopf.

Klein, N. (2014) *This Changes Everything: Capitalism vs. the Climate*, Toronto, ON: Knopf.

Levitin, D. J. (2015) "Why the modern world is bad for your brain", *The Guardian*, January 18, retrieved January 18 2015 from: www.theguardian.com/science/2015/jan/18/modern-world-bad-forbrain-daniel-j-levitin-organized-mind-information-overload.

Levy, N. (2013) "Addiction is not a brain disease (and it matters)", *Frontiers in Psychiatry* 4 (article 24): 1–6.

Lewis, M. (2015) *The Biology of Desire: Why Addiction is Not a Disease*, New York, NY: Public Affairs.

Livingston, W. (2012) "Reflections on abstinence from a controlled drinker", *New Directions in the Study of Alcohol Journal* 35: 35–47.

Mann, C. C. (2011) *1493: Uncovering the New World Columbus Created*, New York, NY: Vintage.

Maté, G. (2008) *In the Realm of Hungry Ghosts: Close Encounters with Addiction*, Toronto, ON: Knopf, pp. 205–207.

McLaren, L. (2014) "All the lonely old people", *Maclean's*, March 17, 127(10): 36.

McWilliams, J. (2015) "Inside big ag: on the dilemma of the meat", *Virginian Quarterly Review*, Spring, retrieved July 29 2015 from: www.vqronline.org/nonfictioncriticism/2015/04/inside-big-ag-dilemma-meat-industry.

Nature editorial (2014) "Animal farm: Europe's policy-makers must not buy animal-rights activists' arguments that addiction is a social, rather than a medical, problem", *Nature*, February 5, 506: 7486, retrieved April 1 2014 from: www.nature.com/news/animal-farm-1.14660.

Nikiforuk, A. (2015) "Fracking industry shakes up northern BC with 231 tremors", *The Tyee*, January 10, retrieved January 17 from: http://thetyee.ca/News/2015/01/10/Fracking_Industry_Shakes_Up_Northern_BC/.

O'Neil, P. (2015) "Canada offers friendly home for illicit cash, report says: Canada ranks among worst at tracking corrupt real estate buys", *Vancouver Sun*, November 13, pp. A1, A14.

Peele, S., Brodsky, A. and Arnold, M. (1992) *The Truth About Addiction and Recovery*, New York, NY: Simon and Schuster.

Pickard, H. (2012) "The purpose in chronic addiction", *AJOB Neuroscience* 3(2): 40 49, doi: 10.1080/21507740.2012.663058.

Polanyi, K. (1944) *The Great Transformation: The Political and Economic Origins of Our Times*, Boston, MA: Beacon.

Pond, M. and Palmer, M. (2016) *Wasted: An Alcoholic Therapist's Fight for Recovery in a Flawed Treatment System*, Vancouver, BC: Greystone Books.

Porter, E. (2016) "We've seen the Trump phenomenon before", *The New York Times*, May 24, retrieved May 26 2016 from: www.nytimes.com/2016/05/25/business/economy/weve-seen-thetrump-phenomenon-before.html?_r=0.

Pryor, W. (2003) *The Survival of the Coolest: an Addiction Memoir*, Bath, UK: Clear Press.

Risen, J. (2014) *Pay any Price: Greed, Power, and Endless War*, New York, NY: Houghton Mifflin, Harcourt.

Rodrik, D. (2016) "The politics of anger", *Social Europe*, March 11, retrieved from www.socialeurope.eu/2016/03/the-politics-of-anger.

Rosin, H. (2015) "The Silicon Valley suicides", *The Atlantic*, December, 316(5): 62–73.

Ryan, D. (2014a) "Hastings Park gallops on: the glory days are gone but Vancouver's racetrack still has its own unique community", *The Vancouver Sun*, April 26, pp. B11–B12.

Ryan, D. (2014b) "Living the life: the mystical beauty of horses and family of racetrack personalities captivate Vancouver author Kevin Chong", *The Vancouver Sun*, April 26, p. B12.

Satel, S. and Lilienfeld, S. C. (2013) *Brainwashed: The Seductive Appeal of Mindless Neuroscience*, New York, NY: Basic Books.

Seelye, K. Q. (2015) "In heroin crisis, white families seek gentler War on Drugs", *The New York Times*, October 30, retrieved November 24 2015 from: www.nytimes.com/2015/10/31/us/heroin-war-on-drugs-parents.html?_r=0W.

Slater, P. (1980) *Wealth Addiction*, New York, NY: Dutton.

Snowden, E. (2014) "How the National Security State kills a free society", *Reader Supported News*, June 7, retrieved June 7 2014 from: http://readersupportednews.org/opinion2/27775/24094-how-the-national-security-state-kills-a-free-society.

Surowiecki, J. (2014) "Real estate goes global", *The New Yorker*, June 26, retrieved June 4 2014 from: www.newyorker.com/talk/fina"ncial/2014/05/26/140526ta_talk_surowiecki.

Sussman, S., Lisha, N. and Griffiths, M. (2011) "Prevalence of the addictions: a problem of the majority or the minority?" *Evaluation and the Health Professions* 34(1): 3–56.

Szasz, T. (1973/1985) *Ceremonial Chemistry: The Ritual Persecution of Drugs, Addicts, and Pushers (Revised Edition)*, Holmes Beach, FL: Learning Publications.

Taylor, C. (1991) *The Malaise of Modernity*, Toronto, ON: Anansi.

Tencer, D. (2015) "5 signs Canada's housing markets are out of control", *Huffington Post*, November 3, retrieved November 4 2015 from: www.huffingtonpost.ca/2015/11/03/canada.housing-market_n_8461888.html?utm_hp_ref=canada-business.

Tharoor, I. (2014) "The haunting poetry of a Chinese factory worker who committed suicide", *The Washington Post*, November 12, retrieved November 16 2014 from: www.washingtonpost.com/blogs/worldviews/wp/2014/11/12/the-haunting-poetry-of-achinese-factory-worker-who-committed-suicide.

Tolman, C.W. (2013) "*Sumus ergo sum*: the psychology of self and how Descartes got it wrong", in W. E. Smythe (ed.), *Toward a Psychology of Persons*, New York, NY: Psychology Press, pp. 3–24.

Vance, A. (2015) *Elon Musk: Tesla, SpaceX, and The Quest for a Fantastic Future*, New York, NY: HarperCollins.

Verhaeghe, P. (2014) "Neoliberal economy brings out worst in us", *The Guardian Weekly*, October 3–9, p. 18.

Volkow, N. D., Koob, G. F. and McLellan, A. T. (2016) "Neurobiologic advances from the brain disease model of addiction", *The New England Journal of Medicine* 374(4): 363–371.

Welch, I. (2015) "American Psycho is a modern classic", *The Guardian*, January 10, retrieved January 12 2015 from: www.theguardian.com/books/2015/jan/10/american-psycho-bret-eastonellis-irvine-welsh.

White, P. (2014) "Solitary: a death sentence", *The Globe and Mail*, December 6: F1, F9.

White, W. L. (1998) *Slaying the Dragon: The History of Alcoholism Treatment and Recovery in America*, Bloomington, IL: Chestnut Health Systems.

White, W. L. (2000) "Toward a new recovery movement: historical reflections on recovery, treatment and advocacy", Recovery Community Support Program Conference, Arlington, Virginia, retrieved February 2016 from: www.morecovery.org/pdf/newrecmove.pdf.

White, W. L. (2011) "Circles of recovery: an interview with Keith Humphreys, PhD", posted at: www.williamwhitepapers.com.

Wolin, S. S. (2008) *Democracy Incorporated: Managed Democracy and the Specter of Inverted Totalitarianism*, Princeton, NJ: Princeton University Press.

Wright, J. (2004) *God's Soldiers: Adventure, Politics, Intrigue, and Power – A History of the Jesuits*, New York, NY: Doubleday.

41
DON'T BE FOOLED BY THE EUPHEMISTIC LANGUAGE ATTESTING TO A GENTLER WAR ON DRUGS

Carl L. Hart

> The resort to euphemism denotes, no doubt, a guilty conscience or—the same thing nowadays—a twinge in the public-relations nerve.
>
> *Mary McCarthy*

Every Friday evening, with sadness and pride, I make the 90-min. journey from Columbia University to Sing Sing Correctional Facility to teach a drugs and behavior course. My students, predominately black and bright, enthusiastically engage the subject matter not least because some of them have a personal stake in the subject. Several are serving time for a drug-related offense, as are hundreds of thousands of other Americans.

The gentler war on drugs isn't new

August 2017, driven largely by public perception that many white Americans are experiencing problems and even dying from opioid use, Donald Trump proclaimed the opioid problem a national emergency. The president's announcement appeared to consolidate a shift in the way we view certain drug users. They are now patients in need of our help and understanding, rather than criminals deserving scorn and incarceration.

This was made abundantly clear in 2014 when the governor of Vermont, Peter Shumlin, devoted his entire State of the State address to the "heroin crisis" (Shumlin, 2014). He urged his overwhelmingly white electorate to deal with addiction "as a public health crisis, providing treatment and support, rather than simply doling out punishment, claiming victory and moving on to our next conviction." Public officials from both parties and throughout the United States have echoed these sentiments, which ultimately spurred the White House to form the task force that urged Trump to declare an emergency in fall 2017.

What looks like a radical shift to a more enlightened drug policy—one that favors treatment over incarceration—has encouraged many to hope that there will be far fewer drug-related arrests than there were in previous decades. I don't count myself among the optimists.

Don't get me wrong, I support this enlightened approach. It's what we should do, certainly as it relates to dealing with individuals who are struggling with drug addiction.[1] But it is not, historically, what we have done for *all* of our citizens.

Recall the so-called crack epidemic of the late 1980s. Can you imagine Gov. George Wallace of Alabama urging his voters to view crack use as a health crisis? I think not. Back then, even northern liberals such as New York's Gov. Mario Cuomo were calling for life sentences for anyone caught selling crack cocaine, at amounts worth as little as $50 (Schmalz, 1986). Public contempt expressed toward those who used or sold crack was intense and visceral. The perceived user and trafficker of crack was black, young, and menacing. But, in reality, the majority of crack users were white (US Department of Health and Human Services, 1995), and most drug users buy their drugs from dealers within their own racial group (Riley, 1997). In 1992, though, more than 90 percent of those sentenced under the harsh crack cocaine laws that were passed in the late 1980s were black (United States Sentencing Commission, 2002). They were required to serve a minimum prison sentence of at least five years for small amounts of crack, even if they were first-time offenders.

To the extent that the use of crack among whites was acknowledged, media reports sympathetically detailed the plight of white middle-class crack users. It was seen as an understandable tool for stressful professional lifestyles. According to this scenario, the users—Wall Street executives, computer programmers and the like—had initially snorted cocaine recreationally after hours but had gradually moved on to smoking the drug during work time. And some developed problematic use.

For white crack abusers, medical experts were quoted extolling the effectiveness of treatment. Any law enforcement perspective was conspicuous by its absence. Public service announcements, geared toward middle-class crack users, encouraged sympathy and not judgment.

This pattern of racial differentiation—one drug policy for white users and another for black users—followed the format of the heroin crisis beginning in the late 1960s. The face of the heroin addict then in the media was black, destitute and engaged in repetitive petty crimes to feed his or her habit. A popular solution was to lock up these users. New York State's infamous 1973 Rockefeller drug laws exemplified this perspective. This legislation created mandatory minimum prison sentences of 15 years to life for possession of small amounts of heroin or other drugs. More than 90 percent of those convicted under the Rockefeller laws were black or Latino, even though they represented a minority of drug users (Drucker, 2002).

This punitive approach to black heroin users coincided with a massive expansion of methadone maintenance programs that benefited large number of white "patients," including addicted soldiers returning from the Vietnam War (Schatz, 1971). Even President Richard Nixon (Nixon 1971) praised methadone "as a useful tool in the work of rehabilitating heroin addicts," one that "ought to be available to those who must do this work."

One key feature of methadone programs that was viewed as a drawback was the requirement that the drug be administered through health clinics or hospitals. This meant that patients had to attend the clinic daily in order to receive the medication. This presented an inconvenience for some patients, especially those with jobs and demanding schedules. Also, the fact that patients were required to stand in line in order to receive the medication was viewed as a form of social shaming, which was stigmatizing. So, in 1971, New York City Mayor John Lindsay picked up their cause pushing for the use of private physicians to distribute methadone to a select group of middle-class and insured patients (Ranzal, 1971).

This characteristically American pattern of cognitive flexibility on drug policy, with harsh penalties for some and sympathetic treatment for others, thus has a long history—one that continues to this day.

Defining racism

In the US, the differential response to drug users based on race did not begin with heroin policies in the 1960s. This type of racism (or racial discrimination) is as American as drug use itself. Before going any further, though, I think it would be instructive to define racism because so many people have misused and diluted the term that its perniciousness gets lost when communicating, although the impact on the victims of racism can continue to fester for a lifetime. When I use the term here, I simply mean an action(s) that results in disproportionate unjust or unfair treatment of persons from a specific racial group. Intent of the perpetrator is not required. What is required is that the treatment must be unjust or unfair, and that such injustice is disproportionately experienced by at least one racial group. Finally, individuals in positions of authority who fail to take actions to abolish or limit racism after being presented with evidence of its existence can be labelled racists in that specific domain.

Brief history of the United States' response to drug use

In the early years of the 20th century, cocaine use by whites was promoted and celebrated. A range of over-the-counter and pharmaceutical products contained the drug and many prominent individuals openly used it, praised it, and recommended it. Sigmund Freud, perhaps the best-known proponent, endorsed cocaine as a general feel-good tonic and addiction cure. Cocaine use by blacks, on the other hand, was condemned. One of the most egregious myths about cocaine was that it made black men unaffected by .32 caliber bullets. Some southern police forces, as a result, switched to the .38 caliber weapon (Williams, 1914). Between 1898 and 1914 numerous articles appeared exaggerating the association of heinous crimes and cocaine use by blacks. In some cases, suspicion of cocaine intoxication by blacks was reason enough to justify lynchings. Mischaracterizations about black cocaine users coincided with the peak period of lynching and restriction of black people's civil liberties and rights.

Around this time, the US Congress was debating whether to pass the Harrison Narcotics Tax Act, one of the country's first forays into national drug legislation. This unprecedented law sought to tax and regulate the production, importation and distribution of opium and coca products. Proponents of the law saw it as a strategy to improve strained trade relations with China by demonstrating a commitment to controlling the opium trade. Opponents, mostly from Southern states, viewed it as an intrusion into states' rights and had prevented passage of previous versions.

By 1914, however, the law's proponents had found an important scapegoat in their quest to get it passed: the mythical "negro cocaine fiend," which prominent newspapers, physicians and politicians readily exploited. Indeed, at congressional hearings, "experts" testified that "most of the attacks upon white women of the South are the direct result of a cocaine-crazed Negro brain." When the Harrison Act became law, proponents could thank the South's fear of blacks for easing its passage. Noted drug-control policy historian David Musto observed that fabrications about blacks and cocaine served as an important tool to preserve white supremacy (Musto, 1987).

Exploitation of white racial fears also played a crucial role in the de facto criminalization of marijuana in 1937, through the Marijuana Tax Act. Law enforcement types like Harry J. Anslinger, commissioner of the Federal Bureau of Narcotics, routinely connected use of the drug with blacks and Mexicans while recounting gruesome stories: "[P]olice found a youth . . . With an ax he had killed his father, mother, two brothers, and a sister . . . he had become crazed from smoking marijuana" (Anslinger and Cooper, 1937). These fabrications were widely disseminated, facilitating draconian policies, racial discrimination, and incalculable human misery.

US current response to opioids misses the mark

With this as background, US drug policy today takes on a sharper focus. Most opioid users are white. Policymakers recognize this fact. That is one reason Republican senators from those states with the largest numbers of opioid-related deaths withheld their support from the recent Senate health care bill—they saw it as underfunding treatment for opioid addiction. But this humane approach is only part of the picture.

Multiple states have drafted or passed legislation that enhances penalties for opioid possession and trafficking. In some states, prosecutors can now charge a drug dealer with murder for selling drugs to someone who died from an overdose. Deaths related to fentanyl, a potent synthetic opioid, have fueled such initiatives. Kentucky's new law makes the sale of fentanyl, in any amount, punishable by a prison sentence of up to 10 years. Sound familiar?

If past drug law enforcement action is predictive of future behavior, most of those convicted of opioid-related crimes will be black and brown. Recent federal data back this up: More than 80 percent of those convicted of heroin trafficking are black or Latino, although all racial groups buy and sell drugs at roughly the same rate. The discretionary nature of drug law enforcement, which continues to focus mostly on black and Latino communities, is basis for this type of racism.

Perhaps what's even worse is that most "official" proposed solutions to deal with the so-called opioid crisis completely miss the mark, mainly because such approaches are not data driven. For example, the overwhelming majority of opioid users do not become addicts. Less than a quarter of heroin users (Anthony et al. 1994) and less than one percent of people prescribed opioids for pain will become addicted (Noble et al. 2010). We know that one's chances of becoming addicted increases if s/he is young, unemployed and/or has co-occurring psychiatric disorders. That is why it is critically important for policies to ensure that people have jobs and affordable access to effective mental health services, rather than exclusively focusing on eliminating drugs from society. If we took this approach the number of people addicted to drugs would be substantially reduced.

The public also seems to be unaware of this fact: for nearly 20 years, the number of Americans who tried heroin for the first time in a particular year has remained between 100,000–200,000 (SAMHSA, 2015). This number translates to about 0.1 percent of Americans aged 12 and older, which is relatively low compared with cannabis and cocaine, for which the numbers are 2.6 million (1 percent) and 1 million (0.4 percent), respectively. Still, the number of new heroin users has not appreciably changed in several years, despite the fact that heroin was banned in 1914. In other words, heroin use specifically, and opioid use in general, isn't going anywhere whether we like it or not. This isn't an endorsement of drug use but rather a realistic appraisal of the empirical evidence with which I deal in my continuing efforts to help keep people safe.

The major concern with illicit opioid use is the potential for overdose and death. It is certainly possible to die from an overdose of an opioid alone, but this accounts for only about a quarter of the thousands of opioid-related deaths (SAMHSA, 2014). Combining an opioid with another sedative such as alcohol, an antihistamine (like promethazine), or a benzodiazepine (like Xanax or Klonopin) causes many of these deaths. Put another way, people are not dying because of opioids, they are dying because of ignorance. Public service announcement campaigns warning users about the real potential dangers of these drug combinations need to be clear and greatly expanded.

Now there is an additional opioid that we have been made to fear—fentanyl. Fentanyl produces a heroin-like high but is considerably more potent, meaning that less of the drug is required to produce an effect, including overdose. To make matters worse, according to some media reports, illicit heroin is sometimes adulterated with fentanyl. This, of course, can be problematic, and even fatal, for unsuspecting heroin users who ingest too much of the substance thinking that it's heroin alone.

One simple solution to this problem is to make available free, anonymous drug-purity testing services. In this way, people who use illegal drugs can submit their drug samples and receive an analysis of the sample's composition and purity. If the sample contains adulterants, the users would be informed. These services already exist in places such as Belgium, Portugal, Spain, and Switzerland, where the first goal is to keep users safe.

In addition, the opioid overdose antidote naloxone should be made more affordable and readily available to opioid users, and their family and friends. Naloxone now comes in multiple forms, including nasal spray, making it easier for family members, friends and caregivers to administer.

Another life-saving piece of information that has been absent from discussions of the opioid problem is acetaminophen toxicity. The most recent data from the National Survey on Drug Use and Health shows that just over 300,000 people reported using heroin at least once in the past 30 days. This number is substantially lower than the number of individuals who reported use of marijuana, which came to about 22 million, prescription opioids, about 4 million, and cocaine, 2 million, over the same period. Most individuals seeking a heroin-like high use prescription opioids recreationally. On the one hand this is a good thing, because the purity of street heroin is often poor due to adulterants added to increase the quantity of the product. Prescription opioids are usually of a higher quality as they are pharmaceutical-grade. But: Popular prescription medications like Percocet, Vicodin and Tylenol 3 contain a relatively low dose of an opioid in combination with a considerably larger dose of acetaminophen—and excessive acetaminophen exposure is the number one cause of liver damage in the United States. Some users may unwittingly risk liver damage by taking too many of these pills. It seems that a responsible society would inform people not to overdo it on opioids containing acetaminophen because it can be more fatal than the low doses of opioids contained in these formulations.

Negative unintended consequences that stem from our approach to dealing with drugs abound. But the fact that many physicians are unwilling or reluctant to prescribe opioid medications for fear of appearing to run "pill mills" has received insufficient discussion, especially in relation to its impact on racism in medical practice. The barrage of media stories attesting to an epidemic of opioid use has left many with the belief that physicians are too quick to distribute opioid medications to patients. Indeed, occasionally media headlines blare with stories about rogue physicians arrested for "pushing pills," that is, dispensing pain medications indiscriminately for quick cash. There are also periodic stories of cunning patients who "doctor shop," deceiving physicians in order to obtain large amounts of opioid medications. The fact that most physicians want to do the right thing and are judicious in their prescribing of pain medications is frequently omitted from sensational accounts of the "drug pusher doctor." These developments, ultimately, make it more difficult for patients to obtain opioids when medically indicated. Importantly, it has been well documented that physicians are less likely to prescribe opioids to blacks than to whites (Singhal et al., 2016). This type of racism will certainly be exacerbated if we continue on our alarmist approach to dealing with opioids.

In terms of treating individuals with opioid addictions, well-run methadone programs, with appropriate adjunctive therapies that address other medical or psychosocial needs of patients with heroin addiction, work to lessen debilitating symptoms associated with the disorder (Greenwald this volume). Moreover, in a growing number of countries, including Switzerland, The Netherlands, Germany and Denmark, effective opioid treatment may include daily injections of heroin, just as the diabetic may receive daily insulin injections. Also, like diabetes treatment, heroin administration is only one aspect of the treatment, which also includes addressing the medical and psychosocial issues of the patient. Some of these programs have been running successfully for more than 20 years. Notably these patients hold jobs, pay taxes, and live long, healthy, productive lives. Yet, in the United States, these programs are not even discussed as an academic exercise, let alone as another tool in our armamentarium used to treat opioid addiction.

Enhancing penalties for opioid possession and trafficking seems to be a less than effective strategy to deal with opioid-related deaths. In addition, the re-imposition by Attorney General Jeff Sessions of more mandatory minimum prison sentences for drug offenses will result in more black and brown people locked away. This is not only anachronistic, mean and racist, but it does not address the real concerns associated with opioid misuse as noted above. Nor does this approach help those afflicted with opioid addiction.

The correction officer's abrupt, loud knock on the thick glass window signaling the end of our class period startled me, just as it did the previous week and the week before that. Immersed completely in my lecture and in the intensely engaged faces of my students, I lost track of time. There are no clocks on the wall; cell phones and other electronics are not permitted. The two hours had imperceptibly passed. My students stoically prepare to go back to being inmates. And I prepare to leave, as I do on every Friday night, with the same sinking feeling that our approach to dealing with drugs hasn't appreciably changed in more than one hundred years. The only thing that changes is the names of the ignorant policy makers and law enforcement officials charged with waging drug wars. Our citizens, including the black and brown ones, deserve better.

Note

1 The terms 'addiction' and 'abuse', as they are used throughout this article, conform to the *Diagnostic and Statistical Manual of Mental Disorders, 5th Edition* (DSM-V) definition of Substance Use Disorder.

References

Anslinger, H. J. and C. R. Cooper (1937) "Marijuana: assassin of youth", *The American Magazine* 124, pp. 19, 153.
Anthony J. C., Warner L. A. and Kessler R. C. (1994) "Comparative epidemiology of dependence on tobacco, alcohol, controlled substances, and inhalants: basic findings from the National Comorbidity Survey," *Experimental and Clinical Psychopharmacology* 2: 244–268.
Drucker, E. (2002) "Population impact of mass incarceration under New York's Rockefeller drug laws: an analysis of years of life lost", *Journal of Urban Health: Bulletin of the New York Academy of Medicine* 79: 1–10.
Greenwald, M. K. (this volume, pp. 478–489) "Opioid substitution treatment and harm minimization approaches".
Musto, D. (1987) *The American Disease: Origins of Narcotic Control*, Expanded Edition, New York, NY: Oxford University Press.
Nixon, R. M. (1971) *Special Message to the Congress on Drug Abuse Prevention and Control*, June 17, www.presidency.ucsb.edu/ws/?pid=3048.
Noble, M., Treadwell, J. R., Tregear, S. J., Coates, V. H., Wiffen, P. J., Akafomo, C. and Schoelles, K. M. (2010) "Long-term opioid management for chronic noncancer pain", *Cochrane Database Systematic Review* 20, pp. CD006605.
Ranzal, E. (1971) "Mayor seeks $9.2-million for methadone", *New York Times*, March 5.
Riley, K. J. (1997) *Crack, Powder Cocaine, and Heroin: Drug Purchase and Use Patterns in Six U.S. Cities*, Washington, DC: US Department of Justice (NCJ #167265), www.ncjrs.gov/pdffiles/167265.pdf.
Schatz, A. (1971) "The war within: portraits of Vietnam Veterans fighting heroin addiction", *Life Magazine*, July, http://time.com/3878718/vietnam-veterans-heroin-addiction-treatment-photos.
Schmalz, J. (1986) "Cuomo plan fuels drug laws debate", *New York Times*, August 18, www.nytimes.com/1986/08/18/nyregion/cuomo-plan-fuels-drug-laws-debate.html?pagewanted=all.

Shumlin, P. (2014) *State of the State Speech*, www.governing.com/topics/politics/gov-vermont-peter-shumlin-state-address.html.

Singhal, A., Tien, Y. Y. and Hsia, R. Y. (2016) "Racial-ethnic disparities in opioid prescriptions at emergency department visits for conditions commonly associated with prescription drug abuse", *PLOS One* Aug 8;11(8): e0159224, doi:10.1371/journal.pone.0159224.

Substance Abuse and Mental Health Services Administration (SAMHSA) (2014) The DAWN Report: Benzodiazepines in Combination with Opioid Pain Relievers or Alcohol: Greater Risk of More Serious ED Visit Outcomes. Rockville, MD: Substance Abuse and Mental Health Services Administration, www.samhsa.gov/data/sites/default/files/DAWN-SR192-BenzoCombos-2014/DAWN-SR192-BenzoCombos-2014.pdf.

Substance Abuse and Mental Health Services Administration (SAMHSA) (2015) *Results from the 2015 National Survey on Drug Use and Health: Detailed Tables*, Rockville, MD: Substance Abuse and Mental Health Services Administration, www.samhsa.gov/data/sites/default/files/NSDUH-DetTabs-2015/NSDUH-DetTabs-2015/NSDUH-DetTabs-2015.htm#tab7-34a.

US Department of Health and Human Services, Public Health Service (1993) *National Household Survey on Drug Abuse, Main Findings*, Rockville, MD: Substance Abuse and Mental Health Services Administration; 1995.

United States Sentencing Commission (USSC) (2002) *Special Report to the Congress – Cocaine and Federal Sentencing Policy*, May, www.ussc.gov/sites/default/files/pdf/news/congressional-testimony-and-reports/drug-topics/200205-rtc-cocaine-sentencing-policy/200205_Cocaine_and_Federal_Sentencing_Policy.pdf.

Williams, E. H. (1914) "Negro cocaine fiends are a new Southern menace", *New York Times*, February 8.

42
DRUG LEGALIZATION AND PUBLIC HEALTH
General issues, and the case of cannabis

Robin Room

Psychoactive substances have been a part of human experience since before recorded history began. They serve many purposes: as sources of pleasure, relief of distress, wakefulness or sleep, energy or tranquillity. They also bring health and social problems, both for the user and for others.

Whether through custom or law, access to and use of most of the substances has often been subject to social control. There is a considerable history of cultural and national prohibitions on use of psychoactive substances. For instance, there was a national prohibition on alcohol in 13 autonomous countries during the first decades of the 20th century (Schrad 2010), and alcohol use is forbidden for observant Moslems and discouraged or forbidden in some strands of other major world religions. Historically, there have also been prohibitions on tobacco and other substances (Austin 1978). For substances that have a substantial effect on behaviour, cultures have generally chosen either to tolerate use but attempt to limit it or to forbid its use (Room and Hall 2013).

Where use of a substance that affects behaviour is not prohibited, there are commonly rules around its availability and use, and often formal regulations and laws. Such laws may specify, on the one hand, who can use and under what circumstances, and on the other, how, where and when the substance can be promoted and sold or served. Where such a substance is legally available in a society with a market economy, controls on the promotion and marketing of the substance become a crucial public health issue – as experience with the difficulty in limiting markets in alcohol and tobacco in the last century makes clear.

In this chapter, our focus is on legalization in the full sense of the term – where a substance is legally produced and supplied, with whatever restrictions, for unrestricted use by the consumer. Thus we are not concerned with decriminalization or depenalization, where penalties are removed or softened for the end user but criminal penalties for producing or supplying the substance remain in place. We give particular attention to legalization of cannabis, since this is the main current arena where issues of legalization and public health are in play.

Public health approaches to legalization

A public health approach to legalization of substances that are attractive enough to carry risks of overuse and of other harms will inevitably be some variation on policies of "permit but discourage", as a relevant book is titled (Bogart 2011). Where private interests are supplying the substance in a market economy, there will always be pressure to expand the market, whether by advertising and other promotion to potential consumers, or by the "ratchet mechanism" (Mäkelä et al. 1981) of pressure to loosen restrictions on availability on policymakers, regulators or enforcers.

A public health approach to legalization thus needs to look in two directions: on the one hand, at loosening the constraints under which a substance has been in a prohibited status, and on the other hand, at installing mechanisms that limit availability and control promotion so as to discourage heavy use (Rehm et al. 2013) or other risky consumption, and which are resistant to commercial pressures for weakening. Concerning the constraints that require prohibition, we pay primary attention here to the international level, since it can be argued that that has been the level which has been overriding until recently, and which is still important. Concerning limiting the market, our attention will be primarily at the national or subnational level (the latter particularly in federal countries), since these are the primary levels for market regulation in the interests of public health. It should be recognized, however, that prohibitionary impulses are not limited to the international level. In the last half-century, many governments have been elected on the basis of platforms of cracking down on illicit drugs, and in many places the weight of popular sentiment is still towards drug prohibition. On the other side, it must also be recognized that the forces seeking to open up and "grow the market" are not only at national and local levels. Large and powerful multinationals dominate the alcohol and tobacco markets, not to mention the markets in medical psychopharmaceuticals, and if unchecked are likely to move to a dominant position in the market in any other substance with wide appeal not long after its legalization. Beyond their marketing expertise, such global corporations tend to be far quicker than regulatory agencies at transferring successful innovations from one jurisdiction to another.

This discussion does not deal with provisions for medical use of a substance under prescription. Where this alternative exists in an otherwise prohibitionary system, the fact of the prohibition may indeed distort the functioning of the prescription system (see Babor et al. 2010b, Chapters 6 and 12). In a further distortion, in some states in the US making "medical marijuana" available as a medication was in part seen as a stepping-stone towards full legalization, and rules around medical availability have been set quite loosely; 5% of adults in California, for instance, had used medical marijuana by 2012 (Ryan-Ibarra et al. 2015). Where cannabis for nonmedical use becomes legally available, medically prescribed cannabis may be expected to lose prominence – although lighter restrictions on availability for medical marijuana, including a lower price because it was not taxed, meant that this was not the case in the months after Colorado's legalization of recreational cannabis (Ghosh et al. 2016). Meanwhile, in the dozen or so countries other than in North America that now provide for prescription of cannabis, the medical availability is often little different from the situation for other prescription psychopharmaceuticals.

Drug control at the international level

Nowadays there is a global system of drug control that includes many psychoactive substances, but by no means all. It is governed by three international treaties, dating from 1961, 1971 and 1988, to which most countries are parties. Under the treaties, governments can permit medical or scientific use of the substances subject to the treaties, but no other use is allowed (Babor et al. 2010b).

Among other goals, the treaties can be regarded as intended to promote public health. A major justification offered in the preamble to the 1961 treaty is that it is aimed against "addiction to narcotic drugs". Among the criteria for including a substance under the control of the 1971 treaty is that there is "evidence that the substance is being or is likely to be abused so as to constitute a public health and social problem warranting the placing of the substance under international control". In another public health-related facet, the treaties also aim to facilitate "adequate provision" of supplies of the substances for medical use.

But the division of the United Nations system that has charge of the drug control treaties, the Office on Drugs and Crime (UNODC), is responsible for international crime control rather than public health. While the system had always been multiply motivated, crime control and the illicit market became its dominant concern in the late 20th century (Carstairs 2005), as epitomized by the combination of the UN drugs and crime offices in 1997. In the Cold War era, drug criminalization and control was a rare international arena in which the two sides could mostly agree. In compliance with the treaties, laws in most countries criminalize not only markets in substances controlled under the treaties, but also a consumer's possession of the substances other than for medical purposes. At national levels, too, criminal justice concerns and agencies are at least as much involved as public health concerns and agencies, and often take priority.

A fourth international treaty, the Framework Convention on Tobacco Control, which took effect in 2005, is unambiguously a public health treaty, under the jurisdiction of the World Health Organization, and with production and distribution remaining legal. Tobacco became ever more identified as a public health menace in the second half of the 20th century, but as a legal substance with politically strong corporate and state producers it seemed inconceivable to prohibit it, and sporadic attempts to bring it under the drug treaties were quickly stifled. Unlike the drug treaties, the tobacco treaty's provisions are primarily "soft law", recommending actions to parties rather than requiring them. Nevertheless, experience in the first decade of the treaty suggests considerable success in motivating parties to adopt measures that it recommends (e.g., Sanders-Jackson *et al.* 2013).

Many psychoactive substances are not covered by the international treaties. Some are in wide traditional use in particular global regions – e.g., khat, betel nut – and some are more local, but use is primarily as a folk custom without highly centralized and capitalized production and distribution. Others are "new psychoactive substances" (NPSs), a mixed category including both newly invented or promoted substances and some longer-established but not listed under the conventions (Seddon 2014).

And then there is alcohol, which is both widely used and highly commercialized, but which is not covered by any international treaty. If it were to be considered for coverage under the 1971 drug treaty, it would clearly qualify for scheduling in terms of the treaty's criteria (Room 2006); but since this would require that all nonmedical use be prohibited, the drug control system has never formally considered it for scheduling. When the present author attempted to get the WHO's Expert Committee on Drug Dependence to review it for consideration for scheduling under the drug treaties, there was resistance to considering the motion, which was then referred for consideration "at some future Expert Committee meeting" (WHO 2012). An alternative approach would be a separate convention on alcohol, for instance adapted from the tobacco convention (Room 2006). But so far there has been no political appetite for this; for elites and staff in the international arena it is "our drug", available for instance in the luncheon cafeteria at the World Health Organization headquarters.

Issues at the intersection between international controls and legalization

In current circumstances, a major reason why international agreements on psychoactive substances matter is the growing scope and power of global and other multilateral trade treaties.

Psychoactive substances have been an important part of international trade since the wine amphorae of ancient times. Legal trade in such substances – for instance, in alcohol and tobacco products – is rarely excluded from trade treaties, and there have been a number of judgements in World Trade Organization (WTO) disputes disallowing national control measures on alcohol and tobacco (Ziegler and Ziegler 2006). Despite formal provisions for exceptions on public health grounds, the narrow wording of the exceptions and the rules and the nature of the expertise drawn on for adjudication of disputes generally weigh against the public health interest (Rehm and Room 2009; Liberman and Mitchell 2010; McGrady 2011). The tobacco control treaty in its current form does not include language addressing conflict with trade treaties, but such provisions are an important issue for future treaties grounded in public health concerns.

Although the three drug treaties also do not specifically address trade treaty provisions, there has been no WTO dispute filed concerning substances under their control (Babor et al. forthcoming); presumably their prohibitionist stance and tight regulatory provisions concerning trade in controlled medications have warded this off. To the extent drugs are legalized, countering the market-opening intentions of the trade treaties becomes an important policy agenda. National or subnational regulations of legalized drugs need to be protected by treaty from attack in trade disputes and lawsuits.

For the substances scheduled under the international drug control treaties, the situation has been changing. There are three primary areas in which there has been movement as of 2016. The first is the setting of a modern precedent by which a country can remove itself from the treaty requirements for prohibition of a particular substance. Coca-leaf chewing is a long-established custom in Andean nations, and specifically in Bolivia, but had been included in the prohibitions of the 1961 treaty. After failing in an attempt to remove coca leaves from coverage by the treaty (just as cannabis leaves had always been excluded), in 2012 Bolivia withdrew from the 1961 convention and in 2013 successfully re-acceded with a reservation concerning the prohibition on coca leaves. There were minor symbolic and fiscal punishments along the way from the US and the European Union, but Bolivia's successful initiative confirmed a path by which a country can use internationally recognized procedures to remove itself from drug treaty coverage of a particular substance (Room 2012b).

The second area has been a more general revulsion in much of Latin America against the "war on drugs" model of drug control. Efforts to eliminate the illicit drug trade, the primary markets for which have been in the US and Europe, have taken a considerable toll on many Latin American countries. In 2009 a Latin American Commission on Drugs and Democracy, headed by ex-presidents of Brazil, Colombia and Mexico, issued a report arguing for decriminalization of cannabis use and to "reframe the strategies of repression against the cultivation of illicit drugs" (Latin American Commission 2009). The push by some Latin American countries for reform of the system has continued, though so far with little effect; for instance, Mexico, Colombia, Guatemala, Ecuador, Uruguay and St Lucia together pushed for opening up the debate at the 2016 United Nations General Assembly Special Session (UNGASS) on drugs, brought forward at the insistence of Latin American countries from an originally scheduled 2019 to April 2016 (IDPC 2014: 10).

The third area is the developing area of cannabis legalization. Other than in Uruguay, the movement has come primarily from civil society rather than governments, and is in the process of disrupting the international status quo, although there has been no change yet in the formal position of the international control system. Whereas the earlier Netherlands "coffee shops" system could be argued to be formally within the treaty limits, accomplishing this left it with the "back door problem" of having no provision for legal production (MacCoun 2011). De-facto legalization of cannabis production as well as retail sale has come in Europe primarily through the growth of cannabis clubs in Spain, Belgium and elsewhere (Kilmer *et al.* 2013; de Corte 2015), with each club locally organized. The formal systems of legalized cannabis production, processing and sale that have emerged in a growing number of US states have been legalized by citizens' ballot initiatives, not through elected legislators. Change from the national political level has begun to emerge, with Uruguay moving first at this level, and Canada committed also to legalize (Room 2014; Rehm *et al.* 2016). Defenders of the international control system are now putting a new emphasis on the system being "flexible and resilient" (IDPC 2014), but there is no question that these systems are outside the bounds set by the treaties.

While there have been a number of suggestions of amendments to the treaties (Room 2012c) or of a new Framework Convention on Cannabis Control to supersede the 1961 treaty with respect to cannabis (Room *et al.* 2010: 162–191), there has as yet been no concrete sign of change in the international drug control system. Full legalization of any drug by the international treaties remains in defiance of the treaties, unless a manoeuver like Bolivia's is undertaken. On the other hand, the legalization of cannabis in states of the US has compromised the moral authority of the country that has had the primary role in building and maintaining the drug prohibition system (Babor *et al.* in press). It is no longer clear that there would be substantial general international reaction against a controlled legalization of cannabis, as is now being undertaken in Uruguay and in Canada, although (as is the case for Uruguay and the US states) neighbouring jurisdictions may well take a strong interest in developments at their borders.

Cannabis does not generally rank high in comparisons of the relative harmfulness of psychoactive substances, and there has now been a half-century in which it has had some popularity among middle-class youth in high-income countries. For most other drugs with substantial illicit use that are under international control there would probably be a stronger international reaction, and legalization seems further off. Thus, primary attention is paid in the remainder of this chapter to cannabis, where legalization already exists or is pending in a number of places.

Legalizing in the shadow of prohibition: a century's experience with building regimes for alcohol and cannabis

There is plenty of precedent for building legal regulatory regimes when prohibition of production and sale of psychoactive substance is the alternative. The US states had to do this in a period of months following the election of November 1932, when it became clear that US federal alcohol prohibition was coming to an end. The states could draw on a considerable literature on relevant experience elsewhere (Catlin 1931; Fosdick and Scott 1933), including experiences with ending prohibition in Canada and Norway, for instance, as well as regulatory regimes elsewhere. There was also the US experience from pre-Prohibition times, less than 15 years before.

Regimes for legalized alcohol

For alcohol, there were two basic choices in societies that had had a strong alcohol temperance history (Levine 1992), so that alcohol was not regarded as just another foodstuff: the state could monopolize the market, or private interests could be licenced under a specific "liquor licencing" system. Government monopolies of alcohol sale were set up at the municipal level in Sweden and other Nordic countries and in the southern US in the latter half of the 19th century (Room 2000), in some places serving alcohol and in others selling it to take away. They often proved highly profitable – which helped them to survive in a number of states, provinces and countries through the era of privatization in recent decades. Currently, the monopoly covers only part of the market, usually for take-away alcohol; in almost all jurisdictions with a monopoly today, there is also a liquor licencing system, for instance for restaurants. In terms of public health interests, alcohol monopolies typically have fewer outlets and shorter sales hours, putting some limits on availability. With a "disinterested management" not driven by a profit motive and staff in secure government jobs, conditions on sales and restrictions on purchasers (e.g., not selling to those under legal age) are more likely to be complied with. An alcohol monopoly also fills a position in the market that, in private hands, would be lobbying to relax regulations so that sales could be increased. Studies of privatizations of alcohol monopolies have found that monopolies have an effect in holding down levels of alcohol sales and of alcohol-related problems (Her et al. 1999; Hahn et al. 2012).

In liquor licencing systems, as they operated until recent decades in societies that had had a strong alcohol temperance history, the number of licences granted was usually restricted, so that the licencee had the advantage of a non-saturated market. In return for this privileged position, the licencee was expected to follow detailed regulations limiting conditions of sale in the interest of public health and order. The systems adopted in this era were much more restrictive than today: in Canada, limits on purchase amounts per visit were common; in Sweden there was an individualized monthly ration of take-away spirits, primarily for males, with one in ten males denied a ration; in the UK and Australia, tavern opening hours were restricted (Room et al. 2006; Room 2012a). Many of these restrictions were abandoned in the course of the second half of the 20th century, in successive deregulatory waves, and substantial increases in alcohol consumption and alcohol-related problems ensued (e.g., Mäkelä et al. 1981; Norström 1987; AMS 2004). The US alcohol control systems that cannabis legalization initiatives have promoted as a model are thus considerably less restrictive than at their inception in the 1930s.

There is a substantial scholarly literature available on the effects of regulatory controls on levels and patterns of alcohol consumption and on rates of alcohol-related problems (Babor et al. 2010a). Reflecting the historical experience since the 1950s, many of the studies are based on what happened when restrictions were removed or loosened (Olsson et al. 2002).

Regimes for legalized cannabis

There are two longstanding systems for legal availability of cannabis: in India and in the Netherlands. India had a historic tradition of consumption of cannabis for religious and secular purposes. Under the 1961 international drug treaty, countries with traditional patterns of use of scheduled drugs were allowed a 25-year period to eliminate this, and the Indian government reluctantly outlawed cannabis preparations forbidden under the treaty in 1985. However, at India's instance, the treaty had specified that it was only the "fruits and flowering tops" of cannabis plants that were forbidden by the treaty. So bhang, an infusion of cannabis leaves, is legally sold at government-licenced shops in at least five states in India (Room et al. 2010: 99–100). However, there seems to be no serious study available in English of the functioning and effects of these Indian state systems, which have operated largely under the radar of the international drug control system and of policy research.

The Dutch system of de-facto legalization of retail sales of cannabis in "coffee shops" was set up after passage of a national drug law reform in 1976. National regulations specify a number of restrictions on the coffee shops: there can be no hard drugs or alcohol on the premises; quantities to be sold are limited (no more than 5 gm per customer per day); no sales or access to persons under 18 are allowed; advertising is forbidden. In 2013 access was limited to residents of the Netherlands only, though enforcement of this is a local matter (van Ooyen-Houben and Kleemans 2015). Additional conditions, including a ban on coffee shops, can be imposed locally. That for many years the Netherlands was known as the only high-income country in which cannabis for recreational use was (de-facto) legally available has resulted in considerable cannabis tourism. In part due to the resulting foreign-policy pressures, the policies have varied considerably over the 40 years of de-facto legalization (van Ooyen-Houben and Kleemans 2015).

In November 2012, voters in the US states of Colorado and Washington voted to legalize cannabis production and sale, in 2014 Alaska and Oregon followed suit, and in November 2016 they were joined by California, Nevada, Maine and Massachusetts, so that recreational cannabis has now been legalized for one-quarter of Americans. As of 2016, the US state systems were still operating in a grey area legally, since cannabis remained legally prohibited under US federal law. The federal government announced in 2013 that it would tolerate state legalization so long as the states had "strong and effective enforcement systems" that conformed to eight federal enforcement priorities (Caulkins *et al.* 2015b). But the continuing legal prohibition at the national level had major effects on how state legalizations proceeded, for instance ruling out the option of a state monopoly on production or sale (Pardo 2014), and so far effectively dissuading large multinational corporations from entering the market, since the prohibition makes it hard to raise capital for expansion and ignoring it would adversely affect their corporate tax rate. The continuing national prohibition also allows aggressive state and local market regulations that would otherwise contravene the US constitution's requirement of unfettered interstate trade.

In early 2012 the President of Uruguay proposed the legalization of cannabis production and use, and late in 2013 the law carrying this into effect was passed (Pardo 2014). By mid-2016, while the new Colorado and Washington systems were fully functional, the Uruguay system was functioning only in part: home-growing and cannabis clubs had been authorized, but sales of cannabis grown under the control of a government monopoly and sold through pharmacies was not yet functioning, in part because of resistance from the pharmacies (Marshall 2016).

Except for the Dutch experience, evaluations of the effects of cannabis legalizations are still very preliminary. In the 40 years since the "Dutch model" was adopted, there have been substantial developments in the policy, responding in part to developments in the market (van Ooyen-Houben and Kleemans 2015). Levels of cannabis use changed modestly, increasing after de-facto legalization, particularly after a proliferation of retail outlets and promotion, but then falling again as the system was somewhat tightened. That production and distribution have not been legalized has kept prices much higher than is now true in the US legalizing states, and thus restrained levels of consumption (MacCoun 2011). For the US experience, reports on the era when medicinal cannabis became available suggest some increase in use, more at lower- than at higher-income levels, and a "professionalization" of distribution – a shift from "gifting toward selling" (Davenport and Caulkins 2016). An early report on the Washington state experience with legalization of recreational cannabis after July 2014 found the expected dramatic drop in prosecutions for cannabis possession, but that replacing illicit supply of cannabis with a regulated market was far from complete; and there was an increase in detected driving under the influence of cannabis (Roffman 2016). As Hall and Weier (2015) note, "it may well be a decade before we can decide whether the legalization of cannabis use [in the US] has increased population cannabis use and harms related to such use".

Cannabis legalization: the public health dimension in designing and regulating legal markets

The systems adopted in Colorado, Washington and other legalizing US states have reflected a number of considerations. All full legalizations in the US so far have been by popular initiative, and distrust of how the political system would implement a proposition worded only in general terms has meant that the regimes proposed in the initiatives are often specified in considerable detail, in over 100 pages of text. The coalitions that put together the proposals have included libertarians, those concerned about arrest rates for those involved in the illicit market, and cannabis aficionados – but also those with a financial interest in a legal market, notably including those already involved in supplying the legal medical marijuana market, which is already well established in the early-legalizing states. On the other hand, the initiatives include counterbalancing provisions to increase their appeal to a hesitant electorate. One appeal to electors has been the prospect of money being raised for worthy causes in the state budget from taxing cannabis in the legal market. Another has been to place cannabis in the same frame as alcohol, a legally available substance: "regulate marijuana like alcohol" has often been the title of US state legalization initiatives. Such a framing suggested a separate state licencing and regulatory regime with civil controls backed up by criminal laws, like the "alcoholic beverage control" authority in each US state. Other provisions seek to prevent problems from cannabis use. Thus both the Washington and Colorado initiatives tightened the standard for the crime of "drugged driving", specifying a relatively low limit of the main psychoactive substance in cannabis, THC, in a driver's blood as the threshold above which driving is a crime. A further discipline on the US systems as they have developed has been the federal enforcement priorities mentioned above. And, as noted, the shaping of the systems has also been influenced, less intentionally, by the continuing fact of illegality at the federal level.

As Caulkins *et al.* (2015b) emphasize, while the emerging US legalized systems differ in a number of details, they all follow the "for-profit commercial (or so-called alcohol) model" in their general architecture, whereas Caulkins *et al.* enumerate a dozen "supply alternatives to status quo prohibition". From a public health perspective, the future results from the model the US states are following seem highly problematic. There are health and social risks from heavy cannabis use, even if they are less serious than from alcohol or tobacco, and a consolidated and eventually multinational legal cannabis industry operating under the for-profit commercial model will offer substantial stumbling-blocks to a public health approach. Already, public health researchers are describing the "legal cannabis industry adopting strategies of the tobacco industry" (Subritzky *et al.* 2016) and proposing approaches aimed at "avoiding a new tobacco industry" (Barry and Glantz 2016b).

The moves toward legalization in North America have stimulated a substantial literature on public health considerations in the design and detailed provisions of regulatory systems for a legalized market. In terms of the alternative architectures described by Caulkins *et al.*, public health-oriented discussions have tended to emphasize a state-monopoly or public authority model, where government-appointed agencies control the supply chain (e.g., Pacula *et al.* 2014; Rehm and Fischer 2015; Barry and Glantz 2016b). This is the primary model in the Uruguayan system (Walsh and Ramsey 2015; Marshall 2016). Caulkins and Kilmer (2016) add the consideration that

> the 'personality' of the regulatory agency may matter more than the specific regulations. Will legalization vest power in an assertive agency that views its mission as reducing health harms ... or a 'good government' agency that merely insists that rules are followed. ... Or, worse yet, does the agency, perhaps over time, end up viewing the industry as its primary constituency?

Experience with government monopolies in gambling, tobacco and alcohol warns us that public health is not necessarily the primary focus of such agencies; in regard to this, where the agency is located in government – whether it is in or reporting to a department with primary responsibility for state revenues, for consumer affairs, or for public health – may well be a crucial decision.

Otherwise, public health-oriented discussions of cannabis legalization have emphasized that "the devil is in the details", as a Canadian contribution is titled (Rehm et al. 2016). Accordingly, public health-oriented analyses have presented and discussed regulatory provisions in substantial detail (e.g., Caulkins et al. 2015b), sometimes with comparative charts of the features of existing cannabis regulatory regimes (Pardo 2014), and sometimes in comparison with a "public health standard" (Barry and Glantz 2016b). An analysis of the two competing 2016 California ballot propositions on legalization compared their provisions with recommendations from a Blue Ribbon Commission and from a public-health-oriented research committee (Barry and Glantz 2016a). Public health concerns have also been expressed around unexpected features of the newly legalized commercial markets, such as the substantial retail promotion of marijuana edibles (MacCoun and Mello 2015) and the proliferation of intake of cannabis by "vaping" (vaporization) (Budney et al. 2015).

The overriding public health issue in legalization: building a durable regime of mild discouragement in the face of vested commercial interests

Cannabis may be less inherently harmful than alcohol (Lachenmeier and Rehm 2015; Nutt et al. 2010), but it is far from harmless (Hall and Weier 2015; Room et al. 2010). There are differences in types and degrees of potential harm between different potencies, cannabinoid composition, and modes of use, as well as issues of potential contamination in the supply chain, and regulatory regimes provide the opportunity for public health-oriented controls and incentives favouring less harmful products that are not available to the state when a market is illegal. The new legal cannabis systems in Colorado, Washington and Uruguay are all strongly committed to regulating these aspects of the market (Pardo 2014). A substantial part of the harm from cannabis is from traffic injuries caused by driving while intoxicated; along with the legalization of cannabis, the new regimes have been updating their legislation and enforcement to deter such driving.

But a central public health issue, discouraging heavy use, is less likely to be tackled. Like other psychoactive substances when widely available and used, the distribution of consumption of cannabis is highly concentrated; for instance, it is estimated from population survey responses in 2012–2013 that the 13% of US cannabis users who used it daily accounted for 56% of the consumption (Davenport and Caulkins 2016). The concentration of consumption means that commercial interests in cannabis sales are inevitably substantially dependent on sales to quite heavy users. Any regulatory regime that effectively discourages regular heavy use will be against their economic interests.

The challenge for a public health approach to legalization of cannabis or any other potentially harmful psychoactive substance is thus the challenge of building a system that provides for availability and use while effectively holding down levels of use. This can be done by generally applicable measures that limit physical or economic availability to all adults; or it can be done by individually-oriented restrictions, such as an effective rationing or licencing system, which limits the supply to those who would otherwise be heavy users, as the "motbok" rationing system in Sweden did concerning alcohol prior to 1955 (Norström 1987). In the current era, such individually-oriented systems, with the labelling, stigmatization and bureaucratization inevitable with the individualization of controls, have proved unsustainable (the medical prescription system can be viewed as an interesting exception to this). The most efficient and sustainable way to hold down levels of use is thus with general limits on availability, such as through excise taxes and limits on places and times of availability.

This is a challenge that is better faced at the point of legalization than at later times. Once legal private interests have been created in a legal market, they will become effective advocates particularly against any impairment of their existing financial opportunities under the system. Thus it will always be politically easier to impose restrictions when the system is initiated than at any later time. In constructing the system, it is also crucial to give attention to insulating it from pressures to loosen controls and expand the market. The strongest argument in the interest of public health for constructing the market to minimize commercial interests (e.g., by setting it up as a government monopoly) is that such arrangements can be more effectively insulated from commercial pressures to "grow the market".

Moving away from the global drug prohibition regime has been a substantial project for many in the 1960s generation, and for that matter in later generations. The multilevel architecture of the regime, constructed particularly in the post-World War II era, and brought to its full fruition in the eras of Nixon and Reagan, has proved durable, and signs of any real change have only become apparent after 2010. Now, as full legalization at least of cannabis has become a reality, public health discussion of drug policy has had to broaden its scope. For years the focus had been particularly on "harm reduction", defined essentially in terms of countering the health harms that accompany the prohibition regime for marginalized heavy users (Room 2010). With legalization, the meaning of harm reduction expands to include the whole range of users. Thus legalization of cannabis is bringing to the fore the issue of how to construct and control legal markets in psychoactive substances in the public interest, so that – against the grain of a neoliberal era that has prioritized free markets – sale and use of the substance is permitted but not promoted. How well public health advocates and others arguing in the public interest succeed in this for cannabis may well set the frame for how calls for legalization of other psychoactive substances will be considered and judged.

Acknowledgements

Thanks to Jonathan Caulkins, Benedikt Fischer and Hanna Pickard for their comments and suggestions, which substantially improved the chapter. They are not responsible, of course, for any remaining faults.

References

Academy of Medical Sciences (AMS) (2004) *Calling Time: The Nation's Drinking as a Major Health Issue*, London: Academy of Medical Sciences.

Austin, G. (1978) *Perspectives on the History of Psychoactive Substance Use: Research Issues No. 24*, Rockville, MD: National Institute on Drug Abuse.

Babor, T., Caetano, R., Casswell, S., Edwards, G., Giesbrecht, N., Graham, K., Grube, J., Hill, L., Holder, H., Homel, R., Livingston, M., Österberg, E., Rehm, J., Room, R. and Rossow, I. (2010a) *Alcohol: No Ordinary Commodity – Research and Public Policy, 2nd Edition*, Oxford, UK: Oxford University Press.

Babor, T., Caulkins, J., Edwards, G., Fischer, B., Foxcroft, D., Humphreys, K., Obot, I., Rehm, J., Reuter, P., Room, R., Rossow, I. and Strang, J. (2010b) *Drug Policy and the Public Good*, Oxford, UK: Oxford University Press.

Babor, T., Caulkins, J., Fischer, B., Foxcroft, D., Humphreys, K., Medina-Mora, M. E., Obot, I., Rehm, J., Reuter, P., Room, R., Rossow, I. and Strang, J. (forthcoming 2018) *Drug Policy and the Public Good, 2nd Edition*, Oxford, UK: Oxford University Press.

Barry, R. A. and Glantz, S. A. (2016a) *A Public Health Analysis of Two Proposed Marijuana Legalization Initiatives for the 2016 Ballot: Creating the New Tobacco industry*, San Francisco, CA: Center for Tobacco Control Research and Education, UC San Francisco.

Barry, R. A. and Glantz, S. (2016b) "A public health framework for legalized retail marijuana based on the US experience: avoiding a new tobacco industry", *PLOS Medicine* 13(9): p.e1002131.

Bogart, W. A. (2011) *Permit but Discourage: Regulating Excessive Consumption*, Oxford, UK: Oxford University Press.

Budney, A. J., Sargent, J. D. and Lee, D. C. (2015) "Vaping cannabis (marijuana): parallel concerns to e-cigs?" *Addiction* 110(11): 1699–1704.

Carstairs, C. (2005) "The stages of the international drug control system", *Drug and Alcohol Review* 24(1): 57–65.

Catlin, G. E. G. (1931) *Liquor Control*, New York, NY: Henry Holt and Co.

Caulkins, J. P. and Kilmer, B. (2016) "The US as an example of how not to legalize marijuana?" *Addiction*, 111(12): 2095–2096.

Caulkins, J. P., Kilmer, B., Kleiman, M. A. R., MacCoun, R. J., Midgette, G., Oglesby, P., Pacula, R. L. and Reuter, P. H. (2015a) *Considering Marijuana Legalization: Insights for Vermont and Other Jurisdictions*, Research Report 864, Santa Monica, CA: RAND Corporation.

Caulkins, J. P., Kilmer, B., Kleiman, M. A. R., MacCoun, R. J., Midgette, G., Oglesby, P., Pacula, R. L. and Reuter, P. H. (2015b) *Options and Issues Regarding Marijuana Legalization*, RAND Perspective 149, Santa Monica, CA: RAND Corporation.

Davenport, S. S. and Caulkins, J. P. (2016) "Evolution of the United States marijuana market in the decade of liberalization before full legalization", *Journal of Drug Issues* 46(4): 411–427.

de Corte, T. (2015) "Cannabis social clubs in Belgium: organizational strengths and weaknesses, and threats to the model", *International Journal of Drug Policy* 26(2): 122–130.

Fosdick, R. L. and Scott, A. L. (1933) *Toward Liquor Control*, New York, NY, and London, UK: Harper and Bros.

Ghosh, T., Van Dyke, M., Whitley, E., Gillim-Ross, L. and Wolk, L. (2016) "The public health framework of legalized marijuana in Colorado", *American Journal of Public Health* 106(1): 21–27.

Hahn, R. A., Middleton, J. C., Elder, R., Brewer, R., Fielding, J., Naimi, T. S., Toomey, T. L., Chattopadhyay, S., Lawrence, B., Campbell, C. A. and Community Preventive Services Task Force (2012) "Effects of alcohol retail privatization on excessive alcohol consumption and related harms: a community guide systematic review", *American Journal of Preventive Medicine* 42(4): 418–427.

Hall, W. and Weier, M. (2015) "Assessing the public health impacts of legalizing recreational cannabis use in the USA", *Clinical Pharmacology and Therapeutics* 97(6): 607–615.

Her, M., Giesbrecht, N., Room, R. and Rehm, J. (1999) "Privatizing alcohol sales and alcohol consumption: evidence and implications", *Addiction* 94(8): 1125–1139.

International Drug Policy Consortium (IDPC) (2014) *The 2014 Commission on Narcotic Drugs and Its High-Level Segment: Report on Proceedings*, London, UK: International Drug Policy Consortium.

Kilmer, B., Kruithof, K., Pardal, M., Caulkins, J. P. and Rubin, J. (2013) *Multinational Overview of Cannabis Production Regimes*, Cambridge, UK: Rand Drug Policy Research Centre.

Lachenmeier, D. W. and Rehm, J. (2015) "Comparative risk assessment of alcohol, tobacco, cannabis and other illicit drugs using the margin of exposure approach", *Science Reports* 5: 8126.

Latin American Commission on Drugs and Democracy (2009) *Drugs and Democracy: Toward a Paradigm Shift*, available from the Commission website: www.drogasedemocracia.org/Arquivos/declaracao_ingles_site.pdf (accessed 19 October, 2016).

Levine, H. G. (1992) "Temperance cultures: concern about alcohol problems in Nordic and English-speaking cultures", in M. Lader, G. Edwards and D. C. Drummond (eds), *The Nature of Alcohol and Drug Related Problems*, Oxford, UK: Oxford University Press, pp. 15–36.

Liberman, J. and Mitchell, A. (2010) "In search of coherence between trade and health: inter-institutional opportunities", *Maryland Journal of International Law* 25: 143–186.

MacCoun, R. J. (2011) "What can we learn from the Dutch cannabis coffeeshop system?" *Addiction* 106(11): 1899–1910.

MacCoun, R. J. and Mello, M. M. (2015) "Half-baked – the retail promotion of marijuana edibles", *New England Journal of Medicine* 372(11): 989–991.

Mäkelä, K., Room, R., Single, E., Sulkunen. P., Walsh, B., et al. (1981) *Alcohol, Society and the State: I. A Comparative Study of Alcohol Control*, Toronto, ON: Addiction Research Foundation.

Marshall, A. (2016) "Uruguay to test world's first state-commissioned recreational cannabis", *The Guardian*, 18 April, www.theguardian.com/society/2016/apr/18/uruguay-first-state-commissioned-recreational-cannibis-marijuana (accessed 20 October, 2016).

McGrady, B. (2011) *Trade and Public Health: The WTO, Tobacco, Alcohol and Diet*, Cambridge, UK: Cambridge University Press.

Norström, T. (1987) "The abolition of the Swedish alcohol rationing system: effects on consumption distribution and cirrhosis mortality", *British Journal of Addiction* 82(6): 633–641.

Nutt, D. J., King, L. A., Phillips, L. D. for the Independent Scientific Committee on Drugs (2010) "Drug harms in the U.K.: a multicriteria decision analysis", *The Lancet* 376(9752): 1558–1565.

Olsson, B., Ólafsdóttir, H. and Room, R. (2002) "Introduction: Nordic traditions of studying the impact of alcohol policies", in R. Room (ed.) *The Effects of Nordic Alcohol Policies: What Happens to Drinking when Alcohol Controls Change?*, Helsinki: NAD, pp. 5–16.

Pacula, R. L., Kilmer, B., Wagenaar, A. C., Chaloupka, F. J. and Caulkins, J. P. (2014) "Developing public health regulations for marijuana: lessons from alcohol and tobacco", *American Journal of Public Health* 104(6): 1021–1028.

Pardo, B. (2014) "Cannabis policy reforms in the Americas: a comparative analysis of Colorado, Washington, and Uruguay", *International Journal of Drug Policy* 25(4): 727–735.

Rehm, J., Crepault, J.-F. and Fischer, B. (2016) "The devil is in the details! On regulating cannabis use in Canada based on public health criteria", *International Journal of Health Policy Management* 6(3): 173.

Rehm, J. and Fischer, B. (2015) "Cannabis legalization with strict regulation, the overall superior policy option for public health", *Clinical Pharmacology and Therapeutics* 97(6): 541–544.

Rehm, J., Marmet, S., Anderson, P., Gual, A., Kraus, L., Nutt, D. J., Room, R., Samokhvalov, A. V., Scafato, E., Trapencieris, M., Wiers, R. W. and Gmel, G. (2013) "Defining substance use disorders: do we really need more than heavy use?" *Alcohol and Alcoholism* 48(6): 633–640.

Rehm, J. and Room, R. (2009) "Why do we need international regulations for alcohol control?" *Nordic Studies on Alcohol and Drugs* 26(4): 447–450.

Roffman, R. (2016) "Legalization of cannabis in Washington State: how is it going?" *Addiction* 111(7): 1139–1140.

Room, R. (2000) "Alcohol monopolies as instruments for alcohol control policies", in E. Österberg (ed.), *International Seminar on Alcohol Retail Monopolies*, Helsinki, Finland: National Research and Development Centre for Welfare and Health, pp. 7–16.

Room, R. (2006) "International control of alcohol: alternative paths forward?" *Drug and Alcohol Review* 35: 581–595.

Room, R. (2010) "The ambiguity of harm reduction: goal or means, and what constitutes harm?" in T. Rhodes and D. Hedrich (eds) *Harm Reduction: Evidence, Impacts and Challenges*, Lisbon: EMCDDA Monograph No. 10, European Monitoring Centre for Drugs and Drug Abuse, pp. 108–111, www.emcdda.europa.eu/publications/monographs/harm-reduction.

Room, R. (2012a) "Individualised control of drinkers: back to the future?" *Contemporary Drug Problems* 39(2): 311–343.

Room, R. (2012b) "Reform by subtraction: the path of denunciation of international drug treaties and reaccession with reservations", *International Journal of Drug Policy* 23(5): 401–406.

Room, R. (2012c) *Roadmaps to Reforming the UN Drug Conventions*, Oxford, UK: Beckley Foundation.

Room, R. (2014) "Legalising a market for cannabis for pleasure: Colorado, Washington, Uruguay and beyond", *Addiction* 109(3): 345–351.

Room, R., Fischer, B., Hall, W., Lenton, S. and Reuter, P. (2010) *Cannabis Policy: Moving Beyond Stalemate*, Oxford, UK: Oxford University Press.

Room, R. and Hall, W. (2013) "Frameworks for understanding drug use and societal responses", in A. Ritter, T. King and M. Hamilton (eds), *Drug Use in Australian Society*, South Melbourne, VC: Oxford University Press, pp. 51–66.

Room, R., Stoduto, G., Demers, A., Ogborne, A. and Giesbrecht, N. (2006) "Alcohol in the Canadian context", in N. Giesbrecht, A. Demers, A. Ogborne, R. Room, G. Stoduto and E. Lindquist (eds), *Sober Reflections: Commerce, Public Health, and the Evolution of Alcohol Policy in Canada, 1980–2000*, Montreal, QC, and Kingston, ON: McGill-Queen's University Press, pp. 14–42.

Ryan-Ibarra, S., Induni, M. and Ewing, D. (2015) "Prevalence of medical marijuana use in California, 2012", *Drug and Alcohol Review* 34(2): 141–146.

Sanders-Jackson, A. N., Song, A. V., Hiilamo, H. and Glantz, S. A. (2013) "Effect of the Framework Convention on Tobacco Control and voluntary industry health warning labels on passage of mandated cigarette warning labels from 1965 to 2012: transition probability and event history analyses", *American Journal of Public Health* 103(11): 2041–2047.

Schrad, M. L. (2010) *The Political Power of Bad Ideas: Networks, Institutions, and the Global Prohibition Wave*, Oxford, UK: Oxford University Press.

Seddon, T. (2014) "Drug policy and global regulatory capitalism: the case of new psychoactive substances (NPS)", *International Journal of Drug Policy* 25(5): 1019–1024.

Subritzky, T., Lenton, S. and Pettigrew, S. (2016) "Legal cannabis industry adopting strategies of the tobacco industry", *Drug and Alcohol Review* 35(5): 511–513.

van Ooyen-Houben, M. and Kleemans, E. (2015) "Drug policy: the 'Dutch model'", *Crime and Justice* 44(1): 165–557.

Walsh, J. and Ramsey, G. (2015) "Uruguay's drug policy: major innovations, major challenges", Washington, DC: Brookings Institution, www.brookings.edu/~/media/Research/Files/Papers/2015/04/global-drug-policy/Walsh--Uruguay-final.pdf. Washington, DC: Washington Office on Latin America (WOLA).

World Health Organization (WHO) (2012) *Expert Committee on Drug Dependence Thirty-Fifth Report, WHO Technical Report Series 973*, Geneva: World Health Organization.

Zeigler, D. W. and Zeigler, D. W. (2006) "International trade agreements challenge tobacco and alcohol control policies", *Drug and Alcohol Review* 25(6): 567–579.

43
ADDICTION AND DRUG (DE)CRIMINALIZATION

Douglas Husak

A dilemma: criminalization or excuse?

A broad consensus exists on all points along the political spectrum that western countries (and the United States in particular) punish too many people with too much severity (Chettiar *et al.* 2015). If we hope to make a dent on both overcriminalization and overpunishment, drug offenses would be a sensible place to begin (Husak 2008). Although the impact of drug prohibition on overpunishment is frequently exaggerated, sentencing experts estimate that a more just and effective drug policy has the potential to reduce the state and federal prison population by approximately 19%.[1] The grounds supporting this recommendation are not merely pragmatic. Even though those who challenge the core of our drug policy are typically asked to provide a compelling rationale for *de*criminalization, I have long argued that *criminalization* presents the more basic challenge. To my mind, a persuasive case for punishing users of *any* illicit drug that actually exists has not been made (Husak 1992, 2002), so these offenses are among the best examples of overcriminalization on our books today. Decriminalization is the default position in the absence of a strong rationale to continue along our present punitive course.

A handful of countries, such as Portugal and Uruguay, rely less on a criminal justice approach to combat the problems caused by users of any drug. Many people think it is reasonable, however, to draw distinctions *within* the class of illicit drugs. What are the characteristics of the kinds of drug that are thought to be the best candidates for criminalization? Among those that should be proscribed, which should be punished with the greatest severity? Labelling some drugs as *hard* and others as *soft* is a conclusion in need of an argument. To be sure, some drugs are more likely than others to be deleterious to health, to be abused by children, and/or to be implicated in violent behavior. In this chapter I will not discuss these familiar rationales for liability and punishment. Instead, I will critically examine how the phenomenon of *addition* might bear on questions about criminalization.

Many people suppose that the case for criminalizing the use of *addictive* drugs is (*ceteris paribus*) more compelling than the case for criminalizing (or for imposing severe punishments) on the use of *non*-addictive drugs. When laypersons are questioned about why they are reluctant to use a given kind of drug such as opiates, either for pain management or for recreational experimentation, they frequently cite their fear that they will become addicted to it. No sensible person welcomes the prospects of drug addiction. But at least three formidable obstacles must be surmounted by those who would proscribe or attach more severe sentences to users of addictive than non-addictive drugs—a policy I call *selective prohibition*. First, we need a set of principles that must be satisfied before given types of conduct should be criminalized. We cannot decide whether users of addictive drugs should be punished without settling on what would count as a persuasive reason to punish anyone for anything (Husak 2008). Second, we need an account of the nature of addiction itself. As several chapters in this Handbook demonstrate, the nature of addiction is highly contested. Third, we need a political theory that attempts to prioritize the extent to which given problems merit a criminal justice solution. Can the relatively few number of drug addicts in the United States (perhaps 4 million in a population of 320 million) justify the cumbersome, expensive, and repressive apparatus of drug prohibition when so many other social problems are not addressed? To avoid these debates, I will make the heroic assumption that progress can be made in assessing the case for selective prohibition on virtually *any* set of political priorities, theory of criminalization or conception of addiction. According to nearly all analyses, addiction is a pathological state of affairs that undermines human welfare or flourishing.[2] In what follows, I loosely characterize the impact of addiction on persons as pathological largely because it causes a diminution of *freedom*.[3] Addicts typically report levels of compulsion that make it difficult for them to bring their conduct into conformity with what they judge to be best in their more reflective moments. Virtually all moral and political traditions count freedom among the most valuable moral goods, and thus assess factors that diminish it as bad. Reducing the likelihood that persons will become addicts seems like a sensible legal objective regardless of whether addition is conceptualized as a defect of will, an impairment of rationality, a loss of control, a disease of the brain, a failing of responsibility, or whatever.[4] Although reasonable minds might differ, the penal law in particular is among the many devices by which this goal might be pursued.

The supposition that criminal laws proscribing drug use are justifiable as a means to prevent persons from becoming addicts and diminishing their freedom construes these laws *paternalistically*—as designed to protect persons from harms they cause to themselves. Of course, the negative effects of addiction typically extend well beyond those suffered by the user—to her family, friends, and entire communities. Still, the main rationale for coercive laws against addictive drugs, in the first instance, is to protect the freedom of the user herself. Theorists steadfastly opposed to criminal paternalism will immediately reject this rationale.[5] Nonetheless, it is worth a serious hearing. If criminal paternalism is ever acceptable, it may be to prevent the diminution of freedom inherent in addiction (see Goodin (1989) as an example). Since I presuppose that addiction undermines freedom, the decision of addicts to continue to use addictive substances is hardly a paradigm of the kind of autonomous choice with which opponents of paternalism are loath to interfere.

How should we formulate a hypothesis about the support that addiction might lend to the rationale for criminalization? This question is more difficult than it may seem. If current law is to serve as even a rough approximation of our guide, no one would believe addiction to be a *necessary* condition for criminalization. Some drugs currently placed on Schedule I, most notably hallucinogens such as LSD, satisfy few of the familiar criteria for addiction. Of course, I do not aspire to justify the status quo; my inquiry aims to provide a normative basis to assess and reform our existing policy. My point is only that a drug might well be too dangerous and harmful to tolerate even though it lacks addictive properties. In any event, it is no more plausible to regard the finding that an alarming percentage of users become addicted to a particular substance as *sufficient* for criminalization. No sensible person proposes that caffeine should be proscribed, even though it satisfies several of the familiar criteria of addiction. Presumably caffeine is exempted from a policy of selective prohibition because its dangers are small or nonexistent, and frequent users report low levels of compulsion. Thus I propose that the hypothesis to be explored is best formulated in terms of *reasons* rather than in terms of necessary or sufficient conditions: Do the addictive properties of a drug provide *a reason* to criminalize its use? Even this latter hypothesis, however, needs further refinement. After all, reasons are ubiquitous, and most are trivial in weight. Therefore, if my hypothesis about selective prohibition is to have any practical salience, evidence that a particular substance produces high rates of addiction among its users must furnish a *substantial* (that is, a not-easily defeated) reason in favor of criminalization. Admittedly, this latest refinement introduces an air of imprecision to the hypothesis I propose to assess. Reasonable minds might agree that a given reason exists while disagreeing about whether its strength rises to the vague threshold of substantiality.

Inevitable imprecision notwithstanding, much can be said against the hypothesis that the phenomenon of addiction lends significant support to the case for selective drug criminalization. I will simply mention one huge problem before focusing on another in a bit more detail. This problem can be introduced by attempting to specify the likelihood that users of given addictive drugs will actually succumb to addiction. Obviously, no informative statistic can be presented in the absence of a clearer conception of addiction itself. But researchers now understand that relatively few users of *any* drug will actually become addicted to it (Heyman 2009). Thus this rationale for criminalization is *overinclusive*—punishing all persons who engage in a given behavior (viz., the use of an addictive drug) because of the harm (viz., the loss of freedom inherent in addiction) that only some of them will suffer. If possible, overinclusive penal laws should be avoided; we should prefer statutes that are more narrowly tailored. The feasibility of crafting such statutes requires us to understand who gets addicted and who does not. We have long known that the prevalence of addiction is correlated with low IQ, abuse during childhood, personality disorders, stress, and socioeconomic status. It is hard to find a principled ground on which users of drugs who do not incur significant risks of becoming addicted should be subjected to punishment because other users do incur significant risks. At the same time, a more narrowly tailored statute would target the most vulnerable members of society. It is equally hard to find a principled ground to confine the application of penal laws to persons who fall into the foregoing categories.

Although problems of overinclusion are potentially fatal to the case for criminalizing the use of addictive drugs, I propose to explore a different difficulty. In short, I attempt to impale those who believe that addiction provides a substantial reason for selective criminalization on the horns of the following dilemma. If the phenomenon of addiction is to provide a substantial reason to criminalize the use of a given drug, the state of affairs of being addicted must be dreadful. Avoidance of a minor loss of freedom cannot provide a weighty reason to prevent persons from engaging in conduct that causes it. But any plausible account of *why* the loss of freedom inherent in addiction is so dreadful would undermine the degree of blame and punishment deserved by addicts. A significant loss of freedom is a familiar ground to *excuse* persons who commit offenses. Thus the case for supposing that addiction provides a powerful reason in favor of criminalization may be self-defeating. Theorists who emphasize the horrors of addiction in their defense of criminalization must simultaneously offer a basis to withhold criminal liability from those who succumb to these horrors. In other words, the very rationale for believing addiction is sufficiently pathological to justify the enactment of penal laws to prevent it also shows why the blame and punishment of those who suffer from addiction is incompatible with standard accounts of the preconditions of criminal responsibility.[6] Theorists who have recognized this dilemma (implicitly or explicitly) have suggested a number of ingenious strategies to try to avoid it. The second half of this chapter identifies and critically assesses the cogency of three of these strategies. Although the dilemma I describe may not provide an insurmountable barrier to supposing that addictive substances are special candidates for criminal punishment, each of the three strategies to avoid it is problematic.

I do not regard the discovery of this dilemma as a novel insight. It is not surprising to notice that theorists tend to portray the horrors of addiction very differently, depending on whether they are discussing the justifiability of criminalizing the use of addictive substances as opposed to assessing the justifiability of recognizing an excuse from penal liability when addicts face punishment. As I have indicated, addiction must be an awful state before the desirability of preventing it could qualify as a substantial basis for punishing persons who use whatever substances cause it. The evils of addiction must be unlike the evils of becoming significantly overweight by eating high-calorie foods or by failing to exercise—even though obesity diminishes longevity and quality of life. I have supposed addiction is special because addicts are significantly less free. But if addicts are significantly less free, it is hard to deny they should be excused from criminal liability. Conditions that undermine freedom are often thought to generate an excuse for persons who suffer from them.

Is addiction really thought to reduce the blame and quantum of punishment deserved by drug addicts in the real world? We might use existing law to answer this question. Anglo-American criminal law does not recognize a general *defense* for drug offenses when the user is addicted.[7] Nor do statutes or sentencing guidelines tend to reduce the severity of punishment when a drug offense is perpetrated by an addict.[8] Legal commentators seldom reserve judgments about responsibility until they learn whether a given instance of drug use was committed by an addict or a non-addict. These facts provide some reason to doubt that addiction is regarded as morally relevant to judgments about the quantum of blame and punishment addicts deserve.

Moreover, if addiction really devastates the capacity for freedom, punishments cannot be very effective in preventing the continued use of addictive substances among those individuals who have already succumbed. The claim that addicts use drugs non-freely suggests that threats of punishment will do little to discourage them from persisting.[9] The rationale for criminalization, then, must be *general deterrence*: punishment can still be effective in dissuading potential addicts, whose capacity for voluntary choice remains intact, from experimenting with addictive drugs and becoming addicted.[10] *If* the state cites its interest in protecting the capacity for free choice to justify its decision to criminalize addictive drug use, however, must it *also* reduce the amount of blame and punishment imposed for acts of drug use committed by addicts? Either (a) addiction does *not* have a major impact on freedom, or (b) addiction *does* undermine freedom significantly. If (a), and the freedom of addicts is mostly intact, the state interest in protecting freedom can hardly justify the enactment of drug proscriptions. But if (b), and addiction seriously undermines freedom, addicts are almost certainly less responsible, and perhaps should be excused altogether, for their use of drugs. In what follows, I will identify some of the possible ways to avoid the dilemma I have just constructed.

Can this dilemma be avoided?

In this section I will critically examine efforts to show that addicts should not be excused from criminal liability even though the state of addiction involves a sufficiently severe impairment of freedom to provide a powerful rationale for prohibiting the use of those substances that cause it. *Some* such effort had better succeed if the state is permitted to enforce selective prohibition while simultaneously withholding an excuse from addicts. Although I am sure that many responses to this dilemma might be devised, I will briefly mention two before settling on the solution I believe is the most promising—even though it too is problematic.

The first two of my three replies to this dilemma are similar in that they purport to rethink exactly what addicts are blamed or punished *for*. The *first* of these replies is probably the most well-known, invoking what might be called a *culpability-in-causing* consideration.[11] This consideration endeavors to withhold an excuse from addicts for their drug use, despite conceding that addictive drug users may lack freedom altogether. According to this first reply, even a totally unfree act is ineligible for an excuse if the agent is somehow blameworthy for causing it, that is, for causing his own subsequent loss of freedom. Such is typically the case (or so the argument continues) in the case of addicts. Culpability-in-causing considerations are not *ad hoc*, but limit the application of a great many excuses from blame and penal liability,[12] including ignorance of law (Husak 2016). Consider duress, for example, typically available to those who commit wrongful acts "under unlawful threats that persons of reasonable firmness would be unable to resist."[13] This excuse is withheld, however, when the defendant "recklessly placed himself in a situation in which it was probable that he would be subjected to duress."[14] Defendants who have recklessly placed themselves in various predicaments—by joining a criminal gang, for example—should have foreseen that they would be likely to be subjected to threats that persons of reasonable firmness would be unable to resist. By parity of reasoning, the argument continues, an excuse should be unavailable to most drug addicts. Even if their freedom has been lost altogether, they are justifiably blamed and punished for using drugs because they have freely performed a prior culpable act (viz., using addictive drugs before the time they became addicted) that they should have foreseen would subsequently result in addiction and deprive them of their freedom.

Despite its popularity, I regard this first reply as unsatisfactory. It is vulnerable to a number of rejoinders that should fuel suspicion about culpability-in-causing considerations generally.[15] Exactly *why* is the defendant denied an excuse from blame when he culpably caused himself to be in the very predicament in which an excuse is needed? Does the blame for the prior free act somehow transfer to or substitute for that of the subsequent unfree act? In addition, a standard involving what a defendant *should* have foreseen involve negligence, even though nearly all of the subsequent offenses for which such persons become eligible require greater amounts of culpability. Moreover, is it really true that first-time users *should* foresee that addiction and the loss of freedom would be the likely results of their experimentation with drugs?[16] Since addiction results infrequently, the probable answer is no. Most importantly, the severity of punishment that can result from denying the excuse seems grossly disproportionate to the gravity of the prior wrongful act the defendant has committed. Return to the case of a person who freely places himself in the company of thugs. The wrongfulness of this act does not seem especially great. Suppose further that the thugs force him to assist in robbing a bank, and his excuse for this serious crime is denied pursuant to a culpability-in-causing consideration. Surely his blame as an accomplice to bank robbery should not be equated either with the blame he merits for joining the gang or with the blame of the garden-variety bank robber who was not subjected to duress at all. Yet the practical import of withholding his excuse produces this very outcome. I hope that these general difficulties indicate that culpability-in-causing rationales for denying an excuse are problematic generally.

The *second* of my three replies is similar to the first in purporting to justify withholding an excuse from addicts while conceding their drug use is wholly unfree. This reply, however, does not purport to blame the addict either for his drug use or for his previous free choice that caused him to become addicted. Instead, it imposes blame *for* a subsequent choice—his decision to forego various options that would allow him to discontinue his use of addictive drugs. All such options might be called *treatments*. Even those addicts whose drug use is wholly unfree enjoy periodic moments of lucidity in which they might be expected to devise a plan to overcome their addiction. According to this reply, addicts become eligible for blame and punishment because of their free decision to forego whatever treatment would enable them to stop using drugs.

The most obvious rejoinder to this second reply is that treatment is often unavailable, and, when available, may not be effective.[17] This point may not be definitive, however, because more promising treatments are on the horizon. Thus a less contingent rejoinder involves what might be called a "truth in blame" (or "truth in labelling") rationale.[18] If addicts are "really" blamed for some act other than using drugs—an act that is subsequent to the time they become addicted—it is misleading to punish them *for* violating statutes proscribing drug use. Instead, they should be blamed for a new offense defined as an omission rather than a positive act: the failure to seek treatment. To be sure, no jurisdiction presently proscribes the failure to seek treatment, but this problem of legality could easily be overcome simply by enacting a new penal statute. Still, implementing this proposal would raise enormous challenges. What modes of treatment would suffice to preclude liability? How recently must treatment have been sought? How sincere and conscientious must the addict have been in his attempt to quit? How would the criminal law deal with persons who sought treatment but continued to use drugs—those whose loss of freedom is most drastic? These questions would have to be answered if we take truth in labelling seriously.[19] An addict who *had* sought treatment, or for whom treatment was unavailable or unsuccessful, would presumably not be liable for this new offense. Since the conduct of such persons would not satisfy the material elements of this statute, the state would be required to bear the burden of proving that a given defendant had not sought treatment before liability could be imposed.

The *third* and final reply to the foregoing dilemma differs from its predecessors in purporting to blame and punish addicts *for* the offense of using drugs—rather than for something *else* that occurs prior to or subsequent to the time they become addicted. It starts from the uncontroversial premise that freedom admits of degrees, and then alleges that addiction reduces but does not preclude blame and criminal responsibility altogether. Although few of those who make this argument develop its implications for the criminal law,[20] its probable consequences are not too hard to discern: addicts become eligible for blame and punishment *for* using drugs, but should be blamed and punished for doing so less severely than non-addicts who commit the same offense. Thus this reply seemingly provides a rationale for *mitigation of sentencing*, rather than for a complete excuse from liability.

Distinct versions of this reply differ about exactly why addicts are less free than non-addicts when they use drugs.[21] Whatever the details, however, any such explanation must be conceptualized roughly as follows if it is to succeed in avoiding the foregoing dilemma. Begin with the oversimplified assumption that the extent to which freedom is exemplified in a particular act can be quantified with precision along a single continuum. Represent fully free conduct as "A", and wholly unfree conduct as "Z". Suppose that the amount of freedom exemplified by the drug use of a given addict is located along this spectrum at point "T". Two distinct questions now arise in confronting my dilemma. First, how much freedom must a type of wrongful choice lack before the state has a substantial reason to enact penal laws to prevent persons from taking whatever drugs deprive them of that amount of freedom? Second, how much freedom must a type of wrongful act lack before a person should be excused for making it? The key to the third solution to the foregoing dilemma is to appreciate that the answers to these two questions may well differ. That is, the amount of freedom that must be absent in order to excuse an act may not be identical to the amount of freedom that must be absent in order to justify the proscription of behaviors that cause it to be lost. Perhaps a choice must only lack freedom to degree

"P" before the state has a substantial reason to enact criminal laws to prevent persons from making it, although a choice must lack freedom to a higher degree "V" before a person should be excused for having made it. The drug use of addicts may lack enough freedom ("T") to justify criminalization ("P"), but possess enough freedom to defeat the claim for an excuse ("V"). If so, the foregoing dilemma is avoided; the state can appeal to its interest in protecting freedom to support its decision to criminalize addictive drug use while simultaneously withholding an excuse from liability for the drug use of addicts.

A few additional premises would build on the foregoing conceptualization to support the basis for mitigation.[22] Insofar as the drug use of addicts ("T") is *less* free than that of non-addicts (which may approximate "A"), the claim for mitigation can be established by an appeal to a version of the principle of proportionality. According to the most familiar such version, the severity of the punishment that should be imposed must be a function of the seriousness of the offense that has been committed. In addition, however, proportionality might further require that (*ceteris paribus*) the severity of the punishment that should be imposed should be proportionate to the amount of freedom the offense manifests. Although addicts are free enough to be punished to some extent, they are not free enough to merit the same quantum of punishment for which non-addicts are eligible.

Although I have indicated that many theorists believe addicts are somewhat but not fully responsible for their acts of drug use, I am unaware of any who have conceptualized their position in quite the way I have presented it. The myth of precision aside, I take this third response seriously, and do not pretend to have a definitive refutation of it. Admittedly, we have long been accustomed to employing different accounts of various legal standards (e.g., voluntariness, maturity, sanity and the like) for different legal purposes (e.g., confessions, capacity to make contracts, stand trial or be executed). We should not simply insist that the same test of freedom should be used to resolve both questions about criminalization and the availability of an excuse. At the same time, it would be a convenient simplifying hypothesis if the amount of freedom that must be lost by addictive drug use to justify the enactment of penal legislation were the same amount that must be lost to warrant the recognition of an excuse for addicts. Is there good reason to believe the former is less than the latter? Most theorists seemingly believe the quantum of justification needed to allow punishment is actually higher than the quantum needed *not* to punish and thus to excuse. Since any form of punishment involves a significant hardship, the diminution of freedom required to justify criminalization may well be *greater* than the amount that would support granting an excuse—if indeed the quanta differ at all.

Although I regard this final alternative as the most promising, I am not wholly persuaded by any of these three responses to the dilemma I have constructed. I am inclined to believe that either the state should excuse addicts for using drugs, or else it lacks a substantial reason to selectively criminalize addictive drug use in the first place. I tend to believe the latter disjunct, reserving judgment about the former. The state interest in protecting the capacity for freedom does not provide a substantial reason to prohibit the use of addictive drugs. In fact, arguments that stress the importance of freedom provide better support for decriminalization than for selective prohibition. But perhaps I am mistaken, and the state can consistently criminalize addictive drug use to protect freedom while simultaneously denying an excuse for addicts. I challenge penal theorists who specialize both in criminalization and addiction to confront this dilemma directly and to defend their own solution. A state that endeavors to justify its decision to criminalize drug use by appealing to its interest in protecting the capacity for freedom is hard-pressed to deny mitigation, or even a complete excuse, to persons who become addicted.

Notes

1 Cautious estimates are provided by Pfaff (2013).
2 Admittedly, my assumption would be rejected by those theorists who deny that addiction is necessarily pathological. Perhaps drug use is a choice that is both purposive and rational given a realistic picture of the social and economic environment and co-morbid mental health problems that most addicts face. See Pickard (2012).
3 I invoke a very general sense of freedom, and might have used the word *autonomy*. I concede that Anglo-American systems of criminal justice have been far more forgiving to defendants who suffer from impairments of rationality than from impairments of volition.
4 See, for example, the several essays in Polland and Graham (2011). My own view is that no single conceptualization of addiction is adequate for all cases.
5 Perhaps they are correct to reject paternalism when enforced by the criminal law. See Husak (2013).
6 For earlier thoughts, see Husak (1999, 2004).
7 The leading case is *U.S. v. Moore*, 486 F.2d 1139 (1973).
8 Exceptions can be found. Connecticut, for example, authorizes a greater penalty for illegal manufacture, distribution, sale, prescription or administration by a non-drug-dependent person. See *Connecticut General Statutes* §21a–278(b) (1992).
9 But see Vaillant (2001).
10 The general deterrent effect of drug prohibitions is controversial. See, for example, Fagan (1994). See also MacCoun and Reuter (2001).
11 I borrow this term from Robinson (1985).
12 But not all. To my knowledge, no one has advanced a culpability-in-causing rationale to ever withhold an excuse of insanity.
13 See *Model Penal Code*, §2.09(2).
14 Id.
15 See the several contributions in the special issue on "Actio Libera in Causa" in (2013) *Criminal Law and Philosophy* 7: 549–636.
16 If the basis for denying a defense to the drug addict is his culpability in causing his own addiction, the excuse should remain available to those persons who lack any prior culpability—as when they become addicted to prescribed medications, for example.
17 Ironically, recovery rates seem higher among addicts who do *not* undergo treatment. See Heyman (2009).
18 For a discussion of the importance of fair labelling, see Ashworth (2006).
19 Drug courts, which make punishment contingent on a failure to comply with a mandated treatment regime, accomplish this result *de facto*. See Husak (2011).
20 See Sinnott-Armstrong (2013), Yaffe (2013), and Holton and Berridge (2013).
21 Different accounts might be given. For one such attempt, see Schroeder and Arpaly (2013).
22 One theorist regards a reduction of the blame deserved by addicts to be a datum that any adequate theory of addiction should explain. Gideon Yaffe writes: "[A]ddicts are rarely thought blameless, but they are often taken to be less at fault than their unaddicted counterparts" (2001: 178).

References

Ashworth, A. (2006) *Principles of Criminal Law*, Oxford, UK: Oxford University Press.
Chettiar, I. M., Waldman, M., Fortier, N. Z. and Finkelman, A. (eds) (2015) *Solutions: American Leaders Speak Out on Criminal Justice*, New York, NY: Brennan Center for Justice.
Fagan, J. (1994) "Do criminal sanctions deter drug offenders?" in D. MacKenzie and C. Uchida (eds), *Drugs and Crime. Evaluating Public Policy Initiatives*, Thousand Oaks, CA: Sage Publications.
Goodin, R. E. (1989) *No Smoking*, Chicago, IL: University of Chicago Press.
Heyman, G. M. (2009) *Addiction: A Disorder of Choice*, Cambridge, MA: Harvard University Press.
Holton, R. and Berridge, K. (2013) "Addiction between compulsion and choice", in N. Levy (ed.), *Addiction and Self-Control*, Oxford, UK: Oxford University Press, pp. 239–268.
Husak, D. (1992) *Drugs and Rights*, Cambridge, UK: Cambridge University Press.
Husak, D. (1999) "Addiction and criminal liability", *Law and Philosophy* 18: 655–684.
Husak, D. (2002) *Legalize This!*, London, UK: Verso.
Husak, D. (2004) "The moral relevance of addiction", *Substance Use and Misuse* 39(3): 399–436.

Husak, D. (2008) *Overcriminalization*, Oxford, UK: Oxford University Press.
Husak, D. (2011) "Retributivism, proportionality, and the challenge of the drug court movement", in M. Tonry (ed.), *Retributivism Has a Past. Has It a Future?*, Oxford, UK: Oxford University Press, pp. 214–233.
Husak, D. (2013) "Penal paternalism", in C. Coons and M. Weber (eds), *Paternalism: Theory and Practice*, Cambridge, UK: Cambridge University Press, pp. 39–55.
Husak, D. (2016) *Ignorance of Law: A Philosophical Inquiry*. New York, NY: Oxford University Press.
MacCoun, R. and Reuter, P. (2001) *Drug War Heresies*, Cambridge, UK: Cambridge University Press, pp. 72–100.
Pfaff, J. F. (2013) "Waylaid by a metaphor: a deeply problematic account of prison growth", *Michigan Law Review* 111(6): 1087–1110.
Pickard, H. (2012) "The purpose in chronic addiction", *AJOB Neuroscience* 3(2): 40–49.
Polland, J. and Graham, G. (eds) (2011) *Addiction and Responsibility*, Cambridge, MA: MIT Press.
Robinson, P. (1985) "Causing the conditions of one's own defense: a study in the limits of criminal law doctrine", *Virginia Law Review* 71: 1–63.
Schroeder, T. and Arpaly, N. (2013) "Addiction and blameworthiness", in N. Levy (ed.), *Addiction and Self-Control*, Oxford, UK: Oxford University Press, pp. 215–240.
Sinnott-Armstrong, W. (2013) "Are addicts responsible?" in N. Levy (ed.), *Addiction and Self-Control*, Oxford, UK: Oxford University Press, pp. 123–144.
Vaillant, G. E. (2001) "If addiction is involuntary, how can punishment help?" in P. B. Heymann and W. N. Brownsberger (eds), *Drug Addiction and Drug Policy: The Struggle to Control Dependence*, Cambridge, MA: Harvard University Press, pp. 144–167.
Yaffe, G. (2001) "Recent work on addiction and responsible agency", *Philosophy and Public Affairs* 30(2): 178–221.
Yaffe, G. (2013) "Are addicts akratic?" in N. Levy (ed.), *Addiction and Self-Control*, Oxford, UK: Oxford University Press, pp. 191–214.

44
CRIMINAL LAW AND ADDICTION

Stephen J. Morse

Introduction

Possessing and sometimes using controlled, potentially addictive substances are crimes in US and English law, although possessing and using legal addictive substances such as ethanol and nicotine are not. Other antisocial behaviors performed in the service of seeking and using drugs, such as burglary, robbery and homicide, are obviously serious crimes. Most users of recreational substances are not addicts and have severely limited or no potential defense to crime based on their use of controlled substances. For those who engage in drug-related criminal activity who are addicts, and therefore more likely to engage in criminal behavior, the question is how this status does and should affect their liability for criminal behavior related to the addiction.

There is a debate among addiction specialists about the degree to which addicts can exert control over seeking and using substances and about other behaviors related to addiction. All agree, as they must, that seeking and using and related actions are actions, but there the agreement largely ends. Some, especially those who believe that addiction is a chronic and relapsing brain disease, think that seeking and using are solely or almost solely signs of a disease and that addicts have little choice about whether to seek and use. In contrast are those who believe that seeking and using are constrained choices but considerably less constrained on average than the first group suggests. This group is also more cautious about, but does not reject, characterizing addiction as a disorder. There is evidence to support both positions. There is a third group who believe that addiction is simply a consequence of moral weakness of will and that addicts simply need to and can pull themselves up by their bootstraps. The empirical evidence for the moralizing third view seems weak, although such attitudes play a role in explaining the limited role the criminal law accords to addiction. The Nobel-prize winning economist Gary Becker famously argued that addiction can be rational (1996). Which view one holds will of course influence how one thinks the criminal law should respond to addiction-related crimes.

This chapter demonstrates that, despite claims to expand the mitigating and excusing force of addiction based on burgeoning scientific research, existing Anglo-American criminal law is most consistent with the choice position. It also argues that this is a defensible approach that is consistent with current science and with traditional justifications of criminal blame and punishment.

The chapter first discusses preliminary issues to avoid potential objections that the discussion adopts an unrealistic view of addiction. It then provides a general explanation of the responsibility criteria of the criminal law and briefly addresses false or distracting claims about lack of responsibility. Then it turns to analysis of the criminal law's doctrines about addiction to confirm that the criminal law primarily adopts a choice model and that addiction per se plays almost no role in responsibility ascriptions. It concludes with a general defense of present doctrine and practice, but briefly suggests beneficial liberalizing reforms.

Preliminary assumptions about addiction

Virtually every factual or normative statement that can be made about addiction is contestable. This section tries to be neutral.

The primary criteria of addiction commonly employed at present are behavioral (mental states and actions), namely, persistent drug seeking and using, especially compulsively or with craving, in the face of negative consequences (without being clear whether these consequences are subjectively recognized or simply objectively exist (Pickard and Ahmed 2016; Morse 2009)). The usual behavioral criteria for compulsion are both subjective and objective. Addicts commonly report feelings of craving or that they have lost control or cannot help themselves. If the agent persists in seeking and using despite ruinous medical, social, and legal consequences and despite an alleged desire to stop, we infer based on common sense that the person must be acting under some type of compulsion. It seems that there is no other way to explain the behavior, but it is not based on rigorous tests of a well-validated concept.

For legal purposes, the following are important considerations. First, seeking and using and related criminal behaviors are actions, not pure mechanisms. By this I mean that the behavioral signs of seeking and using are potentially under the control of reason, including responsive to incentives. Even if addicts have difficulty controlling their behavior, they are not zombies or automatons; they act intentionally to satisfy their desire to seek and to use drugs (Hyman 2007; Morse 2000, 2007a, 2009). Despite current biologizing about the causes of and nature of addiction (e.g., Kasanetz *et al.* 2010; Volkow *et al.* 2016), much of which is genuinely informative, there is still no definitive biological understanding of addiction. There are many findings about the biology and psychology of addicts that differentiate this group from non-addicts, but none of these findings is independently diagnostic. Moreover, the frequent claims that addiction is a chronic and relapsing brain disease and that addicts do not have the capacity to refrain are false (Heyman 2009, 2013; see also Kelly *et al.* 2017). Whether addiction is properly characterized as a disease is itself controversial. Even if it is and seeking and using are signs of this disease, seeking and using are still actions and whether they and other related behaviors should be excused must be evaluated behaviorally (Fingarette and Hasse 1979). Actions can always be morally evaluated. Whether action is compelled is a moral issue because whatever meaning compulsion might have, it is a continuum concept; how much "compulsion" must exist to mitigate or excuse is a normative question. There is no biological or psychologically validated test for compelled action. Indeed, there is no consensually validated, operationalized criterion for the empirical concept of compulsion. In short, terming addiction a disease and characterizing seeking and using as signs does not answer the question of moral and legal responsibility.

A debated question is whether addiction should be limited to substances. Large numbers of people engage persistently and apparently compulsively in various activities, often with negative consequences. Gambling is an example. If there are some activities or non-drug substances that can produce the same "addictive behavior" as drugs, then the criminal law response should perhaps be similar by analogy. I believe that the concept of addiction should be expanded beyond drugs, but for this chapter will confine the analysis to drug-related addictions.

The concept of the person and responsibility in criminal law

This section does not suggest or imply that Anglo-American criminal law is optimal "as is," but it provides a framework for thinking about the role addiction does and *should* play in a fair system of criminal justice.

Criminal law presupposes the "folk psychological" view of the person and behavior. This psychological theory, which has many variants, causally explains behavior in part by mental states such as desires, beliefs, intentions, willings, and plans (Ravenscroft 2010). Biological, sociological and other psychological variables also play a role, but folk psychology considers mental states fundamental to a full explanation of human action. Lawyers, philosophers and scientists argue about the definitions of mental states and theories of action, but that does not undermine the general claim that mental states are fundamental.

For example, the folk psychological explanation for why you are reading this chapter is, roughly, that you desire to understand the relation of addiction to criminal law, you believe that reading the chapter will help fulfill that desire, and thus you formed the intention to read it. This is a "practical" explanation rather than a deductive syllogism.

Folk psychology does not presuppose the truth of free will, it is consistent with the truth of determinism, it does not hold that we have minds that are independent of our bodies (although it, and ordinary speech, sound that way), and it presupposes no particular moral or political view. It does not claim that all mental states are conscious or that people go through a conscious decision-making process each time that they act. It allows for "thoughtless," automatic, and habitual actions and for non-conscious intentions. The definition of folk psychology being used does not depend on any particular bit of folk wisdom about how people are motivated, feel, or act. Any of these bits may be wrong. The definition insists only that human action is in part causally explained by mental states.

Responsibility concepts involve acting people and not social structures, underlying psychological variables, brains, or nervous systems. The latter types of variables may shed light on whether folk psychological responsibility criteria are met, but they must always be translated into the law's folk psychological criteria. For example, demonstrating that an addict has a genetic vulnerability or a neurotransmitter defect tells the law nothing per se about whether an addict is responsible. Such scientific evidence must be probative of the law's criteria and demonstrating this requires showing how it is probative.

The criminal law's criteria for responsibility, like the criteria for addiction, are acts and mental states. Thus, the criminal law is a folk-psychological institution (Sifferd 2006). First, the agent must perform a prohibited intentional act (or omission) in a state of reasonably integrated consciousness (the so-called "act" requirement, sometimes misleadingly termed the "voluntary act"). Second, virtually all serious crimes require that the person had a further mental state, the *mens rea*, regarding the prohibited harm. Lawyers term these definitional criteria for culpability the "elements" of the crime. For example, one definition of murder is the intentional killing of another human being. To be guilty of murder, the person must have intentionally performed some act that kills, such as shooting or knifing, and it must have been his intent to kill when he shot or knifed. If the person does not act at all because his bodily movement is not intentional—for example, a reflex or spasmodic movement—then there is no violation of the "voluntary act" requirement. There is also no violation in cases in which the further mental state required by the definition is lacking. For example, if the defendant's intentionally performed killing action kills only because the defendant was careless, then the defendant may be guilty of some homicide crime, but not of intentional homicide.

Criminal law also provides for "affirmative defenses" that negate responsibility even if the elements have been proven. Affirmative defenses are either justifications or excuses. The former obtain if behavior otherwise unlawful is right or at least permissible under the specific circumstances. For example, intentionally killing someone who is wrongfully trying to kill you, acting in self-defense, is certainly legally permissible and many think it is right. Excuses exist when the defendant has done wrong but is not responsible for his behavior.

Using general descriptive language, the excusing conditions are lack of reasonable capacity for acting rationally and lack of reasonable capacity for self-control (although the latter is more controversial than the former). The so-called cognitive and control tests for legal insanity are examples of these excusing conditions. For example, in the most famous insanity case of all, *M'Naghten's Case* (1843), M'Naghten delusionally believed that the Tory party was conspiring to kill him and that he needed to assassinate the prime minister, Peel, to save his own life. In the event, after carefully planning the homicide, he shot and killed Peel's hapless secretary, Drummond, who was riding in Peel's carriage at the time (see Moran 1981 for a full account, including the correct spelling of the defendant's name). M'Naghten certainly acted—he planned and shot intentionally—and he had the intent to kill the man he thought was Peel, but he was not capable of rationality under the circumstances. Note that these excusing conditions are expressed as capacities. If an agent possessed a legally relevant capacity but simply did not exercise it at the time of committing the crime or was responsible for undermining his capacity, no defense will be allowed. Finally, the defendant will be excused if he was acting under duress involving a do-it-or-else threat by a third party (some theorists would claim the defendant is justified, but excused is the dominant view).

The capacity for self-control or "will power," is conceived of as a relatively stable, enduring trait or congeries of abilities possessed by the individual that can be influenced by external events (Holton 2009). This capacity is at issue in "one-party" cases, such as those involving addiction, in which the agent claims that he could not help himself in the absence of external threat. In some cases, the capacity for control is poor characterologically; in other cases it may be undermined by variables that are not the defendant's fault, such as mental disorder. Many investigators around the world are studying "self-control," but there is no conceptual or empirical consensus. Indeed, such conceptual and operational problems motivated both the American Psychiatric Association (1983) and the American Bar Association (1989) to reject control tests for legal insanity. In cases of lack of control, unlike cases of no action, the person does act intentionally to satisfy the allegedly overpowering desire.

This account of criminal responsibility is most tightly linked to traditional retributive or desert-based justifications of punishment, which hold that punishment is not justified unless the offender morally deserves it because the offender was responsible. No offender should be punished unless he at least deserves such punishment. Even if good consequences might be achieved by punishing non-responsible addicts or by punishing responsible addicts more than they deserve, such punishment would require very weighty justification in a system that takes desert seriously. The account is also consistent with traditional practical justifications for punishment, such as general deterrence. Most Anglo-American punishment theorists are mixed theorists who believe that desert is a necessary and limiting condition, but that consequential concerns should be considered within the limits, so the account of responsibility offered is consistent with both pure retributivism and mixed theories of punishment.

False starts and dangerous distractions

This section considers four false and distracting claims that are sometimes made about the responsibility of addicts (and others): 1) the truth of determinism undermines genuine responsibility; 2) causation, and especially abnormal causation, of behavior entails that the behavior must be excused; 3) causation is the equivalent of compulsion, and 4) addicts are automatons. Space constraints prevent me from doing more than being conclusory about these complex issues in this section, but the interested reader can find the full arguments in other writing (Morse 2016).

The alleged incompatibility of determinism and responsibility is foundational. Determinism is not a continuum concept that applies to various individuals in various degrees. There is no partial or selective determinism. If the universe is deterministic or something quite like it, responsibility is possible or it is not. If human beings are fully subject to the causal laws of the universe, as a thoroughly physicalist, naturalist worldview holds, then many philosophers claim that "ultimate" responsibility is impossible (e.g., Pereboom 2001; Strawson 1989). On the other hand, plausible "compatibilist" theories suggest that responsibility is possible in a deterministic universe (Vihvelin 2013; Wallace 1994).

There seems no resolution to this debate in sight, but our moral and legal practices do not treat everyone or no one as responsible. Determinism cannot be guiding our practices. If one wants to excuse addicts because they are genetically and neurally determined or determined for any other reason to be addicts or to commit crimes related to their addictions, one is committed to negating the possibility of responsibility for everyone.

Our criminal responsibility criteria and practices have nothing to do with determinism or with the necessity of having so-called "free will" (Morse 2007b). Free will, the capacity to cause one's own behavior uncaused by anything other than oneself, is neither a criterion for any criminal law doctrine nor foundational for criminal responsibility. Criminal responsibility involves evaluation of intentional, conscious, and potentially rational human action. The truth of determinism does not entail that actions and non-actions are indistinguishable (except, trivially, that they are both determined although they are different phenomena) and that there is no distinction between rational and non-rational actions or compelled and uncompelled actions. Our current responsibility concepts and practices use criteria consistent with and independent of the truth of determinism because the distinctions it draws, such as between action and non-action, presence or absence of *mens rea*, and presence or absence of an affirmative defense, are clear, even if determinism is true. These distinctions are also consistent with non-consequential and consequential moral theories of responsibility that we have reason to endorse.

A related confusion is that, once a non-intentional causal explanation has been identified for action, the person must per se be excused. This is sometimes called the "causal theory of excuse." Thus, if one identifies genetic, neurophysiological, or other causes for behavior, then allegedly the person is not responsible. In a thoroughly physical world, however, this claim is either identical to the determinist critique of responsibility and furnishes a foundational challenge to all responsibility, or it is simply an error. I term this the "fundamental psycholegal error" because it is erroneous and incoherent as a description of our actual doctrines and practices (Morse 1994). Non-causation of behavior is not and could not be a criterion for responsibility because all behaviors, like all other phenomena, are caused. Causation, even by abnormal physical variables, is not per se an excusing condition. Abnormal physical variables, such as neurotransmitter deficiencies, may cause a genuine excusing condition, such as the lack of rational capacity, but then the lack of rational capacity, not causation, is doing the excusing work.

Third, causation is not the equivalent of lack of self-control capacity or compulsion. All behavior is caused, but only some defendants lack control capacity or act under compulsion.

If causation were the equivalent of lack of self-control or compulsion, no one would be responsible for any criminal behavior. This is clearly not the criminal law's view.

A last confusion is that addicts are automatons whose behavioral signs are not human actions. We have addressed this issue before, but it is worth re-emphasizing that even if compulsive seeking and using substances are the signs of a disease, they are nonetheless human actions and thus distinguishable from purely mechanical signs and symptoms, such as spasms. Moreover, actions can always be evaluated morally (Morse 2007a).

Now, with a description of addiction and responsibility criteria in place and with an understanding of false starts, let us turn to the relation of addiction to criminal responsibility, beginning with the law's doctrines.

Criminal law doctrine and addiction: background

The introduction to this chapter suggested that the law's approach to addiction is most consistent with the choice model. The ancient criminal law treated the "habitual" or "common" drunkard as guilty of a status offense and drunkenness was considered wrong in itself. Although the legal landscape has altered, the choice model is still dominant.

Constitutional background

Let us begin with the two United States Supreme Court cases that have explicitly addressed addiction: *Robinson* v. *California* (1962) and *Powell* v. *Texas* (1968). Although these cases are older, their holdings and reasoning, especially in *Powell*, continue to be robustly emblematic of the criminal law's response to addiction.

Robinson was a needle-injecting drug addict who was convicted of a California statute that made it a crime to "be addicted to the use of narcotics" and he was sentenced to ninety days in jail. Robinson appealed to the Supreme Court on the ground that punishing him for being an addict was a violation of the 8th and 14th Amendment's prohibition of cruel and unusual punishment. The Court agreed that punishing for addiction was unconstitutional, but there were so many different opinions that it is difficult to pinpoint the Court's rationale. One possibility is that being an addict is a status and it is unconstitutional to blame and punish people for a status. The other is that addiction involves involuntary behavior and it is unconstitutional to punish involuntary behavior. Herbert Fingarette and Anne Fingarette Hasse demonstrated conclusively decades ago that the third, disease, rationale collapses into either the status rationale or the involuntariness rationale (1979), so let us assume that only these two were the most promising rationales. The status claim is narrow because it does not address whether the Constitution requires any mitigating or excusing condition to apply to acts and omissions. It simply prohibits punishing for a status, such as having red hair or being an addict. The "involuntariness" claim extensively suggests that punishing people for conditions and actions that they are helpless to prevent is also unconstitutional.

Leroy Powell was a chronic alcoholic who spent all his money on wine and who had been frequently arrested and convicted for public drunkenness. Mr Powell argued that he was afflicted with "the disease of chronic alcoholism, . . . his appearance in public [while drunk] was not of his own volition" (*Powell* v. *Texas* (1968: 517) and it was "part of the pattern of his disease and is occasioned by a compulsion symptomatic of the disease." Consequently, to punish him for this behavior would be a violation of the Eighth Amendment prohibition of cruel and unusual punishment.

Mr Powell's proposed defense was supported by the testimony of an expert psychiatrist, Dr David Wade, who testified that, a "chronic alcoholic" is an "involuntary drinker," who is "powerless not to drink," and who "loses his self-control over his drinking".

Powell v. Texas 1968: 518

Based on his examination of Mr Powell, Dr Wade concluded that Powell was,

a "chronic alcoholic," who "by the time he has reached [the state of intoxication] . . . is not able to control his behavior, and . . . has reached this point because he has an uncontrollable compulsion to drink".

Powell v. Texas 1968: 518

Dr Wade also opined that Powell lacked "the willpower to resist the constant excessive consumption of alcohol." The doctor admitted that Powell's first drink when sober was a "voluntary exercise of will," but qualified this answer by claiming that alcoholics have a compulsion that is a "very strong influence, an exceedingly strong influence," that clouds their judgment. Finally, Dr Wade suggested that jailing Powell without treatment would fail to discourage Powell's consumption of alcohol and related problems. One could not find a more clear expression of the medicalized, disease concept of addiction to ethanol.

Powell himself testified about his undisputed chronic alcoholism. He also testified that he could not stop drinking. Powell's cross-examination concerning the events of the day of his trial is worth quoting in full.

Q: You took that one [drink] at eight o'clock [a.m.] because you wanted to drink?
A: Yes, sir.
Q: And you knew that if you drank it, you could keep on drinking and get drunk?
A: Well, I was supposed to be here on trial, and I didn't take but that one drink.
Q: You knew you had to be here this afternoon, but this morning you took one drink and then you knew that you couldn't afford to drink anymore and come to court; is that right?
A: Yes, sir, that's right.
Q: Because you knew what you would do if you kept drinking, that you would finally pass out or be picked up?
A: Yes, sir.
Q: And you didn't want that to happen to you today?
A: No, sir.
Q: Not today?
A: No, sir.
Q: So you only had one drink today?
A: Yes, sir.

Powell v. Texas 1968: 519–520

On re-direct examination, Powell's attorney elicited further explanation.

Q: Leroy, isn't the real reason why you just had one drink today because you just had enough money to buy one drink?
A: Well, that was just give to me.
Q: In other words, you didn't have any money with which you could buy drinks yourself?
A: No, sir, that was give to me.

Q: And that's really what controlled the amount you drank this morning, isn't it?
A: Yes, sir.
Q: Leroy, when you start drinking, do you have any control over how many drinks you can take?
A: No, sir.

Powell v. Texas *1968: 520*

Powell wanted to drink and had that first drink, but, despite that last answer, his alleged compulsion did *not* cause him to engage in the myriad lawful and unlawful means he might easily have used to obtain more alcohol if his craving was desperately compulsive. Although Powell was a core case of an addict, he could refrain from using if he had a good enough reason to do so.

This is an extremely sympathetic case for a compulsion excuse. The crime was not serious and the criminal behavior, public intoxication, was a typical manifestation of his alcoholism. The Supreme Court rejected Mr Powell's claim for many reasons. Justice Marshall's plurality opinion was skeptical of the involuntariness/compulsion claim and concluded that it went too far on the basis of too little knowledge. It also suggested that it was unclear that providing a defense in such cases would improve the condition of alcoholics. Although this was a sympathetic case, the plurality was simply unwilling to abandon the choice model that guides legal policy and to impose a "one size fits all" constitutionally-required involuntariness/compulsion defense. The case interpreted *Robinson* as barring punishment for status and not as imposing a constitutional involuntariness defense. If the Court had accepted Powell's argument, it would not have created a specific "addiction" defense. Rather, it would have adopted a general compulsion defense in any case in which criminal behavior was a sign allegedly compelled by a defendant's disease, whether the disease was addiction or any other.

Current doctrine and the choice model

This section will first discuss the affirmative defenses, then it will address the use of intoxication to defend against the elements of the crime charged, which is termed "negating" an element, and will finally discuss the role of addiction in sentencing and diversion.

Given that there is still controversy about how much choice addicts have, it is perhaps unsurprising that the conclusion in *Powell* is still regnant. Addiction is not an affirmative defense per se to any crime in the United States, England or Canada. With one limited exception in English homicide law (Ashworth and Horder 2013: 271–272; *R. v. Bunch* 2013; *R. v. Joyce(Trevor)* 2017), it is also not the basis for any other affirmative defense, such as legal insanity. Some United States jurisdictions explicitly exclude addiction (or related terms) as the basis for an insanity defense, thus implicitly adopting the choice model although substance-related disorders are recognized by the American Psychiatric Association (2013). The claim that an intoxicated addict might not have committed the crime if he had not been intoxicated has no legal purchase, although some disagree (Williams 1961: 564). Addiction does not merit an index entry in most Anglo-American criminal law texts, except in the context of the use of alcohol intoxication as a defense in some instances that will be explored below.

The only exception to the bar to using addiction as the basis of an affirmative defense is "settled insanity." If a defendant has become permanently mentally disordered beyond addiction, say, suffers from delirium tremens, as a result of the prolonged use of intoxicants, the defense of legal insanity may be raised.

An enormous number of crimes are committed by people who are under the influence of intoxicating substances. In what follows I shall discuss the use of intoxication to negate the elements of the crime charged, but these doctrines apply generally to addicts and non-addicts alike. Of course, addicts are more likely to be high than non-addicts and thus these rules will disproportionately affect them.

Recall that all crimes require an "act" and most crimes require a *mens rea*, a culpable mental state that accompanies the prohibited conduct. Does evidence of intoxication in fact tend to show that an element was not present? First, the defendant might be so drunk that his consciousness is sufficiently dissociated to negate the act requirement. Second, the defendant's intoxication may be relevant to whether he formed the mental state required by the definition of the crime. For example, imagine a very drunk defendant in the woods with a gun. In the drunken belief that he is shooting at a tree because his perceptions are so altered, he kills a human being wearing camouflage gear. If he really believed that he was shooting at a tree, he simply did not form the intent to kill required for intentional homicide, although he may be guilty of another type of homicide crime based on carelessness.

The point is straightforward. If the defendant did not act or lacked the *mens rea* for the crime charged, how can he be guilty of that crime? Despite this logic, a substantial minority of United States jurisdictions refuse to admit into evidence factually relevant and probative voluntary intoxication evidence proffered to negate *mens rea*. The remaining United States jurisdictions and English law admit it only with substantial restrictions.

The reasons for complete exclusion and for restriction of the admissibility of relevant evidence of voluntary intoxication result, I believe, primarily from the choice model and from fears for public safety. In the case of restricted testimony, the rules are highly technical. Most commonly, evidence of intoxication is admitted to negate the *mens reas* for some crimes, typically those that are more serious and have more complicated mental state requirements, but not for others, even if *mens rea* in the latter case might actually be negated. The defendant will therefore be convicted of those crimes for which intoxication evidence is not admissible even if the defendant lacked *mens rea*. The rules are a compromise between culpability and public safety and the apparent unfairness of convicting a defendant of a crime for which he lacked *mens rea* is in part justified by his own fault in becoming intoxicated, a classic choice model rationale.

Leading precedents in the United States and England adopt choice reasoning explicitly. In *Montana v. Egelhoff* (1996) the United States Supreme Court held that complete exclusion of voluntary intoxication evidence proffered to negate *mens rea* was not unconstitutional. Justice Scalia's plurality opinion provided a number of reasons why a jurisdiction might wish on policy grounds to exclude otherwise relevant, probative evidence. Among these were public safety and juror confusion. But one is a perfect example of the choice model.

> And finally, the rule comports with and implements society's moral perception that one who has voluntarily impaired his own faculties should be responsible for the consequences.
>
> Montana *v.* Egelhoff *1996: 50*

This view is standard in both common law and continental criminal law: a defendant should not benefit from a defense that he has culpably created.

In *D.P.P. v. Majewski* (1977), a unanimous House of Lords upheld one of the technical distinctions alluded to above that permit defendants to introduce intoxication evidence to negate the *mens reas* of only some crimes. Most of the Lords recognized that there was some illogic in the rule, but all upheld it as either a justifiable compromise or as sound in itself and it had a long history. Most striking, however, is one passage from Lord Elwyn-Jones' opinion for the Court. He wrote,

> If a man of his own volition takes a substance which causes him to cast off the restraints of reason and conscience, no wrong is done to him by holding him answerable criminally for any injury he may do while in that condition. His course of conduct in reducing himself by drugs and drink to that condition in my view supplies the evidence of mens rea, of guilty mind certainly sufficient for crimes of basic intent. It is a reckless course of conduct and recklessness is enough to constitute the necessary mens rea in assault cases . . . The drunkenness is itself an intrinsic, an integral part of the crime, the other part being the evidence of the unlawful use of force against the victim. Together they add up to criminal recklessness.
>
> D.P.P. *v.* Majewski *1977: 474–475*

In other words, the culpability in getting drunk—itself not a crime—is the equivalent of actually foreseeing that there might be criminal consequences of one's intoxication. The rationale is that it is common knowledge that intoxication can be disinhibiting or cloud judgment, even if the intoxicated defendant did not actually foresee specifically what those consequences might be. The choice model reigns.

Despite massive academic criticism of the *Majewski* rule and numerous Law Commission reform proposals, it remains the rule and many think it works reasonably well. Some Commonwealth countries, such as Australia, New Zealand and Canada, have the more expansive rule that permits admission of intoxication evidence without restriction and it seems not to have opened the floodgates of alcohol-awash crime (Ashworth and Horder 2013). Apparently, however, juries in those jurisdictions seldom fully acquit, suggesting that the culpability-based-on-choice model is implicitly guiding decision making even if the law is more lenient. Finally, even the Model Penal Code in the United States, which has had major influence on law reform and which strongly emphasizes subjective culpability, adopted a similar rule (American Law Institute 1962, Sec. 2.08(2)). When substances are involved, the choice model seems recalcitrant to change.

The need for completeness compels me at this point to mention involuntary intoxication, that is, intoxication occasioned through no fault of the agent. Examples would be mistakenly consuming an intoxicant, or being duped into or forced to consume one. The law treats such cases more permissively than cases of voluntary intoxication by providing a limited complete defense and the ability to negate all *mens rea*. But it does not apply to intoxication associated with addiction because the law currently treats such states of intoxication as the agent's fault. The law's view of involuntariness in this context could apply to addicts and non-addicts alike. Even addicts could be duped or coerced into becoming intoxicated on a given occasion.

Addiction-related legal practices

There are two United States contexts in which addiction has potential mitigating force: sentencing, particularly capital sentencing, and diversion to specialized drug courts. The same considerations about addiction's impact would apply in both non-capital and capital sentencing, so I shall discuss only the latter.

The United States Supreme Court has repeatedly held that capital defendants can produce virtually any mitigating evidence (*Lockett* v. *Ohio* 1978) and the bar for the admissibility for such evidence is low. Thus, even if addiction is not a statutory mitigating factor, an addicted defendant convicted of capital murder may certainly introduce evidence of his condition for the purpose of showing that addiction diminished his capacity for rationality or self-control or to support any other relevant mitigating theory. Doing so also raises the danger that addiction will be thought to aggravate culpability based on the choice perspective—especially the moralistic strain—and it is possible that it will make the defendant seem more dangerous, which is a statutory aggravating factor in some jurisdictions. Addiction is a knife that could cut both ways in capital and non-capital sentencing.

Drug courts are an increasingly common phenomenon in the United States. The substantive and procedural details vary across jurisdictions, but these courts aim to divert from criminal prosecution to the drug courts addicted criminal defendants charged with non-violent crimes whose addiction played a role in their criminal conduct. If diverted defendants successfully complete the drug-court-imposed regimen of staying clean and in treatment, they are discharged and the criminal charges are dropped. This approach seems eminently sensible and these courts have fervent supporters, but they also have critics on the grounds that they do not afford proper due process and genuinely solid evidence for their cost–benefit-justified efficacy is lacking. Whatever the merits of the debate may be, drug courts are now an entrenched feature of criminal justice in a majority of United States' jurisdictions and they do permit some number of addicts to avoid criminal conviction and punishment.

A defense of current criminal law

Given the profound behavioral effects of addiction, can the criminal law's generally unyielding rules be fair? Although many addicts are responsible for becoming addicted, the following discussion will assume that an addict is not responsible for becoming an addict, say, because he became addicted as a youth or because he was in pathological denial about what was happening. I shall also assume that the rules apply to adults and that juveniles require special treatment.

I believe that there are roughly two accounts for why addicts might not be responsible for addiction-related crimes, including possession and other crimes committed to obtain drugs (Morse 2011). The first is irrationality, a form of cognitive deficit. As a result of various psychological factors the addict simply cannot "think straight," cannot bring to bear the good reasons to refrain (see Pickard 2016, who suggests that cognitive deficit plays a greater role in addiction than is commonly assumed). This assumes that addicts do have good reasons to refrain, but this may not always be true (Burroughs 2013: 144–147). The other account analogizes the addict's subjective state at times of peak craving as akin to the legal excuse of duress. The addict is threatened by such dysphoria if he doesn't use substances that he experiences the situation like a "do it or else" threat of a gun to one's head. Whether one finds these accounts or another convincing, there is surely some plausible theory of excuse or mitigation that would apply to many addicts at the time of criminal behavior. A very attractive case for a more forgiving legal response arises if one believes that once an agent is addicted, he will be in an excusing state at the time of his crimes on some and perhaps most occasions.

There are at least three difficulties with this position, one of which seems relatively decisive. Much is still not understood about the actual choice possibilities of "typical" addicts. Maybe most can in fact think straight at the times of their crimes but choose not to or they are not substantially threatened by dysphoria or, even if they are threatened with severe dysphoria, they retain the capacity not to give in (Pickard 2016; Pickard and Ahmed 2016 (summarizing the evidence for the retention of this capacity)). The criminal law is justified in adopting the more "conservative" approach under such conditions of uncertainty. Unforgiving criminal law doctrines enhance deterrence. The demand for and use of drugs is price elastic for addicts. If it costs more, they will use less. The threat of criminal sanctions might well deter addiction-related criminal behavior.

The third and seemingly most decisive reason is the potential for what is termed "diachronous responsibility" (Kennett 2001) for addicts who do not suffer from settled insanity. Even if they are not responsible at the times of peak craving or when they are acutely intoxicated, at earlier quiescent times they are lucid. They know then from experience that they will again be in a psychological state in which they will find it subjectively very difficult not to use drugs or to engage in other criminal conduct to obtain drugs. In those moments they are responsible and know it is their duty not to permit themselves to be in a situation in which they will find it supremely difficult to refrain from criminal behavior. They then must take whatever steps are necessary to prevent themselves from allowing that state to occur, especially if there is a serious risk of violent, addiction-related crimes such as armed robbery or burglary. If they do not, they will be responsible for any crimes they commit, although they might otherwise qualify for mitigation or an excuse.

For similar reasons, the criminal law is justified in not providing addicts with enhanced ability to negate *mens rea*. Even if the intoxication of addicts is a sign of their disorder, for the reasons addressed just above, when addicts are not intoxicated and not in peak craving states, they know they will become intoxicated again unless they take steps to avoid future intoxication, which they are capable of doing when lucid.

Two counter-arguments may be raised, however. As people slide into addiction—and almost no one becomes an addict after first use—they may well deny to themselves and others that they are on such a perilous path. This suggests that they may not be fully responsible or responsible at all for becoming addicts. Even if denial, anyway a vexed concept in psychiatry, prevents addicts from understanding that they are addicted, if they get into trouble with the law as a result of drug use, they know that they at least have a "problem" resulting from use. At that point, they also know in their lucid moments that they have the duty to take the steps necessary to avoid criminal behavior.

Regrettably, only limited treatment resources are available in many places to addicts who wish to exercise their diachronous responsibility and to refrain from further criminal behavior. We know from spontaneous remission rates that most addicts can apparently quit using permanently without treatment, but typically they do so after numerous failed attempts and only after they have recognized the good reasons to do so, usually involving family obligations, self-esteem or the like (Heyman 2009, 2013). Fear of criminal sanctions appears to be an insufficient reason for many. Thus, especially when the typical addict is young, having trouble quitting and at higher risk for crimes other than possession, it may be too much to ask of such addicts to refrain without outside help. If outside help is unavailable, diachronous responsibility would be unfair. I think that there is much to this counter-argument, although it certainly weakens as the addiction-related crimes become more serious, such as armed robbery or even homicide.

Having offered a principled defense of current legal doctrines concerning addiction, I should now like to add that I think much current legal response to addicted defendants is too harsh and unwise social policy. Space constraints prevent me from offering my suggestions, but interested readers may find them in my other writings (e.g., Morse 2016) and other suggestions in the chapter by Husak in this volume. In brief, however, there is a powerful moral argument for excusing most addicts for the crimes of possession and use, even if not for more serious crimes. In short, the current criminal law response to drugs and addiction is defensible, but it is far from optimum.

References

American Law Institute (1962) *Model Penal Code*, Philadelphia, PA: The American Law Institute.
American Psychiatric Association (2013) *DSM-5: Diagnostic and Statistical Manual of Mental Disorders*, 5th Edition, Arlington, VA: American Psychiatric Publishing.
Ashworth, A. and Horder, J. (2013) *Principles of Criminal Law*, 7th Edition, New York, NY: Oxford University Press.
Becker, G. S. (1996) "A theory of rational addiction", in G. Becker (ed.), *Accounting for Tastes*, Cambridge, MA: Harvard University Press, pp. 50–76.
Burroughs, A. (2013) *This is How: Surviving What You Think You Can't*, New York, NY: Picador.
D.P.P. v. Majewski [1977] AC 443.
Everitt, B. J. and Robbins, T. W. (2005) "Neural systems of reinforcement for drug addiction: from actions to habits to compulsion", *Nature Neuroscience* 8(11): 1481–1489.
Fingarette, H. and Hasse, A. F. (1979) *Mental Disabilities and Criminal Responsibility*, Berkeley, CA: University of California Press.
Heyman, G. (2009) *Addiction: A Disorder of Choice*, Cambridge, MA: Harvard University Press.
Heyman, G. (2013) "Quitting drugs: quantitative and qualitative features", *Annual Review of Clinical Psychology* 9: 29–59.
Holton, R. (2009) *Willing, Wanting, Waiting*, New York, NY: Oxford University Press.
Husak, D. (this volume, pp. 531–539) "Addiction and drug (de)criminalization".
Hyman, S. (2007) "The neurobiology of addiction: implications for the control of voluntary behavior", *American Journal of Bioethics* 7: 8–11.
Kalant, H. (2010) "What neurobiology cannot tell us about addiction", *Addiction* 105(5): 780–789.
Kasanetz, F., Deroche-Gamonet, V. and Berson, N., et al. (2010) "Transition to addiction is associated with a persistent impairment in synaptic plasticity", *Science* 328(5986): 1709–1712.
Kelly, J., Bergman, B. and Hoeppner, B, et al. (2017) "Prevalence and pathways of recovery from drug and alcohol problems in the United States population: implications for practice, research, and policy", *Drug and Alcohol Dependence* 181: 162–169.
Kennett, J. (2001) *Agency and Responsibility: A Common-Sense Moral Psychology*, New York, NY: Oxford University Press.
Lockett v. Ohio, 486 U.S. 586 (1978).
M'Naghten's Case, 8 ER 718 (1843).
Montana v. Egelhoff, 518 U.S. 37 (1996).
Moran, R. (1981) *Knowing Right from Wrong: The Insanity Defense of Daniel McNaughtan*, New York, NY: The Free Press.
Morse, S. J. (1994) "Culpability and control", *University of Pennsylvania Law Review* 142: 1587–1660.
Morse, S. J. (2000) "Hooked on hype: addiction and responsibility", *Law and Philosophy* 19(1): 3–49.
Morse, S. J. (2003) "Diminished rationality, diminished responsibility", *Ohio State Journal of Criminal Law* 1(1): 289–308.
Morse, S. J. (2007a) "Voluntary control of behavior and responsibility", *American Journal of Bioethics* 7(1): 12–13.
Morse, S. J. (2007b) "The non-problem of free will in forensic psychiatry and psychology", *Behavioral Sciences & the Law* 25(2): 203–220.
Morse, S. J. (2009) "Addiction, science and criminal responsibility", in N. Farahany (ed.), *The Impact of the Behavioral Sciences on Criminal Law*, Oxford, UK: Oxford University Press, pp. 241–289.
Morse, S. J. (2011) "Addiction and criminal responsibility", in J. Poland and G. Graham (eds), *Addiction and Responsibility*, Cambridge, MA: MIT Press, pp. 159–199.
Morse, S. J. (2016) "Addiction, choice and criminal law", in N. Heather and G. Segal (eds), *Addiction and Choice*, Oxford: Oxford University Press, pp. 426–445.
Pereboom, D. (2001) *Living Without Free Will*, Cambridge, UK: Cambridge University Press.
Pickard, H. (2016) "Denial in addiction", *Mind & Language* 31(3): 277–299.
Pickard, H. and Ahmed, S. H. (2016) "How do you know you have a drug problem? The role of knowledge of negative consequences in explaining drug choice in humans and rats", in N. Heather and G. Segal (eds), *Addiction and Choice*, Oxford, UK: Oxford University Press, pp. 31–48.
Powell v. Texas, 392 U.S. 514 (1968).
R v. Bunch [2013] EWCA Crim 2498.
R. v. Joyce (Trevor) [2017] 2 Cr. App. R. 16.

Ravenscroft, I. (2010) *Folk Psychology as a Theory*. Available online at: http://plato.stanford.edu/entries/folkpsych-theory, accessed 27 May 2015.
Robinson v. *California*, 370 U.S. 660 (1962).
Sifferd, K. (2006) "In defense of the use of commonsense psychology in the criminal law", *Law and Philosophy* 25(6): 571–612.
Strawson, G. (1989) "Consciousness, free will and the unimportance of determinism", *Inquiry* 32: 3–27.
United States v. *Moore*, 486 F. 2d 1139 (1973).
Vihvelin, K. (2013) *Causes, Laws and Free Will: Why Determinism Doesn't Matter*, New York, NY: Oxford University Press.
Volkow, N. D., Koob, G. F. and McLellan, A. T. (2016) "Neurobiologic advances from the brain disease model of addiction", *New England Journal of Medicine* 374(4): 363–371.
Wallace, R. J. (1994) *Responsibility and the Moral Sentiments*, Cambridge, MA: Harvard University Press.
Williams, G. (1961) *Criminal Law: The General Part, 2nd Edition*, London, UK: Stevens & Sons.
Zinberg, N. (1984) *Drug Set and Setting: The Basis for Controlled Intoxicant Use*, New Haven, CT: Yale University Press.

45
ADDICTION AND MANDATORY TREATMENT

Steve Matthews

Introduction

What is at stake in mandatory treatment for those with substance addictions? We will answer this by addressing several related tasks. The first is conceptual: what exactly is mandatory treatment, and what are the forms that it takes? The second is factual: what is the current legal and policy situation for mandatory treatment across some representative jurisdictions? The third is normative: what is its justification and how does this articulate with the law? A fourth task is to identify a difficulty that arises in its justification: addiction impairments are conative, cyclical and diachronic, and so the standard cognitively flavoured synchronic criteria for judging competence do not readily apply. So in what sense are addicted persons lacking in the competence required to refuse treatment? A final task is to consider whether there are utilitarian grounds for accepting a limited system of mandated treatment. If, on balance, it demonstrably reduces harm could a case be made for it on the grounds of its effectiveness? Public health studies in this area reveal mixed results, with some positive outcomes in reducing recidivism in criminal justice contexts, but the data for cases of civil commitment is less clear.

Mandatory treatment

I will use 'mandatory treatment' to cover both coercive and compulsory treatment modalities. 'Coercion' has a loose and popular sense covering many different cases, however, let me stipulate a use reminiscent of Nozick (1969), in which the outcome desired by the coercing agent is achieved through manipulation of the choices available to a person convicted of an addiction-related crime.[1] So, for example, a person on a drug-related charge might be presented with the option of treatment where refusal would be losses of other supports, such as housing. This can work because it is backed up by the potential enforcement of the punishment-like alternative to choosing treatment; but nevertheless the coerced agent is given the chance to make a choice. Although this person is not directly forced into rehabilitation – they are certainly not physically removed to a hospital or treatment centre – it counts as coercion partly because the choice situation is designed and enforced by a body external to the one coerced.

The choice-making dimension within mandatory treatment arguably makes a difference to moral perceptions of treatment. For example, a rehabilitated person may later feel that they deserve credit for having chosen the best option.[2] But this is by no means universal and a more favourable experience depends on many variables. In this connection Christopher *et al.* (2015: 319) call for more research. As they put it, this research,

> should include simultaneous assessment of the range of additional pressures that may coexist with a commitment order (e.g., urging of family and employers), one's perceptions of such pressures, changes in motivation during the commitment period, the severity and treatment of co-occurring disorders, and the . . . potential social determinants of substance outcomes (e.g., insurance status, financial resources, and social supports).

Finally, in this connection, what should not be assumed is that the experiences and motivations of those who choose treatment under conditions of coercion are always different in kind from two other groups: those who enter a treatment facility voluntarily, or those who enter involuntarily via civil commitment. Rather, there is some degree of crossover, and this is because, *inter alia*, treatment facilities can run differently regardless of how their patients arrived there, and also because these perceived pressures of coercion may come not just from the State, but from family, friends or medical authorities (Stevens *et al.* 2006).

Coercion, then, works through the will of the coerced, and it occurs typically within the criminal justice system. This is contrasted with *compulsory treatment* where the affected persons are simply directed into treatment without regard for their preferences. (Sometimes 'involuntary treatment' or 'forced treatment' are used as synonyms for 'compulsory treatment'.) Compulsory treatment orders can be applied to offenders within criminal justice – this is known as *court-mandated treatment* – or in the health sector, where they can be applied in civil proceedings where no offence has been committed – this is known in most jurisdictions as *civil commitment*.

In civil commitment an order for treatment occurs by appeal to a state-sanctioned authority specifying strict criteria usually involving harm (or posing a 'danger', as it is put in the US) or the loss of the capacity for decision-making. Admittedly this simplification disguises great complexity in the way different jurisdictions legislate. Some assimilate the case of drug dependence to mental health impairments; some impose age restrictions; some distinguish between the impending dangers of harm (to self or others), and the losses in decision-making capacities, whereas others conjoin these elements. In jurisdictions with strong legal safeguards, civil commitment cases are typically rarer than cases of coerced treatment, perhaps due to the liberty-depriving nature of forcibly removing someone who has not committed any offence. For example, in New South Wales (NSW), Australia, a state with a population of just over 7.5 million, typically around 15 persons per year are committed. (A similarly low figure exists for the state of Victoria.) By contrast, the different states of North America vary significantly. For the period 2010–2012, according to Christopher *et al.* (2015: 315),

> Thirty-three of the 51 states [had] . . . civil commitment . . . Of these, 9 states never apply and 4 more very rarely apply their statutes. Of the remaining 20 states [the figures are] . . . Colorado: 150–200 (annual average); Florida: 9,000 (annual average); Hawaii: 83 in 2009; Massachusetts: 4,500 (annual average) around 2011; Missouri: 166 in 2011; Texas: 22 in 2010; and Wisconsin: 260 in 2011 . . . Seven other states reported that commitment occurred regularly or frequently.

Civil commitment type cases and coercion type cases target different groups, and their purposes and the differing outcomes from each practice suggest that it would be a mistake to attempt to draw lessons from one practice in support of the other. In this connection, Sally Satel (1999, and elsewhere) argues that coercive treatment is effective in reducing recidivism because length of treatment is correlated with better outcomes, and since authorities can effectively control the length of these periods, these better outcomes are more easily enabled.[3] But obviously reduced rates of recidivism are not at issue when we come to the justification question for civil commitment.

Mandatory treatment of addiction includes detoxification, counselling, education, and various other forms of rehabilitation. To provide a concrete example, at the Women's Addiction Treatment Center in Massachusetts US, the services there include a detoxification unit (with 24-hour monitoring by doctors, nurses, and counsellors), clinical and transitional support services, family support, access to twelve-step programmes that incorporate counselling, opioid overdose prevention workshops, and informational sessions focusing on relapse and coping strategies. The initial aim of such programmes is containment and safety, followed by restoration of long-term planning capacity to facilitate decisions supporting health and well-being, reduction in social and economic costs and to improve prospects for employment and better relationships.

The state of play

Treatment programmes involving the criminal justice system operate in many countries including Britain, the USA, Canada, many western European nations (such as Germany, Italy, Spain, Sweden), and parts of Oceania.[4] Again, there is great variation in how the programmes are run. For example, whereas in the USA treatment tends in some places to be the upshot of *any* offence that is drug related, in the Netherlands it is those with a *pattern* of drug use who come to the attention of criminal justice authorities who are then directed into treatment (See note 4, ANCD report, p. 2). Civil commitment programmes also vary, and have changed over time in response to evidence-based medicine advances and human rights considerations, e.g., in the states of Victoria, and NSW (Australia), New Zealand, Britain and North America.

To provide a detailed example of the variation and complexity in this area consider the case of England and Wales. Involuntary detention/treatment of addicted persons there could in principle come under The Mental Capacity Act (2005), but *not* the Mental Health Act where addiction is explicitly ruled out as a condition sufficient for detention. Under the Mental Capacity Act provision to deprive a person, 'P', of liberty is permitted in P's best interests and when a reasonable belief exists that such deprivation is necessary to sustain life, prevent serious deterioration, or prevent harm to P. It would seem that addicted persons could be judged to come under this Act in case they are decisionally incapacitated due to an impairment of the mind or brain, and it would seem also that the most relevant part of the Act on this score pertains to the lost ability to weigh information as part of the process of making decisions.

Proper justification for mandatory treatment depends on practices that embed within a normative framework of culture, policy and law. The emphasis in the jurisdictions described above is typically on a degree of voluntariness (in the alternative consequence models), and a variety of safeguards including short time frames, evidence-based treatment protocols, medical care for co-morbid conditions, and a non-punitive rationale governing both the process and the aims of treatment, viz., harm reduction and the restoration of decision-making capacity.

Unfortunately this kind of normative framework is not universally adopted. So, for example, infamously in East and Southeast Asia, compulsory drug detention centres (CDDCs) have in recent times sprung up and indeed proliferated. Kamarulzaman and McBrayer (2014), quoting an Open Society Institute report, claim that around 235,000 people who use drugs (PWUD) are detained in over 1,000 CDDCs. In some jurisdictions PWUD are picked up in police sweeps or from positive urine tests, or even on the basis of mere suspicion, and are then forcibly sent to the centres. Medical treatments, not atypically, are lacking, and stays can be for indefinite periods and where no attention is paid to such things as withdrawal or the possibility of relapse. Human rights abuses in the centres – including physical abuse, torture, forced work, gruelling exercise regimes of a militaristic nature – underlie the fact that many of the centres operate chiefly to punish illicit drug users. International scrutiny of the centres led to a United Nations joint statement in 2012 condemning them and calling for their closure.[5]

These cases demonstrate what can go wrong in the absence of checks and balances. Mandatory treatment is a treatment type of last resort (certainly in civil commitment cases), and it is supposed to be motivated by a concern for the health and social welfare of the drug-dependent person. Given this motivation, and given the extreme liberty-depriving nature of forced removal into a place of detention, it is necessary to have available processes that act as a check on the system at point of entry, within the treatment centre itself, and to exit. Legislation must carefully specify these processes, and the duties associated with the process be bestowed on those with specific *ex officio* roles. Safeguards, transparency including oversight by other bodies, and appeals mechanisms need to accompany all aspects of the process. Those administering, both within (such as doctors, nursing staff, therapists and so on) and outside (such as civil managers of various sorts, magistrates or politicians), ideally should be motivated by the health needs of those committed, and the welfare of their families. Suffice to say: any attempted justification for mandatory treatment that fails to register the need for these things, and to warn of the fragility of the legal and political background, even in cases where a polity is seemingly stable, will fail.[6]

Civil commitment: an example

Consider the Severe Substance Dependence Treatment Act 2010 from Victoria, Australia. The Act provides for,

> the detention and treatment of people with severe substance dependence in a treatment centre where this is necessary as a matter of urgency to save the person's life or prevent serious damage to their health. Detention must be the only means by which treatment can be provided and there must be no less restrictive means reasonably available to ensure the treatment. In addition, the person must be incapable of making decisions about their substance use and personal health, welfare and safety due primarily to their substance dependence. The purpose is to give the person access to medically-assisted withdrawal, time to recover, capacity to make decisions about their substance use, and the opportunity to engage in voluntary treatment. Detention and treatment must always be an option of last resort. Detention and treatment is limited to a maximum of 14 days.[7]

Two elements are important. First, the addicted person must have lost a 'capacity to make decisions about their substance use', and second, detention must be the 'only means by which treatment can be provided'. We note, then, that this lost capacity leads to waiving of what is taken as a norm in medical settings, consent of the patient. We are to presume that dispensing with the need for consent is based on a loss of decision-making capacity. But in what sense is there a loss here? It is supposed to be due to the *addiction* ('severe substance dependence'), not intoxication, a state in which the loss of decision-making capacity might be clear, and which is present in many non-addicted persons. The impairment grounding the claim of incompetence, then, is a disturbance in the will, not a matter of a cognitive failing here and now, such as a delusion generated by a psychosis. The impairment must also be tied to the threat of harm or death, and be the outcome of a *pathological* condition, since there are non-pathological conditions that also give rise to patterns of behaviour leading to harm. Base jumping, rodeo riding or mountain climbing, undertaken in obsessive ways, are all potentially harmful activities, but they do not issue from some kind of pathological fixation. Certainly they can be habit-forming, but civil commitment statutes must be carefully formulated to exclude them.

So, to condense all of this: the justification for civil commitment of addicted persons is that they have sustained a pathological loss of competence with respect to making and implementing decisions over time, and this loss places them (or others) in danger of harm and/or death. This raises two difficulties. The first is that currently used assessment tools for competence are ill-suited to the case of addiction. The second is that testing for competence (however we assess it) either comes too early, because the justificatory preconditions have not yet obtained, or it comes too late when the harm caused by addiction – which compulsory detention might have prevented – has been done. In the next two sections I expound on the first difficulty; in the section following that I attend to the second.

Justifying mandatory treatment in civil commitment cases

The criteria normally cited for involuntary treatment in the area of mental health do not readily apply in cases of addiction. For example, a psychotic person may be detained and treated because their behaviour risks imminent harm to self and/or others. The loss of competence here typically springs from delusional thinking. Sober drug-dependent persons have not lost competence in this sense. They may be in non-imminent danger of harming themselves, but so are people who engage recklessly in extreme sports. Now it would seem that to justify mandatory treatment an ineliminable element in competence criteria is diminished cognitive awareness and reasoning capacity. Normally failures to understand information, to reason and assess risk with respect to one's diagnosis, and to be able to express one's view are sufficient for an intervention, yet this kind of competence loss is not salient in addiction. (I give a detailed analysis of competence criteria in the next section.)

I will take the concept of competence in medical settings to apply to relatively task-specific aspects of decision-making that affect one's health and one's life. Though it comes in degrees it is best to regard competence as reaching a threshold and its application in law must be regarded this way for the purposes of deciding who meets the criteria for competence and who does not. Losses in competence usually appear in acute care settings where diminished decisional capacities may be easily observed. However, in chronic conditions like addiction typically the problem is not the making of a decision, but rather the ongoing implementation of it (Naik *et al.* 2009; Levy 2016). The addicted person cannot effectively play the executive role required to implement the steps needed to adhere to earlier resolutions to stay clean, to remain healthy, and thereby get their life back on track.[8] This is the loss of competence that the standard criteria do not assess. It is the failure to unify one's agency over time, and because the loss of control stretches over time it is difficult to make assessments of it and to design a competence test that accurately captures what has gone wrong.

Competence criteria

When assessing competence in mental health contexts, there is a relatively consistent approach (Craigie 2009: 2; Tan *et al.* 2006: 268; Wild *et al.* 2012: 167). We may take here the MacArthur

Competence Assessment Tool—Treatment (MacCAT-T) as representative. Competence on this model includes:

(1) Understanding information that is relevant to a condition and its treatment.

(2) Reasoning about the potential risks and benefits of a choice.

(3) Appreciating the nature of one's situation and the consequences of making a choice.

(4) Communicating a choice.

Grisso et al. 1997

These criteria are not suitable for judging competence to refuse treatment for non-intoxicated, non-craving addicted persons. Sober addicted individuals retain an understanding of the medical aspects of their condition, and they have relatively clear insights regarding its effects on behaviour. In the clinic they may readily acknowledge the difficulty of their situation. Suffice to say, addicted persons are competent on the basis of the measurements contained in the standard test.[9] If so, then mandatory treatment orders should not be made using the standard test; different test criteria would need to be formulated for cases of addiction. Ideally, these criteria would need to combine an interview assessment with evidence of the clinical history of the presenting individual in order to capture the diachronic nature of the addiction impairment.

To sharpen up this difficulty it is helpful to compare the case of addiction with the case of anorexia where a clinical dilemma arises for treatment professionals. Jillian Craigie (2009: 2) points out that, on the one hand, a requirement of medical ethics is to pay due respect to the autonomous wishes of an anorexic patient who may be refusing treatment; on the other hand, in adhering to their refusal decision, a doctor knows that he or she thereby places that patient at great risk of further harm or death. Standard criteria for competence are insufficient to show that the patient with the anorexic condition is a candidate for involuntary treatment. And yet suppose in response to this situation the criteria for competence were 'gerrymandered' to include a strong desire for thinness. On the face of it this seems sensible but once certain specific preferences are deemed to be a product of a disordered mind, there is a risk that legitimate, unusual values are wrongly rendered as pathological. The history of disease categorisation contains some infamous examples where the process has gone wrong, including homosexuality and masturbation.

The further challenge, then, is to make sure that politically charged or moralistic values do not come to inform criteria governing mandatory detention. These criteria should focus only on cases where a failure of intervention looks seriously unsafe, leading to the question: what harms *would* this failure to treat bring about? This counterfactual test need not be problematic once an assessment is made based on a person's prior medical history. Indeed, this kind of assessment is familiar from prognoses that are commonly made already in mental health contexts. It is probabilistic, to be sure, and the uncertainty is exacerbated because, unlike predictions based on relatively simple acute physical illnesses like pneumonia, predicting outcomes in addiction is beset with the need to factor in a range of non-clinical variables, such as social supports or their lack. But again, these uncertainties are already common in the area of mental health patient care (Matthews 2014: 215–216).

The problem of knowing when to commit

The criteria to civilly commit an addicted person in the Victorian example (and many others) let in very few cases. The addicted person must be in *imminent* danger of severe harm or death where the *only* means available to prevent this is mandatory treatment. The effect of this is that treatment often comes too late when a lot of damage has already been done. But forced detention at an early stage of an addiction returns us to the problem of justification: for who is to say that a person with a nascent heavy drug habit will progress to the stage of severe and dangerous addiction? A barrier to prediction here is that the identification of some internal condition for substance dependence is not sufficient for prognosis when we know that addiction is an externalizing disorder – the maladaptive behaviour conceptually depends on features of the environment.[10] We know also from public health studies that cases of spontaneous remission are very common, suggesting that addiction typically does not progress to the point where intervention is required (see Heyman (2009) for a sustained review of the relevant literature).

Still, could an exception be made in cases of early intervention involving juveniles? Take, say, a teenager whose five-year addiction is severe, whose parents/guardians have tried extraordinarily hard to manage the situation (but to no avail), and where treating professionals and the criminal justice system have become involved, and where all these efforts have led precisely nowhere. In such early-stage treatment-resistant cases is the point of civil commitment made especially salient? Minimising harm and saving lives seems more likely achieved here, whereas in late-stage cases the damage may be advanced to the point of irreversibility. In such cases the point seems lost. The additional rationale in the early intervention juvenile cases is simply that it makes sense *then* to take an addicted adolescent out of the cycle of consumption and withdrawal.

This argument, as it stands, *might* succeed, but for it to succeed the legal provisions, such as are seen in the Victorian example, would need softening. A rationale for intervention there is that urgent action is needed to save the life of the affected individual or to prevent serious damage to their health, and in the case just rehearsed we are assuming that this stage has not yet been reached; indeed, that is the whole point of the example. Nevertheless, someone might argue that for *juvenile* intervention cases the justificatory paradigm is altered anyway, on account of the sub-adult threshold, and indeed it is a norm of medico-legal practice to recognise that the teenage years constitute a grey area where the authority to make decisions may be vested (at least partly) in parents or guardians, or even in the State. For example, in a famous case in NSW in 2013 a Supreme Court judge ordered that a 17-year-old Jehovah's Witness with Hodgkin's Lymphoma undergo a blood transfusion. The court overrode the wishes of both the patient and his parents saying that '[t]he interest of the State is in keeping him alive until [he is an adult], after which he will be free to make his own decisions as to medical treatment' (Olding 2013).

Notwithstanding my guarded support for civil commitment in these cases, one further problem remains, though it does not seem insurmountable. The argument we rehearsed above was based on our single case, but is this sufficient as justification for a *system* of detention? The main problem is that there are risks that attend the introduction of a system of detention that description of the single case omits. For example, there is a risk that any such programme of detention would be transformed into one in which young poor addicted teenagers are punished and/or removed from social life as a *de facto* (perceived) benefit to others. Such a transformation might occur where the institution comes to be embedded within changed external political conditions, where say an authoritarian political regime comes to misuse its power to control what it views as deviant types (Wild *et al.* 2012: 163). Nevertheless, this difficulty is perhaps not sufficient to defeat the argument for juvenile early intervention; rather, it adds to recognition of the very important need for thinking through the consequences of the institutional arrangements in order to provide for safeguards and judicial oversight.

What evidence supports paternalistic civil commitment?

The attempted justification of civil commitment is based on a paternalism argument in which it is claimed that the refusal to be treated can be overridden for the good of the patient. This conclusion assumes that on balance a *net benefit* will arise from intervention. But for the purposes of making sound policy decisions can we *really* make this assumption? What evidence is there in support of it?

The picture is a diverse one. In a recent paper by Werb *et al.* (2016) the results of a systematic review of the effectiveness of compulsory drug treatment showed grounds for pessimism. The authors (2015: 8) wrote that 'there is little evidence that compulsory treatment is effective in promoting abstention from drug use'.[11] In another analysis Wild *et al.* (2012) present a comprehensive summary of the empirical literature concluding pessimistically that the results are 'mixed'. They point out that in the light of all these inconsistent findings 'an impasse' has been reached. Part of the difficulty in making progress, they further point out, is a problem inherent in the attempts to make *consistent* meta-analytic comparisons of the different studies. There is a lack of uniformity with respect to the type of addictive behaviours; a lack of uniformity in the terminology to describe mandatory treatment; a range of different discipline areas that have their own assumptions and approaches; and a lack of agreement over how we might measure and assess effectiveness in the variety of mandated drug and alcohol programmes. Clearly what are needed are measures that allow for much more meaningful comparisons in order to progress our understanding in this area.

Conclusion

Mandatory treatment of addiction can be compulsory or coercive, applied within criminal or civil contexts. There is some evidence of its effectiveness as a coercive mode within *some* criminal contexts (Satel 1999), but the successes seem correlated with many other factors, both internal and external to the practices themselves. Focusing on civil commitment the State must have evidence that competence in diachronic agency has been lost. The compulsive-like pattern of drug consumption only becomes harmful and severely disordered after a period of time within a certain context and so the measure of competence must be attuned to this diachronic dimension. The longitudinal dimension also raises the problem that early-stage addiction is not severe enough to warrant civil commitment (though I offer qualified support for such a possibility in juvenile cases), and yet the main point of intervention would seem to be the prevention of development of addiction into more serious and harmful phases.

Notes

1 This usage of 'coercion' for alternative consequences models is a standard one; see Gerdner and Israelsson (2010). They also make further fine-grained distinctions between the types of manipulative coercion that is available to states, see p. 119.
2 See Olson *et al.* (1997: p. 1318) who register patient attitude toward recovery as one central variable in response to being committed.
3 Satel's analysis relates primarily to the United States; Werb *et al.* (2016: 7), on the other hand, performed a global meta-analysis, and asserted that 'the majority of studies (78%) . . . failed to detect any significant positive impacts on drug use or criminal recidivism over other approaches, with two studies (22%) detecting negative impacts of compulsory treatment on criminal recidivism compared with control arms'. What helps to explain the difference is that the US practices are situated within a relatively favourable environment of culture, policy and law.

4 For a concise and comprehensive description of the current worldwide situation see the report from the Australian National Council on Drugs (ANCD), *Mandatory Treatment*, www.atoda.org.au/wp-content/uploads/Mandatory_Treatment.pdf. For a highly detailed international analysis see Gerdner and Israelsson (2010).
5 Not all have called for their closure. See, for example, Zunyou Wu (2013).
6 For a general overview, see Ronli Sifris (2011).
7 See www.health.vic.gov.au/ssdta/index.htm.
8 See Jeanette Kennett and Steve Matthews (2003: 310).
9 This is a *general* claim, and there are cases of long-term addiction involving cognitive deficits, typically memory impairments. But these tend to be substance-relative, e.g., long-term opiate use is associated with impairments in 'verbal working memory ... [and] cognitive flexibility (verbal fluency)' (see A. Baldacchino *et al.* 2012).
10 Cf. Will Davies (2016).
11 The authors measured success in both compulsory and coercive treatment.

References

Baldacchino, A., Balfour, D. J. K., Passetti, F., Humphris, G. and Matthews, K. (2012) "Neuropsychological consequences of chronic opioid use: a quantitative review and meta-analysis", *Neuroscience & Biobehavioral Reviews* 36(9): 2056–2068.
Christopher, P. P., Pinals, D. A., Stayton, T., Sanders, K. and Blumberg, L. (2015) "Nature and utilization of civil commitment for substance abuse in the United States", *Journal of the American Academy of Psychiatry and the Law*, 43(3): 313–320.
Craigie, J. (2009) "Competence, practical rationality and what a patient values", *Bioethics* 25(6): 326–333.
Davies, W. (2016) "Externalist psychiatry", *Analysis* 76(3): 290–296.
Gerdner, A. and Israelsson, M. (2010) "Compulsory commitment to care of substance misusers: international trends during 25 years", *European Addiction Research* 18(6): 302–321.
Grisso, T., Appelbaum, P. S. and Hill-Fotouhi, C. (1997) "The MacCAT-T: a clinical tool to assess patient's capacities to make treatment decisions", *Psychiatric Services* 48(11): 1415–1419.
Heyman, G. M. (2009) *Addiction: A Disorder of Choice*, Cambridge, MA: Harvard University Press.
Kamarulzaman, A. and McBrayer, J. L. (2014) "Compulsory drug detention centers in east and southeast Asia", *International Journal of Drug Policy* 26(s1): 533–537.
Kennett, J. and Matthews, S. (2003) "The unity and disunity of agency", *Philosophy, Psychiatry & Psychology* 10(4): 302–312.
Levy, N. (2016) "Addiction, autonomy, and informed consent: on and off the garden path", *Journal of Medicine and Philosophy* 41(1): 56–73.
Matthews, S. (2014) "Addiction, competence and coercion", *Journal of Philosophical Research* 39: 199–234.
Naik, A. D., Dyer, C. B., Kunik, M. E. and McCullough, L. B. (2009) "Patient autonomy for the management of chronic conditions: a two-component re-conceptualization", *The American Journal of Bioethics* 9(2): 23–30.
Nozick, R. (1969) "Coercion", in S. Morgenbesser, P. Suppes and M. White (eds), *Philosophy, Science and Method: Essays in Honor of Ernest Nagel*, New York, NY: St Martin's Press.
Olding, R. (2013) "Jehovah's Witness teen loses appeal over life-saving transfusion", *Sydney Morning Herald*, 27 September, smh.com.au.
Olson, D. H., Mylan, M. M., Fletcher, L. A., Nugent, S. M., Lynch, J. W. and Willenbring, M. L. (1997) "A clinical tool for rating response to civil commitment for substance abuse treatment", *Psychiatric Services* 48(10): 1317–1322.
Satel, S. (1999) *Drug Treatment: The Case for Coercion*, Washington, DC: AEI Press.
Sifris, R. (2011) "An international human rights perspective on detention without charge or trial", in B. McSherry and P. Keyzer (eds), *Dangerous People: Policy, Prediction, Practice*, New York, NY: Routledge, pp. 12–23.
Stevens, A., Berto, D., Frick, U., Hunt, N., Kerschl, V., McSweeney, T., Oeuvray, K., Puppo, I., Santa Maria, A., Schaaf, S., Trinkl, B., Uchtenhagen, A. and Werdenich, W. (2006) "The relationship between legal status, perceived pressure and motivation in treatment for drug dependence: results from a European study of quasi-compulsory treatment", *European Addiction Research* 12(4): 197–209.

Tan, J. O. A., Stewart, A., Fitzpatrick, R. and Hope, T. (2006) "Competence to make treatment decisions in anorexia nervosa: thinking processes and values", *Philosophy, Psychiatry and Psychology* 13(4): 267–282.

Werb, D., Kamarulzaman, A., Meacham, M. C., Rafful, C., Fischer, B., Strathdee, S. A. and Wood, E. (2016) "The effectiveness of compulsory drug treatment: a systematic review", *International Journal of Drug Policy* 28: 1–9.

Wild, T. C., Wolfe, J. and Hyshka, E. (2012) "Consent and coercion in addiction treatment", in A. Carter, W. Hall and J. Illes (eds), *Addiction Neuroethics: The Ethics of Addiction Neuroscience Research and Treatment*, London, UK: Academic Press (Elsevier), pp. 27–54.

Wu, Z. (2013) "Arguments in favour of compulsory treatment of opioid dependence", *Bulletin of the World Health Organization* 91(2): 142–145.

INDEX

Figures given in *italics*, Tables in **bold** and Notes by page number, 'n', note number(s).

AA *see* Alcoholics Anonymous
ABM *see* Agent-Based Models
abnormal states of mind 90–101
"abrupt" concept 262
abstinence: actions 54; CM treatment 456, 459–60; cocaine use disorder 383–4; contingency management 187–8; cost of 15–16; defining addiction 128; judgment-based self-control 48; stress responses 307; TSF efficacy 469; TSMHO effectiveness 468, 472–3; withdrawal and 55
abstinence self-efficacy 471–2
"abstinence violation effect" 38
abuse, use of term 88n3
abusive relationships 417
access: to documentary participants 422–3; to heroin 231–3; to treatment 479–80
accident hypothesis 114
acetaldehyde dehydrogenase (ALDH) 256
acetaminophen problem 515
Acquired Immune Deficiency Syndrome (AIDS) 147
"act" requirement, criminal law 542, 548
action-guiding self-concepts 407
action impulsivity 287
action model 64–5
active inner struggle 72–3
acute stress responses 301
AD *see* axial diffusivity
AD (alcohol-dependent individuals) 308
adaptation and dislocation 505–6
adaptive behavior 300–3
adaptive brain circuits 304
addict identity 443

addiction: assumptions about 541–2; characteristics of addict 185; common pathways of 147–8; conceptions of 9–22; consequences of 149–51; definitions 123–31, 132–4, 320–1, 506; deriving 23–33; disciplinary perspectives 1–3; diversity of study in 163; experience of 23–5, 209–10; manifestations 149–51; models of 145–59, 160–72, **162–3**; nature of 351–4, 532; re-interpreting explanations 37–42; reconsidering concept of 146–7; theories of 160–72, *164*; use of term 78, 161; vocabulary of 28–9
Addiction Research Center (ARC) 241–2
Addiction Severity Index (ASI) 140
Addiction Syndrome Model 145–59, *152*
addictive activities, properties of 28
addictive disorders 92, 140–1, 275–85, *277*
addictive drugs: common properties 24, 28; criminalization 531–2; trade in 212–13
"addictive personalities" 243
addictive phenotype 275
ADHD *see* attention deficit hyperactivity disorder
administration of drugs 108, 116n3, 193, 196–201, 306–7, *see also* self-administration
adolescent communities 79, 83, 85
adolescent drug use 111, 344
adverse life events **300**, 345
adversity and vulnerability 303–4
affect 64, 67–9, 242
affective blame 449
affirmative defenses 543–4, 547
age differences, drug use 109–12, *113*
agency 63, 64–5, 441–2
Agent-Based Models (ABM) 263–4

564

agent–exposure trigger 262
agent–host–environment (AHE) models 255–6, 259
AIDS *see* Acquired Immune Deficiency Syndrome
akrasia 46
alcohol addiction *see* alcoholics; alcoholism
alcohol-dependent (AD) individuals 308
alcohol dependence, PET imaging 384–5, 387
alcohol industry 213, 223
alcohol metabolism genes 283
alcohol regulatory regimes 522–5
alcohol use: age differences *111*; decision-making dysfunctions 343–4; drug control system and 520; drugs distinction 60n1; harm reduction 187; opioid use and 150; placebo effects 397–8; public health consequences 188; "social drinking" 405; social gradients 211–12; sociology of 221
alcohol-use disorder (AUD) 177, 464–77, 481
alcoholics: communities 85; identification as 78; judgment-based self-control 46; quitting 31
Alcoholics Anonymous (AA) 431–2, 464–72
alcoholism: brain dysfunction 99; criminal law and 545–7; defining addiction 126–8; gene–environmental interactions 256; treatment of 431–9; use of term 78
ALDH *see* acetaldehyde dehydrogenase
Alexander, Bruce 199–200
allele frequency studies 278–80
Alline, Henry 434–5, 437
allostasis 245
ambiguity 165, 339, 344–6
amphetamine use 129–30, 355, 357
analgesia 241, 393–4
Anglo-American criminal law 540, 543, 547
animal models/research: addiction rates 91; drug addiction 11, 29; food addiction 183–4, 186; history/significance of 192–203; imaging studies 388; reward prediction 56; sensitization 356; stress 304; SUD endophenotypes 292–3
anorexia case 559
Anthony, J. C. 263, 272
anxiety mechanisms 97, 303
appetite: consumption patterns *41*; discount delay 40; personal rules 41
ARC *see* Addiction Research Center
architectural models, mind 72–4
arrogance 433–4, 437
ASI *see* Addiction Severity Index
"assisted self-change" 492, 493
association delay 193, 195, 197, 200
association studies 278–81, 283–4
associative learning 353
attention deficit hyperactivity disorder (ADHD) 290
attentional bias 51–2
attentional control 45
attributional errors 222

AUD *see* alcohol-use disorder
authority, epistemic 431–9
automatic behavior 198, 374
automatic learning processes 325–38
automatic nervous system 302
autotrophs 104
aversiveness, toxins 107–9
axial diffusivity (AD) 372

"balanced placebo design" 396, 398
Barber, B. 492
bargaining 38, 42, *see also* intertemporal bargaining
barrier defenses 106–7, *107*, 108
baseline state, brain 375
Basic Formal Ontology (BFO) 167–71, *167*
Baumol, W. 29
BBB *see* blood–brain barrier
BDMA *see* brain disease model of addiction
BDNF *see* Brain-Derived Neurotrophic Factor
Becker, Gary 540
behavior change 470–2
behavioral addictions 99, 125, 132, 137–41, *139*, 183
behavioral aspects: assumptions about addiction 541; choice 23–4; cocaine use disorder 385–6, **386**; contingency management 456; gambling 173–5; learning mechanisms 325; neuroscience 242–3
behavioral automatism 198
behavioral disorder models 315–24
behavioral sensitization 307
behavioral signs, desire 194
behavioral therapies 178, 481
belief oscillation hypothesis 54–62
benefits/costs, drug addiction 14–15, 19, 24
benzodiazepine treatment 481
Berridge, K. 322
betting 40–2, *see also* gambling disorder
BFO *see* Basic Formal Ontology
biasing effects: dependence syndromes 268; drug-associated stimuli 51–2; epidemiological studies 257–8; gambling disorder 175; male drug use 112, *113*; myths 222
binding, desire/choice systems 98
binding potential (BP), PET 380–1, *380*
binge-eating disorder 187
binging choice frames 26–9
biological aspects: assumptions about addiction 541; dependence 222–3; gambling 176–7
biological design 92, 93, 97–8
biomarkers, cravings 141
biopsychosocial sequelae 149
bitter taste receptors 106, 110, 114, 116
black populations 513, 514
blame 449, 534, 535–6, 538n22
Blinder, A. 29

Index

blood–brain barrier (BBB) 106, 108
"blunt" smoking 272
blunted dopamine release 385–7
Bolivian drugs legalization 521–2
Bolles-Bindra-Toates principles 354
boundaries of mechanisms 317
boundary-blurring, illegal drug markets 229–39
Bourgois, P. 211, 224–5, 423
BP *see* binding potential
brain circuitry 243, 286–99, *292*, 301–4, 321–2
brain connectivity 362–79
brain damage 90–1, 92–4, 95, 98–9
Brain-Derived Neurotrophic Factor (BDNF) 243–4
brain disease 78, 87, 237, 240, 258, 540, 541, *see also* disease
brain disease model of addiction (BDMA) 221–2, 224, 401, 402–4, 436
brain disorders/conditions 90, 125–6, 128–9, 137–9, 145, *see also* mental disorders
brain function, error prediction 57–8
brain pathology 12–13
brain regions: attentional bias 52; choice and 34–5; decision-making 343–4; gambling disorder 176–7; molecular targets 244; pathways of addiction 147–8; resting-state connectivity *364*; reward circuits 198, 243
brain-reward circuitry concept 243
"brain-stimulation reward" (BSR) 242
brain systems/processes 90–101, 317, 320, 354, 401–10
brokering 230, 232–8, 239n11
BSR ("brain-stimulation reward") 242
bundling approach: choices 27, 29–30; rewards 37–8, *38*
buprenorphine treatment 478–9
Burroughs, Augusten 80–1, 87
buyer–seller interactions, heroin markets 230–5

caffeine 110, *112*
cancer cases 258–61, 281
"candidate gene" studies 277, 290
cannabinoid receptors 393
cannabis (marijuana) 30–1, 513; age of use differences *111*; "blunt" smoking 272; decision-making dysfunctions 344–5; genes causing dependence 283; legalization of 518–30; PFC and 322; psychosis onset 260–1; regulatory regimes 522–4; sociology and 220
canonical models 169
capacity 543, 557
capital sentencing 549–50
carbon monoxide poisoning 98–9
care programs, opioid use disorders 481–2
case-control approaches 259, 266–8
case-control association studies 279, *279*
case-crossover approaches 261–4, 272

cash incentives 457
catechol-O-methyltransferase (COMT) 283, 290
categorizing people 18
Caulkins, J. P. 525
causal analysis 468–9, 506–7
causal knowledge 16–17
causal relations models 316
causal relevance, definition 317
"causal theory of excuse" 544
CBT *see* cognitive behavioral therapies
CD (cocaine-dependent) individuals 307–8, 388
CDA *see* Committee on Drug Addiction
CDCV (common disease/common variant) model 276, *277*
central nervous system (CNS) 104, 108
cerebellum 344–5
"chains of inference", smoking 261
change methods/models 214, 494–5, *see also* self-change
chemical defences 104
"chemotherapy" 242
children 81, 109–10, 185, 293
chimpanzee studies 194–7, 199
choice: administration methods 197–200; agent-host-environment models 255; binding systems 98; brain regions and 34–5; chimpanzee studies 195; costs/benefits of addiction 14, 19; criminal law and 540; discounting role 15; framing 26–9; instrumental learning 330–1; loss of freedom and 536–7; mandatory treatment and 554–5; marketplace model 36; motivational mechanisms 92–3; negative mood effects 333; outcome devaluation **331**; pathology and 12; three elementary features of 23–33
choice impulsivity (*CI*) 286–98
choice model, criminal law 547–50
Christopher, P. P. 555
chronic diseases 188
chronic drug use, effects of 307–9
chronic pain 346–7, 482
chronic relapsing disorders 351
chronic stress 303–4
chronic stressors **300**
CI see choice impulsivity
cigar smoking 272
cigarette smoking *see* smoking
CIPAS *see* cumulative incidence proportion among survivors
civil commitment 555–61
Clark, Larry 419, *420*
Clarkes, Lincoln 419–21
"classic genetics" studies 277
clinical abnormality 169
clinical definitions 123–31
clinical interventions 407
clinical research 466
clinician training 459

Index

close relationships 440, 442–6
CM *see* contingency management
CNS *see* central nervous system
co-occurring disorders 177–8, 470
co-occurring manifestations of addiction 150–1
cocaine/cocaine use: administration methods 193, 197; age of use differences *111*; at the synapse 383; BDNF role in abuse 243–4; contingency management and 458; decision-making dysfunctions 345–6; high-affinity states 356; medication assisted treatment 481; placebo effects and 396; quitting 30–1; resting-state connectivity 371, 373; structural connectivity 373–4; white communities' use 512–13
cocaine-dependent (CD) individuals 307–8, 388
cocaine use disorder (CUD) *381*, 382, 383–7
coercive treatment 554–6, 561n1
cognitive aspects: cocaine use disorder 386–7; gambling 175–6
cognitive behavioral therapies (CBT) 224, 406, 469, 471
cognitive blocks 39
cognitive content, motivation 69
cognitive control network 372–3
cognitive deficits 28
cognitive failure 47
cognitive interventions 224, 377, 406, 469, 471
cognitive paradigms 325–6
cognitive wanting 355
"cohort" approaches 259–61
"cold turkey" 28–9
Coleridge, Samuel Taylor 210–11
colonization effects 503
color drug "showcards" 265, *265*
commercial interests 526–7
commercial models, legal markets 525
commitment, civil 555–60
commitments research 240–9
Committee on Drug Addiction (CDA) 241
common disease/common variant (CDCV) model 276, *277*
common usage, defining addiction 123
communities: power and 211–12; of recovery 87; support by 501; of use 79, 83, 85
community-based CM treatments 459, 460–1
community bonds, addicts 13, 18
community identification 78
competence criteria 558–9
compulsion 10–11; behavioral criteria 541; disease and 12–13; *DSM* accounts of 25, 26; as excuse 547; impulse distinction 43n2; intertemporal bargaining 39; learning networks 374
compulsory treatment 555, 561
computationalism 318
COMT *see* catechol-O-methyltransferase

CON see constraint-disinhibition
conceptual aspect, self model 406–7
conditioned responses (CRs) 186, 326–7, 392
conditioned taste aversion 107–8
connectivity in brain 362–79
conscientiousness construct 289
constellation symptoms 132–4, *133*, 135–6
constitutional criminal law 545–7
constraint-disinhibition (*CON*) 286, 288–9, 291
constructivist approach, self-change 490
consumer choice 29
consumption factors 39–40, *41*, 55–6, 58, 492, 526
context and cravings 357–9
contingency management (CM) 11, 15, 187–8, 387, 455–63, 481
continuants, BFO 168
continuous access, heroin 232–3
control: consumption of drugs 55; defining addiction 126, 129–30; degrees of 61n3; drug use 198–9; illusion of 175; impaired 135, 321; inhibitory systems 182; judgment shifts 59–60; lapses 57; loss of 38, 59–60, 61n3, 132–3, 196, 199, 386–7, 491, *see also* impulse control; self-control
"control interval" 262, 272
controlled drugs 519–20
controlled learning processes 325–38
"controlled retrospection" cases 259
coping mechanisms 245, 301
core symptoms 132–4, *133*, 135–9
Corlett, P. R. 57
corpus callosum 373–4
cortical brain regions 301–2, 304
cortico-striatal reward/motivation pathways 306–7
corticotrophin releasing factor (CRF) 302, 305
cortisol effects 305–6
costs/benefits, drug addiction 14–15, 19, 24
counter-narratives 418–19, 428
court-mandated treatment 555
courts of law 438, 550
covert arrogance 433, 437
crack cocaine epidemic 512
Craver, C. 317
cravings 10–11, 12; behavior/brain/environment link 137–9; BFO associations 168; biomarkers 141; characteristics 133–4; chronic drug effects 307–9; context and 357–9; cue reactivity 327; negative mood effects 332; neuroadaptations *309*, 310–11; reward system 365; self-control and 48–9, 52–3; theories of 161; treatment of 437
"creeping" reductionism 402, 407–8
CRF *see* corticotrophin releasing factor
crime control, drug treaties 520
criminal law 540–53, 556
criminal paternalism 532

criminal responsibility 533–4, 542–5
criminalization of drug use 513–14, 520, 531–9
cross-sectional surveys 264–72
cross-sensitization 357
CRs *see* conditioned responses
CUD *see* cocaine use disorder
cue-induced craving 307–8
cue reactivity theories 326–9
cues exposure, food addiction 182, 185–6
culpability-based-on-choice model 549
culpability-in-causing consideration 535, 538n16
cultural issues 494, 495
"culture of terror" 211
Cummins, R. 318–19
cumulative incidence proportion among survivors (CIPAS) 268–70
cumulative stressors 303
Cunningham, J. A. 493
cycle of addiction 506

"damage" 90, *see also* brain damage
DAT *see* dopamine transporter
DD *see* delay discounting
de la Fuente-Fernández, R. 394
dealers 229–30, 232–5, 237–8
deaths, opioid use 514
decision-making: under ambiguity 339, 344–6; causal knowledge 17; CM mechanisms 460; competence for 558; dysfunctions 339–50; paradigms **340–3**; under risk 339; temporally myopic 15–16; under uncertainty 343–4
decriminalization, drug use 531, *see also* criminalization; legalization of drugs
defence mechanisms, toxins 105–7, *107*, 108
defenses in criminal law 543–4, 547
defining "definition" 124
"degrees" of addictiveness 161
delay discounting (DD): appetite and 40; choice impulsivity 287–9, 291–3; decision-making and 339; heroin use 236; implications 36–7, *38*, 42n1
delayed associations 193, 195, 197, 200
delusions 317
demand association 330
demand reduction approaches 229
democratization of authority 432
denial 16–18, 551
Dennett, Daniel 35
dependence: biological processes 222–3; craving and 307–8; cue reactivity 327, 329; demand association 330; drug addiction 11, 24, 30–2; genes causing 283; goal-directed behavior 332; heroin injection 268; performance deficit 332; procedures 331
dependence syndromes 257, 260, 270–1
depletion effects 52
desert condition, punishment 543
desire/choice mechanisms/systems 92–3, 98

desires: action/agency 64; animal models 192; behavioral signs of 194; chimpanzee studies 195; evaluative theory of 68–70; inner struggle 72; inoculation of addiction 196; irresistibility of 10–12, 73–4; judgment-centered account 55; motivation and 66–7, 92–3; peremptory 90, 93; stress responses 307; as weaknesses 86
detention centers 556–7
detention systems 560
determinism 435–6, 491, 495, 544
deterrence, punishment as 534
devaluation of outcome **329**, 330–2, **331**
deviant behavior 490
deviant subculture theory 225
diacetylmorphine 416, 428
diachronic self-control 46
diachronous responsibility 551
diagnosis of addictions 132–44, 153–5
Diagnostic and Statistical Manual of Mental Disorders (DSM) 23–4; 5 134–7; 6+ 140–1; accounts of addiction 25, *26*; defining addiction 124–5, 127, 320; food addiction 184–5; mental disorders 134–5
dichotomy resolution, relationships 448–9
dietary tryptophan cases *254*
"difference markers" 407–8
"differential susceptibility" concept 293
diffusion-weighted imaging 362–3, 372–3, 376
dimensional approach, diagnoses 136
diminished attentional control 45
diminishing marginal returns rule 28
directed attention 51–2
discounting models: choice impulsivity 287–9, 291–3; drug choices 15; habit 39; heroin use 236; reward anticipation 35, *38*; self-control failure 47; "suicide option" 16, *see also* delay discounting
discouragement regimes, commercial interests 526–7
disease: addiction as 42, 78, 87, 221–2, 237, 240–1, 258; brain pathology 12–13; construction of term 169–70; criminal law rationale 545; policy implications 188; reduction of rates 253, *see also* brain disease; medical approach
disease model of addiction 401–10, 436, 490–1
disempowerment 209–10
dislocation 503–6, 507
dislocation theory 501–2
disorders: addiction as 30, 32, 45–53, 90–1, 315–24; definition 169; desire/choice mechanisms 92–3; neuroplasticity as 240; in normal brain 93–8, *see also* medical disorders; mental disorders; psychiatric disorders; substance use disorders
distal influences, syndrome development 153–4
divided motivational architecture 73

Index

Dixon, M. R. 173–4
DLPFC *see* dorsolateral prefrontal cortex
DoC *see* drugs of choice
documentary photography 415–30
domain-specific ontology 167
dopamine receptors *380*, 381–2, **384**, 394
dopamine release 355–6, 359, 385–6, 393, 395–6
dopamine responses: cravings 357–8; drug addiction 12–13, 47–8, 56–7, 60; gambling disorder 176; neuroadaptation 245; neurobiological context 148; pharmacological treatment 151; reward hypothesis 243, 365
dopamine signaling 380–91
dopamine system/pathway: brain disease model 402–3; genes 282–3; personality traits 290; placebo effects 395–6; reward and 394; role in brain 382–3; self model 405; stress effects 304, 305–7, 310
dopamine theory, reinforcement 103
dopamine transmission imaging 387–8
dopamine transporter (DAT) 383
dorsolateral prefrontal cortex (DLPFC) 345
dose-dependent effects 303
dose levels, addictive drugs 24
"dose-response" regularity, cancer 261
doxastic mediation 71
D.P.P. v. Majewski (1977) 548–9
Drinking: A Love Story (Knapp) 84–5
A Drinking Life (Hamill) 82
drive theory, cravings 161
drug addict–user distinction 23–4
drug addiction/abuse: behavioral concepts of 242–3; brain disease model 403–4; control and 60; dopamine receptor genes 282; features of choice 23–33; "gentler" approach to 512; illegal markets 229; impaired self-control 47–9, 53; judgment-centered account 56–7; natural history of 24; negative consequences 9–10, 14, 16–18; orthodox conceptions of 9–22; stress/reward pathways 306–9, *see also* prescription drug addiction
drug-associated stimuli 51–2
drug control *see* control . . .
drug courts 438, 550
drug criminalization *see* criminalization
drug dealers *see* dealers
drug euphoria 352
drug functions: pleasure and 13–14; "self-medication" hypothesis 15
drug genre photography 418–21
drug injection 226, *see also* self-administration
drug laws *see* legal . . .; legislation
drug legalization 518–30
drug-modifiable personality traits 286–98
drug-oriented values 50
drug policy *see* policy . . .
drug rewards *see* reward . . .

drug seeking behavior 386
drug self-administration 193, 196–8, 200–1, 268, 386, **386**
drug toxicity, reward over 102–19
drug use: age differences 109–12, *113*; agent–host–environment models 255; altering stress/reward pathways 306–9; brain disease model 403–4; control of 198–9; decreasing with treatments 481; defining drugs 221; objects of addiction and 146; persistence 263–4; "problematic use" and 225; recreational 505; sex differences 112–14; social gradients 211–12; social identity 225–6; US responses to 513; vulnerability increase 303–4
drug user–addict distinction 23–4
"drugged driving" laws 525
drugs of choice (DoC) 401, 405–7
drunkenness 431, 434–5, *see also* alcohol . . .
Dry (Burroughs) 80–1
DSM *see Diagnostic and Statistical Manual of Mental Disorders*
du Plessis, G. P. 163, *164*
duress as excuse 535
"Dutch model" 523–4
dysfunction analysis 90–101
dysfunctional decision-making 339–50

Earleywine, M. 396–7
early-stage interventions 560
eating behavior disorder 134–5, *see also* food addiction
ecological interpretations 318, 404–5, 408
Ecological Momentary Assessment (EMA) 138
ED ("effort discounting") 288
effective agency 441–2
effectiveness, twelve-step organizations 466–8
efficacy, twelve-step organizations 466, 468–70
"effort discounting" (ED) 288
Elster, J. 320–1
EMA (Ecological Momentary Assessment) 138
embodiment, identity and 226
emotions: relationships and 441; simplified model 64; stress and 300–3
empathy 69
EMS *see* eosinophilia-myalgia syndrome
endogenous rewards 41–2
endophenotypes 286, 290–4, *292*
English law 548, *see also* Anglo-American criminal law
entities concept 168
environmental conditions 96–9, 137–9
environmental non-specificity 148
eosinophilia-myalgia syndrome (EMS) 254, *254*
epidemiological methods/models 253–74
epigenetics 284
epistemic authority 431–9
epistemic injustice 436–7

epistemic virtue 432–4
Erdős, Paul 129–30
error prediction 48, 57–8
escalating rewards technique 456
escape-based functions, disorders 174–5
escape clauses, resolutions 58–9
escape from withdrawal 352
ethics 63, 427–8
ethnographic research 229, 238n1
euphemistic language 511–17
euphoria 352, *see also* "highs", experience of; pleasure
evaluative theory, desire 68–70
Everitt, B. J. 322
evidence-based approaches *see* contingency management
evolution-based analysis 91–3, 95–6, 99, 102–19
excess, definition 132–3
excessive "wanting" 353–4
excuse: causal theory of 544; compulsion as 547; in criminal law 543, 550; criminalization versus 531–7
executive control network *364*, 372–4
"exome sequencing" 280–1, *281*
expectable environmental conditions 96–9
expectations, placebo effects 392, 394–7
experimental approaches 263, 321
explanations, vision of 318
exponential discounting 35–6
exposure-discordant drivers 262–3
extended simple models 64, 66
extra-medical drug use 266, *266*
extraversion trait 289

FA *see* food addiction; fractional anisotropy
faceless drug user trope 419–20
failure model 407, 447
Fairman, B. J. 263, 272
fallibility, self-control 70–2
familial aggregation reports 276
family association studies 279, *279*, 280
family-history-positive, alcohol dependence 384–5
family members/relationships: power framework use 216; reactive attitudes 442–4, 450n2–3; recovery role 494; subordinated by power 210–11
fatalistic stance 447
FCD (Functional Connectivity Density) mapping 365
female . . . *see* women
female smokers, dependence 271
fentanyl use 514–15
first-order desires 66–8, 72
first person narratives 432
Fischer, J. M. 59
fitness benefits, toxin intake 114–15

Flanagan, O. 407
flawed beliefs 175
fluoxetine responses 395
folk psychology 9, 14, 542
followup approaches 259–61, 268–71
food addiction (FA) 182–91, 358
food craving criteria 139
food neophobia 110
food-specific attack rates 269
for-profit commercial models 525
forgiveness 445–6
Foucault, Michel 418
fractional anisotropy (FA) 371–6
fragmentation *502*, 503–4, 506–7
framing choices 26–9
Frankfurt, Harry 66–7, 72
free will 69, 544
freedom, loss of 532–7
Freud, Sigmund 513
friendships 442–4, 446, 450n2–3
frontoparietal control network 372–3
functional accounts: cocaine use disorder 385–7; deviant subculture theory 225; mechanisms 319–20
functional connectivity 362, 371, 374–5
Functional Connectivity Density (FCD) mapping 365
functional hierarchies 96–7
functioning–function differentiation 168–9
"fundamental attribution error" 222
funding barriers, CM treatment 459

gambler's fallacy 175
gambling addiction 40–2, 211–12, 505
gambling disorder 173–81; criteria 134–5, 136–8, 173; incentive sensitization 359; shadow syndromes 151
gambling industry, power of 213, 215
Garcia, J. L. A. 433–5, 438
"garden path" to consumption 56, 58
gender differences 83–4, 470, 472
gene-by-environment interactions 256, 263, 293
gene ontology 160
gene types: decision-making dysfunctions 345; dopamine pathway 282–3; tobacco addiction 281–2
general deterrence, punishment as 534
general models/theories of addiction 160–72
generalist cultures 494
generalizations 16–17
generalized linear modeling (GZLM) 259
genetic analysis, strategies/methods 276–81
genetic factors: food addiction 358; gambling 176; personality traits 286–98
genetic framework 275–85
genetic linkage analysis 277–8, *279*
genetic non-specificity 148

Genome-Wide Association Study (GWAS) 280–1, 283–4, 290–1
genotypes 278–9
"gentler war on drugs" 511–17
global bookkeeping 25–30
global society, problems of 501–10
goal-directed behavior 198, 330–3
goal-tracking 186
Goldberger, Joseph 253
Goldstein, A. 320
Goodman, Aaron, works of *424–6*
government monopolies 522–5
gray matter 177
group efforts, importance of 507
group identification 80–3, 84
guilt 86
gut toxin processing 107–8
Guze, Samuel 94
GWAS *see* Genome-Wide Association Study
GZLM (generalized linear modeling) 259

habit 39–40, 330–2, 352
habit learning 374
Hadley, S. H. 434–5, 437–8
Haenszel, W. 260–1
Hamill, Pete 82
harm: cannabis potential for 526; competence assessment 558; consequences of addiction 506; defining addiction 126, 129–30, 131n2, 321; reducing 187, 527
harm minimization (HM) 212–13, 478–89, **480**
harmful dysfunction analysis 90–101
Harrison Narcotics Tax Act 513
Harvey, W. 316
HAT *see* heroin-assisted treatment
"hazard intervals" 262
health and culture 494
healthcare 482
hedonic importance, rewards 41–2
helplessness, definition 39
heritability, personality traits 289
heritability component, genetics 276
hermeneutical injustice 436
heroin-assisted treatment (HAT) 415–16, 427–8, 437, 479, 483
"heroin crisis", United States 511–12
heroin markets 229–39
heroin use: case-control approaches 266, 268; decision-making dysfunctions 346–7; dependence syndromes 270; fentanyl dangers 515; humanizing users 415–30; neuroscientific research 241; OST/healthcare 482; US figures 514; withdrawal 353
heterogeneous histories 240–9
heterotrophs 104
Heyman, Gene M. 258
hierarchical instrumental learning 329

hierarchical structure, motivation 66–7
high-affinity states 356
high-frequency reinforcers 456
high-magnitude rewards 456
high uncontrollable stress 302–3
"highs", experience of 396, *see also* euphoria
hijack hypothesis 98, 102–4, 108–9
Hikosaka, O. 288
Hine, Lewis 418
history type entities 168
HM *see* harm minimization
holistically oriented care 481–2
hormone dysfunctions 98
host characteristics, choice and 255
hot thinking 34–5
Houston, Kerr 419
HPA axis *see* hypothalamic-pituitary-adrenal axis
human rights abuses 557
humanization, heroin users 415–30
humility 432–5
hyperbolic discounting 15, 36–40, *38*, 42n1, 47, 70–1
hypersensitivity 357–9
hypodermic injections 116n3, 192–4
"hypophoria" 242
hypothalamic-pituitary-adrenal (HPA) axis 245, 302, 305–7

ICD *see International Statistical Classification of Diseases and Related Health Problems*
"ideal type" approach 63–4, 66–7
ideals–norms mismatch 77
identification and identity 77–9, 87
identity 65–70, 77–89, 225–6, 441–3
identity-constitutive addiction 78, 87
identity gain 18
identity loss 18, 19–20n4
ignorance and addiction 40–2
IGT *see* Iowa Gambling Task
illicit/illegal drugs: distinctions within class 531; market boundaries 229–39; photography use 427; quitting 31; trading 213, 521; use patterns 24; vested interests 225
illusion of control 175
imaging studies 380–91, *see also* neuroimaging
immediacy of outcomes 70, 456
impaired control 47–50, 53, 135, 321
imprinting systems 95–8
impulse–compulsion distinction 43n2
impulse control 45, 286
impulsivity 174, 286–98, *see also* reflection impulsivity
incarceration of addicts 257, 483
incentive learning 332–3
incentive salience 322, 353–6, 365, 382
incentive sensitization 161, 185–6, 351–61
Incentive-Sensitization Theory 353–5

incentives, compulsion and 11, *see also* reinforcers; reward . . .
"incidence density" 269–70
incompetence claim 557–8
Indian drugs legalization 523
individual factors: dislocation effects 503–4; gambling disorder 174; power 209–19
individualism 493
infants 110
inhibitory control systems 182
injection drug users (IUDs) 483
injections *see* drug self-administration; hypodermic injections
injustice 436–7, 448
inner struggle 70–4
insanity defense 543, 547
instances, entity class 168
instrumental learning 329, 330–2
instrumental transfer procedure 328–9
integrated physical healthcare 482
integrated psychiatric care 482
"intellectual republic" 432, 435–8
intention/intentionality 55, 64, 72, 542–3
interests: intertemporal bargaining 39; reward paths 36–7
interference procedures 326
international controls 519–22
International Statistical Classification of Diseases and Related Health Problems (ICD) 134, 140–1
internet drug purchases 229, 238n5
intersubjective dimension, self model 405–6
intertemporal bargaining 38, 42
interventionist accounts 407–8
intoxication in criminal law 547–9, 551
intravenous (IV) drug administration 197, 200–1
intrinsic errors, judgment shifts 72
involuntariness concept 545, 547
involuntary intoxication 549
Iowa Gambling Task (IGT) 343–6
irrational "wanting" 354
irrationality of addicts 71, 550
irresistibility 10–11, 12, 73–4
IUDs *see* injection drug users
IV (intravenous) administration 197, 200–1

Jack Barleycorn (London) 82–3
James, William 431–2, 434–7
judgment-based self-control 45–8, 50–1
judgment-centered accounts 55–9
judgment shifts 46–7, 49, 52, 74; control and 59–60; lapses 56, 57; weak will 70–2
justification of punishment 537
juvenile intervention cases 560

Kalivas, P. W. 246
Kandel, Eric 94–5
kappa opioid receptors (KOR) 256

Karkowski, L. M. 148
Kennedy, Charles 211
"kicking the habit" 28–9
Kim, H. F. 288
King, M. P. 491
Knapp, Caroline 84–5
knowledge 16–17, 431–9
Koch, Robert 253
Koob, George 408n2
KOR (kappa opioid receptors) 256
Kourounis, G. 479
Kraipelin, Emil 94, *see also* neo-Kraepelinianism
Kuhn, T. S. 495

labelling 18, 536
lapses 54–5, 57, 59–60, 72
large-scale generalizations 16–17
larger, later (LL) rewards 35–7, *38*, 40
Latin American drugs trade 521
Latino communities 514
law *see* legal . . .; legislation
law of effect 102
LD *see* linkage disequilibrium
learned habits 352
learning: associative 353; motivation and 69; placebo effects 392; toxin avoidance 110
learning networks *364*, 374–5
learning paradigms 325–38
"left-censoring/truncation" bias 257, 268
legal addictive drugs, use patterns 24
legal insanity 543, 547
legal markets 524–6
legal trade in addiction 213
legalization of drugs 518–30, *see also* decriminalization
legislation: cannabis harm potential 526; criminal law 540–53; excuses and 534; "heroin crisis" 512; opioid use 514, 516; overinclusive laws 533; paternalistic 532; psychoactive substances 518; tax on drugs 513
levels and mechanisms 317–18
Levy, Neil 47–9, 71
Lewis, Marc 443
licenses to use, societal 79–82
licensing systems, alcohol 523
Lidstone, S. 395
life plan 169, 210
lifetime frequency, tobacco smoking 260
lifetime prevalence, epidemiology 269–70
"liking" 186, 322, 354, 358
Lilienfeld, A. M. 263
limbic networks 301–4, 321–2
limited disorder, addiction as 30, 32
Lindesmith, Alfred 192–3, 195
lineage communities 79, 83, 85
linkage analysis 277–8, *279*
linkage disequilibrium (LD) 278

liquor licensing systems 523
LL rewards *see* larger, later rewards
local bookkeeping 25, 26–30
local markets, heroin 229–39
Logan, Frank 198–9
London, Jack 82–3
long-term interests, rewards 36–7
long-term potentiation (LTP) 244
longitudinal studies 467–8
Looby, A. 396–7
Lorenz, Konrad 95–6
"loss of control": animal models 196, 199; cocaine use disorder 386–7; excess and 132–3; intertemporal bargaining 38; lapses 59–60; mechanisms of 61n3; self-change and 491
loss of freedom 532–7
"losses disguised as wins" 174
LTP (long-term potentiation) 244
lumpen abuse 222, 224–5
lung cancer 259, 261, 281

McAuliffe, W. E. 353
magnetic resonance imaging (MRI) 364, 372, 373
Majewski rule 548–9
maladaptive behaviors 301
male bias, drug use 112, *113*
management techniques 222
mandatory treatment 554–63
manliness 82–3
Mantel-Haenszel estimator **261**
marijuana *see* cannabis/marijuana
market access, heroin 231–2
market position and drug control 519
market system 229–39, 503, 524–7
marketplace model 35–6
Marr, D. 318
Martin, William R. 241–2
MAT *see* medication assisted treatment
"matching law" 29
Maté, Gabor 427
mathematical modeling 263–4
"maturing out" hypothesis 11, 15, 256–8, 493–4
mechanisms: behavior change studies 472; definition 316–17; functional accounts 319–20; levels and 317–18; reward system as 321–3
mechanistic models 315–24
media dramatization 506, 515
mediation model, TSMHO outcomes 471–2
medical approach/model 65, 315, 319, 501, *see also* disease
medical disorders 90–5
medication assisted treatment (MAT) 438, 481
medications: cannabis as 519; craving and 139
Mele, Alfred 46
Mental Capacity Act (2005) 556
mental disorders 93–8, 134–5, 140, 170, *see also* brain disorders; psychiatric disorders

mental health problems 14, 15, 558–9
mental hygiene 241
mental states 542, 548
mesolimbic dopamine 306–7, 355–6, 394–5, 402–3, 405, *see also* reward system
mesolimbic systems 351, 358
MET *see* motivational enhancement therapy
methadone programs/use 346–7, 416, 458, 481, 512, 515
methamphetamine use 345, 375, 387–8
methylphenidate *381*, 382, 396–7
MHOs (mutual-help organizations) 464–77
midbrain system 56–7, 60, 321–2, 382
mild substance use disorders 124–5
Milner, P. 102
mind–brain distinction 94
mind models 72–4, 90–101
misery and addiction 15
"misery thresholds" concept 435
mitigation of sentencing 536–7, 549–50
M'Naghten's Case 543
models of addiction: challenges 160–3; clarity/diversity 160–3
moderate substance use disorders 124–5
modernity, problems of 501–10
molecular targets 243–6
money: availability of 231–3; as reinforcer 386
monkey studies 193, 195
monopolistic regimes 522–5
Montana v. Egelhoff (1996) 548
mood disorders 303
moral expectations 441–2
moral panics 220
moral psychology 63–76
moral visibility 441
moralistic paradigm 501, 540–1
morphine use/addiction 128–9, 194–6, 199–200, 241
motivation: affect/value in 67–9; agency and 63; cocaine use disorder 386; cognitive blocks of 39; desire/choice mechanisms 92–3; disease of 42; as function 168–9; gambling disorder 174; hyperbolic delay discounting 37; ignorance and 40; moral psychology and 69–70, 73; processes 161; self-administration of drugs 197; self-control and 45–6, 49; shame effects 86; stress effects 304, *309*, 310–11; structure/hierarchy in 66–7; values/identity and 65–70
motivational disorders 93
motivational enhancement therapy (MET) 469
motivational pathways 302–3, 306–7
motivational theory 163
MRI *see* magnetic resonance imaging
MRV (multiple rare variants) model 276, *277*
Muller, Christian 13
multi-dimensionality, impulsivity 287–8
multiple commitments, research 240–9

multiple rare variants (MRV) model 276, *277*
multitudinous self view 404–6, 408
mutual-help organizations (MHOs) 464–77
mutual shaping, relationships 448
myopic processes 34–5
myth of addiction 222–3

NA *see* Narcotics Anonymous
naloxone treatment 483, 515
Narcotics Anonymous (NA) 464, 468, 470, 472
narratives 418–19, 428, 432
National Comorbidity Survey (NCS) 270
National Epidemiologic Survey on Alcohol and Related Conditions (NESARC) 270
National Institute of Mental Health (NIMH) 403
national-level drug control 519–20
National Surveys on Drug Use and Health (NSDUH) 264–6, 272
natural functions 92
natural histories 150
natural rewards 403
naturalistic studies 467–8
nausea 107–8
NCS (National Comorbidity Survey) 270
negative consequences: change over time 24; denial 16–18; self-hatred/-harm 14; significance of 9–10
negative emotionality/neuroticism (*NEM/N*) 286, 289
negative mood 332–3
negative self-evaluations 445
negative social gradients 211–12
NEM/N see negative emotionality/neuroticism
neo-Kraepelinianism 94, 96–7
neophobic food rejection 110
NESARC (National Epidemiologic Survey on Alcohol and Related Conditions) 270
Netherlands, cannabis legalization 523–4
neural correlates, impulsivity 287–8
neural pathways 301–2
neuroadaptations 245, 308, *309*, 310–11, 355–6
neurobiological disease 12–13
neurobiological factors: addiction syndrome 147–9; placebo effects 392–400; stress 302–3
neurobiological theories 102–3
neurocognitive manifestation, addiction syndrome 149
neurocognitive research **340–3**
neuroimaging 141, 362–79, *see also* imaging studies
neuroinflammation effects 376
neuropharmacology 241
neuroplasticity 240, 244–6
neuroscientific definitions 125, 127–30
neuroscientific research 240–9, 315, 320, 408, 436–7
neuroticism trait 289
neurotoxin regulation model *115*, 116

neurotoxins 105–6, 109–12, 114–16, *see also* nicotine
neurotransmitter systems 176, 381
"new psychoactive substances" (NPSs) 520
Next-Generation Sequencing (NGS) 280
nicotine 104–5, 108–10, 115, 375–6, 397, *see also* smoking
nicotinic receptor genes 281–2
NIMH *see* National Institute of Mental Health
non-addictive drugs 531–2
non-specificity 147–8, 150–1
non-toxic activities 27
nonhuman animals *see* animal models/research
nonparametric linkage analysis 277–8
normal brains, dysfunction 90–101
normative failures 407
normative framework, treatment 556
norms 18, 77, 79, 85, 87
NPSs ("new psychoactive substances") 520
NSDUH *see* National Surveys on Drug Use and Health
nucleus accumbens 394–5

obesity 185, 187–8, 358
object non-specificity 150
objective diagnosis methods 155
objective stance 441–3, 446–9, 456
objects of addiction 146, 151, 153
OBO (Open Biological and Biomedical Ontologies) 167
O'Brien, Charles P. 245–6
occasions 40–2, *see also* "special occasions"
occurrents, BFO 168
odds ratio (OR) 258–9, 262, 279
OGMS *see* Ontology for General Medical Science
Olds, J. 102
ontologies 160, 165–71
Ontology for General Medical Science (OGMS) 167, 169–70
Open Biological and Biomedical Ontologies (OBO) 167
openness trait 289
operant conditioning 455
opiate epidemics 223
opioid dependence 271, 283
opioid receptor antagonists 176
opioid substitution treatment (OST) 478–89, **479**, 480–1
opioid system 393
opioid use: AHE models 255–6; alcohol use and 150; decision-making dysfunctions 346–7; dependence syndrome 271; "gentler" approach to 511; humanizing users 415; law enforcement 516; overdose crisis 428, 437; transitioning risk **267**; US current response to 514–16, *see also* heroin use
opioid use disorders (OUD) 478–89

"opium problem" 241
Oppenheim, P. 317
OR *see* odds ratio
oral drug administration 199–200
orbitofrontal cortex 343–7
orthogonality trait 288
Osborne, Lawrence 81
OST *see* opioid substitution treatment
OUD *see* opioid use disorders
Outcasts Project 415–30
outcome devaluation **329**, 330–2, **331**
overdosing, opioid use 428, 437, 481, 483, 514
overinclusive rationale 533
overlearning, habit and 39, 352
OWL ontology 166–7, 170

pain inhibition 394
pain medication 128–9, 266, 271, 346–7, 415
pain responses 393, 482
paradigm cases, defining addiction 127
parametric linkage analysis 277–8
paraphilic disorders 92
parasites, defence against 115
parental failure, feelings of 505–6
parietal executive control region 373
Parkinson's disease 394–5
participant observation 421
participant stance 441–6, 448–9
paternalism 532, 561
pathological conditions 558
pathological gambling *see* gambling disorder
pathology, drug addiction 12–13
Pavlovian conditioning 109, 186, 326–7
Pavlovian to instrumental transfer (PIT) 328–9, **328–9**
PDMP *see* prescription drug monitoring program
peer relationships 229–32, 235–6
Pellens, M. 241
PEM/E *see* positive emotionality/extraversion
penal laws 533–4
peremptory desires 90, 93
performance deficit 332
permissions to use 80–3
persistent use 10, 263–4
"person" concept, criminal law 542–3
person-specific factors *139*, *see also* self model of addiction
personal rules 38–9, 41
personality disorders 178
personality effect, drug abuse 243
personality traits 286–98, *292*, 343, 344
personalized care 482
pesticides evolution 105, 110–12, 114–16
PET *see* Positron Emission Tomography
PFC *see* prefrontal cortex
pharmaco-therapy 139
pharmacological adaptation 136

pharmacological substitution 478–89
pharmacological treatment 151, 436
pharmacophagy hypothesis 115
phasic dopamine 56–7
phenomena: defining addiction 320–1; mechanistic explanations 316–17
phenotypes 275, 278–9
philosophical definitions 126–7, 129–31, 316
photo-elicitation 415–30
photographic methods 421, *see also* documentary photography
physiological pathways 301–2
Pickard, H. 54, 88n1, 88n5, 321, 449
picoeconomics 34–44
"pinching" 234
Pink, Sarah 418–19, 421
PIT (Pavlovian to instrumental transfer) 328–9
placebo analgesia 393–4
placebo effects 392–400
plant drugs 104–6, 109–10, 114, *see also* tobacco
plasticity genes 293
pleasure 13–14, 321–2, *see also* euphoria
policy implications: food addiction 186–8; "gentler war on drugs" 511; racial differentiation 512
political barriers, treatment 460
political conditions, detention systems 560
poly-substance abuse 372, 376
polymorphisms 283–4, 290
"population science" 253–8
population surveys 264–72
positive emotionality/extraversion (*PEM/E*) 286, 289
positive reinforcement 197, 327
Positron Emission Tomography (PET) 380–5, 387
post-traumatic stress disorder (PTSD) 244–5
poverty, heroin and 231
Powell v. Texas (1968) 545–7
power 209–19
powerlessness 71, 214–16
PPR *see* prescription pain relievers
pragmatic approach, defining addiction 123–31
prediction error minimization 48, 57, 58–9
predictors: dopamine responses 56–7; overlearning 352; quitting drugs 30–1, *see also* self-prediction
preference dynamics 25
prefrontal cortex (PFC) 304, 322, 387
prefrontal regions, stress 302, 304, 308
pregnancy 112–13, 483
premorbid deficits 347–8
prescription drug addiction 78, 88n1
prescription drug monitoring program (PDMP) 481
prescription medications 515, 519
prescription pain relievers (PPR) 266, 271
"prevalence" of cases 270
prevention efforts, syndrome model 154
pricing drugs 234
prison sentences 512, 516, *see also* incarceration

private aspects, self model 406
prize-based CM 455–6, 458–9
"problematic use" 225
process capture, addiction 161, **162–3**
professional services 464, 467
prohibitions 518–19, 521–4, 532–4
Project MATCH 469, 471–2
proportionality principle 537
proximal influences, syndromes 153–4
psychiatric disorders 29, 177, 470, 482, *see also* mental disorders
psychiatry and reductionism 402, 407–8
psychoactive substances 108, 149, 518–30
psychological contents, brain disorders 94
psychological dysfunction 96–7
psychological implications: adaptation 505; addictive drugs 24; dislocation 503–4, 507
psychological self-control 49–51
psychological theory 9, 15, 542
psychological visibility 440–1
psychologically describable functions 98
psychology: moral 63–76; self-hatred/-harm 14; social 79
psychomotor sensitization 356
psychopathology 148–9, 241
psychosis 259–61
psychosocial factors 147–9
PTSD *see* post-traumatic stress disorder
public health issues 188, 215, 224, 253–4, 518–30
punishment 534, 536, 537, 543
purity of drugs 108
Putnam, H. 317

quitting drugs 30–2, 129–30

racial differentiation 512
racism 513, 515
radial diffusivity (RD) 372, 374
radioligand imaging 380–1, 383–5
radiotracers 380, 381, *381*
randomized studies 262
Rat Park experiment 199–201
rat studies 197, 199
rational processes 34, 58–9, 65, 93
rationing system, legal drugs 526
Ravizza, M. 59
RD *see* radial diffusivity
RDF *see* Resource Description Framework
RDoC (Research Domain Criteria) initiative 134, 403
reactive attitudes 440–51
"reactive measures" problem 271
"real self" concept 85–6
reason/reasoning 52–3, 59, 533
receptor systems, molecular 243
recidivism 556

recovery: communities of 87; cultural issues 494; definition 491; "intellectual republic" of 435–8; objective stance 447; responsibility for 449; TSMHOs and 464–77
recovery support groups 18–19
recreational drugs 105–6, 108–9, *115*, 116, 505, 524
recursive processes 41–2
recursive self-prediction 37
reductionism 229, 238, 401–3, 407–8
reflection impulsivity 339, 344, 346
regulatory mechanisms 73, 114–16
regulatory regimes 522–6
reinforcement concept 102–3, 197, *see also* reward . . .
reinforcement process 174, 198
reinforcers, contingency management 455–7
relapse: abstinence and 54, 128; choice frames 26–9; cravings link 138–9; incentive sensitization 351–3; OST reducing 483; risk of 307–9
relationships, reactive attitudes 440–51
relative risk (RR) 258–61, 272, 279
relativism 126–8
religion 471–2, 492
remission 26–9, 30, *31*, 469, 560
representation content, desire 68–9
representation-update system 57
representational states 58
Research Domain Criteria (RDoC) initiative 134, 403
research integration 216
resentment 444, 446, 449
resilient coping 301
resistance: desires 10; weakness of 211–12
resolutions 58–9
"resolving power" concept 256
Resource Description Framework (RDF) 165–6
responsibility 99; criminal 533–4, 542–5; determinism and 544; diachronous 551; family members 211; intentional actions 64; resentment/blame 449; selective attribution 214–15; treatments 216
responsible drinking concept 223
resting-state brain connectivity 362–79, *364*, **366–8**
retention improvement, treatment 480–1
"reverse genetics" 276–7
reward anticipation: brain pathology 12; consumption patterns *41*; discounting role 15; dysfunctions 98; intertemporal bargaining 38, 42; marketplace model 35–7
reward-based learning 102–19
reward-based treatments 11, 15, 455–63
reward circuits, brain 198, 305–6
reward contingencies 29
reward cues 13, 47–8, 58, 359
"reward deficiency syndrome" 147
reward versus drug toxicity 102–19
reward hypothesis, dopamine response 243

reward motivation 42
reward paradox 109
reward pathways, drugs altering 306–9
reward prediction systems 56–7
reward-seeking behavior 186
reward system/mechanisms 321–3, *364*, 365–71; baseline state of brain 375; brain disease model 402–3; dopamine imaging 382, 386, 388; incentive salience 354; placebo effects 392–5, 396; resting-state connectivity 371, 374; structural connectivity 371–2, *see also* mesolimbic dopamine
reward theories/models 40, 161–2
reward valuation, ventral striatum 288
Richard, Eugene 419
Riis, Jacob 418
risk factors: dislocation theory 502; opioid use **267**; pre-existing 135–6; psychiatry 407–8; relapse 307–9; social 149; stress 303–6; structural theories 224, *see also* relative risk
risk-taking 339
Robbins, T. W. 322
Roberts, B. W. 289
Roberts, R. 433–4
Robinson, T. 322
Robinson v. California (1962) 545, 547
Rockefeller laws 512
Ross, D. 322–3
Roth, M. 319
RR *see* relative risk

S-S (stimulus-stimulus) structure 326–7
salience network *364*, 365–72, 374–5, *see also* incentive salience
SALOME *see* Study to Assess Long-term Opioid Maintenance Effectiveness
Satel, Sally 556, 561n3
Schaffner, Ken 401–2
Schonberg, J. 224–5, 423
Schultz, Wolfram 47–8
Schumann, Gunter 13
science of addiction 160–72, 241–2, 401–2
Science and Technology Studies (STS) 240
scientific definitions 123–31
SDT *see* social dominance theory
Seabrooke, T. 329
"Second Chance" program 484
second-order desires 66–8
Seevers, M. H. 193
selective attribution 214–15
selective prohibition 532–4
self: "real self" concept 85–6; stability of 46
self-abasement 433–4
self-administration 193, 196–8, 200–1, 306–7, 386, **386**
self-categorizing 18

self-change 490–8
self-concepts 67, 406–7, 443–6
self-control 543–5; animal models 199; definition 45, 51; disorder of 45–53; executive control network 372; failures of 46–7, 51; inner struggle 70–4; motivation and 63; stress and 310–11
self-destructive behavior 197–8, 200–1
self-forgiveness 445
self-harm/-hatred 14
self-identity 13, 18–19
self-injection *see* self-administration
self-knowledge 432, 435, 440–1
self-labelling 18
self-medication hypothesis 15, 332–3
self model of addiction 401–2, 404–8
self-prediction, reward and 37–8
"self-regulation" *see* self-control
"self-serving bias" 222
"self-signaling" theory 38
self-visibility 442, 446
seller–buyer interactions 230–5
Semantic Web 160, 163–6, 170
sensitization 307, 322, 356–9, *see also* incentive sensitization
sensory cues, smoking 397
sentencing mitigation 536–7, 549–50
sequencing genomes 280–1
serotonin 345
Severe Substance Dependence Act (2010) 557
severe substance use disorders 125
severity of addiction 161, 443
severity assessment diagnosis 140
sex differences 112–14, *113*, 308
shadow syndromes 150–1
shame 85–6, 445
shared experiences 149–51
shared psychosocial antecedents 148–9
short-term interests 36, 39
sign-tracking 186
signs, syndromes 146–7, *see also* symptoms
simplified models 64–5, 66, *139*
single nucleotide polymorphisms (SNPs) 278, 281, 283, 290–1
singularity, betting occasions 41–2
Sinnott-Armstrong, W. 321
small-scale generalizations 17
smaller, sooner (SS) rewards 35–7, *38*, 40
Smith, Adam 234
smoking: automatic processes 326; causal knowledge 16–17; incentive learning 332–3; placebos and 397; regulatory mechanisms 115; research examples 258–64; self-control and 48–51; sex differences *113*, 114; sociology of 220; temporal profile 24, *see also* nicotine; tobacco
Snow, John 253, 255

Index

SNPs *see* single nucleotide polymorphisms
social behavior 237, 490–2
social communion, licenses to use 79–82
social decision-making 339–40, 345–6
social dominance theory (SDT) 211, 384
"social drinking" 405
social gradients 211–12
social identity 18, 225–6, 441
social impairment diagnosis 135
social isolation effects 200
social learning 110
social networks 471–2
social psychology 79
social relations, brokers 235
social responsibility 99, 212–13
social risk factors 149
social roles: photo-fieldwork 421; self-knowledge 432, 435
societal fragmentation 503–4, 506
societal problems, theories of 501–10, *502*
socio-economic opportunities 15
sociology 220–8
"sovereign reason" approach 65
spatial relations 316
"special occasions" 27, 28, 58–9
specialist cultures 494
spirituality 471–2, 494
spontaneity 72–3
spontaneous remission 560
Spragg, Sidney 192–7, 199, 201
SRAD (substance-related and addictive disorders) 185
Sripada, C. 72–3
SS rewards *see* smaller, sooner rewards
stability trait 288
state interest, criminalization 537
status rationale, disease 545
Sterelny, K. 318
stereotypes 436
stigma/stigmatization 226, 421, 423, 441, 493
stimulant use/addiction 345–6, 362–79, *364*, **366–8**, 458, *see also* cocaine/cocaine use
stimulus-bound perspective 52–3
stimulus-stimulus (S-S) structure 326–7
storytelling 431–9
Strawson, Peter 440–2, 447
strength-based approaches 492–3
stress 53, 299–311, 357
stress responses 300–2, 307–9
stressors 300, 309
striatal networks/compartments 287–8, 301–4
striatum 382, **386**, 387
Stroop task 51, 326
structural barriers, treatment 479, 481–2
structural brain connectivity 362–79, **369–70**
structural theories 224–5, 490–1, 501–10
STS *see* Science and Technology Studies

Study to Assess Long-term Opioid Maintenance Effectiveness (SALOME) 415–17, 422–3, 425, 427–8
subculture deviance 225
"subject-as-own-control" logic 262–3
"subjugated knowledges" 418
subordination by power 210–11
substance addiction: analysis 91, 99; diagnosis 132, 140–1, *see also* drug addiction
substance-induced deficits 347–8
substance-related and addictive disorders (SRAD) 185
substance use, sociology of 221
substance use disorders (SUD): choice impulsivity 286–7; CM treatment 455–63; co-occurring disorders 177; definition 124–6; diagnostic criteria 134–5, 137, *139*; DSM account 320; endophenotypes 290–4, *292*; food addiction as 183–4; genetic framework 275; impaired control 135; risky use 135–6; TSMHOs recovery programs 464–77
substantive theories 492, 533
subversive implications, powerlessness 215–16
SUD *see* substance use disorders
sugar, definition 123–4
"suicide option" 16
supply chains, illicit drugs 213
supply reduction approaches 229
survival mechanisms 182–3
"sweeping" reductionism 402–3
sympathetic nervous system 302
symptoms: core/constellation 132–4, *133*, 135–6; defining addiction 124–5; syndromes manifestation 146–7
synapse, cocaine at 383
syndromes: addiction as 145–59; definition 146–7
systems of detention 560

taste aversion, toxins 107–8
taste receptors 105–6, 110, 114, 116
Tatum, A. L. 193
tax legislation 513
TCD (tobacco cigarette dependence) 271
TDT (transmission disequilibrium test) 280
Tellegen's personality traits 291
temporal profile, costs/benefits 24
temporally extended aspect, self model 406
temporally myopic decision-making 15–16
Terry, C. E. 241
test criteria, treatment 559
testimonial injustice 436
testing frequency, contingency management 456
theft of drugs 236
"theories", use of term 161
therapies, limitations 39, *see also* treatment . . .
Thomas, Dylan 211

Thorndike, E. L. 102
tissue damage, brain 90, 92–3, 98–9
tobacco: addictive gene types 281–2; age of use differences *111*, *113*; cigar smoking 272; dependence syndromes 270; goal-directed behavior 331; international treaties 520; research examples 258–64; sex use differences *113*; simulation studies 264; toxic effects 104–5, 108, 110, *see also* nicotine; smoking
tobacco cigarette dependence (TCD) 271
tolerance 136, 153
tonic dopamine 365
"top-down computationalism" 318
toxic activities, choice/value 27
toxicity of drugs 102–19
trade: addictive products 212–13; illegal drugs 230
trade treaties 521
trait approach 286–98, 343, 344
transfers, brokering 234–5
transformative change 214
transitioning risk **267**
translational neuroscience 243–6
transmission disequilibrium test (TDT) 280
transversal population surveys 264–72
trauma 53, **300**, 427
trauma-focused research 244–6
treaties: cannabis legalization 523; international drug control 519–22
treatment/treatment programs: amphetamine use 130; attendance 458; entry/engagement 479–80; epistemic authority in 431–9; food addiction 186–8; heroin-assisted 415–16, 427, 437; loss of freedom 535–6; mandatory 554–63; neuroimaging studies 376–7; opioid substitution 478–89; outcomes of 330; responsibility 551; reward-based 11, 15; syndrome model 154; twelve-step recovery 464–77; use of term 216; Veterans Affairs 467–9; voluntary drug use 32, *see also* contingency management; therapies
treatment non-specificity 151
treatment planning 140
treatment providers 215, 492
treatment response, dopamine imaging 387–8
"triggering" hypotheses 262
"triples", models of addiction 165–6, **165**
Trump, Donald 511
"truth in labelling" rationale 536
TSF *see* Twelve-Step Facilitation
TSMHO *see* twelve-step mutual-help organizations
Tucker, J. A. 491
Twelve-Step Facilitation (TSF) 464–6, 468–70
twelve-step mutual-help organizations (TSMHOs) 464–77
twin normative failure model 407
twin studies 259–60, 262–3, 276

UD ("uncertainty discounting") 288
Ugor, Paul 419
UN (United Nations) treaties 520
uncertainty 343–4, 359
"uncertainty discounting" (UD) 288
Uniform Resource Identifiers (URIs) 165–6, 170
Uniform Resource Locators (URLs) 164
United Nations (UN) treaties 520
United States (US): cannabis legalization 524–5; criminal law 548, 549–50; "gentler war on drugs" 511–17; historical responses to drug use 513; NSDUH data 264–6, *see also* Anglo-American criminal law
universals, entity class 168
"unwilling addict" type 66–8, 72
"urges" *see* cravings
URIs *see* Uniform Resource Identifiers
URLs (Uniform Resource Locators) 164
Uruguay, cannabis legalization 521–2, 524–5
US *see* United States

VA (Veterans Affairs) treatment programs 467–9
value-based self-control 45–7, 50–1
values, moral approach 65–70
vanity 433–4
variable number of tandem repeats (VNTR) polymorphisms 282–3
The Varieties of Religious Experience (James) 431–2, 434–5, 437
ventral striatum 176, 287–8, 301–2
ventromedial prefrontal cortex (VmPFC) 176, 301–2
vested interests: commercial 526–7; lumpen abuse 225
Veterans Affairs (VA) treatment programs 467–9
virtues 432–4
visceral thinking 34–5
visibility 440–1, 442, 446
visual methods *see* documentary photography
Vivitrol 437
VmPFC *see* ventromedial prefrontal cortex
VNTR (variable number of tandem repeats) polymorphisms 282–3
volitional failure, self-control 47
Volkow, N. D. 322, 385, 396, 408n2
voluntary drug use 30–2
voluntary intoxication 548–9
vomiting 107–8
voucher-based CM 455–7
voxel-wise methods 365, 374
vulnerability 245–6, 279, 282, 291–3, *292*, 303–4

"wanting" 186, 307, 322, 353–9
"wanton addict" type 66–7

"war on drugs" model 511–17, 521
Watson, Gary 46
weakness: desires 86; of resistance 211–12; of will 37–9, 70–2
Weatherly, J. N. 173–4
weight gain 183
Werb, D. 561
The Wet and the Dry (Osborne) 81
white communities, cocaine use 512–13
white matter 177, 376
WHO (World Health Organization) 137
whole-brain connectivity methods 365
Wikler, A. 327
Wild, T. C. 561
will/will power 37–9, 42, 65, 69–72, 543
"willing addict" type 66
Wilson, Bill 431–2, 465
Winick, Charles 257–8
Wise, R. A. 103
withdrawal 11; abstinence and 55; animal research 195; cravings link 137; escape from 352; food addiction 184; incentive sensitization 352–3; pharmacological adaptation 136; stress responses 307; syndrome model 153
within-network connectivity 373
women 83–4, 112–13
Wood, W. 433–4
World Health Organization (WHO) 137
World Trade Organization (WTO) 521
Worldwide Web (www), effects of 164
WTO *see* World Trade Organization
www (Worldwide Web) 164

xenobiotics 106, 108–9

Yerkes, Robert 192, 194
young people 470, 472

Zinberg, N. E. 149